NATURAL HISTORY
UNIVERSAL LIBRARY

西方博物学大系

主编：江晓原

THE CACTACEAE : DESCRIPTIONS AND ILLUSTRATIONS OF PLANTS OF THE CACTUS FAMILY

仙人掌科

[美] 纳撒尼尔·洛德·布里顿
[美] 约瑟夫·纳尔逊·罗斯 著

华东师范大学出版社

图书在版编目 (CIP) 数据

仙人掌科 = The Cactaceae : descriptions and illustrations
of plants of the cactus family : 英文 / (美) 纳撒尼尔·洛德·
布里顿,(美) 约瑟夫·纳尔逊·罗斯著. — 上海 : 华东师范
大学出版社, 2018
　(寰宇文献)
　ISBN 978-7-5675-7993-4

　Ⅰ.①仙… Ⅱ.①纳… ②约… Ⅲ.①仙人掌科–英文
Ⅳ.①Q949.759.9

　中国版本图书馆CIP数据核字(2018)第179776号

仙人掌科

The Cactaceae : descriptions and illustrations of plants of the cactus family
(美) 纳撒尼尔·洛德·布里顿,(美) 约瑟夫·纳尔逊·罗斯

特约策划　黄曙辉　徐　辰
责任编辑　庞　坚
特约编辑　许　倩
装帧设计　刘怡霖

出版发行　华东师范大学出版社
社　　址　上海市中山北路3663号　邮编 200062
网　　址　www.ecnupress.com.cn
电　　话　021-60821666　行政传真　021-62572105
客服电话　021-62865537
门市（邮购）电话　021-62869887
地　　址　上海市中山北路3663号华东师范大学校内先锋路口
网　　店　http://hdsdcbs.tmall.com/

印 刷 者　虎彩印艺股份有限公司
开　　本　787×1092　16开
印　　张　85.75
版　　次　2018年8月第1版
印　　次　2018年8月第1次
书　　号　ISBN 978-7-5675-7993-4
定　　价　1598.00元 (精装全二册)

出 版 人　王　焰

(如发现本版图书有印订质量问题，请寄回本社客服中心调换或电话021-62865537联系)

总　目

卷一

《西方博物学大系总序》（江晓原）　　　　1

出版说明　　　　1

The Cactaceae　　VOL.I　　　　1

The Cactaceae　　VOL.II　　　　311

卷二

The Cactaceae　　VOL.III　　　　1

The Cactaceae　　VOL.IV　　　　311

《西方博物学大系》总序

江晓原

《西方博物学大系》收录博物学著作超过一百种，时间跨度为 15 世纪至 1919 年，作者分布于 16 个国家，写作语种有英语、法语、拉丁语、德语、弗莱芒语等，涉及对象包括植物、昆虫、软体动物、两栖动物、爬行动物、哺乳动物、鸟类和人类等，西方博物学史上的经典著作大备于此编。

中西方"博物"传统及观念之异同

今天中文里的"博物学"一词，学者们认为对应的英语词汇是 Natural History，考其本义，在中国传统文化中并无现成对应词汇。在中国传统文化中原有"博物"一词，与"自然史"当然并不精确相同，甚至还有着相当大的区别，但是在"搜集自然界的物品"这种最原始的意义上，两者确实也大有相通之处，故以"博物学"对译 Natural History 一词，大体仍属可取，而且已被广泛接受。

已故科学史前辈刘祖慰教授尝言：古代中国人处理知识，如开中药铺，有数十上百小抽屉，将百药分门别类放入其中，即心安矣。刘教授言此，其辞若有憾焉——认为中国人不致力于寻求世界"所以然之理"，故不如西方之分析传统优越。然而古代中国人这种处理知识的风格，正与西方的博物学相通。

与此相对，西方的分析传统致力于探求各种现象和物体之间的相互关系，试图以此解释宇宙运行的原因。自古希腊开始，西方哲人即孜孜不倦建构各种几何模型，欲用以说明宇宙如何运行，其中最典型的代表，即为托勒密（Ptolemy）的宇宙体系。

比较两者，差别即在于：古代中国人主要关心外部世界"如何"运行，而以希腊为源头的西方知识传统（西方并非没有别的知识传统，只是未能光大而已）更关心世界"为何"如此运行。在线性发展无限进步的科学主义观念体系中，我们习惯于认为"为何"是在解决了"如何"之后的更高境界，故西方的分析传统比中国的传统更高明。

然而考之古代实际情形，如此简单的优劣结论未必能够成立。例如以天文学言之，古代东西方世界天文学的终极问题是共同的：给定任意地点和时刻，计算出太阳、月亮和五大行星（七政）的位置。古代中国人虽不致力于建立几何模型去解释七政"为何"如此运行，但他们用抽象的周期叠加（古代巴比伦也使用类似方法），同样能在足够高的精度上计算并预报任意给定地点和时刻的七政位置。而通过持续观察天象变化以统计、收集各种天象周期，同样可视之为富有博物学色彩的活动。

还有一点需要注意：虽然我们已经接受了用"博物学"来对译 Natural History，但中国的博物传统，确实和西方的博物学有一个重大差别——即中国的博物传统是可以容纳怪力乱神的，而西方的博物学基本上没有怪力乱神的位置。

古代中国人的博物传统不限于"多识于鸟兽草木之名"。体现此种传统的典型著作，首推晋代张华《博物志》一书。书名"博物"，其义尽显。此书从内容到分类，无不充分体现它作为中国博物传统的代表资格。

《博物志》中内容，大致可分为五类：一、山川地理知识；二、奇禽异兽描述；三、古代神话材料；四、历史人物传说；五、神仙方伎故事。这五大类，完全符合中国文化中的博物传统，深合中国古代博物传统之旨。第一类，其中涉及宇宙学说，甚至还有"地动"思想，故为科学史家所重视。第二类，其中甚至出现了中国古代长期流传的"守宫砂"传说的早期文献：相传守宫砂点在处女胳膊上，永不褪色，只有性交之后才会自动消失。第三类，古代神话传说，其中甚至包括可猜想为现代"连体人"的记载。第四类，各种著名历史人物，比如三位著名刺客的传说，此三名刺客及所刺对象，历史上皆实有其人。第五类，包括各种古代方术传说，比如中国古代房中养生学说，房中术史上的传说人物之一"青牛道士封君达"等等。前两类与西方的博物学较为接近，但每一类都会带怪力乱神色彩。

"所有的科学不是物理学就是集邮"

在许多人心目中，画画花草图案，做做昆虫标本，拍拍植物照片，这类博物学活动，和精密的数理科学，比如天文学、物理学等等，那是无法同日而语的。博物学显得那么的初级、简单，甚至幼稚。这种观念，实际上是将"数理程度"作为唯一的标尺，用来衡量一切知识。但凡能够使用数学工具来描述的，或能够进行物理实验的，那就是"硬"科学。使用的数学工具越高深越复杂，似乎就越"硬"；物理实验设备越庞大，花费的金钱越多，似乎就越"高端"、越"先进"……

这样的观念，当然带着浓厚的"物理学沙文主义"色彩，在很多情况下是不正确的。而实际上，即使我们暂且同意上述"物理学沙文主义"的观念，博物学的"科学地位"也仍然可以保住。作为一个学天体物理专业出身，因而经常徜徉在"物理学沙文主义"幻影之下的人，我很乐意指出这样一个事实：现代天文学家们的研究工作中，仍然有绘制星图，编制星表，以及为此进行的巡天观测等等活动，这些活动和博物学家"寻花问柳"，绘制植物或昆虫图谱，本质上是完全一致的。

这里我们不妨重温物理学家卢瑟福(Ernest Rutherford)的金句："所有的科学不是物理学就是集邮（All science is either physics or stamp collecting）。"卢瑟福的这个金句堪称"物理学沙文主义"的极致，连天文学也没被他放在眼里。不过，按照中国传统的"博物"理念，集邮毫无疑问应该是博物学的一部分——尽管古代并没有邮票。卢瑟福的金句也可以从另一个角度来解读：既然在卢瑟福眼里天文学和博物学都只是"集邮"，那岂不就可以将博物学和天文学相提并论了？

如果我们摆脱了科学主义的语境，则西方模式的优越性将进一步被消解。例如，按照霍金（Stephen Hawking）在《大设计》（The Grand Design）中的意见，他所认同的是一种"依赖模型的实在论（model-dependent realism）"，即"不存在与图像或理论无关的实在性概念（There is no picture- or theory-independent concept of reality）"。在这样的认识中，我们以前所坚信的外部世界的客观性，已经不复存在。既然几何模型只不过是对外部世界图像的人为建构，则古代中国人干脆放弃这种建构直奔应用（毕竟在实际应用中我们只需要知道七政"如何"运行），又有何不可？

传说中的"神农尝百草"故事，也可以在类似意义下得到新的解读："尝百草"当然是富有博物学色彩的活动，神农通过这一活动，得知哪些草能够治病，哪些不能，然而在这个传说中，神农显然没有致力于解释"为何"某些草能够治病而另一些则不能，更不会去建立"模型"以说明之。

"帝国科学"的原罪

今日学者有倡言"博物学复兴"者，用意可有多种，诸如缓解压力、亲近自然、保护环境、绿色生活、可持续发展、科学主义解毒剂等等，皆属美善。编印《西方博物学大系》也是意欲为"博物学复兴"添一助力。

然而，对于这些博物学著作，有一点似乎从未见学者指出过，而鄙意以为，当我们披阅把玩欣赏这些著作时，意识到这一点是必须的。

这百余种著作的时间跨度为 15 世纪至 1919 年，注意这个时间跨度，正是西方列强"帝国科学"大行其道的时代。遥想当年，帝国的科学家们乘上帝国的军舰——达尔文在皇家海军"小猎犬号"上就是这样的场景之一，前往那些已经成为帝国的殖民地或还未成为殖民地的"未开化"的遥远地方，通常都是踌躇满志、充满优越感的。

作为一个典型的例子，英国学者法拉在（Patricia Fara）《性、植物学与帝国：林奈与班克斯》（*Sex, Botany and Empire, The Story of Carl Linnaeus and Joseph Banks*）一书中讲述了英国植物学家班克斯（Joseph Banks）的故事。1768 年 8 月 15 日，班克斯告别未婚妻，登上了澳大利亚军舰"奋进号"。此次"奋进号"的远航是受英国海军部和皇家学会资助，目的是前往南太平洋的塔希提岛（Tahiti，法属海外自治领，另一个常见的译名是"大溪地"）观测一次比较罕见的金星凌日。舰长库克（James Cook）是西方殖民史上最著名的舰长之一，多次远航探险，开拓海外殖民地。他还被认为是澳大利亚和夏威夷群岛的"发现"者，如今以他命名的群岛、海峡、山峰等不胜枚举。

当"奋进号"停靠塔希提岛时，班克斯一下就被当地美丽的土著女性迷昏了，他在她们的温柔乡里纵情狂欢，连库克舰长都看不下去了，"道德愤怒情绪偷偷溜进了他的日志当中，他发现自己根本不可能不去批评所见到的滥交行为"，而班克斯纵欲到了"连嫖妓都毫无激情"的地步——这是别人讽刺班克斯的说法，因为对于那时常年航行于茫茫大海上的男性来说，上岸嫖妓通常是一项能够唤起"激情"的活动。

而在"帝国科学"的宏大叙事中，科学家的私德是无关紧要的，人们关注的是科学家做出的科学发现。所以，尽管一面是班克斯在塔希提岛纵欲滥交，一面是他留在故乡的未婚妻正泪眼婆娑地"为远去的心上人绣织背心"，这样典型的"渣男"行径要是放在今天，非被互联网上的口水淹死不可，但是"班克斯很快从他们的分离之苦中走了出来，在外近三年，他活得倒十分滋润"。

法拉不无讽刺地指出了"帝国科学"的实质："班克斯接管了当地的女性和植物，而库克则保护了大英帝国在太平洋上的殖民地。"甚至对班克斯的植物学本身也调侃了一番："即使是植物学方面的科学术语也充满了性指涉。……这个体系主要依靠花朵之中雌雄生殖器官的数量来进行分类。"据说"要保护年轻妇女不受植物学教育的浸染，他们严令禁止各种各样的植物采集探险活动。"这简直就是将植物学看成一种"涉黄"的淫秽色情活动了。

在意识形态强烈影响着我们学术话语的时代，上面的故事通常是这样被描述的：库克舰长的"奋进号"军舰对殖民地和尚未成为殖民地的那些地方的所谓"访问"，其实是殖民者耀武扬威的侵略，搭载着达尔文的"小猎犬号"军舰也是同样行径；班克斯和当地女性的纵欲狂欢，当然是殖民者对土著妇女令人发指的蹂躏；即使是他采集当地植物标本的"科学考察"，也可以视为殖民者"窃取当地经济情报"的罪恶行为。

后来改革开放，上面那种意识形态话语被抛弃了，但似乎又走向了另一个极端，完全忘记或有意回避殖民者和帝国主义这个层面，只歌颂这些军舰上的科学家的伟大发现和成就，例如达尔文随着"小猎犬号"的航行，早已成为一曲祥和优美的科学颂歌。

其实达尔文也未能免俗，他在远航中也乐意与土著女性打交道，当然他没有像班克斯那样滥情纵欲。在达尔文为"小猎犬号"远航写的《环球游记》中，我们读到："回程途中我们遇到一群黑人姑娘在聚会，……我们笑着看了很久，还给了她们一些钱，这着实令她们欣喜一番，拿着钱尖声大笑起来，很远还能听到那愉悦的笑声。"

有趣的是，在班克斯在塔希提岛纵欲六十多年后，达尔文随着"小猎犬号"也来到了塔希提岛，岛上的土著女性同样引起了达尔文的注意，在《环球游记》中他写道："我对这里妇女的外貌感到有些失望，然而她们却很爱美，把一朵白花或者红花戴在脑后的鬓髻上……"接着他以居高临下的笔调描述了当地女性的几种发饰。

用今天的眼光来看，这些在别的民族土地上采集植物动物标本、测量地质水文数据等等的"科学考察"行为，有没有合法性问题？有没有侵犯主权的问题？这些行为得到当地人的同意了吗？当地人知道这些行为的性质和意义吗？他们有知情权吗？……这些问题，在今天的国际交往中，确实都是存在的。

也许有人会为这些帝国科学家辩解说：那时当地土著尚在未开化或半开化状态中，他们哪有"国家主权"的意识啊？他们也没有制止帝国科学家的考察活动啊？但是，这样的辩解是无法成立的。

姑不论当地土著当时究竟有没有试图制止帝国科学家的"科学考察"行为，现在早已不得而知，只要殖民者没有记录下来，我们通常就无法知道。况且殖民者有军舰有枪炮，土著就是想制止也无能为力。正如法拉所描述的："在几个塔希提人被杀之后，一套行之有效的易货贸易体制建立了起来。"

即使土著因为无知而没有制止帝国科学家的"科学考察"行为，这事也很像一个成年人闯进别人的家，难道因为那家只有不懂事的小孩子，闯入者就可以随便打探那家的隐私、拿走那家的东西、甚至将那家的房屋土地据为己有吗？事实上，很多情况下殖民者就是这样干的。所以，所谓的"帝国科学"，其实是有着原罪的。

如果沿用上述比喻，现在的局面是，家家户户都不会只有不懂事的孩子了，所以任何外来者要想进行"科学探索"，他也得和这家主人达成共识，得到这家主人的允许才能够进行。即使这种共识的达成依赖于利益的交换，至少也不能单方面强加于人。

博物学在今日中国

博物学在今日中国之复兴，北京大学刘华杰教授提倡之功殊不可没。自刘教授大力提倡之后，各界人士纷纷跟进，仿佛昔日蔡锷在云南起兵反袁之"滇黔首义，薄海同钦，一檄遥传，景从恐后"光景，这当然是和博物学本身特点密切相关的。

无论在西方还是在中国，无论在过去还是在当下，为何博物学在它繁荣时尚的阶段，就会应者云集？深究起来，恐怕和博物学本身的特点有关。博物学没有复杂的理论结构，它的专业训练也相对容易，至少没有天文学、物理学那样的数理"门槛"，所以和一些数理学科相比，博物学可以有更多的自学成才者。这次编印的《西方博物学大系》，卷帙浩繁，蔚为大观，同样说明了这一点。

最后，还有一点明显的差别必须在此处强调指出：用刘华杰教授喜欢的术语来说，《西方博物学大系》所收入的百余种著作，绝大部分属于"一阶"性质的工作，即直接对博物学作出了贡献的著作。事实上，这也是它们被收入《西方博物学大系》的主要理由之一。而在中国国内目前已经相当热的博物学时尚潮流中，绝大部分已经出版的书籍，不是属于"二阶"性质（比如介绍西方的博物学成就），就是文学性的吟风咏月野草闲花。

要寻找中国当代学者在博物学方面的"一阶"著作，如果有之，以笔者之孤陋寡闻，唯有刘华杰教授的《檀岛花事——夏威夷植物日记》三卷，可以当之。这是刘教授在夏威夷群岛实地考察当地植物的成果，不仅属于直接对博物学作出贡献之作，而且至少在形式上将昔日"帝国科学"的逻辑反其道而用之，岂不快哉！

2018 年 6 月 5 日
于上海交通大学
科学史与科学文化研究院

《仙人掌科》是美国植物学家纳撒尼尔·洛德·布里顿（Nathaniel Lord Britton，1859-1934）和约瑟夫·纳尔逊·罗斯（Joseph Nelson Rose，1862-1928）共著的一部博物学著作。

布里顿生于纽约，虽然双亲希望他成为教会的神职人员，但他从小就对自然生态产生了浓厚的兴趣。在哥伦比亚大学修习矿物学后，留校教地质和植物学，后来又加入植物学家约翰·托雷创立的植物学俱乐部。在那里他遇到古巴长大的苔藓学家伊丽莎白·奈特，并于 1885 年与之成婚，自此夫妇合作直至去世。1888 年，布里顿夫妇访问英国的邱园。回国后，伊丽莎白在俱乐部提议也应在纽约设立专门的植物园，并发起募款活动。1891 年，纽约植物园成立，布里顿辞去大学教职，出任园长。在夫妇二人有效的运作下，当时的美国实业巨子卡耐基、摩根等人都拨款投资，植物园也因此蓬勃发展。

罗斯生于印第安纳州，该地是南北战争期间联邦军队的主要兵源地之一，壮年男性在战争期间伤亡惨重，他的父亲也死在战场上。他起初专修古植物学，在农业部当了几年小职员后，被聘为史密森博物馆的助理，专门研究伞形科和仙人掌科植物，并多次前往墨西哥进行田野考察和样本采集，取得的植物样本分送史密森博物馆和纽约植物园。因此，当布里顿决定出版一部专论仙人掌科的著作时，罗斯成了当仁不让的合作者。二人利用闲暇时间赶赴南美等地做了大量田野工作，终于在 1919 至 1923 年出齐了全部四卷《仙人掌科》。在这部洋洋 1400 页的巨著中，作者详尽记述了这种绚丽斑斓却往往遭人忽视的坚韧植物，并配上了 300 余幅单色铜版插图和 100 余幅彩色插图。这些栩栩如生的插图，均由当时供职于纽约植物园的英国博物画大家玛丽·艾米莉·伊顿绘制，堪称精美绝伦。

今据英文原版影印。

A Cactus Desert in Arizona.   Photograph by D. T. McDougal.

THE CACTACEAE

DESCRIPTIONS AND ILLUSTRATIONS OF PLANTS OF THE CACTUS FAMILY

BY

N. L. BRITTON AND J. N. ROSE

Volume I

THE CARNEGIE INSTITUTION OF WASHINGTON
WASHINGTON, 1919

CARNEGIE INSTITUTION OF WASHINGTON
PUBLICATION No. 248, VOLUME I

PRESS OF GIBSON BROTHERS
WASHINGTON

CONTENTS.

	PAGE.
Introduction	3
Order Cactales	8
Family Cactaceae	8
Key to Tribes	8
Tribe Pereskieae	8
Pereskia	8
Key to Species	9
Tribe Opuntieae	24
Key to Genera	24
Pereskiopsis	25
Pterocactus	30
Nopalea	33
Tacinga	39
Maihuenia	40
Opuntia	42
Key to Subgenera and Series	44
Subgenus Cylindropuntia	46
Series Ramosissimae	46
Series Leptocaules	46
Series Thurberianae	52
Series Echinocarpae	56
Series Bigelovianae	58
Series Imbricatae	60
Series Fulgidae	67
Series Vestitae	71
Series Clavarioides	72
Series Salmianae	73
Series Subulatae	75
Series Miquelianae	78
Series Clavatae	79
Subgenus Tephrocactus	84
Series Weberianae	84
Series Floccosae	86
Series Glomeratae	87
Series Pentlandianae	90

	PAGE.
Family Cactaceae—*continued*.	
Tribe Opuntieae—*continued*.	
Opuntia—*continued*.	
Subgenus Platyopuntia	99
Series Pumilae	100
Series Curassavicae	102
Series Aurantiacae	106
Series Tunae	110
Series Basilares	118
Series Inamoenae	125
Series Tortispinae	126
Series Sulphureae	133
Series Strigiles	136
Series Setispinae	136
Series Phaeacanthae	139
Series Elatiores	149
Series Elatae	156
Series Scheerianae	159
Series Dillenianae	159
Series Macdougalianae	169
Series Tomentosae	172
Series Leucotrichae	174
Series Orbiculatae	176
Series Ficus-indicae	177
Series Streptacanthae	181
Series Robustae	191
Series Polyacanthae	193
Series Stenopetalae	200
Series Palmadorae	201
Series Spinosissimae	202
Series Brasilienses	209
Series Ammophilae	211
Series Chaffeyanae	213
Grusonia	215
Appendix	216
Index	227

III

ILLUSTRATIONS.

PLATES

FACING PAGE

PLATE 1. Cactus Desert in Arizona...Frontispiece
PLATE 2. (1) Flowering branch of Pereskia pereskia. (2 and 3) Fruit of Pereskia pereskia. (4) Leafy branch of Pereskia sacharosa (5) Proliferous fruit of Pereskia sacharosa............................ 10
PLATE 3. (1) Flowering branch of Pereskia grandifolia. (2) Leafy branch of Pereskiopsis chapistle. (3) Leafy branch of Pereskiopsis pititache... 20
PLATE 4. (1) Upper part of flowering joint of Nopalea cochenillifera. (2) Upper part of flowering joint of Nopalea auberi. (3) Fruit of Nopalea auberi. (4) Flowering joint of Nopalea dejecta..... 34
PLATE 5. Nopalea auberi as it grows near Mitla, Mexico.. 38
PLATE 6. (1 and 2) Branch of Opuntia mortolensis. (3 and 4) Branch of Opuntia leptocaulis. (5) Flowering branch of Opuntia arbuscula. (6) Flowering branch of Opuntia kleiniae................ 48
PLATE 7. (1) Leafy branch of Opuntia kleiniae. (2) Terminal branch of Opuntia vivipara. (3) Branch of Opuntia parryi. (4) Flowering branch of Opuntia echinocarpa. (5) Fruiting branch of Opuntia versicolor.. 50
PLATE 8. (1) Type plant of Opuntia vivipara, near Tucson, Arizona. (2) A much branched plant of Opuntia versicolor.. 52
PLATE 9. (1) Joint of Opuntia tetracantha. (2, 3, 4, 5) Flowering joint of Opuntia versicolor. (6) Proliferous fruits of Opuntia fulgida.. 54
PLATE 10. (1) Joint of Opuntia tunicata. (2, 3, 4, 5) Joint of Opuntia spinosior........................ 66
PLATE 11. (1) Leafy branch of Opuntia imbricata. (2) Flowering branch of Opuntia prolifera. (3, 4) Form of Opuntia alcahes. (5, 6) Opuntia vestita.. 68
PLATE 12. (1) Clump of plants of Opuntia fulgida. (2) A very open plant of Opuntia spinosior............ 70
PLATE 13. (1) Opuntia exaltata as seen in the highlands of Peru. (2) Clump of Opuntia floccosa as it grows in the valleys of the Andes of eastern Peru.. 76
PLATE 14. (1) Flowering branch of Opuntia burrageana. (2) Opuntia cylindrica. (3, 4) Joint of Opuntia stanlyi. (5) Flowering joint of Opuntia macrorhiza.. 78
PLATE 15. (1, 2) Part of joint of Opuntia exaltata. (3) Upper part of joint of Opuntia macrarthra. (4) Upper part of joint of Opuntia tortispina.. 80
PLATE 16. (1) Top of Opuntia miquelii. (2) Old and young joints of Opuntia invicta. (3) Upper part of joint of Opuntia ignescens.. 98
PLATE 17. (1) Joint of Opuntia pascoensis. (2) Joints of Opuntia taylori. (3, 4) Form of Opuntia repens. (5) Flower of Opuntia repens. (6) Flowering joint of Opuntia drummondii.............. 102
PLATE 18. (1) Two plants of Opuntia drummondii. (2) Joints of Opuntia retrorsa with flower. (3) Joints of Opuntia triacantha. (4) Joint of Opuntia jamaicensis. (5) Section of fruit of Opuntia jamaicensis.. 104
PLATE 19. (1) Plant of Opuntia jamaicensis. (2, 3) Flower of Opuntia jamaicensis. (4) Longitudinal section of flower of Opuntia jamaicensis. (5, 6) Stamen of Opuntia jamaicensis. (7) Style of Opuntia jamaicensis.. 112
PLATE 20. (1) Flowering joint of Opuntia decumbens. (2) Fruiting joint of Opuntia decumbens. (3) Hybrid 116
PLATE 21. Group of hardy Opuntia, mostly Opuntia tortispina, in grounds of New York Botanical Garden.... 126
PLATE 22. (1) Joints of Opuntia microdasys. (2) Flowering joint of Opuntia macrarthra. (3) Fruit of Opuntia macrarthra. (4) Seed of Opuntia macrarthra. (5) Flowering joint of Opuntia opuntia..... 128
PLATE 23. (1) Flowering joint of Opuntia fuscoatra. (2) Upper part of joint of Opuntia sulphurea. (3) Joint of Opuntia tenuispina.. 132
PLATE 24. (1) Plant of Opuntia santa-rita. (2) Plant of Opuntia discata........................ 142
PLATE 25. (1) Flowering joint of Opuntia atrispina. (2) Flowering joint of Opuntia phaeacantha. (3) Upper part of joint of Opuntia engelmannii.. 144
PLATE 26. (1) Flowering joint of Opuntia bergeriana. (2) Flowering joint of Opuntia elatior. (3) Flowering joint of Opuntia boldinghii. (4, 5) Joint of Opuntia elata.............................. 152
PLATE 27. (1) Upper part of fruiting joint of Opuntia schumannii. (2) Flower of Opuntia schumannii. (3) Flowering joint of Opuntia vulgaris. (4) Flowering joint of Opuntia stricta.............. 156
PLATE 28. (1) Flowering joint of Opuntia laevis. (2) Flowering joint of Opuntia dillenii. (3) Upper part of flowering joint of Opuntia aciculata.. 160
PLATE 29. (1) View of Opuntia keyensis. (2) View of Opuntia dillenii 162
PLATE 30. Flowering joint of Opuntia linguiformis.. 164
PLATE 31. Flowering joints of Opuntia lindheimeri. (1) Orange-flowered race. (2) Red-flowered race........ 166
PLATE 32. (1) Upper part of flowering joint of Opuntia leptocarpa. (2) Fruit of Opuntia leptocarpa. (3) Flowering joint of Opuntia velutina. (4) Upper part of joint of Opuntia megacantha........... 172
PLATE 33. (1) Upper part of joint of Opuntia tomentosa. (2) Flowering joint of Opuntia brasiliensis. (3) Flowering branch of Opuntia brasiliensis. (4) Joint of Grusonia bradtiana.............. 174
PLATE 34. (1) Part of joint of Opuntia leucotricha. (2) Part of joint of Opuntia maxima. (3) Joint of Opuntia lasiacantha. (4) Joint of Opuntia robusta.. 180
PLATE 35. (1) Plant of Opuntia fragilis. (2) Flowering branch of Opuntia rhodantha. (3) Flowering joint of Opuntia polyacantha.. 194
PLATE 36. (1) Flowering joint of Opuntia spinosissima. (2, 3) Single flower of Opuntia spinosissima. (4, 5) Longitudinal section of flower of Opuntia spinosissima. (6) Cross-section of ovary of Opuntia spinosissima. (7) Style of Opuntia spinosissima............................ 204

TEXT-FIGURES.

		PAGE.
FIG.	1. Hedge of Pereskia pereskia	7
	2. Tree of Pereskia autumnalis.........	11
	3. Branches of Pereskia autumnalis.......	12
	4. Branch of Pereskia lychnidiflora.......	12
	5. Leafy branch of Pereskia nicoyana......	13
	6. Branch of Pereskia zehntneri..........	13
	7. Cultivated plant of Pereskia zehntneri..	14
	8. Herbarium specimen of Pereskia moorei.	15
	9. Tree of Pereskia guamacho.............	15
	10. Flowering branch of Pereskia guamacho.	16
	11. Leafy branch and flower of Pereskia colombiana	17
	12. Branch and fruit of Pereskia bleo.......	18
	13. Fruit of Pereskia bahiensis............	19
	14. Leafy branch of Pereskia bahiensis......	19
	15. Tree of Pereskia bahiensis.............	20
	16. Hedge containing Pereskia grandifolia...	21
	17. Branch of Pereskia zinniaeflora.........	21
	18. Tree of Pereskia cubensis.............	22
	19. Leafy branch of Pereskia cubensis.......	22
	20. Branch and fruit of Pereskia portulacifolia.............................	23
	21. Potted plant grown from a cutting of Pereskiopsis velutina................	26
	22. Branch of Pereskiopsis diguetii.........	27
	23. Branch of Pereskiopsis opuntiaeflora	27
	24. Branch of Pereskiopsis rotundifolia......	27
	25. Shows a clump of Pereskiopsis rotundifolia...........................	28
	26. Branch of Pereskiopsis porteri..........	28
	27. Branch of Pereskiopsis aquosa..........	29
	28. Leaf of Pereskiopsis kellermanii........	30
	29. Leaf of Pereskiopsis kellermanii........	30
	30. Leaf of Pereskiopsis kellermanii........	30
	31. Seed of Pterocactus hickenii...........	31
	32. Plant of Pterocactus hickenii...........	31
	33. Branch of Pterocactus fischeri..........	31
	34. Seed of Pterocactus fischeri............	31
	35. Seed of Pterocactus pumilus...........	31
	36. Seed of Pterocactus tuberosus..........	31
	37. Plant of Pterocactus tuberosus, showing a very large root....................	32
	38. Potted plant of Pterocactus tuberosus...	33
	39. Joint of Nopalea guatemalensis.........	35
	40. Joint of Nopalea lutea.................	35
	41. Large plant of Nopalea dejecta.........	36
	42. Joints of Nopalea dejecta..............	37
	43. Joints of Nopalea karwinskiana........	37
	44. Joint of Nopalea inaperta.............	37
	45. Flower of Tacinga funalis..............	38
	46. Longitudinal section of flower of Tacinga funalis...........................	38
	47. Section of stem of Tacinga funalis......	38
	48. Tip of young branch of Tacinga funalis.	38
	49. Plant of Tacinga funalis, climbing over bushes.............................	39
	50. Plant of Maihuenia valentinii..........	40
	51. Fruit of Maihuenia poeppigii...........	41
	52. Joint and flower of Maihuenia brachydelphys.............................	41
	53. Plant of Maihuenia tehuelches.........	41
	54. Branch of Opuntia ramossissima........	46
	55. Section of stem of Opuntia ramosissima.	46
	56. Plant of Opuntia leptocaulis...........	48
FIG.	57. Section of stem of Opuntia leptocaulis·..	48
	58. Joint of Opuntia caribaea.............	48
	59. Thicket formed of Opuntia caribaea.....	49
	60. Clump of Opuntia arbuscula...........	50
	61. Plant of Opuntia arbuscula............	51
	62. Fruiting branch of Opuntia arbuscula...	51
	63. Flowering branch of Opuntia thurberi...	53
	64. Branch of Opuntia davisii..............	55
	65. Branch of Opuntia viridiflora..........	55
	66. Branch of Opuntia whipplei............	55
	67. Plant of Opuntia acanthocarpa.........	56
	68. Joint of Opuntia serpentina............	58
	69. Plant of Opuntia bigelovii.............	59
	70. Joint of Opuntia bigelovii..............	59
	71. Potted plant of Opuntia ciribe..........	60
	72. Joint of Opuntia ciribe................	60
	73. Potted plant of Opuntia cholla..........	61
	74. Joint of Opuntia cholla................	62
	75. Proliferous fruits of Opuntia cholla, developing new joints.....................	62
	76. Proliferous fruits of Opuntia cholla, developing new joints..................	62
	77. Joints of Opuntia loydii................	63
	78. Plant of Opuntia lloydii...............	63
	79. Plant of Opuntia imbricata............	64
	80. Potted plant of Opuntia tunicata.......	65
	81. Plant of Opuntia pallida...............	65
	82. Potted plant of Opuntia molesta........	67
	83. Joint of Opuntia prolifera.............	69
	84. Fruiting branch of Opuntia prolifera....	70
	85. Potted plant of Opuntia alcahes........	70
	86. Joint of Opuntia verschaffeltii..........	72
	87. Grafted plants of Opuntia clavarioides...	73
	88. Potted plant of Opuntia salmiana.......	74
	89. Joints of Opuntia salmiana.............	74
	90. Potted plant of Opuntia subulata.......	76
	91. Joint of Opuntia pachypus.............	77
	92. Joints of Opuntia schottii.............	81
	93. Joints of Opuntia clavata..............	81
	94. Joints of Opuntia parishii.............	82
	95. Joints of Opuntia pulchella............	82
	96. Plants of Opuntia vilis.................	83
	97. Joints and cluster of spines of Opuntia bulbispina........................	83
	98. Joints of Opuntia grahamii.............	84
	99. Plants of Opuntia weberi..............	84
	100. Joints of Opuntia weberi..............	85
	101. Potted plant of Opuntia floccosa.......	86
	102. Mound of Opuntia lagopus............	88
	103. Root, joints, and flower of O. australis..	88
	104. Joints of Opuntia glomerata............	89
	105. Joint of Opuntia aoracantha...........	91
	106. Joint of Opuntia rauppiana............	92
	107. Flowering plant and fruit of Opuntia subterranea............................	92
	108. Joints of Opuntia hickenii.............	92
	109. Joint of Opuntia darwinii.............	94
	110. Joints of Opuntia atacamensis..........	94
	111. Joints of Opuntia russellii.............	94
	112. Joints of Opuntia ovata...............	95
	113. Potted plant of Opuntia sphaerica......	96
	114. Joint of Opuntia skottsbergii..........	97
	115. Joint of Opuntia nigrispina............	97
	116. Joint of Opuntia pentlandii......	97

TEXT-FIGURES—continued.

PAGE.

FIG. 117. Joints of Opuntia pentlandii............ 98
118. Joint of Opuntia ignescens............. 98
119. Mound of Opuntia ignescens.......... 98
120. Plant of Opuntia campestris........... 99
121. Joints of Opuntia ignota.............. 99
122. Thicket of Opuntia pumila............. 100
123. Joints of Opuntia pumila.............. 101
124. Joints of Opuntia pubescens........... 101
125. Joints of Opuntia curassavica.......... 102
126. Joints of Opuntia borinquensis......... 104
127. Joints of Opuntia militaris............. 104
128. Joints and flower of Opuntia tracyi..... 105
129. Joints and flowers of Opuntia pusilla.... 106
130. Joints of Opuntia aurantiaca.......... 107
131. Potted plant of Opuntia schickendantzii. 107
132. Plant of Opuntia kiska-loro........... 108
133. Joints of Opuntia canina.............. 108
134. Plant of Opuntia retrorsa............. 109
135. Plant of Opuntia utkilio.............. 110
136. Joints of Opuntia anacantha.......... 110
137. Thicket of Opuntia bella............. 111
138. Joints of Opuntia bella.............. 112
139. Joint of Opuntia bella............... 112
140. Plant of Opuntia triacantha.......... 113
141. Plant of Opuntia tuna............... 114
142. Joints of Opuntia tuna.............. 114
143. Thicket of Opuntia antillana......... 115
144. Joints of Opuntia antillana.......... 115
145. Plant of Opuntia decumbens.......... 117
146. Plant of Opuntia depressa........... 118
147. Joints of Opuntia lubrica............ 119
148. Landscape showing Opuntia treleasei.... 119
149. Joints of Opuntia basilaris........... 120
150. Plant of Opuntia microdasys.......... 121
151. Potted plant of Opuntia, probable hybrid 121
152. Joint of Opuntia macrocalyx.......... 122
153. Plant of Opuntia rufida.............. 122
154. Plant of Opuntia pycnantha.......... 123
155. Potted plant of Opuntia comonduensis.. 124
156. Plant of Opuntia inamoena........... 125
157. Joint of Opuntia inamoena........... 125
158. Joints of Opuntia allairei............ 126
159. Joints of Opuntia pollardii........... 126
160. Plant of Opuntia opuntia............ 128
161. Fruit of Opuntia grandiflora.......... 129
162. Flowering joint of Opuntia grandiflora.. 129
163. Flowering joints of Opuntia austrina.... 130
164. Joints, flower, and fruit of O. plumbea.. 131
165. Fruit of Opuntia stenochila.......... 132
166. Fruit of Opuntia stenochila.......... 132
167. Joint of Opuntia stenochila.......... 132
168. Potted plant of Opuntia delicata....... 133
169. Joint of Opuntia soehrensii.......... 135
170. Joint of Opuntia microdisca.......... 135
171. Joints of Opuntia strigil............. 136
172. Joints of Opuntia ballii............. 137
173. Joints of Opuntia pottsii............ 138
174. Joint of Opuntia setispina........... 138
175. Plant and fruit of Opuntia mackensensii 139
176. Joint of Opuntia macrocentra......... 140
177. Joints of Opuntia tardospina......... 141
178. Cluster of spines of Opuntia gosseliniana 141
179. Joint of Opuntia gosseliniana........ 141
180. Joint of Opuntia angustata.......... 142
181. Plant of Opuntia azurea............. 143
182. Joints of Opuntia azurea............ 143
183. Joint of Opuntia covillei............ 145
184. Joint of Opuntia covillei............ 145

PAGE.

FIG. 185. Joint of Opuntia vaseyi.............. 146
186. Potted plant of Opuntia occidentalis.... 147
187. Joint of Opuntia brunnescens......... 150
188. Fruit of Opuntia brunnescens......... 150
189. Joint of Opuntia galapageia.......... 150
190. Flower of Opuntia galapageia......... 150
191. Joint and cluster of spines of Opuntia
        galapageia...................... 151
192. Flowering joint of Opuntia delaetiana.. 152
193. Joints of Opuntia hanburyana........ 154
194. Joint of Opuntia quitensis........... 154
195. Joint of Opuntia distans............ 155
196. Joint of Opuntia elata.............. 157
197. Joints of Opuntia cardiosperma....... 157
198. Joint of Opuntia scheeri............ 159
199. Plant of Opuntia chlorotica.......... 160
200. Joints of Opuntia chlorotica......... 160
201. Plant of Opuntia dillenii............ 162
202. Joint of Opuntia tapona............ 164
203. Potted plant of Opuntia littoralis...... 164
204. Joints of Opuntia cantabrigiensis...... 167
205. Part of joint and cluster of spines of
        Opuntia procumbens.............. 167
206. Joint of Opuntia cañada............ 167
207. Joint of Opuntia pyriformis.......... 168
208. Joint of Opuntia durangensis......... 169
209. Plant of Opuntia macdougaliana...... 170
210. Potted plant of Opuntia macdougaliana. 171
211. Joint of Opuntia wilcoxii........... 172
212. Plant of Opuntia tomentosa.......... 173
213. Joint of Opuntia tomentella......... 174
214. Potted plant of Opuntia leucotricha.... 175
215. Joints of Opuntia orbiculata......... 176
216. Potted plant of Opuntia pilifera...... 177
217. Plants of Opuntia ficus-indica........ 178
218. Fruit of Opuntia ficus-indica......... 178
219. Plant of Opuntia crassa............. 179
220. Potted plant of Opuntia maxima....... 180
221. Joint of Opuntia spinulifera.......... 182
222. Joint of Opuntia lasiacantha......... 183
223. Joint of Opuntia zacuapanensis....... 183
224. Joint of Opuntia hyptiacantha........ 183
225. Joint of Opuntia streptacantha....... 184
226. Potted plant of Opuntia megacantha.... 185
227. Plants of Opuntia megacantha........ 186
228. Joint of Opuntia megacantha......... 186
229. Joint of Opuntia deamii............ 187
230. Joint of Opuntia eichlamii........... 188
231. Joint of Opuntia inaequilateralis....... 188
232. Joint of Opuntia pittieri............ 189
233. Joint of Opuntia cordobensis......... 189
234. Fruit of Opuntia cordobensis......... 189
235. Joint of Opuntia quimilo............ 190
236. Fruit of Opuntia quimilo............ 190
237. Joint and flowers of Opuntia quimilo.... 191
238. Plant of Opuntia robusta............ 192
239. Plant of Opuntia fragilis............ 194
240. Joints of Opuntia arenaria........... 195
241. Joint of Opuntia trichophora......... 195
242. Plant of Opuntia erinacea............ 196
243. Joint of Opuntia juniperina.......... 197
244. Seed of Opuntia juniperina.......... 197
245. Joint of Opuntia hystricina.......... 197
246. Joint of Opuntia sphaerocarpa........ 198
247. Joints of Opuntia polyacantha........ 199
248. Joint of Opuntia stenopetala.......... 200
249. Upper part of joint and flower of Opuntia
        stenopetala..................... 201

TEXT-FIGURES—continued.

		PAGE.
FIG. 250.	Plants of Opuntia palmadora..........	202
251.	Joints of Opuntia palmadora...........	202
252.	Plants of Opuntia nashii..............	202
253.	Potted plant of Opuntia nashii.........	203
254.	Joint of Opuntia bahamana............	204
255.	Flower of Opuntia bahamana...........	204
256.	Plants of Opuntia macracantha........	204
257.	Potted plant of Opuntia macracantha...	205
258.	Potted plant of Opuntia spinosissima....	205
259.	Plants of Opuntia millspaughii.........	206
260.	Plant of Opuntia moniliformis.........	206
261.	Plant of Opuntia moniliformis.........	207
262.	Plant of Opuntia moniliformis.........	207
263.	Plants of Opuntia rubescens...........	208
264.	Plants of Opuntia rubescens...........	208
265.	Proliferous fruits of Opuntia rubescens..	209
266.	Joint of Opuntia rubescens............	209
267.	Fruit of Opuntia brasiliensis...........	209
268.	Plant of Opuntia brasiliensis..........	210
269.	Branch of Opuntia bahiensis...........	210
270.	Joint and fruit of Opuntia bahiensis.....	210
271.	Plant of Opuntia bahiensis............	211
272.	Plant of Opuntia ammophila...........	211
273.	Fruiting joint of Opuntia ammophila....	211
274.	Flower of Opuntia argentina..........	212
275.	Potted plant of Opuntia chaffeyi.......	212
276.	Plant of Opuntia chaffeyi..............	213

		PAGE.
FIG. 277.	Small joint of Nopalea gaumeri........	216
278.	Elongated joint of Nopalea gaumeri....	216
279.	Plant of Opuntia depauperata.........	216
280.	Joint of Opuntia depauperata.........	217
281.	Plant of Opuntia pestifer.............	217
282.	Plant of Opuntia discolor.............	218
283.	Joints of Opuntia pestifer.............	218
284.	Joint of Opuntia discolor.............	218
285.	Joint of Opuntia guatemalensis........	219
286.	Joint of Opuntia pennellii.............	219
287.	Joints of Opuntia caracasana..........	219
288.	Plant of Opuntia aequatorialis.........	220
289.	Joints of Opuntia aequatorialis........	220
290.	Joints of Opuntia lata...............	220
291.	Fruits of Opuntia lata................	220
292.	Joint with flower of Opuntia macateei..	220
293.	Joint of Opuntia macateei.............	220
294.	Plants of Opuntia soederstromiana......	221
295.	Plants of Opuntia zebrina.............	222
296.	Fruit of Opuntia zebrina..............	222
297.	Plants of Opuntia keyensis............	223
298.	Section of flower of Opuntia keyensis....	223
299.	Flower of Opuntia keyensis............	223
300.	Joint of Opuntia bonplandii...........	224
301.	Plant of Opuntia dobbieana...........	224
302.	Plant of Opuntia dobbieana (without legend)............................	225

THE CACTACEAE

Descriptions and Illustrations of Plants of the Cactus Family

THE CACTACEAE.

INTRODUCTION.

The writers began field, greenhouse, and herbarium studies of the Cactaceae in 1904 and in the years following they made studies and collections over wide areas in the United States, Mexico, and the West Indies. It was first intended that these should be followed by a general description of the North American species only, but a plan for a more complete investigation of the family was proposed by Dr. D. T. MacDougal in January 1911. This was approved by the trustees of the Carnegie Institution of Washington at its next regular meeting and a grant was made to cover the expenses of such an investigation. Dr. Rose was given temporary leave of absence from his position as Associate Curator in charge of the Division of Plants, United States National Museum, and became a Research Associate in the Carnegie Institution of Washington, with William R. Fitch and Paul G. Russell as assistants; Dr. Britton, Director-in-Chief of the New York Botanical Garden, was appointed an honorary Research Associate, while R. S. Williams, of the New York Botanical Garden, was detailed to select and preserve the specimens for illustration. Work under this new arrangement was begun January 15, 1912, and thus several lines of investigation were undertaken in a comprehensive way.

1. *Reexamination of type specimens and of all original descriptions:* This was necessary because descriptions had been incorrectly interpreted, plants had been wrongly identified, and the errors perpetuated; thus the published geographical distribution of many species was faulty and conclusions based on such data were unreliable. Not only had specific names been transferred to plants to which they did not belong, but generic names were interchanged and the laws of priority ignored. Many valid species, too, had dropped out of collections and out of current literature and had to be restored.

2. *Assembling of large collections for greenhouse and herbarium use:* Extensive greenhouse facilities were furnished by the New York Botanical Garden and the United States Department of Agriculture, while the herbaria and libraries of the United States National Museum and of the New York Botanical Garden furnished the bases for the researches. The New York Botanical Garden has also cooperated in contributing funds in aid of the field operations, in clerical work, and a large number of the illustrations used have been made there, the paintings and line drawings mostly by Miss Mary E. Eaton.

3. *Extensive field operations in the arid parts of both Americas:* Many of these deserts are almost inaccessible, while the plants are bulky and if not handled carefully are easily destroyed. Many plants require several years to mature, in some cases many years to flower in cultivation. Through these explorations were obtained the living material for the greenhouse collections and for exchange purposes, as well as herbarium material for permanent preservation. Of much importance, also, were field observations upon the plants as individuals, their form, habit, habitat, and their relations to other species.

3

Early in 1912 Dr. Rose went to Europe to study the collections there and to arrange for exchanges with various botanical institutions having collections of these plants. He spent considerable time at London, mainly at the Royal Botanic Gardens, Kew, where through the courtesy of the Director, Sir David Prain, he was able to examine the greenhouse, illustrative, and herbarium material for which this institution has long been famed. The collection at the British Museum of Natural History and that of the Linnaean Society of London were examined. At Paris he studied the collections at the Natural History Museum, many of which have historic interest; one of his interesting discoveries there was that the *Pereskia bleo*, collected by Baron Friedrich Alexander von Humboldt in Colombia, is a very different species from the plant which for nearly a century has been passing in our collections and literature under that name. He also visited the famous botanical garden of the late Sir Thomas Hanbury, at La Mortola, Italy, and through the courtesy of Lady Hanbury was given every possible facility for the study of this collection; Mr. Alwin Berger, who was then curator in charge, had brought together one of the most extensive representations of this family to be found growing in the open in any place in the world. Here in the delightful climate of the Riviera were grown many species which were apparently just as much at home as they would have been in their desert habitats. Dr. Rose also visited Rome, Naples, Venice, and Florence, where he saw smaller collections in parks and private gardens. At Munich he examined certain types in the Royal Botanical Museum, then under the charge of Dr. L. Radlkofer, and saw some interesting species in the Royal Botanical Garden then being organized by Dr. K. Goebel. At Berlin he examined the herbarium and living specimens in the Berlin Botanical Garden, through the courtesy of Dr. A. Engler, and the West Indian collection through the courtesy of Dr. I. Urban. He then went to Halle and saw L. Quehl's collection of mammillarias; to Erfurt, where he saw the Haage and Schmidt, and Haage Jr. collections; to Darmstadt to see the Botanical Garden under Dr. J. A. Purpus; and to Antwerp to see DeLaet's private collection.

In 1913 Dr. Britton and Dr. Rose visited the West Indies. Dr. Britton, who was accompanied by Mrs. E. G. Britton, Miss D. W. Marble, and Dr. J. A. Shafer, collected on St. Thomas and the other Virgin Islands, Porto Rico, and Curaçao. At the latter island he rediscovered the very rare *Cactus mammillaris*, which had not been in cultivation for many years. Dr. Rose, who was accompanied by William R. Fitch and Paul G. Russell, also stopped at St. Thomas, and collected on St. Croix, St. Christopher, Antigua, and Santo Domingo.

In 1914 and 1915 Dr. Britton again visited Porto Rico and, assisted by Mr. John F. Cowell and Mr. Stewardson Brown, explored the entire southwestern arid coast and the small islands Desecheo, Mona, and Muertos.

In 1914 Dr. Rose went to the west coast of South America, making short stops at Jamaica and Panama. He made extensive collections in central and southern Peru, central Bolivia, and northern and central Chile. At Santiago, Chile, he examined a number of Philippi's types in the National Museum and obtained some rare specimens from the Botanical Garden through the courtesy of Johannes Söhrens.

In 1915 Dr. Rose, accompanied by Paul G. Russell, visited Brazil and Argentina on the east coast of South America, collecting extensively in the semiarid parts of Bahia, Brazil, and in the region about Rio de Janeiro, so rich in epiphytic cacti. In the deserts about Mendoza and Córdoba, in Argentina, collections were also made. Here he also arranged for exchanges with the leading botanists and collectors. The following persons have made valuable contributions from the regions visited: Dr. Leo Zehntner, Joazeiro, Brazil; Dr. Alberto Löfgren, Rio de Janeiro, Brazil; Dr. Carlos Spegazzini, La Plata, Argentina; Dr. Cristóbal M. Hicken, Buenos Aires, Argentina; and Dr. Carlos S. Reed, Mendoza, Argentina.

In October and November 1916, Dr. Rose, accompanied by Mrs. Rose, visited Curaçao and Venezuela, studying especially the cactus deserts about La Guaira and Puerto Cabello. A number of photographs were taken by Mrs. Rose.

While *en route* for Venezuela, arrangements were made with Mr. Harold G. Foss to make a collection of cacti at Coro, Venezuela. Among the specimens obtained were species not found farther east in Venezuela, so far as known.

In 1916 Dr. Britton, assisted by Mr. Percy Wilson, studied the cacti of Havana and Matanzas Provinces and those of the Isle of Pines, Cuba.

In 1918 Dr. Rose, assisted by George Rose, visited Ecuador on behalf of the United States Department of Agriculture, aided by the Gray Herbarium of Harvard University and the New York Botanical Garden; about thirty rare or little-known species were obtained.

Through the expenditure of about $2,400, contributed by Dr. Britton, a very important collection of cacti was made by Dr. J. A. Shafer during a six months' exploration from November 1916 to April 1917 of the desert regions of northwestern Argentina, southeastern Bolivia, northeastern Argentina, and adjacent Uruguay and Paraguay. Fortunately, for the purposes of this work, this collection was brought back to New York by Dr. Shafer in time for the information yielded by it and by his field observations to be used in the manuscript. It has given us first-hand information concerning over 120 species of cacti as to which we have previously known little.

There are still a few cactus regions which ought to be explored, but the following summary will show the wide field from which we have obtained information.

Our field investigations have covered practically all the cactus deserts of Mexico. The most important of these are the vicinities of Tehuacán and Tomellín, the plains of San Luis Potosí, the chalky hills surrounding Ixmiquilpan, the lava fields in the Valley of Mexico and above Cuernavaca, the deserts of Querétaro, the west coast of Mexico extending from the United States border to Acaponeta, and the seacoasts and islands of Lower California. Other regions in Mexico containing cacti, but not in such great abundance as the foregoing, are those about Pachuca, Oaxaca City, Mitla, Jalapa, Iguala, Chihuahua City, and Guadalajara. All the work in Mexico, however, was done prior to 1912, for, owing to political disturbances, no field work there has been feasible since that time.

In the United States our work has extended over the cactus regions of Florida, Texas, New Mexico, Arizona, southern California, western Kansas, and southeastern Colorado.

In the West Indies we have explored all of the Greater Antilles, the Bahamas, the Virgin Islands, St. Christopher, Antigua, Barbados, and Curaçao.

In South America our field study included the most important deserts of Peru, Bolivia, and Chile, and parts of Brazil, Venezuela, Ecuador, and Argentina. The cactus deserts of South America are so extensive and so remote from one another that it was possible to visit only a part of them in the four seasons allowed for their exploration.

Among many enthusiastic volunteers whose contributions of specimens and data have greatly supplemented our own collections and field studies, the following deserve especial mention:

Mr. Henry Pittier has made valuable sendings from Colombia, Venezuela, Panama, Costa Rica, and Mexico; Mr. O. F. Cook, from Guatemala and Peru; Mr. G. N. Collins, the late Federico Eichlam, Mr. R. H. Peters, Mr. C. C. Deam, Mrs. T. D. A. Cockerell, Baron H. von Türckheim, and the late Professor W. A. Kellerman have sent important collections from Guatemala; Mr. A. Tonduz, Mr. Otón Jiménez, Dr. A. Alfaro, Mr. C. Wercklé, and Mr. Alfred Brade, local collectors and naturalists in Costa Rica, have sent much good material from their country; Mr. William R. Maxon has sent new and rare material from Costa Rica, Guatemala, and Cuba; Professor C. Conzatti and his son, Professor Hugo Conzatti, Dr. C. A. Purpus, Dr. Elswood Chaffey, Mrs. Irene Vera, M. Albert de Lautreppe, and the late Mr. E. A. H. Tays have sent us many interesting specimens from Mexico; Mr. W. E. Safford made a valuable collection in Mexico in 1907; E. W. Nelson and E. A. Goldman, who have collected so extensively in Mexico and the Southwest, have obtained many herbarium and living specimens for our use; Mrs. Gaillard, who lived at Panama several years while the late Colonel D. D. Gaillard was a member of the Isthmian Canal Commission, collected interesting cacti, including *Epiphyllum gaillardae;* the late Dr. H. E. Hasse sent specimens from southern California and Arizona; C. R. Orcutt, the well-known cactus fancier, has aided us in many ways besides sending us specimens from his collections; Dr. R. E. Kunze has frequently sent specimens, especially from Arizona; General Timothy E. Wilcox, for whom *Wilcoxia* was named, has sent us specimens from the Southwest, while his son, Dr. G. B. Wilcox, contributed several sendings from the west coast of Mexico and Guatemala; Dr. D. T. MacDougal has sent many specimens from all over the Southwest, especially from Mexico, Arizona, and southern California; he has made several excursions into remote deserts, which have yielded interesting results, and has contributed many excellent photographs, quite a number of which are reproduced in this report (Plate 1, etc.). Professor F. E. Lloyd, while located in Arizona and in Zacatecas, Mexico, made large collections of living, herbarium, and formalin material, often accompanied by valuable field notes, sketches, and photographs. Dr. Forrest Shreve has sent specimens, especially from northern Arizona and Mr. W. H. Long from New Mexico; Mr. S. B. Parish and Mr. W. T. Schaller have furnished interesting specimens and valuable notes on southern California species; Professor J. J. Thornber has made valuable contributions of material and notes from Arizona; Mr. M. E. Jones, Mr. I. Tidestrom, Mr. Thomas H.

Kearney, and Professor A. O. Garrett have all sent specimens from Utah; Professor T. D. A. Cockerell and Mr. Merritt Cary have sent specimens from Colorado; Dr. P. A. Rydberg has brought many specimens from the Rocky Mountain region; Messrs. Paul C. Standley, E. O. Wooton, Vernon Bailey, and H. L. Shantz have sent specimens from the southwestern United States; Brother León, of the Colegio de la Salle, Havana, and Dr. Juan T. Roig, of the Estación Agronómica, Santiago de las Vegas, Cuba, have contributed Cuban specimens, and Dr. J. A. Shafer has collected widely in Cuba; Mr. William Harris, of Hope Gardens, Jamaica, has collected for us in Jamaica; Dr. John K. Small has obtained collections from nearly all over the southeastern United States, aided by Mr. Charles Deering. Dr. Henry H. Rusby and Dr. Francis W. Pennell have contributed plants and specimens from Colombia, collected in 1917 and 1918. Mr. Frederick V. Coville, of the United States Department of Agriculture, has made many valuable suggestions during the progress of the investigation.

In our studies we have also had use of the cacti of the following American collections: Herbarium of the Missouri Botanical Garden at St. Louis; the Gray Herbarium of Harvard University; the Rocky Mountain Herbarium at Laramie, Wyoming; the collection of the United States Department of Agriculture; the herbarium of the University of California, especially the Brandegee collection; and the herbarium of the Field Museum of Natural History.

The types of the new species described in this work are deposited in the herbaria of the New York Botanical Garden and the United States National Museum, unless otherwise indicated.

In greenhouse collections many kinds of cacti grow very slowly, and flower only after many years' cultivation. We have a number of plants of this kind from various parts of America. It is hoped that some of them may bloom during the period of publication of this book and thus enable us to include them in an appendix.

FIG. 1.—Pereskia pereskia. Grown as a hedge.

Order CACTALES.

Perennial, succulent plants, various in habit, mostly very spiny, characterized by specialized organs termed areoles. Leaves usually none, except in *Pereskia* and *Pereskiopsis*, where they are large and flat but fleshy, and in *Opuntia* and its relatives, where they are usually much reduced and mostly caducous, terete, or subulate. Spines very various in size, form, arrangement, and color, sometimes with definite sheaths. The areoles are peculiar and complex organs, situated in the axils of leaves when leaves are present, and bearing the branches, flowers, spines, glochids, hairs, or glands; in some genera two kinds of areoles occur, either distinct or united by a groove. Flowers usually perfect, either regular or irregular, usually solitary but sometimes clustered, sometimes borne in a specialized terminal dense inflorescence called a cephalium; perianth-tube none, or large and long, the limb spreading or erect, short or elongated, the lobes few or numerous, often inter-grading in shape and color, but sometimes sharply differentiated into sepals and petals; stamens commonly numerous, elongated or short, sometimes clustered in series, the filaments usually borne on the throat of the perianth, the small oblong anthers 2-celled; style one, terminal, short or elongated; stigma-lobes 2 to many, usually slender; ovary 1-celled, distinct, or immersed in a branch or forming a part of a branch; ovules numerous. Fruit a berry, often juicy and sometimes edible, sometimes dry, in one species described as capsular and dehiscing by an operculum, in others opening by a basal pore. Seeds various; cotyledons two, accumbent, sometimes minute knobs, often broad or elongated; endosperm little or copious; radicle terete.

The order consists of the following family only:

Family CACTACEAE Lindley, Nat. Syst. ed. 2. 53. 1836.

Characters of the order as given above. The family is composed of three tribes.

KEY TO TRIBES.

Leaves broad, flat; glochids wanting; flowers stalked (sometimes short-stalked), often clustered...... 1. *Pereskieae*
Leaves (except in *Pereskiopsis*) terete or subterete, usually small, often wanting on the vegetative parts;
   flowers sessile.
    Areoles with glochids (except in *Maihuenia*); vegetative parts bearing leaves, which are usually
      small and fugacious; flowers rotate (petals erect in *Nopalea*).......................... 2. *Opuntieae*
    Areoles without glochids; usually no leaves on the vegetative parts (except cotyledonary); flowers
      with definite tubes (except *Rhipsalis*)... 3. *Cereeae*

Tribe 1. PERESKIEAE.

Stems and foliage as in other dicotyledonous plants; inflorescence in some species compound; flowers more or less stalked, their parts all distinct; glochids wanting; ovule with short funicle; testa of seed thin, brittle.

The genus *Pereskia*, the only representative of this tribe, is, on account of its similarity to other woody flowering plants, considered the nearest cactus relative to the other families, but this relationship is in all cases remote.

The nearest generic relatives of *Pereskia* in the cactus family are doubtless the following:

Pereskiopsis, some of whose species were first assigned to the genus *Pereskia*, but they have different foliage and the areoles often bear glochids.

Opuntia, whose species have leaves, though much reduced and usually caducous, otherwise very different; but some of the species of *Opuntia* were first referred to *Pereskia*.

Maihuenia (two of whose species have only recently been taken out of *Pereskia*), whose seeds are similar but the areoles lack glochids, otherwise very different.

This tribe has a wide geographic distribution, but is found wild only in the tropics.

1. PERESKIA (Plumier) Miller, Gard. Dict. Abr. ed. 4. 1754.

Leafy trees, shrubs, or sometimes clambering vines, branching and resembling other woody plants; spines in pairs or in clusters in the axils of the leaves, neither sheathed nor barbed; glochids (found only in the Opuntieae) wanting; leaves alternate, broad, flat, deciduous, or somewhat fleshy; flowers solitary, corymbose, or in panicles, terminal or axillary, wheel-shaped; stamens numerous; style single; stigma-lobes linear; seeds black, glossy, with a brittle shell, the embryo strongly curved; the cotyledons leafy; seedlings without spines.

Type species: *Cactus pereskia* Linnaeus.

In 1898 about 25 names had been proposed in *Pereskia*, but, in his monograph published that year, Karl Schumann accepted only 11 species. Several new ones have been proposed since the publication of Schumann's monograph.

The species are native in Mexico, the West Indies, Central America, and South America. Some of the species are much used as stocks for growing the various forms of *Zygocactus*, *Epiphyllum*, and other cacti requiring this treatment; *P. pereskia* is most used and *P. grandifolia* next. Several species are widely cultivated as ornamentals in tropical regions; they do not flower freely under glass in northern latitudes. All species studied by us in the living state grow readily from cuttings.

The typical species seems to have been first introduced into Europe from the West Indies in the latter part of the sixteenth century. A straight-spined species was first described and figured by L. Plukenet in 1696, who called it a portulaca, and the next year by Commerson as an apple (*Malus*). In 1703 C. Plumier described the genus *Pereskia*, basing it upon a single species. The genus was repeatedly recognized by Linnaeus in his earlier publications, and by some pre-Linnaean botanists, but in 1753 Linnaeus merged it into *Cactus* along with a number of other old and well-established genera; but it was retained by Philip Miller in 1754 in the fourth edition (abridged) of his Gardeners' Dictionary and has since been generally recognized as a genus by botanical and horticultural authors.

The name is variously spelled *Peirescia*, *Peireskia*, *Perescia*, and *Pereskia*.

Named for Nicolas Claude Fabry de Peiresc (1580–1637).

KEY TO SPECIES.

```
Climbing vines, the twigs with a short pair of reflexed spines from each areole, the stem with
                       acicular spines (Series 1. Typicae)........................  1. P. pereskia
Shrub or trees with slender straight spines (Series 2. Grandifoliae).
  Petals toothed or fimbriate.
    Petals somewhat toothed.................................................  2. P. autumnalis
    Petals fimbriate.
      Species from Mexico; ovary turbinate................................  3. P. lychnidiflora
      Species from Costa Rica; ovary pyriform.............................  4. P. nicoyana
  Petals entire, at least not fimbriate.
    Branches and leaves very easily detached.............................  5. P. zehntneri
    Branches and leaves not easily detached.
      Axils of sepals bearing long hairs and bristles.
        Leaves lanceolate................................................  6. P. sacharosa
        Leaves orbicular.................................................  7. P. moorei
      Axils of sepals not bearing long hairs and bristles.
        Flowers white....................................................  8. P. weberiana
        Flowers not white.
          Petals yellow.
            Leaves lanceolate to oblong or obovate.......................  9. P. guamacho
            Leaves orbicular or broadly ovate............................ 10. P. colombiana
          Petals red or purple.
            Spines few or none........................................... 11. P. tampicana
            Very spiny, at least on old branches.
              Flowers terminal.
                Flowers panicled.
                  Fruit naked, broadly truncate........................... 12. P. bleo
                  Fruit leaf-bearing, not truncate.
                    Leaves of ovary cuneate at base....................... 13. P. bahiensis
                    Leaves of ovary broad at base......................... 14. P. grandifolia
                Flowers solitary.......................................... 15. P. zinniaeflora
              Flowers usually axillary and solitary.
                Leaves 1 cm. long or longer, obtuse or acute.
                  Flowers 2 to 5 together, 1 cm. long; South American species.... 16. P. horrida
                  Flowers solitary, 1.5 cm. long; petals elliptic-obovate; Cuban
                       species.............................................. 17. P. cubensis
                Leaves emarginate, 1 cm. long or less, petals obovate............ 18. P. portulacifolia
Affinity unknown........................................................... 19. P. conzattii
```

Series 1. TYPICAE.

Consists of only the typical species, which is widely distributed, and much cultivated throughout tropical America. Schumann regarded it as a subgenus under the name *Eupereskia*.

1. Pereskia pereskia (Linnaeus) Karsten, Deutsch. Flora 888. 1882.

> *Cactus pereskia* Linnaeus, Sp. Pl. 469. 1753.
> *Pereskia aculeata* Miller, Gard. Dict. ed. 8. 1768.
> *Cactus lucidus* Salisbury, Prodr. 349. 1796.
> *Pereskia longispina* Haworth, Syn. Pl. Succ. 178. 1812.
> *Pereskia aculeata longispina* De Candolle, Prodr. 3: 475. 1828.
> *Pereskia fragrans* Lemaire, Hort. Univ. 2:.40. 1841.
> *Pereskia undulata* Lemaire, Illustr. Hort. 5: Misc. 11. 1858.
> *Pereskia foetens* Spegazzini in Weingart, Monatsschr. Kakteenk. 14: 134. 1904.
> *Pereskia godseffiana* Sander, Gard. Chron. III. 43: 257. 1908.

Shrub, at first erect, but the branches often long, clambering, and forming vines 3 to 10
meters long; spines on lower part of stem solitary or 2 or 3 together, slender and straight; spines
in the axils of the leaves paired, rarely in threes, short, recurved; leaves short-petioled, lanceolate
to oblong, or ovate, short-acuminate at the apex, tapering or rounded at base, 7 cm. long or less;
flowers in panicles or corymbs, white, pale yellow, or pinkish, 2.5 to 4.5 cm. broad; ovary leafy and
often spiny; fruit light yellow, 1.5 to 2 cm. in diameter, when mature quite smooth; seeds black,
somewhat flattened, 4 to 5 mm. in diameter; hilum basal, circular, depressed, or crater-shaped.

The plant and fruit have several common names, one of which, blade apple, was in use
as early as 1697. Lemon vine, Barbados gooseberry, and West Indian gooseberry are three
others, with various French and Dutch modifications. In Argentina it is called sacharosa,
according to Sir Joseph Hooker (Curtis's Bot. Mag. 116: pl. 7147), but this name is properly applied only to the *P. sacharosa* of Grisebach, native of Argentina, a distinct species,
which Hooker thought identical with this.

The berries are eaten throughout the West Indies and the leaves are used as a pot herb
in Brazil. The species was in cultivation in the Royal Gardens of Hampton Court in 1696
and has been at Kew ever since its establishment in 1760, but did not flower until 1889.
In Washington we have one plant among a dozen which flowers abundantly each year;
three plants at New York bloom annually.

In tropical America the plant climbs over walls, rocks, and trees, and at flowering time
is covered with showy, fragrant blossoms, followed by beautiful clusters of yellow berries.
In La Plata it is grown sometimes for hedges (see fig. 1), but its strong, almost offensive
odor makes it objectionable for growing near habitations.

Type locality: Tropical America.

Distribution: West Indies and along the east and north coasts of South America; found
also in Florida and Mexico, but perhaps only as an escape; widely grown for its fruit.

This species consists of several races, differing in shape and size of the leaves and in
color of the flowers. One of these races, with ovate-orbicular leaves rounded at the base,
had heretofore been known to us only in cultivation, but in October 1916, while collecting
in Venezuela, Dr. Rose found this broad-leafed form common in the coastal thickets near
Puerto Cabello.

Pereskia lanceolata (Förster, Handb. Cact. 513. 1846), *P. acardia* Parmentier (Pfeiffer,
Enum. Cact. 176. 1837), and *P. brasiliensis* Pfeiffer (Enum. Cact. 176. 1837), usually
referred as synonyms of *P. aculeata*, were not formally published in the places above cited.

The following varieties, based on the shape of the leaves, are recorded under *P. aculeata:*
lanceolata Pfeiffer (Enum. Cact. 176. 1837); *latifolia* Salm-Dyck (Hort. Dyck. 202. 1834,
name only); *rotundifolia* Pfeiffer (Enum. Cact. 176. 1837); *rotunda* (Suppl. Dict. Gard.
Nicholson 589. 1901) is perhaps the same as *rotundifolia*.

Pereskia aculeata rubescens Pfeiffer (Enum. Cact. 176. 1837) is described with glaucous-
green leaves above, tinged with red beneath.

Near the last belongs *Pereskia godseffiana*, described as a sport in the Gardeners'
Chronicle in 1908. It is a very attractive greenhouse plant, often forming a round,

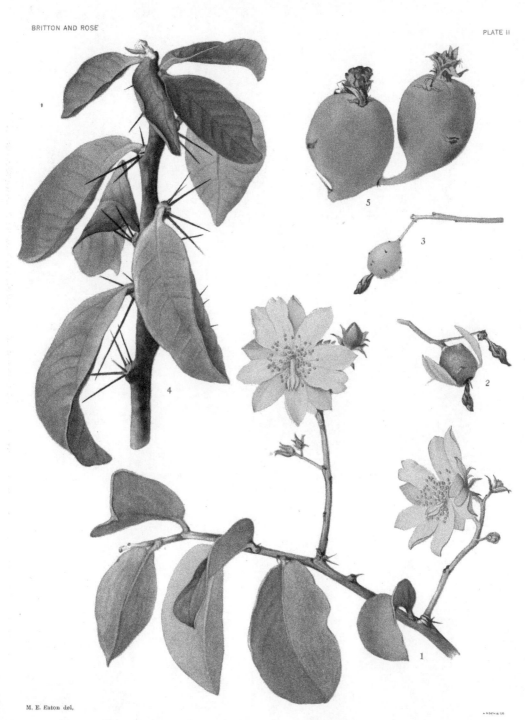

M. E. Eaton del.

1. Flowering branch of *Pereskia pereskia*.   4. Leafy branch of *Pereskia sacharosa*.
2, 3. Fruits of the same.   5. Proliferous fruit of the same.
(All natural size.)

densely branched bush, but is sometimes grown as a climber, as a basket plant, or in the form of a pyramid. It is especially distinguished by the rich coloration of the leaves, which are variously mottled or blotched above with crimson, apricot-yellow, and green, but of a uniform purplish crimson beneath. We have seen this form in the New York Botanical Garden, where it is grown only as a bush. It was exhibited first at Ghent, Belgium, in 1908, and is supposed to have originated in Queensland, Australia.

Illustrations: Stand. Cycl. Hort. Bailey 5: pl. 87; Blühende Kakteen 2: pl. 86; Bot. Reg. 23: pl. 1928; Curtis's Bot. Mag. 116: pl. 7147; Gard. Chron. III. 29: f. 61; Plumier, Nov. Pl. Amer. pl. 26, in part; Safford, Ann. Rep. Smiths. Inst. 1908: f. 10; Schumann, Gesamtb.

Fig. 2.—Pereskia autumnalis.

Kakteen f. 109, all as *P. aculeata.* Descourtilz, Fl. Med. Antill. ed. 2. 4: pl. 294, as Cactier à Fruits Feuilles; Vellozo, Fl. Flum. 5: pl. 26, as *Cactus pereskia;* Gard. Chron. III. 43: f. 114, as *P. godseffiana.*

Plate 11, figure 1, of this volume is a flowering branch of a plant at the New York Botanical Garden obtained from M. Simon, of St. Ouen, Paris, France, in 1901; figure 2, fruit of same plant; figure 3, fruit of another plant. Text-figure 1, from a photograph taken by Paul G. Russell at La Plata, Argentina, in September 1915, shows the plant used as a hedge.

Series 2. GRANDIFOLIAE.

In this series we include 18 species, all tropical American, both continental and insular. Schumann, regarding the series as a subgenus, applied to it the name *Ahoplocarpus.*

2. Pereskia autumnalis (Eichlam) Rose, Contr. U. S. Nat. Herb. 12: 399. 1909.

Pereskiopsis autumnalis Eichlam, Monatsschr. Kakteenk. 19: 22. 1909.

Tree, 6 to 9 meters high, with a large, round, much branched top, the trunk usually very definite and 40 cm. or more in diameter, often covered with a formidable array of spines; young branches cherry-brown, smooth; spines in the axils of the leaves usually solitary, sometimes in threes, long and slender, 3 to 4 cm., rarely 16 cm. long; leaves thickish, oblong to orbicular, 4 to 8 cm. long, rounded or somewhat narrowed at base, mucronately tipped; flowers solitary, near the tops of the branches, short-peduncled; ovary covered with leafy scales; flowers 4 to 5 cm. broad; petals entire, orange-colored; stamens numerous; fruit globular, 4 to 5 cm. in diameter, fleshy, glabrous, bearing small, scattered leaves, these naked in the axils; seeds black, glossy, 4 mm. long.

Type locality: In Guatemala.

Distribution: Widely distributed in Guatemala, usually at an altitude of 120 to 300 meters; but we do not know of its occurrence elsewhere.

Fig. 3.—Pereskia autumnalis. ×0.5.   Fig. 4.—Pereskia lychnidiflora.

The plant, so far as we know, has no common name and no use is made of its fruit.

Illustrations: Contr. U. S. Nat. Herb. **12**: pl. 52 to 54; Safford, Ann. Rep. Smiths. Inst. **1908**: pl. 10, f. 1; Monatsschr. Kakteenk. **21**: 37, the last as *Pereskiopsis autumnalis*.

Text-figures 2 and 3 are copied from the above-cited illustrations. The original photographs were obtained by O. F. Cook in Guatemala.

3. Pereskia lychnidiflora De Candolle, Prodr. 3: 475. 1828.

Evidently a tree or shrub; branches cylindric, woody; leaves large, 4 to 7 cm. long, oval to oblong, pointed, rounded at base, sessile, fleshy, with a prominent midvein; axils of leaves each bearing a stout spine 2 to 5 cm. long and several long hairs; flowers large, 6 cm. broad, solitary, borne at the ends of short, stout branches; petals broadly cuneate, laciniate at the apex; ovary turbinate, bearing small leaves.

Type locality: In Mexico.

Distribution: Mexico.

This species was described by De Candolle from Mociño and Sessé's drawing, but it has never been collected since, so far as we can learn. Its large flowers with laciniate petals must make this a very striking species and it is surprising that it has not been rediscovered. Schumann thought it might be the same as *P. nicoyana* of Costa Rica, but a study of recent Costa Rican collections indicates that the species are distinct. The measurements given in the description are taken from De Candolle's plate, and may require some modification. *Cactus fimbriatus* Mociño and Sessé (De Candolle, Prodr. **3**: 475. 1828) was published only as a synonym of this species.

Illustrations: Mém. Mus. Hist. Nat. Paris **17**: pl. 18; Förster, Handb. Cact. ed. 2. 1003. f. 136; Safford, Ann. Rep. Smiths. Inst. **1908**: 545. f. 11.

Text-figure 4 is copied from the first illustration above cited.

4. Pereskia nicoyana Weber, Bull. Mus. Hist. Nat. Paris **8**: 468. 1902.

Tree, usually about 8 meters high; branches rigid, stout, covered with smooth brown bark; spines wanting or single, long (4 cm. long), stout and porrect; leaves in fascicles on old branches, but alternate on young shoots, lanceolate or oblanceolate, subsessile, the lateral veins almost parallel and some-

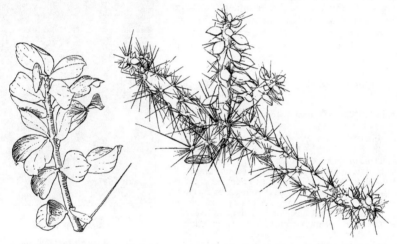

FIG. 5.—Pereskia nicoyana. ×0.6.   FIG. 6.—Pereskia zehntneri. ×0.6.

times seeming to come from the base, acute, bright green, and somewhat shining; axils of the young leaves containing long white hairs; petals reddish yellow, fimbriate; ovary pyriform, bearing 8 to 12 spreading leaves, except the uppermost ones, which are much smaller and connivent.

Type locality: Gulf of Nicoya, Costa Rica.

Distribution: Costa Rica.

The spines, hairs in the axils of the leaves, and fimbriate petals indicate a relationship to the little-known *P. lychnidiflora*.

Mr. H. Pittier informs us that this species is common in the open coastal forests along the Pacific side of Costa Rica. The plant illustrated by Mr. Pittier, referred to below, has a long, slender trunk and is very spiny.

According to Mr. W. E. Safford, it has long, slender spines and the habit of the Osage orange, and is used as a hedge plant in Costa Rica, where it is known as matéare or puipute.

Illustration: Pittier, Pl. Usuales Costa Rica pl. 2.

Text-figure 5 was drawn from a plant obtained by Mr. C. Wercklé at San José, Costa Rica, in 1912.

5. Pereskia (?) zehntneri sp. nov.

Shrub, 2 to 3 meters high, with a central erect trunk, very spiny; branches numerous, horizontal, usually in whorls, sometimes as many as 10 in a whorl; branches terete, green, fleshy, very easily detached from the stem; leaves stiff, fleshy, numerous, small, 2 to 4 cm. long, ovate to orbicular, acute, standing at right angles to the branches; areoles large, filled with short white wool and numerous slender white spines; flowers at tops of branches, large, 7 to 8 cm. broad, bright red, appearing in November; petals broad, retuse; ovary borne in the upper end of the branch, very narrow, 3 to 4 cm. long, bearing the usual leaves, areoles, and spines of the branches.

Collected by Dr. Leo Zehntner (Nos. 567 and 630, type) November 15 and 16, 1912, at Bom Jesus da Lapa, Bahia, Brazil, on the Rio São Francisco.

This is a very rare plant and seen in only one locality, in soil of a peculiar chalky formation.   Living plants were taken by Dr. Zehntner to the Horto Florestal, Joazeiro, Brazil, where they grew well, and whence Dr. Rose obtained specimens in 1915 which were shipped to the United States under No. 19722.

The plant is known in Bahia under the name quiabento.   It is probably not a true *Pereskia;* it suggests in its habit and foliage some of the Mexican species of *Pereskiopsis,* but it may represent a distinct genus.

Text-figures 6 and 7 are from the type plant above cited.

FIG. 7.—Pereskia zehntneri.   Photograph by Paul G. Russell.

6. Pereskia sacharosa Grisebach, Abh. Ges. Wiss. Göttingen **24**: 141.   1879.

> *Pereskia amapola* Weber, Dict. Hort. Bois 938. 1898.
> *Pereskia argentina* Weber, Dict. Hort. Bois 938. 1898.

Small tree or shrub, 6 to 8 meters high; branches green and smooth, but in age becoming yellowish or light brown; leaves lanceolate to oblanceolate, 8 to 12 cm. long, cuneate at base, more or less pointed at apex; young areole with 1 to 3 spines, the longest 5 cm. long, the others when present not over half as long, all acicular and dark in age; older areoles often with 6 or more spines; pedicels sometimes 10 mm. long; flowers in terminal clusters, either white or rose-colored and very showy, 8 cm. broad, open at midday; sepals about 8, 1 or 2 petal-like, the others scale-like, the outer sepals and upper scales bearing long hairs; petals 8, rose-colored, oblanceolate, 3 cm. long; stamens free from the petals, numerous, unequal, erect; filaments, style, and stigma-lobes white; ovules borne on the lower part of ovary; ovary bearing small leaves, their axils filled with short wool and occasionally bearing a spine; fruit hard, 2.5 to 4 cm. in diameter, more or less tapering at base, many-seeded, leafless or nearly so, sometimes proliferous.

Type locality: Cobos, Oran, Argentina.

Distribution: Paraguay and Argentina.

Schumann (Gesamtb. Kakteen 765. 1898) gives *Opuntia sacharosa* Grisebach as a synonym of this species, but erroneously, since it was never taken up by Grisebach as an *Opuntia.*   The Index Kewensis refers this species to *P. aculeata,* doubtless following Hooker's references in Curtis's Botanical Magazine for 1890 in regard to Argentine plants, which even then were little known.

The common name of this plant in Argentina is sacharosa.   It is sometimes used as a hedge plant.

Plate II, figure 4, represents a leafy branch of a plant given to the New York Botanical Garden by Frank Weinberg in 1903; figure 5 shows its fruit.

7. Pereskia moorei sp. nov.

A much branched shrub about 1 meter high, covered with brown bark; branches stout; leaves orbicular or obovate-oblong, rounded or apiculate at the apex, somewhat cuneately narrowed at the base, thick and fleshy, 4 to 8 cm. long, 3.5 to 6 cm. wide; areoles suborbicular, 4.5 mm. in diameter, the wool gray; spines at each areole mostly 2 to 4, very unequal, the longest 7.5 cm. long or less, ashy gray, blackish toward the apex; flowers purplish red, about 4.5 cm. long; ovary few-leafed, its leaves obovate-oblong, 2.5 to 3 cm. long, bearing 1 to 3 black spines about 5 mm. long in the axils; sepals narrowly oblong-obovate, bluntly acute, 2.5 cm. long, bearing long bristles in their axils; petals obovate, obtuse, 3.5 cm. long, rose-colored; stamens about 2 cm. long; areoles on the ovary large, filled with a mass of short, white wool and bearing an occasional short spine; fruit not known.

FIG. 8.—Pereskia moorei.   ✕0.66.   FIG. 9.—Pereskia guamacho.

Described from the specimen preserved in the herbarium of the British Museum of Natural History collected at Corumba, Brazil, by Spencer Moore, No. 955, who has kindly furnished us with data for this description, together with a sketch of the type specimen. Specimens were also collected at Corumba by F. C. Hoehne in 1908, No. 4863, who supposed it to be *P. sacharosa*.

Figure 8 is from a photograph of an herbarium specimen from Matto Grosso, Brazil, received from F. C. Hoehne in 1915.

8. Pereskia weberiana Schumann, Gesamtb. Kakteen 762. 1898.

Shrubby, much branched, glabrous, 2 to 3 meters high, the slender round branches about 3 mm. thick; leaves ovate to elliptic, about 3 cm. long and 2 cm. wide, sessile, acute at the apex, obtuse at the base; areoles circular, slightly elevated, the wool short, whitish, fading brown; spines 3 to 6 at the lower areoles, solitary at the upper, 2 cm. long or less, terete, acicular, yellow or horn-colored; flowers clustered, white, about 1 cm. long or less; ovary about 2 mm. long, bearing a few white, woolly areoles; outer segments of the perianth triangular, acute, woolly at the axils, the inner spatulate to obovate; stamens a little longer than the petals; stigma-lobes 3 or 4, erect.

Type locality: Tunari Mountains,* Bolivia, at 1,400 meters altitude.

Distribution: Bolivia, known only from the type locality.

This species is said to flower in May.

The description has been drawn from a cotype in the herbarium of the New York Botanical Garden, and from Professor Schumann's original account of the species in his Gesamtbeschreibung der Kakteen, p. 762. Dr. Kuntze obtained the specimens during his botanical exploration of Bolivia in 1892. The species was named, but not described, by Professor Schumann in Dr. Kuntze's Revisio Genera Plantarum (3²: 107. 1893).

The material preserved is too imperfect to enable us to give an illustration of this plant.

FIG. 10.—Pereskia guamacho. ×0.8.

9. Pereskia guamacho Weber, Dict. Hort. Bois 938. 1898.

Plant very spiny, usually a small shrub 1 to 3 meters high, but often a tree 10 meters high with a trunk up to 4 dm. in diameter and 3 meters long or more below the much branched top; areoles rather prominent, especially in age often standing out like small knobs on the branches, filled with brown felt, at first usually with only 1 to 4 spines along with a few short accessory ones, but in age often with 20 spines or more; spines somewhat divaricate, rigid, brown, the longer ones often 4 cm. long; leaves on young branches solitary, but on old wood growing in fascicles, acute, lanceolate to ovate or obovate with cuneate bases, usually about 3 cm. long, but sometimes 5 to 9 cm. long by 3 to 6 cm. broad, fleshy; flowers probably solitary, but so thickly set along the branches as to appear almost spicate, sessile, bright yellow, 4 cm. broad; ovary covered with small, lanceolate-acuminate leaves, these hairy in the axils; stamens numerous; fruit globular, about 2 cm. in diameter, becoming naked, said to be orange-colored and edible; seeds black, flattened, 4 mm. broad.

Type locality: Basin of the Orinoco, Venezuela.

Distribution: Venezuela mainland and on Margarita Island.

This plant is very common not only in the flat land along the coast of Venezuela but also in the mountains. It is also widely grown in and about yards, for the leaves are supposed to have medicinal properties, and when properly grown as a hedge it forms a

---

*Tunari Mountains, just northwest of Cochabamba, Bolivia, about at the site of Sacaba.

most formidable protection. In the grazing regions of the country and along railways where wire fencing is employed, the trunks and larger branches are used for posts and smaller branches for intervening supports; these posts and stays, however, do not die, but in time grow to considerable size.

Although the wood, especially the branches, has little strength or endurance, it is used somewhat for making hanging baskets for orchids. It is known everywhere as guamacho, which was taken by Weber as the specific name for the plant.

Figures 9 and 10 are from photographs taken by Mr. H. Pittier at Carácas, Venezuela, in 1913.

10. Pereskia colombiana sp. nov.

A tree, 6 to 11 meters high, or sometimes smaller and shrub-like; main stem covered with clusters of slender spines, 2.5 to 7 cm. long; branches glabrous, either bearing spines or naked, covered with light-brown bark; areoles small, woolly; leaves oblong to obliquely orbicular, short-petioled, unarmed at base, often broad above, usually acute, probably fleshy, glabrous, 4 cm. long or less; flowers bright yellow, opening about midday, borne on the old wood, solitary, sessile, 4 cm. broad; ovary covered with small ovate, acute leaves, these hairy in the axils; sepals oblong, obtuse, about 1 cm. long, entire on the margins; stamens numerous; fruit not known.

Collected by Herbert H. Smith at low altitudes near Santa Marta, Colombia, in April, 1898 to 1905 (No. 1886, type), and from the same locality by Justin Goudot in 1844, and by Francis W. Pennell in 1918 (No. 4765).

Mr. Smith remarks that the leaves are deciduous in March or April, and that the tree is leafless when in bloom in the spring.

Figure 11 is copied from a drawing of an herbarium specimen collected by Herbert H. Smith at Ronda, Santa Marta, Colombia.

11. Pereskia tampicana Weber, Dict. Hort. Bois 939. 1898.

Shrub; branches often without spines or the spines several, needle like, black, 2 to 3 cm. long; areoles globular, appearing as knobs along the stem; leaves about 5 cm. long, petioled; flowers 2.5 cm. long; petals entire, rose-colored.

FIG. 11.—Pereskia colombiana. X0.5.

Type locality: Near Tampico, Mexico.

Distribution: Eastern Mexico, but known only from the type locality.

P. tampicana is not well known and has been reported only from Tampico, Mexico. Dr. E. Palmer made a careful search for it some years ago at the type locality, but in vain. In 1912 Dr. Rose examined the two small specimens of the species preserved in the herbarium of the Royal Botanical Garden of Berlin, and is convinced that it is a *Pereskia* and not a *Pereskiopsis*.

Pereskia rosea A. Dietrich (Allg. Gartenz. **19:** 153. 1851; *Opuntia rosea* Schumann, Gesamtb. Kakteen 764. 1898) is supposed to have come from Mexico, but we have not been able to identify it; Schumann refers to it in a note under *P. tampicana*. Here he also takes up *Pereskia zinniaeflora* De Candolle (Prodr. **3:** 475. 1828). Both these specific names are much older than *P. tampicana*, and should either of them be found identical with it, the name *P. tampicana* would be rejected.

12. Pereskia bleo (HBK.) De Candolle, Prodr. **3:** 475. 1828.

Cactus bleo Humboldt, Bonpland, and Kunth, Nov. Gen. et Sp. **6:** 69. 1823.
Pereskia panamensis Weber, Dict. Hort. Bois 739. 1898.

A tree, sometimes 7 meters high; trunk 10 cm. in diameter or less, when old becoming naked, but young shoots often bear large fascicles of spines (sometimes 25 or more); young branches red, leafy, its spines in fascicles of 5 and 6, but young shoots often bear but 1 to 4, black, acicular, up to 2.5 cm.

long; leaves thin, oblong to oblanceolate, 16 to 21 cm. long, 4 to 5.5 cm. wide, acuminate, cuneate at base, tapering into petioles 2 to 3.5 cm. long; areoles circular, bearing when young a little wool, but soon becoming naked; calyx turbinate, somewhat angled, naked, with linear deciduous sepals; petals 12 to 15, rose-colored, obovate, 3.5 cm. long; style longer than the stamens, red, thick; stigma-lobes 5 to 7; ovary depressed; fruit yellow, truncate, 5 to 6 cm. long; seeds 6 mm. long, black, shining.

Type locality: Near Badillas, on the Magdalena River, Colombia, South America.

Distribution: Northwestern South America and throughout Panama.

This species was collected by Bonpland during Humboldt's trip through the New World and was described and published by Kunth in 1823. Dr. Rose examined two of the original specimens in the herbarium of the Museum of Natural History at Paris, one being the specimen given by Bonpland and the other the specimen in the Kunth Herbarium,

Fig. 12.—Pereskia bleo.

which is kept distinct from the general herbarium. The only other representatives of this species from South America which we have seen are a specimen in the herbarium of the same museum, which was collected by Justin Goudot in Colombia in 1844, and one collected in 1852, by I. F. Holton at San Juancito, Colombia, preserved in the Torrey Herbarium and one recently brought by Francis W. Pennell from Boca Verde, Rio Sinu, Colombia.

Heretofore *Pereskia bleo* has been considered one of the most common species, for many living collections as well as herbaria contain many specimens under that name; the plant which has been known as *P. bleo*, however, is *P. grandifolia* Haworth, now known to be a native of Brazil and not found wild in Colombia.

Since determining that the so-called *Pereskia bleo* of our gardens and of Brazil is not the true *P. bleo* of Humboldt, we have become convinced that *P. panamensis* Weber is the same as *P. bleo;* Mr. Pittier's exhaustive exploration of Panama has strengthened our conclusions, for he has traced this species as far south as the Colombian border. Humboldt's plant came from northern Colombia.

In Panama the plant is known under the name ñajú de Culebra.

Illustrations: All illustrations referred to this species which we have examined are cited under *P. grandifolia.*

Figure 12 is from a photograph taken by Henry Pittier, near Chepo, Panama, October 30, 1911.

13. Pereskia bahiensis Gürke, Monatsschr. Kakteenk. **18**: 86.   1908.

Shrub or tree, sometimes 8 meters high, with a more or less definite trunk, sometimes 1 meter or more long and 1.5 to 2 dm. in diameter, and a large, rounded, much branched top; spines on new growth wanting, but on old wood 5 to 40 at an areole, some of them 5 to 9 cm. long; young branches green; leaves lanceolate, 6 to 9 cm. long, deciduous, somewhat pointed, narrowed at base into short petioles; flowers in small panicles, rose-colored; ovary bearing large leaves with cuneate bases; fruit often proliferous, yellowish when mature, more or less irregularly angled, bearing large leaves 3 to 4 cm. long, which ultimately fall away; seeds black, oblong, 5 mm. long.

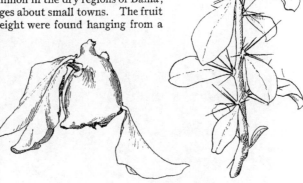

Type locality: In the southeast catinga between Rio Paraguaçu at Tambury and Rio das Contas at Caldeirão, Bahia, Brazil.

Distribution: Dry parts of eastern Brazil.

This species is very common in the dry regions of Bahia; and is often planted for hedges about small towns.   The fruit is proliferous; as many as eight were found hanging from a single peduncle; it is said to be edible, but while half-ripe is very astringent.   The perfect fruits can seldom be found, because the birds peck into them for the large black seeds.

Called   in   Brazil, according   to   Dr.   Leo Zehntner,   Iniabanto, Espinha de São Antonio, and Flor de Cêra.   He

FIGS. 13, 14.—Pereskia bahiensis.   X0.5.

also   says: "I   think Iniabanto is the best and ought to be generalized.   It is derived from Iniabo = Okra = *Hibiscus esculentus,* without doubt because the leaves of the pereskias are sometimes eaten by people, giving a mucilaginous dish like that of the Hibiscus fruit."

Illustration: Monatsschr. Kakteenk. **18**: 87.

Figure 13 is from a specimen, preserved in formalin, collected by J. N. Rose near Machado Portello, Bahia, Brazil, in June 1915; figure 14 is from a plant from the same place; figure 15 is from a photograph obtained by J. N. Rose at Barrinha, Bahia, in June 1915.

14. Pereskia grandifolia Haworth, Suppl. Pl. Succ. 85.   1819.

Cactus rosa Vellozo, Fl. Flum. 206.   1825.
Pereskia ochnacarpa Miquel, Bull. Sci. Phys. Nat. Neerl. 48.   1838.

Tree or shrub, 2 to 5 meters high, usually with a definite, very spiny, woody trunk up to 1 dm. in diameter, the branches fleshy, glabrous, elongated, usually with 1 or 2 acicular spines at the areoles; leaves oblong, obtuse or acute, somewhat narrowed at base, 8 to 15 cm. long; petioles short; inflorescence terminal, usually few-flowered; 3.5 to 4 cm. broad; sepals green; petals rose-colored, sometimes white; filaments red; style and stigma-lobes white; ovary leaf-bearing; fruit described as large, pear-shaped, many-seeded; seeds black.

Type locality: In Brazil.

Distribution: Brazil, widely planted and subspontaneous throughout the West Indies. The plant is extensively used for hedges in tropical America. It is planted by pushing cuttings into the ground, its spiny stems soon forming a capital barrier.

Illustrations: Vellozo, Fl. Flum. **5**: pl. 27, as *Cactus rosa.* Amer. Garden **11**: 462; Blühende Kakteen **3**: 137; Curtis's Bot. Mag. **63**: pl. 3478; Cycl. Amer. Hort. Bailey **1**: f. 309; Dict. Hort. Bois f. 678; Edwards's Bot. Mag. **17**: pl. 1473; Engler and Prantl, Pflanzenfam. **3**[6a]: f. 71; Gard. Chron. III. **20**: f. 75; Karsten, Deutsch. Fl. 887. f. 9; Martius, Fl. Bras. **4**[2]: pl. 63; Pfeiffer and Otto, Abbild. Beschr. Cact. **1**: pl. 30; Reichenbach, Fl. Exot. pl. 328, all as *Pereskia bleo.*

Fig. 15.—Pereskia bahiensis. Photograph by Paul G. Russell.

Plate III, figure 1, represents a flowering branch of a plant obtained by N. L. Britton on St. Christopher in 1901. Figure 16 is from a photograph of the plant used as a hedge near Rio de Janeiro, Brazil.

15. Pereskia zinniaeflora De Candolle, Prodr. **3**: 475. 1828.

Shrub; leaves oval to oblong, 2 to 4 cm. long, acuminate, cuneate at base; spines on young branches 1 or 2 at an areole, on old branches 4 or 5, all short, less than 1 cm. long; flowers broad, 5 cm. wide, rose-red; petals entire, obtuse or retuse; style and stamens very short; ovary truncate, bearing small, stalked leaves.

Type locality: In Mexico.

Distribution: Mexico.

Nicholson associates this species with *Pereskia bleo,* that is, *P. grandifolia,* but the relationship is not close. The measurements of the flower given above are taken from

M. E. Eaton del.

1. Flowering branch of *Pereskia grandifolia*.　　2. Leafy branch of *Pereskiopsis chapistle*.
3. Leafy branch of *Pereskiopsis pititache*.　(All natural size.)

De Candolle's plate cited below, and may not be quite correct. This species, so far as we are aware, has not been again collected.

Cactus zinniaeflora Mociño and Sessé (De Candolle, Prodr. 3: 475. 1828) was given as a synonym.

Illustrations: Förster, Handb. Cact. ed. 2. f. 135; Mém. Mus. Hist. Nat. Paris 17: pl. 17; Rümpler, Sukkulenten f. 127; Suppl. Dict. Gard. Nicholson f. 624.

Figure 17 is a copy of the second illustration above cited.

FIG. 16.—Pereskia grandifolia. Exposed branches are shown above the other foliage.

FIG. 17.—Pereskia zinniaeflora.

16. Pereskia horrida (HBK.) De Candolle, Prodr. 3: 475.   1828.

Cactus horridus Humboldt, Bonpland, and Kunth, Nov. Gen. et Sp. 6: 70.   1823.

Tree, 4 to 6 meters high, with terete slender branches; spines often solitary, sometimes 2 or 3, slender, dark in color, unequal, the longest 2 to 3 cm. long; leaves solitary, alternate, narrowly oblong, 3 cm. long, subsessile, entire, glabrous; flowers 3 to 5 together in upper axils, about 10 mm. long; calyx described as 5-toothed and persistent; petals 5 or 6, red, lanceolate, spreading; fruit fleshy, many-seeded.

Type locality: "Ad flumen Marañon prov. Jaen de Bracamoros." (Schumann says this locality is in Peru.)

Distribution: Northwestern South America.

The above description is compiled from Kunth's original description and from notes made by Dr. Rose upon the type material in the herbarium of the Museum of Natural History at Paris, in which there are specimens from both Bonpland and Kunth. Both of these sheets lack flowers and fruit, and only Kunth's bears leaves. So far as we are aware, no other material of this species has been collected since Humboldt's time except that in 1912 Dr. Weberbauer wrote that he had visited the Marañon, at Humboldt's locality, and had collected a single specimen, which had been sent to the Botanical Museum at Berlin.

17. Pereskia cubensis Britton and Rose, Torreya **12**: 13. 1912.

A tree, 4 meters high, with a trunk 2.5 dm. in diameter and a large, flat, much branched top; bark brownish, rather smooth, marked here and there by black bands (representing the old areoles), these broader than high; young branches slender, smooth, with light-brown bark; spines from young areoles 2 or 3, needle-like, brownish, 2 to 4 cm. long, from old areoles very numerous (25 or more), and much longer (5 cm. or more long); leaves several at each areole, sessile, bright green on both sides, oblanceolate to oblong or obovate, 1 to 4 cm. long, 10 to 12 mm. wide, acute at both ends or obtuse at the apex, fleshy, the midvein broad, the lateral veins very obscure; peduncle very short, jointed near the base, bearing 1 to 3 leaf-like bracts; flowers terminal and also axillary, solitary; sepals 5, obtuse or rounded, ovate-oblong to orbicular, unequal, 7 to 9 mm. long, the larger ones with broad purple margins; petals 8, about 15 mm. long, deep reddish purple, obovate-elliptic, rounded; stamens many, about 6 mm. long; anthers light yellow; ovary turbinate, naked, spineless; fruit not seen.

Fig. 18.—Pereskia cubensis.

Fig. 19.—Pereskia cubensis.
×0.5.

Type locality: In Cuba.

Distribution: Near the southern coast of eastern and central Cuba.

The tree is abundant on the plain between Guantánamo and Caimanera, Oriente, where the type specimens were collected; it also inhabits coastal thickets at Ensenada de Mora, in southwestern Oriente, the plants of this colony bearing leaves with less acute apices than those of the typical ones. A single plant was also observed on La Vigia Hill, at Trinidad, province of Santa Clara, which had shorter and smaller leaves than either of the other two. The description of the flower is from one of a plant collected by N. L. Britton and J. F. Cowell at Ensenada de Mora, southern Oriente, Cuba, in 1912, and brought to the New York Botanical Garden, where it flowered in May 1917.

Dr. Rose finds that the plant in De Candolle's herbarium which represents the *Pereskia portulacifolia* of the Prodromus is undoubtedly *Pereskia cubensis*. It was collected as early as 1821.

Illustration: Journ. N. Y. Bot. Gard. **10**: f. 22.

Figure 18 is from a photograph taken by Dr. M. A. Howe in the colony of this tree at Nuevaliches, near Guantánamo, Cuba, studied by Dr. N. L. Britton in 1909; figure 19 represents a leafy branch of the same plant.

18. Pereskia portulacifolia (Linnaeus) Haworth in De Candolle, Prodr. **3**: 475.   1828.

Cactus portulacifolius Linnaeus, Sp. Pl. 469.   1753.

A tree, 5 to 6.6 meters high, the branches terete, very spiny; spines acicular, sometimes almost bristle-like, 2 cm. long, on old wood in clusters of 7 to 9, but on new growth usually solitary; leaves

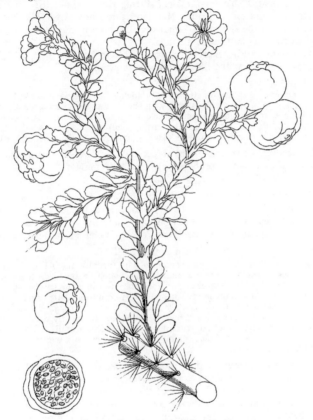

Fig. 20.—Pereskia portulacifolia.   ×0.66.

only 1 cm. long or less, cuneate at base, often retuse at apex; peduncles short but definite, 2 to 5 mm. long, bearing several small spatulate to broadly obovate leaves; flowers rose-colored, about 3 cm. broad; sepals about 3, ovate to shortly oblong, obtuse, fleshy, 8 mm. or less long; petals oblong, about 2 cm. long, thin, obtuse; ovary small, truncate, naked or bearing a single small leaf; immature fruit hard, depressed, 2 cm. long, 2.5 cm. broad, smooth, naked or with a single small leaf 5 to 6 mm. long, with a broad scar at the top 8 to 10 mm. in diameter; fruit globular, naked; seeds large, black.

Type locality: Tropical America, doubtless Hispaniola.
Distribution: Haiti.

The usual reference for the first publication of the plant under *Pereskia* is Haworth's Synopsis (Syn. Pl. Succ. 199. 1812), but it was not here formally transferred from the genus *Cactus*.   His statement is: "*Cactus portulacifolius* is another species of this genus."

Our knowledge of this plant is drawn from the illustration below cited and descriptions, and from a fragmentary specimen collected by W. Buch near Gonaives, Haiti, in 1900, where it grows on dry calcareous rocks, and one obtained by Paul Bartsch at Tomaseau in April 1917. Dr. Bartsch states that the flower reminds one very much of a rose and the fruit is pendent like a green plum.

Lunan in 1814 (Hort. Jam. 2: 236) described a tree nearly a foot in diameter, growing at a residence near Spanish Town, Jamaica, stating that it differed from *Pereskia* by the absence of tufts of leaves on its fruit. His description points to *Pereskia portulacifolia*, but nothing is known of the species in Jamaica at the present day; according to Grisebach, Macfadyen recorded it as cultivated there.

Illustration: Plumier, Pl. Amer. ed. Burmann pl. 197, f. 1.

Figure 20 is copied from the illustration above cited.

19. Pereskia conzattii sp. nov.

Tree, 8 to 10 meters high; bark of stems and branches brown and smooth; leaves orbicular to obovate, acute, 1 to 2.5 cm. long; areoles small, with short white wool and a few long hairs; spines 2 to 6 on young branches, 10 to 20 on main stem, acicular, 2 to 2.5 cm. long, at first yellowish brown, dark brown in age; flowers not known; ovary bearing small scales; fruit naked, pear-shaped, more or less stalked at base, 3 to 4 cm. long; seeds black, glossy, 3 mm. long, with a small white hilum.

Collected at Salina Cruz and Tehuantepec, Oaxaca, Mexico, in February and April, 1913, by Professor C. Conzatti; probably also in Guatemala.

SPECIES UNKNOWN TO US.

Pereskia affinis and *P. haageana* Meinshausen, Wochenschr. Gärtn. Pflanz. 2: 118. 1859.

Pereskia cruenta, *P. grandiflora*, and *P.(?) plantaginea*, the first two given as synonyms and the last merely mentioned by Pfeiffer (Enum. Cact. pp. 176, 177, and 179) can not be identified. The same is true of *P. grandispina* Forbes (Journ. Hort. Tour Germ. 159. 1837).

Tribe 2. OPUNTIEAE.

Plants usually very fleshy, never epiphytic, branched (usually much branched), one to many-jointed; joints diverse in structure, terete, compressed, or much flattened, with irregularly scattered areoles, ribless, except one species; leaves usually caducous, but in some species more or less persistent, small or minute, subulate or cylindric, in one genus broad and flat; areoles usually glochidiferous (except in *Maihuenia*; in *Grusonia* only those of the ovary), mostly spine-bearing; spines usually slender, straight or nearly so, sometimes sheathed; corolla mostly rotate (sepals and petals in *Nopalea* erect); flowers sessile, diurnal, one from an areole; fruit usually a fleshy berry, sometimes dry, rarely capsular; seed white or black, globular, flattened or even winged, with a thin or hard testa; cotyledons large, elongated.

This tribe contains 7 genera and at least 300 species, various in habit, flower, fruit, and seeds. It is more closely related to the *Pereskieae* than to the *Cereeae*. The following characters possessed by some or all genera of the *Opuntieae* are wanting in the *Cereeae*:

Leaves on the stem (see also *Harrisia* and *Hylocereus*); glochids in the areoles; sensitive stamens; sheathed spines; winged, white, and bony-covered seeds; the separation of withering calyx, stamens, and style from the ovary; areoles irregularly distributed over the stem in all the genera except *Grusonia*, in which they are arranged on ribs as in many of the *Cereeae*.

The tribe is distributed throughout the cactus regions of the Americas, but the genera, except *Opuntia*, are localized.

KEY TO GENERA.

```
Leaves broad and flat....................................................................  1. Pereskiopsis
Leaves subulate or cylindric.
    Seeds broadly winged ..............................................................  2. Pterocactus
    Seeds wingless.
        Stamens much longer than the petals.
            Petals erect; joints flat.....................................................  3. Nopalea
            Petals recurved; joints terete...............................................  4. Tacinga
        Stamens shorter than the petals.
            Joints flat to terete, not ribbed.
                Testa of the seed thin, black, shining....................................  5. Maihuenia
                Testa thick, pale, dull...................................................  6. Opuntia
            Joints terete, longitudinally ribbed.........................................  7. Grusonia
```

1. PERESKIOPSIS Britton and Rose, Smiths. Misc. Coll. **50**: 331. 1907.

Trees and shrubs, in habit and foliage similar to *Pereskia;* old trunk forming a solid woody cylinder covered with bark and resembling the ordinary dicotyledonous stem; areoles circular, spine-bearing or sometimes spineless, also bearing hairs, wool, and usually glochids; flowers similar to those of *Opuntia;* ovary sessile (one species described as pedunculate), with leaves at the areoles (except in one species); fruit red, juicy; seeds bony, few, covered with matted hairs.

Type species: *Opuntia brandegeei* Schumann.

The plants are common in hedges and thickets of Mexico and Guatemala.

As to the number of species to be recognized in this genus we are uncertain; about 16 have been described. In our first discussion of the genus (*op. cit.*) we recognized 11 species, including several known only from descriptions. There now seem to be at least 10 species, of which 8 are in cultivation in Washington and New York. Two of the plants were described, as early as 1828, as species of *Pereskia*, and here they remained with 2 later-described species until, in 1898, Dr. A. Weber transferred them to *Opuntia*, placing them in a new subgenus, *Pereskiopuntia*. The same year Dr. Karl Schumann adopted Weber's conclusions, publishing his treatment of the subgenus and assigning 5 species to it.

In its large leaves and woody, spiny stems, this group suggests *Pereskia*, but it has glochids and different flowers, fruit, and seeds; in flowers, fruit, seeds, and glochids it resembles *Opuntia*, but on account of habit and foliage must be excluded from that genus.

In view of these differences, Britton and Rose in 1907 established the genus *Pereskiopsis* and listed 11 species, 4 of which had been originally described as species of *Pereskia* and 5 as species of *Opuntia*. Since then we have grown most of these plants along with the pereskias and opuntias so as to compare them. Unfortunately we are not able to describe all the species fully, for they have never been known to flower in cultivation, although some of the species, at least, bloom freely in the wild state. The leaves on the lower parts of shoots are sometimes broader and shorter than those on the upper parts, and in greenhouse cultivation the leaves of some species are narrower than when the plants are growing under natural conditions.

The generic name is from the Greek and signifies resembling *Pereskia*.

KEY TO SPECIES.

```
Stems, ovary, and often the leaves more or less pubescent.
    Normal leaves long-acuminate, narrow, with narrow cuneate bases.................... 1. P. velutina
    Normal leaves abruptly pointed, somewhat cuneate at base......................... 2. P. diguetii
Stems, ovary, and leaves glabrous.
    Leaves, at least some of them, not much longer than broad.
        Fruit without leaves, at least so figured..................................... 3. P. opuntiaeflora
        Fruit with leaves subtending the areoles.
            Areoles white, with few glochids or none.
                Leaves orbicular or nearly so, rounded or apiculate...................... 4. P. rotundifolia
                Leaves, at least the upper ones, obovate or elliptic, acute at both ends........ 5. P. chapistle
            Areoles dark, filled with numerous brown glochids........................... 6. P. porteri
    Leaves, at least some of them, twice as long as broad or longer.
        Leaves spatulate.......................................................... 7. P. spathulata
        Leaves elliptic to oblong, or obovate.
            Leaves pale green, glaucous............................................. 8. P. pititache
            Leaves bright gr.en, shining.
                Glochids few, yellow................................................. 9. P. aquosa
                Glochids many, brown................................................ 10. P. kellermanii
```

1. Pereskiopsis velutina Rose, Smiths. Misc. Coll. **50**: 333. 1907.

Stems weak and spreading, forming compact bushes 9 to 12 dm. high or sometimes higher; old stems with cherry-brown bark; young branches green, borne nearly at right angles to the old stem, velvety-pubescent; areoles bearing long white hairs and several short spines and some glochids; leaves elliptic to ovate-elliptic, 2 to 6 cm. long by 1.5 to 2.5 cm. broad, acuminate, or acute at both ends, dull green, more or less velvety-puberulent on both surfaces, when very young brighter green; flowers sessile on the second-year branches; ovary obovoid to oblong, pubescent, bearing large leaves and areoles similar to those of the stem; leaves on ovary spreading or ascending and persisting after the flower falls; flower-bud (above the ovary) 2 to 3 cm. long, acute; sepals green or deep red tinged with yellow; petals bright yellow.

Type locality: In hedges about city of Queré-taro, Mexico.

Distribution: Table-lands of central Mexico.

This plant is called by the natives nopaleta and cola de diablo.

Illustration: Smiths. Misc. Coll. **50**: pl. 44.

Figure 21 is from a photograph of type plant.

2. **Pereskiopsis diguetii** (Weber) Britton and Rose, Smiths. Misc. Coll. **50**: 332. 1907.

> *Opuntia diguetii* Weber, Bull. Mus. Hist. Nat. Paris **4**: 166. 1898.

Tall shrub, larger than the preceding species; old stems reddish; branches pubescent; areoles when young filled with long, white, cobwebby hairs, when old large and filled with short black wool; leaves elliptic to obovate, 3 to 5 cm. long, usually abruptly pointed, more or less cuneate at the base; spines usually 1, rarely as many as 4, at first nearly black, in time becoming lighter, sometimes nearly 7 cm. long; glochids brownish, not very abundant; flowers yellow; fruit 3 cm. long, red, pubescent, its areoles often bearing spines as well as glochids; seeds white, 5 mm. broad, covered with matted hairs.

Type locality: Near Guadalajara, Mexico.

Distribution: Central Mexico.

Common in hedges near Guadalajara and Oaxaca, Mexico. According to W. E. Safford, it is known in Guadalajara as tasajillo and alfilerillo.

Figure 22 represents a leafy branch of a plant collected by W. E. Safford in Guadalajara, Mexico, in 1907.

3. **Pereskiopsis opuntiaeflora** (De Candolle) Britton and Rose, Smiths. Misc. Coll. **50**: 332. 1907.

> *Pereskia opuntiaeflora* De Candolle, Prodr. **3**: 475. 1828.
> *Opuntia golziana* Schumann, Gesamtb. Kakteen 654. 1898.

Shrubby, glabrous; leaves obovate, mucronate, often in pairs; spines, when present, solitary, elongated, 2 to 3 times as long as the leaves; flowers subterminal, short-pedunculate; petals numerous, ovate, subacute, reddish yellow, arranged in two series; ovary leafless, bearing areoles filled with glochids.

Type locality: In Mexico.

Distribution: Known only from the original description.

This description is drawn from De Candolle's original description and illustration; otherwise nothing is known of the plant.

Fig. 21.—Pereskiopsis velutina.

This species, as illustrated by De Candolle, is unlike anything we know. In its pedunculate fruit it is like *Pereskia*, but its leafless ovary and its areoles filled with glochids would exclude it from that genus. In a general way the illustration looks more like a *Pereskiopsis*, and we suspect that the delineation is incorrect or that the leaves had fallen away from the specimen drawn.

Cactus opuntiaeflorus Mociño and Sessé (Pfeiffer, Enum. Cact. 178. 1837) was published as a synonym of *Pereskia opuntiaeflora.*

Illustrations: Förster, Handb. Cact. ed. 2. f. 137; Mém. Mus. Hist. Nat. Paris **17**: pl. 19, both as *Pereskia opuntiaeflora.*

Figure 23 is copied from the second illustration above cited.

4. Pereskiopsis rotundifolia (De Candolle) Britton and Rose, Smiths. Misc. Coll. **50**: 333. 1907.

> *Pereskia rotundifolia* De Candolle, Prodr. **3**: 475. 1828.
> *Opuntia rotundifolia* Schumann, Gesamtb. Kakteen 652. 1898.

Stem thick, more or less woody; branches slender, glabrous; leaves nearly orbicular, mucronate; spines elongated, solitary; flowers 3 cm. broad, borne on the second-year branches; petals reddish yellow. broad, with mucronate tips; ovary leafy; fruit obovoid, red, leafy.

FIG. 22.—Pereskiopsis diguetii. ✕0.5.

FIG. 23.—Pereskiopsis opuntiae-flora. ✕0.5.

FIG. 24.—Pereskiopsis rotundifolia. ✕0.5.

Type locality: In Mexico.

Distribution: Known only from the original description and, apparently, from Oaxaca.

Cactus frutescens Mociño and Sessé (Pfeiffer, Enum. Cact. 178. 1837) and *Cactus rotundifolia* Mociño and Sessé (De Candolle, Prodr. **3**: 475. 1828) were given as synonyms of *Pereskia rotundifolia,* but were never published.

Illustrations: Mém. Mus. Hist. Nat. Paris **17**: pl. 20, as *Pereskia rotundifolia;* Schumann, Gesamtb. Kakteen f. 99, as *Opuntia rotundifolia.*

Figure 24 is copied from the first illustration above cited; figure 25 is from a photograph taken by Dr. MacDougal at Oaxaca, Mexico, in 1906.

5. Pereskiopsis chapistle (Weber) Britton and Rose, Smiths. Misc. Coll. **50**: 331. 1907.

> *Opuntia chapistle* Weber in Gosselin, Bull. Mus. Hist. Nat. Paris **10**: 388. 1904.

A large, branching shrub, sometimes 3 to 4 meters high, the branches widely spreading, glabrous; spines single, white, long (6 cm. long), very stout; leaves fleshy, somewhat persistent, obovate to elliptic, sometimes nearly orbicular, 3 to 4 cm. long, glabrous; flowers yellow; fruit red.

Type locality: In Oaxaca.

Distribution: Oaxaca and perhaps Morelos, Mexico.

Illustration: Bull. Soc. Nat. Acclim. France **52**: f. 10, as *Opuntia chapistle.*

Plate III, figure 2, represents a leafy branch of a plant collected by Dr. Rose at Cuernavaca, Mexico, in 1906.

FIG. 25.—Pereskiopsis, apparently P. rotundifolia, with other cacti in the background.

6. Pereskiopsis porteri (Brandegee) Britton and Rose, Smiths. Misc. Coll. **50**: 332. 1907.

Opuntia porteri Brandegee in Weber, Dict. Hort. Bois 899. 1898.
Opuntia brandegeei Schumann, Gesamtb. Kakteen 653. 1898.
Pereskiopsis brandegeei Britton and Rose, Smiths. Coll. **50**: 331. 1907.

Stems stout, woody, branching, 6 to 12 dm. high, 3 cm. in diameter, the old areoles bearing 3 to 8 stout spines 3 to 5 cm. long, but on the trunk often 15 to 20 spines from an areole; first and second year branches usually short, spineless, or with 1 or 2 brown spines, those of the first year green, the second-year brownish; areoles bearing numerous small, brown glochids; leaves sessile, 2 to 3 cm. long, obovate, acute, fleshy, in greenhouse specimens sometimes much narrower; flowers about 4 cm. in diameter; sepals few, spatulate, short; petals few, yellow, broad, entire; fruit joint-like, oblong, 4 to 5 cm. long, orange-colored, with large areoles bearing brown glochids; seeds 1 or few, covered with white deciduous hairs.

Type locality: In Sinaloa, Mexico.
Distribution: Common in the Cape region of Lower California and in the State of Sinaloa, Mexico.

Figure 26 shows a leafy twig of a plant sent in 1904 from the Missouri Botanical Garden to the New York Botanical Garden as *Opuntia brandegeei*, which had been received by the Missouri Botanical Garden from Mrs. Katharine Brandegee in 1901.

7. Pereskiopsis spathulata (Otto) Britton and Rose, Smiths. Misc. Coll. **50**: 333. 1907.

Pereskia spathulata Otto in Pfeiffer, Enum. Cact. 176. 1837.
Opuntia spathulata Weber, Bull. Mus. Hist. Nat. Paris **4**: 165. 1898.

FIG. 26.—Pereskiopsis porteri. ✕0.66.

Branching shrub, 1 to 2 meters high; branches few, glaucescent, deflexed; leaves spatulate, thick, green, 2.5 to 5 cm. long; areoles distant, woolly, hairy when young; spines 1 or 2, rigid, white below, 2.5 cm. long; glochids brown, borne in the upper part of the areoles; flowers red; seeds white.

Type locality: In Mexico.

Distribution: Probably southern Mexico, but no definite locality is known.

There is some confusion in the literature of this species; Schumann describes it as pubescent, while in the original description nothing is said about pubescence; this error is probably due to a misidentification, for Dr. Rose found in the Museum of Paris two specimens collected by Diguet at Guadalajara, Mexico, which were labeled *Opuntia spathulata*, and which have pubescent branches and leaves; these are undoubtedly *O. diguetii*.

Pereskia crassicaulis Zuccarini (Pfeiffer, Enum. Cact. 176. 1837) was never published, simply being given as a synonym of *P. spathulata*.

8. Pereskiopsis pititache (Karwinsky) Britton and Rose, Smiths. Misc. Coll. **50**: 332. 1907.

> *Pereskia pititache* Karwinsky in Pfeiffer, Enum. Cact. 176. 1837.
> *Pereskia calandriniaefolia* Link and Otto in Salm-Dyck, Cact. Hort. Dyck. 1849. 252. 1850. (According to Schumann.)
> *Opuntia pititache* Weber, Bull. Mus. Hist. Nat. Paris **4**: 166. 1898.

Stems rather low and somewhat branching; bark light brownish and flaking off; areoles on main trunk each bearing 1 to 4 slender acicular spines and a small cluster of yellowish glochids; branches, even when several years old, bearing a single long, acicular spine from an areole and no glochids; young and growing branches rather slender and green, their areoles small, black in the center, with long, white hairs from their margins and no spines; leaves obovate or oblong-obovate, 4 cm. long or less, pale green, thin, acute or bluntish at the apex, narrowed at the base.

Type locality: In Mexico.

Distribution: Uncertain, but reported from southern Mexico.

In the original description this species is said to have a very spiny, erect woody trunk, the branches spreading nearly horizontally, the spines unequal, 3 to 6, 25 to 37 mm. long, the leaves fleshy, green, lanceolate to ovate, 37 mm. long, 16 mm. broad. It was named by Baron Wilhelm von Karwinsky and probably collected by him in Mexico, but no definite locality was given; Weber states it is from Tehuantepec, while Schumann gives Tehuacán on a statement of Weber.

Pereskia calandriniaefolia we have referred here, following Schumann, but the original description is somewhat different from that of *P. pititache*, the leaves being described as spatulate to lanceolate, strongly narrowed below, 7.5 cm. long.

Our description is mostly drawn from specimens growing in the New York Botanical Garden obtained from M. Simon, of St. Ouen, Paris, in 1901.

Illustrations: Abh. Bayer. Akad. Wiss. München **2**: pl. 1, sec. 6, f. 1, 2; pl. 2, f. 9, both as *Pereskia pititache*.

Plate III, figure 3, represents a leafy shoot of a plant sent by M. Simon, of St. Ouen, Paris, France, to the New York Botanical Garden in 1901.

9. Pereskiopsis aquosa (Weber) Britton and Rose, Smiths. Misc. Coll. **50**: 331. 1907.

> *Opuntia aquosa* Weber, Bull. Mus. Hist. Nat. Paris **4**: 165. 1898

Shrub, with glabrous, glaucous, green branches, the young shoots with long white hairs at the areoles; leaves bright green, nearly elliptic, acute, about twice as long as wide, narrowed at the base, glabrous; spines usually solitary, standing at right angles to the stem, white; glochids few, yellow; flowers yellow; outer petals blotched with red; fruit pear-shaped, 4 to 5 cm. ong, 2 to 2.5 cm. in diameter, yellowish green.

FIG. 27.—Pereskiopsis aquosa. X0.66.

Type locality: Guadalajara, Mexico.

Distribution: In hedges about Guadalajara, Mexico.

The fruit, called in Mexico tuna de agua and tasajillo, is used in making a cooling drink and for preserves.

Opuntia spathulata aquosa (Bull. Mus. Hist. Nat. Paris 4: 165. 1898) was given as a synonym of this species, but was never published.

Illustration: Safford, Ann. Rep. Smiths. Inst. 1908: pl. 10, f. 2.

Figure 27 represents a leafy shoot of the plant collected by W. E. Safford near Guadalajara, Mexico, in 1907.

10. Pereskiopsis kellermanii Rose, Smiths. Misc. Coll. 50: 332. 1907.

Stem glabrous, herbaceous, weak, and clambering over shrubs to a length of 4 to 5 meters, about 2 cm. in diameter; second-year branches usually at right angles to main stem, with cherry-red bark; old stem bearing several slender, acicular brown spines, sometimes only 1, sometimes wanting, and numerous brown glochids; young branches green, fleshy, their areoles circular, white, filled with long white hairs, brown glochids, and often with several acicular brown spines; spines on wild plants often stout, usually solitary, nearly black, 2 to 3 cm. long; leaves various, shining green, glabrous, thickish, elliptic and two or three times as long as wide, or suborbicular, acute at the apex, narrowed at the base, 5 cm. long or less, 2 to 2.5 cm. broad; flowers not known; fruit red, glabrous, leafy, 3 to 6 cm. long, bearing large areoles filled with brown glochids; seeds covered with matted hairs.

FIGS 28, 29, and 30.—Pereskiopsis keller-manii, showing three leaf forms. X0.5.

Type locality: Trapichite, Guatemala.

Distribution: Guatemala.

Figures 28, 29, and 30 are copied from sketches of the leaf-forms of the type plant, made by W. A. Kellerman in Guatemala in 1908.

2. PTEROCACTUS Schumann, Monatsschr. Kakteenk. 7: 6. 1897.

Stems low, more or less branched above, cylindric, from tuber-like and often greatly enlarged roots; leaves minute, caducous; spines weak, several or many at each areole; glochids small, caducous as in *Opuntia;* flower terminal, regular, without tube; perianth-segments several, erect; filaments and pistil shorter than the petals; ovary nearly turgid, bearing numerous small clusters of spines; fruit dry, capsular, dehiscent; seeds winged, white; embryo curved.

Type species: *Pterocactus kuntzei* Schumann.

Four species have already been described, but three of these we have combined and the fourth is referred to *Opuntia.* Three additional species, however, are here described. The generic name refers to the winged seeds.

This is a remarkable genus, and it is surprising that it remained unrecognized so long, for one of its species was known as long ago as 1837; the fruit and seeds, however, seem not to have been known until about 1897. In habit the plants are nearest some of the anomalous species of *Opuntia,* having large roots and short, weak stems like *Opuntia chaffeyi,* of Mexico; the seeds, however, differ, not only from those of *Opuntia,* but from those of all other cactus genera, in being winged. The fruit, according to Schumann, although we have not been able to confirm his observation definitely, is a capsule with an operculum. Another peculiarity is that the fruit is borne in the end of the stem or branch.

While this genus has good characters, it is no more distinct than many others and does not deserve the relative importance given to it by T. von Post and Otto Kuntze in Lexicon Generum Phanerogamarum, who treat it as one of the only three cactus genera to be conserved, in their view.

Key to Species.

Seeds narrowly winged; spines up to 2 cm. long.. 1. *P. hickenii*
Seeds broadly winged; spines 3 to 10 mm. long.
    Joints strongly tuberculate... 2. *P. fischeri*
    Joints scarcely tuberculate.
        Ovary densely covered with weak spines; wing of seed 1 mm. wide..................... 3. *P. pumilus*
        Ovary loosely covered with stiff spines; wing of seed 2 mm. wide...................... 4. *P. tuberosus*

FIGS. 31, 32.—Pterocactus
hickenii. ×0.7.

FIG. 33.—Pterocactus fischeri. ×1.12. Photograph
by Paul G. Russell.

1. Pterocactus hickenii sp. nov.

Rootstocks moniliform, consisting of at least 4 joints widely separated; joints above ground 2 or 3, 2 to 3 cm. long, almost hidden by the spines; spines from each areole numerous, slender, yellow above, brown at base; glochids numerous; fruit and flower not known; seeds thick, 5 mm. in diameter, with narrow lateral wing.

Collected by Cristóbal M. Hicken (No. 3284) January 10, 1914, near Comodoro Rivadavia, southeastern Chubut, Argentina.

Figures 31 and 32 represent a plant and a seed from the specimen above cited

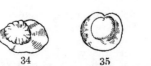

FIGS. 34, 35, 36.—Seeds of three species of Pterocactus.
Natural size.

2. Pterocactus fischeri sp. nov.

Stems low, 1 dm. high or less, spreading or erect, cylindric, 1.5 cm. in diameter, tuberculate; leaves minute, acute; tubercles about as long as broad, arranged in spiral ridges somewhat resembling those of *Opuntia whipplei;* spines numerous, the radials 12 or more, white, setaceous, 6 mm. long, spreading; centrals usually 4, 1 to 1.5 cm. long, brownish, the tips and bases often yellowish;

glochids numerous, yellowish, 3 to 4 mm. long; flowers, in only specimen seen, terminal, almost continuous with the stem; ovary tuberculate and spiny like the stem, deeply umbilicate; seed one, large, flat-winged.

Collected by Walter Fischer in 1914 in the Province of Río Negro, Argentina, and given to Dr. Rose during his visit to Argentina in 1915 by Professor Cristóbal M. Hicken.

While this species resembles some of the species of *Cylindropuntia* of the United States, the spines are not sheathed.

Figure 33 is from a photograph of the specimen above cited; figure 34 shows a seed of the same specimen.

3. Pterocactus pumilus sp. nov.

Plants low, usually prostrate or ascending; joints cylindric, 1 cm. in diameter, covered with weak appressed spines; areoles very woolly; flower terminal; ovary sunk in the apex of the terminal joint, somewhat umbilicate; ovules several; seed flattened, 7 mm. in diameter, very thin.

Collected by Cristóbal M. Hicken (No. 3286), January 8, 1914, at Puerto Piramides, Chubut, Argentina.

Figure 35 shows a seed of above specimen.

4. Pterocactus tuberosus (Pfeiffer).

Opuntia tuberosa Pfeiffer, Enum. Cact. 146. 1837.
Pterocactus kuntzei Schumann, Monatsschr. Kakteenk. **7:** 6. 1897.
Pterocactus kurtzei Schumann in Engler and Prantl, Pflanzenfam. Nachtr. 259. 1897.
Pterocactus decipiens Gürke, Monatsschr. Kakteenk. **17:** 147. 1907.

Roots tuber-like, single or in clusters, usually small but sometimes large and thick, up to 12 cm. long by 8 cm. in diameter, deep-seated, giving off several erect stems, these branching at surface of the ground; terminal branches purplish, turgid, 3 to 40 cm. long, 1 cm. in diameter, more or less clavate; areoles numerous, small, bearing numerous small white appressed spines; flowers terminal, 2 to 3 cm. long; petals long, lanceolate, apiculate, yellow; ovary with numerous areoles bearing long bristles; ovules numerous; fruit dry; seeds large, flat, winged, 10 to 12 mm. in diameter.

Type locality: Near Mendoza, Argentina.
Distribution: Western provinces of Argentina, chiefly in the mountains.

FIG. 37.—Pterocactus tuberosus. Natural size. Photograph by Paul G. Russell.

We have not seen the type of *P. kuntzei*, which is doubtless at Berlin, but we have examined cotypes in the Kurtz Herbarium at Córdoba, Argentina, and at New York.

Opuntia tuberosa, described from Mendoza as long ago as 1837, has long been a puzzle to botanists, who have tried to associate it with various opuntias. Dr. Rose, who visited Mendoza in 1915, found a tuberous-rooted cactus in the mountains above that city, which we are convinced is the plant described by Pfeiffer. There is no doubt, on the other hand, that it is *Pterocactus kuntzei*, from the same region, which was described as new by Schumann in 1897.

Opuntia alpina Gillies (Pfeiffer, Enum. Cact. 146. 1837) was not published, but was given as a synonym of *Opuntia tuberosa*. Schumann referred both names to *Opuntia platyacantha*.

Illustrations: Monatsschr. Kakteenk. **7**: 7; Schumann, Gesamtb. Kakteen f. 107, both as *Pterocactus kuntzei;* Blühende Kakteen **3**: pl. 140, as *P. decipiens*.

Figure 36 shows a seed of a plant, collected by Dr. Rose near Mendoza, Argentina, in 1915; figure 37 is from a photograph of same plant; figure 38 is from a photograph taken by Dr. Carlos Spegazzini.

Fig. 38.—Pterocactus tuberosus.

3. NOPALEA Salm-Dyck, Cact. Hort. Dyck. 1849. 63. 1850.

Much branched cacti with definite cylindric trunks; roots so far as known fibrous; branches or joints flattened, fleshy, often narrow; glochids usually less abundant than in *Opuntia;* spines solitary or in clusters at the areoles, sheathless; leaves small, subterete, soon deciduous; areoles bearing white wool, glochids, and often spines; flowers originating in the areoles usually at or near the edges of the joints; sepals ovate, erect; petals red or pinkish, erect, closely appressed against the numerous stamens and the style; filaments and style slender, much longer than the petals; ovary more or less tuberculate, naked or spiny, with a very deep umbilicus; fruit a juicy berry, red, edible, usually spineless; seeds numerous, flat, covered by a hard bony aril.

Nopalea is closely related to *Opuntia*, with which it is sometimes united; the erect petals and elongated filaments and style are constant in *Nopalea*, however.

Three species were included by Salm-Dyck in this genus when it was described, of which *Opuntia cochenillifera* Linnaeus was the first and is therefore considered the type.

Karl Schumann described five species in his monograph, but since then two species, *N. guatemalensis* and *N. lutea*, have been described by Dr. Rose, and one, *N. inaperta*, by Dr. Griffiths. *N. moniliformis* (Linnaeus) Schumann, based on plate 198 of Plumier, is *Opuntia moniliformis* (Linnaeus) Steudel.

The species are natives of Mexico and Guatemala, and have been accredited to Cuba, although none has recently been observed wild on that island. Some of them are widely cultivated and may be found throughout the warmer parts of the world. Two are of some economic importance and two or three are grown as ornamentals.

The name *Nopalea* is doubtless from nopal, the common name of Mexicans for certain opuntias and nopaleas.

KEY TO SPECIES.

Spineless, or rarely a few short spines on old joints.............................. 1. *N. cochenillifera*
Joints spiny (spines few in *N. auberi*).
    Spines, at least those of young joints, very slender, acicular, several at each areole.
        Spines white.. 2. *N. guatemalensis*
        Spines yellow or becoming brown.
            Joints obovate to oblong, 10 to 22 cm. long, 5 to 10 cm. wide................. 3. *N. lutea*
            Joints linear-oblong to oblong-lanceolate, 6 to 12 cm. long, 2 to 3 cm. wide.......... 3a. *N. gaumeri*
    Spines stouter, subulate.
        Areoles with 1 or 2 spines, or spineless; joints glaucous......................... 4. *N. auberi*
        Areoles with 2 to 4 spines; joints green.
            Joints linear or linear-oblong, 4 to 7 times as long as wide.................... 5. *N. dejecta*
            Joints oblong or oblong-obovate, 2 to 4 times as long as wide.
                Spines 2 to 4; joints not tuberculate..................................... 6. *N. karwinskiana*
                Spines 4 to 12; joints strongly tuberculate............................... 7. *N. inaperta*

1. Nopalea cochenillifera (Linnaeus) Salm-Dyck, Cact. Hort. Dyck. 1849. 64. 1850.

    Cactus cochenillifer Linnaeus, Sp. Pl. 468. 1753.
    Opuntia cochenilifera Miller, Gard. Dict. ed. 8. No. 6. 1768.

Often tall plants, 3 to 4 meters high, with trunks up to 2 dm. thick; branches of ascending or spreading oblong joints, sometimes 5 dm. long, green, bright green at first; spines none or rarely minute ones develop on the older joints; glochids numerous, caducous; leaves small, awl-shaped, soon deciduous; flowers appearing from the tops of the joints, usually in great abundance; flower, from base of ovary to tip of style, 5.5 cm. long; ovary nearly globular, 2 cm. long, with low diamond-shaped tubercles, its areoles bearing many glochids; sepals broadly ovate, acute, scarlet; petals a little longer than the sepals, otherwise similar, persistent; stamens pinkish, exserted 1 to 1.5 cm. beyond the petals; stigma-lobes 6 or 7, greenish, exserted beyond the stamens; style swollen just above its base into a broad disk; fruit red, about 5 cm. long, rarely maturing in greenhouse plants; seeds about 5 mm. long and 3 mm. wide.

Type locality: Jamaica and tropical America.

Distribution: Cultivated in the West Indies and tropical America; its original habitat unknown.

Opuntia magnifolia Noronha (Verhandl, Batav. Genootsch. 5[4]: 22. 1790), published without description, is referred to this species by Schumann and others. The name *Opuntia mexicana*, although it has been used for more than one species, first appeared in Pfeiffer's Enumeratio (p. 150. 1837) as a synonym of *O. cochenillifera*. *Cactus subinermis* Link (Steudel, Nom. ed. 2. 1: 246. 1840) is given as a synonym of *Opuntia cochenillifera*.

The specific name of this plant was given because it is one of the species of cactus from which cochineal was obtained. Cochineal was long supposed to be a vegetable product; it was not until 1703 that, by the aid of the microscope, it was definitely determined to be of insect origin. The cochineal industry is of prehistoric origin. The Spaniards found it well established when they conquered Mexico in 1518, and began at once to export the product. As early as 1523 Cortez was ordered to obtain and send to Spain as much as he possibly could, while during the early colonial days it was one of the chief articles of tribute to the crown. From Mexico and Peru the industry was taken to southern Spain, India, Algiers, South Africa, New Granada (Colombia), Jamaica, and the Canary Islands. The industry grew rapidly and was very profitable. The greatest source of the cochineal was probably the Canary Islands. In about the year 1868 more than 6,000,000 pounds, valued at $4,000,000, were exported from these islands alone, of which the largest part was sent to England.

The cochineal insects were placed on the joints or branches of the cactus plants, where they rapidly multiplied and in about four months were collected by brushing them off into baskets or bags. Then, after being dried in various ways, they became the cochineal of commerce. Two or three such collections were made each year.

PLATE IV

M. E. Eaton del.

1. Upper part of flowering joint of *Nopalea cochenillifera*.   3. Fruit of *Nopalea auberi*.
2. Upper part of flowering joint of *Nopalea auberi*.   4. Flowering joint of *Nopalea dejecta*.

(All natural size.)

The cactuses upon which the cochineal was raised were often grown in large plantations called nopalries, sometimes containing 50,000 plants in rows about 4 feet apart. Since the introduction of the aniline dyes, the cochineal industry has almost disappeared. The cochineal colors, while brilliant and attractive, are not very permanent.

According to J. J. Johnson, this plant was introduced into cultivation in England, in 1688; but according to Ray it was growing in Chelsea before that time.

Illustrations: Hernandez, Nov. Pl. Hist. 78 and 479. f. 1. 1651, as Nopalnochetzli; Andrews, Bot. Rep. **8**: pl. 533; Curtis's Bot. Mag. **54**: pl. 2741, 2742; Descourtilz, Fl. Pict. Antilles **7**: pl. 516, all as *Cactus cochenillifer.* Cycl. Amer. Hort. Bailey **1**: 205. f. 308; Gard. Chron. III. **34**: 92. f. 41; Pfeiffer and Otto, Abbild. Beschr. Cact. **1**: pl. 24, all as *Opuntia cochenillifera;* Förster, Handb. Cact. ed. 2. f. 3, as *Opuntia coccifera;* Dillenius, Hort. Elth. pl. 297, as tuna, etc.; Agr. Gaz. **25**: pls. opp. p. 884; Amer. Garden **11**: 457; Martius, Fl. Bras. 4²: pl. 60. Schumann Gesamtb. Kakteen f. 109, B.

Plate IV, figure 1, shows a plant which flowered in the New York Botanical Garden in 1912.

2. Nopalea guatemalensis Rose, Smiths. Misc. Coll. **50**: 330. 1907.

Tree-like, 5 to 7 meters high, branched, sometimes nearly to the base; joints bluish green, ovate to oblong, 15 to 20 cm. long; areoles numerous, filled with short white wool; spines 5 to 8, unequal, nearly or quite porrect, white or sometimes rose-colored, the longest 2.5 to 3 cm. long; leaves small, linear, reflexed; flower, including ovary, 5 to 6 cm. long; sepals ovate, thickened; petals red; fruit 4 to 5 cm. long, clavate, red, more or less tuberculate, deeply umbilicate, without prominent glochids; seeds irregular, 4 mm. broad.

Type locality: El Rancho, Guatemala.
Distribution: Arid valleys of Guatemala.
Illustrations: Safford, Ann. Rep. Smiths. Inst. **1908**: f. 13, 14; Smiths. Misc. Coll. **50**: pl. 41, 42.

Figure 39 illustrates joints of a plant obtained from Frank Weinberg in 1910.

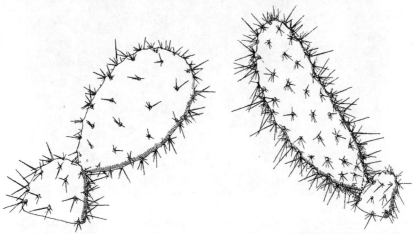

FIG. 39.—Nopalea guatemalensis. ×0.4.   FIG. 40.—Nopalea lutea. ×0.4.

3. Nopalea lutea Rose, Contr. U. S. Nat. Herb. **12**: 405. 1909.

More or less arborescent, 5 meters high or less, with a short, definite trunk and several large, lateral, more or less spreading branches; joints obovate to elliptic or oblong, 10 to 22 cm. long, pale green, slightly glaucous; areoles about 2 cm. apart, large, filled with short brown wool; spines weak, yellow, acicular or bristle-like, the longest 4 cm. long; flowers 5 cm. long; petals red, 2 cm. long;

ovary with numerous prominent areoles filled with yellow bristles; fruit red, 4 cm. long; seeds 4 to 5 mm. in diameter.

Type locality: Near El Rancho, Guatemala.

Distribution: Guatemala, Honduras, and Nicaragua.

This species, although not discovered until 1907, is very common, extending from altitude 300 meters at El Rancho to altitude 1,100 meters near Aguas Calientes. Accord-

Fig. 41.—Nopalea dejecta.

ing to Mr. Charles C. Deam, who has explored extensively in Guatemala, the plant when growing on river sand-bars is low, but in rich soil is tall.

Our reference of this species to Nicaragua is based on a specimen collected by A. S. Oersted in 1845–1848 between Granada and Tipitapa. The joints of this, however, are nearly orbicular or a little longer than broad, with numerous brown spines and glochids. More material may show that this specimen should be referred elsewhere.

Illustration: Contr. U. S. Nat. Herb. 12: pl. 58.

Figure 40 shows a joint of a plant from Guatemala, received from F. Eichlam in 1911.

3*a*. **Nopalea gaumeri** sp. nov.   (See Appendix, p. 216.)

4. **Nopalea auberi** (Pfeiffer) Salm-Dyck, Cact. Hort. Dyck. 1849. 64.   1850.

> *Opuntia auberi* Pfeiffer, Allg. Gartenz. **8**: 282.   1840.

Often 8 to 10 meters high, with a cylindric, jointed trunk, never very spiny, but the areoles bearing tufts of brown glochids; branches often at right angles to the stem; joints narrow, thick, 3 dm. long, bluish green and glaucous; areoles circular, about 2 mm. broad, bearing short white wool and later a tuft of brown glochids; spines, when present, 1 or 2, subulate, the upper one about twice as long as the other, white or nearly so, with brownish tips, the longest one 2 to 3 cm. long; flowers from base of ovary to tip of style about 9 cm. long; petals erect, closely embracing the stamens, rose-pink, ovate-lanceolate, acuminate, 2 to 3.5 cm. long; filaments 12 to 15 mm. longer than the petals, white below, but the exposed parts pinkish; anthers dehiscing before maturing of stigma; style stout, light pink with a large, white, circular disk just above the constricted base; stigma-lobes green; ovary 4 cm. long, with low but very distinct tubercles and a deep umbilicus, its areoles bearing many brown glochids, these sometimes 10 mm. long.

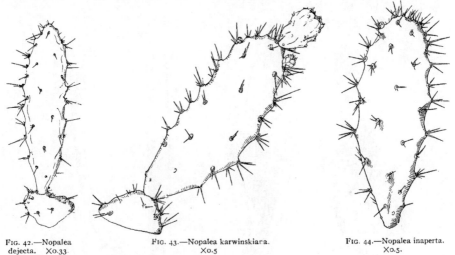

FIG. 42.—Nopalea dejecta. X0.33.

FIG. 43.—Nopalea karwinskiara. X0.5.

FIG. 44.—Nopalea inaperta. X0.5.

Type locality: Erroneously cited as Cuba.

Distribution: Central and southern Mexico.

Illustration: Addisonia **1**: pl. 10.

Plate IV, figure 2, represents a flowering joint of a plant obtained by W. E. Safford at Guadalajara, Mexico, in 1907; figure 3 shows young fruit of the same plant; plate V is from a photograph taken by Dr. MacDougal near Mitla, Mexico, in 1906.

5. **Nopalea dejecta** Salm-Dyck, Cact. Hort. Dyck. 1849. 64.   1850.

> *Opuntia dejecta* Salm-Dyck, Hort. Dyck. 361.   1834.

Plant 1 to 2 m. high, with a definite trunk, very spiny, the old areoles often bearing 6 or 8 spines; joints narrow, 10 to 15 cm. long, only moderately thick, often drooping, bright green even in age, bearing usually two somewhat spreading spines at an areole; spines at first pale yellow or pinkish, in age gray, the longest 4 cm. long; flower, including ovary and style, 5 cm. long; sepals obtuse; petals erect, dark red; stamens long-exserted, dark red.

Type locality: Erroneously cited as Havana, Cuba.

Distribution: Common in cultivation in tropical America; perhaps native in Panama.

Opuntia diffusa and *O. horizontalis* are both given by Pfeiffer (Enum. Cact. 159. 1837) as synonyms of this species.

Illustrations: Agr. Gaz. N. S. W. **25** : pl. opp. p. 138; Roig, Cact. Fl. Cub. pl. [6], this last as *Nopalea auberi*.

Plate IV, figure 4, shows a flowering joint of a plant obtained from Mr. S. F. Curtis in 1897. Figure 41 is from a photograph taken by Dr. Juan T. Roig in the Havana Botanical Garden, Cuba; figure 42 shows a joint of a plant collected by Mr. J. F. Cowell at Panama in 1905.

6. **Nopalea karwinskiana** (Salm-Dyck) Schumann, Gesamtb. Kakteen 752. 1898.

    Opuntia karwinskiana Salm-Dyck, Cact. Hort. Dyck. 1849. 239. 1850.

A tree, 2 meters high or more, with a definite jointed terete spiny trunk; joints oblong, 1.5 to 3 dm. long, light dull green, only slightly glaucous; leaves elongated, acute; areoles distant; spines 3 to 7 from an areole, porrect, 1 to 2 cm. long, pale yellow to nearly white; glochids yellow, numerous, caducous; flowers red, 11 to 12 cm. long; ovary deeply umbilicate, 3 cm. long.

    Type locality: In Mexico.
    Distribution: Mexico.

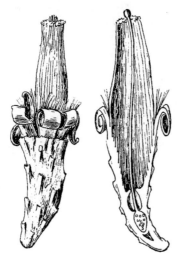
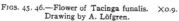
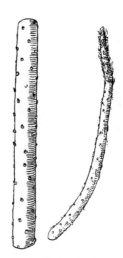

FIGS. 45, 46.—Flower of Tacinga funalis. X0.9.   FIGS. 47, 48.—Tacinga funalis. X0.6.
Drawing by A. Löfgren.

This species was sent from Mexico by Karwinsky, who supposed it was an *Opuntia*. When described by Salm-Dyck in 1850 it had not flowered. It was re-collected by Edmund Kerber near Colima, Mexico, and flowered for the first time in cultivation in 1879.

Our description is drawn chiefly from a plant now in the New York Botanical Garden, obtained from M. Simon, of St. Ouen, Paris, France. In the original description it is stated that the young spines are 2 to 4 and rose-colored, but afterwards 18 to 20, gray and deflexed. *O. nopalilla* Karwinsky (Salm-Dyck, Cact. Hort. Dyck. 1849. 68. 1850) was first given as a synonym of this species.

Figure 43 represents a joint with young fruit, from a plant sent by M. Simon, St. Ouen, Paris, France, in 1901.

7. **Nopalea inaperta** Schott in Griffiths, Monatsschr. Kakteenk. **23**: 139. 1913.

Described as 5 to 7 meters high, but in cultivation much smaller, diffusely branched, often bush-like; trunk very spiny; terminal joints rather small, obovate, 6 to 17 cm. long, strongly tuberculate, bright green; spines usually 3 to 6 at areoles of young joints, more at old ones, yellowish

PLATE V

Nopalea auberi as it grows near Mitla, Mexico.
Photograph by D. T. MacDougal.

brown, 2 cm. long or less; flowers rather small, including ovary and stamens 4 cm. long; filaments numerous, long-exserted; style much longer than the stamens; stigma-lobes 5, green; fruit small, red, 1.5 cm. long.

Type locality: In Yucatán, Mexico.

Distribution: Yucatán.

Dr. Griffiths states that he found this species in the Albert S. White Park, Riverside, California, in 1904. In the Bulletin of the New Mexico Agricultural Experiment Station No. 60 he describes and illustrates it, but without specific name. Later he identified it as the same as one of Schott's specimens from Yucatán, and then published it as above.

Dr. Griffiths compares it with *N. auberi*, but its nearest relative is *N. karwinskiana*, from which it differs in its smaller and more tuberculate joints and much smaller flowers.

Illustration: N. Mex. Agr. Exp. Sta. Bull. **60**: pl. 3, f. 1, as *Nopalea*.

Figure 44 shows a joint from a plant obtained by Dr. David Griffiths at Riverside.

4. TACINGA gen. nov.

Long, clambering or climbing cacti, more or less branched; old stems smooth, brown; branches faintly ribbed, terete; young branches green, each tipped with a tuft of long wool or soft hairs; areoles small but conspicuous, black, the margin giving off long, white, cobwebby hairs; leaves minute, soon deciduous, 3 to 4 mm. long; spines sometimes present, on young joints 2 or 3, reflexed, appressed, brown, 2 to 3 mm. long, not seen on old branches; glochids from the upper parts of the areoles, pale yellow, numerous, caducous, falling in showers at the slightest jarring of the branch; flower-buds acute; flowers usually terminal, opening in the evening or at night; ovary narrow, bearing numerous areoles, the umbilicus very deep; petals few, spreading or recurved; a row of hairs between the petals and the stamens; stamens and style erect, much longer than the petals; fruit oblong, the upper half sterile, bearing areoles but no spines; seeds nearly globular, white, covered with a bony aril.

This genus is intermediate between *Opuntia* and *Nopalea*, having the erect, non-sensitive stamens of the latter, and the areoles, leaves, and glochids of the former. From *Opuntia* it differs in its narrow, green, recurved petals, in having one or possibly more rows of hairs between the stamens and the petals, in the clambering or climbing habit, and its very caducous glochids.

Only one species is known, this a common and characteristic plant of the catinga* in Bahia, Brazil, whence the anagramatic name.

1. Tacinga funalis sp. nov.

At first erect, then climbing over shrubs or through trees, 1 to 12 meters long, somewhat branching; old stems woody, slender; branches usually reddish, the areoles borne on low ribs; glochids short; flower, including ovary, 7 to 8 cm. long; sepals about 10, short, ovate, acute, 5 to 15 mm. long; petals about 7, green, 4 cm. long, acute, revolute;

FIG. 49.—Tacinga funalis. Showing how it climbs over bushes.

stamens erect, connivent, not sensitive; anthers narrow, elongated; style elongated, thread-like, most slender below, a little longer than the stamens, 4.5 cm. long, cream-colored; stigma-lobes 5, green; fruit 4 to 5 cm. long; seeds 3 to 4 mm. broad.

---

* Catinga or caatinga is the common Brazilian name for the thorn-bush desert region in Bahia, Brazil. Dr. Albert Löfgren says that the name (best spelled caatinga) is of Indian origin, meaning caa = wood, forest; tinga = white, clear; a forest in which one can see far.

Common in the dry parts of Bahia, Brazil, where it was collected by Rose and Russell in 1915 (No. 19723, type). Dr. Zehntner thinks there may be a second species, as he has found one with purple flowers; specimens from southern Bahia had purple buds, but the open flowers were not seen. The type comes from Joazeiro, northern Bahia.

Dr. Rose studied this species in the field and believed it to be new. On reaching Rio de Janeiro, he found that Dr. A. Löfgren had also studied it, referring it, however, to *Opuntia*, using the above specific name.

Figures 45 and 46 are copied from drawings of the flowers given to Dr. Rose by Dr. Löfgren; figures 47 and 48 are from twigs of the plant grown at the New York Botanical Garden; figure 49 is from a photograph of the type plant.

5. MAIHUENIA Philippi, Gartenflora 32: 260. 1883.

Plants low, cespitose, often forming small, dense mounds; stems jointed; joints small, globular or short-cylindric; leaves small, usually terete, persistent; leaves of seedlings terete, ascending, with 2 long white bristles in the axils; areoles filled with white wool; spines 3, the central one elongated, the 2 lateral ones small and very short; glochids wanting; flowers large for the size of the plant, yellow or red, usually terminal; petals distinct; flower-tube none; stamens and style much shorter than the petals; fruit juicy (described as dry in one species), oblong to obovoid, bearing small scattered, ovate, persistent leaves; wall of fruit thin; cotyledons linear; seed black, shining, with a brittle testa.

Fig. 50.—Maihuenia valentinii.

Type species: *Opuntia poeppigii* Otto.

There are five species described, rather closely related, natives of the high mountains of Chile and Argentina.

The generic name is derived from maihuen, the native name of the plant.

This is a small, localized genus; it is perhaps nearest *Opuntia*, but is without glochids and has different seeds. The first species was described in 1837, and a second in 1864, both as *Opuntia*. Weber in 1898 transferred them to *Pereskia*, proposing a new subgenus for them, but they are much less like *Pereskia* than *Opuntia*, for, except as to the seeds, they have little in common with *Pereskia;* in habit, leaves, spines, flowers, and fruits they are quite unlike any of the pereskias.

KEY TO SPECIES.

Joints subglobose...........................1. *M. patagonica*
Joints oblong to cylindric.
  Leaves linear, 4 to 6 mm. long............2. *M. poeppigii*
  Leaves ovate to subulate, 2 to 4 mm. long.
    Joints spineless below.................3. *M. brachydelphys*
    Joints spiny all over.
      Leaves on the ovary with white hairs
        in their axils.....................4. *M. valentinii*
      Leaves on the ovary without hairs in
        their axils......................5. *M. tehuelches*

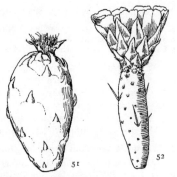

Fig. 51.—Maihuenia poeppigii. ×0.75.
Fig. 52.—Maihuenia brachydelphys. ×0.75.

1. Maihuenia patagonica (Philippi).

> *Opuntia patagonica* Philippi, Linnaea 33: 82. 1864.
> *Pereskia philippii* Weber, Dict. Hort. Bois 939. 1898.
> *Maihuenia philippii* Weber in Schumann, Gesamtb. Kakteen 757. 1898.

Low, much branched, and dense, resembling *Sempervivum tomentosum* in habit; joints subglobose, 1 to 1.5 cm. in diameter; leaves subulate, green; young areoles bearing white hairs; spines weak, hardly pungent, white, the longest 10 to 15 mm. long; flowers 2.8 to 3 cm. long; fruit 8 to 10 mm. long, thicker than long; leaves on the ovary ovate to lanceolate, fleshy, naked in their axils, except some of the upper ones; seeds round, 3 to 4 mm. in diameter.

Type locality: In southern Argentina.

Distribution: Near snow-line on southern mountain ranges of Argentina and Chile.

Opuntia philippii Haage and Schmidt, without description, is given by Weber (Dict. Hort. Bois 939. 1898) as a synonym of this species.

This is called by the natives espina blanca.

2. Maihuenia poeppigii (Otto) Weber in Schumann, Gesamtb. Kakteen 755. 1898.

> *Opuntia poeppigii* Otto in Pfeiffer, Enum. Cact. 174. 1837.
> *Opuntia maihuen* Remy in Gay, Fl. Chilena 3: 29. 1847.
> *Pereskia poeppigii* Salm-Dyck, Cact. Hort. Dyck. 1849. 252. 1850.

Shrubby, much branched, prostrate, forming dense cespitose masses 1 meter broad; joints spiny to the bases, cylindric, 6 cm. long or more, 1.5 cm. in diameter; leaves cylindric, green, 4 to 6 mm. long; spines 3 from each areole, the 2 laterals very short, the central one 1.5 to 2 cm. long; flowers terminal, yellow; fruit oblong to obovoid, about 5 cm. long and 3 cm. thick.

Type locality: In Chile, without definite locality.

Distribution: High mountains of Chile.

Illustrations: Schumann, Gesamtb. Kakteen f. 108, B, C.; Gartenflora 32: pl. 1129, f. 1 to 4, as *Opuntia poeppigii;* Dict. Gard. Nicholson 3: f. 82, as *Pereskia poeppigii.*

Figure 51 is from a fruit obtained by Dr. Rose at the National Museum of Chile, Santiago, in 1914.

Fig. 53.—Maihuenia tehuelches

3. Maihuenia brachydelphys Schumann, Gesamtb. Kakteen 756. 1898.

Cespitose, prostrate; joints cylindric or nearly ellipsoid, naked below, 2 cm. long; spines 2 or 3, one much stouter and longer, yellow except at base and there brown; leaves terete, 2 to 3 mm. long; areoles circular, full of white wool; flowers usually from the tips of joints, red, 3.5 cm. long.

Type locality: Pasco Cruz, Argentina, 34° south latitude, province of Mendoza.
Distribution: Western Argentina.
Opuntia brachydelphys Schumann is mentioned by Kuntze (Rev. Gen. Pl. 3^2: 107. 1898) by name only.
Illustration: Schumann, Gesamtb. Kakteen f. 108, A.
Figure 52 is copied from Schumann's illustration above cited.

4. Maihuenia valentinii Spegazzini, Anal. Mus. Nac. Buenos Aires II. **4**: 289. 1902.

Shrubby, 1 to 2.5 dm. high, dull green; joints cylindric, somewhat clavate, 1 to 3.5 cm. long; leaves ovate, small; spines 3, the central much larger, 2 to 6 cm. long; flowers from near the ends of the branches, 2 cm. broad, the sepals reddish, the petals white to light yellow; stamens indefinite; filaments white; style 6 mm. long, white, longer than the stamens; stigma-lobes 5, short, 2 mm. long, purplish; ovary globular to obconic, 5 to 8 mm. long, bearing numerous triangular fleshy leaves with long white hairs and sometimes 1 or 2 spines in their axils; fruit unknown.

Type locality: Near Trelew, Chubut, Argentina.
Distribution: Territory of Chubut, southern Argentina.
Related to *M. tehuelches* and *M. poeppigii,* but said to be very distinct.
Figure 50 is from a photograph furnished by Dr. Carlos Spegazzini.

5. Maihuenia tehuelches Spegazzini, Anal. Mus. Nac. Buenos Aires II. **4**: 288. 1902.

Shrubby, 2 to 3 dm. high, with many intricate branches, dull green; joints cylindric, ellipsoid to somewhat clavate, 2 to 8 cm. long by 10 to 12 cm. in diameter; leaves ovate, small, 2 to 4 mm. long; spines 3, the central one erect, 2 to 4 cm. long, the 2 lateral ones only 5 to 10 mm. long; flowers at the apex of the branches, 35 to 45 mm. broad, white to yellowish white; fruit globose, naked, dry, 2 cm. in diameter; seeds black, 3 mm. broad.

Type locality: Between San Julián and Río Deseado, Argentina.
Distribution: Dry, rocky deserts, southwestern Argentina.
Figure 53 is from a photograph furnished by Dr. Carlos Spegazzini.

6. OPUNTIA (Tournefort) Miller, Gard. Dict. Abridg. ed. 4. 1754.

Cactodendron Bigelow, Pac. R. Rep. **3**: 102; **4**: 7, 11, iii. 1856.
Consolea Lemaire, Rev. Hort. **1862**: 174. 1862.
Tephrocactus Lemaire, Cact. 88. 1868.
Ficindica St. Lager, Ann. Soc. Bot. Lyon **7**: 70. 1880.

Cacti, sometimes with definite trunks, or more often much branched from the base, the branches often spreading, reclining, or prostrate, sometimes clambering, but never climbing (one species known with annual stems); roots fibrous or rarely tuberous and large and fleshy; ultimate branches (joints or pads) cylindric to globose or flattened, usually very fleshy, sometimes woody; areoles axillary, bearing spines, barbed bristles (glochids), hairs, flowers, and sometimes glands; leaves usually small, terete, mostly early deciduous; spines solitary or in clusters, terete or flattened, naked or sheathed, variously colored; glochids usually numerous, borne above the spines; flowers usually one at an areole; ovary inferior, one-celled, many-ovuled, bearing leaves, the areoles often with spines and glochids; sepals green or more or less colored, usually grading into the petals; petals usually of various shades and combinations of green, yellow, and red (rarely white), widely spreading; stamens much shorter than the petals, sensitive; style single, thick; stigma-lobes short; fruit a berry, dry or juicy, often edible, spiny or naked, globular, ovoid or ellipsoid; seed covered by a hard, bony aril, white, flattened; embryo curved; cotyledons 2, large.

The species grow naturally from Massachusetts to British Columbia south to the Strait of Magellan. Several have been naturalized and have become very abundant locally in the Old World and in Australia.

The type species is *Cactus opuntia* Linnaeus.

Karl Schumann recognized 131 species in his "Gesamtbeschreibung der Kakteen," published during the years 1897 and 1898. Many have been described since this monograph was published.

The name Opuntia was that of a town in Greece, where some cactus-like plant is said to have grown.

The genus is important economically. It furnishes the well-known tuna fruit largely imported into our eastern cities from Italy and which is common in the markets of Mexico. Some species are used for hedges, the branches of others are cooked like spinach, and still others furnish forage for stock.

The species are numerous and very diverse, and have at various times been grouped by authors into several genera, while other species, now referred by us to *Nopalea*, *Maihuenia*, and *Pereskiopsis*, were included in *Opuntia*.

The following genera now referred to *Opuntia* have been regarded as distinct from it:

Consolea was described by Lemaire in 1862. He described five species, of which *C. rubescens* is the first and therefore the type. This group is a striking one, characterized by a pronounced cylindric trunk in old plants, with an unjointed central woody axis, peculiar semaphore-like branches at the top, and very small flowers. There are eight species of this group, described under our series *Spinosissimae*. They are confined to the West Indies, although *C. rubescens*, the spineless race of *Opuntia catacantha*, was originally described as from Brazil—doubtless erroneously.

Tephrocactus was described by Lemaire in 1868, and to it he referred eight species of *Opuntia*. *T. diadematus* is the type species. Schumann included it in *Opuntia* as a subgenus, with 15 species. They are all South American, chiefly in Argentina and Bolivia.

Ficindica was established by St. Lager in 1880, based on *Opuntia ficus-indica*, which is clearly congeneric with *Opuntia opuntia*.

In 1856 the name *Cactodendron* was proposed in an account of Whipple's Expedition, published in volumes 3 and 4 of the Pacific Railroad Reports. It was apparently not intended to be a formal publication, but as a definite species is indicated, the name is published. It will be of interest to record here the evidence upon which we reach this conclusion:

Cactodendron Bigelow Pac. R. Rep. **3**: 102; **4**: 7, 11; Additional Notes and Corrections iii. 1856.

"There are * * * *Opuntia* of many varieties; some with wide leaf-like joints, others of shrubby form and woody fibre, which the botanist proposes to name *Cactodendron*." Pac. R. Rep. **3**: 102.

"Immediately on our entrance into this valley (November 19 [1856]) we found and collected a new species of *Opuntia*, with prostrate, nearly terete joints, entirely devoid of woody fibre; * * *. Lieutenant Whipple discovered the first specimen of our new *Cactodendron*, as we were pleased to call it, to distinguish it from the *O. arborescens*." Pac. R. Rep. **4**: 7.

"The arborescent *Opuntia*, first found near Zuni, which, to distinguish from the true *O. arborescens*, we called *Cacto-dendron*, finds its western limits near the termination of this region." Pac. R. Rep. **4**: 10.

"15. 'New arboresent *Opuntia*,' called also 'our new *Cactodendron*,' pages 7 and 11, is *Opuntia whipplei*, E. & B., new species." Pac. R. Rep. **4**: Additional Notes and Corrections iii.

Opuntias are known under a great variety of names. Among the names for the flat-jointed species, the most common are: prickly pear in the United States; tuna in Mexico; sucker and bullsucker in the Lesser Antilles. For the round-stemmed forms we have: cane cactus, and such Mexican names as cholla and tasajo. Dr. David Griffiths has published a list of names used in Mexico.

The genus *Opuntia*, as understood by us, is composed of at least 250 species, but more than 900 names are to be found in literature. No type specimens of many of the species were preserved by their authors, some have, apparently, been lost, and some, which are probably preserved, we have been unable to study.

The genus shows a great range in stem structure, varying from cylindric to broad and flat. These extremes suggest different generic types, but these characters can not be used except in the most general way, for some species have both rounded and flattened stems. Some with round stems have flowers which suggest a closer relationship with the species with flattened stems.

The habits of some of the species are very characteristic, while others show a wide range of forms. Many of the erect or tree-like forms, when grown from cuttings, develop bushy habits much unlike their normal shapes.

The spines, while somewhat constant in color in some species, vary considerably in others, and the number of spines is rather inconstant. Species which are normally abundantly spined are sometimes naked when cultivated, while species which are normally naked sometimes develop spines in cultivation; cultivated specimens usually have weaker spines and sometimes decidedly different ones from wild plants.

The flowers often vary greatly in color, as is seen especially in *O. versicolor* and *O. lindheimeri*, which show wide ranges of color forms. Some flowers vary in color during the day.

We group the species known to us into 3 subgenera, 46 series, and with the following characteristics:

KEY TO SUBGENERA AND SERIES OF OPUNTIA.

A. Joints all terete, elongated or short, cylindric to globose.
  B. Branches several, many-jointed.......................... Subgenus 1. CYLINDROPUNTIA
    C. Spines with papery sheaths.
      D. Spines, at least some of them, solitary, sometimes several, acicular; ultimate branches slender, rarely more than 1 cm. thick.
        E. Stem and branches conspicuously marked by flattened, diamond-shaped tubercles; fruit dry, covered with long bristle-like spines...................... Series 1. *Ramosissimae* (N. A.)
        EE. Tubercles not flattened nor diamond-shaped; fruit usually a naked berry...................... Series 2. *Leptocaules* (N. A.)
      DD. Spines always more than 1; ultimate branches stouter.
        E. Ultimate branches not over 2 cm. thick.............. Series 3. *Thurberianae* (N. A.)
        EE. Ultimate branches 2 cm. thick or more.
          F. Fruit dry.. Series 4. *Echinocarpae* (N. A.)
          FF. Fruit fleshy.
            G. Tubercles of young joints scarcely longer than broad. Series 5. *Bigelovianae* (N. A.)
            GG. Tubercles distinctly longer than broad.
              H. Tubercles narrow, high, laterally flattened...... Series 6. *Imbricatae* (N. A.)
              HH. Tubercles broad, low.................. Series 7. *Fulgidae* (N. A.)
    CC. Spines without sheaths.
      D. Joints not tuberculate, or with broad or flat tubercles.
        E. Areoles long-woolly or with weak hairs (without hairs in *O. verschaffeltii*)........................ Series 8. *Vestitae* (S. A.)
        EE. Areoles neither long-woolly nor long-hairy.
          F. Joints clavate or crested.................. Series 9. *Clavarioides* (S. A.)
          FF. Joints neither clavate nor crested. \
            G. Low, slender species, scarcely, if at all, tuberculate.. Series 10. *Salmianae* (S. A.)
            GG. Tall, stout species, the tubercles broad or flat; leaves large.............................. Series 11. *Subulatae* (S. A.)
      DD. Joints strongly tuberculate; joints cylindric.
        E. Tall, shrubby species; joints cylindric.............. Series 12. *Miquelianae* (S. A.)
        EE. Low, prostrate species; joints clavate (transition to *Tephrocactus*)........................ Series 13. *Clavatae* (N. A.)
  BB. Branches 1 to few-jointed, the short joints usually clustered.... Subgenus 2. TEPHROCACTUS (S. A.)
    C. Joints, at least some of them, cylindric, tuberculate, the tubercles contiguous (transition to *Cylindropuntia*) Series 1. *Weberianae*
    CC. Joints globose to oblong, mostly little, if at all, tuberculate.
      D. Areoles normally bearing many long white hairs, which often cover the whole plant...................... Series 2. *Floccosae*
      DD. Areoles without hairs.
        E. Spines, when present, at least some of them, modified into flat, papery processes...................... Series 3. *Glomeratae*
        EE. Spines, when present, all subulate or acicular, terete or somewhat flattened............................ Series 4. *Pentlandianae*

KEY TO SUBGENERA AND SERIES OF OPUNTIA—continued.

AA. At least some of the joints flat or compressed.................... Subgenus 3. PLATYOPUNTIA
  B. Stems perennial, stout or slender.
    C. Plants branching from near or at the base, not forming erect,
      cylindric unjointed trunks; flowers mostly large.
      D. Epidermis glabrous or pubescent, not papillose-tuberculate
        when dry.
        E. Flowers perfect; petals obovate to oblong.
          F. Fruit a juicy berry (exceptions in Series 5, *Basilares*).
            G. Joints readily detached.
              H. Joints very readily detached; low, mostly small-
                jointed species.
                I. Joints little flattened, subterete (transition to
                  Cylindropuntia).......................... Series 1. *Pumilae* (N. A.; S. A.)
                II. At least the ultimate joints distinctly flattened.
                  J. Ultimate joints or all joints turgid.......... Series 2. *Curassavicae* (N. A.; S. A.)
                  JJ. Ultimate joints flat and thin.............. Series 3. *Aurantiacae* (S. A.)
              HH. Joints less readily detached; mostly taller and
                larger-jointed species...................... Series 4. *Tunae* (N. A.; S. A.)
            GG. Joints not readily detached, persistent.
              H. Areoles small, 1 to 2 mm. in diameter, not ele-
                vated, mostly close together............... Series 5. *Basilares* (N. A.)
              HH. Areoles larger, mostly distant.
                 I. Prostrate or spreading species; joints relatively
                small. (*O. austrina* suberect.)
                  J. Joints not tuberculate.
                    K. Flowers small, brick-red............... Series 6. *Inamoenae* (S. A.)
                    KK. Flowers large, yellow................. Series 7. *Tortispinae* (N. A.)
                  JJ. Joints strongly tuberculate................ Series 8. *Sulphureae* (S. A.)
               II. Bushy, depressed or tall species.
                 J. Spines, when present, brown or yellow (white
                 in *O. setispina*).
                 K. Spines brown, at least at the base or tip.
                  L. Bushy or depressed species.
                    M. Fruit very small.................. Series 9. *Strigiles* (N. A.)
                    MM. Fruit large.
                      N. Spines acicular.............. Series 10. *Setispinae* (N. A.)
                      NN. Spines subulate.............. Series 11. *Phaeacanthae* (N. A.)
                  LL. Tall species, sometimes with a definite
                  trunk (*O. galapageia* sometimes de-
                  pressed).
                  M. Spines several at each areole........ Series 12. *Elatiores* (N. A.; S. A.)
                  MM. Spines, when present, 1 to few at each
                    areole .... Series 13. *Elatae* (S. A.)
                KK. Spines, if any, yellow, at least partially.
                L. Epidermis glabrous.
                 M. Areoles close together, bearing long
                 brown wool.................. Series 14. *Scheerianae* (N. A.)
                 MM. Areoles distant, without long wool. Series 15. *Dillenianae* (N. A.)
                LL. Epidermis, at least that of the ovary,
                pubescent...................... Series 16. *Macdougalianae* (N. A.)
             JJ. Spines, when present, white (or faintly yellow).
               K. Epidermis pubescent.
               L. Spines, when present, acicular.......... Series 17. *Tomentosae* (N. A.)
               LL. Spines several, setaceous, flexible........ Series 18. *Leucotrichae* (N. A.)
              KK. Epidermis glabrous.
               L. Areoles bearing long, soft hairs......... Series 19. *Orbiculatae* (N. A.)
               LL. Areoles without long hairs.
                M. Joints green or bluish green.
                 N. Spineless, or with few, usually short,
                 spines...................... Series 20. *Ficus-indicae* (N. A.; S. A.)
                 NN. Spiny, at least old joints so........ Series 21. *Streptacanthae* (N. A.)
                MM. Joints blue........................ Series 22. *Robustae* (N. A.)
    FF. Fruit dry, not juicy............................. Series 23. *Polyacanthae* (N. A.)
    EE. Flowers dioecious; petals very narrow .... Series 24. *Stenopetalae* (N. A.)
  DD. Epidermis densely papillose-tuberculate when dry........ Series 25. *Palmadorae* (S. A.)
  CC. Plants with erect, unjointed trunks, the branches with flat
    joints; flowers mostly small.
    D. Flowers small; joints spreading.
      E. Joints all flat, relatively thick.................... Series 26. *Spinosissimae* (N. A.)
      EE. Some joints terete, others flat and very thin.......... Series 27. *Brasilienses* (S. A.)
    DD. Flowers large; joints ascending....................... Series 28. *Ammophilae* (N. A.)
BB. Stems annual, very slender................................ Series 29. *Chaffeyanae* (N. A.)

Subgenus 1. CYLINDROPUNTIA.

Includes the many-jointed species in which none of the joints is at all flattened.

Series 1. RAMOSISSIMAE.

The series consists of a single bushy species, with slender joints, the nearly flat tubercles diamond-shaped and contiguous, the acicular spines, when present, usually only 1 at an areole.

1. Opuntia ramosissima Engelmann, Amer. Journ. Sci. II. **14**: 339. 1852.

Opuntia tessellata Engelmann, Proc. Amer. Acad. **3**: 309. 1856.

Frutescent, bushy, sometimes 2 meters high, the branches gray, often widely spreading, and 9 cm. long; tubercles low, slightly convex, 4-angled to 6-angled, giving the surface an appearance of being covered with diamond-shaped plates; leaves ovoid, 1 to 3 mm. long, acute; areoles on young shoots circular, with white or tawny wool and pale glochids, the upper part in age compressed into the narrow slit between the two adjoining tubercles, the lower part depressed-linear, with a slightly elevated border; spines often wanting, but when present abundant, usually one at each areole, rarely 2, porrect, acicular, sometimes 6 cm. long, usually reddish when young, covered by loose, yellow, papery sheaths; flowers, including ovaries, 3 to 4 cm. long; sepals subulate, similar to the leaves of the ovary, but longer; petals greenish yellow, tinged with red, obovate, aristulate, about 1 cm. long; stamens greenish yellow; anthers orange-colored; style and stigma-lobes cream-colored; ovary narrowly obconic, covered with emarginate tubercles, the areoles bearing wool and long glochids, but no spines; fruit dry, obovate, 2 to 2.5 cm. long, covered with clusters of weak, slender spines, appearing like a bur; seeds few, white, 5 mm. broad.

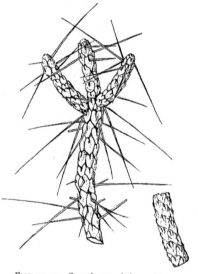

FIGS. 54, 55.—Opuntia ramosissima. X0.75.

Type locality: In California, near the Colorado River.

Distribution: Southern Nevada, western Arizona, southeastern California, northwestern Sonora and probably northeastern Lower California.

The flowers of this species have been described as purple, apparently erroneously.

This species is found in the most arid deserts of the southwestern part of the United States, usually growing on low hills, and is confined chiefly to the lower Colorado; it is here rather inconspicuous and might easily be overlooked. It is one of the least succulent species of the genus, the terminal shoots soon becoming hard, and hence the plant is difficult to propagate from cuttings, and is rarely found in greenhouse collections.

Opuntia tessellata cristata Schumann (Monatsschr. Kakteenk. **8**: 70. 1898) is a striking monstrosity which Schumann has described and figured.

Illustrations: Cact. Journ. **1**: pl. [1]; Cycl. Amer. Hort. Bailey **3**: f. 1549; Pac. R. Rep. **4**: pl. 21; 24, f. 20, all as *Opuntia tessellata*.

Figure 54 represents a spiny branch drawn from a specimen sent by Mr. S. B. Parish from Barstow, California, in 1915; figure 55 shows a portion of an unarmed branch sent by the same collector from the same locality.

Series 2. LEPTOCAULES.

Bushy species, with slender joints, the ultimate ones 4 to 15 mm. thick, often readily detached; the flowers small.

Inhabitants of the southwestern United States, Mexico, northern South America, and one species in Santo Domingo.

KEY TO SPECIES.

Ultimate joints short, usually at right angles to the branches, 4 to 7 mm. thick.
Bushy plants, 1.5 meters high or less; fruit small, fertile.
Branches scarcely if at all tuberculate.
Leaves ovoid to ovoid-subulate; young areoles long-hairy......................... 2. *O. mortolensis*
Leaves linear; areoles not long-hairy... 3. *O. leptocaulis*
Branches long-tuberculate... 4. *O. tesajo*
Elongated plants, up to 2 meters long; fruit larger, sterile.............................. 5. *O. caribaea*
Ultimate joints longer, 8 to 15 mm. thick, usually at an acute angle to the branches.
Joints only slightly tuberculate... 6. *O. arbuscula*
Joints manifestly tuberculate... 7. *O. kleiniae*

2. Opuntia mortolensis sp. nov.

Slender, 6 dm. high or less, dull green, with dark blotches below the areoles, the ultimate twigs short, sometimes only 2 cm. long, 4 to 5 mm. thick, scarcely tuberculate; leaves ovate to ovate-subulate, 2 to 4 mm. long, green, with acute bronze-colored tips; young areoles with numerous, early deciduous, weak white hairs sometimes longer than the leaves, and several brown glochids; areoles of old branches with solitary acicular spines 3 to 5 cm. long, these with tightly fitting brownish sheaths; flowers and fruit unknown.

Described from No. 25360, New York Botanical Garden, received from the garden of Sir Thomas Hanbury, La Mortola, Italy, in 1906. Mr. Berger has referred this specimen to *Opuntia leptocaulis longispina*, but this was considered by Dr. Engelmann as the "usual western form" of *O. leptocaulis*.

An herbarium specimen collected by Rose, Standley, and Russell at Empalme, Sonora, Mexico, March 11, 1910 (No. 12644), appears to be referable to this species.

The short leaves and long-hairy young areoles appear to distinguish this plant from *O. leptocaulis*.

Illustration: Gard. Chron. III. 34: f. 37, as *Opuntia leptocaulis longispina*.

Plate VI, figure 1, represents a branch of a plant sent from La Mortola, Italy, in 1906; figure 2 shows a leafy twig of the same plant.

3. Opuntia leptocaulis De Candolle, Mém. Mus. Hist. Nat. Paris 17: 118. 1828.

Opuntia ramulifera Salm-Dyck, Hort. Dyck. 360. 1834.
Opuntia gracilis Pfeiffer, Enum. Cact. 172. 1837.
Opuntia fragilis frutescens Engelmann, Bost. Journ. Nat. Hist. 5: 245. 1845.
Opuntia virgata Link and Otto in Förster, Handb. Cact. 506. 1846.
Opuntia vaginata Engelmann in Wislizenus, Mem. Tour North. Mex. 100. 1848.
Opuntia frutescens Engelmann, Bost. Journ. Nat. Hist. 6: 208. 1850.
Opuntia frutescens brevispina Engelmann, Proc. Amer. Acad. 3: 309. 1856.
Opuntia frutescens longispina Engelmann, Proc. Amer. Acad. 3: 309. 1856.
Opuntia leptocaulis brevispina S. Watson, Bibl. Index 1: 407. 1878.
Opuntia leptocaulis vaginata S. Watson, Bibl. Index 1: 407. 1878.
Opuntia leptocaulis stipata Coulter, Contr. U. S. Nat. Herb. 3: 456. 1896.
Opuntia leptocaulis longispina Berger, Bot. Jahrb. Engler 36: 450. 1905.

Usually bushy, often compact, 2 to 20 dm. high, but sometimes with a short, definite trunk 5 to 8 cm. in diameter, dull green with darker blotches below the areoles, with slender, cylindric, ascending, hardly tuberculate branches; branches, especially the fruiting ones, thickly set with short, usually spineless joints spreading nearly at right angles to the main branches, very easily detached; leaves green, awl-shaped, 12 mm. long or less, acute; spines usually solitary at young areoles, very slender, white, at areoles of old branches 2 or 3 together, 2 to 5 cm. long or less; sheaths of spines closely fitting or loose and papery, yellowish brown to whitish; areoles with very short white wool; flowers greenish or yellowish, 1.5 to 2 cm. long including the ovary; sepals broadly ovate, acute, or cuspidate; ovary obconic, bearing numerous small woolly brown areoles subtended by small leaves, its glochids brown; fruit small, globular to obovate or even clavate, often proliferous, red or rarely yellow, 10 to 18 mm. long, turgid, slightly fleshy; seeds compressed, 3 to 4 mm. broad, with narrow, often acute, margins.

Type locality: In Mexico.
Distribution: Southwestern United States and Mexico.

This species has a wide distribution for an *Opuntia*, extending from southern United States to Puebla, Mexico.

The great variation in the length of the spines and in the character of the spine sheaths has led to the description of several varieties. These all seem to us to merge into the one species, as above indicated. It sometimes hybridizes with *O. imbricata*. See C. B. Allaire's plant from San Antonio, New Mexico.

The following names, *Opuntia leptocaulis laetevirens* Salm-Dyck (Hort. Dyck. 184. 1834), *O. gracilis subpatens* Salm-Dyck (Cact. Hort. Dyck. 1849. 73. 1850), and *O. leptocaulis major* Toumey (Cycl. Amer. Hort. Bailey 3 : 1152. 1901) are printed but not described.

Illustrations: Bull. Torr. Club 32 : pl. 10, f. 9; Rep. Mo. Bot. Gard. 19 : pl. 21, in part; Safford, Ann. Rep. Smiths. Inst. 1908 : f. 12; Emory, Mil. Reconn. app. 2. f. 12; Pac. R. Rep. 4 : pl. 20, f. 1; pl. 24, f. 13 to 15, all as *Opuntia vaginata*. Cact. Journ. 1 : 154, as *Opuntia*

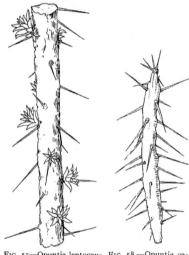

FIG. 56.—Opuntia leptocaulis in the foreground.

FIG. 57.—Opuntia leptocaulis. ×0.4. FIG. 58.—Opuntia caribaea. ×0.66.

frutescens. Pac. R. Rep. 4 : pl. 20, f. 4, 5; pl. 24, f. 19, all as *Opuntia frutescens brevispina.* Pac. R. Rep. 4 : pl. 20, f. 2, 3; pl. 24, f. 16 to 18, all as *Opuntia frutescens longispina.*

Plate VI, figure 3, represents a fruiting branch from a plant collected by Dr. Rose near Sierra Blanca, Texas, in 1913; figure 4 shows a fruiting branch from another Texas plant obtained by the same collector. Figure 56 is from a photograph taken by Dr. MacDougal near Tucson, Arizona, in 1913; figure 57 represents a branch with young leafy shoots, of a specimen collected by Dr. Rose in 1913 at Laredo, Texas.

4. Opuntia tesajo Engelmann in Coulter, Contr. U. S. Nat. Herb. **3** : 448. 1896.

Bushy, 3 dm. broad and high; joints slender, indistinctly tuberculate, 2 to 5 cm. long; areoles 5 to 6 mm. apart; leaves awl-shaped, 2 to 4 mm. long, often red; spines at first 2, small, dark brown, 4 to 8 mm. long, either erect or reflexed; later a long central spine develops, this porrect, 5 cm. long, yellow near the tip; flowers yellow, small, 1.5 to 1.8 cm. long, including the ovary; style whitish; stigma-lobes 5, yellowish.

Type locality: In Lower California.

Distribution: Central part of Lower California.

The type of this little-known species should be in the herbarium of the Missouri Botanical Garden, at St. Louis, but it can not now be found. The species has been in cultivation at La Mortola, Italy, but it does not do well under cultivation. Dr. C. A. Purpus, who has collected the plant in Lower California, regarded it as related to *O. ramosissima*, claiming that the stems have the peculiar marking of that species. This

M. E. Eaton del.

1, 2. Branches of *Opuntia mortolensis*.   5. Flowering branch of *Opuntia arbuscula*.
3, 4. Branches of *Opuntia leptocaulis*.   6. Flowering branch of *Opuntia kleiniae*.
(All natural size.)

6. Opuntia arbuscula Engelmann, Proc. Amer. Acad. **3**: 309. 1856.

Opuntia neoarbuscula Griffiths, Rep. Mo. Bot. Gard. **19**: 260. 1908.

Forming a bush 2 to 3 meters high, often with a rounded, very compact top with numerous short branches; trunk short, 10 to 12 cm. in diameter, with several woody branches; ultimate joints 5 to 7.5 cm. long, 8 mm. in diameter, with low, indistinct tubercles; leaves small; spines usually 1, but sometimes several, especially on old joints, porrect, up to 4 cm. long, covered with loose straw-colored sheaths; flowers greenish yellow tinged with red, 3.5 cm. long; fruit often proliferous, sometimes only one-seeded.

Fig. 60.—Opuntia arbuscula.

Type locality: On the lower Gila near Maricopa village.

Distribution: Arizona and Sonora.

Opuntia congesta Griffiths (Rep. Mo. Bot. Gard. **20**: 88, pl. 2, f. 4, 7; pl. 8; pl. 13, f. 5. 1909), from the description, is near this species and probably a race of it.

Races of the species differ in size, in armament, in the length of the tubercles, and in size and shape of the fruit.

Illustrations: Ariz. Agr. Exp. Sta. Bull. **67**: pl. 6, f. 2; Bull. Torr. Club **32**: pl. 9, f. 3; Plant World **11**[10]: f. 11; Rep. Mo. Bot. Gard. **19**: pl. 22; **19**: pl. 23, in part, this last as *Opuntia neoarbuscula;* Carnegie Inst. Wash. **269**: pl. 11, f. 95.

Plate vi, figure 5, represents a flowering branch from Professor J. W. Toumey's collection at Tucson, Arizona. Figure 60 is from a photograph taken by Dr. MacDougal near Tucson, Arizona, in 1906; figure 61 is from a photograph taken by George B. Sudworth in Santa Rita Mountains, Arizona; figure 62 shows a fruiting branch from the same collection.

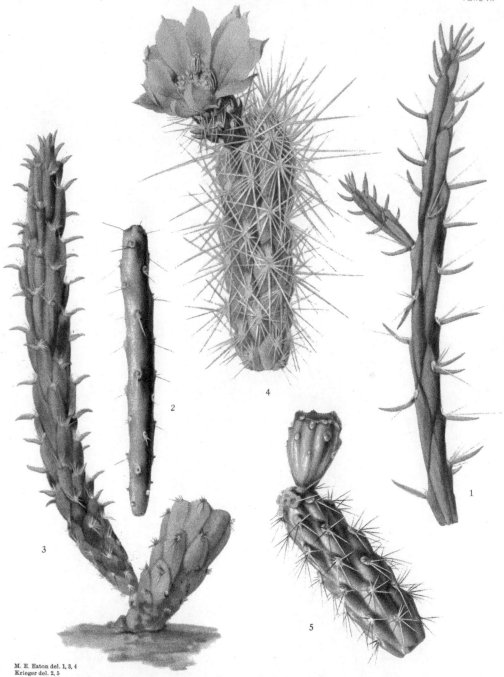

M. E. Eaton del. 1, 3, 4
Krieger del. 2, 5

1. Leafy branch of *Opuntia kleiniae*.
2. Terminal branch of *Opuntia vivipara*.
3. Branch of *Opuntia parryi*.

4. Flowering branch of *Opuntia echinocarpa*.
5. Fruiting branch of *Opuntia versicolor*.

(All natural size.)

7. Opuntia kleiniae De Candolle, Mém. Mus. Hist. Nat. Paris **17**: 118. 1828.

Opuntia wrightii Engelmann, Proc. Amer. Acad. **3**: 308. 1856.
Opuntia caerulescens Griffiths, Rep. Mo. Bot. Gard. **20**: 86. 1909.

Stems pale, glaucous, sometimes 2.5 meters tall, woody at base; tubercles long; areoles large, a little longer than wide, filled with white wool from the very first; spines usually 1, but sometimes more, from the base of the areole, covered with yellow sheaths, on old joints accompanied by several bristle-like spines from the lower margin of the areole; glochids yellow to brown; leaves linear, 15 cm. long, acute; flowers 3 cm. long, purplish; petals broad, rounded at apex; fruit red, 2 to 2.5 cm. long, long persisting; seeds 4 to 5 mm. broad.

Fig. 61.—Opuntia arbuscula. Fig. 62.—Opuntia arbuscula. ×0.75.

Type locality: In Mexico.

Distribution: Texas to central Mexico.

Opuntia kleiniae was originally described as without tubercles on the stems, which has raised the question whether the plant bearing this name is properly referred; in this respect *O. arbuscula* answers the description better, but it is very doubtful whether *O. arbuscula* could have been known at that time.

Opuntia kleiniae has long been in cultivation and is to be seen in most collections.

In 1910 Dr. Rose collected near Alamos, Mexico, an *Opuntia* very similar in habit and joints to *O. kleiniae*, but much more spiny.

Opuntia kleiniae cristata (Cat. Darrah Succ. Manchester 55. 1908) is a garden form. *O. kleiniae laetevirens* Salm-Dyck (Cact. Hort. Dyck. 1849. 73. 1850) is only a name.

Illustrations: Abh. Bayer. Akad. Wiss. München **2**: pl. 1, sec. 7, f. 9; Rep. Mo. Bot. Gard. **19**: pl. 21, in part; **20**: pl. 6, in part, this last as *Opuntia caerulescens.*

Plate VI, figure 6, represents a flowering branch of a specimen obtained from M. Simon, of St. Ouen, Paris, France, in 1901; plate VII, figure 1, represents a leafy branch of a specimen collected by Dr. Rose at Ixmiquilpan, Mexico, in 1905.

Two remarkable opuntias were collected in Lower California by Dr. Rose in 1911, but as they were not in flower or fruit, and have not developed flowers since they were brought into cultivation, we are unable to describe them fully; they are doubtless of this relationship and their characters are given as follows:

OPUNTIA sp.

Stems 1.3 to 2 meters high, rather weak, often clambering over bushes, 10 mm. in diameter, woody below, pale, when dry the white epidermis peeling off; lateral branches numerous, horizontal, short (2 to 6 cm. long); areoles on old stems bearing 3 or 4 long (2 to 4 cm. long) needle-like brownish spines; young areoles usually with a single spine each, filled with brown wool; glochids brown, numerous sheaths on young spines straw-colored, soon deciduous; flowers and fruit unknown.

Description based on field notes and on living and herbarium specimens.

Collected by Dr. Rose on Santa Cruz Island, Gulf of California, April 1, 1911 (No. 16845).

OPUNTIA sp.

Procumbent, forming an indeterminable mass of spiny branches, 3 to 10 dm. in diameter; old stems woody, smooth, brown, and shiny, 2 cm. in diameter; branches 10 to 20 cm. long, bluish green; spines of two kinds; the 2 to 4 principal ones long (2 to 3 cm. long), needle-like, at first covered with thin yellow sheaths, straw-colored when young, becoming purplish, finally fading to gray; secondary spines 4 to 6, radial, inconspicuous; glochids brownish; flowers and fruit unknown.

Description based on field notes and living and herbarium specimens.

Collected by Dr. J. N. Rose on East San Benito Island, off the coast of Lower California, March 9, 1911 (No. 16085). This is, doubtless, the plant referred to by Walton (Cact. Journ. 2: 137. 1899) as *O. ramosissima*, but it is not that species.

Series 3. THURBERIANAE.

Bushy, arborescent, or depressed species, with slender joints, the ultimate ones tuberculate, about 2 cm. thick or less, the areoles bearing several spines. We recognize 8 species, 7 of them natives of the southwestern United States and northern Mexico, and 1 in Lower California.

KEY TO SPECIES.

```
Bushy or arborescent species, 6 dm. high or higher.
   Tubercles narrowly oblong, 1 cm. long or more.
      Joints readily detached.................................................... 8.  O. vivipara
      Joints not readily detached.
         Longer spines 2.5 cm. long or longer.
            Flowers orange to scarlet.......................................... 9.  O. tetracantha
            Flowers purple....................................................10.  O. recondita
         Spines 2 cm. long or less............................................11.  O. thurberi
   Tubercles low, oblong, 6 to 8 mm. long.....................................12.  O. clavellina
Depressed species, 6 dm. high or less.
   Spines yellow or brown; flowers green or tinged with yellow.
      Spines yellow, up to 5 cm. long; petals 1 to 1.5 cm. long...............13.  O. davisii
      Spines brown, 2.5 cm. long or less; petals 2 to 2.5 cm. long............14.  O. viridiflora
   Spines white; flowers yellow..............................................15.  O. whipplei
```

8. Opuntia vivipara Rose, Smiths. Misc. Coll. **52**: 153. 1908.

Plant 2 to 3.5 meters high, usually several strong branches from the base, 8 to 10 cm. in diameter, much branched above, but not compactly so; old stems with rather smooth bark; young branches bluish green, slender, 1 to 2 cm. long, 10 to 12 mm. in diameter; tubercles low, oblong, 15 to 20 mm. long; areoles when young bearing a dense cushion of yellow wool with few or no glochids; spines 1 to 4, 2 cm. long or less, porrect or ascending, covered with straw-colored sheaths; leaves small, terete, acutish, purple; flowers numerous, borne in clusters at the top of the branches, purplish; ovary strongly tuberculate, bearing white deciduous bristles; fruit oblong, 4 to 6 cm. long, smooth, with a somewhat depressed umbilicus, yellowish green, spineless; seeds white, very thick, 5 mm. long.

Type locality: Near Tucson, Arizona.
Distribution: Known only from type locality.

The relationship of this species is doubtful; it resembles certain garden forms of *O. tetracantha*, but differs from typical forms of that species in its much larger fruit and seeds, different armament, and habit. The type grew associated with *O. spinosior* and *O. versicolor*, but there is no indication that it is the result of hybridization of those species.

Illustrations: Smiths. Misc. Coll. **52**: pl. 12; Plant World 11[10]: f. 12.

Plate VII, figure 2, represents a branch drawn by L. C. C. Krieger at the Desert Botanical Laboratory, Tucson, Arizona; plate VIII, figure 1, is from a photograph of the type plant taken by Dr. MacDougal in 1908.

1

2

1. Type plant of *Opuntia vivipara*, near Tucson, Arizona.
2. A much-branched plant of *Opuntia versicolor*.

9. Opuntia tetracantha Toumey, Gard. and For. **9**: 432. 1896.

Low bush, 5 to 15 dm. high, branching; central stem woody, 5 to 8 cm. in diameter; young joints 23 to 30 cm. long, 10 to 15 mm. in diameter, purplish; tubercles at first prominent, elongated, 16 to 22 mm. long; areoles bearing wool, light brown glochids, prominent glands and spines; spines 3 to 6, usually 4, slender, somewhat deflexed, 2 to 3.5 cm. long; flowers greenish purple, 1.5 to 2 cm. broad; fruit 2 to 2.5 cm. ong, yellowish orange to "scarlet," nearly smooth, but rarely bearing a few spines, deeply umbilicate; seeds 3 to 5 cm. broad, with irregular faces and a thick, spongy commissure.

Type locality: Five miles east of Tucson, Arizona.

Distribution: Known only from the region about Tucson, Arizona.

The species was originally compared by Mr. Toumey with *O. thurberi*, with which he thought it to be closely associated, but differing in "its longer, more strongly deflexed spines, smaller and different-colored flowers."

The type specimen was not indicated, but Toumey's own plant, collected in 1895, which was recently purchased by the U. S. National Herbarium, is doubtless the type.

Illustration: Bull. Torr. Club **32**: pl. 9, f. 2.

Plate IX, figure 1, shows a joint painted by L. C. C. Krieger at the Desert Botanical Laboratory, Tucson, Arizona.

10. Opuntia recondita Griffiths, Monatsschr. Kakteenk. **23**: 131. 1913.

"A stout broad-branched shrub, 1 to 1.5 meters in height; trunk cylindric, 4 to 7 cm. in diameter, with constrictions corresponding to each year's growth, with gray bark, and having a few lateral, easily detachable, weakly spined joints about 10 cm. long, the remaining joints being 20 to 30 cm. long, very spiny, in the second year about 2 cm. in diameter, tuberculate; tubercles forming a ridge, flattening out below, above extending precipitously, about 2 to 5 cm. long, 5 to 6 mm. wide, and 4 to 5 mm. high, remaining recognizable three years, and then disappearing; areoles broadly obovate, 5 to 6 mm. in the longest diameter, in age becoming larger and more prominent, forming new wool for several years; glochids yellow, in a thick 3 mm. long cluster on the upper part of the areole, also smaller clusters on the other parts of the areole, mostly at the base of the longest and most central spine; spines first 2 to 4, later 6 to 8 or 10, upr ght, spreading, 2.5 to 5 cm. long, in cross-section weakly circular, gray at the base, becoming deep reddish brown at the tips, surrounded the entire length by a loose, comparatively bright sheath; between the spines are scattered a few dirty-black, sheathless bristles about 6 mm. long; leaves subulate, finely tipped, terete, 12 to 20 mm. long.

Fig. 63.—Opuntia thurberi. Natural size.

"Flowers bright purple, when open about 2.5 cm. in diameter; petals finely and irregularly serrate, inconspicuously but finely irregularly notched; sepals thick, triangular pointed, greenish purple; anthers greenish with purple tinge; pistil greenish at base, with purple tinge above; stigma-lobes 6, white; ovary obovoid, tuberculate, with small areoles, 2 mm. in diameter, short greenish brown glochids 1 to 2 mm. long, and 1, 2, or 3 brown, caducous spines sheathed in part; fruit not deciduous, 3 to 3.5 by 2 to 2.4 cm., large, greenish yellow with a reddish tinge on the outermost side, only weakly tuberculate in the second year, with projecting brownish glochids 3 mm. long; seeds white, thick, mostly flat but often lightly angled with narrowly thickened edges, and often somewhat concave."

Type locality: La Perla, Mexico.

Distribution: Known only from type locality, and, to us, only from the description of which the above is a translation by Mr. Russell.

11. Opuntia thurberi Engelmann, Proc. Amer. Acad. **3**: 308. 1856.

Large bushy plants, 2 to 4 meters high; joints slender, elongated, 1.5 to 2.5 dm. long, 10 to 12 cm. in diameter; tubercles 1.5 to 2 cm. long, flattened laterally; leaves linear, 6 to 8 mm. long, spreading; spines 3 to 5, short (10 to 12 mm. long), spreading, covered with thin, brown, papery sheaths, the lowest one stoutest; flowers 3.5 cm. broad, brownish; fruit 2 cm. to 3 cm. long, spineless; seeds nearly globular, 4 mm. in diameter.

Type locality: Bacuachi, Sonora, Mexico.
Distribution: Western coast of Mexico.

Opuntia thurberi has long been one of our least-known species. The type, which is but a fragment, has not been clearly associated with any recent collections, but we are disposed now to believe that specimens collected on the west coast of Mexico by Dr. Rose in 1910 belong here. If we are correct, it ranges from Sonora to Sinaloa, Mexico. It is sometimes associated with *Opuntia versicolor* in its northern range, but is not so stout and has fewer and longer spines.

Figure 63 is from a photograph of the type specimen.

12. Opuntia clavellina Engelmann in Coulter, Contr. U. S. Nat. Herb. **3**: 444. 1896.

Plant 1 meter high or less, rather openly branched; ultimate joints slender, spreading or ascending, somewhat clavate, 5 to 10 cm. long, a little over 1 cm. in diameter; tubercles prominent, elongated; spines 3 to 6 in a cluster, very long, covered with loose straw-colored or brown sheaths, the central one much longer and porrect; flowers yellow; fruit clavate, short, tuberculate.

Type locality: Near Misión Purísima, Lower California.
Distribution: Interior of central Lower California.

The above description is based on the original one and on the type If the plant illustrated as cited below belongs here, this is a very distinct species, which was referred, however, by Mrs. Brandegee to *Opuntia molesta* Brandegee.

Illustration: Contr. U. S. Nat. Herb. **16**: pl. 129, A.

Of this series there is another peculiar Lower California species, perhaps nearest *O. clavellina*, but of different habit and spines. It also suggests *O. tetracantha* of Arizona. It was obtained first by Dr. Rose in 1911, but was without flowers or fruit. It may be characterized as follows:

OPUNTIA sp.

Stems slender (1 to 1.5 cm. in diameter), weak, often clambering over bushes, pale green in color, terete, pointed, 6 to 7 dm. long; areoles set on low tubercles, circular; chief spines 2 to 6, only slightly spreading, nearly equal, 1.5 to 2.5 cm. long, clothed with loose straw-colored sheaths (rose-colored when very young); accessory spines 3 or 4, almost bristle-like, borne from the lower parts of the areoles; glochids short, greenish when young, yellow in age; flowers and fruit not seen.

Collected by Dr. J. N. Rose on Cerralvo Island, off southern Lower California, April 19, 1911 (No. 16875), and also by Nelson and Goldman on the same island in 1906 (No. 7524).

13. Opuntia davisii Engelmann and Bigelow, Proc. Amer. Acad. **3**: 305. 1856.

Plants low, 3 to 5 dm. high, much branched, their dense covering of straw-colored spines making them conspicuous objects in the landscape; terminal joints slender, 6 to 8 cm. long, about 1 cm. in diameter, strongly tuberculate; spines 6 to 12, unequal, the longest ones 4 to 5 cm. long, acicular, covered with thin sheaths; glochids numerous, yellow; flowers, including ovary, 3.5 cm. long; petals olive-green to yellow, broad, with rounded mucronate tips; ovary with large areoles bearing a few spines each; fruit 3 cm. long, somewhat tuberculate, naked; seeds not known.

Type locality: Upper Canadian, about Tucumcari Hills, near the Llano Estacado.
Distribution: Western Texas and eastern New Mexico.

For many years this plant was not collected and the name was confused with other species, so that at one time it was supposed to extend as far west as California. It is now believed to have a rather circumscribed range. It is first seen going west on the Texas & Pacific Railroad about Colorado, Texas.

The plant was named for Jefferson Davis, who was Secretary of War when Whipple's report was made.

1. Joint of *Opuntia tetracantha*.   2 to 5.   Flowering joints of *Opuntia versicolor*.
6. Proliferous fruits of *Opuntia fulgida*.   (All natural size.)

Illustrations: Curtis's Bot. Mag. **108**: pl. 6652; Pac. R. Rep. **4**: pl. 16.
Figure 64 is copied from the second illustration above cited.

14. Opuntia viridiflora sp. nov.

A low, round, bushy plant 30 to 60 cm. high; terminal joints 5 to 7 cm. long, 1.5 to 2 cm. thick, often quite fragile; areoles prominent, flattened from the sides; areoles circular, filled with short, yellow or dull-gray wool; spines 5 to 7, somewhat spreading, the longest ones 2 cm. long, dark brown in color; glochids numerous, very short, yellow; flowers at tips of branches in clusters of 3 to 8, 3.5 to 4.5 cm. long (including ovary), "green, tinged with red"; fruit strongly tuberculate, except for a few long, deciduous bristles, with a deep umbilicus; seeds smooth, white, 3 mm. broad.

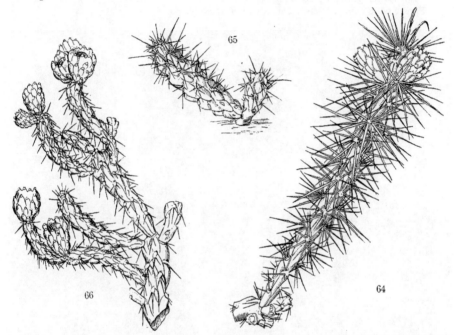

FIG. 64.—Opuntia davisii. ×0.5.   FIG. 65.—Opuntia viridiflora. ×0.5.   FIG. 66.—Opuntia whipplei. ×0.5

Collected in the vicinity of Santa Fé, New Mexico, altitude about 2,225 meters, by Paul C. Standley, July 6, 1911 (No. 6493, type) and at the same locality by T. D. A. Cockerell in 1912, and by J. N. Rose in 1913 (No. 18776). It is quite common on the hills just north of Santa Fé about Fort Marcy, where it is one of the dominant plants, but it was not observed elsewhere in that region.

This species differs from *Opuntia imbricata* with which it is found, in its much lower stature, more bushy habit, in its branches, spines, and smaller, differently colored flowers, different fruit, and smaller seeds.

Figure 65 represents two joints of a specimen collected by Dr. Rose at the type locality in 1913.

15. Opuntia whipplei Engelmann and Bigelow, Proc. Amer. Acad. **3**: 307. 1856.

Opuntia whipplei laevior Engelmann, Proc. Amer. Acad. **3**: 307. 1856.

Low, much branched, with long, fibrous roots; areoles prominent, flattened laterally, 10 to 15 cm. long, circular, filled with light-brown wool; glochids pale yellow, short; spines about 12, the longest about 2 cm. long, dark brown, covered with lighter colored papery sheaths; flowers yellow, small (2 cm. broad); young ovary bearing brown spines in the axils of the leaves; fruit strongly tuberculate, spineless, 2.5 to 4 cm. long, with a deeply depressed umbilicus, sometimes with only one seed but usually many; seeds small, 4 cm. broad, smooth.

Type locality: About Zuni, New Mexico.

Distribution: Northern New Mexico and Arizona to southwestern Colorado and probably southern Utah. Also reported by Coulter in southern California, Lower California, and Sonora, but not to be expected there.

Illustration: Pac. R. Rep. 4 : pl. 24, f. 9, 10.

Figure 66 is copied from the illustration above cited.

Fig. 67.—Opuntia acanthocarpa in the foreground. Photograph by MacDougal.

Series 4. ECHINOCARPAE.

Dry-fruited, rather stout-jointed, bushy or depressed species, the areoles bearing several spines, the flowers red, yellow, or yellowish. Four species, inhabiting the southwestern United States, Sonora, and Lower California.

KEY TO SPECIES.

Tubercles elongated, 2 to 3 times as long as wide.
    Fruit long-spiny, strongly tuberculate...16. *O. acanthocarpa*
    Fruit short-spiny, little tuberculate..17. *O. parryi*
Tubercles short, less than twice as long as wide.
    Spines with white or straw-colored sheaths..18. *O. echinocarpa*
    Spines with yellow-brown sheaths...19. *O. serpentina*

16. Opuntia acanthocarpa Engelmann and Bigelow, Proc. Amer. Acad. **3**: 308. 1856.

Much branched, 1.5 to 2 meters high; branches becoming woody, alternate, making a narrow angle with the trunk; terminal joints 4 to 8 cm. long, strongly tuberculate; tubercles elongated, flattened laterally; spines 8 to 25, acicular, dark brown, covered with thin and lighter colored sheaths, 2 to 3 cm. long; glochids numerous, yellow; flowers large, red to yellow, 5 cm. long, and when fully open nearly as broad; ovary rather short, turbinate, with few prominent tubercles; fruit dry, about 3 cm. long, naked below, tuberculate above, each tubercle crowned by a cluster of 10 to 12 stout spines; umbilicus broad and somewhat depressed; seeds 5 to 6 cm. broad, sharply angular.

Type locality: On the mountains of Cactus Pass, Arizona, about 500 miles west of Santa Fé, New Mexico.

Distribution: Arizona and California; reported also from Utah, Nevada, and Sonora.

Illustrations: N. Amer. Fauna **7**: pl. 7, 8; Pac. R. Rep. **4**: pl. 18, f. 1 to 3; pl. 24, f. 11.

Figure 67 is from a photograph by Dr. MacDougal of a plant near Pictured Rocks, Tucson Mountains, Arizona.

17. Opuntia parryi Engelmann, Amer. Journ. Sci. II. **14**: 339. 1852.

> Opuntia bernardina Engelmann in Parish, Bull. Torr. Club **19**: 92. 1892.

Low and bush-like, 2 to 4 dm. high; joints cylindric, 7 to 30 cm. long by 1.5 to 2 cm. in diameter, strongly tuberculate; tubercles 1 to 1.5 cm. long; areoles rather large, bearing light-brown wool, yellow glochids, and spines; spines about 10, dark brown, the longer ones 3 cm. long, covered with loose sheaths; flowers, several near together at ends of branches, 4 cm. long; sepals greenish or dull red; petals yellow, obtuse; stigma-lobes cream-colored; ovary tuberculate; fruit dry, ovoid, 2 cm. long, strongly umbilicate, when mature and fertile plump, otherwise more or less tuberculate; areoles on the fruit large, filled with wool and glochids, those at top of fruit often with short spines; seeds white, 4 to 6 mm. broad, beaked, the margins channeled.

Type locality: Near San Felipe, eastern slope of California Mountains—San Jacinto Mountains.

Distribution: Interior valleys of southern California.

This is common in some of the interior valleys of southern California, although its range has not been very definitely determined. It was first collected by Dr. C. C. Parry in 1851 and named for him by Dr. Engelmann in 1852; but when the latter again took up this name a few years later, he associated it with a very different species, which most later writers and dealers accepted as the true *Opuntia parryi*. Later on Dr. Engelmann segregated a species which he named *O. bernardina*, including therein Parry's specimen, but this was not published until after his death. We therefore regard *O. bernardina* as a synonym of *O. parryi*, while the *O. parryi* of most collections becomes *O. parishii*. We are under obligation to Mr. C. R. Orcutt for first calling our attention to this confusion.

Mr. Orcutt thinks that this species is near *O. serpentina;* but the former has larger flowers, different spines, much less spiny fruit, and is of different habit.

Opuntia bernardina cristata Schumann (Monatsschr. Kakteenk. **12**: 20. 1902), an abnormal form, has been described.

Plate VII, figure 3, is from a plant collected by W. T. Schaller at Pala, California, showing a leafy joint.

18. Opuntia echinocarpa Engelmann and Bigelow, Proc. Amer. Acad. **3**: 305. 1856.

> Opuntia echinocarpa major Engelmann, Proc. Amer. Acad. **3**: 305. 1856.
> Opuntia echinocarpa nuda Coulter, Contr. U. S. Nat. Herb. **3**: 446. 1896.
> Opuntia echinocarpa parkeri Coulter, Contr. U. S. Nat. Herb. **3**: 446. 1896.
> Opuntia echinocarpa robustior Coulter, Contr. U. S. Nat. Herb. **3**: 446. 1896.
> Opuntia deserta Griffiths, Monatsschr. Kakteenk. **23**: 132. 1913.

Plant usually low, but sometimes 1.5 meters high, much branched and widely spreading, with a short woody trunk 2 to 3 cm. in diameter, in age with nearly smooth bark; joints short, turgid, strongly tuberculate; spines numerous, when young bright yellow, when older brownish, or in age grayish, unequally covered with thin papery sheaths; flowers yellowish, but the sepals often tipped with red; ovary short, turbinate, densely spiny especially in the upper part; fruit dry, very spiny; seeds somewhat angular, 4 mm. broad.

Type locality: In the Colorado Valley near the mouth of Bill Williams River.

Distribution: Nevada, Utah, Arizona, California, and Lower California.

Coulter has described three varieties of this species, none of which is quite typical, but without seeing more specimens we can only refer them all to the species proper. His variety *parkeri* seems more like a very spiny form of *O. parryi*. *O. parkeri* Engelmann (Coulter, Contr. U. S. Nat. Herb. **3** : 446. 1896) was published as a synonym.

Mrs. Brandegee thought *Opuntia echinocarpa nuda* very near *O. alcahes*, if not identical with it (Erythrea **5** : 122).

Illustrations: Pac. R. Rep. **4** : pl. 18, f.5 to 10; pl. 24, f. 8; Monatsschr. Kakteenk. **23** : 132, the last as *Opuntia deserta*.

Plate VII, figure 4, is from a plant collected by Dr. Rose near the Salton Sink, California, showing a flowering joint.

19. Opuntia serpentina Engelmann, Amer. Journ. Sci. II. **14** : 338.   1852.

> *Cereus californicus* Torrey and Gray, Fl. N. Amer. **1** : 555.   1840. Not *Opuntia californica* Engelmann. 1848.
> *Opuntia californica* Coville, Proc. Biol. Soc. Washington **13** : 119.   1899.

Ascending, erect, or prostrate; branches slender, 2 to 2.5 cm. in diameter, bluish green, strongly tuberculate; leaves minute; tubercles elevated, 1 to 1.5 cm. long, longer than broad; spines 7 to 20, brown, covered with yellowish-brown papery sheaths about 1 cm. long; glochids light brown; flowers close together at the top of short branches, about 4 cm. broad, greenish yellow, the outer petals tinged with red; ovary strongly tuberculate, spiny, with a depressed umbilicus; fruit dry, very spiny.

Type locality: Near the seacoast about San Diego, California.

Distribution: Southern California and northern Lower California.

Cactus californicus Nuttall, although given in the Index Kewensis (**1** : 367), was never published by Nuttall, although he did have the name in manuscript, as stated in Torrey and Gray's "Flora" in the place cited above, where it was taken up as a *Cereus*.

Figure 68 is from a plant collected by Mr. G. Sykes near San Diego, California.

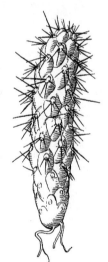

Series 5. BIGELOVIANAE.

We recognize two species in this series, natives of the southwestern United States and Lower California. They are low, bushy plants, with short definite trunks densely covered with short, stout, very spiny branches, the spines white, straw-colored, or yellow, the tubercles, at least those of young shoots, little if any longer than broad, and considerably elevated. Their fruits are fleshy berries.

KEY TO SPECIES.

Larger spines numerous; upper tubercles on fruit larger than lower ones 20. *O. bigelovii*
Larger spines 4 to 6; tubercles on fruit all alike.................... 21. *O. ciribe*

20. Opuntia bigelovii Engelmann, Proc. Amer. Acad. **3** : 307.   1856.

FIG. 68.—Opuntia serpentina.   X0.66.

Usually with a central, erect trunk, 1 meter high or less, with short lateral branches, the upper ones erect; joints usually 5 to 15 cm. long, very turgid, with closely set areoles and almost impenetrable armament; tubercles slightly elevated, pale green, somewhat 4-sided, about as long as broad, 1 cm. broad or less; spines, as well as their papery sheaths, pale yellow; flowers several, borne at the tips of the branches, 4 cm. long including the ovary; sepals orbicular, about 1 cm. in diameter, tinged with red; petals about 1.5 cm. long, pale magenta to crimson; ovary 2 cm. long, its large areoles bearing brown wool and several acicular spines; fruit usually naked, strongly tuberculate, the upper tubercles larger than the lower.

Type locality: Bill Williams River, Arizona.

Distribution: Southern Nevada, Arizona, California, northern Sonora, and northern Lower California.

Illustrations: Ann. Rep. Bur. Amer. Ethn. 26: pl 12; Contr. U. S. Nat. Herb. 16: pl. 128, B; Hornaday, Camp-fires on Desert and Lava, facing p. 154; Journ. N. Y. Bot. Gard. 5: f. 16; Pac. R. Rep. 4: pl. 19; Plant World 11¹⁰: f. 10.

Figure 69 is from a photograph by Dr. MacDougal of a plant in Pima Canyon, Santa Catalina Mountains, Arizona; figure 70 is copied from the Pacific Railroad Report above cited.

21. Opuntia ciribe Engelmann in Coulter, Contr. U. S. Nat. Herb. **3**: 445. 1896.

One meter high or less, with numerous stout branches densely armed; ultimate joints 4 to 5 cm. in diameter, strongly and regularly tuberculate, 3 cm. in diameter; tubercles about as long as broad

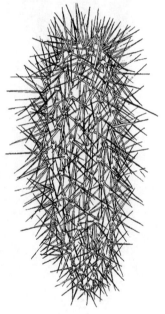

FIG. 69.—Opuntia bigelovii.

FIG. 70.—Opuntia bigelovii.   X0.66.

(5 to 7 cm. broad); larger spines 4 to 6, stout, 2 to 3 cm. long, covered with loose yellow sheaths, accompanied by several bristle-like spines or hairs; glochids numerous; flowers yellow; ovary somewhat bristly; fruit strongly tuberculate, 3 to 4 cm. long, spineless.

Type locality: Comondu and Loreto northward to beyond Rosario, Lower California.
Distribution: Central Lower California.

Opuntia ciribe is near *O. bigelovii*, but differs from it in having less spiny stems and globular, slightly different fruits.

Figure 71 is from a photograph of a plant collected by Dr. Rose at the head of Concepción Bay, Lower California; figure 72 is from a drawing of a joint from the same plant.

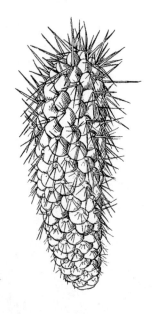

Fig. 71.—Opuntia ciribe.   Fig. 72.—Opuntia ciribe.   ✕0.8.

Series 6. IMBRICATAE.

The typical species are tall, much branched, very spiny. The terminal joints are fleshy and strongly tuberculate, the tubercles large and flattened laterally. The fruit is either smooth or strongly tuberculate. We recognize 8 species, natives of Mexico and southwestern United States.

KEY TO SPECIES.

Joints cylindric; tubercles much flattened laterally.
  Fruit smooth or but slightly tuberculate.
    Branches very stout, 5 cm. thick or more..22. *O. cholla*
    Branches relatively slender, 2 cm. thick or less.
      Plant glaucous; spines 4 at an areole...23. *O. calmalliana*
      Plant not glaucous; spines more than 4 at an areole.........................24. *O. versicolor*
  Fruit manifestly tuberculate.
    Tall species, up to 2 or 4 meters high.
      Flowers small; petals 1.5 cm. long...25. *O. lloydii*
      Flowers large; petals 2 to 3 cm. long...26. *O. imbricata*
    Low species, 6 dm. high or less.
      Flowers yellow...27. *O. tunicata*
      Flowers rose-colored...28. *O. pallida*
Joints clavate; tubercles not much flattened laterally..................................29. *O. molesta*

22. Opuntia cholla Weber, Bull. Mus. Hist. Nat. Paris 1: 320. 1895.

Usually tree-like, 1 to 3 meters high, with a definite trunk 7 to 15 cm. in diameter; trunk very spiny at first and becoming more spiny each year for some time, but in age spineless and developing a smooth, brownish yellow bark; top of plant often dense and broad; joints often in whorls, horizontal, pale, with large compressed tubercles; spines usually numerous, more or less porrect, covered with loose brownish sheaths; glochids numerous, yellow; flowers rather small, 3 cm. broad, deep purple; fruit often 4 to 5 cm. long, usually proliferous, often in long chains of 8 to 12 individuals or forming compound clusters; seeds numerous, very small, often abortive.

Type locality: In Lower California.

Distribution: Lower California.

This is one of the commonest opuntias in southern Lower California and was usually seen by Dr. Rose at every locality visited south of Magdalena Bay on the west coast and on the east coast as far north as Muleje. It is undoubtedly the plant referred to *O. prolifera* by Mr. Brandegee, but it differs in habit and armament from that species; the fruit of *O. prolifera* is nearly or quite devoid of seeds, while this species often has numerous small ones. In this species, as in a few other opuntias, the fruits are quite proliferous, hanging on for a number of years and usually remaining green. They are, however, easily detached, and on falling to the ground, readily take root and start new colonies. Our illustration shows some of the fruits which have already rooted and have developed young joints.

The plant here described is the true "cholla" of the people of Lower California, and is the plant cultivated under that name by A. Berger at La Mortola from a cutting of Weber's type specimen, and by the late Mr. Darrah at Manchester, England.

Illustrations: Contr. U. S. Nat. Herb. 16: pl. 128, A; Karsten and Schenk, Vegetationsbilder 13: pl. 17, B.

Figure 73 is from a photograph of a plant collected by Dr. Rose at Cape San Lucas; figure 74 represents a joint of the same plant; figures 75 and 76 represent its proliferous fruits developing new joints.

23. **Opuntia calmalliana** Coulter, Contr. U. S. Nat. Herb. 3: 453. 1896.

"Habit and height unknown; joints cylindrical, 1 to 2 cm. in diameter, glaucous, with linear-oblong crested (mostly distinct) tubercles 20 to 25 mm. long; pulvini densely covered with yellowish wool, and with a penicillate tuft of whitish bristles at upper edge; spines usually 4, the upper one stout and porrect, reddish with yellowish tip (as are all the spines), 2 to 2.5 cm. long (occasionally 1 to 2 short upper ones added), the usually 3 (sometimes 4) lower ones more slender and sharply deflexed, 1 to 1.5 cm. long (occasionally one of them longer); flowers apparently purple; ovary covered with very prominent woolly pulvini which are more or less bristly and spiny, but ripening into a smooth juicy obovate fruit; seeds discoid and beaked, irregularly angular, with broad commissure, about 4 mm. broad." (Coulter, *l. c.*)

FIG. 73.—Opuntia cholla.

Type locality: Calmalli, Lower California.

Distribution: Lower California.

Type in the Brandegee Herbarium, University of California.

Referred by Mrs. Brandegee (Erythea **5**: 122) to *O. molesta* Brandegee. It is closely related to *O. molesta*, but its spines are different, though on the same general plan, and its seeds are quite different.

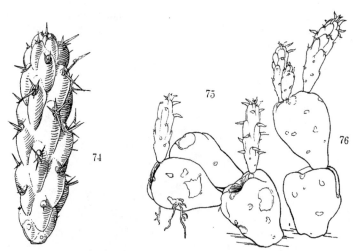

FIGS. 74, 75, 76.—Opuntia cholla. ×0.66.

24. Opuntia versicolor Engelmann in Coulter, Contr. U. S. Nat. Herb. **3**: 452. 1896.

Opuntia arborescens versicolor E. Dams, Monatsschr. Kakteenk. **14**: 3. 1904.

Bush or tree-like, 2 to 4 meters high, with a large, open top sometimes 5 meters broad; trunk and larger stems woody throughout, except the younger branches; terminal joints 10 to 20 cm. long, 2.5 cm. in diameter, variously colored, not strongly tuberculate when living; tubercles 1.5 cm. long; spines 5 to 11, 5 to 25 mm. long, dark colored, with close-fitting sheaths; glochids reddish brown; flowers variously colored, yellow, greenish, reddish, or brown, 3 to 5.5 cm. broad; ovary tuberculate, with large areoles bearing wool, glochids, and long deciduous bristles; fruit persisting for months, sometimes for a year, 2.5 to 4 cm. long, at first somewhat tuberculate, becoming pear-shaped or globose, sometimes proliferous; seeds white, 5 mm. broad.

Type locality: Tucson, Arizona.

Distribution: Arizona and northern Mexico.

This species is common on the lower foothills and is only rarely found on the mesas. It is of slow growth, propagating almost entirely from seed. As the name suggests, it has flowers of many colors; each plant has its own color and the color of the flowers is to a greater or less extent paralleled in that of the branches. The contrast in color shown by a colony of these plants is very striking and one's first impression is that more than one species exists.

Named specimens of this species were distributed by the late Dr. C. G. Pringle in 1881, but the species was not published until 1896.

Illustrations: Ariz. Agr. Exp. Sta. Bull. **67**: pl. 6, f. 1; Bull. Torrey Club **32**: pl. 9, f. 4 to 8; Hornaday, Camp-fires on Desert and Lava, pl. facing p. 18, 116, 320; N. Mex. Agr. Exp. Sta. Bull. **60**: pl. 6, f. 1; Plant World 11[10]: f. 8; Sargent, Man. Trees N. Amer. f. 561.

Plate VII, figure 5, represents a fruiting joint; plate VIII, figure 2, is from a photograph taken by Dr. MacDougal near the base of the Santa Catalina Mountains, Arizona; plate IX, figures 2 to 5, are paintings made at the Desert Laboratory, Tucson, Arizona, by Kako Morita, showing the range in color of the flowers.

25. Opuntia lloydii Rose, Contr. U. S. Nat. Herb. **12**: 292. 1909.

Much branched, 2 to 3 meters high and nearly as broad; joints terete, 2 cm. in diameter; tubercles prominent, oblong; spines few, on last year's joints 3, reddish, 1.5 cm. long; leaves terete, 6 to 8 mm. long; flowers 3 cm. long, opening after midday; petals 15 mm. long, dull purple; filaments olive-green below purplish above; style rose-colored; stigma-lobes white; ovary yellowish, strongly tuberculate, naked; fruit 3 cm. long, yellow to orange, slightly tuberculate.

Type locality: On foot slopes, Hacienda de Cedros, Zacatecas, Mexico.
Distribution: Central Mexico.

According to F. E. Lloyd, for whom this species was named, it is known to the Mexicans as tasajo macho.

We have had this plant in cultivation for several years, but it does not grow well under glass; these specimens have white areoles; no glochids are developed the first

Fig. 77.—Opuntia lloydii.   Fig. 78.—Opuntia lloydii.   Photograph by F. E. Lloyd.

year, but on old branches dark-brown bunches of glochids are developed in the upper edges of the areoles, and the several brownish spines are acicular.

Illustration: Contr. U. S. Nat. Herb. **12**: f. 34; pl. 25.

Figure 77 represents two joints of the type specimen; figure 78 is from a photograph of the type plant.

26. Opuntia imbricata (Haworth) De Candolle, Prodr. **3**: 471. 1828.

> *Cereus imbricatus* Haworth, Rev. Pl. Succ. 70. 1821.
> *Cactus cylindricus* James, Cat. 182. 1825. Not Lamarck. 1783.
> *Cactus bleo* Torrey, Ann. Lyc. N. Y. **2**: 202. 1828. Not Humboldt, Bonpland, and Kunth. 1823.
> *Opuntia rosea* De Candolle, Prodr. **3**: 471. 1828.
> *Opuntia decipiens* De Candolle, Mém. Mus. Hist. Nat. Paris **17**: 118. 1828.
> *Opuntia exuviata* De Candolle, Mém. Mus. Hist. Nat. Paris **17**: 118. 1828.
> *Opuntia exuviata angustior* De Candolle, Mém. Mus. Hist. Nat. Paris **17**: 118. 1828.
> *Opuntia exuviata spinosior* De Candolle, Mém. Mus. Hist. Nat. Paris **17**: 118. 1828.
> *Opuntia exuviata stellata* Lemaire, Cact. Gen. Nov. Sp. 67. 1839.
> *Opuntia exuviata viridior* Salm-Dyck, Cact. Hort. Dyck. 1844. 48. 1845.
> *Opuntia arborescens* Engelmann in Wislizenus, Mem. Tour North. Mex. 90. 1848.
> *Opuntia imbricata crassior* Salm-Dyck, Cact. Hort. Dyck. 1849. 249. 1850.

Opuntia imbricata ramosior Salm-Dyck, Cact. Hort. Dyck. 1849. 73. 1850.
Opuntia imbricata tenuior Salm-Dyck, Cact. Hort. Dyck. 1849. 73. 1850.
Opuntia vexans Griffiths, Rep. Mo. Bot. Gard. **22**: 28. 1912.
Opuntia magna Griffiths, Proc. Biol. Soc. Washington **27**: 23. 1914.
Opuntia spinotecta Griffiths, Proc. Biol. Soc. Washington **27**: 24. 1914.

Tree-like, often 3 meters high or higher, with a more or less definite woody trunk 2.5 cm. in diameter; ultimate joints 2 to 3 cm. in diameter, strongly tuberculate; leaves 8 to 24 mm. long, terete; tubercles 2 to 2.5 cm. long, flattened laterally; spines 8 to 30, 2 to 3 cm. long, brown, covered with papery sheaths; flowers borne at ends of branches, 4 to 6 cm. long, sometimes 8 to 9 cm. broad, purple; ovary tuberculate, bearing a few bristles from some of the upper areoles; fruit naked, yellow, 2.5 to 3 cm. long, strongly tuberculate or, when long persistent, smooth; seeds 2.5 to 3.5 mm. in diameter.

Type locality: Unknown; introduced into England by Loddiges in 1820.

Distribution: Central Colorado to Texas, New Mexico, and central Mexico.

The plant is hardy in southwestern Kansas, and has been recorded as a native of that State; it has existed through three winters out of doors at the New York Botanical Garden, but has made little growth.

We have followed Schumann and Weber in uniting *Opuntia arborescens* and *O. imbricata.* As thus treated, the species has a wide geographic distribution, and in our view con sists of many slightly differing races.   In its northern limits it is much smaller than in its southern range.

Opuntia cristata tenuior Salm-Dyck (Cact. Hort. Dyck. 1844. 49. 1845, name only), *O. decipiens major* Hort. in Salm-Dyck (Cact. Hort. Dyck. 1844. 49. 1845, as synonym), *O. cristata* Salm-Dyck (Cact. Hort. Dyck. 50. 1842), and *O. stellata* Salm-Dyck (Cact. Hort. Dyck. 50. 1842) are unpublished names.   *O. ruthei* is a garden name mentioned by Berger.

Opuntia exuviata major (Salm-Dyck, Cact. Hort. Dyck. 1844. 49. 1845) is an unpublished name.

Opuntia cardenche Griffiths (Rep. Mo. Bot. Gard. **19**: 259. pl. 21, in part. 1908) is described as standing between *Opuntia kleiniae* and *O. imbricata*, being stouter than the

Fig. 79.—Opuntia imbricata.

one and more slender than the other. It resembles very closely specimens collected by Dr. Rose at Ixmiquilpan, Mexico, in 1905, which we have referred to *O. kleiniae*.

Opuntia galeottii de Smet (Miquel, Nederl. Kruidk. Arch. **4**: 337. 1858) and *O. costigera* Miquel (Nederl. Kruidk. Arch. **4**: 338. 1858), if really from Mexico, may belong here, but the descriptions are indefinite. Dr. Schumann did not know them.

Opuntia mendocienses (Cat. Darrah Succ. Manchester **56**. 1908) is said to be "probably only a form of *O. imbricata*."

Opuntia undulata Link and Otto (Verh. Ver. Beförd. Gartenb. **6**: 434. 1830) was not published. According to Pfeiffer, it is the same as *O. exuviata*, which we refer here.

Opuntia decipiens minor (Pfeiffer, Enum. Cact. 172. 1837) is unpublished.

Cactus subquadrifolius Mociño and Sessé (De Candolle, Prodr. **3**: 471. 1828) was given as a synonym of *Opuntia rosea* and therefore belongs here.

FIG. 80.—Opuntia tunicata.   FIG. 81.—Opuntia pallida.

Illustrations: Agr. Gaz. N. S. W. **22**: pl. opp. p. 696; Bull. U. S. Dept. Agr. **31**: pl. 5; pl. 6, f. 1; Cact. Mex. Bound. pl. 73, f. 7, 8 ; Curtis's Bot. Mag. **135**: pl. 8290; N. Mex. Agr. Exp. Sta. Bull. **60**: pl. 7, f. 2; Förster, Handb. Cact. ed. 2. f. 134; Mém. Mus. Hist. Nat. Paris **17**: pl. 15; W. Watson, Cact. Cult. f. 85, the last three as *Opuntia rosea*. W. Watson, Cact. Cult. f. 8, in part, this as *Opuntia decipiens*. Ann. Rep. Bur. Amer. Ethn. **26**: pl. 8, f. *a*; Cact. Mex. Bound. pl. 75, f. 16, 17; Gard. Chron. III. **34**: f. 36; Gard. and For. **9**: f. 1; Illustr. Fl. **2**: f. 2533; ed. 2. **2**: f. 2992; N. Mex. Agr. Exp. Sta. Bull. **78**: pl. [10]; Pac. R. Rep. **4**: pl. 17, f. 5, 6; pl. 18, f. 4; pl. 24, f. 12; Rep. Mo. Bot. Gard. **22**: pl. 7, in part; all as *Opuntia arborescens*. Rep. Mo. Bot. Gard. **22**: pl. 6, 7, in part, these two as *Opuntia vexans*.

Plate XI, figure 1, represents a joint of a plant collected by W. L. Bray in western Texas. Figure 79 is from a photograph taken by Professor F. E. Lloyd in Zacatecas, Mexico, in 1908.

27. Opuntia tunicata (Lehmann) Link and Otto in Pfeiffer, Enum. Cact. 170. 1837.

> *Cactus tunicatus* Lehmann, Ind. Sem. Hort. Hamb. 6. 1827.
> *Opuntia stapeliae* De Candolle, Mém. Mus. Hist. Nat. Paris **17**: 117. 1828.
> *Opuntia hystrix* Grisebach, Cat. Pl. Cub. 117. 1866.
> *Opuntia perrita* Griffiths, Rep. Mo. Bot. Gard. **22**: 33. 1912.

Very variable, sometimes low and spreading from the base and forming broad clumps, at other times 5 to 6 dm. high, with a more or less definite woody stem and numerous lateral branches; joints easily detached, sometimes short and nearly globular to narrowly oblong, 10 to 15 cm. long, strongly tuberculate; spines reddish, normally 6 to 10, elongated, 4 to 5 cm. long, covered with thin, white, papery sheaths; flowers 3 cm. long, yellow; petals obtuse; ovary often bearing long spines at the areoles, but usually naked.

Type locality: In Mexico.

Distribution: Highlands of central Mexico; also in Ecuador, Peru, and northern Chile.

Opuntia stapeliae has long puzzled collectors and students of cacti. We are convinced now that it is only starved or stunted greenhouse specimens of the common *O. tunicata.* When grown in cultivation, *O. tunicata* takes on abnormal shapes, for the joints, which break off easily, rarely grow to their full size. In its native home many small dwarf plants are found everywhere about the larger plants. We have discussed this explanation of *O. stapeliae* with Mr. A. Berger, and he agrees with our conclusion.

No specimens of the type of *O. stapeliae* are preserved in the De Candolle Herbarium. The plant figured as *Opuntia stapeliæ* (?) by Goebel in Pflanzenbiologische (f. 36) does not belong here. It is erect, has strongly tuberculate joints, very short spines and narrow elongated leaves.

Cereus tunicatus (Pfeiffer, Enum. Cact. 170. 1837) is given as a synonym of *Opuntia tunicata,* but has never been formally taken up.

We believe *Opuntia hystrix* Grisebach, collected by C. Wright in Cuba, belongs here, probably being an escape from a garden. Dr. Rose examined the specimens in the Krug and Urban Herbarium in Berlin in 1912; the loose sheaths of the spines of these specimens are now brown, while the flowers seemed a little smaller than those of the Mexican specimens. The flowers were described as red.

Opuntia furiosa Wendland (Pfeiffer, Enum. Cact. 170. 1837) is referred to *O. tunicata* by Pfeiffer, while Salm-Dyck refers it to his variety *O. tunicata laevior* (Cact. Hort. Dyck. 1849. 73. 1850).

Illustrations: Bull. U. S. Dept. Agr. 31 : pl. 4; Cact. Journ. 1 : October; The Garden 62 : 425; Safford, Ann. Rep. Smiths. Inst. 1908: pl. 10, f. 5; Schumann, Gesamtb. Kakteen f. 2; Rep. Mo. Bot. Gard. 22 : pl. 13, 14, these two as *Opuntia perrita.*

Plate x, figure 1, represents a joint of a plant collected by Dr. Rose near Cuzco, Peru. Figure 80 is from a photograph of the same plant.

28. Opuntia pallida Rose, Smiths. Misc. Coll. 50: 507. 1908.

Stems 5 cm. in diameter, about 1 meter high, with widely spreading branches, the whole plant often broader than high; old areoles very spiny, often bearing 20 spines or more, often 3 to 4 cm. long, with white, papery sheaths; young areoles bearing few spines; ovary tuberculate, the areoles either naked or bearing a few bristly spines; flowers pale rose-colored; petals 15 mm. long.

Type locality: Near Tula, Hidalgo, Mexico.

Distribution: State of Hidalgo, Mexico.

This species is known only from near Tula, Mexico, where it was discovered by Dr. J. N. Rose in 1905, and afterwards collected near the same station by Mr. E. W. Nelson. It grows interspersed with *O. imbricata,* but is much lower in stature and has smaller leaves and lighter-colored flowers. It is much like *O. tunicata,* but that species has yellow flowers and is always smaller.

Illustration: Contr. U. S. Nat. Herb. 10 : pl. 17, A.

Figure 81 is from a photograph of the type specimen.

29. Opuntia molesta Brandegee, Proc. Cal. Acad. II. 2 : 164. 1889.

Stems 1 to 2 meters high, or in cultivation only 6 dm. high, with few, long, spreading branches; joints clavate to subcylindric, 10 to 40 cm. long, sometimes as much as 4 cm. in diameter at the top,

M. E. Eaton del. 1, 4, 5
Kako Morita del. 2, 3

A. HOEN & CO.

1. Joint of *Opuntia tunicata*.   2 to 5.   Joints of *Opuntia spinosior*.
(All natural size.)

pale green, with low, broad tubercles, these elongated and often 4 cm. long or more; leaves linear, 10 mm. long or less; spines few, 6 to 10, unequal, the longest ones 2.5 to 5 cm. long, straw-colored, with loose, papery sheaths; flowers purple, 5 cm. in diameter; fruit ovoid, 2.5 cm. long, somewhat spiny or naked; seeds 6 mm. in diameter, irregular in shape.

Type locality: San Ignacio, Lower California.

Distribution: Lower California.

The type of the species is deposited in the Brandegee Herbarium, now a part of the herbarium of the University of California. Living plants have been distributed by A. Berger from La Mortola, Italy, and are now to be found in various collections.

In the Index Kewensis, first supplement, this species is wrongly entered as *Opuntia modesta!*

Figure 82 is from a photograph of a plant sent from La Mortola, Italy, to the New York Botanical Garden in 1913.

Series 7. FULGIDAE.

Much branched, bushy plants, usually with the terminal joints very fleshy, the tubercles broad and low, about as broad as long. The species, of which we recognize five, inhabit the southwestern United States and western Mexico.

KEY TO SPECIES.

```
Joints very readily detached, freely falling ....... 30. O. fulgida
Joints not very readily detached, persistent.
    Spines brown or reddish, at least at base.
        Branches slender; fruit not proliferous.... 31. O. spinosior
        Branches stout; fruit proliferous......... 32. O. prolifera
    Spines white or yellow.
        Spines white; petals greenish yellow, 1 cm.
            long or less ....................... 33. O. alcahes
        Spines yellow; petals red, 2 cm. long...... 34. O. burrageana
```

30. Opuntia fulgida Engelmann, Proc. Amer. Acad. 3: 306. 1856.

> *Opuntia mamillata* Schott in Engelmann, Proc. Amer. Acad. 3: 308. 1856.
> *Opuntia fulgida mamillata* Coulter, Contr. U. S. Nat. Herb. 3: 449. 1896.

Plant sometimes 3 meters high or even more, with a rather definite woody trunk 10 to 20 cm. in diameter, much branched, sometimes almost from the base, and forming a compact flattened crown; terminal joints 10 to 20 cm. long, 3 to 5 cm. in diameter, very succulent, strongly tuberculate, easily breaking off; spines 2 to 12, yellowish to brown, 2.5 to 3.5 cm. long, acicular, covered with loose, papery sheaths; glochids small, whitish to light yellow; flowers light rose, 2.5 to 3 cm. broad; petals few, obtuse;

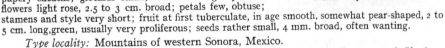
FIG. 82.—Opuntia molesta.

stamens and style very short; fruit at first tuberculate, in age smooth, somewhat pear-shaped, 2 to 5 cm. long, green, usually very proliferous; seeds rather small, 4 mm. broad, often wanting.

Type locality: Mountains of western Sonora, Mexico.

Distribution: Gravelly and sandy situations, southern Arizona, Sonora, and Sinaloa.

We consider *O. mamillata* as synonymous with *O. fulgida;* in herbarium and greenhouse specimens we can find no constant differences. Professor J. J. Thornber, who has long studied this group, says there is no difference between the flowers and fruits, and that there is no difference in distribution (Ariz. Agr. Exp. Sta. Bull. **67**: 501). In the field, however, one can see two rather distinct forms which differ in armament, the typical plant being the more spiny.

This is one of the most characteristic opuntias of southern Arizona, being very abundant on the valley slopes and lower foothills. It often forms dense colonies almost to the exclusion of other cacti, or it may be associated with other species, especially of *Opuntia.*

It is a most troublesome plant to come in contact with, for, as the sharp, barbed spines pierce the flesh, the joints easily break loose from the plant and are detached with difficulty from the unfortunate victim.

The flowering season extends from early spring to September. The fruit is markedly proliferous, often developing in chains, and so persisting for several years, possibly eight or ten years, as suggested by Professor D. S. Johnson. They grow in chains of 8 or 9 fruits (12 to 14 have been reported), several chains hanging from a single joint and forming a large cluster. We have seen as many as 38 fruits (40 to 50 have been reported) in a single cluster, and doubtless under favorable conditions many more would be found. These juicy fruits, usually spineless, are much sought by grazing animals.

According to Professor Johnson, who has studied this species several years, the seeds are not known to germinate in nature. Only by cutting away a part of the hard, bony coat could they be made to germinate in the greenhouse. The species is propagated easily by the terminal joints, which come off readily and are transported far and wide like burs, and soon strike root on reaching the soil. New plants are also started occasionally by the fruits themselves.

This species appears to hybridize with *O. spinosior*.

Illustrations: Ariz. Agr. Exp. Sta. Bull. **67**: pl. 1, f. 2; Bull. Torr. Club **32**: pl. 9, f. 1; Cact. Mex. Bound. pl. 75, f. 18; Gard. and For. **8**: f. 46; Hornaday, Camp-fires on Desert and Lava opp. p. 42, 320; Lumholtz, New Trails in Mex. opp. p. 18; Monatsschr. Kakteenk. **18**: 153; Nat. Geogr. Mag. **21**: 710; N. Mex. Agr. Exp. Sta. Bull. **60**: pl. 6, f. 2; Plant World 11^6: f. 1, in part; 11^{10}: f. 9, in part; Sargent, Man. Trees N. Amer. f. 559; Ariz. Agr. Exp. Sta. Bull. **67**: pl. 5, f. 1; Cact. Mex. Bound. pl. 75, f. 19; Lumholtz, New Trails in Mex. opp. p. 152; Nat. Geogr. Mag. **21**: 710; Plant World 11^6: f. 1, in part; 11^{10}: f. 9, in part, the last six as *Opuntia mamillata*; Carnegie Inst. Wash. **269**: Frontispiece; pl. 1 to 7; pl. 8, f. 76 to 79; pl. 12.

Plate IX, figure 6, represents the proliferous fruit; plate XII, figure 1, is from a photograph taken by Dr. MacDougal near Tucson, Arizona, showing the typical plant to the left and the less spiny plant to the right.

31. Opuntia spinosior (Engelmann) Toumey, Bot. Gaz. **25**: 119. 1898.

Opuntia whipplei spinosior Engelmann, Proc. Amer. Acad. **3**: 307. 1856.

Plants 2 to 4 meters high, tree-like in habit, with a more or less definite, woody trunk, openly branched; ultimate joints 1 to 3 dm. long, 1.5 to 2.5 cm. in diameter, often bright purple, strongly tuberculate; tubercles about 6 to 12 mm. long, longer than broad, more or less flattened laterally; spines 6 to 12, but on old branches sometimes as many as 25, 10 to 15 mm. long, divergent, gray to brownish, covered with thin sheaths; glochids yellowish white; flower-buds short, acute; flowers 5 to 6 cm. broad, purple to pink, yellow, or even white; petals about 10, broad at apex, narrowed at base; style thick, cream-colored or pinkish; ovary tuberculate, bearing small, purple leaves and long, white, easily detached bristles; fruit strongly tuberculate, spineless, yellow, globose to broad'y oblong, 2.5 to 4 cm. long, with a depressed umbilicus; seeds white, 4 mm. broad, smooth, with a very indistinct marginal band.

Type locality: South of the Gila River.

Distribution: Arizona, western New Mexico, and northern Mexico.

Opuntia spinosior neomexicana (Toumey, Bot. Gaz. **25**: 119. 1898) seems to be a yellow-flowered form of this species. Mr. Toumey writes that his original material of this variety came from the low foothills north of the Rillito River near Tucson.

Opuntia spinosior was described by Engelmann in 1856 as a variety of *O. whipplei*, to which it is only remotely related, but it was not separated until 1898, when it was described as distinct by Professor J. W. Toumey.

Illustrations: Ariz. Agr. Exp. Sta. Bull. **67**: pl. 1, f. 1; pl. 5, f. 2; Gard. and For. **9**: f. 1; N. Mex. Agr. Exp. Sta. Bull. **60**: pl. 7, f. 1; Plant World 11^{10}: f. 7; Sargent, Man. Trees N. Amer. f. 560.

PLATE XI

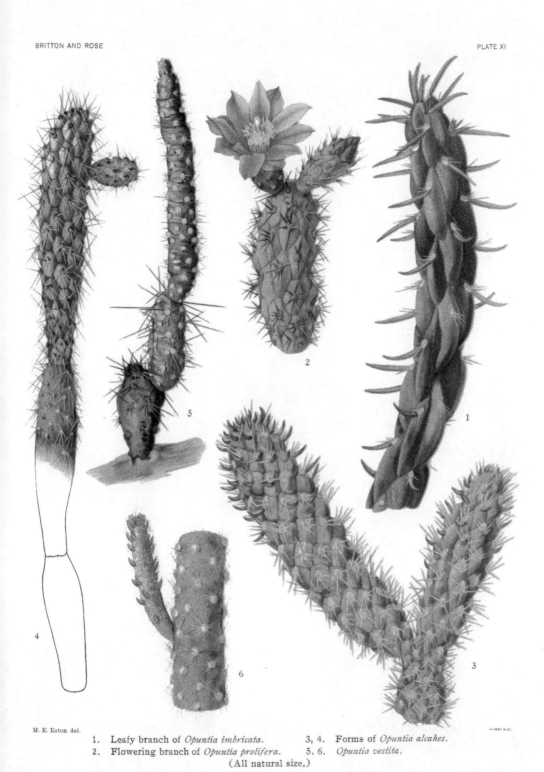

M. E. Eaton del.

1. Leafy branch of *Opuntia imbricata*.   3, 4. Forms of *Opuntia alcahes*.
2. Flowering branch of *Opuntia prolifera*.   5, 6. *Opuntia vestita*.

(All natural size.)

Plate X, figures 2 and 3, are from paintings showing different flower-colors, made at the Desert Laboratory, Tucson, Arizona; figure 4 represents a fruiting joint of a plant collected by F. Gilman at Sacaton, Arizona; and figure 5 represents a leaf-bearing joint of the same plant; plate XII, figure 2, is from a photograph of the plant in the Tucson Mountains, Arizona, by Dr. MacDougal.

32. Opuntia prolifera Engelmann, Amer. Journ. Sci. II. **14**: 338. 1852.

Stems 1 to 2 meters high, the trunk and old branches terete and woody; terminal joints 3 to 12 cm. long, easily breaking off, fleshy, covered with short, more or less turgid tubercles; spines 6 to 12, brown, 10 to 12 mm. long; glochids pale; flowers small; sepals orbicular, obtuse, dark red; petals red; filaments yellow; style stout; stigma-lobes red; ovary 1 cm. long, strongly tuberculate; upper areoles bearing 2 to 6 reddish spines or the joints naked throughout; fruit proliferous, 3 to 3.5 cm. long and often without seeds; seeds, if present, large, regular, 6 mm. broad.

Type locality: Arid hills about San Diego, California.

Distribution: Southern California and coast of Lower California.

The range of this species is not well known. We have referred here, with some doubt, specimens collected by Dr. Rose on Guadalupe Island, off the coast of Lower California, as well as specimens from the south end of Lower California, but we have seen no flowers from these Lower California collections. A peculiar form less than 5 dm. high with bluish-green joints and small seeds, from near Newport, Orange County, California, deserves further study.

This species, although common in southern California, has never been fully and accurately described. It is often confused in collections with *O. serpentina*, with which it grows, although they are very different.

In greenhouse specimens the joints and spines are not well developed.

Illustration: Meehan's Monthly **3**: pl. 1.

Plate XI, figure 2, represents a flowering joint of a plant collected by E. W. Nelson and E. A. Goldman in Lower California, which bloomed at the New York Botanical Garden in April 1914. Figure 83 represents a joint of a plant sent from La Mortola, Italy, in 1912; figure 84 is from a photograph of this plant.

Of this relationship, but of very different habit, is the species collected by Dr. Rose on West San Benito Island in 1911. Unfortunately no flowers or fruits could be obtained, and hence we have not named it here. It may be briefly characterized as follows:

OPUNTIA sp.

Low, much branched plants; joints short (10 cm. long), thick, and fleshy; leaves cylindric, 10 mm. long, acute; areoles distant, circular, bearing brown wool, tawny glochids and numerous spines; spines 6 to 8, often 4 cm. long, slender, reddish brown, inclosed in loose, thin, brownish sheaths. Collected by Dr. J. N. Rose on West San Benito Island, off the west coast of Lower California, March 9, 1911 (No. 16043).

33. Opuntia alcahes Weber, Bull. Mus. Hist. Nat. Paris **1**: 321. 1895.

Plant about 1 meter high, much branched, very spiny, especially when old; branches terete; spines on young joints about 12, short, covered with white or very pale sheaths; tubercles prominent, diamond-shaped; leaves small, 1 cm. long, terete, somewhat bronzed; sepals small, brownish, closely imbricated, hardly spreading at tips; petals sometimes wanting, or, if present, about 1 cm. long, greenish yellow, obtuse; stamens numerous; stigma-lobes very short, 6 to 8, at first exserted beyond the sepals, yellowish; fruit globular, small, becoming turgid in age, yellowish, more or less proliferous, the umbilicus truncated or slightly depressed.

FIG. 83.—Opuntia prolifera.

Type locality: In Lower California.

Distribution: Lower California.

Plate xi, figure 3, represents a leaf-bearing joint of a plant obtained by the same collector on Espíritu Santo Island, Lower California; figure 4 is from a plant sent to the New York Botanical Garden from La Mortola, Italy, in 1906. Figure 85 is from a photograph of a plant collected by Dr. Rose at San Francisquito, Lower California.

34. Opuntia burrageana sp. nov.

Usually low and bushy, rarely 1 meter high; stems slender, 1 to 2 cm. in diameter, densely spiny; leaves small, 2 mm. long, green, early deciduous; old stem and branches terete; young

FIG. 84.—Opuntia prolifera

FIG. 85.—Opuntia alcahes.

joints cylindric to narrow-clavate, 15 cm. long or less; areoles closely set; tubercles rather low, not much broader than long; spines numerous, similar, spreading, rarely 2 cm. long, all covered with thin, bright-yellow sheaths; wool in areoles short, brown; glochids, when present, short, light yellow; flower 3 to 4 cm. broad; petals few, brownish red with green bases; filaments green; stigma-lobes white; ovary very spiny; fruit not proliferous, globular, 2 cm. in diameter, somewhat tuber-culate, probably dry; seeds pale, 4 mm. in diameter.

Common on the hills along the coast of southern Lower California.

1. Plants of *Opuntia fulgida*.　　　2. A very open plant of *Opuntia spinosior*.

The following specimens were collected by Dr. J. N. Rose in 1911: Near Pichilinque Island (No. 16533, type); near San José del Cabo (No. 16468); near Cape San Lucas (No. 16379); on Carmen Island (No. 16630); on San Josef Island (No. 16552).

Plate XIV, figure 1, is from a plant collected by Dr. Rose on San Josef Island, Lower California, in 1911, which flowered the next year at the New York Botanical Garden.

Series 8. VESTITAE.

The series *Vestitae* contains three or perhaps four species, two of which possibly represent greenhouse forms of species of *Tephrocactus*, natives of the high Andes. They are low species with elongated cylindric joints sometimes arising from subglobose ones, and form a connecting link between the true species of *Tephrocactus* and *Cylindropuntia*. *Opuntia vestita* in the field was supposed to be a form of *O. pentlandii*, but in cultivation it has developed quite differently: *O. floccosa*, a *Tephrocactus*, sometimes develops like the *Vestitae;* one specimen which we have grown shows a slender cylindric stem with few long hairs or none. *Opuntia boliviana* and *O. pentlandii*, both from Bolivia and described at the same time by Salm-Dyck, and which we have united, seem to represent two forms of the same species, *O. pentlandii* being the abnormal form. The same condition seems to exist in *O. verschaffeltii* and its variety *digitalis*, the variety being the normal form. Schumann had these species in his series *Teretes* (our series *Subulatae*), but *O. subulata* and *O. cylindrica* are tall woody, much branched plants.

KEY TO SPECIES.

Areoles with hairs; joints not or scarcely tuberculate.
    Joints 1 to 1.5 cm. thick; spines 2.5 cm. long or less; fruit mostly sterile.................... 35. *O. vestita*
    Joints 2.5 to 3 cm. thick; spines up to 5 cm. long; fruit many-seeded .... 36. *O. shaferi*
Areoles without hairs; joints distinctly tuberculate.. 37. *O. verschaffeltii*
Of this series?.. 38. *O. hypsophila*

35. Opuntia vestita Salm-Dyck, Allg. Gartenz. 13: 388. 1845.

Opuntia teres Cels in Weber, Dict. Hort. Bois 898. 1898.

Roots fibrous; stems much branched, weak, forming small clumps 3 dm. broad or less and nearly as high, fragile; joints short or elongated, becoming in greenhouse cultivation 2 dm. long or more, oblong or cylindric, 1 to 1.5 cm. thick, very spiny, easily breaking apart; areoles circular, conspicuous, bearing short wool, spines, and several long hairs; spines about 6 in each cluster, acicular, brownish, 2 to 2.5 cm. long; leaves minute, acute; flowers small, including the ovary; 2 cm. long, deep red; petals 1 cm. long; areoles on ovary conspicuous, filled with white wool and long hairs; fruit red, usually sterile, globular or a little longer than broad, usually naked, generally truncate at apex, often bearing small spiny joints at the areoles.

Type locality: In Bolivia.
Distribution: Common on the sterile hills about La Paz, Bolivia.

Specimens were collected by Miguel Bang some years ago and segregated as a new species by the late Karl Schumann, but this was never published; others were obtained by Dr. H. H. Rusby in 1885, and by R. S. Williams in 1901. It was again collected by Dr. Rose in 1914, and living plants are now growing at the New York Botanical Garden. As seen wild, it is a strange little plant, growing in low clumps, its fragile stems easily breaking apart, especially at the terminal joints. The bright red fruits remain on the parent plant until they produce a number of spiny joints, often as many as five, which, after falling off, strike roots and start new colonies.

Dr. Rose suspected at the time he collected his material that it might be *Opuntia vestita*, and suggested that it should be carefully compared with it. This he was not able definitely to prove in the field, but the living specimens sent to the New York Botanical Garden put out new branches which are long, slender, and cylindric, and are devoid of long acicular spines, quite unlike the wild plants but almost identical with the specimens received from La Mortola, Italy, some years ago as *O. vestita*.

Opuntia teres Cels must belong here, at least in part. Weber states that the flowers are very similar to *O. vestita*, while the fruit is said to be small, red, and proliferous, just as found in *O. vestita*. The leaves are described as 2 cm. long, however, and there is a possibility that *O. exaltata* may be partly represented in the description, as we find herbarium material of both species, from Bolivia, mounted on the same sheet.

Plate XI, figure 5, shows the plant collected by Dr. Rose in 1914; figure 6 is from a plant received from La Mortola, Italy, in 1912.

36. Opuntia shaferi sp. nov.

Plants in clusters of 2 to 4, erect, about 3 dm. high; joints terete, 2.5 to 3.5 cm. in diameter, elongated, very spiny; tubercles low, often indistinct; leaves deciduous, 6 mm. long; areoles 1 cm. apart or less, circular, white-felted; glochids numerous, whitish from the upper margin of the areole; spines about 6 at an areole, brownish, acicular, often 4 to 5 cm. long and associated with long white hairs; flowers not known; fruit globular, about 2 cm. in diameter, bearing numerous large areoles, the areoles white-felted, with glochids and hairs, but no spines; seeds turgid, pointed at base, 4 mm. long.

Collected by J. A. Shafer among stones between Purmamaria and Tumbaya, Argentina, February 6, 1917 (No. 90).

Nearest *O. vestita* but less cespitose, taller and larger, and with fertile fruit.

37. Opuntia verschaffeltii Cels in Weber, Dict. Hort. Bois 898. 1898.

Opuntia verschaffeltii digitalis Weber, Dict. Hort. Bois 898. 1898.

Forms low, in dense clumps, much branched; joints globular to short-cylindric, 1 to 4 cm. long, somewhat tuberculate, pale green; spines 1 to 3, yellowish, weak, and bristle-like, 1 to 3 cm. long; in cultivated plants joints elongated, 6 to 21 cm. long, slender, 1 to 1.5 cm. in diameter, strongly tuberculate, spineless; glochids few, white; areoles narrow, longer than broad, filled with short white wool.

Type locality: In Bolivia.

Distribution: Bolivia.

In 1914 Dr. Rose collected this species on the barren hills about La Paz, Bolivia, and from his observations it seemed to be only a form of *Opuntia pentlandii*. In cultivation, however, it behaves very differently from his specimens of the latter, and in fact has developed a phase very unlike its normal type but identical with other greenhouse specimens sent out by Mr. Berger some years ago under the name of *O. verschaffeltii*.

Opuntia digitalis Weber (Dict. Hort. Bois 898. 1898) was given as a synonym of *O. verschaffeltii digitalis*.

Figure 86 represents an elongated joint, from a greenhouse specimen; this grew from the short normal joint, collected by Dr. Rose near La Paz, Bolivia, in 1914.

38. Opuntia hypsophila Spegazzini, Anal. Mus. Nac. Buenos Aires III. 4:509. 1905.

Cespitose, branching, small, 5 to 10 cm. high, pale green; joints globose to cylindric, 1.5 to 3 cm. long; tubercles depressed; spines 3 to 5, subulate, weak, spreading, white at first, in age brownish; flowers and fruit unknown.

Type locality: In the Province of Salta, Argentina, in the Andes, at an altitude of 2,500 to 4,000 meters.

Distribution: Salta, Argentina.

We do not know this species, but Dr. Spegazzini thought it might be a *Tephrocactus* and associated it with *Opuntia verschaffeltii digitalis*.

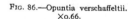

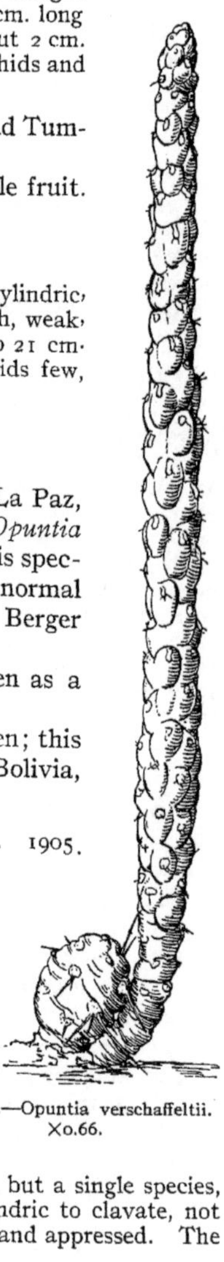

FIG. 86.—Opuntia verschaffeltii.
X0.66.

Series 9. CLAVARIOIDES.

This series is the same as the *Etuberculatae* of Schumann and contains but a single species, recorded as a native of Chile. According to Schumann, the stems are cylindric to clavate, not tuberculate, the leaves are small and caducous, and the spines are very small and appressed. The fruit is said to contain one woolly seed.

39. Opuntia clavarioides Pfeiffer, Enum. Cact. 173. 1837.

Low, much branched, grayish brown, 4 dm. high or less, truncate or cristate at apex; joints not tuberculate, rather fragile, short-cylindric or clavate, 1.5 cm. in diameter; leaves minute, 1.5 mm.

long, reddish, caducous; areoles minute, closely set, filled with wool and minute spines; spines 4 to 10, white, appressed; flowers 6 to 6.5 cm. long; sepals linear, pointed, reddish; petals light brown, narrowly spatulate, slightly crenate; ovary bearing minute leaves with wool and short bristles in their axils; filaments white, shorter than the petals; style white, with 7 stigma-lobes; fruit ellipsoid, 1.5 cm. long, one-seeded.

Type locality: In Chile.

Distribution: Originally described from Chile, but often referred to Mexico.

Very little is known of this species, although it was described as long ago as 1837, and it is rare in collections. We have never seen it in flower and have seen only one record of its flowering in cultivation. The peculiar structure of the stem, narrow petals and single lanate seed, join a combination of characters separating it from other opuntias, and lead Schumann to refer it to a distinct series which he calls *Etuberculatae*. The question has been raised in our own minds if this is a true *Opuntia*. In cultivation the plant is usually grafted on some *Platyopuntia*.

Variety *cristata* is offered in the trade journals.

Opuntia microthele, Cereus clavarioides, and *Cereus sericeus* are usually given as synonyms, but all these were cited by Pfeiffer (Enum. Cact. 173. 1837) as synonyms of this species at the place commonly given as their first publication. The varieties *fasciata* Schumann (Monatsschr. Kakteenk. 10: 159. 1900), *fastigiata* Mundt (Monatsschr. Kakteenk. 3: 30. 1893), and *monstruosa* Monville (Labouret, Monogr. Cact. 489. 1853) are anomalous greenhouse forms.

Illustrations: Gartenflora 44: f. 7; Monatsschr. Kakteenk. 3: 9; 16: 169; Schumann, Gesamtb. Kakteen f. 104; Gard. Chron. III. 30: f. 75, this last as *Opuntia clavarioides cristata*.

Figure 87 is copied from the illustration used by Schumann cited above.

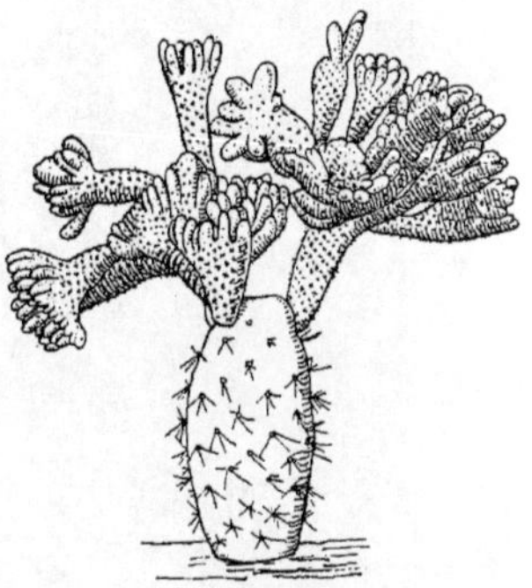

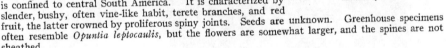

Series 10. SALMIANAE.

This series (*Frutescentes* of Schumann), by some supposed to be composed of five species but here treated as containing but one, is confined to central South America. It is characterized by slender, bushy, often vine-like habit, terete branches, and red fruit, the latter crowned by proliferous spiny joints. Seeds are unknown. Greenhouse specimens often resemble *Opuntia leptocaulis*, but the flowers are somewhat larger, and the spines are not sheathed.

Fig. 87.—Opuntia clavarioides grafted on another species.

40. Opuntia salmiana Parmentier in Pfeiffer, Enum. Cact. 172. 1837.

> *Opuntia spegazzinii* Weber, Dict. Hort. Bois 898. 1898.
> *Opuntia albiflora* Schumann, Gesamtb. Kakteen Nachtr. 152. 1903.

A bushy plant, 3 dm. to 2 meters high, much branched at base; branches often weak, terete, 1.5 cm. in diameter or less, often purplish, etuberculate; areoles small, bearing wool, yellow glochids, and spines; spines sometimes wanting, usually several, 1.5 cm. long or less, white; flowers 2 to 3.5 cm. broad, scattered along the stem; buds pinkish or even scarlet; petals obovate, pale yellow to white, sometimes tinged with pink; stamens and style short; stigma-lobes yellowish green; fruits sterile, clavate, scarlet, with few or no spines.

Type locality: In Brazil.

Distribution: Southern Brazil, Paraguay, and northern Argentina.

After careful consideration, we have combined three species of Schumann's series *Frutescentes* into one. We have examined considerable living material and all the illustrations, but have found no grounds for separating the group into species. All were described as proliferous and sterile. *O. spegazzinii* was supposed to be unarmed, but this character is not constant; flower differences are described, but these are inconstant.   One species, *O. albiflora*, has already been referred to synonymy.

Opuntia salmiana is said to have come from Brazil, but no definite locality is given for it, and it has not been collected there in recent times.   If really from Brazil, and there is no good reason to question this reference, it is doubtless from the southern part, possibly on the border of Paraguay; indeed, *O. albiflora*, one of the three, was described from a Paraguay collection; the other, *O. spegazzinii*, is a native of the deserts of northern Argentina.

Cactus salmianus Lemaire (Cact. 87. 1868, name only) has been referred here as a synonym; as has also *O. floribunda* Lemaire (Cact. Gen. Nov. Sp. 68. 1839).

Opuntia schickendantzii Weber, included by Schumann in this relationship, we refer to our series *Aurantiacae*.

FIG. 88.—Opuntia salmiana.   FIG. 89.—O. salmiana. ×0.6.

Opuntia wagneri Weber in Gosselin (Bull. Mus. Hist. Nat. Paris **10**: 393. 1904), described without flower or fruit, is probably to be referred here; at least Roland-Gosselin believed it to be of this group.   We have not seen any of the specimens from Chaco, Argentina, obtained by M. Emile Wagner in 1902.

Illustrations: Blühende Kakteen **3**: pl. 123; Curtis's Bot. Mag. **76**: pl. 4542; Fl. Serr. **7**: pl. 670; Jard. Fleur. **2**: pl. 194; Loudon, Encycl. pl. ed. 3. f. 19406; Pfeiffer and Otto, Abbild. Beschr. Cact. **1**: pl. 6; Castle, Cactaceous plants f. 15; Blühende Kakteen **2**: pl. 103, this last as *Opuntia spegazzinii*; Hogg, Veg. King. 340. f. 111.

Figure 88 is from a plant in the greenhouses of the United States Department of Agriculture at Washington; figure 89 represents a joint of a plant collected by Dr. Rose at Córdoba, Argentina, in 1915.

OPUNTIA MALDONADENSIS Arechavaleta, Anal. Mus. Nac. Montevideo **5**: 286. 1905.

Cespitose, erect; branches cylindric, entangled or intertwined; joints 3 to 10 cm. long, about 2 cm. in diameter, the terminal ones obovate-spherical, dark green to olive-colored; areoles each surrounded by a violet blotch, small or prominent, orbicular; spines 5 or more, stout, spreading, elongated, unequal, the longest one 2 to 2.5 cm. long, reddish to brown; flowers and fruit unknown.

Type locality: Punta Ballena, near Maldonado, Uruguay.

Distribution: Uruguay.

This species, referred to the subgenus *Cylindropuntia* by Arechavaleta, inhabits the coast of Uruguay and is known to us only from description; we append it to the series *Salmianae,* but its nearest relationship may be elsewhere.

Series 11. SUBULATAE.

This series is confined to South America and represents a very distinct group, differing greatly from the tall cylindric-jointed species of North America. They lack sheaths to the spines, and the typical species has elongated persistent leaves. Although several of the species have long been in cultivation, at least two of them being known only from garden plants; for a long time the flowers were unknown and the plants were as frequently called *Cereus* or *Pereskia* as *Opuntia.*

KEY TO SPECIES.

```
Leaves long-persisting, elongated.
    Leaves up to 12 cm. long; spines yellowish white..................................... 41. O. subulata
    Leaves 1 to 7 cm. long; spines brownish............................................ 42. O. exaltata
Leaves early deciduous, short.
    Stem 1 meter high; leaves 4 mm. long.............................................. 43. O. pachypus
    Stem 3 to 4 meters high; leaves 10 to 13 mm. long ................................ 44. O. cylindrica
```

41. Opuntia subulata (Mühlenpfordt) Engelmann, Gard. Chron. **19**: 627. 1883.

Pereskia subulata Mühlenpfordt, Allg. Gartenz. 13: 347. 1845.
Opuntia ellemeetiana Miquel, Nederl. Krudk. Arch. 4: 337. 1858.*
Opuntia segethii Philippi, Bot. Zeit. 26: 861. 1868.

Either with a simple erect stem or with several main branches from the base, 2 to 4 meters high; trunk 6 to 10 cm. in diameter, the old bark smooth and brown, its areoles bearing clusters of 8 spines or more; branches numerous, more or less clustered but not whorled, at first almost at right angles to main stem but soon erect, bright green; leaves persistent, green, nearly at right angles to branch, straight or somewhat bowed above, nearly terete, pointed, 5 to 12.5 cm. long, grooved on the under side; tubercles large, depressed, becoming obliterated on old branches, arranged in longitudinal or spiral lines, more or less diamond-shaped, but retuse at apex and pointed or attentuate below, 2 to 4 cm. long; areoles in the retuse grooves of the tubercles bearing a few short yellow spines or sometimes spineless, but usually having 1 or 2 slender spines; flowers borne toward the ends of the branches; sepals reddish, minute, 4 to 8 mm. long or less; petals broader than the sepals, orange or greenish yellow; style rose-red except the whitish base, including the stigma-lobes about 3 cm. long, about as long as the longest stamens; stigma-lobes 5 or 6, slender, orange-yellow; fruit oblong, more or less persistent, 6 to 10 cm. long, leafy, with a deep umbilicus, sometimes proliferous; seeds few, 10 to 12 mm. long.

Type locality: Valparaiso, Chile, but doubtless described from cultivated plants.

Distribution: Chile is usually given as the home of this plant, but it is not found wild there. It may be a native in Argentina.

This species has long been in cultivation, it having originally been sent from Valparaiso, but Dr. Rose did not find it wild there or in any other part of Chile. It is rarely seen in cultivation in Chile. For many years it passed as a species of *Pereskia*, but in 1883 Dr. George Engelmann pointed out that it could not be retained in that genus and transferred it to *Opuntia*. The leaves are the largest in the genus, and it has larger seeds than any other *Opuntia*.

We have referred *Opuntia ellemeetiana* (originally described from Chile), a species with very long leaves, to *O. subulata*, although we have never seen specimens. Schumann did not know it and only lists it.

---

*Schumann says 1859.

We have been able more definitely to refer here *Opuntia segethii*, for we saw not only Philippi's type specimens in his herbarium, but also living specimens grown from Philippi's original stock.   The type specimen was from plants cultivated at Santiago, but in a later publication he states that his species grows spontaneously near Arequipa.   A part of this latter material is preserved in his herbarium at Santiago, which Dr. Rose was able to study; he also examined the Arequipa plant alive, and is convinced that it is quite different, being the plant common in Peru and Bolivia described below as *Opuntia exaltata*.

Illustrations: Engler and Prantl, Pflanzenfam. 3[6a]: f. 56, L; Gard. Chron. III. **34**: f. 33, 38; Monatsschr. Kakteenk. **8**: 7; **9**: 183; Schumann, Gesamtb. Kakteen f. 103; Neub. Gart. Mag. **1893**: 291, this as *Pereskia subulata;* Bot. Zeit. **26**: pl. 13, C. f. 1; Gartenflora **32**: pl. 1129, f. 5, the last two as *Opuntia segethii*.

Figure 90 is from a photograph of a plant at the New York Botanical Garden grown from a cutting brought by Mrs. H. L. Britton from the Riviera, Italy, in 1907.

42. Opuntia exaltata Berger, Hort. Mortol. 410. 1912.

Stem 2 to 5 meters high, with a definite trunk 5 to 30 cm. in diameter when well grown, much branched; ultimate joints fleshy, easily detached, somewhat curved upward, clavate, strongly tuberculate; tubercles large, 1.5 to 3 cm. long, more or less diamond-shaped, elevated, and rounded; areoles rounded, filled with short white wool; glochids often wanting, when present brown; leaves fleshy, terete, 1 to 7 cm. long; spines on young joints 1 to 5, mostly 1 to 3, dark yellow or brownish, unequal, the longest ones 5 cm. long; spines on old wood numerous, sometimes 12 or more, often 13 cm. long, brown, with roughened tips; flowers, including ovaries, 7 cm. long; sepals and petals brick-red; outer sepals ovate, small, the inner ones passing into petals; petals 2 cm. long, broadly obovate to broadly spatulate, sometimes nearly truncate at apex; stamens numerous, short, pinkish above, nearly white below; style swollen below, pinkish; stigma-lobes greenish; ovary 4 cm. long, deeply umbilicate, with large flat tubercles; areoles on ovary circular, filled with short brown and white wool, long, loosely attached brown spines, and a few shorter glochids, and subtended by small, tardily deciduous leaves; fruit green, pear-shaped, 9 cm. long, usually sterile; seeds large, irregular, 10 mm. broad.

FIG. 90.—Opuntia subulata.

Type locality: Not cited; described from cultivated plants.

Distribution: Ecuador, Peru, Bolivia, and probably northern Chile.

This *Opuntia* is called pataquisca by the Cuzco and Arequipa Indians, and is also known as espina.

This species was the most widely distributed *Opuntia* seen by Dr. Rose on the west coast of South America; but it is difficult to decide whether it is really native there, for it is widely cultivated as a hedge plant in many places.   It seems to be native along the upper Rimac of central Peru; at least it is well established on the hills.   Although very common in southern Peru and about La Paz, Bolivia, it is probably introduced for it grows only about towns and cultivated fields and seems never to produce fertile fruit.   About Cuzco it is likewise cultivated, but may be a native there also, for the fruit is generally fertile.

1. *Opuntia exaltata* as seen in the highlands of Peru.
2. Clump of *Opuntia floccosa* as it grows in the valleys of the Andes of eastern Peru.

Opuntia maxillare Roezl (Morren, Belg. Hort. **24**: 39. 1874), published without description and probably collected in the high mountains above Lima, may belong here.

Opuntia cumingii, of European gardens, belongs here. It was briefly mentioned in the journal of the Berlin Cactus Society (Monatsschr. Kakteenk. **7**: 160. 1897), but not formally described. Schumann referred it to *O. pentlandii*.

This species is near *Opuntia subulata*, but probably is distinct, although it is not always easy to distinguish them in greenhouse plants. Berger speaks of the similarity of the two as follows:

"This new species stands very close to *O. subulata*, and may be easily mistaken for it, but when grown side by side the differences are quite obvious. *O. exaltata* is a taller plant with generally longer branches, and somewhat glaucous instead of grass-green. The tubercles are more elongated and differently marked. The leaves are shorter, the spines, when young, are not white, but yellowish brown, generally stouter and stiffer. I have not yet seen a flower of it. It is an old inhabitant of our gardens."

Plate XIII, figure 1, is from a photograph taken by Hiram Bingham, July 7, 1912, near Tipon, Cuzco Valley, Peru, showing the plant near the upper left-hand corner; plate XV, figure 1, represents a leaf-bearing joint of a plant sent to the New York Botanical Garden from La Mortola, Italy, in 1915; figure 2 represents the lower part of a fruiting branch obtained by Dr. Rose at Cuzco, Peru, in 1914.

43. Opuntia pachypus Schumann, Monatsschr. Kakteenk. **14**: 26. 1904.

Plant about 1 meter high, much branched and candelabrum-like; branches cylindric, 3 to 5 cm. in diameter, either straight or curved, marked with broad tubercles; leaves subulate, pointed, constricted at the base, 4 mm. long, early deciduous; areoles circular, borne at the upper edges of the tubercles, 4 mm. in diameter, filled with short wool; spines 20 to 30, subulate, 5 to 20 mm. long; glochids yellow; flowers scarlet, 7 cm. long, including the ovary; petals variable, the longest ones 1.4 cm. long; style very thick, 9 mm. long; stigma-lobes 5 mm. long; ovary more or less spiny.

Type locality: Near Santa Clara, Peru.

Distribution: Central Peru, near the coast.

We know this species only from the description and illustrations. In 1914 Dr. Rose made several unsuccessful efforts to find it at Santa Clara, the type locality.

FIG. 91.—Opuntia pachypus.

Illustrations: Engler and Drude, Veg. Erde **12**: pl. 5^b; Monatsschr. Kakteenk. **14**: 27. Figure 91 is copied from the second illustration above cited.

44. Opuntia cylindrica (Lamarck) De Candolle, Prodr. **3**: 471. 1828.

Cactus cylindricus Lamarck, Encycl. **1**: 539. 1783.
Cereus cylindricus Haworth, Syn. Pl. Succ. 183. 1812.

More or less branched, 3 to 4 meters high, the old trunk becoming smooth; joints cylindric, obtuse at apex, green, with slightly elevated tubercles; leaves deciduous, 10 to 13 mm. long, terete, acute; areoles depressed, filled with white wool, bearing some long hairs and at first 2 or 3, afterwards more, short white spines (spines often wanting on greenhouse plants); flowers appearing just below the ends of the terminal branches, small, inconspicuous, about 2.5 cm. broad, scarlet; petals small, erect, obtuse; stamens numerous; style slender, 2.5 cm. long; ovary strongly tuberculate, depressed at apex; fruit about 5 cm. long, yellowish green; seeds more or less angled, 4 to 6 mm. in diameter.

Type locality: In Peru.

Distribution: Highlands of Ecuador and Peru.

The home of this species is usually given by recent writers as Chile, but Lamarck, who described it first in 1783, said it came from Peru. Dr. Rose, who visited Peru and Chile in

1914, was not able to find it wild in either country but found it abundant in Ecuador in 1918. This species was introduced into England in 1799, but flowers were not known until about 1834.

There are two abnormal forms in cultivation which are offered under the names variety *cristata* and *monstruosa*. Several varieties of this species are given in catalogues: *cristata* (Haage and Schmidt, Haupt-Verzeichnis 1908: 228. 1908); *cristata minor* Haage and Schmidt (Verzeichnis Blumenzwiebeln 1913: 37. 1913); and *robustior* (Haage and Schmidt, Haupt-Verzeichnis 1908: 228. 1908).

Illustration: Curtis's Bot. Mag. 61: pl. 3301; Carnegie Inst. Wash. 269: pl. 10, f. 88.

Plate XIV, figure 2, shows a leaf-bearing top of a plant grown at the New York Botanical Garden.

Series 12. MIQUELIANAE.

Bushy plants, with elongated cylindric bluish joints; tubercles large, elevated; leaves minute, early deciduous. The series consists of but one species, confined to the deserts of northern Chile.

45. **Opuntia miquelii** Monville, Hort. Univ. 1: 218. 1840.*

> *Opuntia pulverulenta* Pfeiffer, Allg. Gartenz. 8: 407. 1840.
> *Opuntia pulverulenta miquelii* Salm-Dyck, Cact. Hort. Dyck. 1844. 49. 1845.
> *Opuntia geissei* Philippi, Anal. Univ. Chile 85: 492. 1894.
> *Opuntia rosiflora* Schumann, Gesamtb. Kakteen 686. 1898.

Often growing in colonies 2 to 5 meters broad; stems cylindric, much branched, usually less than 1 meter high, but occasionally 1.5 meters high, with numerous lateral branches; branches rather short, usually only 8 to 20 cm. long, thick (5 to 6 cm. in diameter); old branches bluish green, with low tubercles sometimes 2 cm. long; young joints bright green, with high tubercles flattened laterally; spines tardily developing, but formidable on old branches, very unequal, in clusters of 10 or more, the longest ones nearly 10 cm. long, whitish in age; glochids numerous, brownish, caducous; leaves minute, 2 to 3 mm. long; areoles circular, when young filled with white wool, in age somewhat elevated on the areoles; flowers rather variable in length, 4 to 8 cm. long including the ovary, rose-colored to nearly white; petals broad, apiculate, 2 to 2.5 cm. long; filaments rose-colored; ovary strongly tuberculate; areoles filled with numerous brown glochids and subtended by minute leaves; style white; stigma-lobes green; fruit ovoid to oblong in outline, nearly white; umbilicus truncate; seeds small, 4 mm. broad.

Type locality: In South America, but no definite locality.

Distribution: Province of Atacama, Chile.

Opuntia miquelii and *O. pulverulenta* have long been considered identical. We have not seen the types of either, but are following such authorities as Salm-Dyck (in 1850), Labouret (1853), and Rümpler (1885) in uniting them. They seem to have been published in the same year.

Opuntia geissei, according to a statement made to Dr. Rose by Juan Söhrens, of Santiago, is the same as *O. miquelii*, and this the former was able to verify by later herbarium and field studies.

Opuntia rosiflora Schumann was based on Philippi's unpublished name *O. rosea;* while *O. rosea* was made by Philippi the type of *O. geissei*. This is clearly shown by Philippi's herbarium, where he has erased the name *O. rosea* and substituted *O. geissei*. Dr. Rose also obtained from William Geisse a part of Philippi's original specimen, which came, as the label states, from near Bandurrias, in the valley of Carrizal, in the Province of Atacama. Later on, while making field observations in Atacama, Dr. Rose found this species very common from north of Castillo to Vallenar. This is in the general region of *O. geissei* (*O. rosea* and *O. rosiflora*) and in the river valley of the Huasco. Huasco, the type locality of *O. miquelii*, is 25 miles lower down this valley, and we have no hesitancy in uniting them all.

Although this species is not uncommon in cultivation, it has rarely been seen in flower, and we believe that the fruit has not heretofore been described.

---

*Schumann states that this book was published in 1839.

M. E. Eaton del.

1. Flowering branch of *Opuntia burrageana*.   3, 4. Joints of *Opuntia stanlyi*.
2. *Opuntia cylindrica*.   5. Flowering joint of *Opuntia macrorhiza*.

(All natural size.)

Dr. Rose observed a single plant infested by *Loranthus aphyllus*, the parasite which is so abundant on *Cereus chiloensis*.

Opuntia heteromorpha Philippi (Anal. Mus. Nac. Chile 1891[2]: 28. 1891) we refer here on the authority of Schumann, but we have seen no specimens, the type specimen being missing from the Philippi herbarium in Santiago; it was collected at Chiquito, Tarapaca, Chile.

Dr. Weber thought that *Opuntia segethii* belonged here, but we have referred it to *O. subulata*.

Opuntia carrizalensis Philippi is only mentioned by Schumann (Gesamtb. Kakteen Nachtr. 152. 1903). It is doubtless to be referred here.

Plate XVI, figure 1, is from a plant collected by Dr. Rose at Vallenar, Chile, in 1914.

Series 13. CLAVATAE.

Here we include nine prostrate or spreading, low species, natives of the southwestern United States and Mexico, characterized by clavate joints and by sheathless spines, although rudimentary sheaths have been observed on young spines in some of the species; they appear to form a transition between the subgenus *Cylindropuntia* and the South American subgenus *Tephrocactus*, from which they differ essentially in having clavate joints.

KEY TO SPECIES.

```
Spines flattened.
    Stems very stout.
        Stems hardly clavate; ovary very prickly.......................................  46. O. invicta
        Stems strictly clavate; ovary only slightly prickly............................  47. O. stanlyi
    Stems more slender and weak.
        Spines brown, slender, long (4 to 6 cm. long)................................  48. O. schottii
        Spines stout, white, when old very flat.
            Bristles on ovary and fruit white......................................  49. O. clavata
            Bristles on ovary and fruit brown......................................  50. O. parishii
Spines terete, elongated, and flexible, or the central ones somewhat flattened.
    Flowers pinkish or purple.
        Bristles on ovary numerous, brown........................................  51. O. pulchella
        Bristles on ovary few, white.............................................  52. O. vilis
    Flowers yellow.
        Spines comparatively short, swollen at base...............................  53. O. bulbispina
        Spines long and flexible, not swollen at base.............................  54. O. grahamii
```

46. Opuntia invicta Brandegee, Proc. Calif. Acad. II. **2**: 163. 1889.

Plants growing in large clusters 2 meters in diameter and 2 to 5 dm. high, with many ascending or spreading branches; joints obovoid to clavate, dark green, 8 to 10 cm. long, strongly tuberculate; tubercles large, flattened laterally, 3 to 4 cm. long; areoles large, 1 to 1.5 cm. in diameter; leaves linear, 8 to 14 mm. long, reddish, curved, acute, deciduous; spines very formidable, when young reddish or purple with carmine-red bases, chestnut-brown at tips and grayish between, but in age dull in color; radial spines 6 to 10; central spines 10 to 12, much stouter than the radials, strongly flattened; wool white; glochids few, white, 2 to 4 mm. long; flowers yellow, 5 cm. in diameter; sepals ovate, acuminate; ovary 2 cm. in diameter, almost hidden by the numerous reddish, acicular spines; seeds yellowish, 2 mm. broad.

Type locality: About San Juanico, Lower California.

Distribution: Central Lower California, at low elevations.

Mr. Brandegee has called attention to the strong resemblance in habit of this species to some of the species of *Echinocereus*, and Dr. Rose states that when he first saw it he supposed it to be some strange *Echinocereus*. It grows in great tufted masses and does not suggest in the remotest degree any of our North American opuntias. The species clearly belongs to Engelmann's series *Clavatae*, where it was placed by Schumann, who associated it with *O. cereiformis*, but it is undoubtedly much nearer to *O. stanlyi*. So far as we know, the plant has never been in the trade; it does not succeed well in cultivation. Considerable living material was brought back by the *Albatross* in 1911, most of which was sent to the New York Botanical Garden; but some of the plants were sent to collections at St. Louis, Washington, and Los Angeles.

Illustration: Cact. Journ. **1**: February.

Plate XVI, figure 2, represents a plant collected by Dr. Rose at San Francisquito, Lower California, in 1911.

47. Opuntia stanlyi Engelmann in Emory, Mil. Reconn. 158. 1848.

Opuntia emoryi Engelmann, Proc. Amer. Acad. **3**: 303. 1856.
Opuntia kunzei Rose, Smiths. Misc. Coll. **50**: 505. 1908.

Stems low, usually less than 3 dm. high, much branched, creeping, forming broad, impenetrable masses 2 to 3 meters in diameter; joints 10 to 15 cm. long, clavate, more or less curved, strongly tuberculate; tubercles 3 to 4 cm. long, flattened laterally, 4 to 6 cm. apart; spines numerous, stout, elongated, somewhat roughened, reddish brown, the larger ones strongly flattened, 3.5 to 6 cm. long; flowers yellow, 5 to 6 cm. broad; fruit ovate, clavate at base, yellow, 5 to 6 cm. long, very spiny, with a depressed umbilicus; seeds flattened, 4.5 to 6.5 mm. in diameter.

Type locality: On the del Norte and Gila, New Mexico.

Distribution: Southwestern New Mexico to eastern Arizona and adjacent Mexico.

O. stanlyi was first found October 22, 1846, by W. H. Emory on his first trip across the continent; he reported the plant as abundant on the Del Norte and Gila. There has been much speculation as to what this species is, for no specimens were preserved. Dr. George Engelmann, who named the species, based it upon a sketch made by the artist of the expedition, Mr. J. M. Stanly. By a reference to Emory's itinerary we find that on October 22, 1846, he was in southwestern New Mexico. In 1908 Dr. Rose visited this region where he collected the various species of cacti to be found there. The only plant which we know from that part of New Mexico which could represent *O. stanlyi* is *Opuntia emoryi;* this was the conclusion reached by Wooton and Standley, who, in their Flora of New Mexico, have restored the name *O. stanlyi.*

We have referred *Opuntia kunzei* here because recent specimens sent in by Dr. Kunze have taken on a phase very much like the true *O. stanlyi.* There is a possibility that *O. kunzei* should be maintained, for we are not altogether convinced that certain material we have seen should be merged into *O. stanlyi.* To clear up this point, it is hoped that someone will collect and preserve a full series of specimens showing flowers, fruits, and seeds.

Illustrations: Emory, Mil. Reconn. App. 2. f. 9; Amer. Garden **11**: 531?; Cact. Journ. **1**: 154; Cact. Mex. Bound. pl. 70, 71, these last three as *Opuntia emoryi;* Hornaday, Campfires on Desert and Lava opp. p. 116, this as *Opuntia kunzei.*

Plate XIV, figure 3, represents a plant collected by Dr. R. E. Kunze near Gunsight Mountains, Arizona, in 1912; figure 4 shows a leaf-bearing joint of the same plant.

48. Opuntia schottii Engelmann, Proc. Amer. Acad. **3**: 304. 1856.

Prostrate, rooting from the areoles, forming dense clusters sometimes 2 or 3 meters in diameter; joints clavate, curved, ascending, easily breaking off, 6 to 7 cm. long, 2 cm. in diameter at thickest part, strongly tuberculate; leaves subulate, bronze-colored, 6 to 8 mm. long, acuminate; areoles 1 to 1.5 cm. apart; spines white and sheathed when young, soon brown, the larger ones sometimes as many as 12, very slender, sometimes 6 cm. long, somewhat flattened; wool white when young, turning brown; glochids white when young, turning brown, 4 mm. long or less; flowers yellow, 4 cm. long including ovary; sepals narrow, acuminate; petals acuminate; fruit yellow, narrowly oblong in outline, a little narrowed at base, 4 cm. long, closely set with areoles bearing numerous short spines, bristles, and white wool, the umbilicus depressed; seeds yellow, flattened, 4 mm. in diameter, notched at base.

Type locality: Arid soil near the mouth of the San Pedro and Pecos, western Texas.

Distribution: Southern and western Texas and northern Mexico.

Opuntia schottii greggii Engelmann (Cact. Mex. Bound. 68. pl. 73, f. 4. 1859), which came from near San Luis Potosí, Mexico, where it was collected by Dr. J. Gregg, in December 1848, is much out of the range of the normal form and probably belongs elsewhere; but no specimens have been examined except the type, which is fragmentary. Engelmann at first considered it a distinct species.

Illustration: Cact. Mex. Bound. pl. 73, f. 1 to 3.

Figure 92 represents joints of a plant collected by Dr. Rose at Langtry, Texas, in 1908.

M. E. Eaton del.

1, 2. Parts of joints of *Opuntia exaltata*.   3.   Upper part of joint of *Opuntia macrarthra*.
4.   Upper part of joint of *Opuntia tortispina*.   (All natural size.)

49. **Opuntia clavata** Engelmann in Wislizenus, Mem. Tour North. Mex. 95. 1848.

Plants low, not over 1.5 dm. high, much branched at base, spreading, forming large patches sometimes 2 meters in diameter; joints short, 3 to 7 cm. long, turgid, ascending, clavate; areoles close together; leaves subulate, 4 to 5 mm. long; spines pale, somewhat roughened, the radial ones 6 to 12, slender and acicular, 4 to 16 mm. long; central spines 4 to 7, much longer than the radials, more or less flattened, the largest one dagger-like; glochids numerous, yellowish, 3 to 5 mm. long; flowers yellow, 3.5 to 4 cm. long; fruit 4 to 5 cm. long, with numerous areoles filled with yellow, radiating glochids; seeds white, 5 mm. broad.

Type locality: Albuquerque, New Mexico.

Distribution: New Mexico, chiefly in the central part of the State.

This is one of the most characteristic species of the genus and has no near relative except *O. parishii*, of the deserts of California and Nevada. It is a great pest to grazing stock.

Illustrations: Bull. Agr. Exper. Station N. Mex. **78**: pl. [1, 2], Pac. R. Rep. **4**: pl. 22, f. 1 to 3; pl. 24, f. 6.

Figure 93 represents joints of a plant collected by W. T. H. Long at Albuquerque, New Mexico, in 1915.

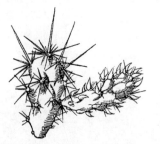 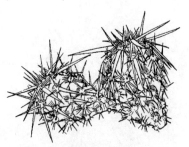

FIG. 92.—Opuntia schottii. ✕0 75.   FIG. 93.—Opuntia clavata. ✕0.75.

50. **Opuntia parishii** Orcutt, West Amer. Sci. **10**: 81. 1896.

Stems low, creeping, rooting along the under surface and forming dense, broad clusters; terminal joints short, clavate, ascending but almost hidden under the dense armament; tubercles prominent but short, 5 to 7 mm. long; spines at first reddish but soon grayish and finally nearly white; radial spines numerous, slender; central spines about 4, strongly angled and more or less flattened, 2 to 4 cm. long; glochids numerous; flowers not known; fruit 5 cm. long, the numerous large areoles bearing many long yellow glochids and short spines forming a radiating band about the margin; seeds dark, 4 mm. broad.

Type locality: Mohave Desert.

Distribution: Southern California and Nevada.

The species here described is the *Opuntia parryi* as described by Engelmann in 1856, although he then suspected it was different from that species. It has been renamed *Opuntia parishii* by Orcutt, who wrote as follows:

"We propose this name for that interesting plant of the Mohave desert region, hitherto called *O. parryi*, and under which it has been well described. The Messrs. Parish have hardly earned this light honor in many laborious trips through these desert regions, and I take pleasure in dedicating this species to them; *Opuntia parryi* (type from San Felipe), along with *bernardina* and *echinocarpa*, and a bewildering host of nameless forms, I unhesitatingly class under *serpentina!*"

Illustrations: Cact. Journ. **1**: 132; N. Amer. Fauna **7**: pl. 10; Pac. R. Rep. **4**: pl. 22, f. 4 to 7; pl. 24, f. 7, all as *Opuntia parryi*.

Figure 94 represents joints of a plant collected by S. B. Parish in southern California.

51. Opuntia pulchella Engelmann, Trans. St. Louis Acad. **2**: 201. 1863.

Low, 10 to 20 cm. high, densely branched, sometimes forming compact heads 6 dm. in diameter; main stem more or less definite, covered with areoles bearing yellow glochids 10 to 12 mm. long; lateral joints 5 to 6 cm. long, narrowly clavate, strongly tuberculate, purplish; areoles 6 to 8 mm. apart, 2 to 3 mm. broad; spines 10 to 16, slender, reddish, the longer ones 5 to 6 cm. long, somewhat flattened; flower 5 cm. long, when open, fully as broad; petals purple, 3 cm. long; ovary 2 cm. long, bearing numerous areoles filled with white wool and purple glochids 10 to 12 mm. long; fruit about 2.5 cm. long; seeds (according to Coulter) thick and round, 4 mm. in diameter, with broad flat commissure.

Type locality: Sandy deserts on Walker River, Nevada.

Distribution: Nevada and Arizona.

The plant was first collected by Henry Engelmann in 1859, and brought to his brother, Dr. George Engelmann. The species does not succeed well in cultivation under glass.

Illustration: Simpson's Rep. pl. 3.

Figure 95 is from an herbarium specimen collected by Thomas H. Means at Fallon, Churchill County, Nevada, in 1909.

 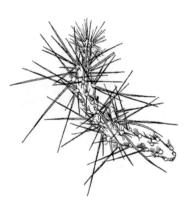

Fig. 94.—Opuntia parishii. ×0.66.   Fig. 95.—Opuntia pulchella. ×0.66.

52. Opuntia vilis Rose, Contr. U. S. Nat. Herb. **12**: 293. 1909.

Low, creeping, often forming mats several meters in diameter and only 10 to 15 cm. high; joints prostrate, becoming erect or ascending, the ultimate vertical ones clavate, 5 cm. long, the others 2 to 4 cm. long, very turgid, pale green, with low tubercles; leaves terete, 2 to 3 mm. long, acute, red; young areoles bearing white wool; radial spines upward of 12, the number increasing with age by the addition of very small whitish ones; central spines on prostrate joints 4, reddish, white-tipped, 1 to 4 cm. long, terete, slightly scabrous, with a sheath 5 mm. long, those of clavate joints white, reddish on the upper surface at the base, and along the whole of the lower surface flattened; flowers 4 cm. long; petals brilliant purplish, 2 cm. long; filaments bright yellow with green bases; style white; stigma-lobes yellow; fruit pale green, blackening in drying, 2 to 2.5 cm. in diameter, 2.5 to 3 cm. long, tuberculate, especially about the margin of the umbilicus, spiny, fluted above, somewhat dry, with large white seeds.

Type locality: Foot-slopes and plains of Zacatecas, Mexico.

Distribution: State of Zacatecas, Mexico.

Illustrations: Contr. U. S. Nat. Herb. **12**: pl. 27; f. 36.

Figure 96 is from a photograph of the type plant taken by F. E. Lloyd in Zacatecas, Mexico, in 1907.

FIG. 96.—Opuntia vilis.

53. Opuntia bulbispina Engelmann, Proc. Amer. Acad. **3**: 304. 1856.

Stems low, forming wide-spreading clumps 6 to 12 dm. broad; joints ovoid in outline, 2 to 2.5 cm. long by 10 to 12 mm. in diameter; tubercles prominent, 6 to 8 mm. long; radial spines 8 to 12, acicular, 3 to 6 mm. long; central spines 4, much stouter than the radials, 8 to 12 mm. long, bulbose at base; flower and fruit not described in original description and as yet unknown.

Type locality: Near Perros Bravos, north of Saltillo, Mexico.

Distribution: Coahuila and probably into Durango, Mexico.

The type of this species was collected by Josiah Gregg in 1848 and it has not with certainty been found since; it has been reported from one or two localities, but doubtless erroneously. At one time we supposed certain plants collected by Dr. Palmer in Chihuahua were to be thus referred. It is possible that specimens collected by Dr. Chaffey near Lerdo, Durango, may be referred here, as they have the short joints of this species, but the central spines are much longer, often reaching 2.5 to 3.5 cm. long. The type is deposited in the Engelmann Herbarium at St. Louis, and although the material is poor, it may yet serve to clear up this species definitely.

As stated by Coulter, this species has been regarded as the same as *O. tunicata*, a plant to which it is very remotely related.

Illustration: Cact. Mex. Bound. pl. 73, f. 5, 6.

Figure 97 is copied from the illustration above cited.

FIG. 97.—Opuntia bulbispina.

54. Opuntia grahamii Engelmann, Proc. Amer. Acad. **3**: 304. 1856.

Roots at first thick and fleshy, becoming woody, 2 cm. thick or more; plant low, much branched, spreading, forming low mounds often half buried in the sand, sometimes giving off roots at the areoles; terminal joints erect, clavate, bright green, 3 to 5 cm. long, with large oblong tubercles; leaves thick, bronze-colored, ovate, acute, 3 to 4 mm. long; areoles about 3 mm. broad; wool white; spines 8 to 15, slender, slightly scabrous, terete or some of the larger ones slightly compressed, white when young, soon reddish, the longest 3.5 to 6 cm. long; glochids numerous, slender, 4 mm. long or less, white, turning brown, persistent on the old stems; flowers yellow, 5 cm. broad; sepals ovate, acute, about 5 mm. long; fruit oblong to ovoid, 3 to 4.5 cm. long, its numerous areoles bearing white glochids and some slender spines; seeds beakless, 5 to 5.5 mm. in diameter, the commissure indistinct, linear.

Type locality: Near El Paso, Texas.

Distribution: Western Texas, New Mexico, and adjacent parts of Mexico.

This species was named for James Duncan Graham, Colonel, Corps of Engineers, United States Army, who died December 28, 1865, at Boston, Massachusetts. Colonel Graham was for a time chief of the scientific corps of the United States and Mexican Boundary Commission, and caused the specimens of this plant to be transmitted to Dr. George Engelmann.

FIG. 98.—Opuntia grahamii. X0.75.

The plant succeeds rather well in cultivation under glass.

Illustrations: Cact. Mex. Bound. pl. 72; Schumann, Gesamtb. Kakteen f. 102.

Figure 98 represents joints of a plant collected by Dr. Rose on hills near Sierra Blanca, Texas, in 1913.

Subgenus 2. TEPHROCACTUS.

Includes all the South American species of *Opuntia* which have short, oblong, or globular joints. It is hardly to be distinguished from the North American series *Clavatae.* Four series are recognized. The plants are confined to Peru, Chile, Bolivia, and Argentina. (See key to series, p. 44.)

Series 1. WEBERIANAE.

Plants low, forming dense clumps; joints subcylindric, strongly tuberculate and bearing numerous spines. This series suggests *Platyopuntia*, while the other series show closer relationship with the *Cylindropuntia.* Only one species known, inhabiting the dry part of northern Argentina.

55. Opuntia weberi Spegazzini, Anal. Mus. Nac. Buenos Aires III. **4**: 509. 1905.

Densely cespitose, forming clumps 2 to 3 dm. in diameter and 10 to 18 cm. high; joints yellowish green, erect, cylindric, strongly tuberculate, 2 to 6 cm. long, 1.5 to 2 cm. in diameter, densely spiny;

FIG. 99.—Opuntia weberi as it grows wild.

leaves described as wanting; tubercles spirally arranged, obtuse, somewhat 4-angled, 5 to 6 mm. broad; areoles somewhat depressed; spines 5 to 7, brown, 3 to 5 cm. long, flexuous, the upper ones erect; flowers borne near the top of the plant, small, solitary; ovary somewhat woolly below and with short spines above; flower rotate, yellow; fruit dry, white, 10 mm. in diameter; seeds somewhat contorted, bony, glabrous.

Type locality: In Sierra Pie de Palo, Province of San Juan, Argentina.

Distribution: Mountains of Provinces of San Juan and Salta, Argentina.

This description, though largely drawn from Dr. Spegazzini's full account of this species, has been amplified from examination made of the type. Dr. Spegazzini refers it

Fig. 100.—Opuntia weberi. Natural size.

to the subgenus *Tephrocactus*, and we have followed him in this; but it differs widely from any other known species of that group and its true affinity may be elsewhere. If the plant is leafless, as Dr. Spegazzini's description implies, this is a most interesting exception to the character of *Opuntia*.

Figure 99 is from a photograph of the plant at Molinos, Argentina; figure 100 is from a photograph of the type specimen in the collection of Dr. Spegazzini, to whom we are indebted for both of these illustrations.

Series 2. FLOCCOSAE.

Low plants, forming dense clumps or mounds; joints short, thick, and fleshy, usually covered with long, white, silky hairs. The two species are common in the high valleys of the Andes of Peru and Bolivia.

KEY TO SPECIES.

Spines yellow, stout.. 56. *O. floccosa*
Spines white, acicular.. 57. *O. lagopus*

56. Opuntia floccosa Salm-Dyck, Allg. Gartenz. **13**: 388. 1845.

Opuntia senilis Roezl in Morren, Belg. Hort. **24**: 39. 1874.
Opuntia floccosa denudata Weber, Dict. Hort. Bois 897. 1898.
Opuntia hempeliana Schumann, Gesamtb. Kakteen 690. 1898.

Plant growing in clumps or low mounds sometimes 1 to 2 meters in diameter, with hundreds of short, erect branches; joints oblong, 5 to 10 cm. long, usually hidden under a mass of long white hairs coming from the areoles; spines usually one from an areole, sometimes as many as three, yellow, 1 to 3 cm. long; leaves minute, green or pinkish; tubercles somewhat elevated, elongated; flowers, small, 3 cm. long, yellow; fruit globular, 3 cm. in diameter: seeds 4 mm. in diameter, with very narrow margins.

FIG. 101.—Opuntia floccosa.

Type locality: Said to be from vicinity of Lima, Peru, but doubtless only from the high mountains east of Lima.

Distribution: High mountain valleys and hills of the Andes from central Peru to central Bolivia.

O. floccosa is one of the most unusual and striking species of all the opuntias. One who is familiar only with the opuntias of North America would not suspect that it belongs to the genus. It does not grow on the hot mesas in the low country, as one would expect, but in the high, cold valleys and hills near the top of the Andes. The following paragraph, taken from John Ball's notes, is interesting in this connection:

Reserving some remarks on the botany of this excursion, there is yet to be mentioned here one plant of the upper region so singular that it must attract the notice of every traveler. As we ascended from Casapalta we noticed patches of white, which from a distance looked like snow. Seen nearer at hand, they had the appearance of large, rounded, flattened cushions, some five or six feet in diameter, and a foot high, covered with dense masses of floss silk that glistened with a silvery lustre. The unwary stranger who should be tempted to use one of these for a seat would suffer from the experiment. The plant is of the cactus family, and the silky covering conceals a host of long, slender, needle-like spines, that penetrate the flesh, easily break, and are most difficult to extract. Unfortunately, the living specimen which I sent to Kew did not survive the journey.

Dr. Rose found the plant very abundant in the Andes from 3,600 to 4,260 meters altitude, while others have reported it as high as 4,570 meters altitude. It is very common, forming everywhere great, conspicuous, usually white mounds. Dr. Rose also found it quite common between Cuzco and Juliaca, in southwestern Peru.

Mr. O. F. Cook, in the Journal of Heredity (8: 113. 1917), who has named this plant the polar bear cactus, wrote of it as follows:

Many exposed slopes on the bleak plateaus of the high Andes are dotted with clumps of pure white cacti that look from a distance like small masses of snow. On closer view, the shaggy white hair of these cacti make them appear like small sheep or poodle-dogs, or like reduced caricatures of the denizens of the arctic regions. We are so accustomed to think of cacti primarily as desert plants, peculiarly adapted to hot, dry deserts, that they seem distinctly out of place on the cold plateaus of the high Andes of southern Peru.

While most of the plants are covered with long white hairs, plants without hairs are not uncommon. These naked plants, which are characteristic of the whole clump or colony, appear at first sight very unlike the other forms, but they grow in the same region and have the same kind of flowers and fruits. In cultivated plants few hairs are developed. The variety *denudata* Weber seems to be only one of these naked forms.

Opuntia involuta Otto (Förster, Handb. Cact. 505. 1846) was not published, but was given as a synonym of this species. It was used the year before (Salm-Dyck, Allg. Gartenz. 13: 388. 1845) as a synonym of *O. vestita*.

Illustrations: Engler and Drude, Veg. Erde 12: pl. 14; Monatsschr. Kakteenk. 11: 41, 44, these last two as *Opuntia hempeliana;* Journ. Heredity 8: f. 3 to 8.

Plate XIII, figure 2, is from a photograph taken by Mr. O. F. Cook in the high mountains of eastern Peru. Figure 101 is from a photograph of a fragment of the plant collected by Dr. Rose in 1914, at Araranca, Peru.

57. Opuntia lagopus Schumann, Gesamtb. Kakteen Nachtr. 151. 1903.

Plants cespitose, growing in compact mounds; joints stout, cylindric, 10 cm. long, 3 to 3.5 cm. in diameter, densely covered with long white hairs; leaves minute, hidden under the wool, 7 mm. long; spines solitary, white, 2 cm. long, slender; glochids white, bristle-like; flowers probably red; fruit not known.

Type locality: Mountains of Bolivia above Arequipa, Peru.
Distribution: On the plains of the high Andes of Peru and Bolivia (altitude 4,000 meters).

This species is related to *O. floccosa*, with which it often grows, but it takes on a very different habit, growing in very dense, peculiar rounded mounds much higher than those formed by *O. floccosa*.

Illustration: Engler and Drude, Veg. Erde 12: pl. 14.

Figure 102 is from a photograph by H. L. Tucker, near Laxsa, Peru, in 1911.

Series 3. GLOMERATAE.

Plants low, composed of globose or oblong joints, the spines, or some of them, modified into flat papery processes. We recognize two species, confined to western Argentina.

KEY TO SPECIES.

Central spines papery; radial spines subulate ....58. *O. australis*
Spines, when present, all developed into long papery processes..............................59. *O. glomerata*

FIG. 102.—Opuntia lagopus, growing in a mound.

58. Opuntia australis Weber, Dict. Hort. Bois 896. 1898.

> *Pterocactus valentinii* Spegazzini, Anal. Soc. Cient. Argentina **48**: 51. 1899.

FIG. 103.—Opuntia australis. Showing large roots, joints, and flower. Natural size.

Plants often with large roots, these 5 to 8 cm. long by 2 to 3 cm. in diameter and larger than the parts above ground; joints described as cucumber-shaped, usually 6 to 8 cm. long by 1 to 2 cm. in diameter, but apparently often much smaller, tuberculate; radial spines 10 to 15, spreading, white, short, 3 to 4 cm. long; central spines 1 or 2, much longer than the radials, 2 cm. long, erect, flattened, and somewhat papery; flowers yellow, 2 to 3 cm. broad; seeds said to be rugose.

Type locality: Between Santa Cruz River and the Strait of Magellan, Argentina.

Distribution: The southernmost parts of Argentina.

We have recently examined three collections of this plant made by Carl Skottsberg in the Territory of Santa Cruz, which in the main agree with Weber's description. We have also seen *Pterocactus valentinii*, which is the same as Skottsberg's plant.

Dr. Spegazzini records this species as being in Santa Cruz, Argentina; but as he regards the plant collected there by him as only a variety of *O. darwinii*, we are inclined to believe he must have collected something else.

This species, which is found at the Strait of Magellan, extends farther south than any other cactus known to us.

Figure 103 is from a photograph of an herbarium specimen collected by Carl Skottsberg in the Territory of Santa Cruz, Patagonia, in 1908.

59. Opuntia glomerata Haworth, Phil. Mag. **7:** 111. 1830.

> *Opuntia articulata* Otto, Allg. Gartenz. 1: 116. 1833.
> *Cereus articulatus* Pfeiffer, Enum. Cact. 103. 1837.
> *Cereus syringacanthus* Pfeiffer, Enum. Cact. 103. 1837.
> *Opuntia platyacantha* Salm-Dyck in Pfeiffer, Allg. Gartenz. **5:** 371. 1837.
> *Opuntia tuberosa spinosa* Pfeiffer, Enum. Cact. 146. 1837.
> *Opuntia andicola* Pfeiffer, Enum. Cact. 145. 1837.
> *Opuntia diademata* Lemaire, Cact. Aliq. Nov. 36. 1838.
> *Opuntia turpinii* Lemaire, Cact. Aliq. Nov. 36. 1838.
> *Opuntia andicola elongata* Lemaire, Cact. Gen. Nov. Sp. 72. 1839.
> *Opuntia andicola fulvispina* Lemaire, Cact. Gen. Nov. Sp. 72. 1839.
> *Opuntia andicola major* Lemaire, Cact. Gen. Nov. Sp. 72. 1839.
> *Opuntia calva* Lemaire, Cact. Gen. Nov. Sp. 73. 1839.
> *Opuntia platyacantha gracilior* Salm-Dyck, Cact. Hort. Dyck. 1844. 43. 1845.
> *Opuntia platyacantha monvillei* Salm-Dyck, Cact. Hort. Dyck. 1849. 71. 1850.
> *Opuntia platyacantha deflexispina* Salm-Dyck, Cact. Hort. Dyck. 1849. 245. 1850.
> *Opuntia papyracantha* Philippi, Gartenflora **21:** 129. 1872.
> *Opuntia syringacantha* Schumann, Monatsschr. Kakteenk. **6:** 156. 1896.
> *Opuntia plumosa nivea* Walton, Cact. Journ. 1: 105. 1898.

Forming low, spreading clumps, the branches either erect or prostrate; joints globular, 3 to 6 cm. in diameter, often in cultivated specimens even smaller, dull grayish brown, hardly tuberculate except in drying; areoles large, bearing numerous long, brown glochids; spines often wanting, when present 1 to 3, long, weak, thin and papery, hardly pungent, either white or brownish, sometimes 10 cm. long; flowers light yellow, small; fruit globose, 1 to 1.5 cm. long, dry; seeds corky.

Type locality: Brazil, according to Haworth, but erroneously.

Distribution: Western Argentina. It has also been referred to Brazil and Chile, but surely not found in Brazil, and we should not expect it to inhabit Chile.

The plant figured by Nicholson (Dict. Gard. 2: f. 755) as *O. platyacantha* hardly belongs here.

O. glomerata, which is common on the dry hills about Mendoza, is very variable, especially as to whether it is spine-bearing or not; while the spines—which are really not spines but thin ribbon-like processes—vary much as to their color, markings, and length. These variations are partly the cause of so many synonyms for the species. Dr. Rose, who visited the region in which this species grows, found wide variation in the size of the joints, as well as in the absence or presence of spines.

Tephrocactus diadematus Lemaire (Cact. 88. 1868), *T. turpinii* Lemaire (Cact. 88. 1868),

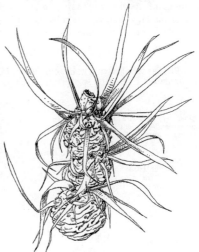

FIG. 104.—Opuntia glomerata. X0.5.

Opuntia polymorpha Pfeiffer (Enum. Cact. 103. 1837), and *Opuntia turpinii polymorpha* Salm-Dyck (Cact. Hort. Dyck. 1849. 71. 1850) are usually given as synonyms of *Opuntia diademata*, but none of them was actually published. *Opuntia polymorpha* Pfeiffer was used by Pfeiffer as a synonym for *Cereus articulatus* Pfeiffer.

Tephrocactus andicolus, *T. calvus*, and *T. platyacanthus*, all of Lemaire (Cact. 88. 1868), without descriptions, are referred here by inference.

Spegazzini (Anal. Mus. Nac. Buenos Aires III. **4:** 511. 1905) describes three varieties of this species under *O. diademata*, from Argentina, as follows: *inermis*, *oligacantha*, and *polyacantha;* while Weber (Dict. Hort. Bois 896. 1898) under the same name describes var. *calva*, but these all seem to be forms of this very variable species.

The following varietal names, under *Opuntia glomerata* var. *albispina* Förster (Handb. Cact. 472. 1846), var. *flavispina* Salm-Dyck (Cact. Hort. Dyck. 1844. 43. 1845), and var. *minor* Salm-Dyck (Cact. Hort. Dyck. 1849. 71. 1850), are mentioned in the places cited, but not described.

Opuntia horizontalis Gillies (Pfeiffer, Enum. Cact. 145. 1837) was used as a synonym of *Opuntia andicola*, and should be referred here.

Opuntia pelaguensis (Salm-Dyck, Cact. Hort. Dyck. 1849. 71. 1850) was published as a synonym of *Opuntia platyacantha deflexispina*.

Opuntia andicola minor, an unpublished variety, is mentioned by name only in Monatsschrift für Kakteenkunde (10: 48. 1900).

Illustrations: Cact. Journ. 1: 100, as *Opuntia andicola*: Engler and Prantl, Pflanzenfam. 3⁶ᵃ: f. 56, K.; Gard. Chron. III. 34: f. 39; Monatsschr. Kakteenk. 13: 23, these three as *Opuntia diademata*. Cact. Journ. 1: February; Dict. Gard. Nicholson Suppl. f. 607; Förster, Handb. Cact. ed. 2. f. 125; Gard. Chron. III. 23: f. 129; 29: f. 63; Gartenflora 21: pl. 721, f. 2, all as *Opuntia papyracantha*; Cact. Journ. 1: 105, as *Opuntia plumosa nivea*.

Figure 104 represents a plant collected by Dr. Rose at Mendoza, Argentina, in 1915.

Opuntia schumannii Spegazzini (Anal. Mus. Nac. Buenos Aires III. 4: 511. 1905, not Berger, 1904) is a homonym, and we hesitate to give it a new name until it is better known. The type comes from Salta, Argentina, from a region where we already have a number of species of *Tephrocactus*. Spegazzini, who described it, says it is related to *O. diademata*, which is now referred to *O. glomerata*, but is very distinct. It is without spines and the flowers are unknown.

Series 4. PENTLANDIANAE.

Plants often growing in large mounds; joints globular to oblong; spines usually slender, acicular to subulate. Seventeen species are here recognized.

KEY TO SPECIES.

Spines very long and stout, up to 15 to 20 cm. long.................................... 60. *O. aoracantha*
Spines slender, 10 cm. long or less.
  Spines appressed to the joints.
    Spines 12 to 20, flexuous; joints 7 cm. long.................................... 61. *O. rauppiana*
    Spines 6 or 7; joints 2 to 4 cm. long.. 62. *O. subterranea*
  Spines straight, not appressed.
    Spines flat or semiterete.
      Spines 7 to 10 cm. long.. 63. *O. hickenii*
      Spines 6 cm. long or less.
        Longer spines 1 to 3.
          Joints ellipsoid, 4 to 5 cm. thick.. 64. *O. darwinii*
          Joints oblong, 1 cm. thick... 65. *O. tarapacana*
        Longer spines 4 or 5.
          Spines gray... 66. *O. atacamensis*
          Spines yellow.. 67. *O. russellii*
    Spines terete.
      Spines white, at least when young.
        Joints tuberculate... 68. *O. corrugata*
        Joints not tuberculate.
          Joints oblong.. 69. *O. ovata*
          Joints globose... 70. *O. sphaerica*
      Spines yellow to brown or nearly black.
        Roots large and woody; spines nearly black............................... 71. *O. skottsbergii*
        Roots fibrous.
          Spines purple-black.. 72. *O. nigrispina*
          Spines yellow to brown.
             Plants forming large clumps.
               Fruit about 2.5 cm. long, nearly unarmed....................... 73. *O. pentlandii*
               Fruit 5 to 6 cm. long, copiously armed with long spines above........ 74. *O. ignescens*
             Plants isolated, not forming clumps.
               Old joints globose; spines acicular............................. 75. *O. campestris*
               Joints all oblong; spines subulate.............................. 76. *O. ignota*

60. Opuntia aoracantha Lemaire, Cact. Aliq. Nov. 34. 1838.

Cereus ovatus Pfeiffer, Enum. Cact. 102. 1837. Not *Opuntia ovata* Pfeiffer, l. c. 144. 1837.
Opuntia formidabilis Walton, Cact. Journ. 1: 105. 1898.

Usually low, cespitose, forming clumps 2 to 5 dm. in diameter and sometimes 1 to 2 dm. high; branches grayish, either erect or prostrate, made up of 5 to 10, perhaps even more, globular joints; joints easily detached, freely rooting and starting new colonies, 5 to 8 cm. in diameter, strongly tuberculate especially when young, the lower part spineless, the upper areoles large, spine-bearing; spines brown or blackish, 1 to 7, the longer ones 13 cm. long, straight, a little flattened, roughish to the touch; flowers white; fruit short-oblong, 3 cm. long, red, weakly tuberculate, bearing numerous areoles, usually naked but sometimes bearing a few short spines near the top, becoming dry; umbilicus of fruit broad and depressed; seeds white, flattened, 4 to 5 mm. broad, the margins thick and corky.

Fig. 105.—Opuntia aoracantha.　X0.66.

Type locality: Not cited, but doubtless from Mendoza.
Distribution: Western provinces of Argentina, from Mendoza to Jujuy.

Opuntia gilliesii Pfeiffer (Enum. Cact. 102. 1837, as synonym) and *Tephrocactus aoracanthus* Lemaire (Cact. 89. 1868) are usually given as synonyms of this species, but they were not described in the places usually cited, and as here given. *Opuntia acracantha* Walpers (Repert. Bot. 2: 354. 1843) is a typographical error.

O. aoracantha, although described nearly 80 years ago, is practically unknown in collections and has been very poorly described. The fruit has heretofore been unknown. Dr. Rose found it in 1915 in great abundance growing on dry, rocky hills west of Mendoza, although in but one locality. A bountiful supply of living material was sent home, several photographs were taken, and fruit and seeds obtained.

Opuntia tuberiformis Philippi (Anal. Mus. Nac. Chile 1891[2]: 28. 1891), referred here by Schumann, doubtless belongs elsewhere. It may possibly belong to some *Platyopuntia*, for it is described as having ovate joints only 5 mm. thick. It comes from the foot of the Andes in the Province of Tarapaca, Chile.

Illustrations: Gard. Chron. III. **34**: f. 40; Monatsschr. Kakteenk. **12**: 172; Cact. Journ. 1: 105, the last as *O. formidabilis.*

Figure 105 represents a joint of a plant collected by Dr. Rose at Mendoza, Argentina, in 1915.

61. Opuntia rauppiana Schumann, Monatsschr. Kakteenk. **9**: 118. 1899.

Joints ellipsoid, rounded at each end, somewhat tuberculate, dark green or becoming grayish green, 7 cm. long by 4 cm. in greatest diameter; glochids yellow, 5 cm. long; spines 12 to 14, sometimes as many as 20, very weak, almost bristle-like, 2 cm. long, hardly pungent.

Type locality: In the Andes.
Distribution: Bolivia, according to Schumann.

Little is known of the habit of this plant, as only one joint is figured and this appears to be a sickly greenhouse specimen. It suggests some of the species which grow in large clumps like the one figured as *Opuntia grata* by Fries.

FIG. 106.—Opuntia rauppiana.

Illustrations: Monatsschr. Kakteenk. **9**: 118; Schumann, Gesamtb. Kakteen Nachtr. f. 36 (same).

Figure 106 is copied from the illustration above cited.

62. Opuntia subterranea R. E. Fries, Nov. Act. Soc. Sci. Upsal. IV. **1**[1]: 122. 1905.

Almost buried in the sand, simple or few-branched from a thick root 7 to 12 cm. deep: joints terete, 2 to 4 cm. long; tubercles low; spines 1 to 7, all radial, short, whitish, recurved, appressed;

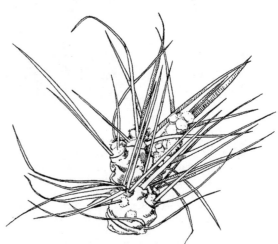

FIG. 107.—Opuntia subterranea.

FIG. 108.—Opuntia hickenii. ✕0.6.

flowers lateral, brownish; ovary small, with a depressed umbilicus, its areoles bearing small glochids and a little wool; fruit 12 to 15 mm. long; seeds 3 mm. broad, irregular.

Type locality: Near Moreno, Jujuy, Argentina.
Distribution: Northern Argentina and adjacent Bolivia.
This peculiar little plant, heretofore known only from the type collection, was obtained by Dr. Shafer on stony plains at Villazón, Bolivia, in February 1917, but was not in bloom.
Illustration: Nov. Act. Soc. Sci. Upsal. IV. 1^1: pl. 8, f. 4 to 8.
Figure 107 is copied from the illustration above cited.

63. Opuntia hickenii sp. nov.

Low, cespitose, forming clusters 1 meter in diameter; joints globular, 3 to 5 cm. in diameter, strongly tuberculate, the lower tubercles usually spineless; areoles rather large, circular; spines 2 to 5, flat and thin, narrow, weak, pungent, 5 to 12 cm. long, silvery-colored but nearly black in age; flowers yellow; fruit not known.

Type in United States National Herbarium, No. 603229, from Puerto Madryn, Chubut, Argentina, collected by Cristóbal M. Hicken.
Common in Chubut and Rio Negro, southern Argentina, where it was collected several times by Dr. Hicken.
Figure 108 represents the type specimen above cited.

A photograph of a plant from San Juan, Argentina, communicated by Dr. Spegazzini, indicates another species of this relationship.

64. Opuntia darwinii Henslow, Mag. Zool. Bot. 1: 466. 1837.

Low, perhaps not more than 2 to 4 cm. high, much branched at base from a more or less elongated woody root; joints normally few, nearly globular, about 3 cm. in diameter, or often nearly cylindric, frequently numerous and small and growing out from the main axis, then only 5 to 10 mm. in diameter; areoles large, filled with wool, the lower ones spineless; spines 1 to 3, nearly erect, the longest one 3 to 3.5 cm. long, yellow or reddish yellow, decidedly flattened; flowers originally described as larger than the joints, but certainly often much smaller; petals yellow, broad, with a truncate or depressed top and usually with a mucronate tip; ovary, in specimens seen, only 2 cm. long, covered with large woolly areoles; styles described as stout, with 9 thick radiating stigma-lobes.

Type locality: Port Desire, Patagonia, latitude 47° south.
Distribution: Southern Argentina.
This species seems to be common in that part of Patagonia known now as the Territory of Santa Cruz, Argentina. We have recently examined four separate collections made in this region, especially one from about Lake Buenos Aires and on the Fenix River by Carl Skottsberg, in 1907–1909.
The plant is in cultivation in Europe and is offered for sale by cactus dealers.
It was first collected by Charles Darwin, but only a single joint was taken, which was described and figured by Rev. J. S. Henslow. The illustration of the flowers seems too large, but otherwise represents fairly well the plant as we know it. The following interesting note is taken from Mr. Henslow's article as it appeared in the Magazine of Zoology and Botany, volume 1, page 467:

I have named this interesting Cactus after my friend C. Darwin, Esq., who has recently returned to England, after a five years' absence on board H. M. S. *Beagle*, whilst she was employed in surveying the southernmost parts of South America. The specimen figured was gathered in the month of January, at Port Desire, lat. 47° S. in Patagonia. He recollects also to have seen the same plant in flower as far south as Port St. Julian in lat. 49° S. It is a small species growing close to the ground on arid gravelly plains, at no great distance from the sea. The flowers had one day arrested his attention by the great irritability which their stamens manifested upon his inserting a piece of straw into the tube, when they immediately collapsed round the pistil, and the segments of the perianth soon after closed also. He had intended to procure fresh specimens on the following day, and returned to the ship with the one now figured, but unfortunately she sailed immediately afterwards, and he was prevented from obtaining any more. The geographical position of this species is beyond the limits hitherto assigned to any of the order, which are not recorded as growing much south of the tropic of Capricorn. The climate is remarkably dry and clear, hot in summer, but with sharp frosts during the winter nights. He found Cacti both abundant and of a large size, a little farther to the north at Rio-Negro in latitude 41° S.

Illustration: Mag. Zool. Bot. 1: pl. 14, f. 1.

Figure 109 is copied from a photograph of an herbarium specimen collected by Carl Skottsberg in Patagonia in 1908.

65. Opuntia tarapacana Philippi, Anal. Mus. Nac. Chile 1891²: 27. 1891.

> *Opuntia rahmeri* Philippi, Anal. Mus. Nac. Chile 1891²: 27. 1891.

Low, cespitose plants; joints small, ovoid, about 2 cm. long by 1 cm. thick, bearing spines from white woolly areoles at tips; spines usually 3, straight, 12 to 15 mm. long, white with yellowish tips; flowers yellow; petals 21 mm. long; ovary elongated, 2 cm. long.

Type locality: Calalaste, Chile.

Distribution: Known only from type locality, although Schumann in his Keys refers this species to Bolivia.

Although the type of this species is preserved in the Museum at Santiago, Chile, it is insufficient to enable us to give a very full description. It seems distinct from the other species of the group.

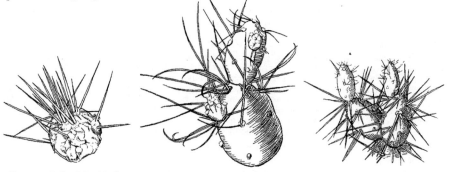

FIG. 109.—O. darwinii. ×0.6.   FIG. 110.—O. atacamensis. ×0.6.   FIG. 111.—O. russellii. ×0.6.

66. Opuntia atacamensis Philippi, Fl. Atac. 24. 1860.

> ? *Pereskia glomerata* Pfeiffer, Enum. Cact. 179. 1837. Not *Opuntia glomerata* Haworth. 1830.

Growing in large, dense clusters sometimes 6 dm. broad and 3 dm. high; joints ovoid, 2.5 cm. long by 2 cm. in diameter; areoles in 5 to 7 series, the lower ones with wool and very short spines; upper areoles each bearing 1 erect central spine 18 to 25 mm. long, yellow or reddish; radial spines 2 to 4, strongly appressed, 2 mm. long; flowers yellow.

Type locality: Profetas, Chile; also Puquios, 23° 50' south latitude.

Distribution: On the high central deserts of northern Chile at an altitude of 2,700 to 3,300 meters.

We have not seen the type of this species, and our reference of *Pereskia glomerata* here may not be correct.

Illustration: Nov. Act. Soc. Sct. Upsal. IV. 1¹: pl. 1, as *Opuntia grata*.

Figure 110 represents a plant obtained by Dr. Rose at the Botanical Garden, Santiago, Chile, in 1914.

67. Opuntia russellii sp. nov.

Forming small, compact clumps 1 to 2 dm. in diameter; joints small, globular to obovoid, dull green to more or less purplish, 2 to 4 cm. long, very spiny near the top; leaves minute, acute, soon falling; prominent spines 3 to 6, yellow, 2 to 3 cm. long, slightly flattened; accessory spines 1 to several, 1 cm. long or less; glochids at first inconspicuous, but in time very abundant, sometimes 2 cm. long, somewhat persistent; flowers not known; fruit globular, 2 to 2.5 cm. in diameter, spineless; seeds pale, 4 mm. broad.

Collected by J. N. Rose and Paul G. Russell on the dry hills at Potrerillos, Mendoza, Argentina, September 2, 1915 (No. 21002).

This is a common species in the foothills of the Andes, in the Province of Mendoza, where it forms low mounds along with other cacti.

Figure 111 represents joints of the type specimen above cited.

68. Opuntia corrugata Salm-Dyck, Hort. Dyck. 360. 1834.

> Opuntia eburnea Lemaire, Cact. Aliq. Nov. 35. 1838.
> Opuntia retrospinosa Lemaire, Cact. Aliq. Nov. 35. 1838.
> Opuntia parmentieri Pfeiffer, Allg. Gartenz. 6: 276. 1838.

More or less cespitose; joints 3.5 cm. long, 8 to 12 mm. in diameter, orbicular to cylindric, often erect, attenuate at both ends, light green, the terminal one often flattened; glochids minute, yellowish; spines 6 to 8, acicular, 8 to 12 mm. long, white; flowers reddish; fruit red.

Type locality: None given.

Distribution: Northwestern Argentina, according to later writers.

Lemaire (Cact. 88. 1868) uses the names *Cactus corrugatus* and *C. eburneus*, both of which Schumann refers here.

Tephrocactus retrospinosus Lemaire (Cact. 88. 1868) is placed by Lemaire in his third section of *Tephrocactus*, but it is without description. It is doubtless the same as *Opuntia retrospinosa* Lemaire, which belongs here.

Opuntia aulacothele Weber (Gosselin, Bull. Mus. Hist. Nat. Paris 10: 392. 1904), which was described without flowers or fruit, may be of this alliance. It comes from San Rafael, Argentina.

Opuntia cornigata, mentioned in Bailey's Standard Cyclopedia of Horticulture (4: 2367. 1916), is a misspelling of this name.

Opuntia corrugata monvillei Salm-Dyck (Cact. Hort. Dyck. 1849. 72. 1850) was not described.

Opuntia longispina Haworth (Phil. Mag. 7: 111. 1830), when first described, was supposed to have come from Brazil; the Index Kewensis refers it to Chile; while Schumann treats it in a note under *O. corrugata* as an Argentine species. It may not be an *Opuntia* but a *Maihuenia*.

69. Opuntia ovata Pfeiffer, Enum. Cact. 144. 1837.

> Opuntia ovallei Remy in Gay, Fl. Chilena 3: 29. 1847.
> Opuntia grata Philippi, Linnaea 30: 211. 1859.
> Opuntia monticola Philippi, Linnaea 33: 82. 1864.

Low, branching, cespitose plants; joints yellowish green, some deep purple when young, subcylindric to ellipsoid, 3 cm. long; spines 5 to 9, 4 to 10 mm. long, when young brownish, in age white; fruit ovoid; umbilicus curved outward.

Type locality: Mendoza, Argentina.

Distribution: Mountains of Argentina and Chile.

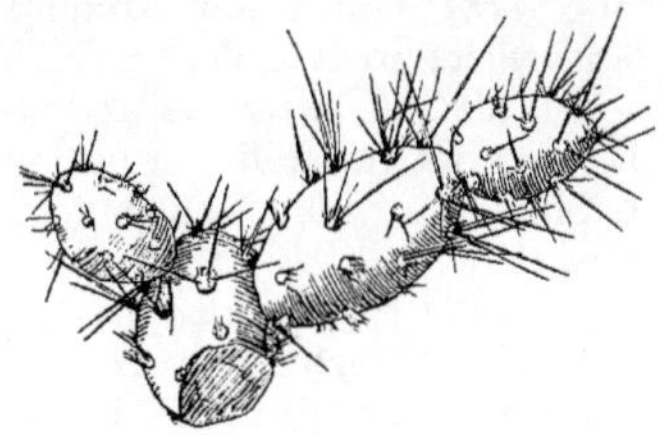

FIG. 112.—Opuntia ovata. X0.5.

Opuntia ovoides Lemaire (Cact. Gen. Nov. Sp. 73. 1839) and *Cactus ovoides* Lemaire (Cact. 88. 1868) are usually cited as synonyms for *Opuntia ovata;* they are unpublished names.

This species forms low clumps, each branch consisting of 2 to 5 joints. Dr. Rose found it abundant in the Andes above Mendoza and it has also been reported from the Chilean side of the Andes. Colonies differ in armament. In cultivation some of the joints are elongated and club-shaped.

Illustration: Schumann Gesamtb. Kakteen f. 105, as *Opuntia grata*.

Figure 112 shows joints of the plant collected by Dr. Rose in 1915 at Potrerillos, Argentina.

70. **Opuntia sphaerica** Förster, Hamb. Gartenz. **17**: 167. 1861.

Opuntia dimorpha Förster, Hamb. Gartenz. **17**: 167. 1861.
Opuntia leonina Haage and Schmidt in Regel and Schmidt, Gartenflora **30**: 413. 1881.
Opuntia leucophaea Philippi, Anal. Mus. Nac. Chile **1891**[2]: 27. 1891.
Opuntia corotilla Schumann in Vaupel, Bot. Jahrb. Engler Beibl. **111**: 28. 1913.

Plants often erect, always low, usually few-branched, often forming large patches; joints usually globular, 12 to 40 cm. in diameter; areoles large, numerous, sometimes nearly hiding the surfaces of the joints with their short brown wool; spines variable as to number, sometimes few, sometimes numerous, brown at first, in age sometimes gray, 1 to 4 cm. long, usually stiff; flowers 4 cm. long, deep orange; petals obtuse; fruit globular, often very spiny; seed globular, white, 4 mm. in diameter, surrounded by a thin, broad band.

Type locality: Near Arequipa, Peru.
Distribution: Central Peru to central Chile.

The three illustrations cited below were made from the same cultivated plant. They look very much like a poor specimen of *Opuntia glomerata*, and, if such it should prove, the name *O. leonina* should be referred to the synonymy of that
species.

We have referred *Opuntia dimorpha* here with some hesitancy.

This plant often passes for
Opuntia ovata and, from herbarium
specimens we have seen, it has been
so identified by Rudolph Philippi.

This species is very common in
sandy places on hills, dry flats, and
in mountain valleys, often covering the ground to the exclusion of
all other plants. The joints readily
break loose and, falling to the
ground, start new colonies. We
found the species very common
both above and below Arequipa,
Peru, where it is called corotilla;
in central Chile it grows at lower
altitudes but in similar situations.

FIG. 113.—Opuntia sphaerica.

In Chile it is called leon or leoncito, which is probably the origin of the name *Opuntia
leonina.*

Opuntia phyllacantha Haage and Schmidt (Regel and Schmidt, Gartenflora **30**: 414.
1881), if it actually came from Chile, as stated, may belong here. The joints are more
elongated, although the habit is somewhat similar. The illustration is poor and has doubtless been made from a greenhouse specimen. This name was given, with Salm-Dyck as
authority, by Förster (Handb. Cact. 508. 1846), but without any description.

Illustrations: Cact. Journ. **1**: 100; Förster, Handb. Cact. ed. 2. f. 133; Gartenflora **30**:
413, all as *Opuntia leonina.*

Figure 113 is from a photograph of joints of the plant collected by Dr. Rose above
Arequipa, Peru, in 1914.

71. **Opuntia skottsbergii** sp. nov.

Roots thick and fleshy, sometimes 10 cm. long, the plant doubtless more or less cespitose; joints,
at least some of them, globular, 3 cm. in diameter, almost hidden by the numerous closely set spines;
areoles close together, small, at times producing long tufts of white wool; spines about 10, black
except the yellowish tips, 1 to 2 cm. long; glochids numerous, elongated; flowers, including the very

spiny ovary, about 6 cm. long; petals about 3 cm. long, drying reddish or reddish green; areoles of the ovary bearing 5 to 7 spines, which are brown or blackish below and with more or less yellowish tips; fruit not known.

Collected near Lake Buenos Aires, Territory of Santa Cruz, Argentina, December 12, 1908, by Carl Skottsberg (No. 675); and again on the Rio Fenix, north of the locality above given, December 10, 1908 (No. 625, type).

This species belongs to the subgenus *Tephrocactus*, but is not closely related to any of the described species. The flower resembles very much the one figured by Henslow as *O. darwinii*, and it is possible that he may have had some of this species in his *O. darwinii;* the plant bodies, however, are so different that one could hardly confuse the two.

Figure 114 is copied from a photograph of the type specimen above cited.

72. Opuntia nigrispina Schumann, Gesamtb. Kakteen 695. 1898.

> *Opuntia purpurea* R. E. Fries, Nov. Act. Soc. Sci. Upsal. IV. 1¹: 123. 1905.

Described as a shrub, 1 to 2 dm. high and much branched, the branches upright; joints dull green or reddish violet, 2 to 4 cm. long, 1 to 2 cm. in diameter, oblong-ellipsoid, terete, when young bearing decurrent, spirally arranged tubercles; areoles 2 to 3 mm. in diameter, bearing abundant

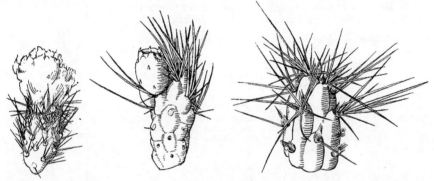

Fig. 114.—Opuntia skottsbergii.   Fig. 115.—Opuntia nigrispina. ✕0.8.   Fig. 116.—Opuntia pentlandii. ✕0.4.

wool and glochids; spines 3 to 5 from upper areoles, 2.5 to 3 cm. long, straight, spreading, subterete, weak, purplish black; flowers small, purple, 2.2 to 2.5 cm. long; petals spatulate, 1.5 cm. long, 6 mm. broad; stigma-lobes 5; ovary 1 cm. long, obovoid, nearly smooth.

Type locality: On the puna of Humahuaca, Bolivia.

Distribution: Rare in stony mountains, altitude 3,500 meters, Jujuy, Argentina, and southern Bolivia.

Figure 115 represents a fruiting joint collected by J. A. Shafer at La Quiaca, Argentina, February 2, 1917 (No. 79).

73. Opuntia pentlandii Salm-Dyck, Allg. Gartenz. **13**: 387. 1845.

> *Opuntia boliviana* Salm-Dyck, Allg. Gartenz. **13**: 388. 1845.
> *Opuntia pyrrhacantha* Schumann, Gesamtb. Kakteen 694. 1898.
> *Opuntia dactylifera* Vaupel, Bot. Jahrb. Engler Beibl. **111**: 29. 1913.
> *Opuntia cucumiformis* Griffiths, Bull. Torr. Club **43**: 524. 1916. (From the description.)

Plant much branched, forming low, rounded, compact mounds sometimes a meter broad with hundreds of short stubby branches; joints obovoid to oblong-cylindric, plump, 2 to 10 cm. long, sometimes 4 dm. in diameter, more or less pointed, pale green or sometimes purplish, tuberculate; areoles small, circular, filled with short wool and yellow glochids, the upper ones sometimes also having spines; spines sometimes wanting, when present mostly from the upper areoles, erect, 2 to 10, usually bright yellow, sometimes brownish becoming dull brown, the longest one 7 cm. long; flowers very variable in color and size, lemon-yellow to deep red, 2 to 3 cm. long, sometimes 5 cm. broad when fully expanded; petals obtuse; filaments short; style thick; stigma-lobes very short; ovary short with few areoles; areoles on ovary subtended by minute leaves, filled with short wool, the upper ones with bristle-like spines; fruit globular to short-oblong, 2 to 3 cm. long, dry; seeds numerous, 4 to 5 mm. long.

Type locality: In Bolivia.

Distribution: Very common on the high pampas of southeastern Peru and Bolivia, and adjacent Argentina.

Cactus pentlandii Lemaire (Cact. 88. 1868), name only, is supposed to apply to this species.

This is one of the most characteristic plants of the high pampas of the Andean region, mostly growing at elevations of 12,000 feet or higher, forming low, broad, compact clumps, sometimes made up of a hundred plants or more.

FIG. 117.—Opuntia pentlandii. ×0.4.

Illustrations: ?Dict. Gard. Nicholson **2**: f. 751; ?Förster, Handb. Cact. ed. 2. f. 124; ?W. Watson, Cact. Cult. f. 77, all as *Opuntia boliviana*; Monatsschr. Kakteenk. **24**: 175, as *Opuntia dactylifera*.

Figure 116 represents a joint of the plant collected in 1914 by Dr. Rose at Comanche, Bolivia; figure 117 shows a flowering joint collected by Dr. Rose in 1914, at Juliaca, Peru.

74. Opuntia ignescens Vaupel, Bot. Jahrb. Engler Beibl. **111**: 30. 1913.

Plants forming clumps 2 dm. high or less, with hundreds of erect or spreading joints; joints bluish green, 8 to 10 cm. long, very fleshy, naked below; upper areoles very spiny; spines 6 to 15 from each areole, nearly equal, 4 to 5 cm. long, erect, acicular, yellow; flowers very showy, deep red; ovary oblong, 3 to 4 cm. long, naked below, but the upper areoles producing numerous spines 4 to 7 cm. long; fruit red, 7 cm. long, spiny and tuberculate above, terete below, with a deep umbilicus; seeds nearly globular, about 5 mm. in diameter.

Type locality: Near Sumbay, southern Peru.

Distribution: On the pampas of southern Peru and northern Chile, at altitude of 3,000 to 3,600 meters.

FIG. 118.—Opuntia ignescens.   ×0.5.

FIG. 119.—Opuntia ignescens forming large mounds.

PLATE XVI

M. E. Eaton del.

1. Top of *Opuntia miquelii*.   2.   Old and young joints of *Opuntia invicta*.
3.   Upper part of joint of *Opuntia ignescens*.   (All natural size.)

Plate xvi, figure 3, represents old and young joints of the plant collected above Ayrampl, Peru, by Dr. Rose in 1914. Figure 118 shows a fruit from the same plant; figure 119 is from a photograph taken by H. L. Tucker at Coropuna, Peru, in 1911.

75. Opuntia campestris sp. nov.

Much branched, often forming low, dense masses, 3 to 6 dm. in diameter; terminal joints readily breaking off; joints globular or a little longer than thick, 3 to 5 cm. long, with numerous prominent areoles, the tubercles conspicuous when young; leaves minute, 1 to 1.5 mm. long, caducous; glochids conspicuous, numerous, yellow; spines usually wanting at the lower areoles, present above, very unequal, 5 to 10, acicular, the longest ones 3.5 cm. long; flowers rosy white to light yellow, 2 to 3 cm. long; ovary naked or spiny; fruit thicker than long, 2.5 cm. long, with deep umbilicus, often very spiny.

Common just below railroad station at Pampa de Arrieros, Peru, where it was collected by Dr. Rose, August 23, 1914 (No. 18957).

Figure 120 represents joints of the type specimen above cited.

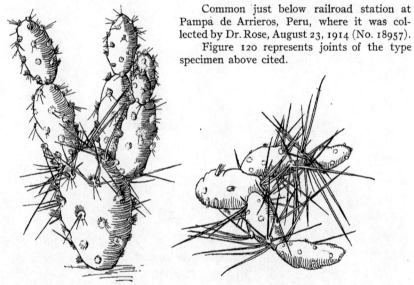

FIG. 120.—Opuntia campestris. ×0.8.

FIG. 121.—Opuntia ignota. ×0.8.

76. Opuntia ignota sp. nov.

Low, much branched, spreading; joints small, narrow, 2 to 3 cm. long, more or less purplish; leaves minute, often purplish; spines 2 to 7 from an areole, brownish, acicular, the longest ones 4 to 5 cm. long; glochids, when present, yellow; areoles large, full of grayish wool; flowers and fruit not seen.

Collected by Dr. Rose on the hills below the railroad station at Pampa de Arrieros, Peru, August 23, 1914 (No. 18974).

Plants grown in greenhouses are dark green and develop few spines or none.

This plant grows in the same region as *O. campestris*, but is quite different from it.

Figure 121 shows joints of the type specimen above cited.

Subgenus 3. PLATYOPUNTIA.

Includes all the species with flattened joints; a few species have nearly terete joints; others have some of the joints terete. Twenty-eight series are recognized. The species are most abundant in North America, but several series are found only in South America, while others have representatives in both Americas. (See Key to the Series, p. 45.)

Series 1. PUMILAE.

Low, spiny species, with slightly flattened, narrowly cylindric or linear-oblong, readily detached ultimate joints, the main stem terete. We know three species, the typical one in Mexico and Guatemala, one from Oaxaca, Mexico, and one Peruvian. In the structure of their joints they form a transitional series between *Cylindropuntia* and *Platyopuntia*, and might be included in either of these subgenera with about equal reason.

KEY TO SPECIES.

Young areoles with only 1 to 3 spines; joints 2 to 3 cm. thick.
    Plant 1 to 5 meters high; joints tubercled; spines yellowish.............................. 77. *O. pumila*
    Plants about 2 dm. high; joints not tubercled; spines reddish to brown................ 77a. *O. depauperata*
Areoles with 3 to 7 spines; plants 1 to 4 dm. high.
    Joints 1 to 1.5 cm. thick; areoles not blotched; spines brownish...................... 78. *O. pubescens*
    Joints 2 to 3 cm. thick; young areoles dark-blotched; spines yellowish................ 79. *O. pascoensis*

FIG. 122.—Opuntia pumila forming low thickets.

77. Opuntia pumila Rose, Smiths. Misc. Coll. **50**: 521.   1908.

Stems low, very much branched, the joints readily falling off when touched, 6 to 20 cm. long, velvety-pubescent, terete or sometimes slightly flattened, turgid, bearing more or less prominent tubercles; areoles small, those of old stems bearing several slender spines, the longer ones 3 cm. long; areoles of young joints usually bearing 2 yellowish spines; ovary pubescent, with few spines or none; petals yellow, tinged with red, 15 mm. long; fruit globular, red, 15 mm. long.

Type locality: Near Oaxaca City, Mexico, on the road to Mitla.
Distribution: Central and southern Mexico.

When this species was described, attention was called to various forms which belonged here or to one or more related species. These we now refer to *O. pubescens*.

Figure 122 is from a photograph of the type; figure 123 represents joints of the same.

77a. **Opuntia depauperata** sp. nov. (See Appendix, p. 216.)
78. **Opuntia pubescens** Wendland* in Pfeiffer, Enum. Cact. 149. 1837.

Opuntia leptarthra Weber in Gosselin, Bull. Mus. Hist. Nat. Paris 10: 393. 1904.

Plants small, usually low, sometimes 4 dm. high, much branched; joints easily becoming detached, nearly terete, glabrous or pubescent, 3 to 7 cm. long; spines numerous, short, brownish; flowers lemon-yellow but drying red; filaments greenish; style white; stigma-lobes cream-colored; fruit small, 2 to 2.5 cm. long, red, a little spiny, with a depressed umbilicus; seeds small, 3 mm. in diameter.

Type locality: In Mexico.

Distribution: Northern Mexico to Guatemala.

This species was sent to the Exposition Universelle at Paris by the Mexican Government in 1889, and was there seen and described by Dr. Weber as *O. leptarthra*. A part of this material finally went to the Hanbury Garden at La Mortola, Italy, whence we obtained specimens in 1913 which prove to be identical with specimens obtained by Dr. Rose and others in Mexico and Guatemala in 1905 to 1909.

This is an insignificant species and hence has generally been overlooked in the region where so many more striking species are found. It is widely distributed, extending from the State of Tamaulipas, in Mexico, to Guatemala, a much greater range than that of most species. Its wide distribution is doubtless due to the fact that the joints, which are covered with barbed spines and are easily detached, fasten themselves to various animals and are scattered like burs over the country; each little joint thus set free starts a new center of distribution.

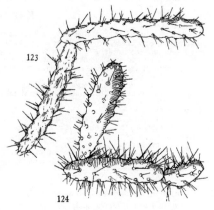

Fig. 123.—Opuntia pumila. X0.4.
Fig. 124.—Opuntia pubescens. X0.75.

This is a difficult plant to grow in greenhouses, for the spreading or hanging branches soon become entangled with other plants and break off in attempts to free or move them; partly for this reason, doubtless, it rarely flowers in cultivation.

Opuntia angusta Meinshausen (Wochenschr. Gärtn. Pflanz. 1: 30. 1858) was unknown to Schumann. It was originally described as similar to the South American species, *O. aurantiaca*, and, if so, it must be near *O. pubescens*, if not identical with it, being a native of Mexico, where it was first collected by Karwinsky.

Figure 124 represents joints of the Guatemalan plant, cultivated in the greenhouses of the United States Department of Agriculture, Washington, obtained in 1907.

79. **Opuntia pascoensis** sp. nov.

Stems erect and rigid, up to 3 dm. high; joints easily breaking apart, erect or ascending, terete or slightly flattened, 3 to 12 cm. long, 1.5 to 4 cm. broad, puberulent, hardly tuberculate but with faint upturned lunate depressions between the dark-blotched areoles; leaves minute; areoles somewhat elevated, filled with brown wool intermixed with longer white cobwebby hairs; spines 4 to 8 on young joints, more on older joints, acicular, yellow, 2 cm. long or less; glochids numerous, short, yellow, tardily developing; fruit globular, 1.5 cm. in diameter, naked below, spiny above. Doubtless of wide distribution, for the joints are easily detached and are distributed like burs, but so far only two collections have been reported.

*Pfeiffer (Enum. Cact. 1837) frequently refers several of Wendland's species to Catal. h. Herrnh. 1835, but we can find no references to Wendland having published a catalogue of the Herrenhausen Garden either in 1835 or about that time. We have therefore cited all of Wendland's species so referred by Pfeiffer to the pages given in his Enumatio.

Collected by Dr. and Mrs. J. N. Rose in central and southern Peru, in 1914, first from just below Matucana (No. 18653), and later at Pasco (No. 18812, type).

Plate XVII, figure 1, represents a joint of the type specimen above cited.

Series 2. CURASSAVICAE.

This series is composed of 10, or perhaps 11, species of low plants, characterized by their fragile branches, the small joints separating and becoming detached very readily, more or less flattened or subterete. They mostly inhabit the southern United States and the West Indies; one is known from Ecuador; the original home of one of the species recognized is unknown.

KEY TO SPECIES.

```
Spines acicular.
  Joints oval, mostly not more than twice as long as wide; plants prostrate, little branched.....80. O. curassavica
  Joints oblong to linear, 2 to 8 times as long as wide; plants ascending or erect, much branched.
    Joints narrowly linear, 1 to 2 cm. wide...........................................81. O. taylori
    Joints oblong to linear-oblong or obovate-oblong, 2 to 4 cm. wide.
      Joints oblong to linear, 4 to 8 times as long as wide; spines 1 to 3 cm. long.
        Joints not tubercled.........................................................82. O. repens
        Joints tubercled, at least when young.......................................82a. O. pestifer
      Joints oblong to obovate-oblong, 2 to 3 times as long as wide; spines 3 to 5 cm. long.....83. O. borinquensis
Spines subulate.
  Spines white.
    Roots fibrous; spines at most of the areoles....................................84. O. militaris
    Roots tuberous; spines only at the upper areoles................................85. O. nemoralis
  Spines brown.
    Joints oval to oblong.
      Joints scarcely repand; plant up to 2 dm....................................86. O. drummondii
      Joints strongly repand; plant 1 dm..........................................87. O. tracyi
    Joints linear-lanceolate...........................................................88. O. pusilla
Affinity uncertain...................................................................89. O. darrahiana
```

80. Opuntia curassavica (Linnaeus) Miller, Gard. Dict. ed. 8. No. 7. 1768.

Cactus curassavicus Linnaeus, Sp. Pl. 469. 1753.

Stems low, 5-jointed, light green, prostrate and creeping or hanging over rocks; joints oval to oblong, decidedly flattened but thick, 2 to 5 cm. long, glabrous; leaves minute, soon withering; areoles small, bearing short wool and longer, white cobwebby hairs; spines 4 to many, acicular, 2.5 cm. long or less, yellowish, becoming white in age; glochids tardily developing.

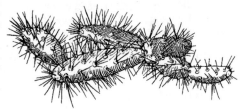

FIG. 125.—Opuntia curassavica. ×0.75.

Type locality: Curaçao Island.

Distribution: Curaçao, Bonaire, and Aruba.

Haworth (Syn. Pl. Succ. 196. 1812) describes three varieties, *major, media,* and *minor,* and later (Rev. Pl. Succ. 71. 1821) also describes the variety *longa.* *O. curassavica elongata* Haworth (Salm-Dyck, Hort. Dyck. 184. 1834), a name only, is supposed to be the same as var. *longa.*

This is one of the oldest species of *Opuntia,* having been described and figured as early as 1696. For a long time it has been unknown, the name having been transferred to a similar species, *O. repens.* In 1913 Dr. Britton visited Curaçao, its native home, and re-collected it. Its flowers have not been described, and several residents informed him that they had never seen it in flower; Dr. Britton did not find it in flower on Curaçao, nor has it flowered with us in cultivation; Haworth, who wrote about it in 1812, speaks of its being a shy bloomer, saying he had seen it in flower but once. In early English books it is called pin pillow, because its turgid joints suggest pincushions filled with pins.

Illustrations: Bradley, Hist. Succ. Pl. ed. 2. pl. 4, as *Opuntia minima americana,* etc.; Commerson Hort. pl. 56, as *Opuntia curassavica minima;* Plukenet, Opera Bot. 3: pl. 281, f. 3, as *Opuntia minor caulescens.*

Figure 125 represents the plant collected on Curaçao by Dr. N. L. Britton and Dr. J. A. Shafer in 1913.

M. E. Eaton del.

1. Joint of *Opuntia pascoensis*.   3, 4. Forms of *Opuntia repens*.   5. Flower of same.
2. Joints of *Opuntia taylori*.   6. Flowering joint of *Opuntia drummondii*.
(All natural size.)

the larger up to 6 cm. long, brown when young, fading white; leaves subulate, acuminate, 1 to 2 mm. long; fruit obovoid, subtruncate, 1.5 cm. long.

Limestone swale, Morillos de Cabo Rojo, Porto Rico (Britton, Cowell, and Brown, No. 4741), growing with *O. repens* Bello, from which it differs by its larger, broader, and flatter joints and much longer spines.

The only locality known for this plant is at the extreme southwestern corner of Porto Rico, where numerous colonies of it were observed. The region is a very dry one, rain falling there only at long intervals; the associated vegetation is of a highly xerophytic character.

Figure 126 represents joints of the type specimen above cited.

84. Opuntia militaris sp. nov.

Stems 3 dm. tall, the branches weak and more or less spreading; joints thick, narrowly oblong to obovate, 5 to 8 cm. long, somewhat shiny when young, easily breaking apart; spines 1 or 2 from an areole, occasionally more, acicular, white, 1 to 2 cm. long; flower-buds pointed; flowers small, 3 cm. long; petals greenish to cream-colored, tinged with pink; ovary small, its small areoles without spines.

Collected by Dr. N. L. Britton, March 17 to 30, 1909, at the U. S. Naval Station, Guantánamo Bay, Oriente, Cuba (No. 1957).

Figure 127 represents joints of the type specimen above cited.

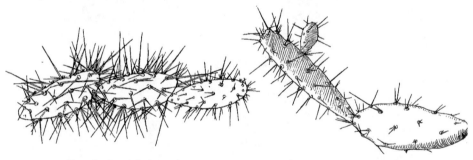

FIG. 126.—Opuntia borinquensis. ×0.5.   FIG. 127.—Opuntia militaris. ×0.5.

85. Opuntia nemoralis Griffiths, Monatsschr. Kakteenk. 23: 133. 1913.

Plants low, usually prostrate, forming clumps 1 meter in diameter, sometimes 3 dm. high; joints ovate to obovate, thick, 7 to 9 cm. long, green, but often with purple blotches about the areoles; spines 1 or 2, only from the upper areoles, 2 to 2.5 cm. long, mostly erect; glochids yellow; flowers yellow; fruit obovoid to pyriform, small, 3 cm. long, light red, truncate.

Type locality: Longview, Texas.
Distribution: Pine woods and fields about Longview, Texas.

This species in habit, joints, and spines suggests the *Tortispinae;* but on account of having easily detached joints we have referred it to the *Curassavicae*, as indicated in the original description, placing it between the Cuban species *O. militaris* and the United States species *O. drummondii.* It is known only from the type specimens.

86. Opuntia drummondii Graham in Maund, Botanist 5: pl. 246. 1846.

Opuntia pes-corvi LeConte in Engelmann, Proc. Amer. Acad. 3: 346. 1856.
Opuntia frustulenta Gibbes, Proc. Elliott Soc. Nat. Hist. 1: 273. 1859.

Plant prostrate or spreading, 2 dm. or less high, from thickened single or sometimes moniliform roots; joints rather variable, narrowly linear to broadly oblong, with entire margins, sometimes 12 cm. long and 5 to 6 cm. broad, usually light green, sometimes darker about the areoles; leaves 2 to

PLATE XVIII

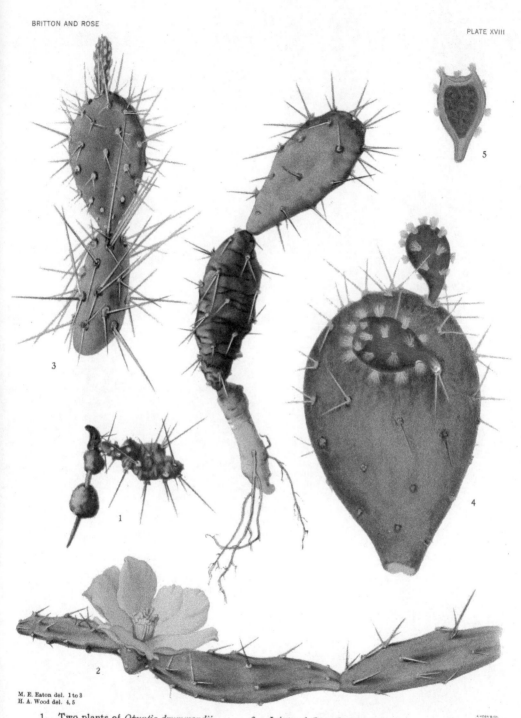

M. E. Eaton del. 1 to 3
H. A. Wood del. 4, 5

1. Two plants of *Opuntia drummondii*.   3. Joints of *Opuntia triacantha*.
2. Joints of *Opuntia retrorsa* with flower.   4, 5. Joint and section of fruit of *Opuntia jamaicensis*.
(All natural size.)

6 mm. long; spines (if present) solitary or 2 to 4, brownish red or gray, 2 to 4 cm. long; flowers yellow, 6 cm. broad; petals obovate; fruit red, juicy but insipid, obovoid to clavate, 22 to 35 mm. long, 15 mm. in diameter at thickest part, bearing few areoles and no spines; umbilicus slightly depressed in the center; seeds 1 to 8, about 4 mm. broad.

Type locality: Apalachicola, Florida.

Distribution: Sandy soil from northern Florida to Pamlico Sound, North Carolina.

In February 1916, Dr. J. K. Small visited the coastal islands near Charleston, South Carolina, for the purpose of collecting Gibbes's *Opuntia frustulenta.* He found this species very common on Folly Island and on the Isle of Palms, where it grows abundantly in the sand, and also very variable as to shape and size of joints. He says the joints break off easily and attach themselves to one's clothing like the sand spur, making progress over these islands difficult and painful. It is the common belief that this species rarely flowers. It usually flowers when first brought into cultivation, but rarely afterward, this doubtless being due to unsuitable greenhouse conditions.

The fruit described was collected by Dr. J. K. Small, December 10, 1917, at Apalachicola, Florida, the type locality.

According to Professor L. R. Gibbes, it is known as dildoes about Charleston.

Illustration: Maund, Botanist 5: pl. 246.

Plate XVII, figure 6, represents flowering joints of a plant sent from La Mortola, Italy, to the New York Botanical Garden in 1912; plate XVIII, figure 1, shows the plant collected by Dr. Small on the Isle of Palms, South Carolina, in 1916.

Herbarium specimens apparently representing a related species, were collected by W. L. McAtee at Cameron, Louisiana, in 1910 (No. 1955).

FIG. 128.—Opuntia tracyi.

87. Opuntia tracyi Britton, Torreya **11**: 152. 1911.

Low, diffusely much branched, pale green, about 2 dm. high or less; older joints oblong to linear-oblong, flat, 6 to 8 cm. long, 1.5 to 2.5 cm. wide, about 1 cm. thick; young joints scarcely flattened or terete, 1 cm. thick; areoles elevated, 5 to 10 mm. apart; spines 1 to 4, acicular, light gray with darker tips, 3.5 cm. long or less; glochids numerous, brownish; corolla pure yellow, 4 cm. broad; ovary 1.5 cm. long, bearing a few triangular acute scales similar to the outermost sepals, which are 2 mm. long; sepals triangular-ovate, 5 to 15 mm. long, the outer green, the inner yellowish with a green blotch; petals obovate, apiculate, 2 to 2.5 cm. long; filaments yellow, 1 cm. long; anthers white.

In sandy soil near the coast, Biloxi, Mississippi.

Figure 128 is from a photograph of the plant collected by S. M. Tracy at Biloxi, Mississippi, in 1911.

88. Opuntia pusilla Haworth, Syn. Pl. Succ. 195. 1812.

> *Cactus pusillus* Haworth, Misc. Nat. 188. 1803.
> *Cactus foliosus* Willdenow, Enum. Pl. Suppl. 35. 1813.
> *Opuntia foliosa* Salm-Dyck in De Candolle, Prodr. 3: 471. 1828.

Low, usually prostrate; joints narrow, more or less flattened, sometimes nearly terete, hardly tuberculate, light green in color; leaves 6 mm. long, linear, early deciduous; areoles remote; spines

1 or 2, subulate, usually brownish when young, in age straw-colored; flowers pale yellow, rather large for the plant; petals few, about 8, spreading, acute.

Type locality: Not cited.

Distribution: Usually assigned to South America, but not known from any definite locality; Schumann, in his Keys, however, says West Indies.

This species has usually passed under the name of *O. foliosa*, although all writers seem to agree that the older name, *O. pusilla*, was given to the same species. It may belong in the series *Aurantiacae* rather than in the *Curassavicae*.

Specimens distributed from European gardens as *O. foliosa* in recent years are not typical, and are probably referable to *O. drummondii*.

Tephrocactus pusillus Lemaire (Cact. 88. 1868), an unpublished name, referred by Lemaire to his third section of *Tephrocactus*, may belong here. The Index Kewensis refers it to *Opuntia pusilla*.

Illustration: Pfeiffer and Otto, Abbild. Beschr. Cact. 1: pl. 18, as *Opuntia foliosa*.

Figure 129 is copied from the illustration above cited.

89. Opuntia darrahiana Weber in Gosselin, Bull. Mus. Hist. Nat. Paris 10: 388. 1904.

Growing in masses, 2 to 2.5 dm. high 3.5 to 4 dm. broad, very much branched, joints 7 to 8 cm. long by 4 to 5 cm. broad, bright green to sea-green; areoles somewhat elevated, especially when young, 1 cm. apart; spines 6, the two uppermost the longest, these 4 to 4.5 cm. long, all suberect, white or grayish white, more or less brownish at tip; glochids said to be wanting; flowers and fruit not known.

Type locality: Turks Islands.

Distribution: Known only from the type locality.

This species is known only from the Turks Islands, a small group at the southeastern end of the Bahaman Archipelago. It was introduced into Europe by the late Charles Darrah.

FIG. 129.—Opuntia pusilla.

We know the plant only from the above-cited description, and, so far as we have been able to learn, it is not now in cultivation, nor have we been able to find any herbarium specimens preserved. The opuntias known to us to inhabit Turks Islands are *O. dillenii*, *O. nashii*, and *O. lucayana*. The description of *O. darrahiana* does not agree with any of these. The species is referred to the series *Curassavicae* with doubt, but as this series has representatives in Florida, Cuba, and Hispaniola, the existence of one in the Bahamas is not improbable.

Series 3. AURANTIACAE.

The species of this series are low plants, mostly with readily detached joints; the main stems are often terete or turgid, the ultimate joints narrow and flat. They inhabit southeastern South America. During the expedition to Brazil and Argentina conducted by Dr. Rose in the summer of 1915, only a few of the species here grouped were found; Dr. Shafer collected several of them in the winter of 1916-17. Dr. Spegazzini has given us photographs of several.

We recognize 8 species, and have appended another, which may belong here.

KEY TO SPECIES.

Joints not conspicuously purple-blotched under the areoles.
Joints linear, elongated.
Stem terete or subterete; branches mostly flat.
Joints dark green, not tubercled.................................... 90. *O. aurantiaca*
Joints tubercled, bluish green when young......................... 91. *O. schickendantzii*
All the joints flat.
Joints elongated, linear...................................... 92. *O. kiska-loro*
Joints linear-oblong... 93. *O. canina*
Joints short, elliptic....................................... 94. *O. montevidensis*
Joints with a long purplish blotch under each areole.
Joints more or less spiny.
Joints flattened.
Joints 2 to 3.5 cm. wide.................................. 95. *O. retrorsa*
Joints 3.5 to 6 cm. wide.................................. 96. *O. utkilio*
Joints subterete, turgid.................................. 96a. *O. discolor*
Joints spineless... 97. *O. anacantha*
Perhaps of this series.. 98. *O. grosseiana*

90. Opuntia aurantiaca Lindley, Edwards's Bot. Reg. **19**: pl. 1606. 1833.

Opuntia aurantiaca extensa Salm-Dyck in Förster, Handb. Cact. 476. 1846.

Low, much branched, and spreading; stem terete or subterete, 1 to 2 cm. thick; joints very fragile, linear, 6 to 8 cm. long, 1.5 to 2.5 cm. broad, almost terete at base, dark green, shining; areoles somewhat elevated, filled with white wool; spines 2 or 3, brownish, 1 to 3 cm. long; flowers yellow, 2.5 cm. broad; fruit 2 to 2.5 cm. long.

Type locality: Chile (in error).
Distribution: Argentina and Uruguay.

Cactus aurantiacus Lemaire (Cact. 87. 1868) is usually cited in synonymy, but Lemaire only mentions the name as a species of *Cactus*. It is in fact Gillies's manuscript name, first published in the Botanical Register in 1833 as a synonym of *O. aurantiaca*.

O. extensa Salm-Dyck (Pfeiffer, Enum. Cact. 147. 1837) is also given as a synonym.

Remy states (Gay, Fl. Chilena **3: 25.** 1847) that it grows in the central provinces of Chile, but he probably had in mind some other plant, as *O. aurantiaca* is not known to be native of Chile by resident botanists.

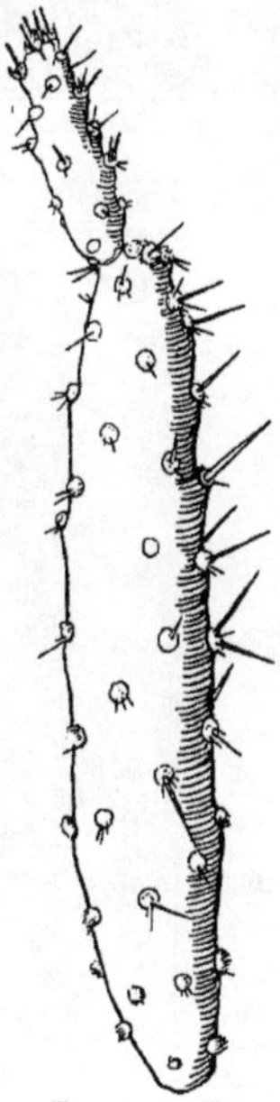

Fig. 130.—O. aurantiaca.

Fig. 131.—O. schickendantzii.

Illustrations: Anal. Mus. Nac. Montevideo **5**: pl. 34; Edwards's Bot. Reg. **19**: pl. 1606.

Figure 130 represents a joint from a plant found by Dr. Rose, in Argentina, in 1915.

91. Opuntia schickendantzii Weber in Schumann, Gesamtb. Kakteen 688. 1898.

Shrub-like, 1 to 2 meters high, much branched, grayish green; branches cylindric or flattened, somewhat tuberculate; leaves minute, 2 mm. long, reddish; spines 1 or 2, subulate, 1 to 2 cm. long; flowers 4 cm. in diameter, yellow; fruit green, sterile.

Type locality: In Tucuman, Argentina.
Distribution: Northern Argentina.

Figure 131 is from a photograph of a plant in Argentina contributed by Dr. Spegazzini.

FIG. 132.—Opuntia kiska-loro.

92. Opuntia kiska-loro Spegazzini, Anal. Mus. Nac. Buenos Aires III. **4:** 516. 1905.

Prostrate, rooting, forming spreading clumps 3 to 6 dm. high; joints flat, at first very narrow, becoming lanceolate, 20 cm. long, 4 cm. broad, shining green; spines 2 to 4, unequal, whitish, 4 to 6 cm. long; flowers orange, rather large, 3 to 6 cm. broad; filaments pale orange; stigma-lobes 6, flesh-colored; fruit 5 cm. long, deep violet-purple without, white within; seeds 5 mm. broad, pubescent.

Type locality: Deserts of La Rioja, Catamarca, Argentina.

Distribution: Northwest Argentina.

Figure 132 is from a photograph of the type plant sent by Dr. Spegazzini.

93. Opuntia canina Spegazzini, Anal. Mus. Nac. Buenos Aires III. **4:** 518. 1905.

At first erect, then decumbent, 1 to 3 meters broad; joints flat, very narrow, attenuate at both ends, 2.5 to 3.5 dm. long, 4.5 cm. broad, shining green; areoles on young joints unarmed; spines of areoles of older joints 1 or 2, sometimes 3, 1.5 to 3.5 cm. long, reflexed, subterete, grayish white with yellowish tips; flowers numerous, medium sized; ovary obovoid; corolla rotate, yellowish orange, 4 to 5 cm. broad; petals obovate; filaments yellow; stigma-lobes 5; fruit obovoid, 2.6 to 2.8 cm. long, red without, white within; seeds 4 mm. broad, white, lanate.

Type locality: Near Pampablanca, Jujuy, Argentina.

FIG. 133.—Opuntia canina.

Distribution: Provinces of Jujuy and Tucuman, Argentina.

Figure 133 is from a photograph sent by Dr. Spegazzini.

94. Opuntia montevidensis Spegazzini, Anal. Mus. Nac. Buenos Aires III. **4:**515. 1905.

Cespitose, the branches 3 to 5 dm. high; joints 5 to 10 cm. long, obovate to elliptic; areoles not very prominent; spines usually 5, 3 longer and stouter, 2 very small, reflexed, and setiform, the 2 or 3 longer ones erect or spreading, 2 to 3 cm. long; flowers 4 to 5 cm. broad, orange-colored; fruit dark purple, clavate, 3.5 to 4 cm. long ; seeds lanate.

Type locality: Cerro de Montevideo, Uruguay.

Distribution: Cerro de Montevideo, and near La Colonia, Uruguay.

95. Opuntia retrorsa Spegazzini, Anal. Mus. Nac. Buenos Aires III. **4:**517. 1905.

(?) *Opuntia platynoda* Griffiths, Bull. Torr. Club **43:** 526. 1916.

Stems prostrate, intricately branched, creeping, rooting at the nodes; joints linear-lanceolate, more or less attenuate at each end, flattened; areoles somewhat prominent, each subtended by a long, dull purplish blotch; spines 1 to 3, reflexed, white below, with pinkish tips; flowers yellowish, 4 to 5 cm. broad; fruit about 2 cm. long, violet-purple on the outside, light rose on the inside; seeds 2 to 2.5 mm. broad, somewhat villous.

FIG. 134.—Opuntia retrorsa.

Type locality: In the Territory of the Chaco, Argentina.

Distribution: Northern Argentina.

Plate XVIII, figure 2, represents a plant from Argentina which flowered at the New York Botanical Garden in 1911. Figure 134 is from a photograph sent by Dr. Spegazzini.

96. Opuntia utkilio Spegazzani, Anal. Mus. Nac. Buenos Aires III. **4:** 516. 1905.

Low, creeping plant, rooting at the joints, with elongated branches 5 to 15 dm. long; joints flat, elliptic-linear, 15 to 30 cm. long, 5 to 6 cm. broad; spines at first 2 or 3, the upper one longer, later more numerous, reflexed; flowers small, 3.5 to 4 cm. broad, yellowish; ovary obovoid, somewhat spiny; fruit small, 3 cm. long, fleshy, insipid, reddish violet both within and without; seeds sub-orbicular, 4 mm. broad, lanate.

Type locality: Province of Tucuman, Argentina.

Distribution: Northern Argentina.

Figure 135 is from a photograph sent by Dr. Spegazzini.

96a. Opuntia discolor sp. nov. (See Appendix, p. 218.)

97. Opuntia anacantha Spegazzini in Gosselin, Bull. Mus. Hist. Nat. Paris **10:** 391. 1904.

Usually decumbent and rooting along the under surface, sometimes ascending and clambering, 1 to 2.5 meters long; joints unarmed, dark green except for purple spots under the areoles, elliptic to lanceolate, narrowed toward each end, 1.5 to 4 dm. long, 3.5 to 7 cm. broad; areoles small; flowers large, numerous, yellowish orange, 4 cm. long, 5 to 6 cm. in diameter; sepals large, reddish, obtuse, emarginate or even 2-lobed; petals 12; style white; stigma-lobes white or rose-colored; fruit 3 cm. long, red, the pulp yellowish or white.

Type locality: In the southern Chaco, Argentina.

Distribution: Northeastern Argentina.

Figure 136 is from a photograph of a part of the type plant, received from Dr. Spegazzini.

98. Opuntia grosseiana Weber in Gosselin, Bull. Mus. Hist. Nat. Paris **10**: 391. 1904.

Described as having joints intermediate between those of *Opuntia elata* and *O. anacantha*, and resembling these species.

Type locality: In Paraguay.
Distribution: Paraguay.

Introduced from Paraguay by Hermann Grosse; known to us only from the description.

Series 4. TUNAE.

Bushy, ascending, depressed, or erect plants, with rather large and more or less readily detached joints, bearing acicular or subulate, often numerous, yellow or white spines. The species inhabit the West Indies, Mexico, Guatemala, and northern South America.

FIG. 135.—Opuntia utkilio.

FIG. 136.—Opuntia anacantha.

KEY TO SPECIES.

```
Joints glabrous.
    Spines slender, acicular.
        Spines white.
            Joints dull.
                Joints dark green, repand; areoles somewhat elevated................ 99.  O. bella
                Joints light green, not repand; areoles not elevated.
                    Spines several at the areoles; plant ascending................. 100. O. triacantha
                    Spines 1 to few at the areoles or often wanting; plant erect....... 101. O. jamaicensis
            Joints shining................................................... 101a. O. guatemalensis
        Spines yellow, at least when young; plant bushy branched.................... 102. O. tuna
    Spines stout, subulate.
        Spines white; joints relatively thick, turgid............................ 102a. O. pennellii
        Spines yellow, at least when young; joints relatively thin.
            Plants low, spreading, 2 dm. high or less.
                Joints repand; spines bright yellow............................ 103. O. antillana
                Joints not repand; spines pale yellow.......................... 103a. O. caracasana
            Plants tall, 1 to 2 meters high.
                Joints obovate or broadly elliptic............................. 104. O. wentiana
                Joints narrowly oblong or oblong-obovate....................... 104a. O. aequatorialis
```

Joints pubescent.
  Areoles surrounded by purplish spots... 105. *O. decumbens*
  Areoles not surrounded by purplish spots....................................... 106. *O. depressa*

99. Opuntia bella sp. nov.

Stems low, 10 to 12 dm. high, forming thickets; joints oblong, repand, 10 to 16 cm. long, dull dark green; areoles 1 to 2 cm. apart, somewhat elevated, small, filled with short brown wool and glochids; leaves minute, 1.5 to 2.5 mm. long; spines white, 2 to 6, unequal, acicular, the longer ones about 2 cm. long; flowers 5 cm. long, "sulphur-yellow turning to orange-red;" petals 20 to 22 mm. long; ovary deeply umbilicate; "fruit small, greenish yellow."

137.—Opuntia bella in the foreground.

Type locality: Venticas del Dagua, Dagua Valley, western cordillera of Colombia.
Distribution: Western Colombia.

The type is based upon plants collected by Mr. Henry Pittier in the State of Cauca, Colombia, in 1906, and grown ever since in Washington and New York. The species is very common in Cauca, forming with other cacti impenetrable thickets.

Figure 137 is from a photograph by Mr. Pittier of the type plant, taken near Cauca, Colombia, in 1906; figure 138 is from a photograph by the same collector, showing flowering and fruiting joints; figure 139 represents a single joint.

100. **Opuntia triacantha** (Willdenow) Sweet, Hort. Brit. 172. 1826.

Cactus triacanthos Willdenow, Enum. Pl. Suppl. 34. 1813.

Stems half procumbent or clambering over rocks, sometimes even erect but always low; joints turgid, oblong, 4 to 8 cm. long, the terminal and often the second and third ones breaking off easily; spines usually 3, white but often drying yellowish, 4 cm. long or less; flowers, including the ovaries, 5 cm. long, brownish yellow to cream-colored, tinged with pink; petals obtuse; filaments and style pale green; fruit 2.5 cm. long, red, spineless.

Type locality: Not cited; cultivated in the Berlin Garden.

Distribution: Desecheo Island, Porto Rico; Lesser Antilles, St. Thomas to Guadeloupe.

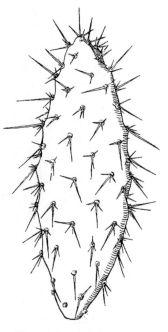

Fig. 138.—Opuntia bella. ×0.75.   Fig. 139.—Opuntia bella. ×0.66.

When published, the origin of this species was uncertain. It has been referred to the South American flora, but if our interpretation is correct it is a West Indian plant. It was introduced into cultivation in 1796.

This species is very common on flats or low hills and, so far as our observation goes, is never found very far inland in the Lesser Antilles.

Professor Schumann's description includes two species, one of which belongs here and one in the *Streptacanthae*, perhaps as Mr. Berger thinks to *O. amyclaea*—and a tall plant, 3.5 meters high, is now grown in Italy under that name. The Index Kewensis refers *O. triacantha* as a synonym of *O. curassavica*, which is erroneous if our interpretation of it is correct.

Plate XVIII, figure 3, represents joints of the plant collected on Antigua by Rose, Fitch, and Russell in 1913. Figure 140 is from a photograph taken on St. Christopher, British West Indies, by Paul G. Russell in 1913.

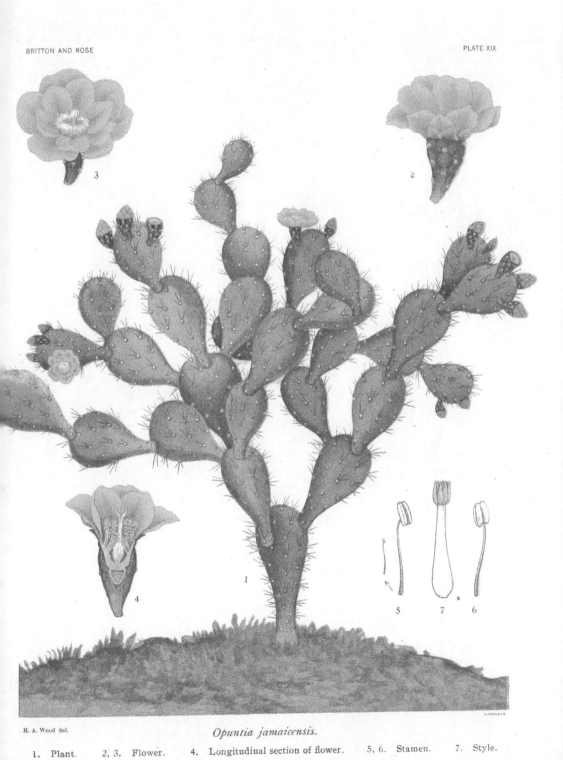

H. A. Wood del.

Opuntia jamaicensis.

1. Plant.   2, 3. Flower.   4. Longitudinal section of flower.   5, 6. Stamen.   7. Style.

An *Opuntia* collected by H. Pittier in Costa Rica and now growing in the cactus house of the U. S. Department of Agriculture has not been identified. It resembles somewhat *Opuntia triacantha*, but is much out of the range of that species and differs from it in some important respects. The joints are small, obovate to oblong, rounded at apex, dark green with purple blotches below the areoles, with low, broad tubercles; margin of the areole bearing short white hairs; spines usually wanting, but cultivated specimens bear a single short spine 6 to 7 mm. long from an areole.

FIG. 140.—Opuntia triacantha.

101. Opuntia jamaicensis Britton and Harris, Torreya 11: 130. 1911.

Erect, 1 meter high, with a short subcylindric trunk; branches several, ascending, joints dull green, obovate, much narrowed at base, flat, rather thin, readily detached, 7 to 13 cm. long, 5 to 7.5 cm. wide; areoles about 2.5 cm. apart; spines 1 to 5, usually only 2, acicular, unequal, white, 2.5 cm. long or less; flowers 4 cm. broad; petals 16 to 18; filaments greenish white; style white; stigma-lobes 7 or 8, creamy white; fruit pyriform, red, 3.5 to 4 cm. long; seeds 4 mm. broad.

Type locality: St. Catherine, Jamaica.
Distribution: Plain south of Spanish Town, Jamaica.
The following figures are from paintings by Miss H. A. Wood:

Plate XVIII, figure 4, shows a fruiting joint; figure 5 is of a section of the fruit; plate XIX, figure 1, shows the type plant about one-third natural size; figures 2, 3, and 4 are of the flowers; figures 5 and 6 show the stamens; figure 7 represents the style.

101a. Opuntia guatemalensis sp. nov. (See Appendix, p. 218.)

102. Opuntia tuna (Linnaeus) Miller, Gard. Dict. ed. 8. No. 3. 1769.

> Cactus tuna Linnaeus, Sp. Pl. 468. 1753.
> Cactus humilis Haworth, Misc. Nat. 187. 1803.
> Opuntia humilis Haworth, Syn. Pl. Succ. 189. 1812.
> Opuntia polyantha Haworth, Syn. Pl. Succ. 190. 1812.
> Cactus polyanthos Sims, Curtis's Bot. Mag. 53: pl. 2691. 1826.
> Opuntia multiflora Nicholson, Dict. Gard. 2: 503. 1885.

Plants 6 to 9 dm. high or less; joints usually small, but sometimes up to 16 cm. long, obovate to oblong, light green, except above the areoles and there brownish; leaves minute, fugacious; areoles large; spines 2 to 6, usually only 3 to 5, slightly spreading, light yellow; glochids yellow; flowers about 5 cm. broad; sepals orbicular, yellowish, with a purple stripe along the center; petals light yellow, slightly tinged with red, oblong, rounded at apex; filaments short, greenish below; style and stigma-lobes cream-colored or yellowish; ovary bright green, narrowed downward; fruit red, obovoid, about 3 cm. long; seeds 3 to 4 mm. broad.

Type locality: Jamaica.

Distribution: Southern side of Jamaica, West Indies.

Opuntia tuna is one of the old *Cactus* species. It was described by Linnaeus as *Cactus tuna* and by Philip Miller as *Opuntia tuna*. In the early part of the Nineteenth Century it was renamed *Opuntia humilis* and also *O. polyantha*, and has long passed under the latter name. *Opuntia tuna*, however, is one of the commonest *Opuntia* names in our botanical literature. This is due partly to the fact that the name was early transferred to *Opuntia dillenii*, one of the most common species, both wild and cultivated, and partly because tuna is the

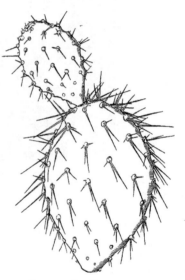

FIG. 141.—Opuntia tuna.   FIG. 142.—Opuntia tuna.   ×0.5.

common Mexican name for opuntias, and many species have therefore been identified as *O. tuna*. So far as our studies indicate, this species is confined to the Jamaica lowlands.

Opuntia multiflora is referred here, although we do not know the plant. It is figured by Nicholson (Dict. Gard. Nicholson 2: f. 754); this figure is republished by Rümpler (Förster, Handb. Cact. ed. 2. f. 130), and by Knippel (Kakteen, pl. 28), both calling it *Opuntia polyantha;* while W. Watson (Cact. Cult. f. 79) uses the same illustration, calling it *O. dillenii*.

Opuntia coccinea (Pfeiffer, Enum. Cact. 161. 1837) is given as a synonym of *O. tuna*, but it was never published; it is doubtless different from *O. coccinea* Rafinesque (Med. Fl. U. S. 2: 247. 1830), also unpublished. The following names seem to belong here, but were not formally published: *Opuntia flexibilis* (Pfeiffer, Enum. Cact. 161. 1837); *O. tuna humilis* Salm-Dyck (Cact. Hort. Dyck. 1844. 46. 1845); *O. tuna laevior* Salm-Dyck (Hort. Dyck. 186. 1834); and *O. tuna orbiculata* Salm-Dyck (Cact. Hort. Dyck. 1844. 47. 1845).

Illustrations: Loudon, Encycl. Pl. ed. 3. f. 6878, as *Cactus tuna;* Wiener Illustr. Gartenz. 10: f. 114, as *Opuntia humilis;* Blühende Kakteen 2: pl. 75; Förster, Handb. Cact.

ed. 2. f. 130; Knippel, Kakteen 2: pl. 28, these three as *Opuntia polyantha;* Curtis's Bot. Mag. 53: pl. 2691, as *Cactus polyanthos;* De Candolle, Pl. Succ. Hist. 2: pl. 138[d], as *Cactus opuntia polyanthos;* Descourtilz, Fl. Med. Antil. pl. 513, as *Cactus opuntia.*

Figure 141 is from a photograph of a plant collected by William Harris, near Kingston, Jamaica, in 1913; figure 142 represents a joint of the same plant.

102*a.* **Opuntia pennellii** sp. nov. (See Appendix, p. 219.)

103. **Opuntia antillana** Britton and Rose, Brooklyn Bot. Gard. Mem. **1**: 74. 1918.

Growing in dense clumps, often 1 meter broad, more or less prostrate; joints usually obovate, 7 to 20 cm. long, narrow and nearly terete at base; terminal joints easily breaking off; leaves conic-subulate, about 2 mm. long; areoles large, 2 to 3 cm. apart, containing soft brown wool; spines stout, terete, 3 to 6 at an areole, unequal, 1 to 6 cm. long, yellow but becoming gray to nearly white in age; glochids numerous, yellow; flowers 5 to 7 cm. long; petals broad, obtuse, yellow, turning reddish in age; fruit reddish purple, 4 cm. long.

FIG. 143.—Opuntia antillana forming thickets.

Type locality: Near Basse Terre, St. Christopher, Rose, Fitch and Russell, No. 3230, February 2, 1913.

Distribution: St. Christopher, St. Croix, Tortola, St. Thomas, Porto Rico, and Hispaniola.

This species is one of the most widely distributed in the West Indies and, on some of the islands on which it occurs, generally the most abundant. This is partly due to the fact that the terminal joints are easily detached and may thus be widely scattered.

The question has frequently been raised in our minds whether this species may not be of hybrid origin. It has some resemblance to *O. dillenii,* but has much smaller joints and these very fragile. What the other parent would be is not so clear. The fragile joints would suggest *O. triacantha* or *O. repens,* but otherwise there is no close alliance with either of these. Owing to the fact that it is more common than any of these species, and is often not associated with any of them, we believe it to be distinct. In the

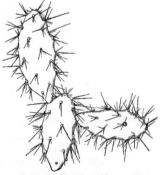

Fig. 144.—Opuntia antillana. X0.33.

desert of Azua, Santo Domingo, this is the dominant cactus, forming dense, impenetrable thickets on the low coastal plain. In the wild state the Azua plant has the joints often bronzed or purple. On Tortola and St. Thomas it occurs with *O. dillenii* and *O. repens*, and is there called bull suckers.

Figure 143 is from a photograph taken by Paul G. Russell in 1913 near Azua, Santo Domingo; figure 144 represents joints of the type plant.

103a. Opuntia caracasana Salm-Dyck. (See Appendix, p. 219.)

104. Opuntia wentiana sp. nov.

> *Opuntia tunoides* Britton and Shafer in Boldingh, Fl. Ned. W. Ind. Eiland 300. 1913. Not *O. tunoidea* Gibbes.

Plant erect, much branched, 1 to 2 meters high· joints obovate to elliptic, rather thin, up to 25 cm. long, usually rounded at apex, pale green, slightly glaucous; terminal joints somewhat fragile; leaves small and subulate; spines on young joints usually 3, afterwards 4 or 5, when young pale yellow but soon white; flowers small, 6 to 7 cm. long including the ovary; petals pale yellow, 3 cm. long, obovate, acute; style cream-colored; fruit small, red.

Type locality: Curaçao.

Distribution: Venezuela, and the neighboring islands, Margarita, Bonaire, Curaçao, and Aruba.

Dr. Rose found this plant repeatedly in Venezuela and writes of it as follows: Very common not only on the savannas along the coast but also on the neighboring hills along with *Lemaireocereus, Cephalocereus*, and other cactus genera; its more or less fragile joints, yellowish spines, bushy stature, and structure of flowers ally it with the *Tunae*.

This species has been confused with the Jamaican *Opuntia tuna* (Linnaeus) Miller, which it resembles. Named in honor of Professor F. A. F. C. Went, distinguished Dutch botanist.

104a. Opuntia aequatorialis sp. nov. (See Appendix, p. 219.)

105. Opuntia decumbens Salm-Dyck, Hort. Dyck. 361. 1834.

> *Opuntia puberula* Pfeiffer, Enum. Cact. 156. 1837.

Stems low, often creeping or trailing, rarely over 4 dm. high; joints 1 to 2 dm. long, oval to oblong, covered with a short, soft pubescence; areole usually small, surrounded by a purple blotch, bearing yellow glochids and wool, the wool cobweb-like on very young joints; spines often wanting, usually solitary but sometimes numerous, slender or rather stout, 4 cm. long and yellow; flowers numerous, small, including the ovary about 4 cm. long; petals dark yellow; fruit deep purple, very juicy; seeds about 4 mm. broad.

Type locality: In Mexico.

Distribution: Guatemala and Mexico as far north as Mazatlan and Tamaulipas.

Opuntia repens Karwinsky (Salm-Dyck, Hort. Dyck. 361. 1834) and *O. irrorata* Martius are usually given as synonyms of this species, but as they were printed without descriptions, they should hardly be referred to synonymy.

The species has long been in cultivation, a colored illustration having been published in Curtis's Botanical Magazine in 1841. It grows luxuriously in greenhouses, flowering profusely in the spring.

We have referred here *Opuntia puberula* Pfeiffer, which seems to be different from the plant now grown in collections under that name. Pfeiffer's original description, based upon sterile plants alone, may be paraphrased as follows: Joints thick, obovate, 7.5 to 12.5 cm. long by 5 to 7.5 cm. broad, puberulent, green; areoles somewhat remote, each surrounded by a red spot, bearing in the upper part a bunch of short glochids and below 2 to 4 slender, white, divergent spines, the longer ones 8 mm. long; leaves 4 mm. long, acute, red at apex.

Labouret's description of 1853, of *O. puberula* Pfeiffer, is very similar to Pfeiffer's, except that he states that the spines are 9 cm. long. Both these descriptions answer very

PLATE XX

M. E. Eaton del.

1, 2. Flowering and fruiting joints of *Opuntia decumbens*.   3. Probable hybrid with fruit and flower.
(All natural size.)

well to the plant which we know as *Opuntia decumbens*, originally described from plants growing in the Botanical Garden in Vienna.

Opuntia decumbens irrorata Forbes (Hort. Tour. Germ. 158. 1837) is doubtless the same as *O. irrorata* Martius (Pfeiffer, Enum. Cact. 154. 1837). These and *O. decumbens longispina* Salm-Dyck (Haage and Schmidt, Haupt-Verzeichnis 1912: 230. 1912) presumably belong here.

Opuntia parvispina Salm-Dyck (Cact. Hort. Dyck. 1849. 238. 1850), described from garden specimens of unknown origin, without flowers, has never been definitely placed. Schumann lists it among his unknown species, but attributes it to Mexico. Salm-Dyck states that it resembles *O. puberula*, but that it is glabrous.

Illustrations: Curtis's Bot. Mag. 68: pl. 3914; Blühende Kakteen 3: pl. 132.

Fig. 145.—Opuntia decumbens.

Plate xx, figure 1, represents a flowering joint of a plant collected by Dr. MacDougal and Dr. Rose at Tehuacán, Mexico, in 1906; figure 2 represents a fruiting joint of a plant collected by William R. Maxon at El Rancho, Guatemala, in 1905. Figure 145 is from a photograph of the plant taken at Tomellín, Mexico, by Dr. MacDougal in 1906.

106. Opuntia depressa Rose, Smiths. Misc. Coll. **50**: 517.   1908.

Low, creeping or spreading plant, sometimes 60 cm. high and forming a patch 3 to 4 meters in diameter; joints of a dark glossy yellowish green color, pubescent, when young, obovate, 20 cm. long, usually with 1 long, somewhat curved spine at each areole, sometimes with 1 to 3 shorter ones, all yellowish; old joints oblong, 30 cm. long, bearing 4 to 6 spines at each areole; flowers red; fruit small, globular, with large clusters of brown glochids, when immature with a broad, deep umbilicus.

Type locality: Near Tehuacán, Mexico.
Distribution: Southern Mexico.

This plant is very common about Tehuacán, growing with species of *Agave, Beaucarnea,* and *Echinocactus.*

Figure 146 is from a photograph taken by Dr. MacDougal of the type plant.

Fig. 146.—Opuntia depressa, in the foreground.

Series 5. BASILARES.

We recognize eight species as forming this series. They are low or bushy, much branched plants, with flat, thin, broad joints, the areoles small, usually numerous and close together.

KEY TO SPECIES.

Joints papillose, not pubescent.
 Fruit juicy, red.. 107. *O. lubrica*
 Fruit dry or nearly dry.. 108. *O. treleasei*
Joints mostly manifestly pubescent.
 Spines none or few.
  Flowers red... 109. *O. basilaris*
  Flowers yellow to orange.
   Joints bright green.
    Glochids long.. 110. *O. microdasys*
    Glochids short... 111. *O. macrocalyx*
   Joints grayish green.. 112. *O. rufida*
 Spines very numerous.
  Areoles close together.. 113. *O. pycnantha*
  Areoles distant... 114. *O. comonduensis*

107. Opuntia lubrica Griffiths, Rep. Mo. Bot. Gard. 21: 169. 1910.

"A low ascending, spreading species very similar in habit to *O. microdasys*, frequently 4½ dm. high and when well developed 10 dm. or more in diameter; joints sub-circular to obovate, about 15 by 20 cm., or in case of last joints of previous year about 12 by 15 cm., bright, glossy, leaf-green, very evidently papillate but scarcely pubescent under a lens; leaves subulate, cuspidate-pointed, 6 to 9 mm. in length; areoles 15 to 22 mm. apart, 4 to 6 mm. in diameter, sub-circular, prominent; spicules prominent, 4 to 5 mm. in length, erect, bushy, in crescentic tufts in upper portion of areoles, becoming much more numerous in age, and at 2 to 4 years completely filling the areole, and, like *O. rufida* and some other species, becoming very abundant and conspicuous by proliferation of areolar tissue into short raised or columnar structures; spines exceedingly variable, sometimes nearly absent, again quite abundant and irregularly distributed, none to many, mostly 1 to 3, becoming more numerous with age and in scattering areoles to as high as 16, mostly about 12 mm. long, but sometimes 2½ cm., yellowish, translucent, bonelike, sometimes darker at base; fruits decidedly acid, light red without with yellowish green rind and red pulp; seed small, thin shelled, about 3 mm. in diameter."

Type locality: Near Alonzo, Mexico.
Distribution: Known only from the type locality.

Our examination of a painting of this plant in the collection made by Dr. Griffiths showed it to have great similarity to *Opuntia rufida*.

Illustration: Rep. Mo. Bot. Gard. **21**: pl. 23.

Figure 147 is copied from the illustration above cited.

108. Opuntia treleasei Coulter, Contr. U. S. Nat. Herb. **3**: 434. 1896.

> *Opuntia basilaris treleasei* Toumey, Cycl. Amer. Hort. Bailey **3**: 1147. 1901.
> *Opuntia treleasei kernii* Griffiths and Hare, N. Mex. Agr. Exp. Sta. Bull. **60**: 81. 1906.

Low, sometimes 3 dm. high, spreading at base, some of the branches of 2 to 4 erect joints; joints obovate, 15 cm. long or more, fleshy, pale bluish green, glabrous, terete at base; areoles numerous, filled with dirty yellow glochids, usually without spines, sometimes quite spiny; flowers rose-colored; fruit dry, subglobose, with large areoles filled with glochids and sometimes bearing spines; seeds large, turgid, 7 cm. in diameter.

Type locality: Caliente, in the Tehachapi Mountains, California.

Distribution: Southern California.

Figure 148 is from a photograph of the plant growing on the mesa southeast of Bakersfield, California, taken by Dr. MacDougal in 1913.

109. Opuntia basilaris Engelmann and Bigelow, Proc. Amer. Acad. **3**: 298. 1856.

> *Opuntia basilaris ramosa* Parish, Bull. Torr. Club **19**: 92. 1892.
> *Opuntia intricata* Griffiths, Proc. Biol. Soc. Washington **29**: 10. 1916.

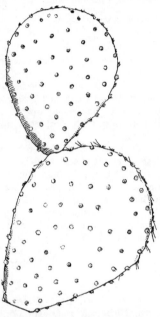

Fig. 147.—Opuntia lubrica.

Fig. 148.—Opuntia treleasei, Southern California.

Stems low, growing in clumps, either prostrate or erect, sometimes 12 dm. high; joints broadly obovate, 8 to 20 cm. long, slightly pubescent or glabrous, usually highly colored; leaves 2 to 5 mm. long, subulate; areoles numerous, filled with white to brown wool and brownish glochids; spines none or rarely a few at upper areoles; flowers large, 6 to 8 cm. long, deep purple or sometimes white; filaments purple; fruit dry, globular to obovoid; seeds large, thick, 6 to 10 mm. broad.

Type locality: From Cactus Pass down the valley of the Bill Williams River.

Distribution: Northern Sonora, western Arizona, southern California, Nevada, and southern Utah.

This is a variable species as to habit, size, pubescence, and color of flowers. The variety *ramosa* described by Mr. Parish is more erect than the ordinary form and glabrous. It has large, handsome flowers, and is a splendid plant for outdoor cultivation where the climate is suitable, but does not live long in greenhouses. It is called beaver-tail in Arizona.

Opuntia humistrata Griffiths (Bull. Torr. Club **43**: 83. 1916) we refer here from the description; it is said to differ from *O. basilaris* "by its much smaller as well as different shaped joints"; it was found in the San Bernardino Mountains, northern California, within the range of *O. basilaris*.

The following varieties are listed, but have not been described: *albiflora, cocrulea, nana,* and *pfersdorffii.*

Opuntia basilaris cordata is a garden plant briefly described by F. Forbes (Monatsschr. Kakteenk. **16**: 46. 1906), of which we have seen no specimens.

Illustrations: Cact. Journ. **1**: 167; Dict. Gard. Nicholson **2**: f. 750; Förster, Handb. Cact. ed. 2. f. 129; Pac. R. Rep. **4**: pl. 13, f. 1 to 5; pl. 23, f. 14; Rümpler, Sukkulenten f. 123; W. Watson, Cact. Cult. f. 76; Cact. Journ. **1**: pl. October, as *Opuntia basilaris* var. *cristata* and var. *nevadensis;* Alverson, Cact. Cat. frontispiece, as *Opuntia basilaris albiflora.*

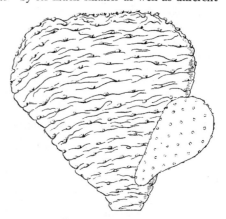

FIG. 149.—Opuntia basilaris.

Figure 149 is copied from Pac. R. Rep. **4**: pl. 13, f. 1, an illustration cited above.

Opuntia brachyclada Griffiths (Proc. Biol. Soc. Washington **27**: 25. 1914) is an anomalous plant with some of the joints terete and others somewhat flattened. It has been suggested that it is a hybrid between a cylindric and a flat-jointed species; but, so far as we know, natural hybrids do not occur between species of these subgenera. It is more likely to be an anomalous form of *Opuntia basilaris,* a form of which is known in the same mountains where it was found.

110. Opuntia microdasys (Lehmann) Pfeiffer, Enum. Cact. 154. 1837.

 Cactus microdasys Lehmann, Ind. Sem. Hamburg. 16. 1827.
 Opuntia pulvinata De Candolle, Mém. Mus. Hist. Nat. Paris **17**: 119. 1828.
 Opuntia microdasys minor Salm-Dyck, Hort. Dyck. 186. 1834.
 Opuntia microdasys laevior Salm-Dyck, Cact. Hort. Dyck. 1849. 241. 1850.

Often low and creeping but sometimes nearly erect and 4 to 6 dm. high; joints oblong to orbicular, 10 to 15 cm. long, soft-velvety, usually pale green, spineless; areoles conspicuous, closely set, filled with numerous yellow or brown glochids; flowers usually produced in abundance, 4 to 5 cm. long, pure yellow or tinged with red; sepals acuminate; petals broad, retuse; filaments and style white; stigma-lobes 6 to 8, green; fruit dark red, juicy, nearly globular; seeds small, 2 to 3 mm. broad.

Type locality: In Mexico, but originally stated by Lehmann as coming from Brazil.
Distribution: Northern Mexico.

In spite of its troublesome glochids, which easily become detached, this species has long been a greenhouse favorite. No cactus collection, however small, lacks one or more pots of this species, which rarely grows large in cultivation.

Opuntia microdasys is usually credited to Lehmann, but he apparently published it as *Cactus microdasys*, and this is the way it is cited in the Index Kewensis. Lehmann soon republished this species (Nov. Act. Nat. Cur. **16:** 317) where it appears as *Cactus (Opuntia) microdasys*. The first use of the name of *Opuntia microdasys* was by Salm-Dyck (Hort. Dyck. 186) in 1834, but was without description or synonymy. Pfeiffer in 1837, however, republishes Lehmann's description under *Opuntia* and is therefore cited as the author of the binomial. Here it is first credited to Mexico, although Lehmann stated definitely that it comes from Brazil; this he does also with regard to *Opuntia tunicata* and *Cactus bradypus*, both Mexican species, while

FIG. 150.—Opuntia microdasys.

FIG. 151.—Opuntia, probable hybrid.

Cactus linkii and *C. ottonis*, both credited to Mexico, are known only from South America. If this *Opuntia* really came originally from Brazil, it might very well be the same as *Opuntia inamoena*.

As shown above (p. 116), *Opuntia puberula* is referred to *O. decumbens*. The *O. puberula* of our gardens, however, is quite a different plant, and in all probability is of hybrid origin. It is almost identical with a hybrid between *O. microdasys* and *O. cantabrigiensis* which Dr. Rose collected in Hidalgo, Mexico, in 1905, and which is now grown in the collection in Washington and in the New York Botanical Garden.

Illustrations: Agr. Gaz. N. S. W. **25:** pl. opp. p. 138; p. 138; Gard. Chron. III. **30:** f. 76; Rep. Mo. Bot. Gard. **19:** pl. 28, in part; **20:** pl. 12, in part; Safford, Ann. Rep. Smiths. Inst. **1908:** pl. 10, f. 4; U. S. Dept. Agr. Bur. Pl. Ind. Bull. **262:** pl. 5, f. 2.

Plate XXII, figure 1, represents joints of the plant grown in a garden at Riverside, California, received by Dr. Rose in 1905. Figure 150 is from a photograph taken by Professor F. E. Lloyd in Zacatecas, Mexico, in 1908.

Plate xx, figure 3, shows a flowering joint of a plant sent to the New York Botanical Garden by M. Simon, of St. Ouen, Paris, France, in 1901, as *Opuntia puberula*. Figure 151 is from a photograph of the plant sent from La Mortola, Italy, to the same institution in 1912, as *Opuntia puberula*.

111. Opuntia macrocalyx Griffiths, Rep. Mo. Bot. Gard. **19**: 268. 1908.

"A profusely, divaricately branched, ascending or erect, spreading plant, 9 to 10 dm. high and about the same in diameter; joints long-obovate, variable but commonly 9 by 22 cm. for last year's growth, gray green, pubescent, velvety to the touch; areoles subcircular, usually 2 to 3 mm. in diameter, very close to 1 cm. apart, slightly sunken; wool tawny, prominent, as long as spicules and occupying lower half of areole; spicules reddish brown, about 1 mm. long, occupying upper half of areole, easily separable and causing fully as much annoyance in handling as those of *O. microdasys*, in age often appearing dirty yellow *in situ* but distinctly reddish brown when removed; strictly spineless; flowers yellow, green outwardly, the leaves on ovary very long subulate and changing gradually into the sepals which are very long subulate, delicately pointed, loosely arranged or often half recurved at apex, giving to the bud a rather ragged appearance; fruit red but both pulp and rind greenish, long obovate to cylindrical, about 2 by 7 cm., with but few rather small seeds, about 3 mm. in diameter."

FIG. 152.—Opuntia macrocalyx. ×0.75.   FIG. 153.—Opuntia rufida.

Type locality: In cultivation at Riverside, California.
Distribution: Known only from cultivated plants; perhaps also from Coahuila, Mexico.
Illustration: Rep. Mo. Bot. Gard. **19**: pl. 28, in part.

Figure 152 is drawn from a joint of the plant collected by Edward Palmer at Saltillo, Mexico, in 1904.

112. Opuntia rufida Engelmann, Proc. Amer. Acad. **3**: 298. 1856.

Opuntia microdasys rufida Schumann, Gesamtb. Kakteen 706. 1898.

More or less erect, 2 to 15 dm. high, with a somewhat definite trunk; joints nearly orbicular, 6 to 25 cm. in diameter, thickish, velvety-tomentose, dull grayish green; leaves subulate, caducous, 4 to 6 cm. long, green with reddish tips; areoles large, filled with numerous brown glochids; flowers yellow to orange, 4 to 5 cm. long including the ovary; petals obovate, 2 to 2.5 cm. long; filaments greenish white, short, 1 cm. long; style 1.5 cm. long, thick, bulbous just above the base; stigma-lobes 5, deep green; ovary globular, 1.5 cm. in diameter, umbilicate, with large areoles; fruit, according to field observation of Dr. Griffiths, bright red.

Type locality: About Presidio del Norte, on the Rio Grande.
Distribution: Texas and northern Mexico.

This species seems much less common than *O. microdasys*, with which it is often confused. The joints are gray or bluish green, and the glochids are brown. It does fairly well under greenhouse conditions.

Illustration: Rep. Mo. Bot. Gard. **20:** pl. 3; Carnegie Inst. Wash. **269:** pl. 11, f. 94.

Figure 153 is from a photograph of a plant brought from Mexico for the New York Botanical Garden in 1896 by Mrs. N. L. Britton.

Fig. 154.—Opuntia pycnantha. Along the coastal plain of Lower California.

113. **Opuntia pycnantha** Engelmann in Coulter, Contr. U. S. Nat. Herb. **3:** 423. 1896.

Opuntia pycnantha margaritana Coulter, Contr. U. S. Nat. Herb. **3:** 424. 1896.

Often low and creeping, but sometimes forming a clump 2 dm. high; joints oblong to orbicular, often 20 cm. long, puberulent or papillose, usually nearly hidden by the thick mass of spines; areoles large and closely set, the upper part filled with yellow or brown glochids, and the lower part with 8 to 12 yellow or brown reflexed spines 2 to 3 cm. long; leaves and flowers unknown; fruit 4 cm. long, very spiny; seeds 2 cm. broad, very thick.

Type locality: Magdalena Bay, Lower California.
Distribution: Southern Lower California.

Coulter's variety *margaritana* is known only from Margarita Island, while the species proper is known only from an adjacent island, Magdalena. They differ only in the color of their spines and glochids. Both have been in cultivation in New York City and Washington, but are not well suited for indoor plants.

This species grows in one of the driest parts of Lower California on islands where there is no surface water and where there is no rain sometimes for five or six years.

Figure 154 is from a photograph taken by Dr. Rose near Santa Maria Bay, Magdalena Island, Lower California, in 1911.

114. Opuntia comonduensis (Coulter) Britton and Rose, Smiths. Misc. Coll. **50**: 519. 1908.

Opuntia angustata comonduensis Coulter, Contr. U. S. Nat. Herb. **3**: 425. 1896.

Low, spreading plants, sometimes 2 dm. high and forming broad clumps; joints obovate to orbicular, 12 to 15 cm. long, softly pubescent; areoles large, filled with brown wool and yellow glochids; lower areoles spineless, the upper ones bearing 1 or 2, rarely 3, or on old stems as many as 10, slender spines, 3 to 5 cm. long or longer, yellow; flowers, including ovary, 6 cm. long, yellow; fruit purple, 4 cm. long, spineless; seeds 4 to 4.5 mm. broad, thick.

Type locality: Comondu, Lower California.
Distribution: Southern Lower California.

As was pointed out by Mrs. K. Brandegee, this plant is not closely related to *Opuntia angustata*.

Fig. 155.—Opuntia comonduensis.

This species has long been known only from herbarium specimens collected by Mr. Brandegee in 1889. In 1911 Dr. Rose collected considerable material both near the town of San José and on Carmen Island which has since been in cultivation in the New York Botanical Garden and in Washington. The above description is based largely on this collection.

This species sometimes grows with *O. tapona*, in fact being confused in the original material; except for its pubescent joints, they are not readily distinguished.

Figure 155 is from a photograph by Mr. T. W. Smillie of a plant collected by Mr. E. W. Nelson and Mr. E. A. Goldman in Lower California in 1906.

FIG. 156.—Opuntia inamoena. A single plant. Photograph by P. H. Dorsett.

Series 6. INAMOENAE.

A single, prostrate or depressed, usually spineless, light-green Brazilian species.

115. Opuntia inamoena Schumann in Martius, Fl. Bras. **4**²: 306. 1890.

Opuntia quipa Weber, Dict. Hort. Bois 894. 1898.

Usually low, often prostrate, forming clumps 2 to 10 dm. broad, or sometimes in sheltered situations 6 cm. high and forming dense, extensive thickets; roots fibrous; joints bluish green, when young bright green, orbicular to oblong, 8 to 16 cm. long, usually quite thick, sometimes 3 cm. thick, usually quite spineless; leaves minute, 2 mm. long; areoles small, when young filled with numerous yellow-ish-brown glochids; glochids unequal, spreading, easily becoming detached; flowers small, brick-red; petals spreading; filaments orange; style yellow; stigma-lobes pale green; fruit globular, yellowish, 2.5 to 3 cm. in diameter.

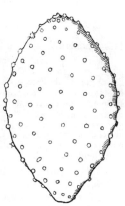

Type locality: Schumann cites Rio de Janeiro in original description.

FIG. 157.—Opuntia inamoena. X0.66.

Distribution: Pernambuco, Bahia, and Minas Geraes, Brazil.

This plant is known as quipa in Bahia, Brazil.

This species is very common in all the dry part of Bahia and, although abundant and mostly spineless, is avoided by all kinds of grazing animals, even when the country is devoid of other suitable forage. It has been suggested that the plant may be bitter, or that the glochids are troublesome; the glochids, however, are usually wanting on old joints.

The plant rarely develops acicular spines up to 3 cm. long on some joints, as shown by specimens collected by Dr. Rose and Mr. Russell near Machado Portello, Brazil.

Figure 156 is from a photograph taken by Mr. P. H. Dorsett near Joazeiro, Brazil, in 1914; figure 157 is from a plant collected by Dr. Rose near Machado Portello, Bahia, Brazil, in 1915.

Series 7. TORTISPINAE.

Prostrate or spreading plants rarely erect, with mostly rather small, persistent, scarcely tuberculate, orbicular or oval joints, and large flowers; natives of the eastern, central, and southern United States.

Plate 21 represents a group of hardy cacti, chiefly species of this series, at the New York Botanical Garden.

KEY TO SPECIES.

Spines none or only 1 or 2 at an areole.
  Joints bluish green; at least when young; roots tuber-like.
    Fruit clavate; joints thin.
      Fruit about 4.5 cm. long................................ 116. *O. allairei*
      Fruit 5 to 7 cm. long .... 116a. *O. lata*
    Fruit obovoid; joints turgid............................ 117. *O. pollardii*
  Joints green; roots not tuberous.
    Flowers 8 cm. broad or less.
      Joints orbicular or little longer than wide.................. 118. *O. opuntia*
      Joints oblong, much longer than wide .... 119. *O. macrarthra*
    Flowers 10 to 12 cm. broad............................. 120. *O. grandiflora*
Spines mostly 2 or more at an areole.
  Ovary obconic, 2 to 4 cm. long.
    Roots tuberous.
      Joints repand; plant suberect........................... 121. *O. austrina*
      Joints scarcely repand; plants nearly prostrate.............. 122. *O. macrorhiza*
    Roots not tuberous.
      Flowers and fruit small............................... 123. *O. plumbea*
      Flowers and fruit large.
        Spines white to light brown, slender.
          Seeds acute-margined............................ 124. *O. tortispina*
          Seeds obtuse-margined.
            Fruit large, 4 to 5 cm. long; spines light colored.... 125. *O. stenochila*
            Fruit small, 2 to 3 cm. long; spines brown......... 126. *O. delicata*
        Spines dark brown, stout............................. 127. *O. fuscoatra*
  Ovary narrowly subcylindric, 5 to 6 cm. long.................. 127a. *O. macateei*

FIG. 158.—Opuntia allairei. ×0.66.

116. Opuntia allairei Griffiths, Rep. Mo. Bot. Gard. **20**: 83. 1909.

A low, spreading, tuberous-rooted, prostrate plant, with some of the joints ascending; joints bluish green, obovate, usually 10 to 15 cm. long, originally described as even longer, with or without spines; spines, if present, 1 to 3, yellowish brown, 2.5 cm. long or less, slender but a little flattened; glochids numerous, especially abundant at very old areoles, yellow; leaves 6 to 8 mm. long; flowers 6 to 7 cm. broad, yellow with a red center; fruit 4 to 5 cm. long, dark red.

Type locality: Mouth of Trinity River, Texas.

Distribution: Southern Texas and western Louisiana.

This species is perhaps nearest *O. macrorhiza*, but differs in the usual absence of spines and in differently colored joints.

Illustrations: Rep. Mo. Bot. Gard. **20**: pl. 2, f. 2; pl. 5; pl. 12, in part.

Figure 158 is copied from the second illustration above cited.

116a. Opuntia lata Small, Journ. N. Y. Bot. Gard. **20**: 26. 1919. (See Appendix, p. 220.)

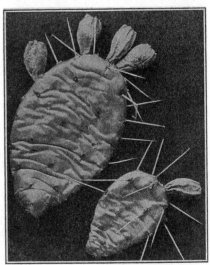

FIG. 159.—Opuntia pollardii. ×0.4.

117. Opuntia pollardii Britton and Rose, Smiths. Misc. Coll. **50**: 523. 1908.

Prostrate, tuberiferous, related to *Opuntia opuntia;* young joints bluish green, glaucous, 5 to 16 cm. long, 1 to 2 cm. thick; areoles 1.5 to 3 cm. apart, bearing numerous yellow glochids 2 to 3 cm.

Group of hardy prickly pears, mostly *Opuntia tortispina*, in the herbaceous grounds of the New York Botanical Garden.

long, those toward the top of the joint each with a single stout, stiff, pungent spine 2.5 to 4 cm. long; fruit short-obovoid, 2.5 cm. long, 1.5 cm. thick, with a few areoles bearing tufts of brownish wool but no spines and but few glochids; flowers yellow, 6 to 8 cm. broad; sepals deltoid to rhombic; fruit 2.5 to 4 cm. long; seeds 4 to 6 mm. wide, much thicker than those of *Opuntia opuntia*.

Type locality: Biloxi, Harrison County, Mississippi.

Distribution: Coastal plain, Church Island, North Carolina, to northern Florida, Alabama, and Mississippi.

Figure 159 is from a photograph of fruiting joints collected by A. H. Howell on Petit Bois Island, Alabama.

118. Opuntia opuntia (Linnaeus) Karsten, Deutsch. Fl. 888. 1882.

> *Cactus opuntia* Linnaeus, Sp. Pl. 468. 1753.
> *Cactus compressus* Salisbury, Prodr. 348. 1796.
> *Cactus opuntia nana* De Candolle, Pl. Succ. Hist. 2: pl. 138. [A]. 1799.
> *Cactus humifusus* Rafinesque, Ann. Nat. 15. 1820.
> *Opuntia vulgaris major* Salm-Dyck, Observ. Bot. 3: 9. 1822.
> *Opuntia vulgaris media** Salm-Dyck, Observ. Bot. 3: 9. 1822.
> *Opuntia humifusa* Rafinesque, Med. Fl. U. S. 2: 247. 1830.
> *Opuntia mesacantha* Rafinesque, Bull. Bot. Seringe 216. 1830.
> *Opuntia caespitosa* Rafinesque, Bull. Bot. Seringe 216. 1830.
> *Opuntia intermedia* Salm-Dyck, Hort. Dyck. 364. 1834.
> *Opuntia nana* Visiani, Fl. Dalmatica 3: 143. 1852.
> *Opuntia rafinesquei†* Engelmann, Proc. Amer. Acad. 3: 295. 1856.
> *Opuntia rafinesquei microsperma* Engelmann, Proc. Amer. Acad. 3: 295. 1856.
> *Opuntia rafinesquei minor* Engelmann and Bigelow, Pac. R. Rep. 4: 55. 1856.
> *Opuntia vulgaris rafinesquei* Gray, Man. Bot. ed. 2. 136. 1856.
> *Opuntia rafinesquei arkansana* Rümpler in Förster, Handb. Cact. ed. 2. 922. 1885.
> *Opuntia mesacantha microsperma* Coulter, Contr. U. S. Nat. Herb. 3: 429. 1896.
> *Opuntia mesacantha parva* Coulter, Contr. U. S. Nat. Herb. 3: 429. 1896.
> *Opuntia vulgaris nana* Schumann, Gesamtb. Kakteen 715. 1898.
> *Opuntia humifusa microsperma* Heller, Cat. N. Amer. Pl. ed. 2. 8. 1900.
> *Opuntia humifusa parva* Heller, Cat. N. Amer. Pl. ed. 2. 8. 1900.

Low, spreading plants, sometimes ascending, with fibrous roots; joints orbicular to oblong, 3 to 13 cm. long, rarely longer, thick, dark green; areoles usually far apart; leaves subulate, appressed or spreading, 4 to 8 mm. long, early deciduous; spines often wanting, when present usually one from an areole, rarely two, 5 cm. long or less, brownish or sometimes nearly white, but on seedlings 5 to 12; glochids numerous, yellow to dark brown; flowers usually bright yellow, sometimes with reddish centers, 5 to 8.5 cm. broad; petals 8 to 10, widely spreading; filaments yellow; stigma-lobes white; fruit obovoid to oblong, red, juicy, 2.5 to 5 cm. long, edible; seeds 4 to 5 mm. broad.

Type locality: In Virginia.

Distribution: Sandy and rocky places from Massachusetts to Virginia, the mountains of Georgia and central Alabama extending north into southern Ontario, Canada (Point Pelee), west in isolated colonies to northern Illinois, eastern Missouri and Tennessee, and long established in the mountains of northern Italy and Switzerland.

Linnaeus undoubtedly had two species in his *Cactus opuntia*, one being the low Virginia plant commonly known as *O. vulgaris*, and the other a tall, branching plant figured by Bauhin (p. 154). Upon Bauhin's illustration Miller based his *Opuntia vulgaris*, a name which was afterwards transferred to the low, procumbent plant of the eastern United States. For this reason Burkill (Rec. Bot. Surv. India 4: 288. 1911) would displace the name *O. vulgaris* and take up the name *O. nana*. We are quite in agreement with him as to the *O. vulgaris* Miller, but we retain for the low plant the specific name *opuntia* Linnaeus. The tall species is *O. monacantha*, which we now call *O. vulgaris*, as suggested by Burkill.

It is to be noted that the southern Atlantic coast specimens of *Opuntia opuntia* have yellow or greenish-yellow glochids, while those in its northern and western range have brown glochids. Its southwestern limit is uncertain. It probably does not extend to Texas, although two varieties have been reported from there; these we are disposed to treat as species under the names *Opuntia macrorhiza* and *O. grandiflora*. It is reported from

---

*Opuntia vulgaris minor (Labouret, Monogr. Cact. 476. 1853) was doubtless intended for this name.
†Sometimes spelled *rafinesquiana*.

eastern Kansas, but the plants found there are not like those found in Illinois and Indiana, having more spines and a glaucous bloom, and are tuberous-rooted, and these are referred by us to *O. macrorhiza*. The published western varieties of *O. humifusa* are specifically distinct; we have referred them to *O. tortispina*.

Some of the joints of this plant elongate under shade conditions, reaching at least 2.5 dm. in length and not more than 5 cm. in width.

Opuntia arkansana (Hirscht, Monatsschr. Kakteenk. 8: 115. 1898) has not been formally described. The name should doubtless be referred here.

Opuntia prostrata Monville and Lemaire (Förster, Handb. Cact. 478. 1846) was given only as a synonym of *O. intermedia*, while *O. intermedia prostrata* Salm-Dyck (Cact. Hort. Dyck 1849. 69. 1850) was based on *O. prostrata*.

O. rafinesquei parva Haage and Schmidt (Verzeichnis Blumenzwiebeln 1915: 29. 1915) is a new name for *O. mesacantha parva* Coulter.

Under *Opuntia vulgaris* Michaele Gandoger in his Flora Europea (9: 145. 1886) has proposed the following new binomials: *O. recedens*, *O. morisii*, *O. cycloidea*, *O. inaequalis*, *O. ligustica*, and *O. mediterranea*. The following varieties cited under *O. humifusa* are in the trade: *cymochila*, *greenei*, *macrorhiza*, *oplocarpa* and *stenochila* (Stand. Cycl. Hort. Bailey 4: 2363. 1916.)

Fig. 160.—Opuntia opuntia in its natural surroundings on Staten Island, New York.

Illustrations: Illustr. Fl. 2: f. 2527; ed. 2. 2: f. 2986; Curtis's Bot. Mag. 50: pl. 2393; Loudon, Encycl. Pl. ed. 3. f. 6884, the last two as *Cactus opuntia;* De Candolle, Pl. Succ. Hist. 2: pl. 138 [A]; DeTussac, Fl. Antill. 2: pl. 30, the last two as *Cactus opuntia nana.* Dept. Agr. N. S. W. Misc. Publ. 253: pl. [1], f. 2; Engler and Prantl, Pflanzenfam. 3⁶ᵃ: f. 57, G; Förster, Handb. Cact. ed. 2. f. 12; Pac. R. Rep. 4: pl.10, f. 1, 2; 4: pl. 23, f. 13; Schumann, Gesamtb. Kakteen Nachtr. f. 1, all as *Opuntia vulgaris.* Standard Cycl. Hort. Bailey 4: f. 2602, in part as *Opuntia humifusa.* Amer. Entom. Bot. 2: f. 160; Amer. Garden 11: 462; Curtis's Bot. Mag. 115: pl. 7041; Dict. Gard. Nicholson 2: f. 756; Fl. Serr. 22: pl. 2328; Förster, Handb. Cact. ed. 2. f. 2; Gard. Mag. 4: 280; Gartenflora 24: 218; Lemaire, Cact. f. 9; Meehan's Monthly 2: pl. 6; 10: 121; Pac. R. Rep. 4: pl. 10, f. 4, 5; pl. 23, f. 7, 8; Rümpler, Sukkulenten f. 125; W. Watson, Cact. Cult. f. 84, all as *Opuntia rafinesquei;* Pac. R. Rep. 4: pl. 11, f. 1, as *Opuntia rafinesquei minor;* Förster, Handb.

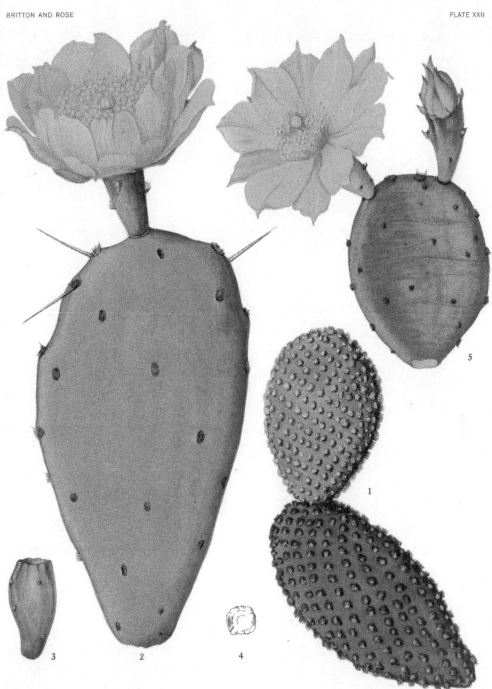

M. E. Eaton del.

1. Joints of *Opuntia microdasys.*
2. Flowering joint of *Opuntia macrarthra.*
3. Fruit of *Opuntia macrarthra.*

4. Seed of same.
5. Flowering joint of *Opuntia opuntia.*
(All natural size except 4.)

Cact. ed. 2. f. 126, as *Opuntia rafinesquei arkansana;* Monatsschr. Kakteenk. 14: 124, as *Opuntia vulgaris nana;* Miller, Fig. Pl. Gard. Dict. 2: pl. 191, as *Opuntia folio minori,* etc., Dict. Hort. Bois f. 638; Rev. Hort. 40: f. 10, 11; 66: f. 59, all as *Opuntia rafinesquiana.* Wiener Illustr. Gartenz. 10: f. 112, as *Opuntia rafinesquiana arkansana.*

Plate XXII, figure 5, represents a flowering joint of the plant which grows naturally on schistose rocks in the New York Botanical Garden. Figure 160 is from a photograph of the plant growing on sand dunes at Crooke's Point, Staten Island, New York, taken by Howard H. Cleaves in 1914.

119. Opuntia macrarthra Gibbes, Proc. Elliott Soc. Nat. Hist. 1: 273. 1859.

Stems prostrate or ascending; joints narrowly oblong to obovate, 12 to 35 cm. long, thick, pale green, somewhat shining; leaves subulate, 10 mm. long, green, sometimes with purplish tips; areoles large, 2 to 3 cm. apart, filled with brown wool; spines wanting, or sometimes 1, up to 2.5 cm. long; glochids when present yellow; flowers not known; fruit narrowly obovoid, red, fleshy, 4 to 6 cm. long.

Type locality: Near Charleston, South Carolina.
Distribution: Coast of South Carolina.

This species, long overlooked, has recently been collected by Dr. J. K. Small in the vicinity of the type locality.

This is doubtless one of the species to which Elliott called attention and which he said he expected to publish, but never did.* The original description long remained unnoticed in the Proceedings of the Elliott Society of Natural History; it is as follows:

"The second, which we will call *Opuntia macrarthra,* falls under the same section with the preceding, and seems to be near *Opuntia angustata,* of Engelmann, from the west of the Rio Grande; a prostrate species, joints from ten to fifteen inches long and three inches wide, one-third of an inch thick; no spines, fruit two and a half inches long, slender, clavate."

Plate XV, figure 3, represents a fruiting joint collected by Dr. Small on James Island, South Carolina, in 1916; plate XXII, figure 2, represents a flowering joint of the plant collected by Dr. Small on the Isle of Palms, near Charleston, South Carolina, in 1916; figure 3 shows a fruit of the same plant and figure 4 a seed, enlarged.

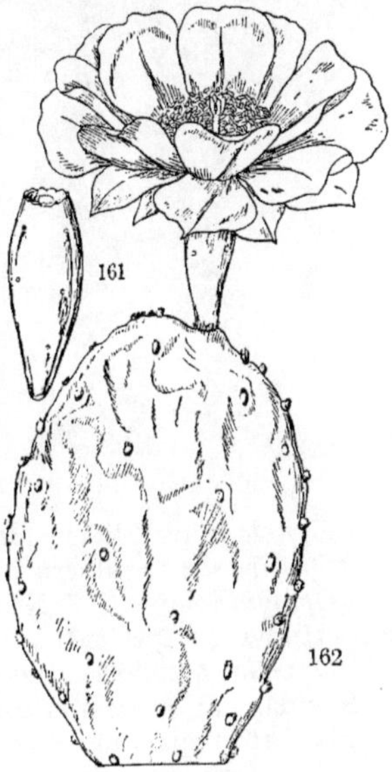

120. Opuntia grandiflora Engelmann, Proc. Amer. Acad. 3: 295. 1856.

FIGS. 161, 162.—Opuntia grandiflora.

Opuntia rafinesquei grandiflora Engelmann, Pac. R. Rep. 4: 55. 1856.
Opuntia mesacantha grandiflora Coulter, Contr. U. S. Nat. Herb. 3: 429. 1896.

Low, with somewhat ascending branches; joints 12.5 to 15 cm. long; areoles 2.5 cm. apart; spines usually wanting; flowers very large, 11 to 12.5 cm. broad, yellow with a red center; petals broad; fruit elongated, 6 cm. long.

Type locality: On the Brazos, Texas.
Distribution: Eastern Texas.

Although Dr. Engelmann formally described this as a species, he introduced it as "probably only a southern variety of *O. rafinesquei.*" A little later he actually used the name as a variety. The position of the plant is still uncertain; if specimens collected by Mr. Wm. R. Maxon at Victoria, Texas, and by Mr. C. V. Piper at Dallas, Texas, belong here, as they appear to, we believe it to be a distinct species.

Illustrations: Pac. R. Rep. 4: pl. 11, f. 2, 3, as *Opuntia rafinesquei grandiflora.*

Figures 161 and 162 are copied from the illustrations above cited.

_____

*Cactus opuntia. "It is probable that there are now three distinct species on the sea coast of the Southern States covered under this name." Elliott, A Sketch of the Botany of South Carolina and Georgia, 1: 537.

121. **Opuntia austrina** Small, Fl. Southeast. U. S. 816. 1903.

    Opuntia youngii C. Z. Nelson, Chicago Examiner. June 13, 1915.

Roots fusiform or tuberous, resembling sweet potatoes, often 4 to 6 cm. in diameter, 5 to 15 cm. long; stems erect or ascending; joints narrowly obovate to oblong-obovate, thick, tuberculate, repand, bright green, 5 to 12 cm. long; leaves soon deciduous, less than 10 mm. long; glochids yellowish; spines usually on the upper half and margin of the joint, often 2, sometimes 1 to 6, from an areole, whitish or pinkish, darker at base and apex, twisted, sometimes wanting; flowers bright yellow, 6 to 7 cm. broad; petals cuneate, truncate or retuse at apex, mucronate; fruit 2.5 to 3 cm. long.

Fig. 163.—Opuntia austrina. X0.5.

Type locality: Miami, Florida.

Distribution: Southern Florida.

Opuntia youngii C. Z. Nelson, published in a Chicago newspaper, we have referred here, after studying a specimen sent by the author.

Opuntia spinalba Rafinesque (Atl. Journ. 1: 147. 1832) was described as from the keys of Florida, and answers in some respects to *O. austrina;* but it is very unlikely that any plants of the region inhabited by *austrina* were known to botanists as early as 1832.

Figure 163 represents a plant collected by Dr. Small at the type locality in 1901.

122. **Opuntia macrorhiza** Engelmann, Bost. Journ. Nat. Hist. **6:** 206. 1850.

    Opuntia fusiformis Engelmann and Bigelow, Proc. Amer. Acad. **3:** 297. 1856.
    Opuntia rafinesquei fusiformis Engelmann, Pac. R. Rep. **4:** 43. 1856.
    Opuntia mesacantha macrorhiza Coulter, Contr. U. S. Nat. Herb. **3:** 430. 1896.
    Opuntia xanthoglochia Griffiths, Rep. Mo. Bot. Gard. **21:** 166. 1910.
    Opuntia roseana Mackensen, Bull. Torr. Club **38:** 142. 1911.

Plant low, usually nearly prostrate, forming a clump 1 meter in diameter, from a cluster of tuber-like roots, these sometimes 5 to 7.5 cm. in diameter; joints orbicular to obovate, dull green, 5 to 16 cm. long, about 1 cm. thick; leaves subulate, 4 to 10 mm. long; areoles rather large, the lower ones and sometimes all of them spineless; glochids numerous, yellow or brown; spines, when present, 1 to 4, unequal, yellow to brown, the longest 2.5 cm. long; flower yellow, with a reddish or purplish center, 7 to 8 cm. broad; fruit narrowly obovoid, 3.5 to 5 cm. long, purple or red, with a depressed umbilicus, not edible; seeds 5 mm. in diameter, with broad margins.

Type locality: Rocky places on the Upper Guadalupe, Texas.

Distribution: Missouri and Kansas to Texas.

Opuntia seguina C. Z. Nelson (Galesburg Register, July 20, 1915), published in a newspaper, and said to have come from San Antonio, Texas, seems to be one of the *Tortispinae,* and is probably referable to *O. macrorhiza.* Through the kindness of Mr. Nelson, we have seen a joint of this species.

Opuntia bulbosa Engelmann (Proc. Amer. Acad. **3**: 297. 1856) was used by Engelmann for *O. fusiformis*, but never described.

Opuntia macrorhiza, originally described by Dr. Engelmann as a species, was afterwards (Proc. Amer. Acad. **3**: 296. 1856) proposed as a subspecies but not formally indicated, so that the reference *O. rafinesquei macrorhiza* Coulter (Contr. U. S. Nat. Herb. **3**: 430. 1896) is the proper designation if it is used as a variety.

Illustrations: Cact. Mex. Bound. pl. 69; Förster, Handb. Cact. ed. 2. f. 11, 127; Pac. R. Rep. **4**: pl. 12, f. 7, 8; pl. 23, f. 6; Suppl. Dict. Gard. Nicholson f. 606; W. Watson, Cact. Cult. f. 82, 83; Rep. Mo. Bot. Gard. **21**: pl. 20, in part, this last as *Opuntia xanthoglochia;* Addisonia 1: pl. 19.

Plate XIV, figure 5, represents a flowering joint of the plant collected at Irving, Dallas County, Texas, by Albert Ruth in 1912.

123. Opuntia plumbea Rose, Smiths. Misc. Coll. **50**: 524. 1908.

Plant low, creeping, 10 cm. high, 20 to 30 cm. broad, few jointed; joints small, nearly orbicular, 3 to 5 cm. in diameter, of a dull lead-color, the surface somewhat wrinkled in dead specimens; areoles rather large for the size of the joints; spines pale brownish, slender, usually porrect, often 3 cm. long, mostly 2 in number, rarely as many as 4, sometimes one or even wanting; flowers very small, red; ovary naked; fruit 1.5 to 2 cm. long with a few small areoles and these simply woolly; seeds small, rather turgid, smooth, and with a shallow obtuse margin.

Type locality: San Carlos Indian Reservation, Arizona.

Distribution: Arizona.

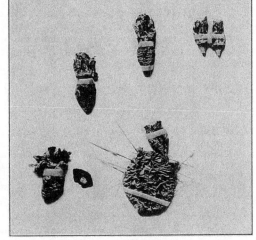

FIG. 164.—Opuntia plumbea.

This is a peculiar little opuntia with very small joints and fruits. It is known only from the original collections made by Mr. F. V. Coville in 1904.

Figure 164 is from a photograph of the type specimen.

124. Opuntia tortispina Engelmann, Proc. Amer. Acad. **3**: 293. 1856.

Opuntia tortisperma Engelmann, Pac. R. Rep. **4**: pl. 23, f. 1 to 5. 1856.
Opuntia cymochila Engelmann, Proc. Amer. Acad. **3**: 295. 1856.
Opuntia rafinesquei cymochila Engelmann, Proc. Amer. Acad. **3**: 295. 1856.
Opuntia rafinesquei cycmochila montana Engelmann and Bigelow, Pac. R. Rep. **4**: 42. 1856.
Opuntia mesacantha cymochila Coulter, Contr. U. S. Nat. Herb. **3**: 430. 1896.
Opuntia mesacantha greenei Coulter, Contr. U. S. Nat. Herb. **3**: 431. 1896.
Opuntia mesacantha oplocarpa Coulter, Contr. U. S. Nat. Herb. **3**: 431. 1896.
Opuntia greenei Engelmann in Britton and Rose, Smiths. Misc. Coll. **50**: 523. 1908.
(?) *Opuntia sanguinocula* Griffiths, Proc. Biol. Soc. Washington **27**: 26. 1914.

Prostrate and creeping; joints ascending, orbicular to obovate, 15 to 20 cm. long; areoles 1.5 to 3 cm. apart; spines several, often 6 to 8, the upper and longer ones 3 to 6 cm. long, either white, yellowish, or brown; on the upper areoles one spine erect, the others spreading or with the lowermost ones deflexed; flowers sulphur-yellow, 6 to 7.5 cm. broad; fruit rather large, 4 to 5 cm. long, 2 to 3 cm. broad; seeds 4 to 6 mm. broad, thick, regular, with a slight indentation at the hilum.

Type locality: On the Camanchica Plains near the Canadian River.

Distribution: Wisconsin to South Dakota, Texas, Kansas, Colorado, and New Mexico.

This has long remained one of our least-understood species. We believe now that it has a wide range, and that it has been referred heretofore to several species. *Opuntia cymochila* does not seem to differ from it, and the two published varieties of *Opuntia mesacantha*, geographically out of harmony with that species, doubtless belong here.

Opuntia oplocarpa Engelmann (Coulter, Contr. U. S. Nat. Herb. **3**: 431. 1896) was published only as a synonym. *Opuntia rafinesquei greenei* (Cat. Darrah Succ. Manchester 58. 1908) is a catalogue name.

The plant is hardy at New York, flowering profusely, and also at Buck Hill Falls, eastern Pennsylvania.

Illustrations: Pac. R. Rep. **4**: pl. 12, f. 1 to 3; pl. 23, f. 10 to 12; Rev. Hort. Belg. **40**: after 186, all as *Opuntia cymochila;* Illustr. Fl. **2**: f. 2528; ed. 2. **2**: f. 2987; Pac. R. Rep. **4**: pl. 10, f. 3; Stand. Cycl. Hort. Bailey **4**: f. 2602, in part, these as *Opuntia humifusa.* Pac. R. Rep. **4**: pl. 8, f. 2, 3; pl. 23, f. 1 to 5, as *O. tortisperma.* Illustr. Fl. **2**: f. 2529; ed. 2. **2**: f. 2988.

Plate xv, figure 4, represents a flowering and fruiting joint of a plant from Colorado, grown at the New York Botanical Garden.

125. Opuntia stenochila Engelmann, Proc. Amer. Acad. **3**: 296. 1856.*

> *Opuntia mesacantha stenochila* Coulter, Contr. U. S. Nat. Herb. **3**: 430. 1896.

Prostrate; joints obovate, 10 cm. long by 7.5 cm. broad; leaves small, 4 to 6 mm. long; spines usually 2, sometimes 3, spreading, 1 long (2.5 to 3 cm. long), and 1 or 2 short and reflexed, usually light-colored, sometimes nearly white; glochids brown; flowers yellow, 6 cm. long; fruit very juicy, 4 to 5 cm. long or more, attenuate at base; seeds thick, quite regular, with very narrow obtuse edges.

Type locality: Canyon of Zuni, New Mexico.

Distribution: Western New Mexico and Arizona.

This species has not been well understood. It has usually passed as a variety of the common species of the eastern Mississippi Valley States, but it grows in a very different region. It is the common low, spreading *Opuntia* of northwestern New Mexico and Arizona.

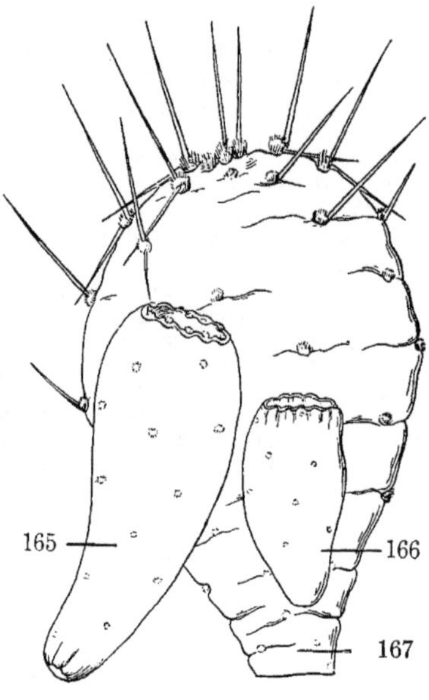

Opuntia stenochila.
FIGS. 165, 166.—Fruits.   FIG. 167.—Joint.

Illustrations: Pac. R. Rep. **4**: pl. 12, f. 4 to 6; pl. 23, f. 9.

Figures 165, 166, and 167 are copied from the first illustration above cited.

126. Opuntia delicata Rose, Contr. U. S. Nat. Herb. **13**: 310. 1911.

A small, procumbent plant with rather thin, ovate, bluish, slightly glaucous joints, often only 4 to 9 cm. in diameter; areoles prominent, bearing conspicuous brown glochids; lower areoles spineless, the upper ones bearing 1 or 2 very slender brownish spines, the longer one 3 to 4 cm. long; flowers yellow, 5 cm. long, 5 to 6 cm. broad; fruit oblong, spineless, 2 to 3 cm. long; seeds small, about 4 mm. in diameter, nearly smooth.

Type locality: Calabasas, Arizona.

Distribution: Southeastern Arizona.

Figure 168 is from a photograph of the type plant.

---

*Although formally published as a species, Engelmann states that it is a form or subspecies, and hence Coulter (Contr. U. S. Nat. Herb. **3**: 430. 1896) uses the synonym *O. rafinesquei stenochila* Engelmann.

M. E. Eaton del.

1. Flowering joint of *Opuntia fuscoatra*.   2. Upper part of joint of *Opuntia sulphurea*.
3. Joint of *Opuntia tenuispina*.   (All natural size.)

FIG. 168.—*Opuntia delicata.*

127. Opuntia fuscoatra Engelmann, Proc. Amer. Acad. **3**: 297. 1856.

Diffuse prostrate plants; joints orbicular to obovate, somewhat tuberculate, 5 to 8 cm. long, areoles 12 to 20 mm. apart, very large for the group; spines single or in twos or threes, one rather stout, sometimes a little flattened, 2.5 to 3 cm. long, yellow to dark brown or even nearly black; usually from the lower areoles; glochids numerous, brown; flowers 7.5 cm. broad, yellow; petals very broad; stigma-lobes 5; ovary 2.5 cm. long, slender; fruit 4 to 5 cm. long, red; seeds 4 mm. broad.

Type locality: Sterile places of prairies west of Houston, Texas.
Distribution: Eastern Texas.
Illustrations: Pac. R. Rep. **4**: pl. 11, f. 4.

Plate XXIII, figure 1, represents a flowering joint of the plant collected by W. L. Mc-Atee at Rockport, Texas, in 1911.

127 a. Opuntia macateei sp. nov. (See Appendix, p. 221.)

OPUNTIA RUBIFLORA Griffiths, Bull. Torr. Club **43**: 529. 1916.

Described as a spreading plant 3 to 4.5 dm. high and a meter broad, with obovate, green joints 12 to 18 cm. long, few white spines up to 5 cm. long with brown or straw-colored bases, and pink flowers. The species is based on cuttings received from European collections, and its origin is unknown.

We have received a similar if not identical plant from Haage and Schmidt of Erfurt, Germany, and we suspect it to be a hybrid, having one of the *Tortispinae* as one of its parents.

The specific name *rubiflora* was used by Davidson a few months earlier than by Griffiths for another plant.

Series 8. SULPHUREAE.

Low or prostrate species, with rather thick, flat, tuberculate joints; fruit small, nearly globular. Three species, natives of central and southern South America.

KEY TO SPECIES.

Flowers yellow.
  Spines stout, subulate.. 128. *O. sulphurea*
  Spines slender, acicular... 129. *O. soehrensii*
Flowers red.. 130. *O. microdisca*

128. Opuntia sulphurea G. Don in Loudon, Hort. Brit. 196. 1830.

Opuntia maculacantha Förster, Handb. Gartenz. **17:** 166. 1861.
Opuntia pampeana Spegazzini, Contr. Fl. Ventana 30. 1896.
Opuntia vulpina Weber, Dict. Hort. Bois 895. 1898.

Plants low and spreading, forming broad clumps 1 to 2 meters in diameter, 3 dm. high or less; joints flattened, oblong to obovate, 12 to 25 cm. long, thick, strongly tuberculate, usually green but sometimes purplish; terminal joints easily detached; leaves conic, about 2 mm. long; spines 2 to 8, generally straight but sometimes curved and twisted, spreading, 3 to 10 cm. long, brownish to red, but sometimes quite pale at first; flowers about 4 cm. long, yellow; fruit with a deep umbilicus, short, about 1 cm. long.

Type locality: Cited as Chile, but doubtless wrong.

Distribution: Dry parts of western Argentina; recorded also from Chile, and perhaps occurring in Bolivia.

This species was not seen in Chile by Dr. Rose, and we are doubtful in considering the Bolivian material to be *O. sulphurea;* the joints, as shown by Dr. Rose's specimens, collected at La Paz (No. 18860), while thick, are not conspicuously tuberculate; the spines are rather short and stiff, white at first, but somewhat yellowish or horn-colored in age.

The name *Cactus sulphureus* Gillies was published by G. Don at the place cited above as a synonym of this species.

Opuntia maculacantha was first described from specimens from Buenos Aires, which had doubtless been sent down from the desert regions to the west or northwest. Schumann in his Monograph referred this species to Mexico, but in his Nachtrag corrects this statement. Dr. Weber, with whom we are in agreement, refers the species to *O. sulphurea.* It is the only species we know with such large tubercles on the joints.

Several varieties of this species, some of which have been described, are given, such as *laevior, major, minor,* and *pallidior.*

Here probably belongs *Opuntia sericea* G. Don (Salm-Dyck, Hort. Dyck. 363. 1834), also reported from Chile, but doubtless from Argentina. *Cactus sericeus* Gillies (Loudon, Hort. Brit. 196. 1830) is the same. There are several varieties of *O. sericea* which we would put with it: *longispina* Salm-Dyck (Hort. Dyck. 363. 1834); *coerulea* Forbes (Hort. Tour Germ. 159. 1837) which is probably *O. coerulea* Gillies (Pfeiffer, Enum. Cact. 155. 1837); *maelenii* Salm-Dyck (Cact. Hort. Dyck. 1844. 46. 1845) which is *O. maelenii* (Salm-Dyck, Cact. Hort. Dyck. 1844. 46. 1845). *Opuntia tweediei* (Schumann, Gesamtb. Kakteen 745. 1898) is given as a synonym of this species by Schumann. *Opuntia albisetosa* Hildmann, a name only, belongs here according to Hirscht (Monatsschr. Kakteenk. **10:** 48. 1900).

Illustrations: Blühende Kakteen **3:** pl. 136; Monatsschr. Kakteenk. **8:** 121; Schumann, Gesamtb. Kakteen f. 106, all as *Opuntia maculacantha.*

Plate XXIII, figure 2, represents a flowering joint of the plant collected by Dr. Rose near Córdoba, Argentina, in 1915.

129. Opuntia soehrensii sp. nov.

Prostrate, in masses usually 1 meter in diameter or less; joints at first erect or ascending, finally prostrate and rooting and forming new colonies, flattened, rather thin, somewhat tuberculate, very spiny, orbicular, 4 to 6 cm. in diameter, often purplish; spines slender, rather variable in color, usually yellow or brown, several from each areole, sometimes as many as eight, the longest ones 5 cm. long, erect; flowers light yellow, 3 cm. long; sepals brown; filaments yellow; style white; stigma-lobes green; fruit naked, 3 cm. long; seeds 3 to 3.5 mm. broad, ovate, thickish, with narrow margin and roughened sides.

Highlands of southern Peru, Bolivia, and northern Argentina. Type collected by Dr. and Mrs. J. N. Rose below Pampa de Arrieros, Peru, August 23, 1914 (No. 18967).

This species is very common in its region, but as it is cultivated somewhat for its seeds as well as used as a protection for gardens and yards, its natural distribution is difficult to determine. On the barren hills below La Paz, Bolivia, the species is well established and grows as if native; on some of these hills it is the dominant and sometimes exclusive plant. In the same general region, however, one finds the plant about the houses, especially on walls, where it has undoubtedly been planted. At Oruro, Bolivia, it was seen only in the wild state, while at several stations along the railroad between Juliaca and Cuzco, Peru, especially at Combatata and Tinta, Peru, it has been planted on top of many of the mud walls about the yards. On the hills below Pampa de Arrieros, Peru, the species is extremely common and undoubtedly native.

The plant is known everywhere by the natives as ayrampo. The seeds are collected in great quantity and dried, and may be bought in the market places, especially in Arequipa. Indeed, there must have been a time when they were shipped by freight, for the name Ayrampo has always appeared on the printed freight classification of the Southern Railroad of Peru. The assistant superintendent of the road, Mr. Brown, states that, so far as he knows, there are few or no shipments made now. One of the places in Peru where Dr. Rose found the plant very abundant is named Ayrampal.

The dry seeds, when placed in water, yield a red substance which is used for coloring jellies and gelatine and, according to some, for coloring wines. In former days the Indians also used this substance in some of their carnival ceremonies. The coloring matter does not come from the seeds themselves, but from the red juice of the fruit which has dried on the surfaces.

Figure 169 represents a joint of this species collected by Dr. Rose at Oruro, Bolivia, in 1914.

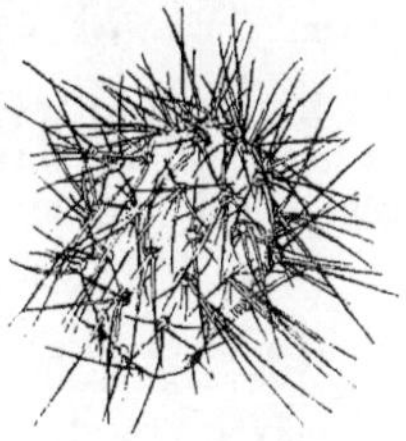

FIG. 169.—Opuntia soehrensii. X0.4.

130. Opuntia microdisca Weber, Dict. Hort. Bois 896. 1898.

Forming small clumps, very much branched, prostrate; joints mostly obovate to oblong, 4 to 8 cm. long, usually much flattened, but sometimes nearly cylindric, grayish green; leaves minute, purple, soon dropping off; areoles numerous, 5 to 6 mm. apart, rather large, when young densely white-felted; spines 10 to 15, white to reddish, unequal, some of the centrals 1.5 to 2.5 cm. long; glochids numerous, yellow; flower-buds red; flowers 2.5 cm. long, bright red; filaments purple; style white; stigma-lobes 6 to 8, short; ovary turbinate, 16 mm. long, bearing numerous areoles subtended by narrow red leaves; areoles on ovary densely felted and bristly; fruit red.

Type locality: In Catamarca, Argentina.
Distribution: Northern Argentina.

Schumann refers this species to *Platyopuntia*, while Weber referred it to *Tephrocactus*. It evidently belongs to our *Sulphureae*, being nearest our *O. soehrensii*.

Our description is drawn chiefly from specimens obtained by J. A. Shafer between Andalgala and Concepción, Argentina, in 1916, supplemented by a living specimen obtained by Dr. Spegazzini in 1915. In Argentina this species also is known as ayrampo.

Figure 170 represents a joint of the plant collected by J. A. Shafer between Andalgala and Concepción, Argentina, December 28, 1916 (No. 24).

To this relationship may belong the following species:

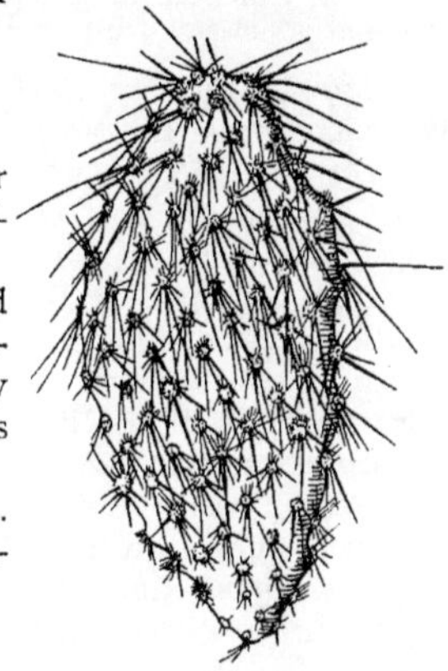

FIG. 170.—Joint of Opuntia microdisca. X0.7.

OPUNTIA PENICILLIGERA Spegazzini, Anal. Mus. Nac. Buenos Aires II. **4:**291. 1902.

Low, nearly prostrate; joints flattened, orbicular to broadly obovate, 10 to 12 cm. long, 7 to 10 cm. broad, dull green; spines slender, twisted, one elongated and 1 to 5 cm. long, the others much shorter, all white; glochids brownish; flowers from the lateral and marginal areoles, citron-yellow;

ovary 3 to 3.5 cm. long, with very many areoles bearing numerous glochids; style thick; stigma-lobes 8 to 10, greenish white; fruit reddish, clavate, 4.5 cm. long, with a depressed umbilicus; seeds small, 3 to 3.5 mm. broad.

Type locality: Argentina, between Río Negro and Río Colorado.
Distribution: Southern Argentina.

According to Dr. Spegazzini, this species is not near to any of the known South American species, but resembles somewhat the North American species *O. microdasys* and *O. basilaris*. We know it only from the description.

OPUNTIA CALANTHA Griffiths, Bull. Torr. Club **43**: 524. 1916.

A low, creeping, prostrate plant 15 cm. high, one meter in diameter; joints obovate, narrowed above and below, inequilateral, 11 cm. long, 4 cm. broad, tuberculate-wrinkled, mostly deep green; areoles 1 to 1.5 mm. long, obovate, at first tawny, turning gray; leaves small, subulate, cuspidate, red. 1 mm. long; glochids yellow; spines 5 to 10, up to 5 mm. long; flowers carmine; fruit globular, 1.5 cm. in diameter.

Recorded as probably of South American origin and usually distributed as *Opuntia microdisca*, but from which it is said to differ very much. The plant is known to us only from the description of cultivated specimens.

Series 9. STRIGILES.

The series consists of a single species, native of Texas. It is a low, bushy plant with large joints bearing many areoles, these close together, each with several acicular, reddish brown spines; the fruit is small.

131. Opuntia strigil Engelmann, Proc. Amer. Acad. **3**: 290. 1856.

Suberect, 6 dm. high; joints orbicular to obovate, 10 to 12.5 cm. long; areoles close together, prominent; spines 5 to 8, spreading, many of them appressed to the joint and deflexed, red to reddish brown with lighter tips, the longer ones 2.5 cm. long; glochids numerous; flowers unknown; fruits small, nearly globular, 12 mm. in diameter, truncate, red; areoles on fruit very small; seeds 3 mm. broad.

FIG. 171.—Opuntia strigil. X0.4.

Type locality: In crevices of limestone rock, between the Pecos River and El Paso, Texas.
Distribution: Texas.

A rare plant, first collected by Charles Wright in 1851. Engelmann says in the Mexican Boundary Report that it was also collected by Wright and Bigelow, but there is no mention of it in his report on Bigelow's plants, nor do we find specimens in the Engelmann herbarium, so that it would appear that this reference to Bigelow was a mistake. Bigelow, it is true, crossed the River Pecos, on which the type was found, but it was well up in New Mexico and not in Texas, where it was crossed by Charles Wright. It was more recently collected by Nealley somewhere in Texas. The place of collection by Wright and the later one by Nealley are very indefinitely indicated on the labels accompanying the specimens.

Illustration: Cact. Mex. Bound. pl. 67.

Figure 171 is copied from the illustration above cited.

Series 10. SETISPINAE.

Bushy or depressed species, with tuberous or thickened roots, broad, flat, thin joints, and elongated, acicular, brown spines which fade whitish; their fruits are large and juicy. We recognize six species, natives of the south central and southwestern United States and northern Mexico. They approach the *Tortispinae* on the one hand and the *Phaeacanthae* on the other.

KEY TO SPECIES.

Joints elongated.. 132. *O. megarhiza*
Joints obovate to orbicular.
  Fruit small, 2 cm. long or less... 133. *O. ballii*
  Fruit large 2.5 to 6 cm. long.
    Flowers red to purple.. 134. *O. pottsii*
    Flowers yellow.
      Areoles large, more or less elevated on old joints; joints glaucous, purplish about the
        areoles... 135. *O. setispina*
      Areoles small; mature joints green throughout.
        Joints usually orbicular; seeds 5 to 6 mm. broad...................... 136. *O. mackensenii*
        Joints obovate; seeds 4 mm. broad or less........................... 137. *O. tenuispina*

132. Opuntia megarhiza Rose, Contr. U. S. Nat. Herb. **10**: 126. 1906.

Roots long and thickened, sometimes 3 to 6 dm. long, 5 to 6 cm. in diameter; stems low, 2 to 3 dm. high, much branched; lower joints elongated, 2 to 3 dm. long, cuneate below, thin, 3 cm. broad; lateral joints appearing along the margins of the older joints and often, if not always, in the same plane; spines 2 to 4, acicular, 1 to 2.5 cm. long, brown; leaves minute; flowers lemon-yellow, often tinged with rose, 5 cm. broad; petals about 13, obovate, mucronately tipped; stigma-lobes 7, greenish; ovary clavate, 3 cm. long; fruit and seeds unknown.

Type locality: Alvarez, Mexico.
Distribution: San Luis Potosí, Mexico.
This species is not very closely related to the other species of this series, but it is referred here on account of its very slender spines.

133. Opuntia ballii Rose, Contr. U. S. Nat. Herb. **13**: 309. 1911.

Plants low, spreading; joints obovate, 6 to 10 cm. long, thickish, pale green, glaucous; spines 2 to 4, brownish, a little flattened, usually ascending or erect, the larger ones 4 to 7 cm. long; glochids conspicuous; fruit small, about 2 cm. long, clavate, glaucous, spineless; seeds thick, 3.5 mm. broad.

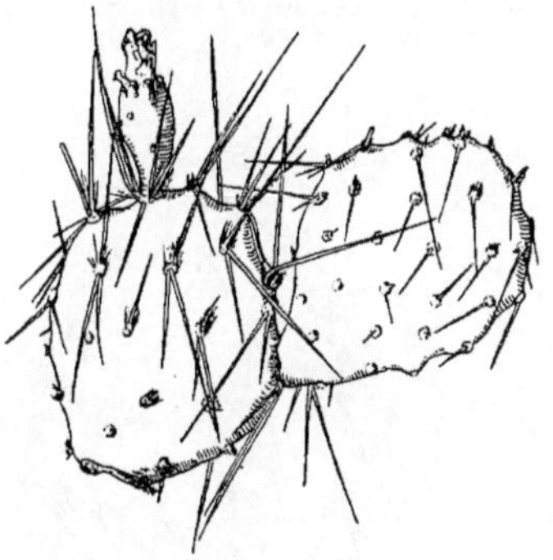

FIG. 172.—O. ballii. Part of type. X0.5.

Type locality: Pecos, Reeves County, Texas.
Distribution: Western Texas.
Wooton and Standley in their Flora of New Mexico refer this species to *Opuntia fili-pendula*, but *O. ballii* grows in a different habitat, has smaller fruit, stouter and erect spines, and different areoles; it grows on the dry mesa beyond Pecos, Texas.
Illustrations: Contr. U. S. Nat. Herb. **13**: pl. 64.
Figure 172 is copied from the illustration above cited.

134. Opuntia pottsii Salm-Dyck, Cact. Hort. Dyck. 1849. 236. 1850.

    Opuntia filipendula Engelmann, Proc. Amer. Acad. **3**: 294. 1856.

Low, spreading plant, 3 dm. high or less, from thickened tuberous roots 2 to 3 cm. in diameter, these sometimes moniliform; joints broadly obovate, 3.5 to 12 cm. long, pale green to bluish; areoles few, either small or large; spines confined to the upper and marginal areoles, 1 or 2, slender, 2 to 4 cm. long, usually white but sometimes purplish; glochids yellow, usually few but sometimes abundant; flowers large, 6 to 7 cm. broad, deep purple; ovary slender, 3 to 3.5 cm. long, with only a few scattered areoles; fruit spineless.

Type locality: Near Chihuahua City, Mexico.
Distribution: Central Chihuahua, Mexico, to Texas and New Mexico.
This species was described by Prince Salm-Dyck in 1850 from material collected by John Potts, who was manager of the mint at Chihuahua and who sent many cacti to F. Scheer at Kew between 1842 and 1850. No types of his species seem to have been retained.

In 1885 C. G. Pringle again collected this species near Chihuahua City and it was distributed as *O. filipendula*, and there Coulter leaves Pringle's specimen (Cont. Nat. Herb. 3: 428). Dr. E. Palmer collected an abundance of material in 1908 which enabled us to reestablish *O. pottsii*, which Coulter omits and Schumann lists under unknown species.

If these Chihuahua specimens are the same as the Texas plants, as Coulter believed and as we regard them, then *Opuntia filipendula* must give place to the older name of Salm-Dyck.

Illustrations: Cact. Mex. Bound. pl. 68; Förster, Handb. Cact. ed. 2. f. 10, 131; Suppl. Dict. Gard. Nicholson 2: f. 605; W. Watson, Cact. Cult. f. 81, all as *Opuntia filipendula*.

Figure 173 shows a joint of a plant collected by Dr. Rose in the valley of the Rio Grande below El Paso, Texas, in 1913.

135. Opuntia setispina Engelmann in Salm-Dyck, Cact. Hort. Dyck. 1849. 239. 1850.

Stem branching and spreading, sometimes 9 to 12 dm. broad, with some of the branches composed of 3 or 4 joints, erect and 6 dm. high; joints deep bluish green, somewhat glaucous, often purplish at the areoles, sometimes more or less tinged with purple throughout, obovate to orbicular, 5 to 15 cm. in diameter; leaves minute, subulate; spines 1 to 6 from an areole, white, 2 to 3 cm. long; glochids yellow, very conspicuous on old joints; flowers yellow; fruit purplish, about 4 cm. long.

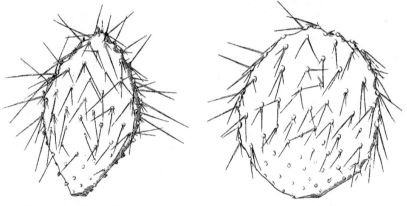

FIG. 173.—Opuntia pottsii. X0.4.   FIG. 174.—Opuntia setispina. X0.4.

Type locality: Pine woods in the mountains west of Chihuahua, Mexico (*fide* Engelmann, Proc. Amer. Acad. 3: 294. 1856).

Distribution: Western Chihuahua, Mexico.

This species has long been known only from the type specimens; but in 1908 Dr. Rose visited western Chihuahua, where this species is quite common; our description is based largely upon the specimens he then collected.

Figure 174 represents a joint of the plant collected by Dr. Rose near Miñaca, Chihuahua, in 1908.

136. Opuntia mackensenii Rose, Contr. U. S. Nat. Herb. 13: 310. 1911.

Plants low, with thick, tuberous roots, spreading, usually resting on the edges of the joints, but some of the branches often erect; joints orbicular to obovate, 10 to 20 cm. long, rarely broader than long, pale and glaucous when young, deep green when older; areoles small, the lower ones without spines, the upper ones with 1 to 4 spines; spines white or brown, or brown at base and white above, somewhat flattened and twisted, slender, 5 cm. long or less; glochids brown; flowers of medium size, 7 to 8 cm. broad, yellow with a reddish brown center; stigma-lobes 7 to 9, white; fruit spineless, 4 to 6 cm. long, truncate or nearly so at apex, rose-purple; seeds suborbicular, 5 to 6 mm. broad, acute on the margin.

Type locality: Near Kerrville, Texas.
Distribution: Kerr County, Texas.
Illustrations: Contr. U. S. Nat. Herb. **13**: pl. 67; Plant World **19**: 142. f. 1; 143. f. 2, the last as *O. macrorhiza.*

Figure 175 is from a photograph of the type plant from near Kerrville, Texas.

FIG. 175.—Opuntia mackensensii.

137. Opuntia tenuispina Engelmann, Proc. Amer. Acad. **3**: 294. 1856.

Opuntia minor C. Mueller in Walpers, Ann. Bot. **5**: 50. 1858.

Low and spreading, but becoming 3 dm. high; joints obovate, attenuate at base, 7 to 15 cm. long, light green; leaves very slender, 4 mm. long or less; spines 1 to 3 from an areole, slender, usually white but sometimes brownish, 3 to 5 cm. long, the upper spines erect or spreading; glochids brown; flowers yellow, 6 to 7.5 cm. broad; ovary with numerous areoles filled with brown wool and brown glochids; fruit oblong, 2.5 to 4 cm. long, with a deep umbilicus; seeds 4 mm. broad or less, very irregular.

Type locality: Sand hills near El Paso, Texas.

Distribution: Southwestern Texas and adjacent parts of Mexico and New Mexico, apparently extending to Arizona.

Engelmann says that this plant grows with *O. phaeacantha,* but is readily distinguished from the latter by its spines and fruit. Cultivated plants and herbarium specimens closely resemble *O. phaeacantha.*

Illustrations: Cact. Mex. Bound. pl. 75, f. 14; N. Mex. Agr. Exp. Sta. Bull. **78**: pl. [15].

Plate XXIII, figure 3, represents a joint of the plant collected by Dr. Rose near El Paso, Texas, in 1913.

Series 11. PHAEACANTHAE.

Bushy or depressed species, with relatively large, flat, persistent joints, the subulate, usually stout spines brown at least at the base, or in some species nearly white. The series is composed of about fifteen species, natives of the south central and southwestern United States, northern and central Mexico.

KEY TO SPECIES.

More or less bushy plants.
  Joints thin; spines, when present, very long and confined to the upper and middle areoles.
    Spines dark brown, stout, rigid.
      Plant pale green to purplish; spines up to 12 cm. long.......................... 138. *O. macrocentra*
      Plant dull dark green; spines 6 cm. long or less............................... 139. *O. tardospina*
    Spines pale brown, flexible or subulate.
      Usually abundantly spiny.. 140. *O. gosseliniana*
      Usually spineless or some areoles with 1 setaceous deflexed spine.............. 141. *O. santa-rita*
  Joints thick; spines not confined to the upper and middle areoles.
    Joints relatively small, seldom over 15 cm. broad; plants relatively low.
      Joints narrowly obovate, about twice as long as ≥ide.......................... 142. *O. angustata*
      Joints broadly obovate to orbicular.
        Flowers yellow.
          Spines subulate, brown at least in part.
            Plant light green.. 143. *O. atrispina*
          Plant bluish green or grayish green.
            Plant erect, 2 meters high or less............................... 144. *O. azurea*
            Plant bushy, rarely over 1 meter high........................... 145. *O. phaeacantha*
            Plant prostrate.. 146. *O. mojavensis*
         Spines acicular, nearly white.. 147. *O. covillei*
        Flowers magenta.. 148. *O. vaseyi*
    Joints relatively large, mostly over 15 cm. broad; plants relatively tall.
      Spines clear brown nearly throughout.. 149. *O. occidentalis*
      Spines nearly white above or throughout.
        Spines with dark brown bases... 150. *O. engelmannii*
        Spines whitish throughout.. 151. *O. discata*
Small creeping plants.. 152. *O. rastrera*

138. Opuntia macrocentra Engelmann, Proc. Amer. Acad. **3**: 292. 1856.

Somewhat bushy, with ascending branches, 6 to 9 dm. high; joints orbicular to oblong, or sometimes broader than long, 10 to 20 cm. long, often bluish or purplish, sometimes spineless but usually bearing spines at the uppermost areoles; spines 1 or 2, rarely 3 together, usually brownish or black but sometimes white above, slender, erect or porrect, 4 to 7 cm. long; flowers yellow, often drying red, 7.5 cm. broad; sepals ovate, acuminate; ovary with few areoles, these bearing brown glochids; filaments very short; fruit 3 to 6 cm. long, purple; seeds 4 to 4.5 mm. broad.

Type locality: Sand hills on the Rio Grande near El Paso, Texas.

Distribution: Western Texas to Eastern Arizona and Chihuahua, Mexico.

This species, especially the forms that have bluish and purplish joints, are very showy. Seedlings sometimes produce long, silky hairs from the areoles, in this respect resembling the *Criniferae*.

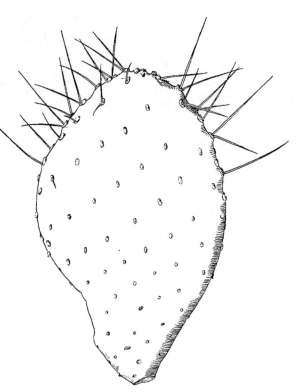

FIG. 176.--Opuntia macrocentra.   X0.5.

Illustrations: Cact. Mex. Bound. pl. 75, f. 8; N. Mex. Agr. Exp. Sta. Bull. **78**: pl. [8].

Figure 176 represents a joint of the plant collected by Dr. Rose near the Rio Grande in New Mexico, northwest of El Paso, Texas, in 1913.

139. Opuntia tardospina Griffiths, Rep. Mo. Bot. Gard. **22**: 34. 1912.

Roots fibrous; low, spreading plant, the joints usually resting on the ground; joints orbicular to obovate, 16 to 24 cm. long; areoles large, usually distant, often 4 cm. apart; spines usually wanting except from the upper areoles and along the upper margin, usually single, sometimes 2 from an areole, 4 to 5 cm. long, brown, but lighter towards the apex; glochids numerous, brown, persistent; fruit red, 6 cm. long; seeds 5 mm. broad, acute on the margin.

Type locality: Near Lampasas, Texas.
Distribution: Eastern Texas.
Illustrations: Rep. Mo. Bot. Gard. **22**: pl. 11, in part; pl. 15.

Figure 177 represents a joint of the plant collected by Albert Ruth in 1912, north of Dallas, Texas.

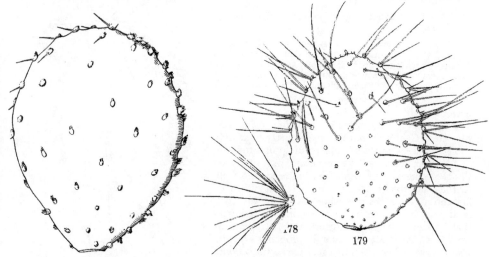

FIG. 177.—O. tardospina. ×0.5.   FIGS. 178, 179.—Cluster of spines and joint of O. gosseliniana. ×0.4.

140. Opuntia gosseliniana Weber, Bull. Soc. Acclim. France **49**: 83. 1902.

One meter or more high, branching from the base, the old trunk often bearing numerous, long, acicular spines; joints usually red or purplish, usually very thin, as broad as or broader than long, sometimes 2 dm. broad; lower and sometimes all the areoles without spines; spines porrect or nearly so, generally 1, sometimes 2, rarely 3 from an areole, 4 to 5 or even 10 cm. long, brown, usually weak; glochids brown, numerous, forming on old joints very large clusters; fruit 4 cm. long, without spines but bearing numerous brown glochids at the areoles, with a depressed umbilicus.

Type locality: Coast of Sonora on the Gulf of California.
Distribution: Sonora and Lower California, Mexico.

This species was placed tentatively in the *Pubescentes* by Schumann, although always glabrous; but it belongs better in the *Phaeacanthae*. In some of its phases it resembles *O. macrocentra*. It is a very showy species and worthy of a place in any collection.

Illustrations: Monatsschr. Kakteenk. **17**: 69.

Figure 179 represents a joint of the plant collected at Hermosillo in Sonora, by Rose, Standley, and Russell in 1910; figure 178 shows a cluster of spines from a trunk areole.

141. **Opuntia santa-rita** (Griffiths and Hare) Rose, Smiths. Misc. Coll. **52**: 195.   1908.

    Opuntia chlorotica santa-rita Griffiths and Hare, N. Mex. Agr. Exp. Sta. Bull. **60**: 64.   1906.
    Opuntia shreveana C. Z. Nelson, Galesburg Register, July 20, 1915.

    Compact plant, 6 to 14 dm. high, with a very short trunk; joints orbicular or a little broader than long, bluish green but deep purple about the areoles and margins; areoles 1.5 cm. apart, bearing numerous chestnut-brown glochids and occasionally a brown spine; flowers very handsome, deep yellow, 6 to 7 cm. broad; ovary purple, oblong.

    Type locality: Selero Mountains, Arizona.

    Distribution: Southeastern Arizona.

    This species is one of the most ornamental of the opuntias, and although it does not grow well in greenhouse cultivation, it would doubtless flourish in the Southwest, where it could be given conditions similar to its wild surroundings.

    Illustrations: Smiths. Misc. Coll. **52**: pl. 15; Plant World **11**[10]: f. 6, this last as *Opuntia chlorotica;* Journ. Inter. Gard. Club **3**: facing page 5, as *O. chlorotica santa-rita.*

    Plate XXIV, figure 1, is from a photograph taken by Dr. MacDougal of a plant near Surritas, Arizona, in 1906.

142. **Opuntia angustata** Engelmann, Proc. Amer. Acad. **3**: 292.   1856.

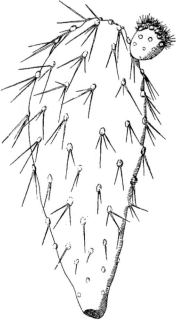

    Ascending to erect; joints narrow, 15 to 25 cm. long, rounded above, gradually narrowing downward; areoles distant, often 2.5 cm. apart, large, oblong; spines sharply angled, straw-colored or whitish but with brown bases, 2.5 to 3.5 cm. long; glochids brown; fruit obovoid, 2.5 to 3 cm. long.

    Type locality: Bottoms, Bill Williams Fork, Arizona.

    Distribution: Recorded as extending from New Mexico to California, but known definitely to us only from central Arizona, perhaps extending north to Utah.

    Engelmann's *Opuntia angustata* was based on three specimens, one from New Mexico, one from Arizona, and one from California.   He stated that the first and last were prostrate, while the second was erect. A study of his specimens and descriptions indicates that he had three species before him.   The first is from Zuni, New Mexico, and is probably *Opuntia phaeacantha.*   The California specimen is the *Opuntia magenta* Griffiths, which is probably the same as *O. vaseyi*, while the suberect plant from the bottoms of the Bill Williams River we have allowed to stand for *O. angustata.*   Wooton and Standley (Contr. U. S. Nat. Herb. **19**: 447.   1915) suggest that the two fruits illustrated by Engelmann in connection with this species may belong to two species of *Cylindropuntia.*

FIG. 180.—Opuntia angustata.

    This plant was first collected by J. M. Bigelow, February 4, 1854.

    Illustrations: Pac. R. Rep. **4**: pl. 7, f. 3, 4.

    Figure 180 is copied from figure 3 of the illustrations above cited.

143. **Opuntia atrispina** Griffiths, Rep. Mo. Bot. Gard. **21**: 172.   1910.

    Usually low and spreading, sometimes 2 meters in diameter, but sometimes the central branches nearly erect and 6 dm. high; joints rather small, nearly orbicular, 10 to 15 cm. in diameter, light green, sometimes a little glaucous; lower areoles spineless; spines from the upper areoles 2 to 4, the principal ones spreading, flattened, dark brown, almost black at base, much lighter above; glochids at first yellow or yellowish, but soon changing to brown; flowers described as yellow, changing to orange; fruit reddish purple.

1. Plant of *Opuntia santa-rita*.   2. Plant of *Opuntia discata*.

Type locality: Near Devil's River, Texas.

Distribution: Type locality and vicinity.

This plant is abundant between Del Rio, Texas, and Devil's River, being one of the two commonest species in that region.

Illustrations: Rep. Mo. Bot. Gard. **21**: pl. 26, in part.

Plate XXV, figure 1, represents a flowering joint of the plant collected near Devil's River, Texas, by Dr. Rose in 1913.

FIG. 181.—Opuntia azurea, Zacatecas, Mexico.

144. Opuntia azurea Rose, Contr. U. S. Nat. Herb. **12**: 291. 1909.

Compact, upright, with a single trunk, or branching from the base and more or less spreading; joints orbicular to obovate, 10 to 15 cm. in diameter, pale bluish green, glaucous; areoles about 2 cm. apart, the lower ones spineless, the upper ones with 1 to 3 rather stout spines; spines, at least when old, almost black, unequal, the longer ones 2 to 3 cm. long, more or less reflexed; glochids numerous, brown; petals 3 cm. long, deep yellow, with crimson claw, but in age pink throughout; filaments greenish or almost white; stigma-lobes pale green; fruit dull crimson, subglobose to ovoid, spineless, truncate, juicy, edible.

Type locality: Northeastern Zacatecas, Mexico.

Distribution: Zacatecas and probably Durango.

Illustrations: Contr. U. S. Nat. Herb. **12**: pl. 24; also f. 33.

Figure 181 is from a photograph by F. E. Lloyd of the type plant; figure 182 represents joints of the plant collected by Albert de Lautreppe near Zacatecas, Mexico, in 1904.

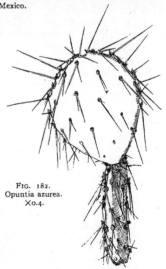

FIG. 182.
Opuntia azurea.
×0.4.

145. **Opuntia phaeacantha** Engelmann in Gray, Mem. Amer. Acad. **4**: 52.   1849.

Opuntia phaeacantha brunnea Engelmann, Proc. Amer. Acad. **3**: 293.   1856.
Opuntia phaeacantha major Engelmann, Proc. Amer. Acad. **3**: 293.   1856.
Opuntia phaeacantha nigricans Engelmann, Proc. Amer. Acad. **3**: 293.   1856.
Opuntia camanchica Engelmann and Bigelow, Proc. Amer. Acad. **3**: 293.   1856.
Opuntia chihuahuensis Rose, Contr. U. S. Nat. Herb. **12**: 291.   1909.
Opuntia toumeyi Rose, Contr. U. S. Nat. Herb. **12**: 402.   1909.
Opuntia blakeana Rose, Contr. U. S. Nat. Herb. **12**: 402.   1909.
Opuntia zuniensis Griffiths, Bull. Torr. Club **43**: 86.   1916. (From the description.)

Low, usually prostrate, with some branches ascending; joints usually longer than broad, 10 to 15 cm. long; areoles rather remote, the lower ones often spineless; spines 1 to 4, those on the sides of the joints more or less reflexed, somewhat flattened, usually rather stout, brown, sometimes darker at base, or often nearly white throughout, the longer ones 5 to 6 cm. long; glochids numerous, yellow to brown; flowers 5 cm. broad, yellow; ovary short; fruit 30 to 35 mm. long, much contracted at base.

Type locality: About Santa Fé and on the Rio Grande, New Mexico.

Distribution: Texas to Arizona and Chihuahua.

We have referred to *Opuntia phaeacantha* the common low, bushy *Opuntia* with small joints, brown spines, and yellow flowers of the Southwest; we formerly regarded it as composed of several species, and others have followed our lead; but we are unable to draw any distinct lines after a study of much additional herbarium and greenhouse material.   Dr. Rose has collected a large series of specimens from the Southwest, especially from the type localities, but his specimens seem to bridge over differences which before seemed tangible; cited differences appear to be racial rather than specific.

Opuntia blakeana, which is found west of the Rocky Mountains, one would expect to be different.   It is characterized by small obovate joints, rather short spines, small yellow flowers purple at center.

Opuntia chihuahuensis, which was first described from Mexican specimens, if it belongs here, is in the southern range of *O. phaeacantha*.   It, too, has yellow flowers with red centers, rather large joints, and long, slender spines.   Mr. Wooton is of the opinion that to *O. chihuahuensis* is to be referred the common, low, brown species from El Paso, which includes the specimens of G. R. Vasey, which Coulter called *Opuntia mesacantha oplocarpa.* This long-spined form extends north throughout eastern New Mexico to southeastern Colorado.   With the latter form Mr. Wooton believes *Opuntia camanchica* belongs.   If we take this broad view of the limits of this species we are forced to include *Opuntia toumeyi*, although it is much larger than *O. blakeana*, and was considered by Dr. Rose to be different.

Opuntia mesacantha sphaerocarpa Wooton and Standley (Contr. U. S. Nat. Herb. **19**: 446. 1915) is a mistake, *O. mesacantha oplocarpa* being intended.

Opuntia rubrifolia Engelmann in Coulter (Contr. U. S. Nat. Herb. **3**: 424. 1896), from St. George, Utah, belongs in this series if E. W. Nelson's No. 156, from the same place has been properly determined as such.   The type specimen of *O. rubrifolia* has, apparently, been lost.

The following varieties of *Opuntia camanchica* have been offered by Haage and Schmidt in their catalogues: *albispina* (Trade Seed Cat. 104. 1911–1912); *orbicularis*, *rubra,* and *salmonea* (all in Haupt-Verzeichnis **1908**: 228. 1908).   Under *O. camanchica* has been mentioned also variety *luteo-staminea* (Cat. Darrah Succ. Manchester 53. 1908).

Opuntia eocarpa Griffiths (Proc. Biol. Soc. Washington **29**: 11. 1916), also *O. recurvospina* Griffiths (Proc. Biol. Soc. Washington **29**: 12. 1916) and possibly *O. superbospina* Griffiths (Proc. Biol. Soc. Washington **29**: 13. 1916) and *O. caesia* Griffiths (Proc. Biol. Soc. Washington **29**: 13. 1916) are of this relationship.

*Opuntia microcarpa** Engelmann (Emory, Mil. Reconn. 158. f. 7. 1848) and *O. violacea* Engelmann (Emory, Mil. Reconn. 158. f. 8. 1848) were described from drawings brought

---

*Since the above was written Dr. Griffiths (Bull. Torr. Club, **43**: 527) has published a detailed account of this species, which he regards as distinct; it inhabits southern Arizona.

M. E. Eaton del.

1. Flowering joints of *Opuntia atrispina*.   2. Flowering joint of *Opuntia phaeacantha*.
3. Upper part of joint of *Opuntia engelmannii*.   (All natural size.)

back from the Southwest by W. H. Emory. They can never be critically identified, but are probably of this relationship.

Illustrations: Engler and Prantl, Pflanzenfam. 3⁶ᵃ: f. 57, C; Förster, Handb. Cact. ed. 2. f. 141; Illustr. Fl. 2: f. 2530; ed. 2. 2: f. 2989; Pac. R. Rep. 4: pl. 9, f. 1 to 5; pl. 22, f. 12 to 15; Wiener Illustr. Gartenz. 10: f. 115, all as *Opuntia camanchica;* N. Mex. Agr. Exp. Sta. Bull. 78: pl. [7], as *Opuntia chihuahuensis;* Contr. U. S. Nat. Herb. 12: pl. 55, as *Opuntia blakeana;* Cact. Mex. Bound. pl. 75, f. 9 to 13.

Plate xxv, figure 2, represents a flowering joint of a plant sent from Tucson, Arizona, in 1916, by Dr. Mac-Dougal.

146. Opuntia mojavensis Engelmann, Proc. Amer. Acad. 3: 293. 1856.

Prostrate, with suborbicular joints; pulvini remote, with large yellow bristles; spines 2 to 6, stout and annulate, acutely angular and compressed, more or less curved, reddish brown, paler toward tip, 2.5 to 6 cm. long, 1 to 3 smaller, slenderer, pale ones added below; fruit oblong, 4.5 cm. long.

Type locality: On the Mojave, west of the Colorado, California.

Distribution: Known only from the type locality.

FIG. 183.—Opuntia covillei. X0.4.

The fragmentary type specimen has been examined; we have been unable to refer any other specimens to this species, which is thus very imperfectly understood.

Illustration: Pac. R. Rep. 4: pl. 9, f. 6 to 8.

147. Opuntia covillei Britton and Rose, Smiths. Misc. Coll. 50: 532. 1908.

> *Opuntia megacarpa* Griffiths, Rep. Mo. Bot. Gard. 20: 91. 1909.
> *Opuntia rugosa* Griffiths, Proc. Biol. Soc. Washington 27: 27. 1914.

Bushy plants, usually growing in dense thickets; joints orbicular to obovate, 10 to 20 cm. long or more, pale green, sometimes purplish, slightly glaucous; areoles 2 to 4 cm. apart; spines several from an areole, slender, unequal, the longest ones 6 cm. long, white when young, brownish when old; flowers large, yellow.

Type locality: San Bernardino, California.

Distribution: Interior valleys of southern California.

FIG 184.—Opuntia covillei.

Opuntia covillei and *O. vaseyi* grow in the same valleys, often in adjoining colonies, and while hybrids may occur, the two species could easily be distinguished. When growing in conjunction, *O. covillei* is considerably taller, has joints of different color, and has yellow flowers. It has doubtless generally passed as *Opuntia occidentalis*, but that is a much larger, stouter plant, with strong, more or less flattened spines, and is common along the coast.

Figure 183 represents a joint of the plant sent by Dr. MacDougal from Elsinore, California, in 1913; figure 184 is from a photograph of a specimen collected by Mr. S. B. Parish from near the type locality in 1916.

148. Opuntia vaseyi (Coulter) Britton and Rose, Smiths. Misc. Coll. **50**: 532. 1908.

> *Opuntia mesacantha vaseyi* Coulter, Contr. U. S. Nat. Herb. **3**: 431.   1896.
> *Opuntia rafinesquei vaseyi* Schumann, Gesamtb. Kakteen 717.   1898.
> *Opuntia humifusa vaseyi* Heller, Cat. N. Amer. Pl. ed. 2. 8.   1900.
> *Opuntia magenta* Griffiths, Rep. Mo. Bot. Gard. **19**: 268.   1908.
> *Opuntia rubiflora* Davidson, Bull. South. Calif. Acad. **15**: 33.   1916.

Plants low, the lower stems spreading at base, but some of the branches erect and 4 to 7 joints high; joints thick, small (usually 10 to 12 cm. long), ovate, pale green, somewhat glaucous; areoles rather large, 2 to 3 cm. apart, bearing 1 to 3 spines; spines porrect, usually short (rarely 2 cm. long), grayish brown or bright brown, whitish or yellowish towards the tips, somewhat flattened; young joints bright green, thickish, bearing short purplish leaves and a single brownish spine from an areole; flowers deep salmon, almost a red-salmon, from the very first; ovary globular to shortly oblong; areoles few, mostly towards the top of the ovary, spineless but with a few brown glochids; fruit globular to shortly oblong, 4 to 5 cm. long, deep purple, truncate at apex, with few areoles, the pulp sweetish but hardly edible; umbilicus broadly depressed.

Type locality: Cited as Yuma, Arizona, presumably erroneously.

Distribution: San Bernardino and Orange Counties, southern California.

Even from a moving train this species is distinguishable from its relatives by the color of its flowers. It forms great thickets along the Southern Pacific Railroad north of Los Angeles, either alone or interspersed with one or more other species, and it is also common in the San Bernardino Valley toward the Cajon Pass where it forms great thickets either alone or with *Opuntia covillei*. Considerable quantities were seen also on hills near Riverside, and it was found cultivated in the cactus garden at Riverside and in the Soldiers' Home Grounds at Santa Monica.

Illustration: Bull. South. Calif. Acad. **15**: 32, as *Opuntia rubiflora*.

Figure 185 represents a joint of the plant collected by Dr. Rose at Fernando, California, in 1908.

149. Opuntia occidentalis Engelmann, and Bigelow, Proc. Amer. Acad. 3: 291.   1856.

> *Opuntia lindheimeri occidentalis* Coulter, Contr. U. S. Nat. Herb. **3**: 421.   1896.
> *Opuntia engelmannii occidentalis* Engelmann in Brewer and Watson, Bot. Calif. **1**: 248.   1876.*
> *Opuntia demissa* Griffiths, Rep. Mo. Bot. Gard. **22**: 29.   1912.

Erect or spreading, often 1 meter high or more, forming large thickets; joints large, obovate to oblong, 2 to 3 dm. long; areoles remote; spines 2 to 7, stout, unequal, the longest ones 4 to 5 cm. long, more or less flattened, brown or nearly white, sometimes wanting; shorter spines often white; glochids often prominent, brown; flowers yellow, large, including the ovary often 10 to 11 cm. long; fruit large, purple.

Type locality: Western slopes of the California Mountains, between San Diego and Los Angeles.

Distribution: Southwestern California and northern Lower California and adjacent islands.

Fig. 185.—Opuntia vaseyi.
×0.5.

In their description of this species, Engelmann and Bigelow state that it was found on the western slope of the California Mountains near San Diego and Los Angeles. In the

---

*Coulter refers this name to Pac. R. Rep. **4**: errata, 3, 1856, but no formal name is published there.

Engelmann herbarium are the two original sheets. One of these comes from the "Mountain Valleys of San Pasquel and Santa Isabel," northeast of San Diego. This consists of a single flower and a small piece of a joint containing three bunches of spines; we doubt if this can be identified. The other comes from near Los Angeles and consists of a large pad and fruit with seeds. The spines are dark brown or nearly black. This specimen appears to be the one figured in the Pacific Railroad Report and may very properly be taken as the type of the species.

There is much uncertainty regarding the range of this species, some referring it to the interior valleys of California. An examination, however, of the type material, and a study of the living plants in southern California by Dr. Rose, convince us that the coastal opuntias can not all be referred to *O. littoralis* as is sometimes done, but a part belongs to *O. occidentalis*. The limits of the latter species, and its distribution, are not well defined.

Of this relationship is to be considered *Opuntia semispinosa* Griffiths (Bull. Torr. Club **43**: 89. 1916), which the author describes as a common, conspicuous species in the coastal region of California.

Illustrations: N. Mex. Agr. Exp. Sta. Bull. **60**: pl. 3, f. 2; Pac. R. Rep. **4**: pl. 7, f. 1, 2; pl. 22, f. 10; Rep. Mo. Bot. Gard. **22**: pl. 8, this last as *Opuntia demissa*.

Figure 186 is from a plant collected on Santa Catalina Island, California, by Mr. S. B. Parish in 1916.

FIG. 186.—Opuntia occidentalis.

150. **Opuntia engelmannii** Salm-Dyck in Engelmann, Bost. Journ. Nat. Hist. **6**: 207. 1850.

> Opuntia engelmannii cyclodes Engelmann, Proc. Amer. Acad. **3**: 291. 1856.
> Opuntia lindheimeri cyclodes Coulter, Contr. U. S. Nat. Herb. **3**: 422. 1896.
> Opuntia dillei Griffiths, Rep. Mo. Bot. Gard. **20**: 82. 1909.
> Opuntia arizonica Griffiths, Rep. Mo. Bot. Gard. **20**: 93. 1909.
> Opuntia wootonii Griffiths, Rep. Mo. Bot. Gard. **21**: 171. 1910.
> Opuntia cyclodes Rose, Contr. U. S. Nat. Herb. **13**: 309. 1911.
> Opuntia gregoriana Griffiths, Rep. Mo. Bot. Gard. **22**: 26. 1912.
> Opuntia valida Griffiths, Proc. Biol. Soc. Washington **27**: 24. 1914.
> Opuntia confusa Griffiths, Proc. Biol. Soc. Washington **27**: 28. 1914.
> Opuntia magnarenensis Griffiths, Proc. Biol. Soc. Washington **29**: 9. 1916.
> Opuntia expansa Griffiths, Proc. Biol. Soc. Washington **29**: 14. 1916.

Originally described as erect and up to 2 meters high, but more properly a widely spreading bush, usually without a definite trunk; joints oblong to orbicular, 2 to 3 dm. long, thick, pale green; areoles distant, becoming large and bulging; spines usually more or less white, with dark red or brownish bases and sometimes with black tips, usually 3 or 4, sometimes only 1, or entirely wanting from the lower areoles, but on old joints 10 or more, usually somewhat porrect or a little spreading, but never reflexed, the larger ones much flattened, the longest one 5 cm. long; leaves subulate, about 15 mm. long; glochids numerous, brown with yellowish tips; flowers large, yellow; fruit 3.5 to 4 cm. long, red; seeds small, 3 to 4 mm. broad.

Type locality: From El Paso to Chihuahua.

Distribution: Chihuahua, Durango, Sonora, Arizona, New Mexico, and Texas.

An examination of the plant collected by Wislizenus (No. 223) north of Chihuahua, now in the herbarium of the Missouri Botanical Garden and labeled by Dr. Engelmann as *O. engelmannii* Salm-Dyck, shows that this species is of Schumann's series *Fulvispinosae* (our series *Phaeacanthae*) rather than series *Tunae*.

Opuntia engelmannii has been more confused than any other species of *Opuntia*. Salm-Dyck, who first studied the species, doubtless had but a single specimen before him, and this or a duplicate is now in the herbarium of the Missouri Botanical Garden. This type specimen came from near Chihuahua City, from which place Dr. Rose has collected identical material. Dr. Engelmann, who published Salm-Dyck's name, described the plant as erect and 5 to 6 feet high, giving its range from Chihuahua City to Texas. These remarks of his were doubtless based on notes of Dr. Wislizenus, who collected the type, and must have included more than one species; as Engelmann says it is both cultivated and wild, the cultivated plants doubtless referring to some of the many forms grown about towns and ranches. In 1852 Engelmann extends the distribution of the species westward to the Pacific Ocean, referring especially to a San Diego specimen. In 1856 he refers here his previously described species *O. lindheimeri*, and extends the range eastward to the mouth of the Rio Grande and to lower Mexico. Coulter brought all this material together under *O. lindheimeri* and four varieties.

An examination of herbarium and greenhouse specimens shows that at least half a dozen species have been passing under the name of *O. engelmannii*. While certain varieties and specimens are evidently to be excluded from the species, we are still uncertain as to its specific limits. It is quite common about Chihuahua City and extends to Monterey and Saltillo or is represented there by a near ally, while Mr. E. O. Wooton would refer here plants of southern New Mexico, and we are including large, bushy opuntias from Arizona.

Dr. Rose was inclined at one time to separate the Tucson plant, which seems to have some just claims for specific recognition, but there is a mass of herbarium material which seems to connect this with the true *O. engelmannii*.

Opuntia engelmannii monstrosa (Cat. Darrah Succ. Manchester 54. 1908) is doubtless one of the abnormal forms so common among the flat-jointed opuntias.

Opuntia cyclodes, first found by Bigelow near Anton Chico, New Mexico, is certainly of this relationship. The characters of orbicular joints, of small fruit and of stout, usually solitary spines, originally assigned to it, are not constant, for it often has obovate to oblong joints bearing as many as four slender spines and large fruit.

In 1913 Dr. Rose explored the upper Pecos, especially about Anton Chico, near the type locality, where he collected specimens similar to the Bigelow plant, but these grade into more spiny forms, some bearing as many as five spines at an areole, usually yellow, especially distally, and more slender than in typical *O. engelmannii*. From the same type locality, and associated with *O. cyclodes*, is *O. expansa* Griffiths, which has more and whiter spines than the typical form, although they are sometimes yellowish with brown bases. *O. dillei* Griffiths is also related to *O. cyclodes*, but the spines are fewer; Dr. Griffiths states, however, that more spines develop on cultivated plants.

Illustrations: Pac. R. Rep. **4**: pl. 8, f. 1; pl. 22, f. 8, 9, all as *Opuntia engelmannii cyclodes;* Rep. Mo. Bot. Gard. **20**: pl. 4, in part, as *Opuntia dillei*. Ariz. Agr. Exp. Sta. Bull. **67**: pl. 7, f. 1; Rep. Mo. Bot. Gard. **20**: pl. 10; Safford, Ann. Rep. Smiths. Inst. **1908**: pl. 10, f. 3, 6, all as *Opuntia arizonica*. Rep. Mo. Bot. Gard. **21**: pl. 26, in part, 27, both as *Opuntia wootonii*. Rep. Mo. Bot. Gard. **22**: pl. 3, this last as *Opuntia gregoriana*. Standley, Ann. Rep. Smiths. Inst. **1911**: pl. 2; Bull. Torr. Club **32**: pl. 10, f. 10 to 13; Cact. Journ. **2**: 147; Cact. Mex. Bound. pl. 75, f. 1 to 4; Cycl. Amer. Hort. Bailey **3**: f. 1547; Gard. Chron. III. **30**: f. 123; N. Mex. Agr. Exp. Sta. Bull. **78**: pl. [5, 6].

Plate xxv, figure 3, represents a flowering joint of a plant sent from Arizona by Dr. MacDougal in 1902.

151. Opuntia discata Griffiths, Rep. Mo. Bot. Gard. **19**: 266. 1908.

Opuntia gilvescens Griffiths, Rep. Mo. Bot. Gard. **20**: 87. 1909.
Opuntia riparia Griffiths, Proc. Biol. Soc. Washington **27**: 26. 1914.

Plants bushy, spreading, sometimes 15 dm. high; joints thick, orbicular to broadly obovate, 2.5 dm. in diameter or less, pale bluish green, somewhat glaucous; areoles rather few, distant, in age becoming very large, hemispheric, filled with short brown wool; spines usually 2 to 4, sometimes 7 or more in old areoles, 2 cm. long or more, grayish with dark bases, somewhat flattened; flowers large, 9 to 10 cm. broad, light yellow, darker near the center; style white; stigma-lobes green; fruit magenta, pyriform, 6 to 7 cm. long.

Type locality: Foothills of Santa Rita Mountains, Arizona.
Distribution: Foothills and high mesas of southern Arizona and northern Sonora.
Illustrations: Rep. Mo. Bot. Gard. **20**: pl. 2, f. 5; pl. 7; pl. 13, f. 6, all as *Opuntia gil-vescens;* Amer. Garden **11**: 469, this last as *Opuntia angustata.* Ariz. Agr. Exp. Sta. Bull. **67**: pl. 1, f. 2; Bull U. S. Dept. Agr. **31**: pl. 3, f. 2; Rep. Mo. Bot. Gard. **19**: pl. 27, in part.

Plate XXIV, figure 2, is from a photograph taken by Dr. MacDougal in the Tortolita Mountains, Arizona, in 1916; *Opuntia discata* is the plant shown in left foreground.

152. Opuntia rastrera Weber, Dict. Hort. Bois 896. 1898.

?Opuntia lucens Griffiths, Rep. Mo. Bot. Gard. **19**: 269. 1908.

Creeping plant; joints circular to obovate, the largest 2 dm. in diameter; spines white, several from an areole, the longest 4 cm. long; glochids yellow; flowers yellow; fruit purple, acid, obovoid.

Type locality: San Luis Potosí, Mexico.
Distribution: The type locality and vicinity.

This species was very briefly described in 1898 by Dr. Weber, who states that it is quite distinct from *O. tuna*, the Jamaican species. Schumann, who treats it in a note under *O. tuna*, states that it is a well-differentiated species from Mexico.

From descriptions we are referring here *O. lucens* Griffiths, also described from San Luis Potosí specimens. Dr. Griffiths states that his *O. lucens* is related to *O. engelmannii*, but has a different habit; he says it is called cuija by the Mexicans, but that it is very different from *Opuntia cuija*.

Series 12. ELATIORES.

Tall species, with flat, broad, persistent joints, the areoles bearing acicular, setaceous, or subulate brown spines, or some species spineless. We know about twelve species, most of them South American, with one in Florida (see Appendix p. 222), possibly one (*O. fuliginosa*) in Mexico.

KEY TO SPECIES.

Joints very spiny.
  Spines not banded.
    Areoles surrounded by a purple blotch.. 153. *O. brunnescens*
    Areoles not surrounded by a purple blotch.
      Spines setaceous; petals yellow.. 154. *O. galapageia*
      Spines, when present, acicular or subulate; petals mostly red or orange.
        Joints strongly undulate; spines short, stout............................ 155. *O. delaetiana*
        Joints not undulate or scarcely undulate.
          Joints bluish green, glaucous...................................... 156. *O. bergeriana*
          Joints bright green, not glaucous or slightly glaucous.
            Spines, at least on young joints, acicular, slender.
              Spines, when present, dark brown or blackish; joints dull.......... 157. *O. elatior*
              Spines light brown to straw-colored.
                Spines up to 5 cm. long; joints shining...................... 158. *O. hanburyana*
                Spines 3 cm. long or less; joints dull.
                  Flowers 12 to 15 mm. wide; spines 1 to 4 at an areole or
                    wanting... 159. *O. quitensis*
                  Flowers 5 to 6 cm. wide; spines up to 10 at an areole....... 159a. *O. soederstromiana*
              Spines subulate, stout; joints shining.............................. 160. *O. schumannii*
      Spines acicular; petals yellow; joints shining [in this series?].................. 161. *O. fuliginosa*
      Spines distinctly banded; joints dark green, obscurely glaucous.................... 161a. *O. zebrina*
Joints usually spineless.
  Bushy, 1 to 2 meters high; flowers rose... 162. *O. boldinghii*
  Erect, 3 to 4 meters high; flowers orange-red.................................... 163. *O. distans*

153. Opuntia brunnescens sp. nov.

Usually low and prostrate, sometimes 1 meter high, without a definite trunk, usually forming a bushy clump; joints oblong to orbicular, 15 to 30 cm. long, smooth, dull green, except the purple blotches about the prominent areoles; spines 2 to 5, brownish, porrect or pointing forward, up to 4.5 cm. long, stout, sometimes twisted.

Hills about the city of Córdoba, Argentina, where it was collected by Rose and Russell, September 8, 1915 (No. 21029).

This species is very common on the dry hills about Córdoba, where it is often associated with *Opuntia sulphurea*. It apparently extends northward into Jujuy.

Figure 187 represents a joint of the type specimen above cited; figure 188 shows its fruit collected by Dr. Shafer (No. 78).

154. Opuntia galapageia Henslow, Mag. Zool. and Bot. **1**: 467. 1837.

> *Opuntia myriacantha* Weber, Dict. Hort. Bois 894. 1898.
> *Opuntia helleri* Schumann in Robinson, Proc. Amer. Acad. **38**: 180. 1902.
> *Opuntia insularis* Stewart, Proc. Calif. Acad. IV. **1**: 113. 1911.

Sometimes low and creeping, but often becoming very large, 5 to 10 meters high, with a large top either open or very compact and rounded; trunk at first very spiny and made up of flat joints set end to end, with the short axis of each joint at right angles to that of the adjacent joint, in time becoming terete, and when old nearly naked, 3 to 13 dm. in diameter; bark of old trunks smooth, brown, peeling off in thin layers; joints oblong to orbicular, usually very large, 1.5 to 3.5 dm. long, very spiny; areoles large, often prominent on the trunk, there especially forming knobs bearing numerous spines; spines extremely variable, but nearly all yellowish brown; areoles on young, vigorous plants very stout and rigid, very unequal, the longest 7 to 8 cm. long; joints of old plants bearing more or less pungent bristles or sometimes very weak soft hairs instead of spines, while the spines from the trunks often are very stout and sometimes 40 in a cluster; flowers yellow, 7.5 cm. broad; ovary more or less tuberculate; fruit greenish, sometimes borne in the ends of joints, more or less spiny; seeds large, 5 to 6 mm. broad, white, covered with soft hairs.

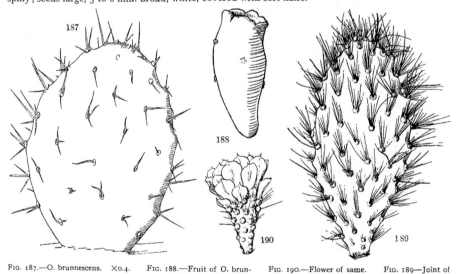

FIG. 187.—O. brunnescens. X0.4.   FIG. 188.—Fruit of O. brun-   FIG. 190.—Flower of same.   FIG. 189—Joint of
                                          nescens. X0.9.   X0.75   O. galapageia.

Type locality: Galápagos Islands.

Distribution: Very common, often forming forests, on the Galápagos Islands.

We have here combined the four species reported from the Galápagos Islands, while Alban Stewart, in his admirable paper on the botany of these islands, not only recognizes four species, but describes a fifth without specific name.   He also has fourteen full-page illustrations showing fine habit views of the Galápagos *Opuntia*. The early descriptions of this species were very inaccurate and, as pointed out by Mr. Stewart, the characters assigned to its fruit are those of a Cereus-like plant. Mr. Stewart visited the Galápagos Islands in 1905–1906 and brought back a remarkable series of photographs and specimens. Through the kindness of Miss Alice Eastwood, Curator of Botany in the California Academy of Sciences, we have been able to study this material. It consists of about forty sheets of well-preserved joints with a few flowers and fruits. These, in connection with the pub-

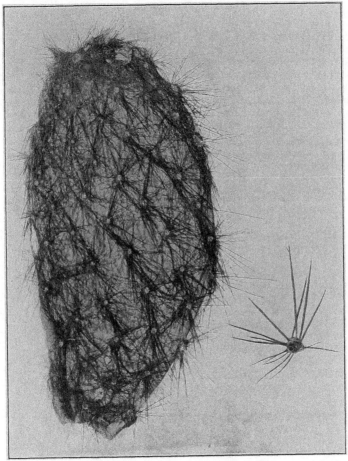

FIG. 191.—Opuntia galapageia. X0.75.

lished illustrations, show a great range of variation in habit, armament of joints, and character of spines. While these differences are very marked, they are similar to what is sometimes met with in other opuntias, such as *O. gosseliniana* and *O. leucotricha*, or in certain Peruvian and Chilean types of *Cereus* relatives; indeed, in a number of cacti which live under intense desert influences, most diverse forms in the same species are often produced. The habit-character in this species seems to be of little value, according to Mr.

Stewart himself, for he calls attention to procumbent and arborescent forms of *O. galapageia*, while the greatest range of spine characters is shown between the young plants and old ones and between the trunk and the joints. The specimen which Mr. Stewart has made the type of his *Opuntia insularis* is quite different from all the others, and yet one can easily believe that intergrades could be found; his species is described without flowers or fruit. Mr. Stewart states that this *Opuntia* forms the principal article of food for the Galápagos land tortoise. Its trunk becomes thicker than that of any other known species of the genus.

Illustrations: Gard. Chron. III. 24: f. 75; Mag. Zool. and Bot. 1: pl. 14, f. 2; Proc. Calif. Acad. IV. 1: pl. 7, f. 2; pl. 8; pl. 9, f. 2; pl. 10 to 12. Gard. Chron. Ser. III. 27: f. 56; Proc. Calif. Acad. IV. 1: pl. 7, f. 1; pl. 13, f. 2; pl. 16 to 18, all as *Opuntia myriacantha*. Proc. Calif. Acad. IV. 1: pl. 13, f. 1; pl. 14, the last two as *Opuntia helleri*. Proc. Calif. Acad. IV. 1: pl. 9, f. 1; pl. 15, the last two as *Opuntia insularis*.

Figure 189 represents a joint of the plant collected by Robert E. Snodgrass and Edmund Heller on Wenman Island, Galápagos, on the Hopkins-Stanford Expedition (type of *Opuntia helleri* Schumann), drawn from the herbarium specimen in the Gray Herbarium; figure 190 is of a flower of the same plant; figure 191 is from a photograph of an herbarium specimen collected by Alban Stewart.

155. Opuntia delaetiana Weber in Vaupel, Blühende Kakteen **3**: pl. 148. 1913.

    Opuntia elata delaetiana Weber in Gosselin, Bull. Mus. Hist. Nat. Paris **10**: 392. 1904.

Joints oblong, 25 cm. long, 8 cm. broad, bright green, at first thin and spineless, the margin strongly undulate; areoles large, bearing 3 to 5 straight, rose-colored or yellowish brown spines up to 4 cm. long; leaves subulate, about 4 mm. long; glochids wanting in young areoles, later appearing numerous and brown; flower-buds rounded at the apex; outer sepals orbicular, obtuse, red; flower rotate, 5 to 7 cm. broad, orange-colored; stigma-lobes white; fruit oblong or pyriform, red, 5 to 7 cm. long, 3 to 5 cm. in thickness.

Type locality: Paraguay.

Distribution: Paraguay and northeastern Argentina.

The plant was collected by Dr. Thomas Morong at Asunción, Paraguay, in 1888, and referred in his list of plants collected in Paraguay (Annal. N. Y. Acad. Sci. **7**: 121. 1892) to *O. nigricans* Haworth; Dr. Shafer found it in 1917 in waste places and in hedge-rows about Concordia and Posados, Argentina. This species may more properly belong in our series *Elatae* than in *Elatiores*.

Illustration: Blühende Kakteen **3**: pl. 148.

Figure 192 is copied from the illustration above cited.

156. Opuntia bergeriana Weber in Berger, Gard. Chron. III. **35**: 34. 1904.

Growing singly or in dense thickets, often 1 to 3.5 meters high and having a trunk 3 to 4 dm. in diameter, with a large, spreading top, or clambering over walls and rocks; joints narrowly oblong, sometimes 2.5 cm.

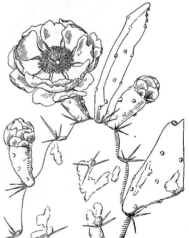

Fig. 192.—Opuntia delaetiana.

long, when young often quite narrow, bright green, but becoming dull and somewhat glaucous; areoles rather distant, on old joints 2 to 4 cm. apart, filled with short gray wool; spines 2 or 3, rarely 5, unequal, the longest one 3 to 4 cm. long and somewhat flattened, more or less brownish at base, sometimes yellowish, porrect, or somewhat turned downward; leaves 2 to 3 mm. long, fugacious; glochids yellow but sometimes turning brown, rather prominent, forming a half circle in the upper part of the areole; areoles circular, when young filled with light brown wool in the center and white in the outer region; flowers numerous, showy, deep red;

M. E. Eaton del.

1. Flowering joint of *Opuntia bergeriana*.
2. Flowering joint of *Opuntia elatior*.
3. Flowering joint of *Opuntia boldinghii*.
4, 5. Joints of *Opuntia elata*.
(All natural size.)

some joints bearing 20 or more; petals 2.5 cm. long, mucronate; filaments numerous, scarlet-rose; stigma-lobes 6, green; fruit small, 3 to 4 cm. long, red, not edible; seeds few, flattened, 5 mm. broad.

Type locality: Described from cultivated specimens.

Distribution: Not known in the wild state, but now very common on the Riviera, northern Italy, forming large thickets.

Mr. Berger would place this species next to *O. nigricans*, which we now call *O. elatior*. This species was named for Alwin Berger, formerly curator of the Hanbury Garden at La Mortola, Italy, who sent material to the late Dr. Weber, from which the species was described. The species is quite common on the Riviera and has run wild in many places, especially about Bordighera, Italy. It produces a great abundance of flowers in May, but blooms more or less throughout the year.

Opuntia ledienii (Berger, Hort. Mortol. 233. 1912), unpublished, is referred here.

Illustrations: Gard. Chron. III. 35: f. 14; Monatsschr. Kakteenk. 16: 156.

Plate XXVI, figure 1, represents a flowering joint of a plant sent from La Mortola, Italy, to the New York Botanical Garden in 1906.

157. Opuntia elatior Miller, Gard. Dict. ed. 8. No. 4. 1768.

> *Cactus nigricans* Haworth, Misc. Nat. 187. 1803.
> *Opuntia nigricans* Haworth, Syn. Pl. Succ. 189. 1812.
> *Cactus elatior* Willdenow, Enum. Hort. Berol. Suppl. 34. 1813.
> *Cactus tuna nigricans* Sims, Curtis's Bot. Mag. 38: pl. 1557. 1813.
> *Cactus tuna elatior* Sims, Curtis's Bot. Mag. 38: under pl. 1557. 1813.
> *Cactus pseudococcinellifer* Bertoloni, Excerpta Herb. Bonon. 11. 1820.

Plants densely bushy-branched, up to 5 meters high; joints obovate to oblong or suborbicular, olive-green, 1 to 2 dm. or even 4 dm. long; leaves 4 mm. long, green with reddish tips; areoles 2 to 4 cm. apart; spines 2 to 8, acicular, mostly terete, dark brown, 2 to 4 cm. or even 7 cm. long; flowers about 5 cm. broad; petals dark yellow striped with red or sometimes salmon-rose, with mucronate tips; filaments numerous, pink or red; style nearly white; stigma-lobes 5, green; ovary ovoid, deeply umbilicate, its areoles either with or without spines; fruit obovoid, truncate when mature, reddish, the pulp dark red; seeds about 4 mm. broad.

Type locality: Unknown.

Distribution: Common or frequent in Curaçao, Venezuela, Colombia, and Panama, escaped from cultivation in Australia. *O. nigricans* has been referred to Mexico, but doubtless wrongly, unless cultivated there. Plants brought by Dr. Howe from Tobogilla Island, Panama, have narrowly obovate joints.

The early history of this species and its various synonyms are rather confusing. Dillenius figured *Opuntia elatior* and this name was taken up by Miller in 1768. There is some doubt as to its native home, but it probably came from northern South America, or possibly Curaçao. *Opuntia nigricans*, also referred here, was described by Haworth from cultivated specimens; plate 1557 of Curtis's Botanical Magazine was made from Haworth's specimen and may be considered typical.

Introduced into cultivation in Europe about 1793.

Illustrations: Loudon, Encycl. pl. ed. 3. f. 6877, as *Cactus elatior;* Curtis's Bot. Mag. 38: pl. 1557, this last as *Cactus tuna nigricans;* Dillenius, Hort. Elth. pl. 294, this as *Tuna elatior,* etc.; Agr. Gaz. N. S. W. 23: pl. opp. 208; pl. opp. 210, both these as *Opuntia nigricans;* Journ. Hort. Home Farm. III. 60: 30, this as *Opuntia occidentalis.*

Plate XXVI, figure 2, shows a flowering joint of a specimen obtained by Dr. Britton and Dr. Shafer in Curaçao in 1913.

158. Opuntia hanburyana Weber in Berger, Gard. Chron. III. 35: 34. 1904.

Bushy, 1 to 2 meters high, somewhat straggling; joints narrowly oblong, about 3 dm. long, bright green; leaves subulate, 4 to 5 mm. long; areoles closely set, filled with brown or blackish wool; spines several, spreading, acicular, somewhat flattened and twisted, yellowish brown, the longest 3 cm. long; flowers widely spreading, rather small; fruit small.

Type locality: Described from cultivated plants.

Distribution: Not known in the wild state.

The species commemorates Sir Thomas Hanbury, who, through his extensive garden at La Mortola, Italy, contributed much to botany and horticulture.

Illustration: Gard. Chron. III. **35**: f. 15.

Figure 193 represents joints of the plant sent from La Mortola, Italy, in 1913.

159. Opuntia quitensis Weber, Dict. Hort. Bois 894. 1898.

Bushy, sometimes 2 meters high; joints obovate, 1 to 4 dm. long, 8 to 9 cm. broad; areoles small, distant, 2 cm. apart, bearing some white tomentum and short glochids; spines wanting, or 1 to 3, sometimes as many as 4 on old joints, straight, yellowish brown, or nearly white when young, acicular, somewhat flexuous, 2 to 3 cm. long; leaves green, minute, acute; flowers red, 12 to 15 mm. broad; petals erect, obtuse; anthers white; style white, short and thick; stigma-lobes 13, white, about as long as the style; fruit obovoid, red, nearly spineless, about 2 cm. long; seeds about 3 cm. broad.

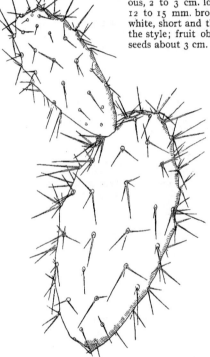

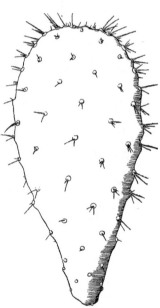

FIG. 193.—O. hanburyana. ×0.5.

FIG. 194.—O. quitensis. ×0.5.

Type locality: Near Quito, Ecuador.

Distribution: Ecuador.

As observed by Dr. Rose in Ecuador in 1918, this species is very variable in habit, for when grown in the open it is low and bushy with rather small joints, but when growing in thickets it becomes tall and has large joints. About Huigra, where it is very common, it is often spineless, and when the spines are present they are few and weak. In southern Ecuador there is a plant which has small, red flowers like this species, but the joints have stout subulate spines.

Figure 194 represents a joint of a plant obtained in 1901 for the New York Botanical Garden from M. Simon, of St. Ouen, Paris, France.

159a. Opuntia soederstromiana sp. nov. (See Appendix, p. 221.)

160. Opuntia schumannii Weber in Berger, Gard. Chron. III. **35**: 34. 1904.

Bushy, 1 to 2 meters high; joints obovate to oblong, 1.5 to 2.5 cm. long, dull dark green; areoles distant, medium sized; spines 2 to 10, slightly spreading, very unequal, the longest ones 4.5 cm. long, more or less twisted, flattened, dark brown; glochids few, soon disappearing; flowers 6 cm. long, yellowish to orange, turning in age to dull red; ovary tuberculate, spineless, deeply umbilicate; fruit dark purple, turgid, juicy, deeply umbilicate, 5 cm. long.

Type locality: Not cited.
Distribution: Northern South America; sometimes assigned to Argentina.
Opuntia schumannii is described by Berger as being intermediate between *Opuntia* and *Nopalea*, and according to him, it has long stamens and upright petals; otherwise it has little in common with *Nopalea;* a plant from Santa Clara, Colombia, which agrees with plants of *O. schumannii* from La Mortola, Italy, has a normal *Opuntia* flower.
Illustration: Gard. Chron. III. **35**: f. 16.
Plate XXVII, figure 1, represents a fruiting joint of the plant collected by John G. Sinclair at Santa Clara, Colombia, in 1913; figure 2 shows a flower of the same plant.

161. Opuntia fuliginosa Griffiths, Rep. Mo. Bot. Gard. **19**: 262. 1908.

Tall, tree-like, 4 meters high or more, much branched; joints orbicular to oblong, 3 dm. long or less, shining; leaves subulate, 8 to 12 mm. long; areoles distant; spines few, rarely as many as 6, dull brown or horn-colored, the longest ones 4 cm. long, slightly twisted; glochids yellow to brown; flowers at first yellow but in age red, 5 to 6 cm. long including the ovary; stigma-lobes yellowish green; fruit pyriform to short-oblong, 3 to 4 cm. long, red; seeds 5 mm. broad.

Type locality: Near Guadalajara, Mexico.
Distribution: Central Mexico.
We refer this species to our series *Elatiores* with hesitation.
Illustration: Rep. Mo. Bot. Gard. **19**: pl. 25.

161a. Opuntia zebrina Small, Journ. N. Y. Bot. Gard. **20**: 35. 1919. (See Appendix, p. 222.)

162. Opuntia boldinghii sp. nov.

Bushy, 2 meters high; joints dull green, somewhat glaucous, obovate, 2 cm. long, spineless or with very short brown spines; leaves conic, red, 2 to 3 mm. long; areoles large, elevated, filled with short brown wool; flowers rose-colored, 5 cm. long; petals obtuse; filaments pink, much shorter than the petals; style nearly white; stigma-lobes yellowish; fruit obovate, 4 cm. long, spineless; seeds 4 mm. in diameter.

Collected by Dr. N. L. Britton and Dr. J. A. Shafer, March 1913, in cultivation on Curaçao (No. 2905, type); also collected by H. Pittier around El Palito, Venezuela, July 2, 1913 (No. 6450), and by Dr. Rose in a hedge at Valencia, Venezuela, October 27, 1916 (No. 21842).

This species is named in honor of Dr. I. Boldingh, a Dutch botanist, author of a valuable descriptive flora of the Dutch West Indian islands.

Plate XXVI, figure 3, shows a flowering joint of a specimen obtained by Dr. Britton and Dr. Shafer in Curaçao in 1913.

163. Opuntia distans sp. nov.

Erect, densely much branched, 3 to 4 meters tall, with a short trunk 1.5 dm. in diameter; joints flat, bluish green when young, grayish green when old, obovate, 2 to 2.5 dm. long, about 1.5 dm. wide and nearly 2 cm. thick, rounded above, narrowed

Fig. 195—Joint of O. distans. ×0.4.

at the base, glabrous; areoles few, only about 12 on each side of a joint, distant, large, nearly circular, 8 to 10 mm. broad, slightly elevated, bearing many short glochids, but quite spineless; leaves subulate, about 3 mm. long; ovary obconic, 3 to 4 cm. long, bearing a few small areoles; sepals broadly triangular, acute, 6 to 10 mm. long; petals broad, rounded, 1 to 2 cm. long, orange-red.

Distribution: Sandy places, Andalgala, Catamarca, Argentina, J. A. Shafer, December 15, 1916 (No. 7).

A spineless species noteworthy for its few, large, distant areoles. We append it to the series *Elatiores,* but are uncertain as to its real affinity. The large distant areoles forbid associating it with the *Ficus-indicae* or the *Streptacanthae.*

Figure 195 represents a joint of the type specimen.

Series 13. ELATAE.

Erect, tall species, natives of South America, with oblong or oval joints, the brown or white spines, when present, only one or few at each areole, except on the trunk and old joints.

KEY TO SPECIES.

Joints ovate to broadly oblong or obovate.
  Joints thin, lustrous, light green.. 164. *O. vulgaris*
  Joints turgid, dull green.
    Leaves purplish, rigid; joints dark green............................... 165. *O. elata*
    Leaves green, not rigid; joints pale green.
      Spines slender, terete... 166. *O. cardiosperma*
      Spines stout, angled, elongated.................................. 167. *O. arechavaletai*
Joints narrowly oblong to linear or spatulate.
  Joints oblong to linear; flowers brick-red............................... 168. *O. mieckleyi*
  Joints spatulate; flowers orange....................................... 169. *O. bonaerensis*

164. Opuntia vulgaris Miller, Gard. Dict. ed. 8. No. 1. 1768.

    Cactus monacanthos Willdenow, Enum. Pl. Suppl. 33. 1813.
    Opuntia monacantha Haworth, Suppl. Pl. Succ. 81. 1819.
    Cactus urumbeba Vellozo, Fl. Flum. 207. 1825.
    Cactus indicus Roxburgh, Fl. Indica 2: 475. 1832.
    Cactus chinensis Roxburgh, Fl. Indica 2: 476. 1832.
    Opuntia monacantha gracilior Lemaire, Cact. Gen. Nov. Sp. 68. 1839.
    Opuntia umbrella Steudel, Nom. ed. 2. 2: 222. 1841.
    Opuntia roxburghiana Voigt, Hort. Suburb. Calcutt. 62. 1845.
    Opuntia monacantha deflexa Salm-Dyck, Cact. Hort. Dyck. 1849. 66. 1850.
    Opuntia lemaireana Console in Weber, Dict. Hort. Bois 894. 1898.

Plant 2 to 4 or even 6 meters high, often with a definite trunk, usually with a large much branched top; trunk cylindric, 1.5 dm. in diameter, either spiny or smooth; joints ovate to oblong, narrowed at base, 1 to 3 dm. long, bright shining green; leaves subulate, 2 to 3 mm. long; areoles filled with short wool; glochids brownish; spines 1 or 2, sometimes more (on the trunk often 10 or more) from an areole, erect, 1 to 4 cm. long, yellowish brown to dark reddish brown; flowers yellow or reddish, 7.5 cm. broad; sepals broad, each with a broad red stripe down the middle; petals golden-yellow, widely spreading; filaments greenish; style white; stigma-lobes 6, white; ovary spineless, 3.5 cm. long; fruit obovoid, 5 to 7.5 cm. long, reddish purple, long-persisting, sometimes proliferous.

Type locality: Type based on an illustration, the origin unknown.

Distribution: Coast and islands of Brazil, Uruguay, and Argentina; in the interior to Paraguay; an escape in Cuba, India, and south Africa and naturalized in Australia; frequently cultivated. According to J. H. Maiden it is found in every state of Australia, but it is not inclined to spread and become a pest.

As has been recently pointed out by Burkill, the *Opuntia vulgaris* of Miller is the same as *O. monacantha* Haworth. *O. vulgaris* was based on Bauhin's figure (Hist. Pl. 1: 154. 1650), which was taken from Lobelius (Icones 2: 241. 1591), and is a tall, branching plant. This species is not to be confused with the low, spreading species of the eastern United States, long known as *O. vulgaris.* (See p. 127.)

This species is said by Burkill to be distributed over the earth more widely than any other, but our observation in America is that *O. ficus-indica* is by far the most widely spread species.

M. E. Eaton del.

1. Upper part of fruiting joint of *Opuntia schumannii*.   3. Flowering joint of *Opuntia vulgaris*.
2. Flower of *Opuntia schumannii*.   4. Flowering joint of *Opuntia stricta*.

(All natural size.)

O. vulgaris was one of those most commonly used in the nopalries of India and South Africa in the cochineal industry.

We have referred both of Roxburgh's species here, although Burkill was inclined to refer *Cactus chinensis* to *O. decumana*, which in his sense is *O. ficus-indica.*

Opuntia monacantha variegata (listed in Cat. Darrah Succ. Manchester 57. 1908) is common in cultivation. Some of the joints are normally green; others are more or less blotched with white or yellow, while others may be entirely white or yellow; the leaves are bright red and though small are conspicuous.

Opuntia urumbella Steudel (Nom. ed. 2. 1: 246. 1840), given as a synonym of *Cactus urumbella*, is doubtless a name for this species.

Opuntia deflexa Lemaire (Cact. Gen. Nov. Sp. 68. 1839) was given as a synonym of *O. monacantha gracilior;* while the latter was given as a synonym of *O. elatior deflexa* Salm-Dyck (Cact. Hort. Dyck. 1844. 47. 1845).

Illustrations: Rev. Hort. **41:** f. 37; **66:** f. 58; Bauhin, Hist. Pl. **1:** 154 [=Lobelius, Icones **2:** 241], this last as *Opuntia vulgo,* etc. Anal. Mus. Nac. Montevideo **5:** pl. 32; Curtis's Bot. Mag. **68:** pl. 3911; Dept. Agr. N. S. W. Misc. Publ. **253:** pl. [3], [4]; Agr. Gaz. N. S. W. **24:** facing p. 864; Edwards's Bot. Reg. **20:** pl. 1726; Gard. Chron. III. **30:** f. 122, in part; **34:** f. 35; Journ. Dept. Agr. Vict. **6:** pl. 25; Martius, Fl. Bras. **4²:** pl. 62; Weeds, Pois. Pl. Nat. Al. Vict. pt. 1. pl. [10], [32], all as *Opuntia monacantha;* Amer. Garden **11:** 529; Cact. Journ. **1:** 167, these last two as *Opuntia monacantha variegata;* Vellozo, Fl. Flum. **5:** pl. 32, as *Cactus urumbeba;* De Candolle, Pl. Succ. Hist. **2:** pl. 138 [B]; De Tussac, Fl. Antill. **2:** pl. 31, these last two as *Cactus opuntia tuna;* Gard. Chron. III. **47:** f. 174, this as *Opuntia ficus-indica;* Rümpler, Sukkulenten f. 122, this as *Opuntia tuna;* Addisonia **1:** pl. 38.

Plate XXVII, figure 3, represents a flowering joint of a plant presented to the New York Botanical Garden by Mr. Gustav Rix in 1900.

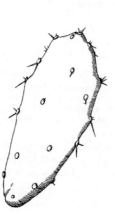 

Fig. 196.—O. elata. X0.4.   Fig. 197.—O. cardiosperma. X0.4.

165. Opuntia elata Link and Otto in Salm-Dyck, Hort. Dyck. 361. 1834.

An erect plant, 1 meter high or more; joints thick, dark green, oblong, 5 to 25 cm. long, half as broad as long; leaves minute, caducous; areoles remote, large (7 mm. in diameter), filled with short white wool, usually spineless; spines if present only 1 to 3, except on old stems and there more, horn-colored, stiff, sometimes 3.5 cm. long; glochids very tardy in appearing, long persistent; flowers about 5 cm. broad, orange-yellow; petals obtuse, broad; filaments short; stigma-lobes white; fruit oblong, 6 cm. long, spineless, with a truncate umbilicus; seeds 6 mm. broad.

Type locality: In Brazil.

Distribution: Paraguay, but according to Salm-Dyck and Pfeiffer, from Brazil and probably Curaçao; our exploration of Curaçao failed to prove its existence there. It is grown for ornament in Cuba and has there escaped from cultivation in gardens to roadsides and waste grounds.

Schumann did not know where to place this species, but we believe it is most nearly related to *Opuntia vulgaris.*

Plate XXVI, figure 4, represents a flowering joint of a plant given to the New York Botanical Garden by Frank Weinberg in 1903; figure 5 represents another joint of the same plant. Figure 196 represents a joint of a plant obtained by Professor Carlos de la Torre at Punta de los Molinos, Cuba, in 1912.

166. Opuntia cardiosperma Schumann, Monatsschr. Kakteenk. **9**: 150. 1899.

About 2 meters high, erect, branching; joints narrowly oblong to obovate; rounded at apex, 10 to 15 cm. long or smaller in greenhouse specimens, easily breaking apart, pale green, more or less tuberculate; leaves minute, subulate; areoles large, 1 to 2 cm. apart, with white wool, when young having conspicuous secreting glands; spines, when present, 1 to 4, acicular, stiff, more commonly 1 or 2 from an areole, short, 1 to 2 cm. long, brownish at first but nearly white when old, porrect or ascending; glochids tardily developing, never conspicuous, brownish; flowers unknown; fruit elongated, pear-shaped, 7.5 cm. long; seeds 6 mm. broad, 2.5 to 3 mm. thick, cordate, gray, with broad yellow margins, woolly on the sides.

Type locality: At Recoleta, near Asunción, Paraguay.
Distribution: Paraguay.
Figure 197 represents joints of the plant sent to the New York Botanical Garden from La Mortola, Italy, in 1913.

167. Opuntia arechavaletai Spegazzini, Anal. Mus. Nac. Buenos Aires. III. **4**: 520. 1905.

Plants tall, 1 to 3 meters high, much branched; joints flattened, oblong to obovate, 25 to 30 cm. long, green; spines, usually 1, sometimes 3, elongated, porrect, up to 9 cm. long, white, flattened; flowers 4.5 cm. long, yellow; stamens and style white; fruit violet-purple, 7 cm. long.

Type locality: Near Montevideo, Uruguay.
Distribution: Argentina and Uruguay.
Illustration: Anal. Mus. Nac. Montevideo **5**: pl. 35.

168. Opuntia mieckleyi Schumann, Blühende Kakteen **1**: pl. 44. 1903.

Plant erect, much branched; joints narrowly oblong, 15 to 25 cm. long, 4 to 6 cm. broad, glabrous, dark green, darker below the areoles; tubercles rather prominent; leaves small; areoles large, filled with white wool; spines, when present, 1 or 2, very short (5 mm. long), dark-colored; flower-buds obtuse; flowers brick-red, 6 cm. broad; petals irregularly notched; ovary spineless.

Type locality: In Paraguay.
Distribution: Paraguay; Estancia Loma, in San Salvador.
Named for W. Mieckley, gardener in the Berlin Botanical Garden.
Illustration: Blühende Kakteen **1**: pl. 44.

169. Opuntia bonaerensis Spegazzini, Contr. Fl. Tandil 18. 1904.

   Opuntia chakensis Spegazzini, Anal. Mus. Nac. III. **4**: 519. 1905.

Two meters high, very much branched; joints spatulate to elliptic-spatulate, 15 to 25 cm. long, green; spines wanting or one, short; flowers orange, large, 4 cm. long; fruit obconic, 6 to 7 cm. long, dull purple; seeds 5 to 6 mm. long, subglobose.

Type locality: Sierra de Curamalal, Argentina.
Distribution: Argentina and perhaps Paraguay.
Opuntia paraguayensis Schumann (Monatsschr. Kakteenk. **9**: 149. 1899) according to Spegazzini, and if so this name would supplant *O. bonaerensis.*
Illustration: Anal. Mus. Nac. Montevideo **5**: pl. 23.

The three following, known to us only from descriptions, may belong to this series.

OPUNTIA STENARTHRA Schumann, Monatsschr. Kakteenk. **9**: 149. 1899.

Shrubby, erect or decumbent, creeping over stones or ascending trees; joints thin, narrow, yellowish green, oblong to lanceolate, rounded at base, glabrous; spines either wanting or 1 to 3 from an areole, stoutish, subangular; flowers yellow; seeds woolly.

Type locality: Estancia Tagatiya, Paraguay.
Distribution: Paraguay.

OPUNTIA ASSUMPTIONIS Schumann, Monatsschr. Kakteenk. **9**: 153. 1899.

Erect, 1 meter high; joints obovate, narrowed at base, thickish; spines at areoles on the faces of the joints none, but on the edges 1 or 2, stout, subulate, the upper one stouter, 3 to 4.5 cm. long; flower 3.5 cm. long, lemon-yellow; fruit pear-shaped, with a deep umbilicus; seeds densely villous.

Type locality: Ascunción, Paraguay.
Distribution: Known only from the type locality.

OPUNTIA CANTERAI Arechavaleta, Anal. Mus. Nac. Montevideo **5**: 278. 1905.

Stems erect, branching, 5 to 10 dm. high; joints elongated, shining green, attenuate below, 15 to 20 cm. long, 4 to 6 cm. broad; areoles orbicular, when young each surrounded by a violet spot, mostly spineless, about 4 cm. apart; spines, when present, 1 or 2 from an areole, 1.5 to 2 cm. long, whitish, with brownish tips; flowers orange-colored, 4 to 4.5 cm. broad; stigma-lobes 6 or 7, light flesh-colored; fruit somewhat pear-shaped, 5 cm. long; seeds flattened.

Type locality: In Uruguay.
Distribution: Along the coast of Uruguay.

In Uruguay this species flowers in January and February.

Series 14. SCHEERIANAE.

A single bushy species, with broad, thin, persistent joints, the areoles close together, each bearing several yellow, acicular spines and long white or yellow hairs. Its home is unknown.

170. Opuntia scheeri Weber, Dict. Hort. Bois 895. 1898.

About 1 meter high, branching at base, the lower branches sprawling over the ground; joints oblong to orbicular, 1.5 to 3 dm. long, bluish green; areoles circular, elevated, filled with short brown wool; spines 10 to 12, yellow, acicular, each surrounded by a row of long white or yellow hairs; flowers large, pale yellow, but in age salmon-colored; stigma-lobes deep green; fruit globular, red, juicy, truncate; seed small, 4 mm. broad, with a broad irregular margin.

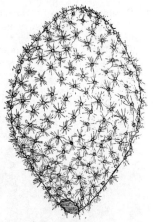

FIG. 198.—Opuntia scheeri. ×0.5.

Type locality: In Mexico.
Distribution: Mexico, but known to us only from cultivated specimens.

This is a very beautiful species, covered as it is by yellow spines and long hairs. A fine plant is growing in the open at La Mortola, Italy. The seedlings produce a long mass of soft white hairs almost covering the joints and giving an appearance very different from the adult plant. In this young stage, according to Mr. Alwin Berger, they readily pass for *Opuntia senilis* (*O. crinifera*).

Figure 198 represents a joint from a specimen sent from La Mortola, Italy, in 1912.

Series 15. DILLENIANAE.

Mostly bushy or tall species, with large, flat, persistent joints, and yellow spines which are sometimes brown at base, some species spineless or nearly so. We recognize thirteen species as composing the series, but many more have been described. The plants inhabit the southern United States, the West Indies, Mexico, and northern South America.

KEY TO SPECIES.

Spines nearly setaceous, most of them reflexed....................................... 171. *O. chlorotica*
Spines, when present, acicular to subulate.
    Joints spineless, or with only 1 or 2 spines at some of the areoles, or spines very short.
        Corolla rotate; petals yellow.
            Plant tall; spines, when present, 2 cm. long or less........................... 172. *O. laevis*
            Plant depressed, bushy or spreading; spines, when present, up to 7 cm. long .... 173. *O. stricta*
        Corolla cup-shaped; petals salmon .... 173*a*. *O. keyensis*
    Joints usually manifestly spiny; spines mostly 2 or more at the areoles.
        Spines mostly stout, commonly flattened .... 174. *O. dillenii*
        Spines acicular to subulate, terete, or slightly flattened at the base.

Joints elongated-lanceolate or oblong, several times longer than wide..........175. *O. linguiformis*
Joints obovate to suborbicular.
    Spines long.
        Areoles mostly 1.5 to 2 cm. apart.
            Spines subulate, up to 7.5 cm. long.............................176. *O. tapona*
            Spines acicular, 4 cm. long or less.
                Spines nearly clear yellow, short..........................177. *O. littoralis*
                Spines brown at base, long and slender......................178. *O. aciculata*
        Areoles mostly 2.5 to 4 cm. apart.
            Bushy species.
                Spines yellow or yellowish brown..........................179. *O. lindheimeri*
                Spines pale yellow or whitish...............................180. *O. cantabrigiensis*
            Depressed or procumbent plant...............................181. *O. procumbens*
    Spines only 1.5 cm. long or less, or becoming longer on old joints.
        Plant 1 meter high or less; joints thin.............................182. *O. cañada*
        Plant 3 to 5 meters high; joints very thick.
            Spines reflexed; flowers yellow...............................183. *O. pyriformis*
            Spines spreading, deciduous; flowers orange-red..................183a. *O. bonplandii*

171. Opuntia chlorotica Engelmann and Bigelow, Proc. Amer. Acad. **3**: 291. 1856.

    Opuntia tidballii Bigelow, Pac. R. Rep. **4**: 11. 1856.
    Opuntia curvospina Griffiths, Bull. Torr. Club **43**: 88. 1916.

    Erect bushy, sometimes 2 meters high or more, with a definite trunk; main branches nearly erect; joints ovate to orbicular, sometimes broader than long, 15 to 20 cm. long, more or less glaucous, bluish green; leaves subulate, small, reddish at tip; areoles closely set, prominent; spines yellow, several, most of them usually appressed and reflexed, setaceous, 3 to 4 cm. long; glochids yellow, numerous, elongated, persistent; flowers yellow, 6 to 7.5 cm. broad; filaments white; fruit purple without, green within, 4 cm. long; seeds small.

FIG. 199.—Opuntia chlorotica.   FIG. 200.—Opuntia chlorotica. ×0.4.

    Type locality: On both sides of the Colorado from San Francisco Mountains to headwaters of Bill Williams River.
    Distribution: Sonora and New Mexico to Nevada, California, and Lower California.
    This species is of wide distribution, but is chiefly confined to mountain canyons, being rarely found on the open mesas.
    Illustrations: Bull. Torr. Club **43**: pl. 3; Pac. R. Rep. **4**: pl. 6, f. 1 to 3; Bull. Torr. Club **43**: pl. 2, this last as *Opuntia curvospina.*

1. View of *Opuntia keyensis.*
2. View of *Opuntia dillenii.*

M. E. Eaton del.

1. Flowering joint of *Opuntia laevis*.   2. Flowering joint of *Opuntia dillenii*.
        3. Upper part of flowering joint of *Opuntia aciculata*.   (All natural size.)

Figure 199 is from a photograph of a plant with narrow joints, in McCleary's Canyon, Santa Rita Mountains, Arizona, taken by Dr. MacDougal; figure 200 represents a joint of a plant from the collection made by Professor J. W. Toumey at Tucson, Arizona, obtained by Dr. MacDougal in 1902.

OPUNTIA PALMERI Engelmann in Coulter, Contr. U. S. Nat. Herb. **3**: 423. 1896.

This plant has not been again collected and is still a doubtful species; it came from St. George, southwestern Utah. In 1909 E. W. Nelson made a collection for us in this region, but the only shrubby, juicy-fruited species which he collected has brown spines and brown glochids, which would seem to exclude it from *O. palmeri*. It is not at all unlikely that *O. palmeri* should be referred to *O. chlorotica*, a widely dispersed species, but of which we have not seen any specimens from Utah.

172. Opuntia laevis Coulter, Contr. U. S. Nat. Herb. **3**: 419. 1896.

Loosely few-branched, 1 to 2 meters high, but in cultivation often forming a low, dense bush; joints obovate to oblong, 1.5 to 3 dm. long, light green, often spineless but usually with a few (1 to 3) short spines 1 cm. long or less at the areoles of the upper part of the joint; areoles rather distant, small; flower large, 6 to 7 cm. broad; petals lemon-yellow, sometimes tinged with red, broad, and obtuse or retuse; filaments and style short, pale yellow; stigma-lobes green; ovary turbinate, more or less tuberculate, at first leafy, often bristly at top; fruit obovoid, 5 to 7 cm. long; seeds 4 to 5 mm. broad.

Type locality: In Arizona.

Distribution: In the mountains about Tucson, Arizona.

Referred by Professor Schumann to *O. inermis* (*O. stricta*), but it is not that species.

Illustrations: Ariz. Agr. Exp. Sta. Bull. **67**: pl. 8, f. 1; N. Mex. Agr. Exp. Sta. Bull. **72**: pl. 1; Plant World **11**[10]: f. 5.

Plate XXVIII, figure 1, represents a flowering joint of a plant brought by Dr. MacDougal from Tucson, Arizona, in 1902, to the New York Botanical Garden.

173. Opuntia stricta Haworth, Syn. Pl. Succ. 191. 1812.

Cactus opuntia inermis De Candolle, Pl. Succ. Hist. **2**: pl. 138 [C]. 1799.[*]
Cactus strictus Haworth, Misc. Nat. 188. 1803.
Opuntia inermis De Candolle, Prodr. **3**: 473. 1828.
Opuntia airampo Philippi, Anal. Univ. Chile **85**: 492. 1894.
Opuntia parva Berger, Hort. Mortol. 411. 1912.
Opuntia bentonii Griffiths, Rep. Mo. Bot. Gard. **22**: 25. 1912.
Opuntia longiclada Griffiths, Bull. Torr. Club **43**: 525. 1916 (according to description and illustration).

Bushy, low, spreading plants, sometimes forming large clumps, seldom over 8 dm. high; joints obovate to oblong, usually 8 to 15 cm. long, but sometimes much elongated and then 30 cm. long or more, green or bluish green, glabrous, often spineless especially in greenhouse specimens, sometimes but a spine or two on a joint, at other times spines more abundant; leaves stout, subulate, 3 to 4 mm. long; areoles distant, the wool brownish, the glochids short; spines, when present, usually 1 or 2 from an areole, stiff, terete, yellow, 1 to 4 cm. long; flowers 6 to 7 cm. long; petals yellow, broad, obtuse, apiculate; filaments yellow to greenish; style usually white; stigma-lobes usually white but sometimes greenish; fruit purple, usually broadest at top, tapering to a slender base, 4 to 6 cm. long, with a more or less depressed umbilicus.

Type locality: Not given.

Distribution: Western Cuba; Florida to southern Texas.

Opuntia vulgaris balearica Weber (Dict. Hort. Bois 894. 1898) is given by Weber as a synonym of *O. inermis; Opuntia balearica* Weber (Hirscht, Monatsschr. Kakteenk. **8**: 175. 1898) has also been used, but not described, and Hirscht says it belongs here.

This species is often cultivated on the west coast of South America. It was there given the name *O. airampo* by Dr. Philippi, who supposed it to be the airampo of the Peruvians, a native species, quite different from this one.

This species is the pest pear of New South Wales and Queensland. It has now run wild over thousands of acres of the best agricultural and grazing land of the interior of

*Berger (Hort. Mortol. 411. 1912) gives the date 1797.

Australia.   J. H. Maiden says: "The growth of this *Opuntia* is one of the wonders of the world, and the spread of few plants in any country can be compared with it."

Illustrations: Dept. Agr. N. S. W. Misc. Publ. **253**: pl. [5]; Gard. Chron. III. **34**: f. 32; Gartenflora **31**: pl. 1082, f. d, e, f; De Candolle, Pl. Succ. Hist. **2**: pl. 138 [C]; De Tussac, Fl. Antill. **2**: pl. 34, the last two as *Cactus opuntia inermis;* Agr. Gaz. N. S. W. **23**: pl. opp. 713; pl. opp. 714; pl. opp. 716; Blühende Kakteen **2**: pl. 108, all these as *Opuntia inermis.*

Plate XXVII, figure 4, represents a flowering joint of the plant collected by Dr. Britton and John F. Cowell on limestone rocks near Pinar del Río, Cuba, in 1911.

173*a.* **Opuntia keyensis** Britton.   (See Appendix, p. 222.)

174. **Opuntia dillenii** (Ker-Gawler) Haworth, Suppl. Pl. Succ. 79.   1819.

> *Cactus dillenii* Ker-Gawler, Edwards's Bot. Reg. **3**: pl. 255. 1818.
> *Opuntia horrida* Salm-Dyck in De Candolle, Prodr. **3**: 472. 1828.
> *Opuntia maritima* Rafinesque, Atl. Journ. 146. 1832.
> *Opuntia tunoidea* Gibbes, Proc. Elliott Soc. Nat. Hist. **1**: 272. 1859.

Fig. 201.—Opuntia dillenii, Antigua, West Indies.

Low, spreading bushes growing in broad clumps and often forming dense thickets, or tall and much branched, 2 to 3 meters high, sometimes with definite terete trunks; joints obovate to oblong, 7 to 40 cm. long, the margin more or less undulate, bluish green, somewhat glaucous, but bright green when young, the areoles somewhat elevated; leaves subulate, curved backward, 5 mm. long; areoles often large, filled with short brown or white wool when young, usually few and remote, on old joints 10 to 12 mm. in diameter; spines often 10 from an areole on first-year joints, very variable, usually more or less flattened and curved, sometimes terete and straight, yellow, more or less brown-banded, or mottled, often brownish in age, sometimes 7 cm. long, but usually shorter, sometimes few or none; glochids numerous, yellow; wool in areoles short, sometimes brown, sometimes white; flowers in the typical form lemon-yellow, in some forms red from the first, 7 to 8 cm. long; petals broadly obovate, 4 to 5 cm. long; filaments greenish yellow; style thick, white; stigma-lobes white; fruit pear-shaped to subglobose, narrowed at base, 5 to 7.5 cm. long, purplish, spineless, juicy.

Type locality: Based on Dillenius's illustration.

Distribution: Coasts of South Carolina, Florida, Bermuda, the West Indies, east coast of Mexico, and northern South America; extending inland in Cuba.

This species is composed of many races varying greatly in habit, character and number of spines, shape of joints, and color of flowers. Brother León has sent us specimens of several individually quite different plants which inhabit hilltops in Cuba.

Opuntia lucayana Britton (Bull. N. Y. Bot. Gard. 4: 141. 1906), inhabiting Grand Turk Island, Bahamas, differs in having elongated, often narrowly oblong joints 2 to 4 times as long as wide and many elongated, little-flattened spines. It grows with *Opuntia dillenii* and *O. nashii*, and is believed to be a hybrid with these species as parents. A closely similar plant was observed on Buck Island, St. Thomas, Danish West Indies, growing immediately with *O. dillenii* and *O. rubescens*, the hybrid nature of which was unmistakable, and similar plants were seen also on Antigua, British West Indies.

Opuntia cubensis Britton and Rose (Torreya 12: 14. 1912), observed in a valley near the southern coast of Cuba at Guantánamo Bay, differs in having narrower joints, rather readily separable and smaller flowers, its stout spines little flattened. It grows near colonies of *Opuntia dillenii* and *O. militaris*, and is probably a hybrid between them.

Reference has already been made to the possible hybrid origin of *Opuntia antillana*, with *O. dillenii* as one of its parents. (See p. 115).

Two varieties of *Opuntia dillenii* are given by name only; *minor* Salm-Dyck (Hort. Dyck. 185. 1834); *orbiculata* Salm-Dyck (Cact. Hort. Dyck. 1849. 67. 1850).

Opuntia gilva Berger (Hort. Mortol. 233. 1912) is unpublished. The name was applied to a specimen collected by Carl F. Baker in Cuba in 1907, and has been distributed under this name. It is only a form of this very variable species.

The plant is hardy on the Gulf coast of the United States and in southern California. It is widely distributed through cultivation in the warmer parts of the Old World, being a "pest pear" in southern India and in Australia; it is used for hedges in Teneriffe, and is common along the sea on Grand Canary Island. On Bermuda, when growing in shade, the plant is often spineless, and its joints elongate sometimes to a length of 3 dm., while only 6 or 7 cm. wide. This elongation of the joints also appears in plants from Florida.

Illustrations: Edwards's Bot. Reg. 3: pl. 255, as *Cactus dillenii;* Rep. Mo. Bot. Gard. 22: pl. 1, 2, both these as *Opuntia bentonii;* Dillenius, Hort. Elth. 2: pl. 296, this as *Tuna major*, etc.; Amer. Journ. Pharm. 68: pl. opp. 169, as *Opuntia vulgaris;* Descourtilz, Fl. Med. Antill. 7: pl. 513, this as *Cactus opuntia*. Abh. Bayer. Akad. Wiss. München 2: pl. 3, f. 7 (?); Amer. Garden 11: 473 (?); Cycl. Amer. Hort. Bailey 3: f. 1545, 1546; Cact. Journ. 1: 154 (?); Dept. Agr. N. S. W. Misc. Publ. 253: pl. [2]; Dict. Gard. Nicholson 2: f. 757; W. Watson, Cact. Cult. f. 86, all these as *Opuntia tuna;* Journ. N. Y. Bot. Gard. 10: f. 26, this as *Opuntia inermis;* Loudon, Encycl. Pl. ed. 3. f. 6878, this as *Cactus tuna;* Britton, Fl. Bermuda 255.

Plate XXVIII, figure 2, represents a flowering joint of a plant collected in 1901 by N. L. Britton and J. F. Cowell on the Island St. Martin, West Indies; plate XXIX, figure 1, is from a photograph of the related *Opuntia keyensis* growing on Boot Key, Florida, taken by Marshall A. Howe in 1909; figure 2 is from a photograph of the plant on Bermuda, obtained by Dr. Britton in 1912. Figure 201 is from a photograph of the plant growing on Antigua, British West Indies, taken by Paul G. Russell in 1913.

175. Opuntia linguiformis Griffiths, Rep. Mo. Bot. Gard. 19: 270. 1908.

A bushy plant, 1 meter high or more; joints elongated, oblong to ovate-oblong or lanceolate, 2 to 5 dm. long or even more, often several times longer than wide, pale green and slightly glaucous; leaves 6 mm. long, terete; spines yellow, very slender, terete or nearly so; areoles filled with brown wool; flowers yellow, 7 to 8 cm. broad; petals broad; filaments white or greenish at base; stigma-lobes 9, green; ovary bearing numerous long glochids at the upper areoles; fruit reddish purple; seeds 3 or 4 mm. broad, acute on the back.

Type locality: Near San Antonio, Texas.
Distribution: Southern Texas, in the vicinity of San Antonio.

This plant is rather common in cultivation in the Southwest and is now found in most cactus collections. According to Dr. Griffiths, it is occasionally found wild near San Antonio. We have seen somewhat similar plants from near Brownsville, Texas, probably referable to one of the races of *Opuntia lindheimeri*.

On account of the shape of the joints, this species is commonly called cow's tongue or lengua de vaca.

Illustration: Rep. Mo. Bot. Gard. **19**: pl. 27, lower figure.

Plate xxx represents a flowering joint of a plant obtained by Dr. MacDougal from the collection of Professor J. W. Toumey at Tucson, Arizona, for the New York Botanical Garden in 1902.

176. Opuntia tapona Engelmann in Coulter, Contr. U. S. Nat. Herb. **3**: 423. 1896.

Low, spreading plants rarely over 6 dm. high; joints glabrous, orbicular to obovate, 20 to 25 cm in diameter, turgid, pale green; spines 2 to 4, yellow, one much longer, 5 to 7 cm. long, slender, porrect or sometimes curved downward; glochids brownish; fruit 4 to 6 cm. long, clavate, dark purple without, red within, spineless.

Type locality: Near Loreto, Lower California.

Distribution: Southern part of Lower California.

Figure 202 represents a joint of the plant collected by Dr. Rose on Pichilinque Island, Lower California, in 1911.

Related to *O. tapona*, but probably specifically distinct from it, is a plant growing in the mountains of Cedros Island, Lower California; it was recorded from this island by Dr. E. L. Greene as *O. engelmannii*, and a specimen was brought to Washington by Dr. Rose in 1911. This plant may be described as follows: About 1 meter high; joints oblong, large, 20 cm. long or more, smooth; areoles 3 cm. apart or more, very large, filled with brown wool; spines usually about 7, pale yellow, slender, terete, the longest ones 3 cm. long; glochids yellow. (Rose No. 16170.)

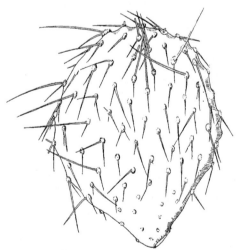

FIG. 202.—Opuntia tapona.   ×0.4.

FIG. 203.—Opuntia littoralis.

PLATE XXX

M. E. Eaton del.

Flowering joint of *Opuntia linguiformis.*
(Natural size.)

177. Opuntia littoralis (Engelmann) Cockerell, Bull. South. Calif. Acad. **4:** 15. 1905.

Opuntia engelmannii littoralis Engelmann in Brewer and Watson, Bot. Calif. **1:** 248. 1876.
Opuntia lindheimeri littoralis Coulter, Contr. U. S. Nat. Herb. **3:** 422. 1896.

Bushy plants, low and spreading; joints thick, orbicular to oblong, 15 cm. long or more, usually smaller in greenhouse plants, dull green; areoles rather closely set, large, often elevated on old joints; spines numerous, yellow, rather short on young joints (1 to 2 cm. long), but on old joints much longer, in age more or less flattened; wool of the areoles brown; flowers large, yellow, 8 to 12 cm. broad; sepals broad, apiculate; petals retuse; ovary with many areoles; fruit red, juicy; seeds 4 to 5 mm. in diameter.

Type locality: Coast from Santa Barbara to San Diego, California.
Distribution: Along and near the coast of southern California.

This species was very briefly described as a variety of *Opuntia engelmannii* in 1876. No definite locality was given for it, and the original material preserved is so poor that its identification is doubtful. We have taken as our representative of this species the low, bushy plant with rather thick joints, large and closely set areoles and yellow spines.

Opuntia littoralis often grows in proximity to *O. occidentalis* in southern California, and hybrids of the two may exist.

Figure 203 represents joints of the plant collected at Elsinore, California, by Dr. MacDougal in 1913.

178. Opuntia aciculata Griffiths, Proc. Biol. Soc. Washington **29:** 10. 1916.

Low, bushy plant, 1 meter high or more, often 3 meters broad or more, the lower branches decumbent and sending up erect branches; joints obovate, 12 to 20 cm. long, rounded at apex, dull dark green, somewhat glaucous, bearing large, closely set areoles, these often spineless; leaves subulate, 7 mm. long; spines several in a cluster, acicular, slender, 3 to 5.5 cm. long, often reflexed, brownish at base, with yellow tips, seemingly deciduous; glochids numerous, from all parts of the areoles, long, persisting for several years; flower golden yellow, sometimes with a greenish center, large, 8 to 10 cm. broad; petals broad, rounded or retuse; filaments yellowish; style dull yellowish green; stigma-lobes 8 to 10, green; fruit pyriform, purple.

Type locality: Near Laredo, Texas.
Distribution: On high gravelly ground at type locality and vicinity.

This species is not very common about Laredo, Texas, but grows in small colonies usually to the exclusion of all other plants. It can easily be distinguished from related species, and is usually restricted to dry hills. Our description is based on specimens obtained by Dr. Rose at Laredo in 1913. Since then it has been grown both in Washington and New York.

Plate xxviii, figure 3, represents a flowering joint of the plant collected by Dr. Rose near the type locality in 1913.

179. Opuntia lindheimeri Engelmann, Bost. Journ. Nat. Hist. **6:** 207. 1850.

Opuntia dulcis Engelmann, Proc. Amer. Acad. **3:** 291. 1856.
Opuntia lindheimeri dulcis Coulter, Contr. U. S. Nat. Herb. **3:** 421. 1896.
Opuntia engelmannii dulcis Schumann, Gesamtb. Kakteen 725. 1898.
Opuntia cacanapa Griffiths and Hare, N. Mex. Agr. Exp. Sta. Bull. **60:** 47. 1906.
Opuntia ferruginispina Griffiths, Rep. Mo. Bot. Gard. **19:** 267. 1908.
Opuntia tricolor Griffiths, Rep. Mo. Bot. Gard. **20:** 85. 1909.
Opuntia texana Griffiths, Rep. Mo. Bot. Gard. **20:** 92. 1909.
Opuntia subarmata Griffiths, Rep. Mo. Bot. Gard. **20:** 94. 1909.
Opuntia alta Griffiths, Rep. Mo. Bot. Gard. **21:** 165. 1910.
Opuntia gomei Griffiths, Rep. Mo. Bot. Gard. **21:** 167. 1910.
Opuntia sinclairii Griffiths, Rep. Mo. Bot. Gard. **21:** 173. 1910.
Opuntia cyanella Griffiths, Rep. Mo. Bot. Gard. **22:** 30. 1912.
Opuntia gilvoalba Griffiths, Rep. Mo. Bot. Gard. **22:** 35. 1912.
Opuntia convexa Mackensen, Bull. Torr. Club **39:** 290. 1912.
Opuntia griffithsiana Mackensen, Bull. Torr. Club **39:** 291. 1912.
Opuntia reflexa Mackensen, Bull. Torr. Club **39:** 292. 1912.
Opuntia deltica Griffiths, Bull. Torr. Club **43:** 84. 1916.
Opuntia laxiflora Griffiths, Bull. Torr. Club **43:** 85. 1916.
Opuntia flexospina Griffiths, Bull. Torr. Club **43:** 87. 1916.
Opuntia squarrosa Griffiths, Bull. Torr. Club **43:** 91. 1916.

Usually erect, 2 to 4 meters high, with a more or less definite trunk, but at times much lower and spreading; joints green or bluish green, somewhat glaucous, orbicular to obovate, up to 25 dm. long; leaves subulate, 3 to 4 mm. long, somewhat flattened, pointed; areoles distant, often 6 cm. apart; spines usually 1 to 6, often only 2, one porrect and 4 cm. long or more, the others somewhat shorter and only slightly spreading, pale yellow to nearly white, sometimes brownish or blackish at base, or some plants spineless; glochids yellow or sometimes brownish, usually prominent; petals yellow to dark red; stigma-lobes usually green; fruit purple, pyriform to oblong, 3.5 to 5.5 cm. long.

Type locality: About New Braunfels, Texas.

Distribution: Southwestern Louisiana, southeastern Texas, and Tamaulipas, Mexico.

Opuntia lindheimeri is an extremely variable species, composed of many races, differing in armament, color of flowers, size and shape of joints and of fruit. Certain forms have been described which in cultivation we have been able to recognize as possibly distinct; but in the field they seem to intergrade with other forms, indicating that they are at most only races of a very variable species. In the delta of the Rio Grande this is especially true, and from this region a number of species has been described. In fact, all the plants described as species which are cited above in the synonymy grow within a relatively small distributional area. Dr. Rose has examined all this region and is of the opinion that only one species of this series exists there, and this we believe is to be referred to *Opuntia lindheimeri*. It is very common about Brownsville and Corpus Christi, where it forms thickets covering thousands of acres of land. It is very variable in habit, being either low and widely spreading or becoming tall and tree-like, sometimes 3 meters high, with a definite cylindric trunk. Plants from these two extremes, if studied apart from the field, might be considered as different species, but in the field one sees innumerable intergrading forms. The low, prostrate forms gradually pass into others with more or less erect or ascending branches, while the large tree-like forms often bear large lateral branches which lie prostrate on the ground, indicating that they have developed from the prostrate ones. Decided differences in the flower colors have been pointed out in the original descriptions, and we have observed them in greenhouse specimens, but they do not correlate with other characters.

Opuntia ellisiana Griffiths (Rep. Mo. Bot. Gard. **21:** 170. pl. 25. 1910), an unarmed species, is known only from cultivated plants. Dr. Griffiths states that it is quite different from the *Ficus-indicae* series, which it much resembles, and is quite hardy in southern Texas. It may be a spineless race of the common *O. lindheimeri* of this region.

Opuntia pyrocarpa Griffiths (Bull. Torr. Club **43:** 90. 1916) we do not know; in its long pyriform fruit it suggests this plant; the type comes from Marble Falls, Texas.

O. winteriana Berger and *O. haematocarpa* Berger (Bot. Jahrb. Engler **36:** 455 and 456. 1905) are of this relationship, but have browner spines than is usual in the species.

Opuntia leptocarpa Mackensen (Bull. Torr. Club **38:** 141. 1911), characterized by its low, bushy habit and elongated, almost abnormal fruits, suggests a natural hybrid between *O. lindheimeri* and *O. macrorhiza*. Indeed, Mr. Mackensen described the species as intermediate between these two, and all three species are often found growing together. *O. leptocarpa* originally came from San Antonio, Texas.

Illustrations: Ann. Rep. Smiths. Inst. **1911:** pl. 3, 4, B; Cact. Mex. Bound. pl. 75, f. 5 to 7; Karsten, Deutsch. Fl. f. 501. 13, 13ᵃ, 13ᵇ; N. Mex. Agr. Exp. Sta. Bull. **78:** pl. [13, 14], all as *Opuntia dulcis*. Bull. U. S. Dept. Agr. **31:** pl. 3, f. 1, this as *Opuntia cacanapa;* Rep. Mo. Bot. Gard. **20:** pl. 4, in part, this as *Opuntia tricolor;* Rep. Mo. Bot. Gard. **20:** pl. 9; pl. 13, f. 1, these two as *Opuntia texana*. Bull. U. S. Dept. Agr. **31:** pl. 2, f. 1; Rep. Mo. Bot. Gard. **20:** pl. 2, f. 1; pl. 11; pl. 13, f. 4, all these as *Opuntia subarmata*. Rep. Mo. Bot. Gard. **21:** pl. 19; pl. 20, in part, these two as *Opuntia alta*. Rep. Mo. Bot. Gard. **21:** pl. 21; pl. 22, in part, these two as *Opuntia gomei*. Rep. Mo. Bot. Gard. **21:** pl. 28, this as *Opuntia sinclairii*. Rep. Mo. Bot. Gard. **22:** pl. 9, in part; pl. 10; Journ. Agr. Research **4.** pl. f., these three as *Opuntia cyanella*. Rep. Mo. Bot. Gard. **22:** pl. 9, in part; pl. 16, 17, these three as *Opuntia gilvoalba*. Bull. U. S. Dept. Agr. **31:** f. 1.

M. E. Eaton del.

Flowering joints of *Opuntia lindheimeri*.

1. Orange-flowered race.   2. Red-flowered race.
(Natural size.)

Plate XXXI, figure 1, represents a flowering joint of a plant collected near Brownsville, Texas, by Dr. Rose in 1913; figure 2 represents a flowering joint of a plant obtained by the same collector at the same locality; plate XXXII, figure 1, represents a flowering joint of a plant sent by Mr. M. Mackensen from the type locality of *O. leptocarpa* in 1910; figure 2 shows the fruit of the same.

180. Opuntia cantabrigiensis Lynch, Gard. Chron. III. **33**: 98. 1903.

> *Opuntia engelmannii cuija* Griffiths and Hare, N. Mex. Agr. Exp. Sta. Bull. **60**: 44. 1906.
> *Opuntia cuija* Britton and Rose, Smiths. Misc. Coll. **50**: 529. 1908.

Rounded bushy plant, 1 to 2 meters high; joints orbicular to obovate, 12 to 20 cm. long, rather pale bluish green; areoles remote, large, filled with brown wool; spines usually 3 to 6 but sometimes more, somewhat spreading, acicular, yellow with brown or reddish bases, 1.5 to 4 cm. long; glochids numerous, large, 1 cm. long or more, yellowish, not forming a brush; flowers 5 to 6 cm. long, yellowish with reddish centers; upper areoles on the ovary bearing long bristles; stigma-lobes green; fruit, globular, about 4 cm. in diameter, purple; seeds numerous, small, 4 mm. in diamter.

Type locality: Described from specimen in Cambridge Botanic Garden, England.

Distribution: Very common in the States of San Luis Potosí, Querétaro, and Hidalgo, Mexico.

Opuntia chrysacantha (Berger, Hort. Mortol. 231. 1912, name only), an undescribed species, probably belongs here.

Our determination of the identity of *O. cantabrigiensis* and *O. cuija* is based on a living plant of the former received from Mr. Lynch.

Illustrations: N. Mex. Agr. Exp. Sta. Bull. **60**: pl. 2, as *Opuntia engelmannii cuija;* Gard. Chron. III. **30**: f. 123, as *Opuntia engelmannii.*

Figure 204 represents joints of a plant collected by Dr. Rose near Ixmiquilpan, Hidalgo, Mexico, in 1905.

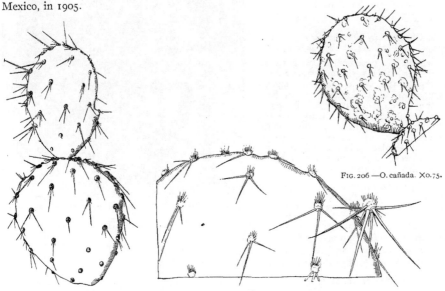

FIG. 206 —O. cañada. X0.75.

FIG. 204.—O. cantabrigiensis.   X0.4.

FIG. 205.—O. procumbens.   X0.5.

181. Opuntia procumbens Engelmann, Proc. Amer. Acad. **3**: 292. 1856.

Stems low and spreading, forming broad masses; joints "always edgewise," orbicular, 2 to 4 or even 5 dm. in diameter, yellowish green, somewhat glaucous; areoles distant (3 to 5 cm. apart), large, bearing long yellow glochids; spines 1 to 5, spreading, 2.5 to 5 cm. long, yellow, lighter above, flattened; flowers said to be yellow; fruit red, juicy.

Type locality: San Francisco Mountains to Cactus Pass, Arizona.

Distribution: Northern Arizona.

This species has long been wanting or poorly represented in our great herbaria. Dr. Rose collected it near Flagstaff, Arizona, and the above description is largely drawn from his notes; but his material was lost. In 1913 it was again collected by Mr. E. A. Goldman.

Illustration: Pac. R. Rep. 4: pl. 6, f. 4, 5.

Figure 205 is copied from the illustration above cited.

182. Opuntia cañada Griffiths, Rep. Mo. Bot. Gard. **20**: 90. 1909.

Plant about 1 meter high, with many erect or ascending branches, forming a broad top; joints ovate to obovate, 16 to 22 cm. long, smooth, and shining; leaves subulate, 1 cm. long; spines various, white to yellow, flattened, sometimes twisted; glochids few on young joints, very abundant on old ones; flowers yellow with red or orange centers; style white to reddish; stigma-lobes green; fruit red.

Type locality: Foothills of the Santa Rita Mountains, Arizona.

Distribution: Southeastern Arizona.

Dr. Griffiths comments on the close relationship of this plant to *O. laevis.*

Illustrations: Rep. Mo. Bot. Gard. 20: pl. 2, f. 6; pl. 6, in part; pl. 13, f. 2, 12.

Figure 206 is copied from the second illustration above cited.

183. Opuntia pyriformis Rose, Contr. U. S. Nat. Herb. **12**: 292. 1909.

Plant 3 to 5 meters high, with widely spreading branches, the lower ones almost resting on the ground and 3 to 5 meters long; joints obovate, thick, 18 cm. long or more; areoles closely set, small; spines 1 or 2, on old joints more, usually reflexed, slender, weak, yellow, 10 to 22 mm. long; flowers yellow; fruit 4 cm. long, somewhat tuberculate, spineless, its large areoles crowded with brown hairs forming hemispherical cushions.

Type locality: Hacienda de Cedros, Zacatecas, Mexico.

Distribution: Zacatecas, Mexico.

FIG. 207.—Opuntia pyriformis.   ×0.5.

The type of this species is in the U. S. National Herbarium. It is known only from the original collection of Professor F. E. Lloyd, made in 1908.

Illustrations: Contr. U. S. Nat. Herb. **12**: f. 35; pl. 26.

Figure 207 is copied from the second illustration above cited.

183 a. Opuntia bonplandii (HBK.) Weber. (See Appendix, p. 223.)

The three following described species may belong to this series:

OPUNTIA BECKERIANA Schumann, Gesamtb. Kakteen 722. 1898.

The plant on which this species is based was sent to Dr. Schumann from a garden at Bordighera, Italy, and its origin is unknown; Dr. Schumann thought that it might have come from Mexico. From the description it may belong to our series *Dillenianae*, but we are unable to associate it with any species known to us; the ovary is described as compressed and tubercled.

OPUNTIA ANAHUACENSIS Griffiths, Bull. Torr. Club **43**: 92. 1916.

A low, reclining or prostrate plant, up to 5 dm. high, 1.5 meters broad; joints obovate, glossy, yellowish green, 27 cm. long, 13 cm. broad; spines yellow or becoming white, 1 or 2, porrect, flattened, twisted, 2 or 3 cm. long; flowers yellow; style white; stigma-lobes 6, white; fruit dark purplish red, pyriform, 7 cm. long.

Type locality: Anahuac, Texas.

Distribution: Known only from the type locality, at the mouth of Trinity River, eastern Texas.

OPUNTIA MEGALANTHA Griffiths, Bull. Torr. Club **43**: 530. 1916.

A tall, erect, open-branching plant, 2 meters high or more; joints obovate, glaucous, grayish green, 21 cm. long, 14 cm. broad; spines yellow, 1 to 3, or even 5 or 6 on old wood, the longest often 4 to 5 cm. long; flowers yellow, 10 to 11 cm. in diameter; petals 5 cm. long, obovate; style white; stigma-lobes 8 or 9, white or tinged with green; fruit dark red.

Known only from cultivated plants received from the Berlin Botanical Garden, where it was grown as *Opuntia bergeriana*.

Series 16. MACDOUGALIANAE.

Erect, mostly tall species, with flat, broad, and thin, persistent joints, the epidermis, at least that of the ovary, pubescent or puberulent. The spines, when present, yellow. There are about half a dozen species, natives of central and southern Mexico.

KEY TO SPECIES.

FIG. 208.—Opuntia durangensis. X0.4.

Joints merely finely puberulent or
  glabrous; spines 1.5 cm. long or
  less; ovary velvety .... 184. *O. durangensis*
Joints distinctly pubescent; spines
  2 to 3 cm. long.
  Petals red.
    Style shorter than the petals 185. *O. atropes*
    Style as long as the petals. 186. *O. affinis*
  Petals yellow.
    Spines acicular, at first yel-
      low, soon white......... 187. *O. macdougaliana*
    Spines subulate.
      Petals retuse; areoles of
      ovary many, approxi-
      mate.............. 188. *O. velutina*
      Petals mucronate; are-
      oles of ovary few, dis-
      tant.............. 189. *O. wilcoxii*

184. Opuntia durangensis Britton and Rose, Smiths. Misc. Coll. **50**: 518. 1908.

Joints broadly obovate, about 20 cm. long, 16 cm. broad, pale green, glabrous or minutely puberulent, bearing numerous areoles; areoles 1 to 2 cm. apart, elevated; spines 3 to 5 at an areole, short, 1.5 cm. long or less, pungent, spreading, yellow, but in age becoming darker; glochids brown, 2 to 3 mm. long; flowers yellow, 5 cm. long; petals broad, apiculate; ovary 3 to 4 cm. long, finely pubescent, bearing many areoles with numerous glochids and a few spines; fruit white or red; seeds about 3 mm. broad.

Type locality: Near Durango, Mexico.

Distribution: Central Mexico.

This species was collected by the late Dr. E. Palmer in 1912, but he did not record the size and habit of the plant. The joints suggest a large, bushy species.

Figure 208 represents a joint of the type specimen.

185. Opuntia atropes Rose, Smiths. Misc. Coll. **50:** 518. 1908.

Plant 1 to 3 meters high, much branched; joints oblong to obovate, 20 to 30 cm. long, deep green, softly pubescent; young joints somewhat glossy, leaves 4 to 5 mm. long, acuminate, pubescent, standing almost at right angles to the joints, the tips reddish, areoles circular, filled with short tawny wool; young spines white or yellowish; old spines 3 to 6 cm. long, somewhat angled, standing almost at right angles to the joints, dark yellow or brown at the base, much lighter, often white above; glochids numerous, long, yellow; petals reddish; ovary pubescent, covered with large cushion-like areoles bearing long glochids near the top but with few spines or none, truncate at apex.

Type locality: Lava beds near Yautepec, Morelos, Mexico.
Distribution: Central Mexico.

186. Opuntia affinis Griffiths, Proc. Biol. Soc. Washington **27:** 27. 1914.

"A low, arborescent species, from 125 cm. high with us at 4 years of age to 2 m. or more in its natural habitat; joints obovate, 13 by 35 cm., broadly rounded above and gradually narrowed below, densely silky, villous to the touch, and villous nature plainly visible when viewed in proper light, slightly raised at areoles, the tubercles being surrounded by a sunken dark-green line; leaves small, subulate, pointed, scarcely 2 mm. in length; areole small, obovate, 3 mm. long, 25 to 30 mm. apart, white to gray; spicules light straw-colored, at first not conspicuous but rather in a connivent tuft, 3 mm. long; spines absent below and 1 to 5 in upper five-sixths of joint, straw-colored, becoming white the second year, the longest 3 cm. and others much shorter, increasing in age in both length

FIG. 209.—Opuntia macdougaliana, Tehuacán, Mexico.

and numbers, at 3 years often 10 in number and some 6 cm. long, divergent, flattened, angular, twisted; flowers dull dark-red in bud, with stigma protruding the day before the petals spread, small, about 3 cm. in diameter when opened, petals 20 to 25 mm. long, slightly, when at all, recurved, ribs of petals red and wings orange, filaments greenish below and pink above, style bright-glossy red, stigma dull greenish red, 4-parted, equaling the petals in length; ovary small, subglobose, deeply pitted, 15 to 17 mm. in diameter, with small subcircular to slightly transversely elongated, dirty brown areoles, 4 mm. apart; fruit small, subglobose, red."

Type locality: State of Oaxaca, Mexico.

Distribution: Known only from type locality.

Our examination of the type specimen of this species showed that it is closely related to *Opuntia macdougaliana*, differing in the color of its petals, which may not be a specific character.

Fig. 210.—Opuntia macdougaliana

187. Opuntia macdougaliana Rose, Smiths. Misc. Coll. **50:** 516. 1908.

Opuntia micrarthra Griffiths, Monatsschr. Kakteenk. **23:** 130. 1913.

Plant about 4 meters high, with a distinct cylindric trunk branching from near the base; joints oblong, 30 cm. long by 8 to 10 cm. broad, softly pubescent; areoles distinct, small; spines generally 4, one much longer (2.5 to 4 cm. long), somewhat flattened, yellowish, becoming whitish in age; glochids short, numerous, yellow; fruit globular to oblong, 5 cm. long, the surface divided into diamond-shaped plates, red, with a broad deep cup at apex, the numerous small rounded areoles filled with clumps of yellow glochids, very rarely with one or two spines.

Type locality: Near Tehuacán, Mexico.

Distribution: Southern Mexico.

Figure 209 is from a photograph of the type plant taken by Dr. MacDougal at El Riego, Tehuacán, Mexico, in 1906; figure 210 represents a plant grown from a cutting of the type.

188. Opuntia velutina Weber in Gosselin, Bull. Mus. Hist. Nat. Paris **10**: 389. 1904.

Opuntia nelsonii Rose, Smiths. Misc. Coll. **50**: 516. 1908.

Stems 1 to 4 meters high; joints flattened, oblong to pear-shaped in outline, 15 to 20 cm. long by 10 to 15 cm. broad near the top, pubescent, pale yellowish green in herbarium specimens; areoles 2 to 3 cm. apart; spines 2 to 6, yellow, becoming white in age, very unequal, the longer ones 3 to 4 cm. long; bristles many, yellow, becoming brownish; flowers rather small; petals yellow, 1 to 3 cm. long; ovary pubescent, bearing many yellowish brown bristles; filaments red; stigma-lobes pale green; fruit "dark red."

Type locality: In Guerrero, Mexico.

Distribution: Southern Mexico.

Plate XXXII, figure 3, represents a flowering joint of a plant collected at Tehuacán, Mexico, by Dr. MacDougal and Dr. Rose in 1906.

189. Opuntia wilcoxii sp. nov.

A tall, bushy plant, 1 to 2 meters high, very much branched; joints oblong, thinnish, large, 2 cm. long, dark green, more or less purplish about the large areoles, finely puberulent; glochids numerous, long, yellow; spines 1 to 3, one very long (5 to 6 cm. long), porrect, white or somewhat yellowish; flower, including ovary, 6 cm. long, yellow; petals oblong, mucronate; ovary bearing few large areoles, these filled with brown wool and yellow glochids; filaments short; style thick, 2 cm. long, with 10 stigma-lobes; fruit pubescent, 4 cm. long.

Very common on the hills in the coastal plain of western Mexico from southern Sonora to southern Sinaloa, Mexico, where it was frequently collected by Rose, Standley, and Russell in 1910; their No. 13546, with flower, from Fuerte, Sinaloa, is selected as the type of the species. It is named for Dr. Glover B. Wilcox, who first sent in living specimens in 1909.

Figure 211 represents a joint of the type specimen.

To this series belong two plants which we have not been able to identify but are here briefly characterized:

The first, a very peculiar species, collected by Rose, Standley, and Russell, March 14, 1910 (No. 12853), on the dry hills near Alamos, Sonora, Mexico, is unlike any of the described species. It is living both in Washington and New York, but it has not done well in cultivation. It may be described as follows:

Bushy; joints oblong, thickish, pale green in color, with very short puberulence, nearly or quite spineless; glochids yellowish or greenish, numerous; young areoles brown in the center, white-woolly in the margin; flowers and fruit not known.

FIG. 211.—Opuntia wilcoxii.   X0.4.

Dr. H. H. Rusby collected the second species on the Balsas River, southern Mexico. It comes from the region of *O. velutina*, but we do not know its flowers. It may be described as follows:

Joints oblong, 18 cm. long, but cultivated specimens smaller, usually obovate, dark green; spines few, short, at first white; young areoles large, bordered with white wool, bearing the spines and glochids from the center.

Living specimens are growing in the New York Botanical Garden under No. 32811.

Series 17. TOMENTOSAE.

Tall, erect, pubescent or puberulent species, with flat persistent joints, the spines, when present, white. We know three species, natives of Mexico and Guatemala.

M. E. Eaton del.

A. HOEN & CO

1. Upper part of flowering joint of *Opuntia leptocarpa*.
2. Fruit of the same.
3. Flowering joint of *Opuntia velutina*.
4. Upper part of joint of *Opuntia megacantha*.

(All natural size.)

KEY TO SPECIES.

Joints narrowly obovate.
    Joints grayish green, densely velvety.. 190. *O. tomentosa*
    Joints bright green, minutely puberulent.. 191. *O. tomentella*
Joints broadly obovate.. 192. *O. guilanchi*

190. Opuntia tomentosa Salm-Dyck, Observ. Bot. 3: 8. 1822.

    Cactus tomentosus Link, Enum. Hort. Berol. **2:** 24. 1822.
    Opuntia oblongata Wendland in Pfeiffer, Enum. Cact. 161. 1837.
    Opuntia icterica Griffiths, Monatsschr. Kakteenk. **23:** 138. 1913.

Becoming 3 to 6 meters high or more, with a broad top and a smooth trunk 10 to 30 cm. in diameter; joints oblong to narrowly obovate, 10 to 20 cm. long, velvety pubescent, somewhat tuberculate when young; glochids yellow; spines usually wanting but sometimes 1 or more appear; flowers orange-colored, 4 to 5 cm. long; filaments white or rose-colored; style dark carmine, longer than the stamens; stigma-lobes 5 or 6, white; fruit ovoid, red, sweetish; seeds 4 mm. broad.

FIG. 212.—Opuntia tomentosa.

Type locality: Not cited; doubtless Mexico.
Distribution: Central Mexico and as an escape in Australia.
This species was first described from cultivated plants and has long been a favorite. When grown out of doors, as it is in Bermuda, it forms a large and conspicuous plant. It is usually nearly or quite spineless, but plants which come from the Valley of Mexico are often spiny.
According to J. H. Maiden, this plant had been sent to him under the unpublished name *Opuntia lurida*, and as *O. pubescens*.
Illustrations: Agr. Gaz. N. S. W. **23:** pl. opp. 1028; Monatsschr. Kakteenk. **16:** 121; De Candolle, Pl. Succ. Hist. **2:** pl. 137 [A, B], this last as *Cactus cochenillifer* (*fide* Berger).

Plate XXXIII, figure 1, represents a fruiting joint of a plant raised from seeds received by the United States Department of Agriculture. Figure 212 is from a photograph of a plant near St. Georges, Bermuda, taken by Stewardson Brown in 1912.

191. Opuntia tomentella Berger, Monatsschr. Kakteenk. **22:** 147. 1912.

Bushy; joints obovate to oblong, 20 to 30 cm. long, 9 to 15 cm. broad, light green, somewhat shining, finely puberulent; areoles about 3 cm. apart; spines 1 or 2, acicular, white, short (7 to 10 mm. long), porrect, sometimes wanting; glochids few; flowers numerous, 5 to 6 cm. long; petals obovate, reddish yellow; filaments yellowish green; style rose-colored; stigma-lobes white; ovary tomentose, armed with numerous black glochids; fruit oblong, red, sour.

Type locality: In Guatemala.
Distribution: Guatemala.

This species was distributed by the late F. Eichlam, who sent plants both to Washington and to La Mortola, those sent to La Mortola being used by Mr. Berger for his description. The species is perhaps near the common Mexican species *O. tomentosa,* but does not grow so tall, and the tomentum is not so dense nor so soft.

Figure 213 represents a joint of a plant collected in Guatemala by F. Eichlam in 1909.

192. Opuntia guilanchi Griffiths, Rep. Mo. Bot. Gard. **19:** 265. 1908.

Becoming 1.5 to 2 meters high, often with a distinct trunk 1.5 to 2.5 cm. in diameter; joints broadly obovate, 14 to 16 cm. wide, 20 to 24 cm. long, minutely pubescent; spines at first white, slightly flattened, the longest 2 cm. long; glochids light yellow; fruit subglobose, 4 cm. in diameter, pubescent, variously colored, aromatic.

Type locality: Near the city of Zacatecas, Mexico.
Distribution: Zacatecas, Mexico.

Series 18. LEUCOTRICHAE.

This series is restricted to a single species. Schumann grouped as *Chaetophorae, O. leucotricha* with *O. ursina,* the latter a species with similar long bristles on the stem but otherwise very different, it being dry-fruited. *Opuntia leucotricha* is characterized by its long, weak, hair-like or bristle-like spines on many of the joints, especially the stem and very old joints. The fruit of this plant is very different from that of related series in that the pulp is fragrant and does not come free from the rind when mature.

Fig. 213.—Opuntia tomentella.
X0.4.

193. Opuntia leucotricha De Candolle, Mém. Mus. Hist. Nat. Paris **17:** 119. 1828.

Opuntia fulvispina Salm-Dyck in Pfeiffer, Enum. Cact. 164. 1837.
Opuntia leucotricha fulvispina Weber in Schumann, Gesamtb. Kakteen Nachtr. 157. 1903.

Often 3 to 5 meters high, with a large top; trunk as well as the older joints covered with long white bristles; joints oblong to orbicular, 1 to 2 cm. long, pubescent; areoles closely set, the upper part filled with yellow glochids, the lower part at first with only 1 to 3 weak white spines; flowers, including ovary, 4 to 5 cm. long; petals yellow, broad; ovary with numerous areoles, the upper ones bearing long, bristly glochids (1 cm. long); style red; stigma-lobes green; fruit variable, 4 to 6 cm. long, white or red, the rind not easily coming off from the pulp, aromatic, edible.

Type locality: In Mexico.
Distribution: Central Mexico.

Opuntia erythrocentron Lemaire (Förster, Handb. Cact. 492. 1846) was given as a synonym of *O. fulvispina.*

Opuntia leucosticta Wendland (Pfeiffer, Enum. Cact. 167, 1837) probably belongs here.

Opuntia leucacantha Link and Otto (Salm-Dyck, Hort. Dyck. 362. 1834), published first in 1834—although the name occurs in literature as early as 1830 (Verh. Ver. Beförd.

M. E. Eaton del.

1. Upper part of joint of *Opuntia tomentosa*.   2, 3.   Flowering joint and branch of *Opuntia brasiliensis*.
4.   Joint of *Grusonia bradtiana*.   (All natural size.)

Gartenb. 6: 434. 1830)—which was later taken up as *Consolea leucacantha* by Lemaire (Rev. Hort. 1862: 174. 1862), seems to belong here rather than to *O. spinosissima*. If it came from Mexico, as reported, it could not be *O. spinosissima* or any of its relatives, for none of them is known from Mexico.

Opuntia subferox Schott (Pfeiffer, Enum. Cact. 167. 1837) was given as a synonym of this species, while *O. leucacantha laevior* Salm-Dyck (Cact. Hort. Dyck. 1844. 47. 1845) and *O. leucacantha subferox* Salm-Dyck (Förster, Handb. Cact. 497. 1846) were supposed to be based on *O. subferox*.

Opuntia leucantha (De Candolle, Prodr. 3: 474. 1828), unpublished, is doubtless the same as *O. leucacantha*.

Fig. 214.—Opuntia leucotricha.

Opuntia fulvispina laevior Salm-Dyck (Pfeiffer, Enum. Cact. 164. 1837) and *O. fulvispina badia* Salm-Dyck (Cact. Hort. Dyck. 1849. 65. 1850) are given as synonyms of *O. leucotricha;* while *O. rufescens* Salm-Dyck (Förster, Handb. Cact. 493. 1846) is given as a synonym of *fulvispina laevior;* all these seem to belong here.

This is called durasnilla in Mexico. It is grown in Bermuda under the name of Aaron's Beard.

Illustrations: Engler and Prantl, Pflanzenfam. 3⁶ᵃ: f. 56, J; N. Mex. Agr. Exp. Sta. Bull. 60: pl. 4, f. 1, 2.

Plate xxxiv, figure 1, represents a flowering joint of a plant in the collection of the New York Botanical Garden. Figure 214 is from a photograph of a plant grown from a cutting received from the collection of M. Simon, St. Ouen, Paris, France, in 1901.

Series 19. ORBICULATAE.

We have retained the series *Criniferae*, although changing its name to *Orbiculatae*, but we have excluded *O. scheeri*, which was placed here by Schumann.   The species are characterized by long hairs produced from the areoles.   The species retained in the series are not closely related; while others, like *O. macrocentra*, in other series, sometimes produce long hairs from the areoles in the seedling stage, and *O. hyptiacantha* and some other species have a few hairs at the areoles of mature joints.

KEY TO SPECIES.

Hairs from the areoles of young plants long and white, long-persistent; plant low. .   194.   *O. orbiculata*
Hairs from the areoles of young joints of old plants early deciduous; plant tall. . .   195.   *O. pilifera*

194. Opuntia orbiculata Salm-Dyck in Pfeiffer, Enum. Cact. 156.   1837.

> *Opuntia crinifera* Salm-Dyck in Pfeiffer, Enum. Cact. 157.   1837.
> *Opuntia crinifera lanigera* Pfeiffer, Enum. Cact. 157.   1837.
> *Opuntia lanigera* Salm-Dyck, Cact. Hort. Dyck. 1849. 65.   1850.

A plant without a very definite trunk, about 1 meter high, often broader than high; joints green or bluish green, orbicular to obovate, sometimes spatulate, about 15 cm. long; leaves subulate, 2 to 3 mm. long; areoles small, in seedlings and young plants producing long white hairs or wool long-persistent; spines acicular, several, yellow; flowers yellow.

Type locality: Cited as Brazil, but undoubtedly by error.
Distribution: Northern Mexico.
Opuntia senilis Parmenteer is given by Pfeiffer (Enum. Cact. 157. 1837) as a synonym of *O. crinifera*, and *O. pintadera* by Salm-Dyck (Cact. Hort. Dyck. 1844. 47. 1845) as a synonym of *O. lanigera*.   They doubtless both belong here.
Opuntia metternichii Piccioli (Salm-Dyck, Cact. Hort. Dyck. 1844. 46. 1845) and *O. orbiculata metternichii* Salm-Dyck (Cact. Hort. Dyck. 1849. 68. 1850), names without descriptions, doubtless belong here.
We have studied living plants sent from the Berlin Botanical Garden as *O. crinifera* and from the Botanical Garden of Santiago, Chile, as *O. orbiculata;* the plant is not native in Chile.

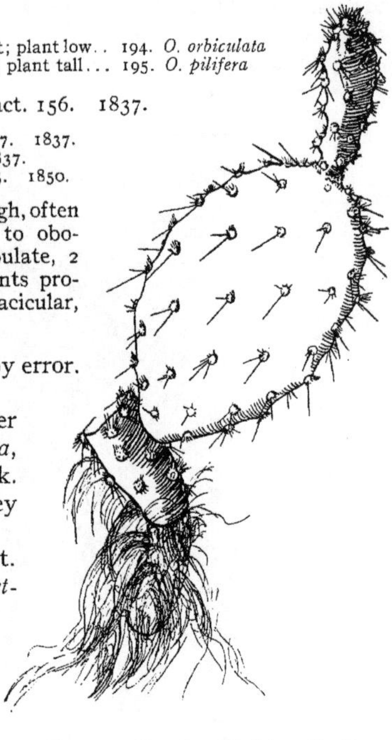

FIG. 215.—Opuntia orbiculata.   X0.66.

Illustration: Monatsschr. Kakteenk. 11: 155, as *Opuntia lanigera.*
Figure 215 represents joints of a plant sent from the Berlin Botanical Garden in 1902.

195. Opuntia pilifera Weber, Dict. Hort. Bois 894.   1898.

Becoming 4 to 5 meters high, with a definite, thick, woody, cylindric trunk and a broad, rounded top; joints oblong to orbicular, 1 to 3 dm. long, obtuse at apex, pale green; leaves subulate, about 5 mm. long; areoles 2 to 3 cm. apart, scarcely elevated; spines 2 to 9, white, slightly spreading, acicular; the outer part of the areole filled with nearly white, more or less deciduous hairs 2 to 3 cm. long; flowers large, red; areoles on the ovary bearing brown glochids and deciduous hairs, the latter especially abundant towards the top of the ovary; fruit red, juicy.

Type locality: In Mexico.
Distribution: Puebla, Mexico.
No definite locality was given for this species when it was first described, and apparently no type material was preserved; living specimens identified by Weber are still grown at La Mortola, Italy.   The species is common about Tehuacán, Mexico, being one of the large forms occurring in that region.   It is common in all large greenhouse collections.

Figure 216 is from a photograph of a plant in the collection of the New York Botanical Garden grown from a cutting brought by Dr. MacDougal and Dr. Rose from Tehuacán, Mexico, in 1906.

Series 20. FICUS-INDICAE.

Large plants, usually with large, nearly spineless green joints; spines, when present, few, small, white; flowers large, usually orange to yellow. None of the species is definitely known in the wild state, but all doubtless originated from tropical American ancestors, and they may all represent spineless races of plants here included in our series *Streptacanthae*. Some of them are cultivated for their fruit and others for forage.

KEY TO SPECIES.

Joints obovate to elliptic, comparatively
   broad, more or less glaucous.
   Joints dull.
      Joints thin, up to 5 dm. long.. 196. *O. ficus-indica*
      Joints thick, 15 cm. long or less. 197. *O. crassa*
   Joints glossy.................. 198. *O. undulata*
Joints elongated, comparatively narrow.
   Flowers yellow; joints somewhat
     tuberculate................... 199. *O. lanceolata*
   Flowers orange-red; joints not tuberculate................... 200. *O. maxima*

196. Opuntia ficus-indica (Linnaeus) Miller, Gard. Dict. ed. 8. No. 2. 1768.

> *Cactus ficus-indica* Linnaeus, Sp. Pl. 468. 1753
> *Cactus opuntia* Gussone, Fl. Sic. Prodr. 559 1827–8. Not Linnaeus.
> *Opuntia vulgaris* Tenore, Syll. Fl. Neap. 239. 1831. Not Miller.
> *Opuntia ficus-barbarica* Berger, Monatsschr. Kakteenk. **22**: 181. 1912.

FIG. 216.—Opuntia pilifera.

Large and bushy or sometimes erect and tree-like and then with a definite woody trunk up to 5 meters high, usually with a large top; joints oblong to spatulate-oblong, usually 3 to 5 cm. long, sometimes even larger; areoles small, usually spineless; glochids yellow, numerous, soon dropping off; leaves subulate, green, 3 mm. long; flowers large, normally bright yellow, 7 to 10 cm. broad; ovary 5 cm. long; fruit normally red, edible, 5 to 9 cm. long, with a low, depressed umbilicus.

Type localtiy: Tropical America.
Distribution: Native home not known, but now found all over the tropics and subtropics either as cultivated plants or as escapes. It is hardy in Bermuda and Florida.

This cactus is widely cultivated in all tropical and subtropical countries, where it is grown for its fruits and for forage. It has run wild in many waste places along the Mediterranean Sea, about the Red Sea, in southern Africa, and in Mexico.

We have not attempted to list the many named garden varieties of this species, which are sometimes Latin and sometimes English in form.

Opuntia amyclaea ficus-indica (Berger, Monatsschr. Kakteenk. **15**: 154. 1905) has never been described.

The origin of this common, cultivated species doubtless dates back to prehistoric times. We have long been convinced that it is a close relative of the *Streptacanthae*, and have kept it out of that series as only a matter of convenience. Mr. A. Berger believed it to be a spineless form of *O. amyclaea*, which is now a well-established species in certain parts of Italy. Dr. Griffiths has recently figured a reversion which appeared on the common spineless form which points very definitely to *O. megacantha* as the origin of this form.   (See Reversion in Prickly Pears, Journ. Hered. **5**: 222.   1914.)·

Illustrations: Amer. Garden ‵**11**: 471; Bull. U. S. Dept. Agr. **31**: pl. 1; pl. 2, f. 1; Cycl. Amer. Hort. Bailey **3**: f. 1543; Dept. Agr. N. S. W. Misc. Publ. **253**: pl. [1], f. 1, 3; Dict. Gard. Nicholson **2**: f. 753; Dodon. Pempt. f. 10, 11; Lemaire, Cact. f. 10; Meehan's Monthly **10**: 28; Mem. Acad. Neap. **6**: pl. 1, 2; Monatsschr. Kakteenk. **15**: 151; W. Watson, Cact. Cult. f. 8, in part; f. 80.

FIG 217.—Opuntia ficus-indica, Córdoba, Argentina.

Figure 217 is from a photograph of the plant growing at Córdoba, Argentina, taken by Paul G. Russell in 1915; figure 218 represents the fruit, obtained in Bermuda by Dr. Britton in 1913.

197. Opuntia crassa Haworth, Suppl. Pl. Succ. 81.   1819.

> *Opuntia parvula* Salm-Dyck, Hort. Dyck. 364.   1834.
> *Opuntia crassa major* Pfeiffer, Enum. Cact. 153.   1837.
> *Opuntia glauca* Forbes, Hort. Tour Germ. 158.   1837.

Plant 1 to 2 meters high, somewhat branched; joints ovate to oblong, 8 to 12.5 cm. long, thick, bluish green, glaucous; areoles bearing brown wool and brown glochids; spines wanting or sometimes 1 or 2, acicular, 2.5 cm. long or less; flowers and fruit unknown.

Type locality: Described from cultivated specimens supposed to have come from Mexico.

Distribution: Unknown in the wild state; locally found in cultivation in tropical America.

Haworth, who first described this species, thought it to be near *O. stricta.*

Pfeiffer (Enum. Cact. 153. 1837) uses *O. glaberrima* Hort. Berol. as a synonym of *O. crassa major.*

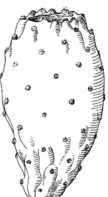

FIG. 218.—Fruit of Opuntia ficus-indica. ×0.66.

Opuntia parvula, when first published, was supposed to be native of Chile, but this was a mistake.   Salm-Dyck compared the species with *O. crassa* and *O. spinulifera,* but says it is thrice smaller than either. Schumann refers *O. parvula* directly to *O. crassa,* which disposition we follow.

Figure 219 is from a photograph of a plant in the Organ Mountains, Rio de Janeiro, Brazil, taken by Paul G. Russell in 1915.

198. Opuntia undulata Griffiths, Rep. Mo. Bot. Gard. **22**: 32. 1912.

Opuntia undosa Griffiths, Mönatsschr. Kaktenk. **23**: 139. 1913.

"Plant tall, large, stout, open-branching, with cylindrical trunk, often 30 cm. or more in diameter; joints very large, obovate, broadly rounded above, widest above middle, commonly 35 by 55 cm., firm, hard, quite fibrous, dished, wavy or flat, glossy light yellowish green at first, but changing through a darker green with a slight touch of glaucous to scurfy brown on old trunks; leaves subcircular in section, subulate, pointed, usually tinged with red at the tip, about 4 mm. long, upon a prominent tubercle and subtending a prominent dark-brown areole; areoles subcircular to ellipsoid or obovate, about 3.5 by 4.5 mm., gray, 5 to 6 cm. apart; spicules yellow in a short, compact tuft in upper part of areole, about 1 mm. long, soon becoming dirty and inconspicuous; spines white, few, short, erect, flattened, straight or twisted, 10 to 15 mm. long, 1 to 3 or 4, mostly one or none; fruit large, 4 to 5 by 9 to 10 cm., dull red to slightly tinged with orange and pulp streaked with red and orange when rind is removed."

FIG. 219.—Opuntia crassa.

Type locality: Described from cultivated plant obtained at Aguascalientes, Mexico.

Distribution: Mexico.

Illustrations: Rep. Mo. Bot. Gard. **22**: pl. 11, in part; pl. 12.

We have doubtfully referred to this species plants collected by Dr. Rose on the west coast of Mexico, where they were growing wild; this is some distance from the place where the type was obtained from cultivated plants. These specimens are like this species in having quite glossy joints with few spines. The plants were not in bloom when seen by Dr. Rose in the spring of 1910.

Dr. Griffiths has changed his first name, *O. undulata*, on account of the use of that name at an earlier time, which was not accompanied, however, by description.

199. Opuntia lanceolata Haworth, Syn. Pl. Succ. 192. 1812.

Cactus lanceolatus Haworth, Misc. Nat. 188. 1803.
Cactus elongatus Willdenow, Enum. Pl. Suppl. 34. 1813.
Opuntia elongata Haworth, Suppl. Pl. Succ. 81. 1819.

Plants tall, much branched; joints elongated, 3.5 cm. long, dull green, somewhat tuberculate; areoles distant, small; spines if present few, small, white, 1 cm. long or less; glochids yellow; flowers large, yellow.

Type locality: In South America.

Distribution: Known only in cultivation.

We have combined *O. lanceolata* and *O. elongata*, although there is a possibility of their being different. *O. lanceolata* was first described essentially as follows: Joints flattened, suberect, subnaked, with leaves 3 lines long; stems at first erect; joints lanceolate, green, when young with many leaves; spines (spicules?) in fascicles, the shortest of all species (except *Cactus coccinellifer*); leaves longer than in other species.

The species was received by Haworth from W. Anderson; no habitat given. In 1812 Haworth calls it the spear-shaped *Opuntia*. He says it probably came from South America, and flowers in July. It had been in cultivation before 1796; it flowered in 1808 with Haworth and was described as follows: Flowers shiny yellow; filaments yellow, half as long as petals; style longer than stamens; stigmas 5, thick, obtuse, 2 lines long, sulphur-colored.

De Candolle says the flowers are 4 inches in diameter.

Pfeiffer states the joints are 5 to 6 inches long by 1 to 1.5 inches broad; that the leaves are red and the spicules yellow.

Opuntia elongata laevior Salm-Dyck (Cact. Hort. Dyck. 1849. 242. 1850) may or may not belong here.

Fig. 220.—Opuntia maxima.

200. **Opuntia maxima** Miller, Gard. Dict. ed. 8. No. 5.   1768.

    Cactus decumanus Willdenow, Enum. Pl. Suppl. 34. 1813.
    Opuntia decumana Haworth, Rev. Pl. Succ. 71. 1821.
    Opuntia gymnocarpa Weber, Dict. Hort. Bois 893. 1898.
    Opuntia labouretiana Console* in Schumann, Gesamtb. Kakteen 717. 1898.
    Opuntia ficus-indica decumana Spegazzini, Anal. Mus. Nac. Buenos Aires III. 4: 512. 1905.
    Opuntia ficus-indica gymnocarpa Spegazzini, Anal. Mus. Nac. Buenos Aires III. 4: 512. 1905.

Forming large, much branched plants; joints elongated, more or less spatulate, 35 cm. long or more, 10 to 12 cm. broad, rounded at apex, somewhat cuneate at base, pale green, not at all tuberculate; areoles small, distant; spines sometimes wanting or sometimes 1 or 2, short, white; glochids yellow (brown in some specimens referred here); flowers conspicuous, 8 cm. broad, orange-red; ovary elongated, 7 to 8 cm. long, bearing numerous large glochids.

---

*Berger (Hort. Mortol. 409. 1912) says this is known as *O. labouretiana* Console.

PLATE XXXIV

M. E. Eaton del.

1. Part of joint of *Opuntia leucotricha*.   3. Joint of *Opuntia lasiacantha*.
2. Part of joint of *Opuntia maxima*.   4. Joint of *Opuntia robusta*.

(All natural size.)

Type locality: In America.

Distribution: Known only in cultivation.

Opuntia maxima Miller was described as the largest of all the opuntias and as the name is older than any of those here cited, it is taken up for this species. Haworth was uncertain whether or not his *O. decumana* is distinct from Miller's *O. maxima*, although in the Index Kewensis the two are considered the same; Burkill considered them distinct, but his idea of *O. decumana* is the *O. ficus-indica* type. Mr. Berger, on the other hand, states that it is evidently of the *O. dillenii* group, but this is hardly warranted by the description. Berger is convinced that *O. elongata* is distinct from *O. decumana*.

Opuntia labouretiana macrocarpa (Cat. Darrah Succ. Manchester 55. 1908) is only a garden name.

Plate xxxiv, figure 2, represents a flowering joint of a plant presented to the New York Botanical Garden by Frank Weinberg in 1901, which bloomed in May 1916. Figure 220 is from a photograph of the same plant.

Opuntia bartramii Rafinesque (Atl. Journ. 1:146. 1832) is based on Bartram's description (Travels p. 163. 1790), in which he states that the plant is 7 to 8 feet high; joints very large, bright green, glossy; spines none; glochids numerous; flowers large, yellow; fruit pear-shaped, purple. It was found about 6 miles from Lake George, northern Florida, associated with *Zamia pumila* and *Erythrina*. We do not know of any *Opuntia* answering the description, growing in Florida at the present time. Dr. Small visited the type locality in 1918 but failed to find any plant answering Rafinesque's description.

Opuntia hernandezii De Candolle (Mém. Hist. Nat. Paris 17: 69. pl. 16. 1828) is a complex. The reference to Hernandez applies to *Nopalea cochenillifera*. Schumann was not able to identify the plant illustrated by De Candolle, but thought it might be referable to *Opuntia ficus-indica*, in which we agree.

Series 21. STREPTACANTHAE.

Tall, branched, glabrous, green species with white or faintly yellow, acicular or subulate spines, large yellow or red flowers, and fleshy fruits, natives of Mexico and Central and South America. We recognize twelve species. The fruits, known as tunas, are mostly edible and are sold in large quantities in Mexican markets, a practice which probably dates from prehistoric time. The long-continued selection of plants for their fruit has perpetuated many slightly differing races.

KEY TO SPECIES.

```
Spines short, 5 mm. to 8 cm. long.
    Joints scarcely if at all tuberculate.
        Joints obovate to elliptic, mostly not more than twice as long as wide.
            Areoles close together, sunken...........................................  201. O. spinulifera
            Areoles not close together, not sunken.
                Joints dull.
                    Spines acicular.................................................  202. O. lasiacantha
                    Spines subulate.
                        Areoles with 2 or more short reflexed hairs or bristles at the lower part
                              of the areoles.
                            Spines strongly depressed; areoles with several hairs..............  203. O. hyptiacantha
                            Spines not strongly depressed; areoles with 1 or 2 hairs.
                                Joints obovate...........................................  204. O. streptacantha
                                Joints oblong............................................  205. O. amyclaea
                        Areoles without reflexed hairs or bristles.
                            Spines clear white, terete or nearly so; fruit spineless, 6 to 8 cm. long,
                                  yellow, edible...................................  206. O. megacantha
                            Spines white to dull yellow, somewhat flattened; fruit 6 cm. long or
                                  less bearing a few spines near the top, red, not edible.
                                Plant with a definite trunk; petals reddish; fruit spiny only at top  207. O. deamii
                                Plant bushy; petals chocolate-colored; fruit spiny all over.... 207a. O dobbieana
                Joints shining............................................................  208. O. eichlamii
        Joints oblong to oblanceolate, some of them much longer than wide.
            Joints shining; wool of young areoles white; petals yellow.................  209. O. inaequilateralis
            Joints dull; wool of young areoles brown; petals deep orange to scarlet.......  210. O. pittieri
    Joints strongly tuberculate.............................................................  211. O. cordobensis
Spines elongated, 10 to 14 cm. long.......................................................  212. O. quimilo
```

201. Opuntia spinulifera Salm-Dyck, Hort. Dyck. 364. 1834.

Opuntia candelabriformis Martius in Pfeiffer, Enum. Cact. 159. 1837.
Opuntia oligacantha Salm-Dyck, Cact. Hort. Dyck. 1849. 241. 1850.

Tall, much branched plants; joints orbicular to oblong, sometimes obovate, 20 to 30 cm. long, glabrous, a little glaucous; leaves small, red, 4 to 6 mm. long; areoles on young joints usually small, sometimes longer than broad, the margin at first bordered with cobwebby hairs, afterwards short white hairs, either spineless or with short white bristle-like spines; areoles on old joints more or less sunken, rather close together; spines on old joints 1 to 3, 1 to 2 cm. long, subulate, bone-colored.

Type locality: In Mexico.
Distribution: Mexico.

We have seen no wild specimens of this species, but Mr. Berger has grown it at La Mortola, Italy, and has distributed specimens now growing in New York and Washington.

So-called *Opuntia candelabriformis* and *O. oligacantha* are also in cultivation; but the original descriptions indicate that these two species should be merged into *O. spinulifera*, and plants so determined in European collections support this view. In so far as we have been able to ascertain, no type specimens of any of the three supposed species are extant. Schumann (Gesamtb. Kakteen 740. 1898) describes the flowers of *O. candelabriformis* as purple, 6 to 7 cm. broad. *Opuntia candelabriformis rigidior* Salm-Dyck (Cact. Hort. Dyck. 1849. 68. 1850), an unpublished variety, may belong here.

Figure 221 represents a joint of a plant presented to the New York Botanical Garden by Mrs. George Such in 1900.

202. Opuntia lasiacantha Pfeiffer, Enum. Cact. 160. 1837.

Opuntia megacantha lasiacantha Berger, Bot. Jahrb. Engler **36**: 453. 1905.

FIG. 221.—Opuntia spinulifera. ×0.4.

A tall plant, with a more or less definite trunk; joints obovate to oblong, 20 to 30 cm. long; leaves short, red; areoles small, 2 to 3 cm. apart; spines usually 1 to 3, acicular, white, 2 to 4 cm. long, slightly spreading; glochids numerous, prominent, dirty yellow to brown; flowers large, yellow or deep orange, 6 to 8 cm. broad; ovary bearing long, brown, deciduous bristles, especially from the upper areoles; style pinkish; stigma-lobes pale green.

Type locality: In Mexico.
Distribution: Central Mexico.

Schumann refers *O. lasiacantha* to *O. robusta*, but wrongly, as Berger states, and as living plants show. Pfeiffer said it is near *O. candelabriformis*, here taken up under *O. spinulifera*.

This species is very variable and, while it seems distinct from *O. megacantha*, it is to be noted that Mr. Berger referred it to that species as a variety.

Opuntia chaetocarpa Griffiths (Proc. Biol. Soc. Washington **27**: 25. 1914), in its few long white spines, resembles plants collected by Dr. Rose in southern Mexico which we have referred to this species.

Illustration: Addisonia **3**: pl. 90.

Plate XXXIV, figure 3, represents a flowering joint of a plant collected by Dr. Rose near the City of Mexico in 1906. Figure 222 represents a joint of a plant collected by Dr. MacDougal and Dr. Rose at Tehuacán, Mexico, in 1906.

OPUNTIA ZACUAPANENSIS Berger, Hort. Mortol. 413. 1912.

"A fine new species with bright-orange flowers. We received this plant a few years ago from M. L. Puteaux, Versailles, as *Opuntia spec.* from Zacuapan.* Joints 13 to 20 cm. long and 9.5 cm. broad, obovate, smooth, glossy green, areoles 15 to 25 mm. distant, slightly elevated, small, roundish or obovate. Spicules yellow, short, not numerous. Spines generally two, white, with yellowish points and base, terete, the lower deflexed shorter, the upper one spreading (2–) 3 cm. long. Flowers numerous from the top of the joint, 7.5 cm. long and 6.5 cm. broad, ovary obovate turbinate, 3.5 to 4 cm. long and 22 mm. broad, areoles somewhat elevated, prickly; petals obovate lanceolate, acute and aristate, orange-yellow, with a more reddish-brown hue along the midrib on the back and as well on the shorter obtuse outer petals; stamens yellow, style yellowish, thickened or clavate above the base, stigmata (6–) 7, dirty rose-coloured."

We have studied a plant, sent from La Mortola to the New York Botanical Garden in 1913, which has not flowered; it appears to be related to *O. lasiacantha*.

Figure 223 represents a joint from the plant received from La Mortola, Italy, in 1913.

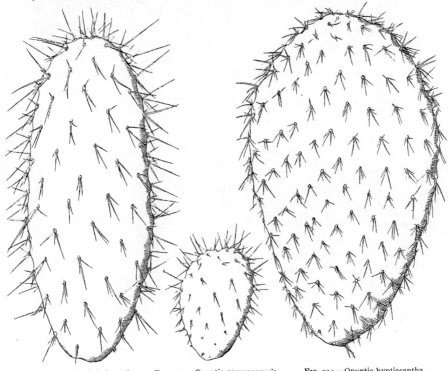

FIG. 222.—Opuntia lasiacantha. ×0.4. FIG. 223.—Opuntia zacuapanensis. ×0.4. FIG. 224.—Opuntia hyptiacantha. ×0.5.

203. **Opuntia hyptiacantha** Weber, Dict. Hort. Bois 894. 1898.

Opuntia nigrita Griffiths, Rep. Mo. Bot. Gard. **21**: 169. 1910.
? *Opuntia cretochaeta* Griffiths, Proc. Biol. Soc. Washington **29**: 11. 1916.

A tall, much branched plant, but in cultivation often only 1 meter high; joints oblong to obovate, 20 to 30 cm. long, pale green, but when young bright green; spines on young joints single, porrect, and accompanied by 2 or 3, sometimes many, white, slightly pungent hairs; spines on old joints

*Perhaps Zacualpan, in Vera Cruz, Mexico.

4 to 6 (in the original description 8 to 10), somewhat spreading or appressed, 1 to 2 cm. long; glochids few, brownish; areoles small, 1.5 cm. apart; leaves small, brownish; flowers red; fruit globular, yellowish, its areoles filled with long, weak glochids; umbilicus broad, only slightly depressed.

Type locality: In Mexico.

Distribution: Oaxaca, Mexico.

This species is very near *Opuntia streptacantha*, and in many cases it is difficult to separate them. It is also near *O. pilifera*, but the areoles are not so hairy. Weber, who first described it, gives no definite locality for the species; but Dr. Rose has examined, at La Mortola, Italy, a living plant sent by Weber which seems to be the same as one of the large opuntias from Tehuacán, Mexico.

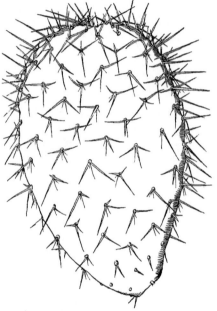

Opuntia chavena Griffiths (Rep. Mo. Bot. Gard. **19**: 264. pl. 23, in part. 1908) is a near relative of *O. hyptiacantha* or not distinct from it.

Illustration: Rep. Mo. Bot. Gard. **21**: pl. 24, as *Opuntia nigrita*.

Figure 224 represents a joint of a plant obtained for the New York Botanical Garden from the collection of M. Simon, St. Ouen, Paris, France, in 1901.

204. Opuntia streptacantha Lemaire, Cact. Gen. Nov. Sp. 62. 1839.

Much branched, up to 5 meters high, sometimes with a trunk 45 cm. in diameter; joints obovate to orbicular, 25 to 30 cm. long, dark green; areoles small, rather close together for this group; spines numerous, spreading or some of them appressed, white; glochids reddish brown, very short; flowers 7 to 9 cm. broad, yellow to orange, the sepals reddish; filaments greenish or reddish; stigma-lobes 8 to 12, green; fruit globular, 5 cm. in diameter, dull red or sometimes yellow, both within and without.

Type locality: Not cited.

Distribution: Very common on the Mexican table-lands, especially on the deserts of San Luis Potosí.

Fig. 225.—Opuntia streptacantha. ×0.5.

This species is known as tuna cardona or nopal cardón, and is one of the most important economic opuntias in Mexico. It has many forms and seems to grade into some of the species which we have here recognized.

Opuntia cardona Weber (Dict. Hort. Bois 895. 1898) and *O. coindettii* Weber (Dict. Hort. Bois 895. 1898) are two names given as synonyms of the species by Weber, but they were never published. *O. diplacantha* (Berger, Hort. Mortol. 232. 1912) must be referred here, but of this, so far as we know, there is no published description. Berger has distributed living specimens which we are inclined to refer here.

Opuntia pachona Griffiths (Rep. Mo. Bot. Gard. **21**: 168. pl. 22. 1910) is closely related to *O. streptacantha*, if not a race of that species. *Opuntia megacantha tenuispina* Salm-Dyck (Cact. Hort. Dyck. 1844. 45. 1845) was given as a new name for *O. lasiacantha*, but was never described.

Illustrations: N. Mex. Agr. Exp. Sta. Bull. **60**: pl. 1; Safford, Ann. Rep. Smiths. Inst. 1908: pl. 9, f. 6; U. S. Dept. Agr. Bur. Pl. Ind. Bull. **102**[1]: pl. 1; **116**: pl. 1, this last as tuna cardona; Engler and Prantl, Pflanzenfam. 3^[6a]: f. 70, this last as *Opuntia pseudotuna*.

Figure 225 represents a joint of a plant received from C. Wercklé in 1902 as *O. cardona*.

205. Opuntia amyclaea Tenore, Fl. Neap. Prodr. App. **5:** 15. 1826.

Opuntia ficus-indica amyclaea Berger, Hort. Mortol. 411. 1912.

Erect; joints oblong to elliptic, 3 to 4 dm. long, about twice as long as broad, thick, dull green, a little glaucous; leaves 4 mm. long, acute, red; areoles small, with 1 or 2 short bristles from the lower parts of areoles; spines 1 to 4, stiff, nearly porrect, usually less than 3 cm. long, white or horn-colored, the stoutest angled; glochids brown, soon disappearing; flowers yellow; fruit yellowish red, not very juicy.

Type locality: Described from specimens grown in Italy.

Distribution: Doubtless Mexico, but not known in the wild state.

Our description is based on the original description and a specimen collected by A. Berger near Palermo, where it is grown as a hedge plant. Berger's plant suggests very much *O. streptacantha*, but is not quite so spiny; it does not suggest very much *O. ficus-indica*, where Berger has placed it. Our description of the spines is taken from Berger's plant, while the original description states that the spines are 3 to 8, stout, spreading, unequal, white, the longest 35 mm. long.

O. alfagayucca (Salm-Dyck, Cact. Hort. Dyck. 1849. 68. 1850) and *O. alfayucca* (Rümpler in Förster, Handb. Cact. ed. 2. 938. 1885) were given as synonyms of *O. amyclaea.*

FIG. 226.—Opuntia megacantha.

206. Opuntia megacantha Salm-Dyck, Hort. Dyck. 363. 1834.

Opuntia castillae Griffiths, Rep. Mo. Bot. Gard. **19:** 261. 1908.

? *Opuntia incarnadilla* Griffiths, Rep. Mo. Bot. Gard. **22:** 27. 1912.

Plant tall, 4 to 5 meters high or more, with a more or less definite woody trunk; joints of large plants obovate to oblong, often oblique, sometimes 40 to 60 cm. long or more, but in greenhouse specimens often much smaller, pale dull green, slightly glaucous; leaves minute, often only 3 mm. long, green or purplish; areoles rather small, on large joints often 4 to 5 cm. apart, when young bearing brown wool; spines white, usually 1 to 5, slightly spreading, sometimes nearly porrect, usually only 2 to 3 cm. long, sometimes few and confined to the upper areoles; glochids few, yellow, caducous, sometimes appearing again on old joints; flowers yellow to orange, about 8 cm. broad; ovary spiny or spineless, obovoid; fruit 7 to 8 cm. long.

FIG. 227—Opuntia megacantha on Lanai, Hawaiian Islands.

Type locality: In Mexico.

Distribution: Much cultivated in Mexico; grown also in Jamaica and southern California and escaped from cultivation in Hawaii.

This species was originally described by Salm-Dyck essentially as follows: Erect and of the size of *O. decumana;* joints 17.5 cm. long by 7.5 cm. broad and 2.5 cm. or more thick; areoles close together, filled with gray wool; glochids brownish, becoming blackish; spines 7 to 10, white, unequal, acicular, somewhat radiating, the longest one deflexed, 5 cm. long; flowers not known; leaves small, reddish.

Opuntia megacantha trichacantha Salm-Dyck was given as a synonym of this species by Förster (Handb. Cact. 486. 1846), but was never published.

Opuntia tribuloides Griffiths (Monatsschr. Kakteenk. **23:** 137. 1913), according to the description, is of this relationship.

This is the chief Mission cactus. It is the one from which the best varieties of edible tunas are obtained and is one of the commonest cultivated opuntias in Mexico, having numerous forms, many of them bearing local names.

Illustrations: Ariz. Agr. Exp. Sta. Bull. **67:** pl. 8, f. 2; Rep. Mo. Bot. Gard. **19:** pl. 24, both as *Opuntia castillae.* Rep. Mo. Bot. Gard. **22:** pl. 4, 5, these two as *Opuntia incarnadilla;* Amer. Journ. Bot. **4:** 572. f. 6.

Plate XXXII, figure 4, represents a flowering joint of a plant in the same collection received from Fairmount Park, Philadelphia, in 1905. Figure 226 is from a photograph of a plant in the collection of the New York Botanical Garden;

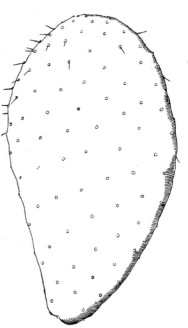

FIG 228.—Opuntia megacantha.   X0.4.

figure 227 is from a photograph taken by A. S. Hitchcock on Lanai in 1916; figure 228 represents a joint of a plant obtained by Dr. MacDougal near Mount Wilson, California, in 1906, a nearly spineless form.

207. Opuntia deamii Rose, Contr. U. S. Nat. Herb. **13**: 309. 1911.

One meter or so high, with a definite cylindric trunk, branching a short distance above the base; branches few, ascending; joints erect or spreading, very large, obovate to oblanceolate, 25 to 30 cm. long, at first bright leaf-green, in age dark green, glabrous; areoles remote, often 4 cm. apart, rather small; spines 2 to 6, usually 4, white or dull yellow, stout, somewhat flattened, spreading or porrect, 3 to 5.5 cm. long; flowers 7 cm. long, reddish; fruit oblong, 6 cm. long, naked, except for a few spines near the top, wine-red both within and without, not edible; seeds small, 3 mm. broad.

Type locality: Fiscal, Guatemala.
Distribution: Fiscal to San José de Golfo and Sanarata, Guatemala.
Illustration: Contr. U. S. Nat. Herb. **13**: pl. 65.

Figure 229 represents a joint of the type specimen.

A tall, white-spined *Opuntia*, closely resembling the Mexican *O. macracantha*, was obtained by Dr. Rose in 1918 (No. 22390) along roadsides at Ambato, Ecuador, presumably escaped from cultivation; its fruit is edible.

207a. **Opuntia dobbieana** sp. nov. (See Appendix, p. 225.)

208. Opuntia eichlamii Rose, Contr. U. S. Nat. Herb. **13**: 310. 1911.

Tree-like, 5 to 6 meters high, the main branches nearly erect; joints obovate to orbicular, 15 to 20 cm. long, more or less glaucous, especially in dried specimens; leaves minute, caducous; areoles small, 3 to 3.5 cm. apart; spines 4 to 6, very unequal, 2 cm. long or less, rose-colored at first, soon becoming white, spreading, the larger ones flattened; glochids brown; flower 3.5 cm. long; petals carmine; style red; stigma-lobes 8 to 11, bright green; fruit 4 cm. long, strongly tuberculate, not edible.

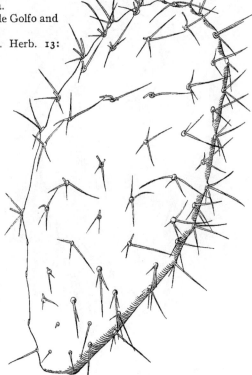

FIG. 229.—Opuntia deamii.

Type locality: Near Guatemala City.
Distribution: Suburbs of Guatemala City, Guatemala.
Illustration: Contr. U. S. Nat. Herb. **13**: pl. 66.
Figure 230 represents a joint of the type specimen.

209. Opuntia inaequilateralis Berger, Bot. Jahrb. Engler **36**: 453. 1905.

About 12 dm. high, with spreading branches; joints oblique, narrowly ovate to subrhomboid, 40 to 70 cm. long, 2 to 4 times as long as wide, narrowed at base, obtuse at apex, with somewhat sinuate margins, green, shining; young joints bright green, not at all glaucous, oblanceolate to narrowly oblong, rounded at apex; leaves reddish, subulate, 2 to 3 mm. long; areoles small, circular, filled with white wool when young, and having white, somewhat cobwebby hairs on the outer edge; glochids brown, in a dense cluster; spines 3 to 7, acicular on young joints, but finally 10 to 15, stout, 3 to 4 cm. long, at first yellowish, becoming white, somewhat spreading but not appressed to the joint; flowers large, borne at the apex of the joints; petals yellow, broadly obovate, retuse with crenulate margins; stigma-lobes green; fruit oblong, truncate, reddish, juicy, sweet.

Type locality: Described from cultivated specimens grown at La Mortola, Italy.
Distribution: Known only from cultivated specimens, their origin unknown.
Illustration: Figure 231 shows a joint of a plant sent from La Mortola, Italy, in 1913.

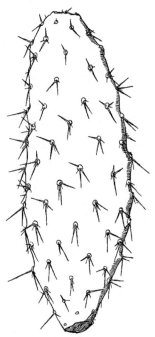

FIG. 230.—Opuntia eichlamii. ×0.5.   FIG. 231.—Opuntia inaequilateralis. ×0.5.

210. Opuntia pittieri sp. nov.

Plant up to 5 meters high, with a rather definite cylindric spiny trunk; joints large, 25 to 50 cm. long, 2 to 4 times as long as wide, narrowly oblong, green; leaves subulate, with purple tips; wool in young areoles dark brown to purple; areoles elevated, rather large, 2 to 3 cm. apart; spines 3 to 6, slightly spreading, acicular, white, the longest 2 to 2.5 cm. long; glochids tardily developing, few, often wanting; flowers deep orange, turning to scarlet; ovary nearly globular, more or less spiny, nearly truncate at apex.

Collected at Venticas del Dagua, Dagua Valley, western cordillera of Colombia, February 1906, by H. Pittier, and since grown in Washington and New York.

Opuntia pittieri differs from *O. inaequilateralis* in having the young joints thinner, somewhat tuberculate, and with longer leaves; the areoles, too, are filled with brown or purple wool, while the glochids develop more slowly or never appear.

Figure 232 represents a joint of the type plant.

211. Opuntia cordobensis Spegazzini, Anal. Mus. Nac. Buenos Aires III. **4**: 513. 1905.

Much branched, the trunk 1 to 2 meters long, 20 cm. in diameter, very spiny; joints large, 3 dm. long or more, broadly oblong to obovate; areoles prominent, numerous; spines 1 to 6, white, somewhat spreading, a little flattened and twisted; flowers usually on the margins of the joints; petals about 12, yellow; fruit pyriform, yellowish both within and without, 8 cm. long; seeds about 3 mm. long.

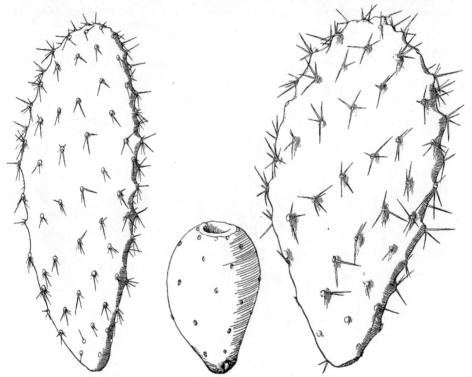

FIG. 232.—O. pittieri. ×0.4.   FIG. 234.—Fruit of O. cordobensis. ×0.7.   FIG. 233.—O. cordobensis. ×0.4.

Type locality: Near Córdoba, Argentina.

Distribution: Northern Argentina.

The only white-spined species observed by Dr. Rose in 1915 about Córdoba were *O. ficus-indica*, in cultivation, and what we have taken to be *O. cordobensis*. The latter is very abundant, growing on the hills about the city, and sometimes planted as hedges. Dr. Spegazzini states that it has the habit of *O. labouretiana*.

Figure 233 represents a joint of the plant collected by Dr. Rose near Córdoba, Argentina, in 1915; figure 234 represents the fruit as collected by J. A. Shafer at Calilegua, Argentina, in 1917 (No. 197).

212. Opuntia quimilo Schumann, Gesamtb. Kakteen 746. 1898.

Much branched, about 4 meters high; joints large, elliptic or obovate, 5 dm. long by 2.5 dm. broad, 2 to 3 cm. thick, grayish green; spines very long, usually 1, sometimes 2 or 3 from an areole, twisted, 7 to 14.5 cm. long; flowers red, 7 cm. broad; fruit pear-shaped to globular, 5 to 7 cm. long, greenish yellow; seeds 8 mm. broad, 1.5 to 2 mm. thick, with broad, thick, white margins.

Type locality: La Banda, Santiago del Estero, Argentina.
Distribution: Northern Argentina.
This plant is known to the natives as quimilo.

Dr. Rose obtained a good photograph of it from Dr. J. A. Dominguez, and seed and a photograph from Dr. Spegazzini. While the volume was going through the press a fine specimen in fruit with the long spines so characteristic of this species was obtained by H. M. Curran at Quilino, Córdoba, Argentina. Dr. Shafer's specimens collected at Río Piedras, show that the trunk-areoles sometimes bear as many as eight spines.

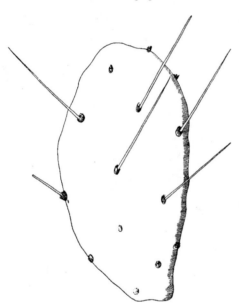

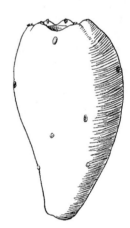

FIG. 235.—Joint of Opuntia quimilo. ×0.3.   FIG. 236.—Fruit of Opuntia quimilo. ×0.3.

Figure 235 represents a joint obtained by Dr. Shafer at Río Piedras, Salta, Argentina, January 4, 1917 (No. 34); figure 236 represents the fruit from the same plant; figure 237 is from a photograph of a flowering joint of the plant, contributed by Dr. Spegazzini.

The following may belong to this series:

OPUNTIA ITHYPETALA Griffiths, Bull. Torr. Club **43**: 529. 1916.

Tall, erect plant, 2 meters or more high; joints large, obovate, 26 to 45 cm. long, 14 to 19 cm. broad, much contracted below, bright dark green, somewhat tuberculate at the areoles; subulate, 5 to 6 mm. long; areoles large, often 1 cm. in diameter, 4 to 5 cm. apart; spines white at least on second year's growth, 3 to 5; central spine largest, porrect, 3 to 4 cm. long; flowers yellow, fading to rose-purplish; petals erect, 3 cm. in diameter; style white; stigma-lobes 6, light green.

Known only from cultivated plants received from the Berlin Botanical Garden.

Series 22. ROBUSTAE.

Tall or large plants with blue or bluish green joints, the spines, when present, white or yellowish. Two of the species are widely distributed in warm regions through cultivation for their edible fruits; the other is known in cultivation only in central Mexico. All are presumably Mexican in origin.

KEY TO SPECIES.

Joints orbicular to broadly obovate or
  elliptic.
  Fruit deep red, 7 to 9 cm. in diameter. 213. *O. robusta*
  Fruit greenish white, 4 to 5 cm. in
    diameter...................... 214. *O. guerrana*
Joints oblong, narrowed at both ends.. 215. *O. fusicaulis*

213. Opuntia robusta Wendland in Pfeiffer, Enum. Cact. 165. 1837.

> *Opuntia flavicans* Lemaire, Cact. Gen. Nov·
>   Sp. 61. 1839.
> *Opuntia larreyi* Weber in Coulter, Contr. U. S.
>   Nat. Herb. **3**: 423. 1896.
> *Opuntia gorda* Griffiths, Monatsschr. Kakteenk.
>   **23**: 134. 1913.

FIG. 237.—Opuntia quimilo.

Often erect, sometimes 5 meters high, usually much branched; joints orbicular to oblong, 20 to 25 cm. long by 10 to 12.5 cm. broad, very thick, bluish green, glaucous; leaves 4 mm. long, reddish, acute; spines 8 to 12, stout, very diverse, brown or yellowish at base, white above, up to 5 cm. long, but often wanting on greenhouse specimens; flowers 5 cm. broad, yellow; stigma-lobes green; fruit globular to ellipsoid, at first more or less tuberculate, deep red, 7 to 9 cm. long.

Type locality: In Mexico.

Distribution: Central Mexico; cultivated in Argentina.

This is one of the few species of *Opuntia* of which we have not been able to verify the original publication. It was redescribed by Pfeiffer in 1837.

Opuntia camuessa Weber (Dict. Hort. Bois 895. 1898) was given as a synonym of *O. robusta*, but was never described; and the same is true of *O. piccolominiana* Parlatore (Schumann, Gesamtb. Kakteen 741. 1898).

The variety *Opuntia robusta viridior* Salm-Dyck (Förster, Handb. Cact. 487. 1846) was never described.

Opuntia albicans Salm-Dyck (Hort. Dyck. 361. 1834) we do not know, but A. Berger, who has grown a plant under that name at La Mortola, says it is closely related to *O. robusta*, while in the New York Botanical Garden are specimens labeled *O. albicans* which are difficult to distinguish from *O. ficus-indica*. Here belong the following: *O. prate* Sabine (Pfeiffer, Enum. Cact. 155. 1837), given as a synonym of *O. albicans; O. albicans laevior* Salm-Dyck (Cact. Hort. Dyck. 1849. 67. 1850), name only; and *O. pruinosa* Salm-Dyck (Cact. Hort. Dyck. 1849. 67. 1850) given as a synonym of *O. albicans laevior.*

Opuntia larreyi, a manuscript name of Weber, which was published by Coulter in 1896, is based on the plant known to the Mexicans as camuessa. Weber gave it the name of *O. camuessa*, as shown above, but did not publish it; it is usually considered to be only a race of *O. robusta*, but Dr. Griffiths considers it a distinct species, even referring it to a different series, the *Ficus-indicae* (N. Mex. Agr. Exp. Sta. Bull. **64**: 56. 1907).

Berger remarks that this species is very variable, but that it can not well be divided even into varieties.

Opuntia megalarthra Rose (Smiths. Misc. Coll. **50:** 529. 1908), in its very spiny joints, yellow spines, and small fruits, seems very different from the common cultivated *O. robusta;* yet when grown in the greenhouse for several years it takes on much the appearance of *O. robusta.*   If this view is correct, *O. megalarthra* represents the wild form of the species.

Opuntia cochinera Griffiths (Rep. Mo. Bot. Gard. **19:** 263. pl. 26. 1908) from Zacatecas, Mexico, is, perhaps, a hybrid between *Opuntia robusta* and one of the *Streptacanthae.*

Illustrations: N. Mex. Agr. Exp. Sta. Bull. **60:** pl. 5, f. 1; Monatsschr. Kakteenk. **23:** 135; Journ. Inter. Gard. Club **3:** 14, the last two as *Opuntia gorda;* U. S. Dept. Agr. Bur. Pl. Ind. Bull. **74:** pl. 5, as Tapuna pear.   ? N. Mex. Agr. Exp. Sta. Bull. **64:** pl. 1.

Plate xxxiv, figure 4, represents a joint of the plant collected by Dr. Rose in Hidalgo, Mexico, in 1905, and described by him as *Opuntia megalarthra.*   Figure 238 is from a photograph taken in Zacatecas, Mexico, by Professor F. E. Lloyd in 1908.

Fig. 238.—Opuntia robusta.

214. Opuntia guerrana Griffiths, Rep. Mo. Bot. Gard. **19:** 266.   1908.

Plant 9 to 12 dm. high, with an open, branching top; joints oblong to orbicular, 15 to 25 cm. long, thick, glaucous; areoles 5 mm. in diameter, filled with tawny wool; spines white to yellow, 1 to 6, flattened, twisted; petals yellow; filaments greenish white; stigma-lobes green; fruit globose greenish white, 4 to 5 cm. in diameter.

Type locality: Near Dublán, Hidalgo, Mexico.
Distribution: Known only from type locality.

Except in size and color of fruit this species is very much like the common *Opuntia robusta* of this part of Mexico.

215. Opuntia fusicaulis Griffiths, Rep. Mo. Bot. Gard. **19:** 271.   1908.

Plant 5 meters high or less, the branches erect or spreading; joints oblong, elongated, 4 dm. long or less, much longer than wide, glaucous, bluish green, spineless, narrowed at both ends; glochids often wanting; areoles small, filled with tawny wool; fruit greenish white.

Type locality: Described from cultivated plants.
Distribution: Known only from cultivated specimens.
Illustration: Rep. Mo. Bot. Gard. **19:** pl. 23, in part.

The following may be referable to this series:

OPUNTIA CRYSTALENIA Griffiths, Bull. Torr. Club **43:** 528. 1916.

Erect, 2 to 2.5 meters high; joints broadly obovate, 25 cm. long, 18 cm. wide, glaucous, bluish green, becoming yellowish in age; leaves 4 mm. long, subulate; spines white, porrect, only on the upper parts of the joints, 1 to 4, usually only 2, the longest 1 to 1.5 cm. long; glochids yellow; flowers yellow; stigma-lobes 10, dark green; fruit subglobose, 4 to 4.5 cm. in diameter.

Type locality: Cardenas, Mexico.

Distribution: Highlands of Mexico, where it is also cultivated.

Series 23. POLYACANTHAE.

This series is confined chiefly to plains of the western United States. The species are all low, creeping plants, very spiny, with dry fruits. On account of the dry fruit this series forms a natural group, although some species in the series *Basilares* also have dry fruits. One species of series *Polyacanthae* has fragile branches, in this respect resembling the *Curassavicae.* The species hybridize with those of the *Tortispinae.*

KEY TO SPECIES.

Joints readily detached, turgid, some of them subterete or subglobose........................ 216. *O. fragilis*
Joints not readily detached, usually flat and thin, or in *O. arenaria* sometimes turgid and
    nearly terete.
    Joints turgid, usually small.. 217. *O. arenaria*
    Joints thinner than the last, mostly flat, larger.
        Spines, or some of them, very long, flexible and bristle-like.
            Flowers 4 to 5 cm. long..................................... 218. *O. trichophora*
            Flowers 5 to 6 cm. long..................................... 219. *O. erinacea*
        Spines stiff, acicular or subulate; areoles distant.
            Spines subulate.
                Fruit naked....................................... 220. *O. juniperina*
                Fruit spiny.
                    Flowers yellow............................... 221. *O. hystricina*
                    Flowers red................................... 222. *O. rhodantha*
            Spines acicular, slender; areoles close together.
                Ovary and fruit without spines........................... 223. *O. sphaerocarpa*
                Ovary and fruit with spines.............................. 224. *O. polyacantha*

216. Opuntia fragilis (Nuttall) Haworth, Suppl. Pl. Succ. 82. 1819.

    Cactus fragilis Nuttall, Gen. Pl. **1:** 296. 1818.
    Opuntia brachyarthra Engelmann and Bigelow, Proc. Amer. Acad. **3:** 302. 1856.
    Opuntia fragilis brachyarthra Coulter, Contr. U. S. Nat. Herb. **3:** 440. 1896.
    Opuntia fragilis caespitosa and *tuberiformis* Hortus, Stand. Cycl. Hort. Bailey **4:** 2363. 1916.
    (?) *Opuntia columbiana* Griffiths, Bull. Torr. Club **43:** 523. 1916.

Usually low and spreading, small and inconspicuous, but sometimes forming mounds 2 dm high in the center and 4 dm. in diameter, with hundreds of joints; joints fragile (the terminal ones especially breaking off at the slightest touch), often nearly globular but sometimes decidedly flattened, usually dark green, 1 to 4 cm. long; areoles closely set, small, filled with white wool; spines 5 to 7, brown or only with brown tips and lighter below, 1 to 3 cm. long; glochids yellowish; flowers pale yellow, about 5 cm. broad; fruit dry, spiny, 1.5 to 2 cm. long, with a truncate or slightly depressed umbilicus; seeds large, 5 to 7 mm. broad.

Type locality: "From the Mandans to the mountains, in sterile but moist situations."

Distribution: Wisconsin to central Kansas and northwestern Texas, westward to Arizona, Oregon, Washington, and British Columbia.

Dr. Engelmann says "it is rarely found in flower and still more rarely seen in fruit." The only fruit we have seen was collected by Dr. Rose near Liberal, Kansas, in 1912.

Opuntia brachyarthra, sometimes regarded as a variety of *O. fragilis,* we regard as not specifically separable from that species. An examination of the type material now preserved in the Missouri Botanical Garden does not warrant a separation of any kind.

This species is of wide distribution and is especially common on the plains. It usually grows low, often being hidden by the grass. In the grazing country it is a most troublesome weed, for the joints easily break off and become attached by their spines to passing objects, thus greatly annoying and pestering all animals on the range, even frightening horses. The wide distribution of the species is doubtless largely due to the fact that the joints are so easily scattered. A hybrid with *O. tortispina* has been found in Kansas (Rose, No. 17132).

The plant is of especial interest as the most northern in distribution of the opuntias. It is stated that *Opuntia cervicornis* Späth (Cat. 156. 1906–7) is "probably a hybrid of which *O. fragilis* is a parent" (Kew Bull. Misc. Inf. 1907: App. 74. 1907). *O. sabinii* (Pfeiffer, Enum. Cact. 147. 1837) was given as a synonym of *O. fragilis*.

Illustrations: Cact. Journ. 1: 100; Dict. Gard. Nicholson 2: f. 752; Förster, Handb. Cact. ed. 2. f. 132; Gartenflora 30: 413; Pac. R. Rep. 4: pl. 12, f. 9; Rümpler, Sukkulenten f. 126; W. Watson, Cact. Cult. f. 78; Wiener Illustr. Gartenz. 10: f. 113, all as *Opuntia brachyarthra*. Illustr. Fl. 2: f. 2532; ed. 2. 2: f. 2991; Pac. R. Rep. 4: pl. 24, f. 5.

Plate xxxv, figure 1, shows old and young joints of the plant collected by C. Birdseye at Florence, Montana, in 1910. Figure 239 is from a photograph of the plant taken by E. R. Warren at San Acacio, Colorado, in 1912.

Fig. 239.—Opuntia fragilis.

217. **Opuntia arenaria** Engelmann, Proc. Amer. Acad. 3: 301. 1856.

Roots in clusters of 10 to 15, spindle-form, somewhat fleshy; stem prostrate, 2 to 3 dm. long, much branched; joints during growing season quite turgid, afterwards much thinner, 4 to 8 cm. long, half as broad as long; areoles large, numerous, filled with brown wool, glochids, and spines; spines 5 to 8 from an areole, 2 or 3 much longer than the others, sometimes 4 cm. long; flowers red, 7 cm. broad; fruit dry, spiny, 3 cm. long; seeds large, 7 cm. broad.

Type locality: Sandy bottoms of the Rio Grande near El Paso.
Distribution: Texas and southern New Mexico.

This species is very rare and has been reported only a few times. Dr. Rose, who has repeatedly collected at El Paso, was never able to find it until October 1913, and then but a single plant about 8 miles above El Paso on the New Mexican side of the Rio Grande. It grows in nearly pure sand not far above the level of the river.

PLATE XXXV

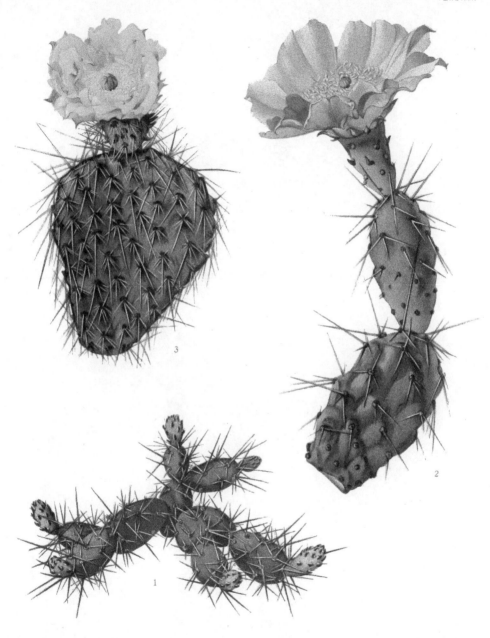

M. E. Eaton del.

1. Plant of *Opuntia fragilis*. 2. Flowering branch of *Opuntia rhodantha*.
3. Flowering joint of *Opuntia polyacantha*. (All natural size.)

Illustration: Cact. Mex. Bound. pl. 75, f. 15.

Figure 240 is from a drawing of the plant collected by Dr. Rose near El Paso, Texas, in 1913.

218. Opuntia trichophora (Engelmann) Britton and Rose, Smiths. Misc. Coll. **50**: 535. 1908.

> *Opuntia missouriensis trichophora* Engelmann, Proc. Amer. Acad. **3**: 300. 1856.
> *Opuntia polyacantha trichophora* Coulter, Contr. U. S. Nat. Herb. **3**: 437. 1896.

A low, spreading plant, often forming small clumps 6 to 10 dm. in diameter; joints orbicular to obovate, 6 to 10 cm. in diameter; areoles closely set; spines numerous, very unequal, the longer one 4 cm. long or so, acicular, pale, often white, but on old joints developing into long, weak hairlike bristles; flowers yellow, the sepals tinged with red; ovary with numerous areoles, these bearing weak, pale bristles; fruit unknown.

Type locality: Mountains near Albuquerque, New Mexico.

Distribution: New Mexico, Texas, and Oklahoma.

This plant, while closely related to *Opuntia polyacantha*, seems worthy of specific rank, its long weak spines being apparently characteristic. Its northern extension into Oklahoma has recently been determined from plants collected by G. W. Stevens.

Illustrations: Pac. R. Rep. **4**: pl. 15, f. 1 to 4; pl. 23, f. 19, all as *Opuntia missouriensis trichophora*.

Figure 241 is copied from the first illustration above cited.

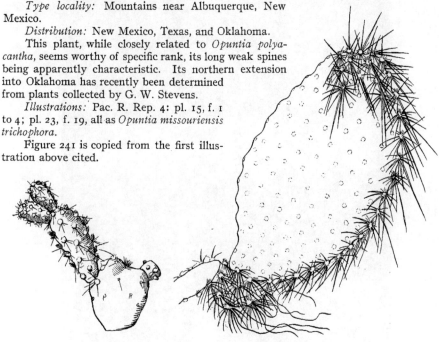

Fig. 240.—Opuntia arenaria. ×0.75.   Fig. 241.—Opuntia trichophora. ×0.75.

219. Opuntia erinacea Engelmann, Proc. Amer. Acad. **3**: 301. 1856.

> *Opuntia ursina* Weber, Dict. Hort. Bois 896. 1898.

Growing in small, low clumps, the branches ascending or erect; joints ovate to oblong, flattened or thick, sometimes nearly terete, 8 to 12 cm. long; areoles somewhat tuberculate, large, numerous, closely set, 4 to 10 mm. apart; spines numerous, usually white or sometimes brownish or with brown tips, slender, often 5 cm., sometimes 12 cm. long or even more, stiff, often developing on the old joints as long hairs or bristles; glochids numerous; flowers rather large, 6 to 7 cm. long, either red or yellow; ovary and fruit very spiny; seeds large, rather regular.

Type locality: On Mojave Creek, California.

Distribution: Northwestern Arizona, southern Utah, southern Nevada, and eastern California.

This species has long been passing under the name of *Opuntia rutila* Nuttall (Torrey and Gray, Fl. N. Amer. 1: 555. 1840). Dr. Engelmann referred it there in the Report of Simpson's Expedition (page 442), and again in the Botany of California, with the remark that "this plant seems to be Nuttall's long lost *O. rutila*." And while it is true that the identification of Nuttall's plant is still doubtful, it seems improbable that this reference is correct,

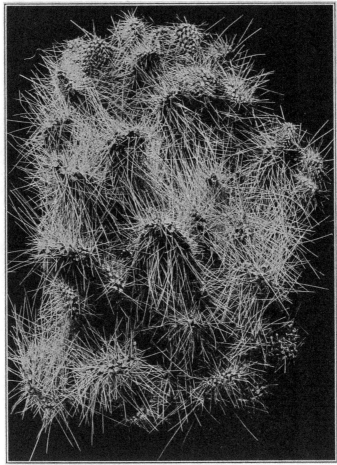

FIG. 242.—Opuntia erinacea.

for the description does not agree with that of the above, and the original station of *O. rutila* in Wyoming is far removed from the other; keen collectors like A. Nelson and V. Bailey, who have searched for the plant for us, have failed to find it in Wyoming. We suspect that *O. rutila* will prove to be *O. polyacantha*.

Opuntia ursina, which comes from the Mojave Desert, seems to be only a slender form with long weak spines. This is known in the trade as the California grizzly bear cactus. Alverson has described it as follows: "This curious plant is covered with tawny white hairs or flexuous spines, some of which are from 3 to 6 inches long, and I have some extra fine specimens with the spines or hairs 9 and 12 inches long.''

Illustrations: Alverson, Cact. Cat. 9 as *Opuntia ursina;* Pac. R. Rep. 4: pl. 13, f. 8 to 11; pl. 24, f. 4.

Figure 242 is from a photograph of the plant taken by F. B. Headley at a point about 29 miles east of Fallon, Nevada, in 1910.

220. Opuntia juniperina sp. nov.

Somewhat of the habit of *Opuntia polyacantha*, but not so procumbent, stouter, and with fewer and stouter spines; joints obovate, 10 to 12 cm. long, broad, rounded at top; areoles small, all below the middle of the joint naked, the upper ones each bearing one stout spine and 1 to several very short accessory ones; the longer spine very stout, 3 to 4 cm. long, brown; flowers not known; fruit dry, oblong, 3 cm. long, spineless, with a shallow, flat umbilicus; seeds large, irregular, 6 to 8 mm. broad.

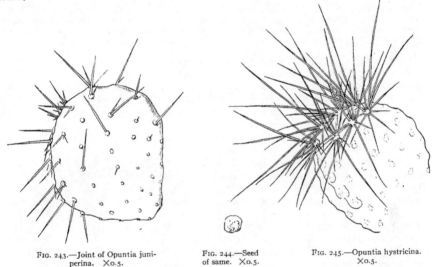

Fig. 243.—Joint of Opuntia juni-   Fig. 244.—Seed   Fig. 245.—Opuntia hystricina.
perina. ✕0.5.   of same. ✕0.5.   ✕0.5.

On dry hills among junipers in vicinity of Cedar Hill, San Juan County, New Mexico, altitude about 1,900 meters, August 17, 1911, Paul C. Standley (No. 8051).

This species is nearest *Opuntia rhodantha*, but has stouter joints and much larger seeds. Figure 243 represents a joint of the type specimen; figure 244 represents a seed.

221. Opuntia hystricina Engelmann and Bigelow, Proc. Amer. Acad. 3: 299. 1856.

More or less diffuse; joints obovate to orbicular, 8 to 20 cm. long; areoles numerous, 10 to 15 mm. apart, rather large; spines numerous, pale brown to white, the longer ones 5 to 10 cm. long, stout, flattish, often reflexed; glochids yellow; flowers 6 cm. long; petals broad, yellow; ovary nearly glob-ular; fruit oblong to obovoid, 2.5 to 3 cm. long, spiny above, dry, with a compressed umbilicus; seeds 7 mm. broad.

Type locality: Colorado Chiquito and on San Francisco Mountains.

Distribution: New Mexico to Arizona and Nevada.

Although this species has a wide range, it is not very well understood; it approaches *O. rhodantha* in some of its forms. We have referred here a very remarkable form collected by E. W. Nelson at Lee's Ferry, Arizona, in 1909. This plant has thick, obovate joints 17 to 22 cm. long, strongly tuberculate, with some of the spines very strong, flattened, and reflexed; the fruit is very spiny; the seeds are 8 mm. broad, angled, with margins thin and acute. This may be the plant listed in Weinberg's catalogue, also from the Grand Canyon, under the name of *Opuntia hochderfferi.*

Opuntia xerocarpa Griffiths (Proc. Biol. Soc. Washington **29**: 15. 1916), from Kingman, Arizona, is of this relationship, described as "readily distinguished from other species of its dry-fruited allies by its spines, shape of joints and color of plant body."

Illustrations: Pac. R. Rep. **4**: pl. 15, f. 5 to 7; pl. 23, f. 15.

Figure 245 is copied from the first illustration above cited.

222. Opuntia rhodantha Schumann, La Semaine Hort. 1897.

> *Opuntia xanthostemma* Schumann, Gesamtb. Kakteen 735. 1898.
> *Opuntia utahensis* J. A. Purpus, Monatsschr. Kakteenk. **19**: 133. 1909.

Joints obovate to oblong, 5 to 12 cm. long; areoles distant, 10 mm. apart or more; spines rather stout, 3 or 4, 2 to 3 cm. long, brownish, with 2 or 3 short accessory ones; lower areoles usually naked; glochids brown; flowers, including ovaries, 5 to 6 cm. long, 8 cm. broad; petals red or pink to salmon-colored, obovate, apiculate; stamens red or yellow; fruit spiny; seeds small, 5 mm. in diameter.

Type locality: Colorado, at 2,000 to 2,300 meters altitude.

Distribution: Western Nebraska, Colorado, and Utah.

After a careful examination of living plants of both *O. rhodantha* and *O. xanthostemma,* we feel convinced that the latter is only a form of the other. The color of the stamens in the opuntias does not furnish a constant character. It is hardy in cultivation at New York and highly ornamental when in bloom.

Haage and Schmidt, in their 1915 catalogue, list several varieties of this species: *brevispina, flavispina, pisciformis,* and *schumanniana;* and under *Opuntia xanthostemma* in the same place they list the following varieties: *elegans, fulgens, gracilis, orbicularis,* and *rosea.*

Illustrations: Meehan's Monthly **7**: 133; Gartenwelt **1**: 83, this last as *Opuntia xanthostemma;* Monatsschr. Kakteenk. **19**: 135, this last as *Opuntia utahensis.*

Plate xxxv, figure 2, represents a flowering plant received by the New York Botanical Garden from Haage and Schmidt, of Erfurt, Germany, in 1913.

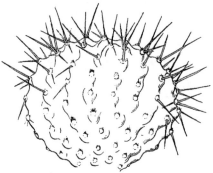

FIG. 246.—Opuntia sphaerocarpa.   X0.66.

223. Opuntia sphaerocarpa Engelmann and Bigelow, Proc. Amer. Acad. **3**: 300. 1856.

Small, spreading plants; joints orbicular, 6 to 7 cm. broad, thickish, strongly tuberculate, wrinkled in drying, light green or becoming more or less purple; areoles 8 to 10 mm. apart, mostly spineless or the upper and marginal ones bearing short acicular spines, the longest ones about 2 cm. long; glochids yellow; flowers not known; fruit naked, 18 mm. in diameter, with a truncate umbilicus; seeds 5 mm. broad, very irregular.

Type locality: Mountains near Albuquerque, New Mexico.

Distribution: Known only from type locality.

We have not, with certainty, identified any recently collected plants with this species, although some New Mexican specimens appear to be referable to it.

Illustrations: Pac. R. Rep. **4**: pl. 13, f. 6, 7; pl. 24, f. 3.

Figure 246 is copied from the first illustration above cited.

224. Opuntia polyacantha Haworth, Suppl. Pl. Succ. 82. 1819.

Cactus ferox Nuttall, Gen. Pl. 1: 296. 1818. Not Willdenow. 1813.
Opuntia media Haworth, Suppl. Pl. Succ. 82. 1819.
Opuntia missouriensis De Candolle, Prodr. 3: 472. 1828.
Opuntia splendens Pfeiffer, Enum. Cact. 159. 1837.
Opuntia missouriensis albispina Engelmann and Bigelow, Proc. Amer. Acad. 3: 300. 1856.
Opuntia missouriensis microsperma Engelmann and Bigelow, Proc. Amer. Acad. 3: 300. 1856. Not
    O. rafinesquei microsperma Engelmann, Proc. Amer. Acad. 3: 295. 1856.
Opuntia missouriensis platycarpa Engelmann, Proc. Amer. Acad. 3: 300. 1856.
Opuntia missouriensis rufispina Engelmann and Bigelow, Proc. Amer. Acad. 3: 300. 1856.
Opuntia missouriensis subinermis Engelmann, Proc. Amer. Acad. 3: 300. 1856.
Opuntia polyacantha albispina Coulter, Contr. U. S. Nat. Herb. 3: 437. 1896.
Opuntia polyacantha borealis Coulter, Contr. U. S. Nat. Herb. 3: 436. 1896.
Opuntia polyacantha platycarpa Coulter, Contr. U. S. Nat. Herb. 3: 436. 1896.
Opuntia polyacantha watsonii Coulter, Contr. U. S. Nat. Herb. 3: 437. 1896.
Opuntia schweriniana Schumann, Monatsschr. Kakteenk. 9: 148. 1899.

FIG. 247.—Opuntia polyacantha.

Low, spreading plants, with fibrous roots, usually forming small clumps; joints not very thick, orbicular, usually less than 10 cm. in diameter, generally light green; areoles small, closely set, usually less than 1 cm. apart, all spiny; spines numerous, often 9, those from the sides mostly short, appressed, and white, but often 1 or 2 of these elongated and like those from the upper and marginal areoles, dark brown, with lighter tips and about 3 cm. long; glochids yellow; flowers small, 4 to 5 cm. long, including the ovary; sepals tinged with red; petals lemon-yellow; stigma-lobes green; fruit dry, oblong, 2 cm. long, bearing small clusters of white, acicular spines at the areoles; seeds white, 6 mm. long, acute on the margin.

Type locality: Arid situations on the plains of the Missouri.

Distribution: North Dakota to Nebraska, Texas, and Arizona to Utah, Washington, and Alberta.

Opuntia sphaerocarpa utahensis Engelmann (Trans. St. Louis Acad. 2: 199. 1863) can not be referred to *O. sphaerocarpa*, where Dr. Engelmann only provisionally placed it when he first described it. On account of its yellow flowers we have referred it here. *Opuntia polyacantha microsperma* and *O. polyacantha rufispina*, mentioned in Bailey's Standard Cyclopedia of Horticulture (3: 2363. 1916), belong here.

Opuntia polyacantha was one of the first of our western opuntias to be collected and described. It was first collected by Thomas Nuttall on his memorable trip to the Upper Missouri. He described it in 1818 as *Cactus ferox*, a name which had been previously used by Willdenow, which led A. H. Haworth in 1819 to rename Nuttall's plant, calling it *Opuntia polyacantha*. At the same place Haworth published a second name, *Opuntia media*, undoubtedly based on a less spiny form of *O. polyacantha*. In 1828 Nuttall's plant was again renamed, this time by A. De Candolle, who called it *Opuntia missouriensis*, under which name it was known for many years. In 1896 Dr. John M. Coulter very properly restored Haworth's name *O. polyacantha*.

This species has a wide distribution laterally and altitudinally. It is properly a plains' species, but is found in mountain valleys and on dry hills, usually in the open, but sometimes in sparse pine woods. In a species of such wide distribution and growing under such diverse circumstances, a wide range of forms is to be expected and a number of varieties have been proposed for the various races, some of which may perhaps have red flowers. The plant is hardy at New York, flowering freely in June.

Illustrations: Curtis's Bot. Mag. **115**: pl. 7046; Illustr. Fl. **2**: f. 2531; ed. 2. **2**: f. 2990; N. Mex. Agr. Exp. Sta. Bull. **78**: pl. [3]; Cact. Journ. **1**: 167; Gard. Chron. **50**: 340, the last two as *Opuntia missouriensis;* Pac. R. Rep. **4**: pl. 14, f. 8 to 10; pl. 23, f. 18, the last two as *Opuntia missouriensis albispina;* Pac. R. Rep. **4**: pl. 14, f. 5 to 7; pl. 24, f. 1, 2, the last two as *Opuntia missouriensis microsperma;* Pac. R. Rep. **4**: pl. 14, f. 4; pl. 23, f. 17, these last two as *Opuntia missouriensis platycarpa;* Pac. R. Rep. **4**: pl. 14, f. 1 to 3; pl. 23, f. 16, these last two as *Opuntia missouriensis rufispina;* Monatsschr. Kakteenk. **9**: 148, this last as *Opuntia schweriniana.*

Plate xxxv, figure 3, represents a flowering joint of the plant collected by Dr. Rose in western Kansas in 1912. Figure 247 represents joints of the plant from Colorado, photographed by T. W. Smillie.

Series 24. STENOPETALAE.

This is an anomalous group in *Opuntia*, since the flowers are diœcious and the petals are linear and more or less erect. It contains three species which are very different in habit and color of spines, but which were all united into a single species by Professor Schumann. Dr. Engelmann was so much impressed by the peculiar structure of the flowers of this group that he proposed for it a new subgenus, *Stenopuntia.*

KEY TO SPECIES.

Spines dark; plants low,
  prostrate.............. 225. *O. stenopetala*
Spines white; plants erect.
  Joints narrow; spines
    acicular.......... 226. *O. glaucescens*
  Joints broader; spines
    stouter........... 227. *O. grandis*

Fig. 248.—Opuntia stenopetala.

225. Opuntia stenopetala Engelmann, Proc. Amer. Acad. **3**: 289. 1856.

Low bushy plant, often forming thickets, the main branches procumbent and resting on the edges of the joints; joints obovate to orbicular, 1 to 2 dm. long, grayish green, but often more or less purplish, very spiny; areoles often remote, 1 to 3 cm. apart, the lower ones often without spines, bearing white wool when young; leaves only on young joints, spreading, dark red, about 2 mm. long; spines usually reddish brown to black, but sometimes becoming pale, usually 2 to 4, the longest ones 5 cm. long, the larger ones somewhat flattened; glochids very abundant on young joints, brown; flowers diœcious, small, including the ovary only 3 cm. long; petals orange-red, very narrow, 10 to 12 mm. long, with long acuminate tips; filaments short; style very thick in the middle, the male flowers with an abortive, pointed style, but female flowers with 8 or 9 yellow stigma-lobes on style; ovary leafy, the upper leaves similar to the sepals; fruit globular, 3 cm. in diameter, acid, naked or spiny; seeds small, smooth, 3 mm. in diameter, with broad, rounded margins.

Type locality: On battlefield of Buena Vista, south of Saltillo, Mexico.

Distribution: In States of Coahuila to Querétaro and Hidalgo, central Mexico.

Referred by Schumann to *O. glaucescens*, but surely a distinct species, as indicated by Berger (Monatsschr. Kakteenk. 14: 171. 1904).

Although in its habit this *Opuntia* is much like many others, its flowers are unique, the petals being very narrow and erect; it is a very beautiful plant, and at flowering time is covered with numerous, small, beautiful flowers. Dr. Griffiths states that it is one of the most valuable ornamental opuntias, and that it is hardy in southern California.

Illustrations: Cact. Mex. Bound. pl. 66; Monatsschr. Kakteenk. 14: 172. f. 1.

Figure 248 is from a photograph of a fruiting joint of a specimen collected by Dr. Edward Palmer near Saltillo, Mexico, in 1905; figure 249 is copied from the illustration first above cited.

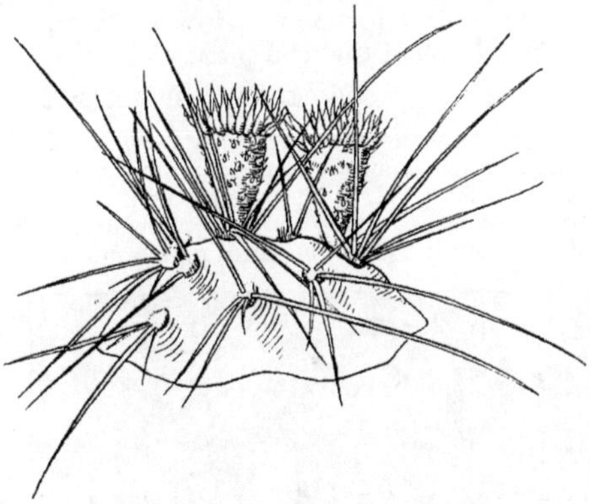

226. Opuntia glaucescens Salm-Dyck, Hort. Dyck. 362. 1834.

Probably erect; joints erect, oblong-obovate, 12 to 15 cm. long, 5 cm. broad, sometimes narrowed at both ends, pale green, glaucous, usually purplish around the areoles; leaves small, reddish when young; areoles filled with gray wool; spines 1 to 4, elongated, acicular, white, 2.5 cm. long; glochids brownish to rose-colored.

Type locality: In Mexico.

Distribution: Mexico, but range unknown.

FIG. 249.—*Opuntia stenopetala.*

The flowers were not known when the species was first described and we do not know that they have since been observed. It has long been in cultivation, but specimens grown under glass at New York have not flowered.

227. Opuntia grandis Pfeiffer, Enum. Cact. 155. 1837.

More or less erect, 6 dm. high or more; joints oblong, 12 to 18 cm. long, erect, when young reddish, glaucous; leaves rose-colored; spines few, white; flowers small, a little open, 2 cm. broad; petals few, narrowly lanceolate, 12 mm. long; filaments reddish; style shorter than the stamens, rose-colored; stigma-lobes 2 or 3, acute.

Type locality: In Mexico.

Distribution: Mexico, but range unknown.

Referred by Schumann to *O. glaucescens*, but doubtless distinct, as indicated by Berger.

Illustration: Monatsschr. Kakteenk. 14: 172. f. 2.

Series 25. PALMADORAE.

An erect plant with narrow flat joints, small, brick-red flowers, and apparently erect stamens; the epidermis densely papillose-tuberculate when dry. The flowers suggest those of the *Spinosissimae*, but otherwise the plant is quite different. The series consists of a single species, from the catinga region of eastern Brazil.

228. Opuntia palmadora sp. nov.

Plant often 3 meters high, sometimes even 5, but often low; trunk sometimes 9 cm. in diameter, sometimes with brown, smooth bark, but usually very spiny; branches numerous, usually erect, at times forming a compact, almost globular top, at other times quite open; joints unusually thin and narrow, 1 to 1.5 dm. long, generally erect, very spiny; leaves subulate, minute, 3 to 4 mm. long, green with reddish tips, found only on very young joints; areoles filled with white wool; spines usually 1 to 4, sometimes 6, from an areole, all yellow at first, in age white, the largest one porrect, 3 cm. long; petals erect or only slightly spreading, brick-red in color; stamens short, erect; filaments orange-colored; style cream-colored; stigma-lobes white; ovary broadly turbinate, 2 cm. long, tuberculate; fruit small.

Collected by Rose and Russell at Barrinha, Bahia, Brazil, June 7, 8, 1915 (No. 19787).

This plant is common in the semiarid parts of Bahia, where it is known as palmadora or palmatoria. Johnston and Tryon describe it briefly without giving it a name in their Report of the Prickly-Pear Travelling Commission, 104. 1914.

Figure 250 represents joints of the type plant; figure 251 is from a photograph of the wild plant from which the above was taken.

Series 26. SPINOSISSIMAE.

Erect species, mostly tall, with terete, continuous, unjointed, usually densely spiny trunks, the ultimate branches spreading or divaricate, flat, usually elongated, spiny or sometimes unarmed; flowers small, yellow, orange or red, or changing from yellow to red; fruit fleshy. We recognize seven species, all natives of the West Indies. The series represents the genus *Consolea* of Lemaire.

FIG. 250.—O. palmadora.

FIG. 251.—Opuntia palmadora. A thicket in Bahia, Brazil.

FIG. 252.—Opuntia nashii.

KEY TO SPECIES.

Areoles of the joints distant, 2 to 4 cm. apart.
  Spines few, 3 cm. long or less, or none.
    Areoles elevated, bearing 2 to 5 grayish spines 3 to 6 cm. long.....................229. *O. nashii*
    Areoles scarcely elevated, spineless or with 1 to 4 weak yellow spines 1 to 2 cm. long...230. *O. bahamana*
  Spines, when present, many, the older up to 12 cm. long..............................231. *O. macracantha*
Areoles of the joints closer together, 1 to 1.5 cm. apart.
  Spines of the trunk-areoles, or most of them, deflexed.
    Young spines straw-colored or whitish; plant up to 5 m. tall.......................232. *O. spinosissima*
    Young spines purple; plant 6 dm. high or less...................................233. *O. millspaughii*
  Spines of the trunk-areoles, when present, spreading.
    Joints distinctly reticulate-areolate, light green; ovary prominently tuberculate.......234. *O. moniliformis*
    Joints indistinctly reticulate-areolate, mostly dark green or reddish; ovary low-tuberculate 235. *O. rubescens*

229. Opuntia nashii Britton, Bull. N. Y. Bot. Gard. **3**: 446. 1905.

Tree-like, or sometimes bushy, dull green; main axis round, 1 to 4 meters high, 5 to 12 cm. in diameter, spiny; branches flat or becoming round below, the principal ones continuous, 1 meter long or more, 6 cm. wide or less, crenate, blunt; lateral branches opposite or alternate, oblong to linear-oblong, often 3 dm. long, and 8 cm. wide, only about 6 mm. thick, blunt, crenate; areoles 1 to 3 cm. apart, slightly elevated; spines mostly 5 at each areole (2 to 5), divergent, slender, straight, light gray, pungent, the longer 3 to 6 cm. long; glochids very small, brownish; ovary 3 cm. long, 1.5 cm. thick, somewhat clavate, tubercled, the tubercles bearing areoles and spines similar to those of the joints, but the spines somewhat shorter; flowers 1.5 cm. broad when expanded, red; petals broadly oval to obovate, blunt, about 8 mm. long, much longer than the stamens.

Type locality: Inagua, Bahamas.

Distribution: Andros, Crooked Island, Fortune Island, Atwood Cay, Caicos Islands, Turks Islands, Ship Channel Cay, and Inagua, Bahamas.

Figure 252 is from a photograph of a plant at Matthew Town, Inagua, Bahamas, taken by George V. Nash, in 1904; figure 253 is from a photograph of a plant from the same place in the collection of the New York Botanical Garden.

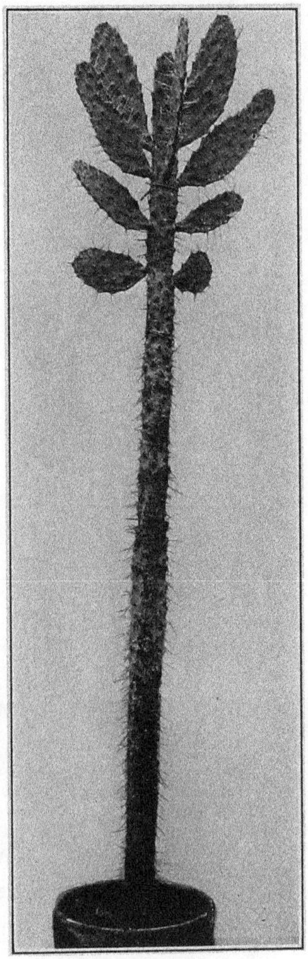

230. Opuntia bahamana sp. nov.

Branched from near the base, bushy, about 1.5 m. high; joints oblong to lanceolate, flat, and thin, 1 to 5 dm. long, 4 to 10 cm. wide, dull green, obtuse, scarcely undulate; leaves red, subulate, 3 cm. long; areoles 1.5 to 3 cm. apart, scarcely elevated, about 2 mm. in diameter, spineless, or bearing 1 to 4 acicular yellow spines 2 cm. long or less when young; glochids few and short; flower about 6 cm. broad; petals obovate, rose-tinted below, yellowish rose above; sepals dark rose, whitish margined.

Distribution: Rocky slopes, The Bright, Cat Island, Bahamas, collected by N. L. Britton and C. F. Millspaugh, March 1907, No. 5794.

This plant was tentatively referred by us (Smiths. Misc. Coll. **50**: 525. 1908) to *Opuntia lanceolata* Haworth. It has been grown under glass at New York ever since, but does not respond well to greenhouse conditions.

FIG. 253.—Opuntia nashii.

It is here included in the series *Spinosissimae*, but with hesitation, its bushy habit and larger flowers being anomalous in this group.

Figure 254 represents a joint of the type specimen above cited; figure 255 is copied from a sketch of a flower made by Dr. Millspaugh on Cat Island, when the plant was discovered.

231. Opuntia macracantha Grisebach, Cat. Pl. Cub. 116. 1866.

Erect, the trunk up to 15 cm. in diameter, its areoles 1 to 2 cm. broad, bearing many brownish glochids and several divergent spines 15 cm. long or less; upper portion of the trunk, and the ultimate, oblong, or oblong-ovate, spreading branches flat, green, faintly shining, the areoles 2 to 3 cm. apart, scarcely elevated, the numerous glochids brown; spines 1 to 4, up to 15 cm. long, nearly white, stout, subulate, or wanting; flowers often numerous; ovary 2.5 to 3 cm. long, densely beset with glochid-bearing areoles; petals orange-yellow, 1 to 1.3 cm. long.

Type locality: Cuba, in maritime depressions.

Distribution: Southern coast of eastern Cuba and adjacent plains.

Specimens of the plant were erroneously referred by Grisebach to *O. triacantha*. It is a picturesque feature of the flora of its native habitat.

Figure 256 is from a photograph of the plant on the United States Naval Station, Guantánamo Bay, Cuba, taken by Marshall A. Howe in 1909; figure 257 is from a photograph of a plant from the same place, grown at the New York Botanical Garden.

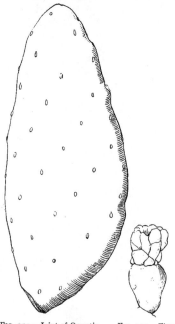

FIG. 254.—Joint of Opuntia   FIG. 255.—Flower   FIG. 256.—Opuntia macracantha.
        bahamana.   of the same.

232. Opuntia spinosissima Miller, Gard. Dict. ed. 8. No. 8.   1768.

Cactus spinosissimus Martyn, Cat. Hort. Cant. 88. 1771.
Consolea spinosissima Lemaire, Rev. Hort. 1862: 174. 1862.

Erect, up to 5 m. high, the trunk sometimes 8 cm. in diameter, densely clothed with areoles bearing many long brownish glochids and acicular, deflexed or spreading spines up to 8 cm. long; ultimate branches flat, dull green, narrowly oblong, 2 to 4 times as long as wide, their areoles 1 to 1.5 cm. apart, slightly or not at all elevated, bearing brown glochids and 1 to 3 acicular, straw-colored or whitish spines 8 cm. long or less, or spineless; ovary 3 to 8 cm. long, often flattened, its areoles bearing short glochids; petals about 1 cm. long, oblong-obovate, rounded at the apex, at first yellow, turning dull red.

Type locality: Jamaica.

Distribution: Southern coast of Jamaica.

Plate xxxvi, from a painting by Miss H. A. Wood at Hope Gardens, Jamaica, sent by William Harris in 1907. Figure 258 is from a photograph of a plant obtained by Professor John F. Cowell in Jamaica and sent from the Buffalo Botanical Garden to the New York Botanical Garden in 1904.

233. Opuntia millspaughii Britton, Smiths. Misc. Coll. 50: 513. 1908.

Trunk terete, 7 cm. thick at base, 5 cm. thick at top, 6 dm. high or less, branching at the summit, the branches divaricate-ascending, narrowly oblong, much compressed, 40 cm. long or less, 5 to

PLATE XXXVI

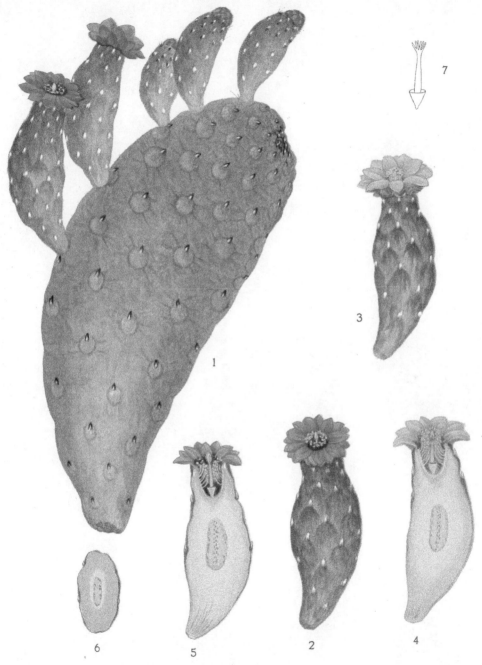

H. A. Wood del.

Opuntia spinosissima.

1. Flowering joint.
2, 3. Single flowers.
4, 5. Longitudinal section of flower.
6. Cross-section of ovary.
7. Style.

10 cm. wide, 1 to 1.5 cm. thick, light green; branchlets obliquely lanceolate, obtuse, as wide as the branches, but shorter, 1 cm. thick or less, floriferous at and near the apex; areoles of the older branches pitted, about 1 cm. apart, those of very young shoots slightly elevated, the glochids very short, yellowish brown; spines of the trunk 15 cm. long or less, very numerous and densely clothing the trunk, very slender, gray, mostly strongly reflexed, pungent, those of the branches and branchlets restricted to the areoles on their edges, shorter than those of the trunk but similar, purple when

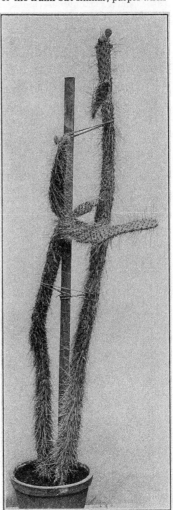

FIG. 257.—Opuntia macracantha.

FIG. 258.—Opuntia spinosissima.

young, those of the fruit yellowish gray, 2 cm. long or less; flowers cupulate, crimson-lake, 1 cm. wide; sepals fleshy, ovate, acute, 4 mm. long and wide; petals erect-ascending, obovate, mucronulate, about 4 mm. wide; stamens half as long as the corolla; style about as long as the corolla; stigma-lobes oblong, yellowish crimson; fruit compressed-obovoid, 2 cm. long, 1.5 cm. thick, bearing one or two spines at most of the areoles.

Type locality: Rock Sound, Eleuthera Island, Bahamas.

Distribution: Eleuthera and Great Ragged Island, Bahamas; Cayo Paredón Grande, Cuba.

Figure 259 is from a photograph of the type plant taken at the type locality by Dr. C. F. Millspaugh, February 22, 1907.

234. **Opuntia moniliformis** (Linnaeus) Haworth in Steudel, Nom. ed. 2. **2**: 221. 1841.

> *Cactus moniliformis* Linnaeus, Sp. Pl. 468. 1753.
> *Cactus ferox* Willdenow, Enum. Pl. Suppl. 35. 1813.
> *Opuntia ferox* Haworth, Suppl. Pl. Succ. 82. 1819.
> *Cereus moniliformis* De Candolle, Prodr. **3**: 470. 1828.
> *Consolea ferox* Lemaire, Rev. Hort. **1862**: 174. 1862.
> *Opuntia microcarpa* Schumann, Gesamtb. Kakteen 714. 1898. Not Engelmann. 1848.
> *Nopalea moniliformis* Schumann, Gesamtb. Kakteen 750. 1898.
> *Opuntia testudinis-crus* Weber in Gosselin, Bull. Mus. Hist. Nat. Paris **10**: 389. 1904.
> *Opuntia haitiensis* Britton, Smiths. Misc. Coll. **50**: 513. 1908.

Fig. 259.—Opuntia millspaughii.

Trunk somewhat flattened above, 3 to 4 m. high, branching at the top, densely armed with acicular, yellowish or gray spines 12 cm. long or less, their bases clothed with yellowish-white wool 1 to 2 cm. long; joints obliquely linear-oblong to obovate, 1 to 3 dm. long, 13 cm. wide or less, about 1 cm. thick, obtuse, distinctly areolate-reticulate, the areoles somewhat elevated, 1 to 1.5 cm. apart, those of young joints bearing near the edges 3 to 6 acicular spines 1 to 2.5 cm. long, those on the sides of the young joints often spineless or with 1 to 3 yellowish

Fig. 260.—Opuntia moniliformis on the plain at Azua, Santo Domingo.

spines, and with small tufts of grayish wool; older joints bearing at all areoles 5 to 8 yellowish spines similar to those of the trunk, and brown glochids 6 or 8 mm. long; flowers about 2.5 cm. broad; sepals as broad as long, or broader, apiculate; petals yellow to orange, ovate, apiculate, spreading; stamens much shorter than the petals; ovary cylindric to obovoid-cylindric, terete or nearly so, 4 to 5 cm. long, its distinctly elevated areoles close together, only 5 or 6 mm. apart, bearing brown glochids 2 mm. long, but no spines; fruit oblong-obovoid, about 6 cm. long.

Type locality: Hispaniola.

Distribution: Hispaniola; Desecheo Island, Porto Rico.

The ovaries, fruits and small joints of this species are readily detached and on falling to the ground strike root and proliferate, forming masses of subglobose or turgid joints entirely different in aspect from the fully developed, tree-like plant. It was on this stage of the organism that the *Cactus moniliformis* of Linnaeus, founded on Plumier's conventionalized plate above cited, was based; this illustration is, however, apparently erroneous in showing the style as long-exserted.

The names *Opuntia dolabriformis* and *Opuntia cruciata* were published by Pfeiffer (Enum. Cact. 167. 1837) as synonyms of *O. ferox*. Some of the joints and, perhaps, some whole plants of this species are nearly or quite spineless.

Illustrations: Descourtilz, Fl. Med. Antill. ed. 2. **7**: pl. 514, as Cactier moniliforme; Plumier, Pl. Amer. ed. Burmann. pl. 198, as *Cactus*, etc.

Fig. 261.—Opuntia moniliformis. The same species as 260, but showing a different mode of growth.

Fig. 262.—Opuntia moniliformis. ×0.66.

Figure 260 is from a photograph of a plant at Azua, Santo Domingo, taken by Paul G. Russell in 1913; figure 261 is from a photograph taken by Frank E. Lutz on Desecheo Island, Mona Passage, Porto Rico, in 1914, showing a mass of proliferating sterile ovaries or small joints below and the mature stage of the plant above; figure 262 represents several of the small joints of the Desecheo plant.

235. **Opuntia rubescens** Salm-Dyck in De Candolle, Prodr. **3**: 474. 1828.

Opuntia catacantha Link and Otto in Pfeiffer, Enum. Cact. 166. 1837.
Consolea rubescens Lemaire, Rev. Hort. 1862: 174. 1862.
Consolea catacantha Lemaire, Rev. Hort. 1862: 174. 1862.
Opuntia guanicana Schumann in Gürke, Monatsschr. Kakteenk. 18: 180. 1908.

Trunk erect, nearly cylindric below, flattened above, 3 to 6 meters high, sometimes 1.5 dm. in diameter, branching above, its areoles bearing several or many acicular spines up to 8 cm. long or more, or spineless: ultimate joints thin and flat, mostly dark green or reddish green, not reticulate-areolate except when young, oblong to oblong-obovate, 2.5 dm. long or less, mostly 2 to 4 times as long as wide, the terminal ones often much smaller; areoles 1 to 1.5 cm. apart, bearing several acicular nearly white spines 1 to 6 cm. long, or spineless; flowers yellow, orange or red, about 2 cm. broad; ovary long-tuberculate, 4 to 5 cm. long, about 1.5 cm. in diameter; petals obovate, apiculate; stamens about half as long as the petals; fruit reddish, obovoid or subglobose, 5 to 8 cm. in diameter, spiny or spineless; seeds suborbicular, 6 to 8 mm. in diameter.

FIGS. 263, 264.—Opuntia rubescens.

Type locality: Cited as Brazil, but erroneously.
Distribution: Mona and Porto Rico to Tortola, St. Croix, and Guadeloupe.

Culebra, St. Thomas, St. Jan, and Montserrat plants agree with the description of *Opuntia rubescens*, which clearly belongs with the *Spinosissimae (Cruciformes)*, as pointed out by Berger, rather than with the South American series *Inarmatae*, where it was placed by Schumann; it is a spineless state of *O. catacantha*, as was conclusively proven by us through field observations in the Virgin Islands, and greenhouse plants of *O. rubescens* develop spines.

Both the spiny and spineless races exhibit remarkable proliferation of the ovaries, these often forming chains of several joints while attached to the plant; these, falling to the ground, strike root and form many small, flattened joints 2 to 4 cm. long, as in *Opuntia moniliformis*, to which this species is otherwise closely related.

Illustration: Journ. N. Y. Bot. Gard. 7: f. 6, as *Opuntia*.

Figure 263 is from a photograph of the plant taken by Professor John F. Cowell at Guanica, Porto Rico, in 1915; figure 264 is from a photograph taken by Professor Cowell at the same time and place, showing in the foreground a mass of young plants arisen from proliferating joints, and a mature plant behind; figure 265 represents proliferating joints of a plant grown at Nisky, St. Thomas, collected by Dr. Britton and Dr. Rose in 1913; figure 266 represents a fruit, collected by Dr. Britton and Dr. Shafer on Buck Island, St. Thomas, in 1913.

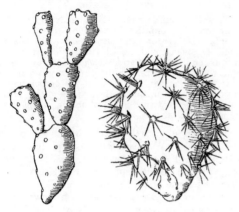

FIGS. 265, 266.—Opuntia rubescens. X0.66.   FIG. 267.—Opuntia brasiliensis. X0.75.

Series 27. BRASILIENSES.

This series represents one of the five subgenera described by Dr. Schumann, which he called *Brasiliopuntia*. It perhaps should be recognized as a distinct genus. We recognize three species in the series, which may be races of a single one, characterized by an erect cylindric trunk with cylindric, horizontal branches terminating in a series of flattened, thin, leaf-like branches. The leaves are small and caducous. The spines are few on the young growth, but large clusters are developed on the old stem and trunk. The flowers are small, the fruit is juicy, and the seeds are large and covered with a dense mass of wool. Unlike most species of *Opuntia*, these grow in the moist tropical forests, forming tall, slender, tree-like plants.

KEY TO SPECIES.

Fruit globular, yellow... 236. *O. brasiliensis*
Fruit clavate to oblong, red.
   Fruit oblong, 3 to 4 cm. long... 237. *O. bahiensis*
   Fruit clavate, 5 cm. long.. 238. *O. argentina*

236. Opuntia brasiliensis (Willdenow) Haworth, Suppl. Pl. Succ. 79. 1819.

    Cactus brasiliensis Willdenow, Enum. Pl. Suppl. 33. 1813.
    Cactus paradoxus Hornemann, Hort. Hafn. **2**: 443. 1815.
    Cactus arboreus Vellozo, Fl. Flum. 207. 1825.
    Opuntia arborea Steudel, Nom. ed. 2. **2**: 220. 1841.
    Cereus paradoxus Steudel, Nom. ed. 2. **1**: 335. 1841.

Becoming 4 meters high, with a cylindric woody trunk and a small rounded top; old trunk either naked or spiny; branches dimorphic, the lateral ones horizontal, terete; the terminal joints flat and leaf-like, many of these in time dropping off; flowers 5 to 5.5 cm. long; petals yellow, oblong, obtuse; filaments very short; fruit yellow, globular, 3 to 4 cm. in diameter, with a low or nearly truncate umbilicus, bearing large areoles; seed usually one, very woolly, 10 mm. broad.

Type locality: Near Rio de Janeiro, Brazil.

Distribution: Southern Brazil, Paraguay, Argentina, and central Bolivia. Naturalized in southern Florida.

A number of varieties of this species appear in literature, of which we may mention the following: *minor* Pfeiffer (Enum. Cact. 169. 1837); *schomburgkii* Salm-Dyck (Cact. Hort. Dyck. 1849. 74. 1850); *spinosior* Salm-Dyck (Hort. Dyck. 184. 1834); *tenuifolia* Forbes (Hort. Tour Germ. 159. 1837); and *tenuior* Salm-Dyck (Hort. Dyck. 376. 1834).

Dr. John H. Barnhart recently called our attention to a number of cactus names published by St. Hilaire which have been overlooked by later writers. One of these, *Cactus heterocladus* St. Hilaire (Voy. Rio de Janeiro and Minas Geraes 2: 103. 1830) seems to belong here, as the following free translation would indicate:

"Another cactus, which I have already seen near Rio de Janeiro, raised its branches in the midst of tortuous lianas; its trunk, which grows more slender from the base to the summit, is covered with fascicles of spines arranged in a quincunx, and it shows various stages of verticillate, horizontal, rounded branches, to the number of seven in each whorl; these branches, like those of the spruce tree, grow shorter toward the summit of the plant, and they bear secondary branches, flattened and oval-oblong, which may in a certain sense be taken as leaves."

Illustrations: Curtis's Bot. Mag. 61: pl. 3293; Dept. Agr. N. S. W. Misc. Publ. 253: pl. [6]; Martius, Fl. Bras. 4²: pl. 61; Pfeiffer and Otto, Abbild. Beschr. Cact. 1: pl. 29; Schumann, Gesamtb. Kakteen f. 100; Vellozo, Fl. Flum. 5: pl. 28, this last as *Cactus arboreus*.

Plate xxx, figure 2, represents a flowering joint taken from a specimen in the New York Botanical

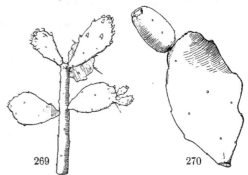

FIG. 268.—Opuntia brasiliensis.   FIGS. 269, 270.—Opuntia bahiensis.   ×0.5.

Garden; figure 3 is from the same plant, showing terete and flat joints. Figure 267 represents a fruit collected by Dr. Rose near Iguaba Grande, Brazil, in 1915; figure 268 is from a photograph taken by Paul G. Russell in a public park in Bahia, Brazil.

237. Opuntia bahiensis sp. nov.

Trunk 3 to 15 meters high, cylindric, 20 to 25 cm. in diameter, tapering gradually upward; the center of trunk pithy, hollow in age, surrounded by an open woody cylinder; lateral joints terete, the terminal ones flat and thin, ovate to oblong; leaves small, 2 to 3 mm. long, turgid; spines on terminal joints, if present, 1 or 2, slender, red at first, then brown; spines on old trunk forming large clusters at all the areoles; flowers not seen; fruit deep red both within and without, oblong, 3 to 4 cm. long; its small areoles with brown glochids; seeds 1 to 5, mostly 1 or 2 in each fruit, very hairy, thick, 8 mm. broad.

Collected in the vicinity of Toca da Onca, Bahia, Brazil, by Rose and Russell, June 27 to 29, 1915 (No. 20068).

Figure 269 represents joints of the type plant above cited; figure 270 represents a joint with fruit; figure 271 is from a photograph of the type specimen.

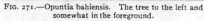
FIG. 271.—Opuntia bahiensis. The tree to the left and somewhat in the foreground.

FIG. 272.—Opuntia ammophila.

238. Opuntia argentina Grisebach, Abh. Ges. Wiss. Göttingen **24**: 140. 1879.

Opuntia hieronymi Grisebach, Abh. Ges. Wiss. Göttingen **24**: 140. 1879.

Erect, 5 to 15 meters high, branching at the top, the lateral branches subverticillate, teretes terminal branches flat, 5 to 12 cm. long, 3 to 8 cm. broad; ovary 2 to 2.5 cm. long; petals elliptic to spatulate, 1.8 cm. long, 8 mm. broad, greenish yellow; filaments white; style white; stigma-lobe; yellowish green; ovary flattened, tuberculate, deeply umbilicate; fruit clavate, 5 cm. long, dull purplish violet, with wine-colored pulp; seeds lens-shaped, 5 to 6 mm. long, 2.5 to 3 mm. broad.

Type locality: Near San Andrés, Oran, Argentina.

Distribution: Northern Argentina.

This species was considered identical with *O. brasiliensis* by Schumann, but they separate on very good fruit characters.

Figure 274 is from a photograph of a flowering branch furnished by Dr. C. Spegazzini.

Series 28. AMMOPHILAE.

One peculiar species, native of Florida, constitutes this series, characterized by a continuous erect subterete trunk, flat, spiny branches, and large, yellow flowers.

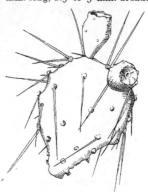

239. Opuntia ammophila Small, Journ. N. Y. Bot. Gard. **20**: 29. 1919.

FIG. 273.—Opuntia ammophila.

Plant erect, more or less branched throughout or ultimately with a stem 1 to 2 meters tall or more, becoming 2 to 2.5 dm. in diameter, bearing several spreading branches near the top, thus tree-like, tuberous at the base; joints various, those of the main stem elongate, ultimately fused on the ends and subcylindric, those of the branches typically obovate or cuneate, varying to elliptic or oval, thickish, 5 to 17 cm. long, becoming grayish green; leaves stout-subulate, 6 to 10 mm. long, green; areoles relatively numerous, conspicuous on account of the densely crowded long bristles, especially on the older joints, the marginal ones, at least, armed; spines very slender, solitary or 2 together, reddish or red, at maturity gray, mostly 2 to 6 cm. long, nearly terete, scarcely spirally twisted; flowers several on a joint; sepals lanceolate, acute or slightly acuminate; buds sharply pointed; corolla bright yellow, 5 to 8 cm. wide; petals obovate, cuneate, notched, and prominently apiculate, 3 cm. long, scarcely erose; stigma-lobes cream-color; berries obovoid, 2 to 3 cm. long, more or less flushed with reddish purple, many-seeded; seeds about 4 mm. in diameter.

FIG. 274.—Opuntia argentina.

FIG. 275.—Opuntia chaffeyi. Photograph by Señor Don Teodoro Chairez.

Type locality: Fort Pierce, Florida.

Distribution: Inland sand-dunes (scrub), peninsular Florida.

The plant was first collected by Dr. Small near Fort Pierce, Florida, in 1917, and again studied by him in its more northern range west of St. George in 1918. He describes it as the most conspicuous native prickly pear of Florida, always viciously armed and with a characteristically unjointed trunk. In spite of its many slender spines, cattle browse upon it.

Illustration: Journ. N. Y. Bot. Gard. **20**: pl. 224.

Figure 272 is from a photograph of the plant taken by Dr. Small near Fort Pierce, Florida; figure 273 shows a fruiting joint of the type specimen.

Series 29. CHAFFEYANAE.

This series contains a single Mexican species, differing from all the other opuntias in having an annual stem which arises from a large, fleshy root or rootstock. The joints, which are elongated and nearly terete, resemble somewhat those of *O. leptocaulis*, but are more fleshy, while the flowers and fruit are like those of the platyopuntias.

240. Opuntia chaffeyi Britton and Rose, Contr. U. S. Nat. Herb. **16**: 241. 1913.

Perennial by a large, fleshy, deep-seated root or rootstock often 35 cm. long by 4 cm. in diameter; stems normally annual, 5 to 15 cm. long, sometimes in cultivated specimens 25 cm. long, much branched, often weak and prostrate; joints elongated, 3 to 5 cm. long, 6 to 7 mm. broad, slightly

FIG. 276.—Opuntia chaffeyi.

flattened, glabrous, pale bluish green or sometimes purplish; leaves minute, caducous; areoles small, circular, with white wool in the lower parts and brown wool in the upper parts; spines 1, rarely 2 or 3, acicular, 2 to 3 cm. long, whitish or pale yellow; glochids numerous, pale yellow; flower-buds, including ovary, 8 cm. long; flower opening at 10 a. m., closing at 2 p. m., 6 cm. broad; sepals few, small, ovate to oblong, greenish; petals few, 7 to 9, pale lemon-yellow, but slightly pinkish on the outside; filaments numerous, about 1 cm. long; style slender, extending beyond the stamens, about 22 mm. long, somewhat swollen at base; ovary deeply umbilicate, somewhat club-shaped, 4 to 5 cm. long, bearing flattened tubercles and large areoles filled with white wool; upper areoles on ovary bearing also white bristly spines; ovules numerous, borne in the upper third of the ovary; fruit and seeds still unknown.

Type locality: Hacienda de Cedros, near Mazapil, Zacatecas, Mexico.
Distribution: State of Zacatecas, Mexico.
Illustration: Contr. U. S. Nat. Herb. **16**: pl. 72.

Figure 275 is from a photograph of part of the original collection as grown by Dr. E. Chaffey, taken and contributed by Señor Don Teodoro Chairez, of Ciudad Lerdo, Mexico; figure 276 is from a photograph of the type showing the large root and the young shoot.

As stated in the original description, this is a remarkable *Opuntia*, being the only one known which has an annual stem. In cultivation, where the plant is grown under abnormal conditions, the stem persists for more than a year; but Dr. Chaffey assures us that in the desert, where the species grows naturally, the stem dies down to the ground in the dry season. We have had it in cultivation since 1910, but it does not do well, and is gradually dying out. It has not been found in flower in a wild state, but it flowered with Dr. Chaffey at Ciudad Lerdo, Durango, Mexico, in 1915. Dr. Chaffey, who has been studying this species for several years, has made a number of interesting observations; he states that the large base, which usually is found 15 to 20 cm. beneath the surface of the ground, when allowed to grow above the ground develops clusters of spines like those on the normal stems, and finds that the plant is easily started from cuttings which soon develop the normal, large, underground part. He further states that the desert turtle eats this plant. It is well known that the Galápagos turtles feed upon the native opuntias of those islands.

The native name of this plant is sacacil.

The following described Opuntias we have been unable to refer to any of the species otherwise mentioned in this work:

> *Opuntia bicolor* Philippi, Linnaea 33: 83. 1864.
>   *glaucophylla* Wendland, Cat. Hort. Herrenh. 1835.
>   *laevior* Salm-Dyck, Cact. Hort. Dyck. 1844. 46. 1845.
>   *longiglochia* C. Z. Nelson, Galesburg Register. July 20, 1915.
>   *lucida* Hortus, Wiener Illustr. Gartenz. 14: 146. 1889.
>   *prostrata spinosior* Schumann, Gesamtb. Kakteen 723. 1898.
>   *spinaurea* Karwinsky in Salm-Dyck, Cact. Hort. Dyck. 1844. 46. 1845. As synonym for *O.*
>   *pseudotuna elongata* Salm-Dyck.
>   *tuberculata* Haworth, Suppl. Pl. Succ. 80. 1819, first described as *Cactus tuberculatus* (Enum.
>   Hort. Berol. Suppl. 34. 1813).

The following names of *Opuntia* are chiefly found in catalogues or in lists, or have been so briefly described that we have not been able to identify them, and it does not seem worth while even to cite the places where they first occur in literature:

Opuntia alpicola Schumann
  americana Forbes
  attulica Forbes
  barbata K. Brandegee
  barbata gracillima K. Brandegee
  bernhardinii Hildmann
  betancourt Murillo
  calacantha
  calacantha rubra
  carolina Forbes
  ciliosa Forbes
  consoleana Todaro
  consolei Haage and Schmidt
  demorenia Forbes
  demoriana Förster
  deppei Wendland
  dichotoma Förster
  eborina Förster
  erecta Schumann
  festiva Sencke
  ficus-indica albispina Haage and Schmidt
  flavispina Förster
  hevernickii Hildmann
  hitchenii Forbes
  italica Tenore
  joconostle Haage and Schmidt
  jussieuii Haage
  leucostata Forbes
  macrophylla Haage and Schmidt

Opuntia missouriensis elongata Salm-Dyck
  erythrostemma Haage and Schmidt
  salmonea Haage and Schmidt
  montana Sencke
  morenoi Schumann
  myriacantha Link and Otto. Not Weber
  ottonis Salm-Dyck
  pachyarthra flava Haage and Schmidt
  pachyclada rosea Haage and Schmidt
    spaethiana Haage and Schmidt
  parote Forbes
  piccolomini Hort.
  platyclada Haworth
  praecox Forbes
  protracta Lemaire
    elongata Salm-Dyck
  pseudococcinellifer Bertoloni
  pseudotuna Salm-Dyck
    elongata Salm-Dyck
    spinosior Salm-Dyck
  pulverata Förster
  reptans Karwinsky
  salmii Forbes
  schomburgkii Salm-Dyck
  speciosa Steudel
  spinuliflora Salm-Dyck
  spinulosa Salm-Dyck
  straminea Sencke
  stricta spinulescens Salm-Dyck
  subinermis Link

Opuntia clavata Philippi (Anal. Univ. Chile **41**: 722. 1872), *O. ottonis* G. Don (Hist. Dichl. Pl. **3**: 172. 1834), *O. phyllanthus* Miller (Gard. Dict. ed. 8. No. 9. 1768), *O. salicornioides* Sprengel (Pfeiffer, Enum. Cact. 141. 1837), and *O. spiniflora* Philippi (Linnaea **30**: 211. 1859) are of the tribe *Cereeae.*

7. GRUSONIA F. Reichenbach in Schumann, Monatsschr. Kakteenk. **6**: 177. 1896.

A low, much branched cactus, the branches terete, jointed, and ribbed; areoles borne on the tops of the ribs, very spiny, but all except the flowering ones without glochids, subtended by small deciduous leaves; corolla rotate, yellow; fruit baccate.

This was first described as a *Cereus* from specimens collected by Mrs. Anna B. Nickels in 1895, then as a new genus *Grusonia*, and lastly as an *Opuntia*. It clearly is not *Cereus*, but when growing might easily be mistaken by its habit for *Echinocereus*. The leaves, glochids, flowers, and fruit are those of *Opuntia*, but its ribbed stem is unlike that of any known species of that genus.

1. Grusonia bradtiana (Coulter).

Cereus bradtianus Coulter, Contr. U. S. Nat. Herb. **3**: 406. 1896 (April).
Grusonia cereiformis F. Reichenbach in Schumann, Monatsschr. Kakteenk. **6**: 177. 1896 (December).
Opuntia bradtiana K. Brandegee, Erythea **5**: 121. 1897.
Opuntia cereiformis Weber, Dict. Hort. Bois 897. 1898.

Forming dense, often impenetrable thickets 2 meters high or less, very spiny; stems light green, 4 to 7 cm. thick, with 8 to 10 low, longitudinal, somewhat tuberculate ribs; areoles 1 to 1.5 cm. apart, 3 to 5 mm. in diameter; leaves linear, fleshy, green, 8 mm. long, early deciduous; spines 15 to 25, yellowish brown when young, soon becoming white, acicular, terete or slightly compressed, 1 to 3 cm. long, not sheathed, some of the longer ones reflexed; wool white, turning brown, early disappearing; corolla rotate, opening in bright sunlight, 3 to 4 cm. broad; sepals ovate, acute, fleshy; petals bright yellow, spatulate, fringed; filaments brownish yellow; stigma-lobes 8, yellow; areoles of the ovary with long, yellow, weak spines, white wool, and yellow glochids; berry (according to Schumann) ellipsoid, deeply umbilicate; seeds not seen.

Type locality: Plains of Coahuila, Mexico.
Distribution: Coahuila, Mexico.
This species first appeared in print in the catalogue of Johannes Nicolai under the name of *Grusonia cereiformis*, but we are informed that there was no description and therefore it was not technically published. The same name next appears in the Monatsschrift für Kakteenkunde for 1894. Here Dr. Schumann wrote a long article about the name, especially condemning the loose manner in vogue of publishing new names without descriptions, but giving no characters of the plant, and as a matter of fact he did not then know it. Two months later this name again appears in this same publication, but without description. Two years later Dr. Schumann records seeing this plant and describes it briefly, although he does not approve of the name *Grusonia*. If the name is to be considered published, it should not date earlier than this (December 1896), although Dalla Torre and Harms accept the date of 1894. In 1898 Weber transferred the name to *Opuntia*, publishing it as *Opuntia cereiformis;* in the meantime Coulter (in 1896) published the name *Cereus bradtianus* for the plant and Mrs. Brandegee (in 1897) transferred it to *Opuntia*, calling it *Opuntia bradtiana*.

Illustrations: Monatsschr. Kakteenk. **21**: 121, as *Opuntia bradtiana;* Schumann, Gesamtb. Kakteen f. 101, as *Opuntia cereiformis.*

Plate XXXIII, figure 4, represents a joint of the plant collected by C. A. Purpus at Cerro de Cypriano, near Morano, Mexico, in 1910.

APPENDIX.

3 a. Nopalea gaumeri sp. nov. (See page 37, *ante.*)

About 3 meters high, much branched; joints small, linear-oblong or oblong-oblanceolate, 6 to 12 cm. long, 2 to 3 cm. broad, rather thin; areoles small, 1 to 2 cm. apart; spines very unequal, 5 to 20 mm. long, acicular, 4 to 12, yellowish when young; flower small, including ovary and stamens about 4 cm. long; sepals ovate, acute; petals oblong, 12 mm. long; stamens long-exserted; style longer than the stamens; stigma-lobes 6, greenish; fruit red, darker within, obovoid, 3 cm. long, its numerous areoles bearing spines and yellow glochids; umbilicus prominent, 1 cm. deep; seeds about 4 mm. broad, with a very narrow margin and a very thin testa.

FIGS. 277 and 278.—Nopalea gaumeri. ✕0.8.

Collected by George F. Gaumer and sons near Sisal, Yucatan, March 1916 (No. 23250, type); also by Dr. Gaumer from Port Silam, 1895 (No. 647).

Dr. Gaumer's field note is as follows: "A coastal cactus, 10 feet high, much branched, small-jointed and of slight build, not of robust build like the interior species. It blooms from February to June. The birds are very fond of the fruit and consume it as fast as it ripens."

Figures 277 and 278 show joints of the type-specimen.

77 a. Opuntia depauperata sp. nov. (See page 101, *ante.*)

Plant 1 to 2 dm. high, with a flattened, much branched top; joints dark green, readily detached, terete or slightly flattened, 3 to 12 cm. long, 2 to 3 cm. thick, puberulent; spines on young joints 2 or 3, on old joints sometimes 6 at each areole, reddish to pale brown, acicular, 1 to 2.5 cm. long, nearly porrect; glochids tardily developing, conspicuous on old joints, yellow; ovary with a deep umbilicus.

Collected by Dr. and Mrs. J. N. Rose north of the station of Zig Zag, along the railroad above Carácas, Venezuela, October 17, 1916 (No. 21751). This little cactus is very inconspicuous and only a few specimens were observed. The station is near the top of the mountains which separate the valley, in which Carácas lies, from the sea. The region here is not so dry as it is farther down on the seaward side of the mountains, but there are several other species of cacti associated with it.

Figure 279 is from a photograph of type plant taken by Mrs. Rose; figure 280 shows a joint.

FIG. 279.—Opuntia depauperata.

A plant, apparently of this relationship, was collected by Dr. H. H. Rusby in 1917 on granite rocks, narrows of Magdalena River, Colombia. The joints, however, are glabrous, only 2 to 3 cm. long, the young joints have numerous brown spines and the young areoles produce long white wool.

216

82*a*. **Opuntia pestifer** nom. nov. (See page 103, *ante*.)

Cactus *nanus* Humboldt, Bonpland, and Kunth, Nov. Gen. et Sp. **6**: 68. 1823.
Cereus *nanus* De Candolle, Prodr. **3**: 470. 1828.

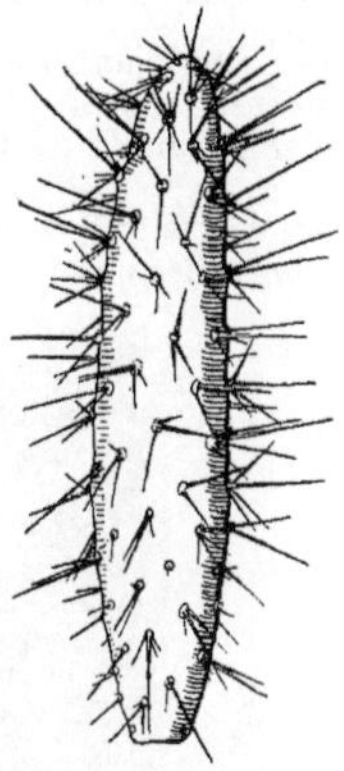

Low and nearly prostrate but sometimes 2 dm. high, much branched; the joints very fragile, glabrous; young joints 2 to 5 cm. long, or when old up to 8 cm. long, nearly terete, 1 to 3 cm. in diameter, or when young flattened and 2 to 3 cm. broad, very spiny; spines 2 to 5 at each areole, acicular, brownish, 1 to 3 cm. long; glochids numerous, yellow; flowers and fruit unknown.

Type locality: Near Sondorello and Guancabamba. In Humboldt's time these places were in southern Ecuador, but they are now in northern Peru.

Distribution: Northern Peru to central Ecuador.

Dr. Rose observed the plant in various places in Ecuador, usually at an altitude ranging from 1,000 to 1,500 meters. The following collections were made: at Huigra (No. 22306); at Sibambe (No. 22433); and west of San Pedro, Province of Loja (No. 23352).

This plant, although widely distributed and very common, has never been seen by botanists in flower or fruit. The joints, which come loose easily, are freely distributed by animals. It is so small that, growing half-hidden in the grass, it is easily overlooked but very annoying when one comes upon it unawares. Humboldt speaks of its being troublesome to men and dogs.

Fig. 280.—Opuntia depauperata. ×0.5.

Fig. 281.—Opuntia pestifer. ×0.5.

Kunth who described it as *Cactus nanus* referred it with hesitancy to the Section *Cereus*. De Candolle transferred it from *Cactus* to *Cereus* placing it in a new subgenus *Opuntiacei* along with *C. moniliformis* (which we know now is an *Opuntia*) and *C. serpens*. He thought these might represent a genus between *Opuntia* and *Cereus*.

Schumann (Gesamtb. Kakteen 166) considered it an *Opuntia* but did not formally refer it to that genus.

This name should not be confused with *Opuntia nana* (Fl. Damatica **3**: 143. 1852) which is *Opuntia opuntia*.

Figure 281 is from a photograph taken by George Rose at Sibambe, Ecuador, in 1918; figure 283 shows the joints of the same plant (Rose, No. 22433.)

96 *a.* **Opuntia discolor** sp. nov.   (See page 109, *ante.*)

A low plant, forming small dense clumps; joints slender, 4 to 12 cm. long, 1.5 to 2.5 cm. in diameter, turgid, glabrous, dark green with dark purple blotches extending downward from the under margin of the areoles; spines 1 to 6, acicular, nearly porrect, somewhat variegated but mostly brown, 3 cm. long or less; glochids tardily developing but conspicuous on old branches, dark brown; flowers light yellow to orange-yellow, only 3 cm. long including the ovary; filaments white; style and stigma-lobes nearly white; fruit evidently very small, bright red.

FIG. 282.—Opuntia discolor.

This species is represented by two collections made by Dr. J. A. Shafer in 1917 which slightly differ from each other.   They are No. 111, from sandy thickets, Santiago del Estaro, Argentina, February 23 (type), and No. 95, from gravelly hills near Tapia, Tucuman, February 9.

Apparently common in dry sandy thickets, growing best under bushes where it is least disturbed.   The joints easily become detached, sticking readily to any disturbing object.

The species differs from *Opuntia retrorsa* in its more nearly terete joints and spreading spines.

Figure 282 is from a photograph of the type plant; figure 284 represents a joint of the plant from near Tapia, Tucuman.

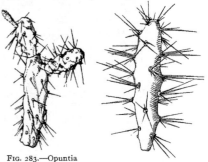

FIG. 283.—Opuntia pestifer. ✕0.5.

FIG. 284.—Opuntia discolor.

101 *a.* **Opuntia guatemalensis** sp. nov.   (See page 113, *ante.*)

Low, spreading plant, resembling *O. decumbens,* but joints glabrous and shining; joints deep green, sometimes with dark blotches below the areoles; areoles small, filled with brown wool, subtended by small leaves; spines 1 to 3 at the areoles, terete, acicular, shining white with blackish tips when young, soon gray, mostly deflexed, somewhat spreading; flower-buds reddish; flowers much smaller than those of *O. decumbens;* petals lemon-yellow, 2.5 cm. long; stigma-lobes cream-colored.

Collected by Dr. Glover B. Wilcox in 1909 while acting as surgeon on a ship plying between Guatemala and San Francisco.   Living specimens were sent directly to Washington and flowered there in April 1915.

Figure 285 represents a joint of the type specimen.

102 *a*. Opuntia pennellii sp. nov.   (See page 115, *ante*.)

Plant low; joints 1 to 1.5 cm. long, obovate, turgid, bright green; spines 1 or 2 at each areole, nearly porrect, subulate, 3.5 cm. long or less, white with dark tips; glochids not very conspicuous, yellowish.

Collected near Magangue, coastal plain of Colombia, Department of Bolivar, at about 100 meters altitude, by Francis W. Pennell in 1918.

Figure 286 shows a joint of the type plant.

Here may belong herbarium specimens which we have seen from northern Colombia but with the material at hand it is impossible to determine them definitely.   One of these was collected by William R. Maxon, April 10, 1906 (No. 3849) at Puerto Colombia.   This plant is described as consisting of 3 to 6 joints, branching at the third or fourth joint, the joints all being in one place.   The flowers are yellow and small, only about 4 cm. long, including the ovary.   Another was collected by H. H. Smith near Bonda in 1898–1899 (No. 2728); this has joints very similar to those of Dr. Pennell's plant.   It is said to be from 2 to 4 feet high.

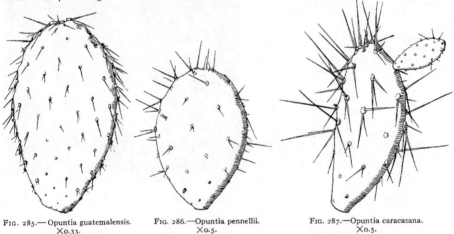

FIG. 285.—Opuntia guatemalensis. ×0.33.

FIG. 286.—Opuntia pennellii. ×0.5.

FIG. 287.—Opuntia caracasana. ×0.5.

103 *a*. Opuntia caracasana Salm-Dyck, Cact. Hort. Dyck. 1849. 238. 1850.   (See page 116, *ante*.)

Stems low, bushy, 4 to 12 dm. high; joints oblong, 10 to 12.5 cm. long, turgid, pale green, "leaves squamiform, minute"; spines 2 to 4, unequal, 2.5 to 4 cm. long or less, pale yellow; flowers and fruit unknown.

Type locality: Near Carácas, Venezuela.
Distribution: Mountains about Carácas, Venezuela.

The type specimens were collected near Carácas by E. Otto, prior to 1849.   Dr. Rose found the plant abundant above Carácas in 1916.   It usually grows on exposed hillsides near the top of the divide which separates Carácas from the coast, and it was especially common along the railroad just below the little station of Zig Zag.   Several other cacti are to be found in this neighborhood, among which are *O. elatior* and *O. depauperata*.

Figure 287 shows a joint of the plant collected by Dr. Rose above Carácas in 1916.

104 *a*. Opuntia aequatorialis sp. nov.   (See page 116, *ante*.)

Bushy, much branched; 1 to 1.5 meters high; the branches spreading or recurved; joints narrowly oblong to obovate, 1.5 to 2 dm. long, 3 to 8 cm. broad, easily becoming detached; spines pale yellow, at first only 2 to 4 but more in age, subulate, 2.5 to 6 cm. long; flower-buds ovoid, acute, red; petals few, 8 to 10, orange-red, spatulate; filaments and style red; stigma-lobes cream-colored.

Collected in thickets on dry hills near Sibambe, Province of Chimborazo, Ecuador, by J. N. Rose and George Rose, August 29, 1918 (No. 22432).

The locality at which this species is found is semiarid and a number of other cacti are associated with it, among which is the little *O. pestifer*, described on a preceding page. *O. aequatorialis* was not so common as some of the other species and was usually found growing up through open-branched bushes and was in this way more or less protected.

Figure 288 is from a photograph of the type plant taken by George Rose; figure 289 shows one of its joints.

116 *a*. **Opuntia lata** Small, Journ. N. Y. Bot. Gard. **20**: 26. 1919. (See page 126, *ante*.)

Plant prostrate, often radially branched, sometimes forming mats nearly a meter in width, the tip of the branches sometimes assurgent, with elongate cord-like roots; joints elliptic to narrowly obovate, often narrowly so, thick, 4 to 15 cm. long, deep green, sometimes glaucous, especially when young; leaves subulate, 6 to 11 mm. long, green or purple-tinged; areoles scattered, often conspicuous, sometimes very prominent and densely bristly, the marginal ones, at least, armed; spines slender, solitary or 2 together, pink, turning red or red-banded, at maturity gray or nearly white, nearly

FIG. 288.—Opuntia aequatorialis.

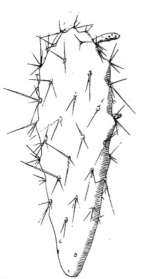

FIG. 289.—O. aequatorialis. ×0.4.

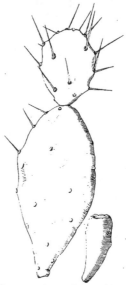

FIGS. 290 and 291.—O. lata. ×0.4.

FIGS. 292 and 293.— Opuntia macateei. ×0.4.

terete, slightly spirally twisted; flowers usually several on a joint, conspicuous; sepals subulate to lanceolate, acute; corolla yellow, 7 to 9 cm. wide; petals numerous, the inner ones broadly obovate to flabellate, erose at the broad minutely mucronate apex; berries clavate, 5 to 6.5 cm. long, red or reddish purple, many-seeded; seeds about 5 mm. in diameter.

Type locality: Twelve miles west of Gainesville, Florida.
Distribution: Pinelands, northern peninsular Florida.

It was first observed by Dr. Small near Gainesville, Florida, in 1917, and plants were taken to Mr. Charles Deering's cactus garden at Buena Vista, Miami, where it has grown luxuriantly, flowering and fruiting freely alongside of *O. pollardii* which it resembles in habit, but differs from in its long clavate berries and more numerous petals.

Figure 290 shows joints of the plant; figure 291 shows its fruit.

127 *a*. Opuntia macateei sp. nov. (See page 133, *ante*.)

Small prostrate plant; joints 2.5 to 6 cm. long, orbicular to obovate, glabrous, dull green, in age somewhat tuberculate; leaves linear, 10 mm. long or less, green; spines 1 to 3, brownish, the longer ones up to 2.5 cm. long; flowers, including the ovary, 8 to 10 cm. long, 7 to 8 cm. broad, yellow with a red center; ovary subcylindric, 5 to 6 cm. long, bearing conspicuous leaves, sometimes 12 mm. long.

Differs from related species by its small joints and slender, elongated, leafy ovaries. Collected by W. L. MacAtee at Rockport, Texas, December 28, 1910 (No. 1992).

Figures 292 and 293 represent the joints and flower of the plant.

159 *a*. Opuntia soederstromiana sp. nov. (See page 154, *ante*.)

Sometimes spreading and bushy, but usually erect, 6 to 10 dm. high, very spiny; joints obovate, 2 to 4 dm. long, bright green when young, or sometimes slightly glaucous, grayish green in age; leaves subulate, small, reddish at top; spines at first 2 to 5, but in age 10 or more, when young reddish or pinkish at base and paler above, soon gray throughout, unequal, subulate, 4 cm. long or less; flowers at first yellow but soon orange to brick-red, rather large, 5 to 6 cm. long; petals few, about 10, oblong, retuse; filaments and style reddish; stigma-lobes pale green; fruit obovate to oblong, 4 to 5 cm. long, usually spiny, red, juicy, with a depressed umbilicus.

Collected at San Antonio, Province of Quito, Ecuador, by J. N. Rose and George Rose, October 29, 1918 (No. 23559).

This plant was first collected for us by Ludovic Söderstrom of Quito, at the request of the President of the Central and South American Cable Company. Although

FIG. 294.—Opuntia soederstromiana.

great care was taken in shipping the plants they all died in transit. In 1918 Dr. Rose visited Mr. Söderstrom's locality and collected herbarium, living, and formalin material which has enabled us to describe the plant fully. The illustration here used was made at the same time.

Figure 294 is from a photograph of the type plant taken by George Rose.

161 *a*. **Opuntia zebrina** Small, Journ. N. Y. Bot. Gard. **20**: 35. 1919. (See page 155, *ante*.)

Plant erect, more or less branched throughout, fully 1 meter tall or less, the roots fibrous; joints oval or obovate, thickish, mostly 1 to 2 dm. long, deep green, sometimes obscurely glaucous; leaves ovoid, 2 to 3 mm. long, bright green; areloes scattered, some of them, usually the lower ones, unarmed, the upper ones irregularly armed; spines slender, solitary or 2, 3, or 4, together, red-brown, finely banded, nearly terete, closely spirally twisted; flowers few on a joint, or solitary; sepals deltoid to deltoid-reniform or nearly reniform; corolla yellow, rotate, 6 to 7 cm. wide; petals rather numerous, the inner ones broadly obovate, undulate, minutely mucronate or notched at the apex; berries obovoid, not constricted at the base, 3.5 to 4.5 cm. long, red-purple; seeds many, 6 to 7 mm. in diameter.

Fig. 295.—Opuntia zebrina.

Type locality: Middle Cape Sable, Florida.

Distribution: Coastal sand-dunes, Cape Sable, Florida, and the lower Florida Keys.

The plant was first discovered by Dr. Britton on Boot Key, Florida, in 1909, and this is the most northern locality yet known for it. The species is interesting not only from its strikingly banded spines but also as being the only known member of the series *Elatiores* growing wild within the United States. In habit it resembles *O. dillenii*, and on Key West the two species were observed growing close together.

Fig. 296.—Fruit of O. zebrina. ×0.5.

Illustration: Journ. N. Y. Bot. Gard. **20**: pl. 226.

Figure 295 is from a photograph of the plant on Cape Sable, Florida, in cultivation at Buena Vista, Miami, Florida; figure 296 shows a fruit collected by Dr. Rose on Key West, Florida, in 1918.

173 *a*. **Opuntia keyensis** Britton in Small, Journ. N. Y. Bot. Gard. **20**: 31. 1919. (See p. 162, *ante*.)

Plant erect, much branched, sometimes forming clumps 3 meters tall, with long fibrous roots; joints elliptic, oval, obovate, or spatulate, thick, 1 to 3 dm. long, bright green; leaves ovoid, 2 to 3 mm. long, green; areoles rather conspicuous, often relatively large and prominent, apparently unarmed; spines stout, 4 to 13 together, very short, mostly hidden in the bristles; at first pink, at maturity salmon-colored, slightly flattened; flowers solitary or 2 or 3 on a joint; sepals deltoid to subreniform, acute or acutish; corolla salmon-colored, cup-like, or short-campanulate, 3 to 3.5 cm. wide; petals rather few, thinner ones broadly obovate or orbicular-obovate, undulate, scarcely, if at all, mucronate; berries obovoid, 4 to 6 cm. long, purple; seeds numerous.

FIG. 297.—Opuntia keyensis.

Type locality: Boot Key, Florida.

Distribution: Hammocks, Florida Keys and Cape Sable.

Opuntia keyensis was first collected by Dr. Britton in 1909 on Boot Key, Florida. Plants brought subsequently by Dr. Small from the Keys to Buena Vista, Miami, and there observed by him under cultivation show the species to be distinct from either *O. dillenii* or *O. stricta*, with both of which it has been associated.

Illustration: Journ. N. Y. Bot. Gard. **20**: pl. 225.

Figure 297 is from a photograph of the plant in cultivation at Buena Vista, Miami, Florida; figures 298 and 299 show its flowers, collected by Dr. Small on Key Largo, Florida, in 1909. See also plate xxx, figure 1.

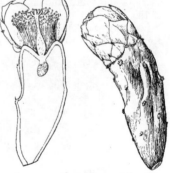

FIGS. 298 and 299.—Flower of Opuntia keyensis. ✕0.5.

183 *a*. **Opuntia bonplandii** (HBK.) Weber, Dict. Hort. Bois 894. 1898. (See page 168, *ante*.)

> *Cactus bonplandii* Humboldt, Bonpland, and Kunth, Nov. Gen. et Sp. **6**: 69. 1823.

Plants tall, 2 to 4 meters high, open-branching; joints ovate to obovate, 2 to 3 dm. long, dull green; spines at first 2 to 7, pale yellow, acicular, 1 to 1.5 cm. long but soon falling off; flowers orange-colored, about 6 cm. long and nearly as broad when fully expanded; petals obtuse; stamens short.

Type locality: Cuenca, Ecuador.

Distribution: Ecuador.

This species was collected by Humboldt and Bonpland at Cuenca, Ecuador, and was first described as *Cactus (Opuntia) bonplandii*. Apparently the type was not pre-

served as Dr. Rose did not find it either at Berlin or Paris in 1912.   Schumann mentions
it only in a note under *O. quitensis* following Weber who associates the two.   Dr. Rose,
while in Ecuador in 1918, spent about a week at Cuenca collecting plants in all direc-
tions from the town.   The only *Opuntia* in this whole region is the one above described
which grows in hedges and along the roadsides.   It may be an introduced species which
has escaped from gardens but we know nothing in cultivation just like it.   It resembles

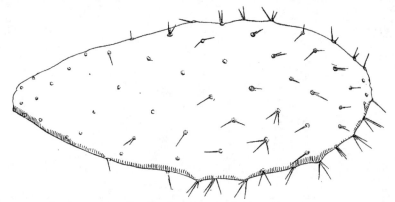

FIG. 300.—Opuntia bonplandii.   ×0.5.

somewhat the Nopal de Castilla, so common in Mexico and the southwestern states
Humboldt compared it with the tuna de Espana which may be the same.   Bonpland
seems to have called his plant *Cactus coccinellifer* which it very much resembles in the
shape of the joints and in being spineless in age.   If we are right in our interpretation
of this species it has no close alliance with *O. quitensis* which Dr. Rose collected also;
it has very small flowers with erect petals which are not readily affected by the sun as
are those of *O. bonplandii* and most of the other species.

FIG. 301.—Opuntia dobbieana.

Figure 300 shows a joint collected by Dr. Rose at Cuenca, Ecuador, in 1918.

207 a. Opuntia dobbieana sp. nov. (See page 187, *ante*.)

Usually low and bushy, forming dense thickets, but sometimes tall and then 3 to 4 meters high; joints orbicular to short-oblong or obovate, 1 to 2.5 dm. long, pale green in color, very spiny; leaves minute, 1 to 2 mm. long, green, spreading; areoles small, closely set; spines white, 5 to 12, usually acicular but on old joints subulate, 1 to 3 dm. long, accompanied by 2 to 4 reflexed hairs from the lower side of the areole; flower, including ovary, 5 to 6 cm. long; petals chocolate-colored, oblong, 2 cm. long; filaments and style pinkish; stigma-lobes dull green; ovary strongly tubercled, leafy, very spiny, especially towards the top; fruit juicy, red, at first spiny, 3 to 5 cm. long.

Common in dry places from Huigra to Sibambe, Province of Chimborazo, Ecuador.

Collected by J. N. Rose and George Rose, August to November 1918, at Huigra (No. 22201, type); at Sibambe, August 29 (No. 22434).

This species, on account of its white spines, is referred to the *Streptacanthae*, although it usually is more bushy than these species generally are. So far as we could learn, the fruit is not used by the Ecuadoreans; the plant was never seen cultivated, and there is every reason to believe it is native to Ecuador.

The species is named for John Dobbie, general manager of the Guayaquil and Quito Railway, whose courtesies and assistance added greatly to the success of Dr. Rose's visit to Ecuador in 1918.

Figures 301 and 302 (the latter at the bottom of this page) are from photographs of the type plant, taken by George Rose.

INDEX.

(Pages of principal entries in heavy-face type.)

Aaron's beard, 175
Agave, 117
Ahoplocarpus, 11
Airampo, 161
Alfilerillo, 26
Ammophilae, 45, **211**
Apple, 9
Aurantiacae, 45, 74, **106**
Ayrampo, 135
Barbados gooseberry, 10
Basilares, 45, **118**, 193
Beaucarnea, 117
Beaver-tail, 120
Bigelovianae, 44, **58**
Blade apple, 10
Brasilienses, 45, **209**
Brasiliopuntia, 209
Bullsucker, 43, 116
Cactaceae, 3, **8**
Cactales, **8**
Cacti, 5, 6, 7, 9, 28, 33, 39, 42, 49, 66, 67, 80, 87, 94, 95, 111, 126, 137, 151, 216, 219, 220
Cactodendron, 42, 43
Cactus, 5, 6, 8, 9, 23, 30, 32, 34, 35, 43, 49, 87, 88, 93, 107, 113, 114, 116, 120, 121, 146, 164, 177, 186, 207, 210, 215, 216, 217, 221
Cactus, arboreus, 209, 210
    aurantiacus, 107
    bleo, 17, 63
    bonplandii, 223
    bradypus, 121
    brasiliensis, 209
    californicus, 58
    chinensis, 156, 157
    coccinellifer, 179, 224
    cochenillifer, 34, 35, 173
    compressus, 127
    corrugatus, 95
    curassavicus, 102
    cylindricus, 63, 77
    decumanus, 180
    dillenii, 162, 163
    eburneus, 95
    elatior, 153
    elongatus, 179
    ferox, 199, 200, 206
    ficus-indica, 177
    fimbriatus, 13
    foliosus, 105
    fragilis, 193
    frutescens, 27
    heterocladus, 210
    horridus, 21
    humifusus, 127
    humilis, 113
    indicus, 156
    lanceolatus, 179
    linkii, 121
    lucidus, 10
    mammillaris, 4
    microdasys, 120, 121
    monocanthos, 156

Cactus—*continued.*
    moniliformis, 206, 207
    nanus, 217
    nigricans, 153
    opuntia, 43, 115, 127, 128, 129, 163, 177
    opuntia inermis, 161, 162
    opuntia nana, 127, 128
    opuntia polyanthos, 115
    opuntia tuna, 157
    opuntiaeflorus, 27
    ottonis, 121
    ovoides, 95
    paradoxus, 209
    pentlandii, 98
    pereskia, 9, 10, 11
    polyanthos, 113, 115
    portulacifolius, 23
    pseudococcinellifer, 153
    pusillus, 105
    rosa, 19, 20
    rotundifolia, 27
    salmianus, 74
    sericeus, 134
    spinosissimus, 204
    strictus, 161
    subinermis, 34
    subquadrifolius, 65
    sulphureus, 134
    tomentosus, 173
    triacanthos, 112
    tuberculatus, 214
    tuna, 113, 114, 163 ·
    tuna elatior, 153
    tuna nigricans, 153
    tunicatus, 65
    urumbeba, 156, 157
    urumbella, 157
    zinniaeflora, 21
Camuessa, 191
Cane cactus, 43
Cephalocereus, 116
Cereeae, 8, 24, 215
Cereus, 58, 75, 151, 215, 217
    articulatus, 89
    bradtianus, 215
    californicus, 58
    chiloensis, 79
    clavarioides, 73
    cylindricus, 77
    imbricatus, 63
    moniliformis, 206, 217
    nanus, 217
    ovatus, 91
    paradoxus, 209
    sericeus, 73
    serpens, 217
    syringacanthus, 89
    tunicatus, 66
Chaetophorae, 174
Chaffeyanae, 45, **213**
Cholla, 43, 61
Clavarioides, 44, **72**

227

- 301 -

Clavatae, 44, **79**, 84
Cola de diablo, 26
Consolea, 42, 43, **202**
    catacantha, 208
    ferox, 206
    leucacantha, 175
    rubescens, 43, 208
    spinosissima, 204
Corotilla, 96
Cow's tongue, 164
Criniferae, 140, 176
Cruciformes, 208
Cuija, 149
Curassavicae, 45, **102**, 104, 106, 193
Cylindropuntia, 32, 44, 45, **46**, 71, 75, 79, 84, 100, 142
Dildoes, 105
Dillenianae, 45, **159**, 169
Durasnilla, 175
Echinocactus, 117
Echinocarpae, 44, **56**
Echinocereus, 79, 215
Elatae, 45, 152, **156**
Elatiores, 45, **149**, 152, 155, 156, 222
Epiphyllum, 9
    gaillardae, 6
Erythrina, 181
Espina, 76
Espina blanca, 41
Espinha de São Antonio, 19
Etuberculatae, 72, 73
Eupereskia, 10
Ficindica, 42, 43
Ficus-indicae, 45, 156, 166, **177**, 191
Floccosae, 44, **86**
Flor de Cêra, 19
Frutescentes, 73
Fulgidae, 44, **67**
Fulvispinosae, 148
Glomeratae, 44, **87**
Grandifoliae, 9, 11
Grizzly bear cactus, 196
Grusonia, 24, **215**
Grusonia bradtiana, **215**
    cereiformis, 215
Guamacho, 17
Harrisia, 24
Hibiscus, 19
Hibiscus esculentus, 19
Hylocereus, 24
Imbricatae, 44, **60**
Inamoenae, 45, **125**
Inarmatae, 208
Iniabanto, 19
Lemaireocereus, 116
Lemon vine, 10
Lengua de vaca, 164
Leon, 96
Leoncito, 96
Leptocaules, 44, **46**, 49
Leucotrichae, 45, **174**
Loranthus aphyllus, 79
Macdougalianae, 45, **169**
Maihuen, 40
Maihuenia, 8, 24, **40–42**, 43, 95
    brachydelphys, 41, **42**
    patagonica, 41
    philippii, 41
    poeppigii, 41, 42

Maihuenia—*continued.*
    tehuelches, 41, **42**
    valentinii, 40, 41, **42**
Malus, 9
Mammillaria, 4
Matéare, 13
Miquelianae, 44, **78**
Mission cactus, 186
Najú de Culebra, 19
Nopal, 34
Nopal cardón, 184
Nopal de Castilla, 224
Nopalea, 8, 24, **33–39**, 43, 155, **216**
    auberi, 34, **37**, 38, 39
    cochenillifera, **34**, 181
    dejecta, 34, 36, **37**
    gaumeri, 34, 37, **216**
    guatemalensis, 33, 34, **35**
    inaperta, 33, 34, 37, **38**
    karwinskiana, 34, 37, 38, 39
    lutea, 33, 34, **35**
    moniliformis, 33, 206
Nopaleta, 26
Nopalnochetzli, 35
Ohulago, 103
Olago, 103
Opuntia, 8, 14, 24, 25, 30, 32, 33, 34, 38, 39, 40, **42–215**,
    217–224
    acanthocarpa, 56, **57**
    aciculata, 160, **165**
    acracantha, 91
    aequatorialis, 110, 116, **219,** 220
    affinis, 169, **170**
    airampo, 161
    albicans, 191
    albicans laevior, 191
    albiflora, 73, 74
    albisetosa, 134
    alcahes, 58, 67, **69**, 70
    alfagayucca, 185
    alfayucca, 185
    allairei, **126**
    alpicola, 214
    alpina, 33
    alta, 165, 166
    americana, 214
    ammophila, **211**
    amyclaea, 112, 177, 181, **185**
    amyclaea ficus-indica, 177
    anacantha, 107, **109**, 110
    anahuacensis, **169**
    andicola, 89, 90
    andicola elongata, 89
    andicola fulvispina, 89
    andicola major, 89
    andicola minor, 90
    angusta, **101**
    angustata, 124, 129, 140, **142**, 149
    angustata comonduensis, 124
    antillana, 110, **115**, 163
    aoracantha, 90, **91**
    aquosa, 29
    arborea, 209
    arborescens, 43, 63, 64, 65
    arborescens versicolor, 62
    arbuscula, 47, **50**, 51
    arechavaletai, 156, **158**
    arenaria, 193, **194**, 195

Opuntia—*continued.*
  argentina, 209, 211, 212
  arizonica, 147, 148
  arkansana, 128
  articulata, 89
  assumptionis, 159
  atacamensis, 90, 94
  atrispina, 140, 142
  atropes, 169, 170
  attulica, 214
  auberi, 37
  aulacothele, 95
  aurantiaca, 101, 107
  aurantiaca extensa, 107
  australis, 87, 88
  austrina, 45, 126, 130
  azurea, 140, 143
  bahamana, 202, 203, 204
  bahiensis, 209, 210, 211
  balearica, 161
  ballii, 137
  barbata, 214
  barbata gracillima, 214
  bartramii, 181
  basilaris, 118, 119, 120, 136
  basilaris albiflora, 120
  basilaris coerulea, 120
  basilaris cordata, 120
  basilaris cristata, 120
  basilaris nana, 120
  basilaris nevadensis, 120
  basilaris pfersdorffii, 120
  basilaris ramosa, 119, 120
  basilaris treleasei, 119
  beckeriana, 168
  bella, 110, 111, 112
  bentonii, 161, 163
  bergeriana, 149, 152, 169
  bernardina, 57, 81
  bernardina cristata, 57
  bernhardinii, 214
  betancourt, 214
  bicolor, 214
  bigelovii, 58, 59
  blakeana, 144, 145
  boldinghii, 149, 155
  boliviana, 71, 97, 98
  bonaerensis, 156, 158
  bonplandii, 160, 168, 223, 224
  borinquensis, 102, 103, 104
  brachyarthra, 193, 194
  brachyclada, 120
  brachydelphis, 42
  bradtiana, 215
  brandegeei, 25, 28
  brasiliensis, 209, 210, 211
  brasiliensis minor, 210
  brasiliensis schomburgkii, 210
  brasiliensis spinosior, 210
  brasiliensis tenuifolia, 210
  brasiliensis tenuior, 210
  brunnescens, 149, 150
  bulbispina, 79, 83
  bulbosa, 131
  burrageana, 67, 70
  cacanapa, 165, 166
  caerulescens, 51
  caesia, 144

Opuntia—*continued.*
  caespitosa, 127
  calacantha, 214
  calacantha rubra, 214
  calantha, 136
  californica, 58
  calmalliana, 60, 61
  calva, 89
  camanchica, 144, 145
  camanchica albispina, 144
  camanchica luteo-staminea, 144
  camanchica orbicularis, 144
  camanchica rubra, 144
  camanchica salmonea, 144
  campestris, 90, 99
  camuessa, 191
  cañada, 160, 167, 168
  candelabriformis, 182
  candelabriformis rigidior, 182
  canina, 107, 108
  cantabrigiensis, 121, 160, 167
  canterai, 159
  caracasana, 110, 116, 219
  cardenche, 64
  cardiosperma, 156, 157, 158
  cardona, 184
  caribaea, 47, 48, 49
  carolina, 214
  carrizalensis, 79
  castillae, 185, 186
  catacantha, 43, 208
  cereiformis, 79, 215
  cervicornis, 194
  chaetocarpa, 182
  chaffeyi, 30, 212, 213
  chakensis, 158
  chapistle, 27
  chavena, 184
  chihuahuensis, 144, 145
  chlorotica, 142, 159, 160, 161
  chlorotica santa-rita, 142
  cholla, 60, 61, 62
  chrysacantha, 167
  ciliosa, 214
  ciribe, 58, 59, 60
  clavarioides, 72, 73
  clavarioides cristata, 73
  clavarioides fasciata, 73
  clavarioides fastigiata, 73
  clavarioides monstruosa, 73
  clavata, 79, 81, 215
  clavellina, 52, 54
  coccifera, 35
  coccinea, 114
  cochenillifera, 33, 34, 35
  cochinelifera, 34
  cochinera, 192
  coerulea, 134
  coindettii, 184
  columbiana, 193
  comonduensis, 118, 124
  confusa, 147
  congesta, 50
  consoleana, 214
  consolei, 214
  convexa, 165
  cordobensis, 181, 189
  cornigata, 95

Opuntia—*continued.*
  corotilla, 96
  corrugata, 90, 95
  corrugata monvillei, 95
  costigera, 65
  covillei, 140, **145**, 146
  crassa, **178**, 179
  crassa major, 178
  cretochaeta, 183
  crinifera, 159, 176
  crinifera lanigera, 175
  cristata, 64
  cristata tenuior, 64
  cruciata, 207
  crystalenia, **193**
  cubensis, 163
  cucumiformis, 97
  cuija, 149, 167
  cumingii, 77
  curassavica, **102**, 103, 112
  curassavica elongata, 102
  curassavica longa, 102
  curassavica major, 102
  curassavica media, 102
  curassavica minima, 102
  curassavica minor, 102
  carassavica taylori, 103
  curvospina, 160
  cyanella, 165, 166
  cyclodes, 147, 148
  cycloidea, 128
  cylindrica, 71, 75, **77**
  cylindrica cristata, 78
  cylindrica cristata minor, 78
  cylindrica monstruosa, 78
  cylindrica robustior, 78
  cymochila, 131, 132
  dactylifera, 97, 98
  darrahiana, 102, **106**
  darwinii, 88, 90, **93**, 94, 97
  davisii, 52, **54**, 55
  deamii, 181, **187**
  decipiens, 63, 65
  decipiens major, 64
  decipiens minor, 65
  decumana, 157, 180, 181, 186
  decumbens, 111, **116**, 117, 121, 218
  decumbens irrorata, 117
  decumbens longispina, 117
  deflexa, 157
  dejecta, 37
  delaetiana, 149, **152**
  delicata, 126, **132**, 133
  deltica, 165
  demissa, 146, 147
  demorenia, 214
  demoriana, 214
  depauperata, 100, 101, **216**, 217, 219
  deppei, 214
  depressa, 111, **117**, 118
  deserta, 57, 58
  diademata, 89, 90
  diademata calva, 89
  diademata inermis, 89
  diademata oligacantha, 89
  diademata polyacantha, 89
  dichotoma, 214
  diffusa, 37

Opuntia—*continued.*
  digitalis, 72
  diguetii, 26, 29
  dillei, 147, 148
  dillenii, 106, 114, 115, 116, 159, **162**, 163, 181, 222, 223
  dillenii minor, 163
  dillenii orbiculata, 163
  dimorpha, 96
  diplacantha, 184
  discata, 140, **149**
  discolor, 107, 109, **218**, 219
  distans, 149, **155**
  dobbieana, 181, 187, 224, **225**
  dolabriformis, 207
  drummondii, 102, **104**, 106
  dulcis, 165, 166
  durangensis, **169**
  eborina, 214
  eburnea, 95
  echinocarpa, 56, **57**, 81
  echinocarpa major, 57
  echinocarpa nuda, 57, 58
  echinocarpa parkeri, 57, 58
  echinocarpa robustior, 57
  eichlamii, 181, **187**, 188
  elata, 110, 156, **157**
  elata delaetiana, 152
  elatior, 149, **153**, 219
  elatior deflexa, 157
  ellemeetiana, 75
  ellisiana, 166
  elongata, 179, 181
  elongata laevior, 180
  emoryi, 80
  engelmannii, 140, **147**, 148, 149, 164, 165, 167
  engelmannii cuija, 167
  engelmannii cyclodes, 147, 148
  engelmannii dulcis, 165
  engelmannii littoralis, 165
  engelmannii monstrosa, 148
  engelmannii occidentalis, 146
  eocarpa, 144
  erecta, 214
  erinacea, 193, **195**, 196
  erythrocentron, 174
  exaltata, 71, 75, **76**, 77
  expansa, 147, 148
  extensa, 107
  exuviata, 63, 65
  exuviata angustior, 63
  exuviata major, 64
  exuviata spinosior, 63
  exuviata stellata, 63
  exuviata viridior, 63
  ferox, 206, 207
  ferruginispina, 165
  festiva, 214
  ficus-barbica, 177
  ficus-indica, 43, 156, 157, **177**, 178, 181, 185, 189, 191
  ficus-indica albispina, 214
  ficus-indica amyclaea, 185
  ficus-incida decumana, 180
  ficus-indica gymnocarpa, 180
  filipendula, 137, 138
  flavicans, 191
  flavispina, 214
  flexibilis, 114
  flexospina, 165

Opuntia—*continued.*
  floccosa, 71, **86**, 87
  floccosa denudata, 86, 87
  floribunda, 74
  foliosa, 105, 106
  formidabilis, 91, 92
  fragilis, **193**, 194
  fragilis brachyarthra, 193
  fragilis caespitosa, 193
  fragilis frutescens,
  fragilis tuberiformis, 193
  frustulenta, 104, 105
  frutescens, 47, 48
  frutescens brevispina, 47, 48
  frutescens longispina, 47, 48
  fulgida, **67**
  fulgida mamillata, 67
  fuliginosa, 149, **155**
  fulvispina, 174
  fulvispina badia, 175
  fulvispina laevior, 175
  furiosa, 66
  fuscoatra, 126, **133**
  fusicaulis, 191, **192**
  fusiformis, 130, 131
  galapageia, 45, 149, **150**, 151, 152
  galeottii, 65
  geissei, 78
  gilliesii, 91
  gilva, 163
  gilvescens, 149
  gilvoalba, 165, 166
  glaberrima, 178
  glauca, 178
  glaucescens, 200, **201**
  glaucophylla, 214
  glaucophylla laevior, 214
  glomerata, 87, **89**, 90, 94, 96
  glomerata albispina, 90
  glomerata flavispina, 90
  glomerata minor, 90
  golziana, 26
  gomei, 165, 166
  gorda, 191, 192
  gosseliniana, 140, **141**, 151
  gracilis, 47
  gracilis subpatens, 48
  grahamii, 79, **83**, 84
  grandiflora, 126, 127, **129**
  grandis, 200, **201**
  grata, 92, 94, 95
  greenei, 131
  gregoriana, 147, 148
  griffithsiana, 165
  grosseiana, 107, **110**
  guanicana, 208
  guatemalensis, 110, 113, **218**, 219
  guerrana, 191, **192**
  guilanchi, 173, **174**
  gymnocarpa, 180
  haematocarpa, 166
  haitiensis, 206
  hanburyana, 149, **153**, 154
  hattoniana, 103
  helleri, 150, 152
  hempeliana, 86, 87
  hernandezii, 181
  heteromorpha, 79

Opuntia—*continued.*
  hevernickii, 214
  hickenii, 90, 92, **93**
  hieronymi, 211
  hitchenii, 214
  hochderfferi, 198
  horizontalis, 37, 90
  horrida, 162
  humifusa, 127, 128, 132
  humifusa cymochila, 128
  humifusa greenei, 128
  humifusa macrorhiza, 128
  humifusa microsperma, 127
  humifusa oplocarpa, 128
  humifusa parva, 127
  humifusa stenochila, 128
  humifusa vaseyi, 146
  humilis, 113, 114
  humistrata, 120
  hypsophila, 71, **72**
  hyptiacantha, 176, 181, **183**, 184
  hystricina, 193, **197**
  hystrix, 65, 66
  icterica, 173
  ignescens, 90, **98**
  ignota, 90, **99**
  imbricata, 48, 55, 60, **63**, 64, 65, 66
  imbricata crassior, 63
  imbricata ramosior, 64
  imbricata tenuior, 64
  inaequalis, 128
  inaequilateralis, 181, **187**, 188, 189
  inamoena, 121, **125**
  incarnadilla, 185, 186
  inermis, 161, 162, 163
  insularis, 150, 152
  intermedia, 127, 128
  intermedia prostrata, 128
  intricata, 119
  invicta, **79**
  involuta, 87
  irrorata, 116, 117,
  italica, 214
  ithypetala, **190**
  jamaicensis, 110, **113**
  joconostle, 214
  juniperina, 193, **197**
  jussieuii, 214
  karwinskiana, 38
  keyensis, 159, 162, 163, **222**, 223
  kiska-loro, 107, **108**
  kleiniae, 47, **51**, 64, 65
  kleiniae cristata, 51
  kleiniae laetevirens, 51
  kunzei, 80
  laboureiana, 180, 189
  laboureiana macrocarpa, 181
  laevis, 159, **161**, 168
  lagopus, 86, **87**, 88
  lanceolata, 177, **179**, 203
  lanigera, 176
  larreyi, 191
  lasiacantha, 181, **182**, 183, 184
  lata, 126, **220**
  laxiflora, 165
  ledienii, 153
  lemaireana, 156
  leonina, 96

Opuntia—*continued.*
leptarthra, 101
leptocarpa, 166, 167
leptocaulis, **47**, 48, 49, 73, 213
leptocaulis brevispina, 47
leptocaulis laetevirens, 48
leptocaulis longispina, 47
leptocaulis major, 48
leptocaulis stipata, 47
leptocaulis vaginata, 47
leucacantha, 174, 175
leucacantha laevior, 175
leucacantha subferox, 175
leucantha, 175
leucophaea, 96
leucostata, 214
leucosticta, 174
leucotricha, 151, **174**, 175
leucotricha fulvispina, 174
ligustica, 128
lindheimeri, 44, 148, 160, 164, **165**, 166
lindheimeri cyclodes, 147
lindheimeri dulcis, 165
lindheimeri littoralis, 165
lindheimeri occidentalis, 146
linguiformis, 160, **163**
littoralis, 147, 160, 164, **165**
lloydii, 60, **63**
longiclada, 161
longiglochia, 214
longispina, 95
lubrica, **118**, 119
lucayana, 106, 163
lucens, 149
lucida, 214
lurida, 173
macateei. 126, 133, 220, **221**
macdougaliana, 169, 170, **171**
mackensenii, 137, **138**, 139
macracantha, 187, 202, **203**, 204, 205
macrarthra, 126, **129**
macrocalyx, 118, **122**
macrocentra, **140**, 141, 176
macrophylla, 214
macrorhiza, 126, 127, 128, **130**, 131, 139, 166
maculacantha, 134
maelenii, 134
magenta, 142, 146
magna, 64
magnarenensis, 147
magnifolia, 34
maihuen, 41
maldonadensis, **75**
mamillata, 67, 68
maritima, 162
maxillare, 77
maxima, 177, **180**, 181
media, 199, 200
mediterranea, 128
megacantha, 178, 181, 182, **185**, 186
megacantha lasiacantha, 182
megacantha tenuispina, 184
megacantha trichacantha, 186
megacarpa, 145
megalantha, **169**
megalarthra, 192
megarhiza, **137**
mendocienses, 65

Opuntia—*continued.*
mesacantha, **127**, 132
mesacantha cymochila, '131
mesacantha grandiflora, 129
mesacantha greenei, 131
mesacantha macrorhiza, 130
mesacantha microsperma, 127
mesacantha oplocarpa, 131, 144
mesacantha parva, 127, 128
mesacantha sphaerocarpa, 144
mesacantha stenochila, 132
mesacantha vaseyi, 146
metternichii, 176
mexicana, 34
micrarthra, 171
microcarpa, 144, 206
microdasys, 118, **120**, 121, 122, 123, 136
microdasys laevior, 120
microdasys minor, 120
microdasys rufida, 122
microdisca, 133, **135**, 136
microthele, 73
mieckleyi, 156, **158**
militaris, 102, **104**, 163
millspaughii, 202, **204**, 206
minima americana, 102
minor, 139
minor caulescens, 102
miquelii, **78**
missouriensis, 199, 200
missouriensis albispina, 199, 200
missouriensis elongata, 214
missouriensis erythrostemma, 214
missouriensis microsperma, 199, 200
missouriensis platycarpa, 199, 200
missouriensis rufispina, 199, 200
missouriensis salmonea, 214
missouriensis subinermis, 199
missouriensis trichophora, 195
modesta, 67
mojavensis, 140, **145**
molesta, 54, 60, 62, **66**, 67
monacantha, 127, 156, 157
monacantha deflexa, 156
monacantha gracilior, 156, 157
monacantha variegata, 157
moniliformis, 33, 202, **206**, 207, 208
montana, 214
montevidensis, 107, **109**
monticola, 95
morenoi, 214
morisii, 128
mortolensis, **47**
multiflora, 113, 114
myriacantha, 150, 152, 214
nana, 127, 217
nashii, 106, 163, 202, **203**
nelsonii, 172
nemoralis, 102, **104**
neoarbuscula, 50
nigricans, 152, 153
nigrispina, 90, **97**
nigrita, 183, 184
nopalilla, 38
oblongata, 173
occidentalis, 140, **146**, 147, 153, 163, 165
oligacantha, 89, 182
oplocarpa, 132

Opuntia—*continued.*
opuntia, 43, 126, **127**, 128, 217
orbiculata, **176**
orbiculata metternichii, 176
ottonis, 214, 215
ovallei, 95, 214
ovata, 90, **95**, 96
ovoides, 95
pachona, 184
pachyarthra flava, 214
pachyclada rosea, 214
pachyclada spaethiana, 214
pachypus, 75, **77**
pallida, 60, 65, **66**
palmadora, **202**
palmeri, **161**
pampeana, 134
papyracantha, 89, 90
paraguayensis, 158
parishii, 57, 79, **81**, 82
parkeri, 58
parmentieri, 95
parote, 214
parryi, 56, **57**, 58, 81
parva, 161
parvispina, 117
parvula, 178
pascoensis, 100, **101**
patagonica, 41
pelaguensis, 90
penicilligera, **135**
pennellii, 110, 115, **219**
pentlandii, 71, 72, 77, 90, **97**, 98
perrita, 65, 66
pes-corvi, 104
pestifer, 102, 103, **217**, 218, 220
phaeacantha, 139, 140, 142, **144**
phaeacantha brunnea, 144
phaeacantha major, 144
phaeacantha nigricans, 144
philippii, 41
phyllacantha, 96
phyllanthus, 215
piccolomini, 214
piccolominiana, 191
pilifera, **176**, 177, 184
pintadera, 176
pititache, 29
pittieri, 181, **188**, 189
platyacantha, 33, 89
platyacantha deflexispina, 89, 90
platyacantha gracilior, 89
platyacantha monvillei, 89
platyclada, 214
platynoda, 109
plumbea, 126, **131**
plumosa nivea, 89, 90
poeppigii, 40, 41
pollardii, **126**, 221
polyacantha, 193, 195, 196, 197, **199**, 200
polyacantha albispina, 199
polyacantha borealis, 199
polyacantha microsperma, 200
polyacantha rufispina, 200
polyacantha trichophora, 195
polyacantha platycarpa, 199
polyacantha watsonii, 199
polyantha, 113, 114, 115

Opuntia—*continued.*
polymorpha, 89
porteri, 28
pottsii, **137**, 138
praecox, 214
prate, 191
procumbens, 160, **167**
prolifera, 61, 67, **69**, 70
prostrata, 128
prostrata spinosior, 214
protracta, 214
protracta elongata, 214
pruinosa, 191
pseudococcinellifer, 153
pseudotuna, 184, 214
pseudotuna elongata, 214
pseudotuna spinosior, 214
puberula, 116, 117, 121, 122
pubescens, 100, **101**, 173
pulchella, 79, **82**
pulverata, 214
pulverulenta, 78
pulverulenta miquelii, 78
pulvinata, 120
pumila, **100**, 101
purpurea, 97
pusilla, 102, **105**, 106
pycnantha, 118, **123**
pycnantha margaritana, 123
pyriformis, 160, **168**
pyrocarpa, 166
pyrrhacantha, 97
quimilo, 181, **190**, 191
quipa, 125
quitensis, 149, **154**, 224
rafinesquei, 127, 128, 129
rafinesquei arkansana, 127, 129
rafinesquei cymochila, 131
rafinesquei cymochila montana, 131
rafinesquei fusiformis, 130
rafinesquei grandiflora, 129
rafinesquei greenei, 132
rafinesquei macrorhiza, 131
rafinesquei microsperma, 127, 199
rafinesquei minor, 127, 128
rafinesquei parva, 128
rafinesquei stenochila, 132
rafinesquei vaseyi, 146
rafinesquiana, 127, 129
rafinesquiana arkansana, 129
rahmeri, 94
ramosissima, **46**, 48, 52
ramulifera, 47
rastrera, 140, **149**
rauppiana, 90, **92**
recedens, 128
recondita, 52, **53**
recurvospina, 144
reflexa, 165
repens, 102, **103**, 104, 115, 116
reptans, 214
retrorsa, 107, **109**, 218
retrospinosa, 95
rhodantha, 193, 197, **198**
rhodantha brevispina, 198
rhodantha flavispina, 198
rhodantha pisciformis, 198
rhodantha schumanniana, 198

Opuntia—*continued.*
riparia, 149
robusta, 182, **191**, 192
robusta viridior, 191
rosea, 17, 63, 65, 78
roseana, 130
rosiflora, 78
rotundifolia, 27
roxburghiana, 156
rubescens, 163, 202, **208**, 209
rubiflora, **133**, 146
rubrifolia, 144
rufescens, 175
rufida, 118, 119, **122**
rugosa, 145
russellii, 90, **94**
ruthei, 64
rutila, 196
sabinii, 194
sacharosa, 14,
salicornioides, 215
salmiana, **73** 74
salmii, 214
sanguinocula, 131
santa-rita, 140, **142**
scheeri, **159**, 176
schickendantzii, 74, **107**
schomburgkii, 214
schottii, 79, **80**, 81
schottii greggii, 80
schumannii, 90, 149, **155**
schweriniana, 199, 200
segethii, 75, 76, 79
seguina, 130
semispinosa, 147
senilis, 86, 159, 176
sericea, 134
sericea coerulea, 134
sericea longispina, 134
sericea maelenii, 134
serpentina, 56, 57, **58**, 69, 81
setispina, 45, 137, **138**
shaferi, 71, **72**
shreveana, 142
sinclairii, 165, 166
skottsbergii, 90, **96**, 97
soederstromiana, 149, 154, **221**
soehrensii, 133, **134**, 135
spathulata, 28, 29
spathulata aquosa, 30
speciosa, 214
spegazzinii, 73, 74
sphaerica, 90, 96
sphaerocarpa, 193, **198**, 199
sphaerocarpa utahensis, 199
spinalba, 130
spinaurea, 214
spiniflora, 215
spinosior, 52, 67, **68**
spinosior neomexicana, 68
spinosissima, 103, 175, 202, **204**, 205
spinotecta, 64
spinulifera, 178, 181, **182**
spinuliflora, 214
spinulosa, 214
splendens, 199
squarrosa, 165
stanlyi, 79, **80**

Opuntia—*continued.*
stapelia, 66
stapeliae, 65, 66
stellata, 64
stenarthra, **158**
stenochila, 126, **132**
stenopetala, **200**, 201
straminea, 214
streptacantha, 181, **184**, 185
stricta, 159, **161**, 178, 223
stricta spinulescens, 214
strigil, **136**
subarmata, 165, 166
subferox, 175
subinermis, 214
subterranea, 90, **92**
subulata, 71, **75**, 76, 77, 79
sulphurea, 133, **134**, 150
sulphurea laevior, 134
sulphurea major, 134
sulphurea minor, 134
sulphurea pallidior, 134
superbospina, 144
syringacantha, 89
tapona, 124, 160, **164**
tarapacana, 90, **94**
tardospina, 140, **141**
taylori, 102, **103**
tenajo, 49
tenuispina, 137, **139**
teres, 71
tesajo, 47, **48**, 49
tessellata, 46
tessellata cristata, 46
testudinis-crus, 206
tetracantha, 52, **53**, 54
texana, 165, 166
thurberi, 52, **53**, 54
tidballii, 160
tomentella, 173, **174**
tomentosa, **173**, 174
tortisperma, 131, 132
tortispina, 126, 128, **131**, 194
toumeyi, 144
tracyi, 102, **105**
treleasei, 118, **119**
treleasei kernii, 119
triacantha, 110, **112**, 113, 115, 204
tribuloides, 186
tricolor, 165, 166
trichophora, 193, **195**
tuberculata, 214
tuberiformis, 92
tuberosa, 32, 33
tuberosa spinosa, 89
tuna, 110, **113**, 114, 116, 149, 157, 163
tuna humilis, 114
tuna laevior, 114
tuna orbiculata, 114
tunicata, 60, **65**, 66, 83, 121
tunicata laevior, 66
tunoidea, 116, 162
tunoides, 116
turpinii, 89
turpinii polymorpha, 89
tweediei, 134
umbrella, 156
undosa, 177, 179

Opuntia—*continued.*
  undulata, 65, 177, **179**
  ursina, 174, 195, 196, 197
  urumbella, 157
  utahensis, 198
  utkilio, 107, **109**, 110
  vaginata, 47, 48
  valida, 147
  vaseyi, 140, 142, 145, **146**
  velutina, 169, **172**
  verschaffeltii, 44, 71, **72**
  verschaffeltii digitalis, 71, 72
  versicolor, 44, 52, 54, 60, **62**
  vestita, **71**, 72, 87
  vexans, 64, 65
  vilis, 79, **82**, 83
  violacea, 144
  virgata, 47
  viridiflora, 52, **55**
  vivipara, **52**
  vulgaris, 127, 128, **156**, 157, 163, 177
  vulgaris balearica, 161
  vulgaris major, 127
  vulgaris media, 127
  vulgaris minor, 127
  vulgaris nana, 127, 129
  vulgaris rafinesquei, 127
  vulgo, 157
  vulpina, 134
  wagneri, 74
  weberi, **84**, 85
  wentiana, 110, **116**
  whipplei, 31, 43, 52, **55**, 68
  whipplei laevior, 55
  whipplei spinosior, 68
  wilcoxii, 169, **172**
  winteriana, 166
  wootonii, 147, 148
  wrightii, 51
  xanthoglochia, 130, 131
  xanthostemma, 198
  xanthostemma elegans, 198
  xanthostemma fulgens, 198
  xanthostemma gracilis, 198
  xanthostemma orbicularis, 198
  xanthostemma rosea, 198
  xerocarpa, 198
  youngii, 130
  zebrina, 149, 155, **222**
  zacuapanensis, **183**
  zuniensis, 144
Opuntiacei, 217
Opuntieae, 8, **24**
Orbiculatae, 45, **176**
Palmadora, 202
Palmadorae, 45, **201**
Palmatoria, 202
Pataquisca, 76
Peirescia, 9
Peireskia, 9
Pentlandianae, 44, **90**
Perescia, 9
Pereskia, **8–24**, 25, 26, 40, 75
  acardia, 10
  aculeata, 10, 11, 14
  aculeata lanceolata, 10
  aculeata latifolia, 10
  aculeata longispina, 10

Pereskia—*continued.*
  aculeata rotunda, 10
  aculeata rotundifolia, 10
  aculeata rubescens, 10
  affinis, 24
  amapola, 14
  argentina, 14
  autumnalis, 9, **11**, 12
  bahiensis, 9, **19**, 20
  bleo, 4, 9, **17**, 18, 20
  brasiliensis, 10
  calandriniaefolia, 29
  colombiana, 9, **17**
  conzattii, 9, **24**
  crassicaulis, 29
  cruenta, 24
  cubensis, 9, **22**
  foetens, 10
  fragrans, 10
  glomerata, 94
  godseffiana, 10, 11
  grandiflora, 24
  grandifolia, 9, 18, **19**, 20, 21
  grandispina, 24
  guamacho, 9, 15, **16**
  haageana, 24
  horrida, 9, **21**
  lanceolata, 10
  longispina, 10
  lychnidiflora, 9, **12**, 13
  moorei, 9, **15**
  nicoyana, 9, **13**
  ochnacarpa, 19
  opuntiaeflora, 26, 27
  panamensis, 17, 18
  pereskia, 7, 9, 10
  philippii, 41
  pititache, 29
  plantaginea, 24
  poeppigii, 41
  portulacifolia, 9, 22, **23**, 24
  rosea, 17
  rotundifolia, 27
  sacharosa, 9, 10, **14**, 15
  spathulata, 28, 29
  subulata, 75, 76
  tampicana, 9, **17**
  undulata, 10
  weberiana, 9, **15**
  zehntneri, 9, 13, **14**
  zinniaeflora, 9, **17**, 20, 21
Pereskieae, 8, 24
Pereskiopsis, 8, 14, 17, 24, **25–30**, 43
  aquosa, 25, 29
  autumnalis, 11, 12
  brandegeei, 28
  chapistle, 25, **27**
  diguetii, 25, **26**, 27
  kellermanii, 25, **30**
  opuntiaeflora, 25, **26**, 27
  pititache, 25, **29**
  porteri, 25, **28**
  rotundifolia, 25, **27**, 28
  spathulata, 25, **28**
  velutina, **25**, 26
Pereskiopuntia, 25
Pest pear, 161, 163
Phaeacanthae, 45, 136, **139**, 141, 148

Pin pillow, 102
Platyopuntia, 45, 73, 84, 92, 99, 100, 135, 211
Polar bear cactus, 87
Polyacanthae, 45, 193
Portulaca, 9
Prickly pear, 43, 212
Pterocactus, 24, 30–33
    decipiens, 32, 33
    fischeri, 31
    hickenii, 31
    kuntzei, 30, 32, 33
    kurtzei, 32
    pumilus, 31, 32
    tuberosus, 31, 32, 33
    valentinii, 88
Pubescentes, 141
Puipute, 13
Pumilae, 45, 100
Quiabento, 14
Quimilo, 190
Quipa, 125
Ramosissimae, 44, 46
Rhipsalis, 8
Robustae, 45, 191
Sacacil, 214
Sacharosa, 10, 14
Salmianae, 44, 73, 75
Scheerianae, 45. 159
Sempervivum tomentosum, 41
Setispinae, 45, 136
Spear-shaped opuntia, 179
Spinosissimae, 43, 45, 201, 202, 203, 208
Stenopetalae, 45, 200
Stenopuntia, 200
Streptacanthae, 45, 112, 156, 177, 181, 192, 225
Strigiles, 45, 136

Subulatae, 44, 71, 75
Sucker, 43, 103
Sulphureae, 45, 133, 135
Tacinga, 24, 39, 40
    funalis, 38, 39
Tapuna pear, 192
Tasajillo, 26, 30
Tasajo, 43
    macho, 63
Tephrocactus, 42, 43, 44, 71, 72, 79, 84, 85, 90, 95, 97,
        106, 135
    andicolus, 89
    aoracanthus, 91
    calvus, 89
    diadematus, 43, 89
    platyacanthus, 89
    pusillus, 106
    retrospinosus, 95
    turpinii, 89
Teretes, 71
Thurberianae, 44, 52
Tomentosae, 45, 172
Tortispinae, 45, 104, 126, 130, 133, 136, 193
Tuna, 35, 43, 114, 181, 186
Tuna cardona, 184
Tuna elatior, 153
Tuna de agua, 30
Tuna de Espana, 224
Tuna major, 163
Tunae, 45, 110, 116, 148
Vestitae, 44, 71
Weberianae, 44, 84
West Indian gooseberry, 10
Wilcoxia, 6
Zamia pumila, 181
Zygocactus, 9

PLATE I

A group of plants of *Cephalocereus macrocephalus* on a forest-covered hillside near
Tehuacán, Mexico.

THE CACTACEAE

DESCRIPTIONS AND ILLUSTRATIONS OF PLANTS OF THE CACTUS FAMILY

BY

N. L. BRITTON AND J. N. ROSE

Volume II

The Carnegie Institution of Washington
Washington, 1920

CARNEGIE INSTITUTION OF WASHINGTON
PUBLICATION No. 248, VOLUME II

PRESS OF GIBSON BROTHERS
WASHINGTON

CONTENTS.

	PAGE.
Tribe Cereeae	1
Key to Subtribes	1
Subtribe Cereanae	1
Key to Genera	1
Cereus	3
Monvillea	21
Cephalocereus	25
Espostoa	60
Browningia	63
Stetsonia	64
Escontria	65
Corryocactus	66
Pachycereus	68
Leptocereus	77
Eulychnia	82
Lemaireocereus	85
Erdisia	104
Bergerocactus	107
Leocereus	108
Wilcoxia	110
Peniocereus	112
Dendrocereus	113
Machaerocereus	114
Nyctocereus	117
Brachycereus	120
Acanthocereus	121
Heliocereus	127
Trichocereus	130

	PAGE.
Tribe Cereeae—continued.	
Subtribe Cereanae—continued.	
Jasminocereus	146
Harrisia	147
Borzicactus	159
Carnegiea	164
Binghamia	167
Rathbunia	169
Arrojadoa	170
Oreocereus	171
Facheiroa	173
Cleistocactus	173
Zehntnerella	176
Lophocereus	177
Myrtillocactus	178
Neoraimondia	181
Subtribe Hylocereanae	183
Hylocereus	183
Wilmattea	195
Selenicereus	196
Mediocactus	210
Deamia	212
Weberocereus	214
Werckleocereus	216
Aporocactus	217
Strophocactus	221
Appendix	223
Index	227

III

ILLUSTRATIONS.

PLATES.

FACING PAGE.

PLATE 1. Group of plants of Cephalocereus macrocephalus on hillside near Tehuacán.............Frontispiece.
PLATE 2. (1) Top of flowering stem of Cereus alacriportanus. (2) Top of stem of Cereus peruvianus. (3) Flower of Cereus peruvianus.. 6
PLATE 3. (1) Top of stem of Cereus validus. (2) Top of flowering stem of Cereus validus. (3) Top of flowering branch of Monvillea cavendishii. (4) Top of branch of M. cavendishii, with fruit. 22
PLATE 4. (1) Flowering stem of Cephalocereus pentaedrophorus. (2) Top of stem of Cephalocereus gounellei. (3) Top of stem of Cephalocereus bahamensis, with flower. (4) Fruit of Cephalocereus deeringii.. 32
PLATE 5. A clump of plants of Cephalocereus deeringii on Lower Matecumbe Key, Florida............. 38
PLATE 6. (1) Top of flowering stem of Cephalocereus arrabidae. (2) Top of flowering stem of Cephalocereus nobilis. (3) Top of flowering stem of Cephalocereus barbadensis.............. 42
PLATE 7. (1) Plants of Cephalocereus polygonus. (2) Large plant of Cephalocereus chrysacanthus...... 48
PLATE 8. (1) Top of flowering stem of Cephalocereus brooksianus. (2) Top of stem of Cephalocereus catingicola. (3) Top of stem of Cephalocereus phaeacanthus. (4) Flowering branch of Leptocereus assurgens... 56
PLATE 9. A large plant of Stetsonia coryne in the desert of northern Argentina...................... 64
PLATE 10. A plant of Escontria chiotilla, near Tehuacán, Mexico................................. 66
PLATE 11. Top of flowering plant of Pachycereus chrysomallus................................. 72
PLATE 12. A mountain-side along Tomellín Canyon, Mexico, covered with Pachycereus columna-trajani.... 76
PLATE 13. (1) Top of flowering branch of Leptocereus arboreus. (2) Top of stem of Lemaireocereus griseus. (3) Fruiting branch of Mediocactus coccineus............................... 80
PLATE 14. (1) Part of branch of Dendrocereus nudiflorus. (2) Flowering branch of Dendrocereus nudiflorus. (3) Flowering branch of Nyctocereus guatemalensis............................ 114
PLATE 15. (1) Top of branch of Eulychnia iquiquensis. (2) Top of stem of Lemaireocereus dumortieri. (3) Part of flowering stem of Nyctocereus serpentinus.............................. 118
PLATE 16. (1) Top of flowering branch of Acanthocereus pentagonus. (2) Top of flowering branch of Acanthocereus subinermis. (3) Top of a fruiting branch of Acanthocereus subinermis.... 124
PLATE 17. (1) End of flowering branch of Heliocereus elegantissimus. (2) End of flowering branch of Heliocereus speciosus. (3) A tip of a fruiting branch of Harrisia portoricensis........... 128
PLATE 18. (1) Tip of a flowering branch of Harrisia eriophora. (2) Fruiting branch of Harrisia eriophora.. 148
PLATE 19. (1) Top of flowering branch of Harrisia fragrans. (2) Top of fruiting joint of Harrisia fragrans. (3) Fruiting branch of Harrisia martinii... 150
PLATE 20. (1) Part of fruiting branch of Harrisia gracilis. (2) Top of flowering branch of Harrisia martinii. 152
PLATE 21. (1) Flowering branch of Harrisia tortuosa. (2) Fruiting branch of Harrisia tortuosa.......... 156
PLATE 22. Top of flowering plant of Carnegiea gigantea....................................... 164
PLATE 23. The giant cactus, Carnegiea gigantea, near Tucson, Arizona........................... 166
PLATE 24. (1) Top of flowering branch of Harrisia fernowi. (2) Flowering branch of Harrisia bonplandii. (3) Top of branch of Binghamia melanostele.................................... 168
PLATE 25. (1) Flowering branch of Rathbunia alamosensis. (2) Flowering branch of Rathbunia alamosensis. (3) Top of flowering branch of Borzicactus acanthurus. (4) Top of stem of Arrojadoa rhodantha.. 170
PLATE 26. (1) Myrtillocactus geometrizans, Tehuacán, Mexico. (2) Myrtillocactus schenckii, near Mitla, Mexico... 180
PLATE 27. (1) End of fruiting branch of Arrojadoa rhodantha. (2) Top of plant of Cleistocactus haumannii. (3) Flower on branch of Hylocereus stenopterus.................................. 184
PLATE 28. Flower on end of branch of Hylocereus ocamponis.................................. 186
PLATE 29. Flower on end of branch of Hylocereus monacanthus............................... 188
PLATE 30. Flower near end of branch of Hylocereus undatus.................................. 190
PLATE 31. Flower on short branch of Hylocereus lemairei..................................... 192
PLATE 32. (1) Fruit of Hylocereus undatus. (2) Flowering branch of Wilmattea minutiflora. (3) Longitudinal section of fruit of Selenicereus grandiflorus................................ 196
PLATE 33. (1) Flowering branch of Selenicereus grandiflorus. (2) Tip of branch of Selenicereus grandiflorus. (3) Fruit of Selenicereus grandiflorus..................................... 198
PLATE 34. Flowering branch of Selenicereus urbanianus...................................... 200
PLATE 35. Flower on branch of Selenicereus coniflorus....................................... 202
PLATE 36. (1) Fruit of Hylocereus trigonus. (2) Flower of Selenicereus boeckmannii. (3) Fruit of Selenicereus boeckmannii.. 204
PLATE 37. Flower of Mediocactus coccineus... 212
PLATE 38. (1) Fruiting branch of Selenicereus pteranthus. (2) Flowering branch of Selenicereus spinulosus. (3) Flowering branch of Weberocereus panamensis............................ 214
PLATE 39. (1) Flowering branch of Weberocereus tunilla. (2) Flowering branch of Weberocereus biolleyi. (3) Flowering branch of Werckleocereus tonduzii. (4) Flower of Werckleocereus glaber. 216
PLATE 40. (1) Flowering plant of Aporocactus leptophis. (2) Flowering plant of A. flagelliformis........ 218

IV

TEXT-FIGURES.

	PAGE.
FIG. 1. Plant of Cereus hexagonus in garden....	5
2. Flower of Cereus hexagonus..........	5
3. Longitudinal section of flower of Cereus hexagonus......................	5
4. Fruit of Cereus hexagonus.............	5
5. Cultivated plants of Cereus hildmannianus.............................	6
6. Cultivated plants of Cereus hildmannianus.............................	6
7. Potted plant of Cereus validus.........	7
8. Potted plant of Cereus tetragonus......	7
9. Plant of Cereus jamacaru.....	8
10. Hedge of Cereus stenogonus...........	10
11. Plant of Cereus dayamii...............	11
12. Cultivated specimen of Cereus argentinensis................................	12
13. Fruit of Cereus peruvianus............	13
14. Plant of Cereus pernambucensis........	14
15. Potted plant of Cereus obtusus.........	16
16. Plant of Cereus aethiops.............	16
17. Fruiting branch of Cereus aethiops.....	18
18. Fruit of Cereus repandus.............	18
19. Plant of Cereus repandus.............	18
20. Plant of Monvillea cavendishii........	22
21. Flower of Monvillea insularis..........	22
22. Potted plant of Monvillea spegazzinii...	23
23. Flower of Monvillea diffusa............	24
24. Potted plant of Cephalocereus senilis....	28
25. Potted plant of Cephalocereus purpureus.	28
26. Fruit of Cephalocereus fluminensis......	29
27. Flower of Cephalocereus purpureus.....	29
28. Cluster of spines of Cephalocereus purpureus...........................	29
29. Plants of Cephalocereus fluminensis...	29
30. Thicket of Cephalocereus dybowskii....	30
31. Plant of Cephalocereus pentaedrophorus.	31
32. Fruit of Cephalocereus pentaedrophorus.	31
33. Flower of Cephalocereus pentaedrophorus.	31
34. Top of plant of Cephalocereus polylophus.	32
35. Potted plant of Cephalocereus euphorbioides	33
36. Flower of Cephalocereus russelianus....	33
37. Fruit of Cephalocereus russelianus......	33
38. Plants of Cephalocereus russelianus.....	34
39. End of branch of Cephalocereus russelianus................................	34
40. Plant of Cephalocereus gounellei.......	35
41. Flower of Cephalocereus zehntneri......	35
42. Potted plant of Cephalocereus leucostele.	36
43. Potted plant of Cephalocereus smithianus...........................	36
44. Flower of Cephalocereus leucostele......	37
45. Fruit of Cephalocereus leucostele.......	37
46. Flower of Cephalocereus smithianus.....	37
47. Fruit of Cephalocereus smithianus......	37
48. Plant of Cephalocereus bahamensis.....	38
49. Plant of Cephalocereus bahamensis.....	38
50. Flower of Cephalocereus deeringii.....	39
51. Fruit of Cephalocereus deeringii........	39
52. Flower of Cephalocereus robinii........	39
53. Fruit of Cephalocereus robinii.........	39
54. Plant of Cephalocereus robinii.........	39
55. Plant of Cephalocereus keyensis.......	40
56. Flower of Cephalocereus keyensis.......	40

	PAGE.
FIG. 57. Fruit of Cephalocereus keyensis........	40
58. Flower of Cephalocereus monoclonos....	40
59. Flower of Cephalocereus moritzianus....	42
60. Fruit of Cephalocereus moritzianus.....	42
61. Plants of Cephalocereus moritzianus.....	42
62. Fruit of Cephalocereus arrabidae.......	43
63. Plant of Cephalocereus arrabidae.......	43
64. Plant of Cephalocereus nobilis.........	45
65. Potted plant of Cephalocereus barbadensis.............................	45
66. Plant of Cephalocereus barbadensis.....	46
67. Plant of Cephalocereus millspaughii....	46
68. Fruit of Cephalocereus millspaughii....	46
69. Flower of Cephalocereus royenii........	46
70. Plant of Cephalocereus swartzii........	47
71. Plant of Cephalocereus maxonii........	48
72. Plant of Cephalocereus piauhyensis.....	48
73. Plant of Cephalocereus lanuginosus.....	50
74. Plant of Cephalocereus royenii.........	50
75. Potted plant of Cephalocereus robustus.	51
76. Potted plant of Cephalocereus cometes.	51
77. Plant of Cephalocereus leucocephalus...	53
78. Plant of Cephalocereus tweedyanus.....	54
79. Plant of Cephalocereus tweedyanus......	54
80. Flower of Cephalocereus tweedyanus....	55
81. Fruit of Cephalocereus tweedyanus.....	55
82. Plant of Cephalocereus colombianus.....	55
83. Stem, flower, and flower-bud of Cephalocereus colombianus...............	56
84. Plant of Cephalocereus brasiliensis......	57
85. Flower of Cephalocereus phaeacanthus.	57
86. Fruit of Cephalocereus phaeacanthus...	57
87. Plant of Espostoa lanata..............	61
88. Plant of Espostoa lanata..............	61
89. Flower of Espostoa lanata.............	61
90. Fruit of Espostoa lanata..............	61
91. Potted plant of Espostoa lanata........	62
92. Flower of Browningia candelaris........	64
93. Young fruit of Browningia candelaris...	64
94. Plant of Browningia candelaris.........	64
95. Flower of Stetsonia coryne............	65
96. Ends of branches of Stetsonia coryne....	65
97. Flower of Escontria chiotilla..........	66
98. Fruit of Escontria chiotilla............	66
99. Plant of Corryocactus brevistylus......	67
100. Plant of Corryocactus brachypetalus....	67
101. Flower of Corryocactus brevistylus.....	68
102. Flower of Corryocactus brachypetalus...	68
103. Fruit of Corryocactus brachypetalus....	68
104. Plant of Pachycereus pringlei..........	69
105. Plant of Pachycereus pecten-aboriginum.	71
106. Fruit of Pachycereus pecten-aboriginum.	72
107. Plant of Pachycereus chrysomallus.....	73
108. Flower of Pachycereus chrysomallus.....	74
109. Longitudinal section of Pachycereus chrysomallus....................	74
110. Flower of Pachycereus marginatus......	74
111. Hedge of Pachycereus marginatus......	75
112. Part of branch of Leptocereus weingartianus..........................	77
113. Plant of Leptocereus leonii............	78
114. Plant of Leptocereus assurgens.........	79
115. Branch of Leptocereus maxonii.........	80
116. Fruit of Leptocereus arboreus.........	80

TEXT-FIGURES—continued.

PAGE.

FIG. 117. Fruit of Leptocereus sylvestris   80
     118. Top of branch of Leptocereus sylvestris.   81
     119. Plant of Leptocereus quadricostatus . . .   81
     120. Fruit of Leptocereus quadricostatus   81
     121. Flower of Leptocereus quadricostatus . .   81
     122. Potted plant of Eulychnia spinibarbis.   83
     123. Flower of Eulychnia acida 84
     124. Flower of Eulychnia castanea   84
     125. Hedge of Lemaireocereus hollianus   86
     126. Plants of Lemaireocereus hystrix   87
     127. Flower of Lemaireocereus hystrix   87
     128. Fruit of Lemaireocereus hystrix   87
     129. Plants of Lemaireocereus griseus   88
     130. Fruit of Lemaireocereus pruinosus   89
     131. Potted plant of Lemaireocereus long-
          ispinus 90
     132. Potted plant of Lemaireocereus eichlamii.   90
     133. Plant of Lemaireocereus chende   91
     134. Plant of Lemaireocereus godingianus . . .   92
     135. Plants of Lemaireocereus aragonii   93
     136. Plants of Lemaireocereus stellatus   94
     137. Plants of Lemaireocereus treleasei   95
     138. Plant of Lemaireocereus deficiens   96
     139. Plant of Lemaireocereus weberi 96
     140. Cluster of spines of Lemaireocereus we-
          beri 97
     141. Fruit of Lemaireocereus weberi 97
     142. Part of rib, showing spine-clusters of
          Lemaireocereus queretaroensis   97
     143. Plant of Lemaireocereus thurberi   98
     144. a. Flower of Lemaireocereus thurberi . .   98
          b. Fruit of Lemaireocereus thurberi   98
     145. Plant of Lemaireocereus laetus 99
     146. Plant of Lemaireocereus laetus 99
     147. Flower of Lemaireocereus laetus 100
     148. Fruit of Lemaireocereus laetus 100
     149. Plants of Lemaireocereus humilis   100
     150. Flowering branch of Lemaireocereus hu-
          milis 101
     151. Cross-section of stem, longitudinal sec-
          tion of rib, spine-cluster, flower,
          and fruit of Lemaireocereus humilis.   101
     152. Fruit of Lemaireocereus dumortieri   102
     153. Plant of Lemaireocereus dumortieri   102
     154. Plant of Erdisia squarrosa 104
     155. Flower, fruit, and stem of Erdisia squar-
          rosa 105
     156. Branch of Erdisia meyenii 106
     157. Plant of Erdisia spiniflora 106
     158. Group of plants of Bergerocactus
          emoryi 107
     159. Flower of Bergerocactus emoryi 108
     160. Potted plant of Leocereus bahiensis   109
     161. Flower of Leocereus bahiensis 109
     162. Flower of Leocereus melanurus 109
     163. Sections of stem of Wilcoxia viperina . . .   110
     164. Potted plant of Wilcoxia poselgeri 110
     165. Cluster of tuberous roots of Wilcoxia
          poselgeri 111
     166. Group of plants of Peniocereus greggii . .   112
     167. Flower of Peniocereus greggii 112
     168. Fruit of Peniocereus greggii 112
     169. Plant of Dendrocereus nudiflorus   114
     170. Fruit of Dendrocereus nudiflorus   114
     171. Plants of Machaerocereus eruca   115
     172. Joint of Machaerocereus eruca 116

PAGE.

FIG. 173. Potted plant of Machaerocereus gum-
          mosus 116
     174. Plant of Machaerocereus gummosus   117
     175. Flower of Machaerocereus gummosus . .   117
     176. Fruit of Nyctocereus serpentinus   118
     177. Flower of Nyctocereus serpentinus   118
     178. Part of flowering plant of Nyctocereus
          guatemalensis   120
     179. Flower of Brachycereus thouarsii   121
     180. Fruit of Brachycereus thouarsii   121
     181. Top of joint of Acanthocereus horridus.   122
     182. Plant of Acanthocereus pentagonus   123
     183. Fruit and withering perianth of Acan-
          thocereus pentagonus   123
     184. Plants of Acanthocereus pentagonus in
          cactus plantation   124
     185. End of joint of Acanthocereus occiden-
          talis 125
     186. Plant of Acanthocereus brasiliensis   126
     187. Plant of Acanthocereus albicaulis   126
     188. Plant of Trichocereus thelegonus   131
     189. Plant of Trichocereus thelegonus   131
     190. Potted plant of Trichocereus spachianus.   132
     191. Plants of Trichocereus pasacana   132
     192. Ends of flowering plants of Trichocereus
          lamprochlorus   133
     193. Flower of Trichocereus pasacana   134
     194. Fruit of Trichocereus pasacana 134
     195. Flower of Trichocereus candicans   134
     196. Plants of Trichocereus pachanoi   135
     197. Plant of Trichocereus peruvianus   136
     198. Plants of Trichocereus chiloensis   137
     199. Potted plant of Trichocereus chiloensis.   137
     200. Flower of Trichocereus chiloensis   138
     201. Potted plant of Trichocereus coquimba-
          nus 138
     202. Plant of Trichocereus coquimbanus. . . .   139
     203. a, flower of Trichocereus terscheckii . . .   140
          b, fruit of Trichocereus terscheckii   140
     204. Plant of Trichocereus terscheckii   140
     205. Plant of Trichocereus fascicularis   141
     206. Flower of Trichocereus fascicularis   141
     207. Fruit of Trichocereus fascicularis   141
     208. Flower of Trichocereus huascha   141
     209. Fruit of Trichocereus huascha 141
     210. Plant of Trichocereus huascha   142
     211. Plants of Trichocereus strigosus   144
     212. Plants of Jasminocereus galapagensis . . .   146
     213. Flower of Jasminocereus galapagensis . .   147
     214. Flower of Jasminocereus galapagensis . .   147
     215. Plant of Harrisia eriophora 148
     216. Plantation of Harrisia fragrans 149
     217. Plant of Harrisia portoricensis 150
     218. Plant of Harrisia nashii 150
     219. Fruit of Harrisia brookii 151
     220. Flower-bud of Harrisia brookii 151
     221. Plant of Harrisia gracilis 152
     222. Flower of Harrisia gracilis 152
     223. Plant of Harrisia simpsonii 153
     224. Plant of Harrisia taylori 153
     225. Part of plant of Harrisia pomanensis . . .   156
     226. Plant of Harrisia adscendens 156
     227. Plant of Harrisia bonplandii 157
     228. Potted plant of Harrisia guelichii   158
     229. Top of plant of Borzicactus sepium   160
     230. Top of plant of Borzicactus morleyanus.   161

TEXT-FIGURES—continued.

PAGE.

FIG. 231. Clump of plants of Borzicactus morleyanus.............................. 161
232. Plant of Borzicactus decumbens........ 162
233. Flower of Borzicactus decumbens...... 162
234. Plant of Carnegiea gigantea........... 165
235. Fruit of Carnegiea gigantea........... 166
236. Plants of Binghamia melanostele...... 167
237. Plants of Binghamia acrantha......... 167
238. Fruit of Binghamia melanostele........ 168
239. Flower of Binghamia acrantha........ 168
240. Fruit of Binghamia acrantha.......... 168
241. Flower of Rathbunia alamosensis...... 169
242. Flower, cut open, of Rathbunia alamosensis........................... 169
243. Plant of Arrojadoa penicillata......... 171
244. Plants of Oreocereus celsianus........ 172
245. Potted plant of Oreocereus celsianus... 172
246. Flower of Oreocereus celsianus........ 172
247. Fruit of Oreocereus celsianus......... 172
248. Potted plant of Cleistocactus smaragdiflorus.......................... 174
249. Plant of Zehntnerella squamulosa...... 176
250. Flower of Zehntnerella squamulosa..... 177
251. Plants of Lophocereus schottii........ 178
252. Cross-section of stem of Lophocereus
schottii.......................... 179
253. Flower of Lophocereus schottii........ 179
254. Section of rib of Myrtillocactus geometrizans with fruit at areoles..... 179
255. Flower of Myrtillocactus geometrizans.. 179
256. Flower and fruits of Myrtillocactus
eichlamii......................... 181
257. Plant of Neoraimondia macrostibas..... 181
258. Flower of Neoraimondia macrostibas... 181
259. Cluster of spines of Neoraimondia macrostibas......................... 182
260. Potted plant of Neoraimondia macrostibas........................... 182
261. Tip of joint of Hylocereus guatemalensis............................ 184
262. Ovary of Hylocereus costaricensis, transformed into a branch............. 186
263. Plant of Hylocereus undatus.......... 187
264. Hedge of Hylocereus undatus......... 188
265. Part of branch of Hylocereus cubensis.. 188
266. Stigma-lobes of Hylocereus lemairei.... 189
267. Flowering branch of Hylocereus stenopterus.......................... 190
268. Plant of Hylocereus trigonus.......... 192

PAGE.

FIG. 269. Joint of Hylocereus triangularis....... 193
270. Plant of Hylocereus antiguensis....... 194
271. Joint of Hylocereus calcaratus........ 194
272. Flowering branch of Wilmattea minutiflora........................... 196
273. Joint of Selenicereus coniflorus........ 198
274. Fruit of Selenicereus coniflorus........ 198
275. Tip of branch of Selenicereus hondurensis.............................. 199
276. Part of branch of Selenicereus donkelaarii............................ 200
277. Part of branch of Selenicereus kunthianus............................ 201
278. Tip of branch of Selenicereus brevispinus. 202
279. Tip of branch of Selenicereus macdonaldiae............................ 202
280. Flower of Selenicereus macdonaldiae... 203
281. Fruit of Selenicereus macdonaldiae.... 203
282. Flower of Selenicereus hamatus....... 204
283. Part of branch of Selenicereus hamatus. 204
284. Flower of Selenicereus vagans........ 205
285. a and b, branches of Selenicereus vagans. 206
c and d, branches of Selenicereus murrillii. 206
286. Top of branch of Selenicereus spinulosus............................ 207
287. Top of branch of Selenicereus inermis... 208
288. Branches of Selenicereus wercklei...... 208
289. Flowering plant of Selenicereus wercklei. 209
290. Plant of Mediocactus coccineus....... 211
291. Fruiting branch, cross-section, and
spines of Mediocactus coccineus..... 211
292. Plant of Mediocactus megalanthus.... 213
293. Plant of Deamia testudo.............. 213
294. Branches of Deamia testudo.......... 214
295. Fruiting branch of Weberocereus panamensis........................... 215
296. Flowering plant of Werckleocereus tonduzii............................ 217
297. Flower of Aporocactus leptophis....... 218
298. Flower of Aporocactus flagriformis..... 218
299. Parts of plant of Aporocactus conzattii. 220
300. Flower of Aporocactus conzattii...... 221
301. Flower of Aporocactus martianus...... 221
302. Parts of plant of Strophocactus wittii.. 222
303. Plant of Cereus grenadensis.......... 223
304. Section of flowering branch of Cereus
grenadensis....................... 223
305. Flower of Selenicereus vagans (without
legend)........................... 239

THE CACTACEAE

Descriptions and Illustrations of Plants of the Cactus
Family

DESCRIPTIONS AND ILLUSTRATIONS OF PLANTS OF THE CACTUS FAMILY.

Tribe 3. CEREEAE.

Plants more or less fleshy, terrestrial or epiphytic, simple and 1-jointed or much branched and many-jointed, the joints globular, oblong, cylindric, columnar or flattened, and winged or leaf-like, often strongly ribbed, angled, or tubercled; leaves* usually wanting on the joints (in a few cases developing as scales) but usually developing as scales on the ovary or perianth-tube; areoles never producing glochids; spines usually present (rare or wanting in most epiphytic genera and in a few species of other genera), various in color, structure, arrangement, and size, never sheathed; flowers sessile, mostly with a definite tube, various in size and shape in different genera, usually solitary at areoles, opening at various times of the day; perianth campanulate, funnelform or rotate; fruit usually a fleshy berry, but sometimes dry and dehiscing by a basal pore (in 1 species by an operculum); seeds usually small, brown or black, with a thin, more or less brittle testa; cotyledons usually minute knobs.

This tribe contains most of the genera and three-fourths or more of the species of *Cactaceae*. It has a wider range in structure of stems and flowers than is exhibited by the other tribes, the species being grouped in many genera. The first two subtribes are treated in this volume.

KEY TO SUBTRIBES.

Perianth funnelform, salverform, tubular, or campanulate; segments several or many.
  Areoles mostly spine-bearing; joints ribbed, angled, or tubercled, very rarely flat; mostly
    terrestrial cacti.
    Flowers and spines borne at the same areoles.
      Several-jointed to many-jointed cacti, the joints long.
        Erect, bushy, arching, or diffuse cacti..................................1. *Cereanae*
        Vine-like cacti, with aërial roots..2. *Hylocereanae*
      One-jointed or few-jointed cacti, the joints usually short, sometimes clustered, ribbed,
        or rarely tubercled.
        Flowers at lateral areoles...3. *Echinocereanae*
        Flowers at central areoles (See *Gymnocalycium*)............................4. *Echinocactanae*
    Flowers and spines borne at different areoles; short, one-jointed cacti.
      Flowering areoles forming a central terminal cephalium.........................5. *Cactanae*
      Flowering areoles at the bases or on the sides of the tubercles..................6. *Coryphanthanae*
  Areoles mostly spineless; joints many, long, flat; perianth mostly funnelform; epiphytic cacti.7. *Epiphyllanae*
Perianth rotate, or nearly so; segments few; mostly spineless, epiphytic, slender, many-jointed cacti.8. *Rhipsalidanae*

Subtribe 1. CEREANAE.

Erect, bushy or sometimes diffuse, stout or slender cacti, the stems and branches several-jointed to many-jointed, usually very spiny, none epiphytic but species of 2 or 3 genera giving off a few roots when the branches touch the ground; flowers 1 or rarely several from the upper part of old areoles; in some genera the flowering areoles and their spines greatly modified; flowers either diurnal or nocturnal, various in size, color, and shape; stamens numerous, borne on the flower-tube; fruit smooth or spiny, usually fleshy, often edible; seeds various.

We group the species known to us in 38 genera.

KEY TO GENERA.

A. Flowers solitary at the areoles, mostly large.
  B. Perianth funnelform, salverform, pyriform, or campanulate; limb relatively large.
    C. Ovary naked, or rarely bearing a few scales, which sometimes subtend tufts
      of short hairs.
      Perianth funnelform, elongated.
        Columnar cacti, or with columnar branches; perianth falling away by
          abscission...1. *Cereus* (p. 3)
        Slender, elongated cacti; perianth withering-persistent................ 2. *Monvillea* (p. 21)
        Perianth short-campanulate or short-funnelform to pyriform; columnar
          cacti... 3. *Cephalocereus* (p. 25)

---

*Plants of the tribe *Cereeae* are usually said to be without leaves. Ganong, however, reports leaves in *Cactus*, *Echinocactus*, and *Cereus*, but we have never seen leaves on any plants of *Cereus* proper. However, they are easily observed on young growth of various species of *Harrisia*, *Acanthocereus*, *Nyctocereus*, *Selenicereus*, *Hylocereus*, and some other genera.

1

Key to Genera—continued.

CC. Ovary squamiferous, often also laniferous, setiferous, or spiniferous.
  Flowers in a large lateral pseudocephalium; columnar cacti............. 4. *Espostoa* (p. 60)
  Plants without a pseudocephalium.
    Ovary squamiferous only; columnar cacti.
      Scales of the ovary fleshy.
        Flower short-funnelform; scales of ovary and flower-tube acute..... 5. *Browningia* (p. 63)
        Flower long-funnelform; scales of ovary and flower-tube broad,
          abruptly cuspidate........................... 6. *Stetsonia* (p. 64)
      Scales of the ovary papery..................................... 7. *Escontria* (p. 65)
    Ovary squamiferous and also laniferous, felted, or spiniferous.
      Perianth short-campanulate or short-funnelform, its tube short or thick.
        Plants mostly stout, columnar, and erect, ribbed or angled; a few
          species spreading or prostrate; rootstocks not tuberous.
          Corolla short-campanulate.
            Corolla falling away by abscission, yellow; columnar cacti.... 8. *Corryocactus* (p. 66)
            Corolla withering-peristent; flowers not yellow.
              Fruit dry; columnar cacti........................... 9. *Pachycereus* (p. 68)
              Fruit a fleshy berry.
                Tree-like, or bushy cacti............................10. *Leptocereus* (p. 77)
                Columnar cacti.....................................11. *Eulychnia* (p. 82)
          Corolla short-funnelform; fruit fleshy.
            Mostly columnar cacti with stout stems, the white to pink
              flowers not widely expanded.....................12. *Lemaireocereus* (p. 85)
            Slender or low cacti, with bright red, scarlet, or yellow, widely
              expanded flowers.
              Branches slender, few to several-ribbed..................13. *Erdisia* (p. 104)
              Branches stout, closely many-ribbed....................14. *Bergerocactus* (p 107)
          Stems very slender, nearly terete or with many low ribs.
            Inner perianth-segments much shorter than tube...............15. *Leocereus* (p. 108)
            Inner perianth-segments as long as tube; rootstocks tuberous...16. *Wilcoxia* (p. 110)
      Perianth funnelform, funnelform-campanulate, or salverform.
        Areoles of the ovary spinuliferous or setiferous (see *Harrisia*).
          Slender cacti, with an enormous fleshy root; flower salverform..17. *Peniocereus* (p. 112)
          Stout or slender cacti, without a large fleshy root; flower funnel-
            form.
            Tree-like cacti; fruit with a thick woody rind; ovary few-spined.18. *Dendrocereus* (p. 113)
            Prostrate or bushy or vine-like cacti; fruit fleshy.
              Stout, bushy or prostrate cacti, the spines dagger-like, flat..19. *Machaerocereus* (p. 114)
              Slender or weak cacti, the spines acicular or subulate.
                Perianth-tube as long as the limb or longer; elongated
                cacti with white flowers.
                Joints ribbed.
                  Perianth-segments and filaments elongated........20. *Nyctocereus* (p. 117)
                  Perianth-segments and filaments short..............21. *Brachycereus* (p. 120)
                Joints angled....................................22. *Acanthocereus* (p. 121)
              Perianth-tube mostly shorter than the limb; bushy cacti
                usually with scarlet flowers......................23. *Heliocereus* (p. 127)
         Areoles of the ovary laniferous or felted (also setiferous in some
          species of *Harrisia*).
         Perianth-limb regular.
          Perianth funnelform or salverform; tube mostly longer than
           limb.
           Stout, upright cacti, columnar or with columnar branches.
             Perianth-tube bearing areoles to top; perianth-segments
               broad..24. *Trichocereus* (p. 130)
            Perianth-tube slender, with few areoles or none; perianth-
              segments narrow.............................25. *Jasminocereus* (p. 146)
           Slender, arching, vine-like or bushy cacti.
             Arching or vine-like cacti...........................26. *Harrisia* (p. 147)
             Low, bushy cacti..................................27. *Borzicactus* (p. 159)
         Perianth funnelform-campanulate, the tube stout.
          Gigantic, columnar cacti; scales of flower broad...........28. *Carnegiea* (p. 164)
          Stout, bushy cacti; scales of flower narrow...............29. *Binghamia* (p. 167)
          Perianth-limb oblique; erect or bushy cacti with scarlet flowers..30. *Rathbunia* (p. 169)
BB. Perianth subcylindric, the limb short or none.
  Scales when present on the ovary and flower-tube naked in their axils........31. *Arrojadoa* (p. 170)
  Scales on the ovary and flower-tube laniferous in their axils.
    Flowers borne from a lateral pseudocephalium.
      Flower-tube elongated; fruit dry......................32. *Oreocereus* (p. 171)
      Flower-tube very short; fruit not dry.................33. *Facheiroa* (p. 173)
    Flowers not borne from a lateral pseudocephalium.
      Perianth-tube elongated, slender; stamens exserted.................34. *Cleistocactus* (p. 173)
      Perianth-tube very short; stamens included..........................35. *Zehntnerella* (p. 176)

KEY TO GENERA—continued.

AA. Flowers 2 to several at an areole; columnar cacti, or with columnar branches;
  flowers small.
  Flowers without wool; areoles small.
    Flowering areoles bearing many long bristles..........................36. *Lophocereus* (p. 177)
    Flowering areoles without bristles...37. *Myrtillocactus* (p. 178)
  Flowers densely woolly; flowering areoles enormously developed..............38. *Neoraimondia* (p. 181)

1. CEREUS (Hermann) Miller,* Gard. Dict. Abridg. ed. 4. 1754.

Piptanthocereus Riccobono, Boll. R. Ort. Bot. Palermo 8: 225. 1909.

Stems mostly upright and tall, but sometimes low and spreading or even prostrate, generally much branched, the branches strongly angled or ribbed; areoles spiny, more or less short-woolly but never producing long silky hairs; flowers nocturnal, elongated, funnelform, the upper part, except the style, falling away from the ovary by abscission soon after anthesis; tube of flower cylindric, expanding above into the swollen throat, nearly naked without; outer perianth-segments obtuse, thick, green or dull colored, the inner thin, petaloid, so far as known white, except in one species and in that red; stamens numerous, varying greatly in length, slender and weak, included; style slender, elongated but often included; stigma-lobes linear; ovary bearing a few scales naked in their axils; fruit fleshy, red, rarely yellow, naked, splitting down one side when mature, often edible; seeds black.

Type species: *Cactus hexagonus* Linnaeus, this being the first species cited by Miller in his Gardeners' Dictionary, 8th edition, 1768, where he described 12 species of *Cereus* (in the 4th edition, abridged, 1754, he described 14 species), which we now know belong to several genera.

The genus *Cereus* has been understood by authors at one time or another since Philip Miller's time as containing species of nearly all the genera of cacti, including even *Rhipsalis* and *Opuntia*. Schumann, in his monograph, recognized 104 species, to which he afterward added 36 in his supplement. His treatment of the genus is artificial and complex; Berger's treatment (Rep. Mo. Bot. Gard. 16: 57 to 86. 1905) is much more natural but more inclusive, for he added *Echinopsis, Pilocereus, Cephalocereus*, and *Echinocereus*, and even suggested the possible transfer here of *Phyllocactus;* he divided the genus into 18 subgenera, most of which we believe require generic recognition (Contr. U. S. Nat. Herb. 12: 413 to 437. 1909), as also indicated by Riccobono (Boll. R. Ort. Bot. Palermo 8: 215 to 266. 1909). From some of Berger's conclusions we differ, but chiefly in cases where he knew the plants only from herbarium specimens or from literature. In his treatment of *Cereus* Berger referred the species which we include in it to his series *Piptanthocereus*, while he took up for the *Eucereus* a different series, but he indicated no type species. Our treatment includes all the species of Schumann's series *Compresso-costati, Formosi*, and *Coerulescentes*, and the two species, *C. tetragonus* and *C. hankeanus* of *Oligogoni*. It corresponds to Berger's subgenus *Piptanthocereus*, but is not so inclusive. We recognize 24 species, which have similar flowers, fruit, spines, and branches; these extend from the southern West Indies through eastern South America to Argentina. The fruits of several species are edible.

The number of published *Cereus* binomials involved is about 900, exceeded in this family only by *Mammillaria* and perhaps by *Opuntia*.

The name *Cereus* is from the Greek, also from the Latin, signifying a torch, with reference to the candelabrum-like branching of the first species known. It was used by Tabernaemontanus on page 386 of the second part of his Kreuterbuch, published in 1625, a plant called *Cereus peruvianus* being there illustrated; this figure represents a tall, columnar, branching species, perhaps the same as the one to which the name *peruvianus* has been applied by modern authors.

---

*Philip Miller credits the genus *Cereus* to P. Hermann (Par. Botavus 112. 1698) although the name *Cereus* had then been in use more than seventy years.

KEY TO SPECIES.

```
A. Flowers large, 10 to 20 cm. long.
    B. Species tall, columnar (except C. pachyrhizus), the joints very thick.
        Ribs 4 to 6, very high, flat or nearly so (Series 1. Hexagonae).
            Young joints glaucous, blue or bluish green.
                Spines of young joints short or none.
                    Ribs usually 4; young joints light blue.............................. 1. C. hexagonus
                    Ribs usually 6; young joints dark blue.............................. 2. C. hildmannianus
                All joints manifestly spiny.
                    Young spines bright yellow......................................... 3. C. alacriportanus
                    Young spines not yellow.
                        Flowers red without............................................ 4. C. validus
                        Flowers green without.......................................... 5. C. jamacaru
            Young joints not glaucous, green, or sometimes glaucous in No. 7.
                Inner perianth-segments red........................................... 6. C. tetragonus
                Inner perianth-segments white (unknown in C. xanthocarpus).
                    Outer perianth-segments red.
                        Spines 1 to 3, short or wanting or elongated in No. 7; seeds dull.
                            Berry red or orange, unpleasant to the taste.............. 7. C. stenogonus
                            Berry yellow, edible..................................... 8. C. xanthocarpus
                        Spines 8 to 13, up to 4 cm. long; seeds shining.
                            Tree-like, 6 to 8 meters high, not densely spiny......... 9. C. lamprospermus
                            Lower, 1 to 3 meters high, densely spiny................10. C. pachyrhizus
                    Outer perianth-segments green or brownish.
                        Spines few, short or wanting...............................11. C. dayamii
                        Spines 6 to 10, rarely 4, up to 10 cm. long................12. C. argentinensis
        Ribs 6 to 9, rarely 4, thicker and lower; outer perianth-segments brownish
            (Series 2. Peruvianae).........................................13. C. peruvianus
    BB. Species lower, prostrate, or bushy, the joints mostly not as stout (C. chalybaeus tall).
        Joints green (Series 3. Obtusae).
            Ribs only 4 to 6 mm. high; plants shining.............................14. C. perlucens
            Ribs much higher; plants dull.
                Spines subulate.....................................................15. C. variabilis
                    Flower 20 to 24 cm. long.
                    Flower 12 to 16 cm. long........................................16. C. pernambucensis
                Spines acicular.
                    Radial spines 5 to 7; central spine 1..........................17. C. obtusus
                    Radial spines 8 to 10; central spines 4 to 7...................18. C. caesius
        Joints glaucous blue; species slender (Series 4. Azureae).
            Ribs strongly sinuate................................................19. C. azureus
            Ribs not strongly sinuate.
                Tree-like; areoles distant.......................................20. C. chalybaeus
                Bush-like; areoles close together................................21. C. aethiops
AA. Flowers small, 8 cm. long or less; plants columnar (Series 5. Repandae).
    Flowers 7 to 8 cm. long; spines straight, acicular.
        Spines up to 5 cm. long, acicular; flowers green; branches constricted..........22. C. repandus
        Spines 2 cm. long or less; flowers purple; branches continuous.................23. C. grenadensis
    Flowers 5 to 6 cm. long; spines curved, subulate..............................24. C. margaritensis
```

1. **Cereus hexagonus** (Linnaeus) Miller, Gard. Dict. ed. 8. No. 1.   1768.

> Cactus hexagonus Linnaeus, Sp. Pl. 466. 1753.
> Cactus octogonus Page in Steudel, Nom. ed. 2. 1: 246. 1840.
> Cereus northumberlandianus* Lambert in Loudon, Gard. Mag. 17: 91. 1841 (February).
> Cereus perrottetianus Lemaire, Icon. Cact. pl. 8. 1841 to 1847.
> Cereus lepidotus Salm-Dyck, Cact. Hort. Dyck. 1849. 207. 1850.

Plant up to 15 meters high, usually branching near the base, with a trunk 4 dm. in diameter; branches usually strict and erect, but in old plants more spreading, made up of short joints 12 cm. in diameter or more, glaucescent or light green, usually 6-angled but sometimes only 4 or 5-angled, occasionally 7; ribs thin, 3 to 5 cm. high, the margins undulate; areoles about 2 cm. apart, small, felted; spines on young branches wanting or few, very short (2 to 3 mm. long), but on old branches often 8 to 10 or perhaps more in a cluster, very unequal, the longest ones 5 to 6 cm. long, when young brown, but lighter in age; flower 20 to 25 cm. long, its tube slender, 10 cm. long; uppermost scales green, short; outer perianth-segments lanceolate to oblong-lanceolate, 6 to 7 cm. long, short-apiculate, tinged with purple; inner perianth-segments much thinner than the outer ones, white,

---

*The name was published in Loudon's Gardener's Magazine first as Cereus northumberlandia with a suggestion by the editor that Cereus northumberlandianus was the preferred spelling but later in the same year (Hort. Univ. 2: 318. 1841) Cereus northumberlandianus was adopted. A re-examination of the description of Linnaeus's Cactus hexagonus, which came from Surinam, leads us to believe that it is the same species and as the name is older than either C. northumberlandianus or C. lepidotus we here use it.

oblong-lanceolate, 7 to 8 cm. long; stamens very numerous; style green; fruit ovoid, 5.5 to 13 cm. long, somewhat oblique, truncate or a little depressed at apex, pale red, a little glaucous, bearing small scattered areoles; rind thick; pulp white or pinkish, edible; seeds black.

Type locality: Surinam.

Distribution: Southern West Indies and northern South America. Often cultivated in the West Indies and South America. Reported from Brazil, but doubtless in error. Also cultivated in the Philippines.

This cactus is a great favorite in the West Indies, where it is much cultivated in yards and parks, and blooms abundantly, the flowers appearing all along the side of the stem. It is sometimes confused with *Cereus jamacaru*, and has long passed under the name of *Cereus lepidotus*. The plant was introduced into England from Tobago Island about 1840 by M. Nightingale, and was then supposed to be the largest cactus ever brought into Europe. Recently Mr. W. E. Broadway has sent us both living and herbarium specimens from Tobago which are identical with the so-called *Cereus lepidotus*. The original specimens of *Cereus lepidotus* came from La Guayra, Venezuela, a floral region similar to Tobago, while the *Cactus hexagonus* type locality was Surinam.

FIG. 1.—Cereus hexagonus.

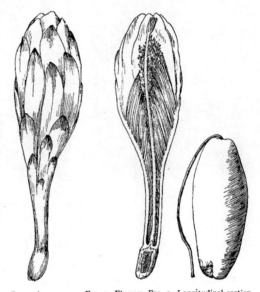

Cereus hexagonus.—FIG. 2, Flower; FIG. 3, Longitudinal section of flower; FIG. 4, Fruit. All ×0.4.

It was introduced into England, according to Salm-Dyck, as *Cereus karstenii*.

In our earlier treatment of this species we combined it with *C. peruvianus* which we now believe was an error. *Cereus hexagonus* is confined to northern South America and the West Indies while *C. peruvianus* is restricted to southeastern South America.

We have seen no Colombian specimens of this species unless we should refer here flowers collected by Dr. Francis W. Pennell from the Sabana of Bolivar (No. 4782).

Cereus horridus Otto (Pfeiffer, Allg. Gartenz. **5**:370. 1837) and *C. thalassinus* Otto and Dietrich (Allg. Gartenz. **6**:34. 1838), referred to *C. jamacaru* by Schumann, probably belong here. Both are from La Guayra, Venezuela. *Cereus thalassinus quadrangularis* (Förster, Handb. Cact. 399. 1846) was used as a synonym of *C. thalassinus*.

Illustrations: Contr. U. S. Nat. Herb. **12**: pl. 61, as *Cereus jamacaru;* Lemaire, Ic. Cact. pl. 8*, as *Cereus perrottetianus;* Maza and Roig, Fl. Cuba pl. 23, as *Cereus lepidotus.*

Text-figure 1 is from a photograph of the plant taken by Marshall A. Howe at Santurce, Porto Rico; text-figure 2 shows a flower and text-figure 3 a longitudinal section of the same drawn by Miss H. A. Wood at Hope Gardens, Jamaica; text-figure 4 shows a fruit collected by Dr. Rose near Carácas, Venezuela, in 1916.

2. Cereus hildmannianus Schumann in Martius, Fl. Bras. **4²**: 202. 1890.

Plant tall, up to 5 meters high, often much branched; ribs 5 or 6, high, thin, rounded, green or often with large yellow patches along the sides; areoles distant, large, at first without spines, afterward a few developing; flower elongated, funnelform, 20 to 23 cm. long; inner perianth-segments white, broad and obtuse; ovary naked, 2.5 to 3 cm. long.

Type locality: State of Rio de Janeiro, Brazil.
Distribution: Eastern Brazil.

FIG. 5.—Cereus hildmannianus.   FIG. 6.—Cereus hildmannianus.

Although this species seems to be a common yard and park plant in Bahia and Rio de Janeiro, it has never been well understood. It there forms bushy plants and is usually without spines. It is probably quite distinct from *Cereus jamacaru,* to which it has been referred by some authors; it grows in moister regions.

Illustrations: Martius, Fl. Bras. **4²**: pl. 41, f. 1; Monatsschr. Kakteenk. **2**: 57.

Text-figure 5 is from a photograph taken by Paul G. Russell near Rio de Janeiro, Brazil, in 1915; text-figure 6 is from a photograph taken by Dr. J. N. Mills at Rio de Janeiro in 1916.

3. Cereus alacriportanus Pfeiffer, Enum. Cact. 87. 1837.

  Cereus peruvianus alacriportanus Schumann, Gesamtb. Kakteen 115. 1897.
  Cereus paraguayensis Schumann in Chodat and Hassler, Bull. Herb. Boiss. II. **3**: 249. 1903.

Stems up to 2 meters high; ribs mostly 5, strongly compressed, 3 cm. high, separated by deep sharp intervals, rounded on the edge; areoles 2 to 2.5 cm. apart, when young filled with white wool; spines 6 to 9, all spreading, when young golden yellow, but gray when older, red at the bases, subulate, 2.5 cm. long; flowers 21 to 22 cm. long, 10 cm. broad at mouth; outer perianth-segments narrow, 1 cm. wide or less; inner perianth-segments spatulate, obtuse to acute, fringed or entire, white with a rosy tinge; stigma-lobes 13, yellowish green; ovary cylindric, naked.

Type locality: Porto Alegre, Brazil.
Distribution: Southern Brazil and Paraguay.

---

  * Lemaire's plates are not numbered and there is more or less uncertainty as to their order. We have followed Schumann in referring this species to plate 8. In the only copy which we have examined it is plate 11.

PLATE II

M. E. Eaton del.

1. Top of flowering stem of *Cereus alacriportanus*.
2. Top of stem of *Cereus peruvianus*.
3. Flower of the same plant.
(Natural size.)

This species has long been in cultivation in the New York Botanical Garden under the name of *Cereus alacriportanus*, where it has frequently flowered.　It differs somewhat from the description of *C. paraguayensis* by Schumann in the color of the spines and closeness of the areoles.

Cereus bonariensis is referred here by Förster (Handb. Cact. 388. 1846) as a synonym.　Sweet also used the name (Hort. Brit. ed. 3. 283. 1839) but does not associate it with this species.

Illustrations: Chodat, Veg. Paraguay 1: f. 90, as *C. paraguayensis;* Karsten, Deutsche Fl. f. 501, No. 7.

Plate 11, figure 1, shows the plant in the New York Botanical Garden above referred to, which flowered in April 1915.

FIG. 7.—Cereus validus.　　　　　　　　　　FIG. 8.—Cereus tetragonus.

4. **Cereus validus** Haworth, Phil. Mag. **10**: 420.　1831.

> *Cereus forbesii* Otto in Förster, Handb. Cact. 398.　1846.*
> *Cereus hankeanus* Weber in Schumann, Gesamtb. Kakteen 88.　1897.
> *Piptanthocereus forbesii* Riccobono, Boll. R. Ort. Bot. Palermo **8**: 228.　1909.
> *Piptanthocereus hankeanus* Riccobono, Boll. R. Ort. Bot. Palermo **8**: 229.　1909.
> *Piptanthocereus labouretianus* Riccobono, Boll. R. Ort. Bot. Palermo **8**: 231.　1909.
> *Piptanthocereus validus* Riccobono, Boll. R. Ort. Bot. Palermo **8**: 234.　1909.

Shrubby, 2 meters high or more, somewhat branched, the branches 5 to 8 cm. thick, glaucous when young; ribs 4 to 8, compressed, obtuse; radial spines 5, short, stout, 1 to 2 cm. long, mostly

*The date of publication of this name is usually given as 1845; this reference, however, is only to the use of the name, without a description, in a publication of that date.

from the lower part of the areole; central spine single or rarely 2 or 3, stouter than the radials, sometimes 16 cm. long; flowers funnelform; outer perianth-segments reddish, obtuse, the inner white or reddish; style green below; stigma-lobes about 16.

Type locality: Not cited.

Distribution: Provinces of Córdoba, Catamarca, and Tucuman, Argentina.

Cereus labouretianus Martius and *C. haematuricus* Weber, mentioned by Schumann, are only catalogue names and should not go into the published synonymy of this species.

Illustration: Blühende Kakteen 2: pl. 114, as *C. hankeanus.*

Plate III, figure 1, shows the top of a plant in the New York Botanical Garden, received from Kew in 1911; figure 2 shows a joint and a flower of a plant received from La Mortola as *Cereus hankeanus.* Text-figure 7 is from a photograph of a plant in the same collection, received from the Missouri Botanical Garden in 1904.

5. **Cereus jamacaru** De Candolle, Prodr. 3: 467. 1828.

> *Cereus glaucus* Salm-Dyck, Hort. Dyck. 335. 1834.
> *Cereus laetevirens* Salm-Dyck, Hort. Dyck. 336. 1834.
> *Cereus lividus* Pfeiffer, Allg. Gartenz. 3: 380. 1835.
> *Cactus jamacaru* Kosteletzky, Allg. Med. Pharm. Fl. 4: 1393. 1835.
> *Cereus horribarbis* Otto in Salm-Dyck, Cact. Hort. Dyck. 1849. 205. 1850.
> *Cereus cauchinii* Rebut in Schumann, Gesamtb. Kakteen 113. 1897.
> *Piptanthocereus jamacaru* Riccobono, Boll. R. Ort. Bot. Palermo 8: 229. 1909.
> *Piptanthocereus jamacaru cyaneus* Riccobono, Boll. R. Ort. Bot. Palermo 8: 230. 1909.
> *Piptanthocereus jamacaru glaucus* Riccobono, Boll. R. Ort. Bot. Palermo 8: 231. 1909.

Plant up to 10 meters high, with a short, thick, woody trunk, very much branched, the branches usually erect, numerous, often forming a compact top, when young often quite blue, with few (4 to 6) ribs; ribs of young branches thin, high, more or less undulate; areoles large, 2 to 3 cm. apart; spines various, on old stems and branches numerous, at first yellow, often very long, 20 to 30 cm. long; flowers nocturnal, large, 30 cm. long, white; ovary purplish, bearing a few minute brown scales; stigma-lobes numerous, 2 cm. long; fruit large, sometimes 12 cm. long by 8 cm. in diameter, bright red, splitting down on one side showing the white edible pulp; seeds 3 mm. long, dull, roughened with blunt tubercles.

Type locality: Brazil.

Distribution: Brazil. Planted in the West Indies; perhaps naturalized on some islands.

Cereus jamacaru is one of the commonest cacti in Bahia and is found in all kinds of situations from the coast to the inland desert. It is always large, 10 meters tall or more, usually much branched. When living in dense forests it has a simple stem or only a few branches, growing tall and erect, the branches have few ribs, but these are high and at first very blue, covered with formidable spines said to be 30 cm. long at times, although we have seen none

Fig. 9.—Cereus jamacaru.

which measured more than 19 cm. in length. The flowers are large and white, opening at night; the perianth cuts off early from the ovary, leaving the style, which is persistent. The woody trunk may be 6 dm. in diameter, and boards suitable for boxes, picture frames, etc., are sawed from it. In most of the smaller houses in the country the cross pieces upon which the tile roofing is laid are from this cactus, which is called mandacaru and mandacaru de boi. The specific name *jamacaru*, said by some writers to be the vulgar name of the plant in Brazil, is doubtless a corruption of mandacaru. It is sometimes planted about country houses, often as a kind of hedge. In times of great drought the farmers cut off the young branches from these cacti to feed to their cattle.

Cereus horridus Otto (Pfeiffer, Allg. Gartenz. 5: 370. 1837) and *C. thalassinus* Otto and Dietrich (Allg. Gartenz. 6: 34. 1838), referred to *C. jamacaru* by Schumann, belong elsewhere; both are from La Guayra, Venezuela.

Cereus lividus was based upon a Brazilian species. Two years after it was described, Pfeiffer redescribed it, referring to it as a synonym *C. perotetti* (Pfeiffer, Enum. Cact. 98), and giving the distribution as Brazil and La Guayra, Venezuela. The plant from La Guayra is doubtless *C. hexagonus*.

Cereus lividus glaucior (Labouret, Monogr. Cact. 359. 1853), given as a synonym of *C. lividus*, may belong here.

Cereus jamacaru glaucus (Ladenberg, Monatsschr. Kakteenk. 3: 70. 1893) is only a name.

Illustrations: Karsten, Deutsche Fl. f. 501, No. 8; Pison, Hist. Nat. Bras. 100. f. 1; Schumann, Gesamtb. Kakteen f. 25; Curtis's Bot. Mag. 95: pl. 5775, this last as *Cereus lividus*.

Figure 9 is from a photograph taken by Mr. P. H. Dorsett near Joazeiro, Bahia, Brazil, in 1914.

6. **Cereus tetragonus** (Linnaeus) Miller, Gard. Dict. ed 8. No. 2   1768.

    Cactus tetragonus Linnaeus, Sp. Pl. 466. 1753.

Plant upright, 1 to 2 meters high, freely branching; branches green, erect, forming a narrow compact top; ribs mostly 4, rarely 5, at first high, separated by acute intervals, compressed, obtuse; areoles close together, white-felted; spines brown to nearly black, usually acicular to subulate; radial spines 5 or 6, 6 to 8 mm. long; central spines solitary or several, a little stouter than the radials; flower funnelform, 13 cm. long; all the perianth-segments reddish; ovary bearing small scales, glabrous.

Type locality: Curaçao, according to Linnaeus, but not known there now.

Distribution: Rio de Janeiro, Brazil, according to Schumann.

Our description is drawn partly from living specimens in the New York Botanical Garden.

Cereus tetragonus ramosior Link and Otto (Verh. Ver. Beförd. Gartenb. 6: 432. 1830) is given by name only; *C. tetragonus major* Salm-Dyck (Walpers, Repert. Bot. 2: 277. 1843) is given as a synonym for *C. tetragonus*.

Illustration: Monatsschr. Kakteenk. 12: 158.

Figure 8 is from a photograph of a plant in the New York Botanical Garden, received from Mr. Frank Weinberg in 1901.

7. **Cereus stenogonus** Schumann, Monatsschr. Kakteenk. 9: 165.   1899.

Tree-like, up to 6 to 8 meters high, much branched or nearly simple, bluish green to yellowish green; ribs 4 or 5, very narrow, high; spines 2 or 3, short, conic, the longest 6 to 7 mm. long or subulate and the longer up to 4.5 cm. long; flowers large, 20 to 22 cm. long, funnelform, the tube long and slender; outer perianth-segments narrow, 7 to 8 cm. long, mucronate, rose-colored or with rose-colored margins; fruit large, 10 cm. long or less, red or orange without, with white or carmine flesh; seeds dull.

Type locality: Paso la Cruz, Paraguay.

Distribution: Paraguay and northeastern Argentina.

We know the species only from description, from a flower collected by Dr. E. Hassler from the region of the type locality, and from living plants and specimens collected by Dr. Shafer at Posadas, Argentina. It is now grown in the Hanbury Garden at La Mortola, Italy.

Figure 10 is from a photograph taken by Dr. Shafer at Posadas, Argentina, in 1917.

8. Cereus xanthocarpus Schumann, Gesamtb. Kakteen Nachtr. 32. 1903.

Tall, tree-like, up to 6 meters high, somewhat branched, very spiny at apex; ribs of branches 4 to 6, high, very narrow; areoles 3 to 4 cm. apart, white-woolly; spines 3 or 4, short, conic, dark brown; flowers opening at night; flower-tube 12.5 cm. long, yellowish green below, whitish green above; outer perianth-segments oblong to lanceolate, 4 to 12 cm. long, whitish green; inner perianth-segments white; fruit yellow, oblong, 6.5 to 7 cm. long, the flesh white; seeds 2 mm. long, kidney-shaped.

FIG. 10.—Cereus stenogonus.

Type locality: Calle Manora, Paraguay.

Distribution: Paraguay.

We have not seen this species; in its yellow fruit it differs from most other known members of this genus.

All we know about *Cereus coracare* Gosselin is that Hirscht (Monatsschr. Kakteenk. 9: 159. 1899) states that Mr. Roland-Gosselin is to be thanked for a splendid fruit of *Cereus coracare*, which in form and size resembles an apple, is of a beautiful color and of excellent taste to eat, and a note of Graebener (Monatsschr. Kakteenk. 12: 174. 1902) that *Cereus coracare* was from Paraguay and was then 19 cm. high. It may belong here.

The status of this and the following two species, all from Paraguay, can be determined only by further observations in that region.

9. Cereus lamprospermus Schumann, Monatsschr. Kakteenk. 9: 166. 1899.

Tree-like, 6 to 8 meters high, very much branched; branches green, soon erect; ribs 6 to 8, thickish and obtuse, separated by rounded intervals; spines 8 to 11, hardly distinguished as radials and centrals; areoles 2 to 2.5 cm. apart, subulate; flower 15 to 16 cm. long; outer perianth-segments green with reddish tips; stigma-lobes 13; ovary nearly naked; seeds black, shining.

Type locality: Fuerte Olympo, Paraguay.

Distribution: Paraguay.

10. Cereus pachyrhizus Schumann, Gesamtb. Kakteen Nachtr. 33. 1903.

Plant upright, 1, or at the most, 3 meters high, with swollen tuberous roots; branches or stem up to 10 cm. thick, rounded at the apex, terminated by large and numerous spines; older joints yellowish brown, younger ones yellowish green, subglaucous; ribs 6, very strongly compressed laterally, up to 1 cm. thick and 5 cm. high, separated by sharp, deep furrows, subsinuate; areoles 2.5 to 3 cm. apart, circular, 5 to 6 mm. in diameter, with short felt, which is not curly even when young; spines 10 to 13, poorly differentiated into radial and central ones, one of the latter being longest and up to 3 cm. long; all spines subulate and very sharp; fruit ellipsoid, 5 cm. long, 3 to 4 cm. in diameter, naked, smooth; seeds 2.5 mm. long, subcompressed, shining.

Type locality: Cerro Noaga, Paraguay.

Distribution: Paraguay.

This species is unknown to us, except from the original description. It is recorded as growing on bare, granitic rocks at 350 meters altitude.

11. Cereus dayamii Spegazzini, Anal. Mus. Nac. Buenos Aires III. **4:** 480. 1905.

Tree-like, 10 to 25 meters high, with a cylindric trunk; branches 5-ribbed or 6-ribbed; ribs 3 cm. high, pale green; areoles orbicular, large, 5 to 6 mm. in diameter; spines few or wanting, when present 4 to 12 mm. long, brown with a yellowish base; flowers funnelform, large, glabrous, up to 25 cm. long; inner perianth-segments white; fruit oblong, glabrous, red without, 6 to 8 cm. long; pulp white, edible; seeds black.

Type locality: Near Colony of Resistencia, Chaco, Argentina.
Distribution: Southern Chaco, Argentina.
Figure 11 is from a photograph given to Dr. Rose by Dr. Spegazzini.

12. Cereus argentinensis nom. nov.

Cereus platygonus Spegazzini, Anal. Mus. Nac. Buenos Aires III. **4:** 481. 1905. Not Otto. 1850.

Erect, 8 to 12 meters high, with a definite trunk; branches numerous, stout, curved at base but soon erect, 10 to 15 cm. in diameter; ribs 4 or 5, 4 to 5 cm. high, thin in section, separated by wide intervals; radial spines 5 to 8, brownish, 3 to 5 cm. long; central spines 1 or 2, 10 cm. long; flowers funnelform, large, 17 to 22 cm. long, inodorous; outer perianth-segments green or reddish at tips; inner perianth-segments white; fruit glabrous, smooth.

Type locality: Central Chaco, Argentina.
Distribution: Territory of the Chaco, Argentina.
This species must be close to *C. stenogonus,* as suggested by Berger, although Spegazzini says it is distinct; it must also be closely related to *C. dayamii.*
Figure 12 is from a photograph of a plant of *C. platygonus* Spegazzini, in Dr. Spegazzini's garden at La Plata, Argentina.

13. Cereus peruvianus (Linnaeus) Miller, Gard. Dict. ed. 8. No. 4. 1768.

Cactus peruvianus Linnaeus, Sp. Pl. 467. 1753.
?*Cereus calvescens* De Candolle, Mém. Mus. Hist. Nat. Paris **17:** 116. 1828.
?*Cereus spinosissimus* Förster, Hamb. Gartenz. **17:** 165. 1861.

Usually tall, said to reach 16 meters in height, tree-like, with a large much branched top; branches 10 to 20 cm. in diameter, usually green, sometimes glaucous, with 6 to 9 ribs, sometimes as few as 4; spines acicular, 5 to 10, brown to black, 1 to 3 cm. long; flower rather large, about 15

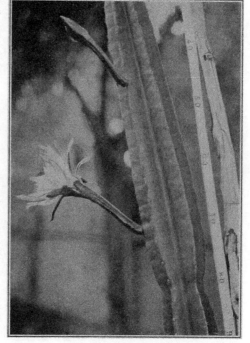

FIG. 11.—Cereus dayamii.

cm. long, with a thick tube; upper scales and outer perianth-segments obtuse, red or brownish; inner perianth-segments oblong, white; fruit subglobose, orange-yellow, somewhat glaucous, about 4 cm. in diameter; seeds black, 2 mm. broad, rough.

Type locality: Uncertain. Linnaeus says it is from Jamaica and the arid coast of Peru. No native *Cereus* is known either from Jamaica or Peru. It was called *Cereus peruvianus* by Bauhin in 1623 but no station was given. Our description applies to the plant from the southeastern coast of South America for which the name *Cereus peruvianus* has been used by most recent authors.

Distribution: Southeastern South America; widely planted in tropical America.

Cereus peruvianus tortuosus (Salm-Dyck, Cact. Hort. Dyck. 1844. 30. 1845) and *C. peruvianus tortus* (Salm-Dyck, Cact. Hort. Dyck. 1849. 46. 1850) are names only.

Cereus peruvianus monstrosus is a common garden form first described as a variety by De Candolle (Prodr. 3: 464. 1828). It is similar to the typical form except that the ribs are often broken into irregular tubercles or are unevenly sulcate. This has also been

FIG. 12.—A cultivated specimen of Cereus argentinensis.

taken up as *Cereus monstrosus* (Steudel, Nom. ed. 2. 1: 334. 1840), as *Cereus monstrosus minor* (Monatsschr. Kakteenk. 1: 163. 1891) and as *C. monstruosus* Schumann (Engler and Prantl, Pflanzenfam. 3⁶ᵃ: 178. 1894). It seems to be the same as *Cactus abnormis* Willdenow (Enum. Pl. Suppl. 31. 1813).* *Cereus peruvianus monstruosus nanus* is a somewhat similar form mentioned by Schumann (Gesamtb. Kakteen 115. 1897) perhaps intended

---

*Taken up later as *Cereus abnormis* by Sweet (Hort. Brit. 171. 1826). Another abnormal form is *C. peruvianus cristatus* (Graebener, Monatsschr. Kakteenk. 11: 29. 1901).

for *C. peruvianus monstrosus minor* (Salm-Dyck, Cact. Hort. Dyck. 1849. 46. 1850). *C. peruvianus brasiliensis* (Förster, Handb. Cact. 390. 1846) probably does not apply to this species.

Cereus surinamensis Trew (Ephem. Nat. Cur. 3: 394. pl. 7, 8, 1733) is referred here by Förster, but to *C. monoclonos* by Pfeiffer. The illustrations, though poor, indicate that it is a *Cereus* and not a *Cephalocereus*. From the name we should expect it to be referable to *Cereus hexagonus*.

Förster (Handb. Cact. 389. 1846) states that Pfeiffer has called this plant *Cereus decandollii*, but Förster doubts the correctness of this.

This species has long been known under the name of *Cereus peruvianus*, and is probably the most widely cultivated *Cereus*. In conservatories it is rarely found more than 2 meters in height.

Illustrations: Anal. Mus. Nac. Montevideo 5: pl. 1, 28 to 31; Blühende Kakteen 3: pl. 131; Cact. Journ. 2: March; Hist. Acad. Paris 1741: pl. 4, 5; Monatsschr. Kakteenk. 10: 7; Pfeiffer, Abbild. Beschr. Cact. 2: pl. 5; Rep. Mo. Bot. Gard. 5: pl. opp. 12; Stand. Cycl. Hort. Bailey 2: f. 884, all as *Cereus peruvianus;* Cact. Journ. 1: 79; October; Gard. Chron. 1873: f. 31; III. 24: f. 46; Home Farm. Gard. III. 60: 145; Mém. Mus. Hist. Nat. Paris 17: pl. 11, all as *Cereus peruvianus monstrosus;* Gerarde, Herball ed. 1. 1015; ed. 2 and 3, 1179, all as *Cereus peruvianus spinosus;* Bradley, Hist. Succ. Pl. ed. 2. pl. 1, as *Cereus erectus maximus* etc.; DeTussac, Fl. Antill. 2: pl. 33; Loudon, Encycl. Pl. 410. f. 6855, as *Cactus peruvianus.*

Fig. 13.—C. peruvianus. X0.8.

Plate 11, figure 2, represents the top of a plant in the collection of the New York Botanical Garden; figure 3 shows the flower of the same plant. Figure 13 shows a fruit of a plant in the same collection.

14. Cereus perlucens Schumann, Monatsschr. Kakteenk. 10: 173. 1900.

Columnar, erect, not very high; ribs 6 or 7, thin but obtuse, 4 to 6 mm. high, bright green or more or less bluish green or even violet when young, somewhat pruinose; areoles about 1 cm. apart, circular, bearing curly woolly hairs; radial spines 8 to 10, chestnut-brown, spreading, acicular, 1 cm. long; central spine solitary, stouter than the radials.

Type locality: Along the Amazon, Manaos, Brazil.
Distribution: Brazil.

We know this species only from a cutting received from the Berlin Botanical Garden; Dr. Schumann referred it to his series *Formosi*, between his *C. pitajaya* (*C. obtusus*) and *C. caesius;* he did not know the flower, however, and was not certain but that it might not belong to his genus *Pilocereus;* the cutting appears to us to represent a *Cereus*.

15. Cereus variabilis Pfeiffer, Enum. Cact. 105. 1837.

Creeping over rocks or clambering into trees, up to 4 meters high; stem made up of short thick joints, 18 to 30 cm. long by 6 to 9 cm. in diameter; ribs 3 to 5, stout, when old strongly crenate, obtuse, with strong indentations on the sides extending from the areoles with a broad upward bend to the bottoms of the ribs; spines about 8, yellowish, the longest about 5.5 cm. long; flower 20 to 27 cm. long, with a very slender, green, and somewhat angled tube; scales of the ovary and lower part of the flower-tube ovate, acute; outer perianth-segments green or yellowish green, linear, acute; inner perianth-segments white.

Type locality: Not definitely cited.
Distribution: Coast of central Brazil.

The plant collected by Dr. Rose on Ilha Grande, near Rio de Janeiro, flowered in the New York Botanical Garden, August 9, 1916, the flower being unusually large. We feel convinced that this is the plant illustrated as below cited. Here, too, is perhaps to be referred plate 4084 of Curtis's Botanical Magazine with the name *Cereus pitajaya*, although its flowers are smaller and the inner perianth-segments are more serrate.

While this species is somewhat similar to the common low *Cereus pernambucensis* of the Brazilian coast, it is stouter, often reaching a height of 4 meters, and has much larger flowers.

Cereus glaucus speciosus (Pfeiffer, Enum. Cact. 106. 1837) is referred to *Cereus variabilis* by both Pfeiffer and Rümpler. *C. brandii* (Salm-Dyck, Cact. Hort. Dyck. 1849. 49. 1850) and *C. colvillii* (Rümpler in Förster, Handb. Cact. ed. 2. 736. 1885) of English gardens are also referred here.

C. variabilis glaucescens Salm-Dyck, var. *laetevirens* Salm-Dyck, var. *micracanthus* Salm-Dyck, var. *salm-dyckianus*, and var. *obtusus* are all given by Walpers (Repert. Bot. 2: 277. 1843) as synonyms of this species. The last name probably should be referred to *Cereus obtusus*. The varieties *gracilior* and *ramosior* (Salm-Dyck, Cact. Hort. Dyck. 1849. 49. 1850) are only names. Of this relationship is *Cereus grandis* Haworth (Suppl. Pl. Succ. 76. 1819) and its two varieties *gracilior* Salm-Dyck and *ramosior* Salm-Dyck (Labouret, Monogr. Cact. 376. 1853).

Cereus prismatiformis, C. hexangularis, and *C. affinis* (Pfeiffer, Enum. Cact. 106. 1837) were all given as synonyms of *Cereus variabilis*.

Illustrations: Pfeiffer, Abbild. Beschr. Cact. 2: pl. 15, as *Cereus variabilis;* Vellozo, Fl. Flum. 5: pl. 23, as *Cactus tetragonus.*

FIG. 14.—Cereus pernambucensis.

16. **Cereus pernambucensis*** Lemaire, Cact. Gen. Nov. Sp. 58. 1839.

?*Cereus tetragonus minor* Salm-Dyck, Hort. Dyck. 337. 1834.
Cereus formosus Förster, Handb. Cact. 404. 1846.

Plant various in habit, often growing in clumps and then sometimes 4 to 5 meters broad, creeping and sprawling, usually 2 to 4 dm. high, perhaps much higher; branches usually short, with 3, 4, or 5 ribs, pale green, sometimes nearly white; ribs prominent, often strongly crenate and very thick; areoles large, 1.5 to 2 cm. apart, at first brown-woolly, afterwards with short white wool; intervals between ribs of young shoots acute, deep, but on old shoots broad and shallow; spines 4

---

*Originally, but erroneously, spelled *Cereus fernambucensis*.

to 10, acicular, yellowish brown to bright yellow, the longest ones 5 cm. long; flower-buds purplish, erect, 16 cm. long, pointed; scales on ovary and lower part of flower-tube minute, deep red, naked in their axils; flowers white; fruit narrowly oblong, 6 to 7 cm. long, purplish red, when mature splitting on one side exposing the white edible pulp and black seeds; style persisting after the perianth falls; seeds shining, 2 mm. long.

Type locality: Not cited.

Distribution: Coast of Brazil and Uruguay.

This species of the coast of Brazil is what Schumann described as *Cereus pitajaya*, but an examination of the original description of *Cactus pitajaya* Jacquin shows that this plant came from the coast of Colombia and is evidently an *Acanthocereus*.

Cereus pernambucensis is common along the seacoast of Brazil. Dr. Rose observed it at Bahia, Rio de Janeiro, Cabo Frio, and at Santos, but it is reported from both north and south of those regions. It is very common in the sand just back of the ocean beach, and on rocks near the sea, where it is usually low, often prostrate, growing in clumps. At times it grows much taller, unless we have associated another species with it. The taller plants suggest a small form of *C. jamacaru*, which is normally an interior desert species, while *C. pernambucensis* is to be found only on the coast; besides the differences in size of flowers and fruits, *C. pernambucensis* has shining seeds, which in the other species are dull.

Illustration: Velloz0, Fl. Flum. 5: pl. 22, as *Cactus pentagonus*.

Figure 14 is from a photograph taken by Paul G. Russell at Bahia, Brazil, in 1915.

17. Cereus obtusus Haworth, Rev. Pl. Succ. 70. 1821.

Low, branching at base, dull green slightly glaucous; branches at first strongly ribbed, but in age simply angled; ribs on young growth separated by deep intervals, obtuse, 2 to 2.5 cm. high, with long grooves running down from the areoles; areoles 1 to 2 cm. apart; spines acicular, yellowish; radial spines usually 5 to 7; central spine 1; flower and fruit unknown.

Type locality: Not cited.

Distribution: South America, presumably Brazil.

The above description is drawn from a plant sent from the Edinburgh Botanical Garden to New York in 1902.

Figure 15 is from a photograph of the specimen above mentioned.

18. Cereus caesius Salm-Dyck in Pfeiffer, Enum. Cact. 89. 1837.

Cereus jamacaru caesius Salm-Dyck in Fobe, Monatsschr. Kakteenk. **18:** 90. 1908.
Piptanthocereus jamacaru caesius Riccobono, Boll. R. Ort. Bot. Palermo **8:** 230. 1909.

Branching at base; branches strongly angled; ribs 5 to 7, high, somewhat acute, repand; areoles 1.5 to 2.5 cm. apart; spines acicular, brown, the radials 8 to 10; central spines 4 to 7, similar to the radials, 12 mm. long or less; flowers and fruit unknown.

Type locality: Not cited.

Distribution: Probably Brazil.

This species was described from greenhouse plants of unknown origin; later these were supposed to have come from South America, probably from Brazil. We have studied a cutting received from the Berlin Botanical Garden.

Cereus glaucus (Pfeiffer, Enum. Cact. 89. 1837) was published as a synonym of *C. caesius*. *Cereus laetevirens caesius* (Förster, Handb. Cact. 400. 1846), published as a synonym only, doubtless applies to this species.

19. Cereus azureus Parmentier in Pfeiffer, Enum. Cact. 86. 1837.

Cereus seidelii Lehmann in Salm-Dyck, Cact. Hort. Dyck. 1849. 200. 1850.
Cereus azureus seidelii E. Dams, Monatsschr. Kakteenk. **14:** 157. 1904.
Piptanthocereus azureus Riccobono, Boll. R. Ort. Bot. Palermo **8:** 225. 1909.

Probably branching at base, bluish pruinose; branches elongated, slender, flexuous; ribs 6 or 7, obtuse, repand; areoles remote, with brown tomentum and grayish wool; radial spines 8 to 12, white, with black tips; central spines 1 to 3, brown, stouter than the radials; flowers nocturnal, 10 to 12 cm. long; inner perianth-segments white, lanceolate, acuminate, 10 cm. long, the margins dentate; stamens numerous, green; style longer than the stamens, green; stigma-lobes 14, spreading, linear; ovary glabrous, bearing a few scales; fruit not known.

Type locality: Brazil.
Distribution: Brazil.

The illustration of Schumann, here cited, resembles the species of Argentina more than those of Brazil. *Cereus azureus* is reported growing in the Hanbury Garden at La Mortola, Italy, and plants are now to be seen in the New York Botanical Garden, where one flowered in 1915.

Illustration: Schumann, Gesamtb. Kakteen f. 26.

FIG. 15.—Cereus obtusus.   FIG. 16.—Cereus aethiops.

20. Cereus chalybaeus Otto in Förster, Handb. Cact. 382. 1846.

Piptanthocereus chalybaeus Riccobono, Boll. R. Ort. Bot. Palermo **8**: 227. 1909.

Stems 2 to 3 meters high, with few ascending branches; ribs 6, very high on the young parts of the stems and there separated by wide intervals, more or less purplish; radial spines usually 7, but on old stems much more numerous; central spines several, a little longer than the radials, all dark brown; perianth large, about 2 dm. long and about as broad when fully expanded; flower-tube about 1 dm. long, purplish, bearing long tubercles crowned by minute scales; outer perianth-segments pinkish, narrowly oblong, the inner white, acute, sometimes toothed; filaments numerous, long-exserted beyond the throat, but shorter than the perianth-segments; style elongated, much longer than the filaments, weak; stigma-lobes many; fruit spherical, smooth, yellow.

Type locality: Not cited.
Distribution: Northern Argentina.

This species is similar to the so-called *Cereus coerulescens*, of Argentina, which was taken up as *Cereus landbeckii* by Philippi, but the former has different stems, is stouter, and usually has shorter spines.

Cereus chalybaeus was described from a plant grown in the Botanical Garden at Berlin in 1846, which we do not know; but we are accepting as this species the plant so identified and figured by T. Gürke as below cited. Our description of the flower is drawn from this illustration.

Dr. Schumann states that the species comes from near Córdoba, Argentina, and there Dr. Rose collected specimens in 1915 which have been used for this description.

Walpers (Repert. Bot. 2:340. 1843) referred this species to *C. polychaetus*, an older species which seems to have been overlooked by recent writers.

Illustrations: Blühende Kakteen 3: pl. 135; Schumann, Gesamtb. Kakteen f. 27.

21. Cereus aethiops Haworth, Phil. Mag. 7: 109. 1830.

Cereus coerulescens Salm-Dyck, Hort. Dyck. 335. 1834.
Cereus landbeckii Philippi in Regel, Gartenflora 24: 162. 1875.
Cereus coerulescens landbeckii Schumann, Gesamtb. Kakteen 122. 1897.
Cereus coerulescens melanacanthus Schumann, Gesamtb. Kakteen 122. 1897.

Stems bluish green to purplish, 1 to 2 meters high, usually much branched; joints 3 dm. long or more, somewhat tapering toward the apex; ribs 7 or 8, low, somewhat tuberculate, obtuse or rounded, separated by acute intervals; areoles large, black; radial spines about 9 or even more, black, at least at bases and tips; central spines usually solitary, a little stouter than the radials, ascending; flower long, tubular, 22 cm. long, with a limb 12 cm. in diameter; outer perianth-segments linear-lanceolate, rose-colored; inner perianth-segments white; filaments and style included, the former attached all along the inner surface of the long tube; fruit ovoid to oblong-ovoid, more or less brownish when mature, truncate at apex, with a thick rind, smooth, somewhat glaucous, 6 cm. long; seeds black, 2 mm. long, coarsely tuberculate above, finely tuberculate at base, with a large depressed hilum.

Type locality: Brazil.

Distribution: Western border of Argentina to Brazil.

Cereus mendory Hortus (Pfeiffer, Enum. Cact. 85. 1837), *C. melanacanthus* Hortus (Schumann, Gesamtb. Kakteen 122. 1897), and *C. nigrispinus* Labouret (Schumann, Gesamtb. Kakteen 122. 1897), usually cited as synonyms of this species, are unpublished. *Cereus coerulescens fulvispinus* (Graebener, Monatsschr. Kakteenk. 19: 137. 1909) and *C. coerulescens longispinus* (Weingart, Monatsschr. Kakteenk. 16: 93. 1906) are referred here, but they have not been described.

Cereus coeruleus Lemaire (Cact. Gen. Nov. Sp. 80. 1839) was supposed to be a variety of the above species when first described but was said to be twice as large with stouter, longer spines.

We have followed Schumann and others in combining the plants from Brazil and western Argentina under one name, although there are indications that the specimens from Mendoza, Argentina, which were taken up by Philippi as *C. landbeckii*, are distinct.

Illustrations: Curtis's Bot. Mag. 68: pl. 3922; Pfeiffer, Abbild. Beschr. Cact. 2: pl. 24; Schumann, Gesamtb. Kakteen f. 28, all three as *Cereus coerulescens;* Gartenflora 24: pl. 832, as *Cereus landbeckii;* Blühende Kakteen 3: pl. 127, as *Cereus coerulescens melanacanthus.*

Figure 16 is from a photograph taken at Alto Pencoso, San Luis, Argentina, by C. Bruch in 1914; figure 17 shows a fruiting branch of *C. aethiops* from Mendoza, Argentina, brought by Dr. Rose to the New York Botanical Garden in 1915.

22. Cereus repandus (Linnaeus) Miller, Gard. Dict. ed. 8. No. 5. 1768.

Cactus repandus Linnaeus, Sp. Pl. 467. 1753.
Cereus hermannianus Suringar, Versl. Med. Akad. Wetensch. III. 2: 194. 1886.
Pilocereus repandus Schumann in Engler and Prantl, Pflanzenfam. 3⁶ᵃ: 181. 1894, as to name.

Tall, tree-like plant, up to 10 meters high, with a much branched top; trunk 4 dm. in diameter; branches grayish green, usually upright or somewhat curved below, bearing numerous constrictions about 2 dm. apart; ribs usually 9 or 10, rather low for this genus, about 1 cm. high; areoles 5 to 15 cm. apart, small; spines numerous, gray, acicular, the longest ones 5 cm. long; flowers nocturnal, narrowly funnelform, 7 to 8 cm. long, the limb 2.5 to 3 cm. broad, dark green except tips of inner perianth-segments; ovary bearing a few small ovate scales with a little felt in their axils; fruit dark red (occasionally white), oblong, 3 to 4 cm. long, with white flesh; seeds dull black, tuberculate

Type locality: Tropical America.

Distribution: Curaçao, Aruba, and Bonaire.

Schumann (Engler and Prantl, Pflanzenfam. 3^6a: 181) has confused this species with *Cephalocereus lanuginosus* and has published it under *Pilocereus repandus.*

Common on Curaçao, where it often grows in thickets, sometimes forming the dominant feature of the landscape and there known as kadoesji and breebee.

Figure 18 shows a fruit of a plant on Curaçao; figure 19 is from a photograph of the same plant taken by Dr. Britton and Dr. Shafer in 1913.

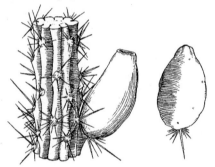

Fig. 17.—Fruiting branch   Fig. 18.—Fruit of Cereus
of Cereus aethiops. ×0.6.   repandus. ×0.6.

Fig. 19.—Cereus repandus.

23. Cereus grenadensis sp. nov.   (See Appendix, p. 223.)

24. Cereus margaritensis Johnston, Proc. Amer. Acad. **40**: 693. 1905.

Stem columnar, erect, 5 to 8 meters high, with a trunk 1 to 2 meters long; branches ascending, gray; ribs usually 8; areoles 1 cm. apart or less; spines 11 to 15, somewhat swollen at base; radial spines about 10, acicular, 5 to 10 cm. long, spreading or reflexed; central spines 1 to 3, stouter and twice as long as the radials, porrect or reflexed; flower-bud obtuse; flowers 5 to 6 cm. long; fruit oblong, 4 cm. long; seeds black, covered with blunt tubercles.

Type locality: El Valle, Margarita Island, Venezuela.

Distribution: Known only from Margarita Island.

OTHER SPECIES DESCRIBED AS BELONGING TO THE GENUS CEREUS.

The following species have been described under *Cereus,* but their flowers are unknown or incompletely described:

CEREUS BENECKEI Ehrenberg, Bot. Zeit. **2**: 835. 1844.

> *Cereus farinosus* Haage in Salm-Dyck, Allg. Gartenz. **13**: 355. 1845.
> *Cereus beneckei farinosus* Salm-Dyck, Cact. Hort. Dyck. 1849. 48.   1850.
> *Piptanthocereus beneckei* Riccobono, Boll. R. Ort. Bot. Palermo **8**: 226. 1909.

Plants 4 to 5 meters high, much branched; branches 6 to 7 cm. in diameter, the growing tips glaucous; ribs 8, strongly tuberculate, obtuse, separated by narrow intervals; areoles small, borne on the upper side of the tubercle, black-felted; spines 1 to 5, acicular, about 1 cm. long, brownish; flowering areoles without wool; flowers small, less than 4 cm. long, greenish brown, night-blooming; inner perianth-segments rose-colored; fruit small, spineless.

Type locality: Mexico, on red lava beds.

Distribution: Central Mexico.

This species is reported by Dr. Purpus from near Tehuacán, Mexico, while Dr. Rose collected it at Iguala Canyon, Guerrero, Mexico, in 1905. This latter specimen is now growing in the New York Botanical Garden, but has never flowered. It is not a true *Cereus* nor is it referable to any genus which we know. It is characterized by its peculiar tuberculate ribs and small flowers. It was named for A. Benecke, a dealer in succulents, at Birkenwerder near Berlin. *Echinocactus farinosus* (Förster, Handb. Cact. 396. 1846) is a synonym.

Illustration: Schumann, Gesamtb. Kakteen f. 22.

CEREUS GRACILIS Haworth, Phil. Mag. 1: 126. 1827. Not Miller, 1768.

Slender, green, nearly erect, terete, simple or with few branches; spines white, at first 2 to several but in age solitary, long; flowers and fruit unknown.

Type locality: "In America calidiore."

According to Haworth this species has the habit of *Euphorbia hystrix* but is less spiny and the spines are half as long. According to Haworth and De Candolle, this species is related to *Cereus nanus* (*Opuntia pestifer*), but a careful study of the descriptions does not suggest a very close relationship.

CEREUS TENUIS Pfeiffer, Allg. Gartenz. 8: 407. 1840.

Described as erect, slender, green, shining, with 8 angles; intervals between the ribs acute, narrow; areoles close together, small, bearing white felt, white wool, and straight, acicular yellow spines; radial spines 8, the central solitary; flowers and origin unknown.

Cereus subintortus, C. *subintortus flavispinus* Salm-Dyck, and *C. haageanus* Salm-Dyck (Förster, Handb. Cact. 381. 1846) are, according to Förster, of this relationship.

CEREUS TRIGONODENDRON Schumann, Bot. Jahrb. Engler 40: 413. 1908.

Tall, 15 meters high, with simple or few-branched stems; ribs 3, prominent; radial spines 6; central spine 1, about 6 mm. long; flowers described as about 10 cm. long and red.

This species was very briefly described by Schumann. Vaupel (Monatsschr. Kakteenk. 23: 184. 1913) has described the species at more length but not in sufficient detail to enable us to place it. It is very tall with few strict branches and only 3-angled stems, and with red flowers. It probably is not a *Cereus* nor is it like any other Peruvian cacti.

Type locality: Department of Loreto, Peru.

Distribution: Northeastern Peru.

E. C. Erdis, in 1915, collected at Pumachaca, at an altitude of about 1,500 meters, a very peculiar cactus which may be referable here. The small plant which he sent in had only 4 thin wing-like ribs, but the newer growth has 5 ribs; the spines are 6 to 9, dark brown, acicular. A small live plant is in the collection at Washington.

Illustration: Bot. Jahrb. Engler 40: pl. 10.

CEREUS MULTANGULARIS (Willdenow) Haworth, Suppl. Pl. Succ. 75. 1819.

> *Cactus multangularis* Willdenow, Enum. Pl. Suppl. 33. 1813.
> ?*Cereus multangularis pallidior* Pfeiffer, Enum. Cact. 78. 1837.
> *Echinocereus multangularis* Rümpler in Förster, Handb. Cact. ed. 2. 825. 1885.
> *Echinocereus multangularis pallidior* Rümpler in Förster, Handb. Cact. ed. 2. 825. 1885.

Cactus multangularis, when first described, was not sufficiently characterized for identification. Schumann associated the name *multangularis* with a Peruvian plant and referred considerable synonymy to it. We know no plant of Peru which answers his description.

To this species Schumann refers *Cereus flavescens* (Pfeiffer, Enum. Cact. 79. 1837) and with it should be referred *Echinocereus flavescens* (Rümpler in Förster, Handb. Cact. ed. 2. 826. 1885). *Cereus multangularis* var. *albispinus* and var. *prolifer* Salm-Dyck (Hort. Dyck. 62. 1834) and var. *rufispinus* Fobe (Monatsschr. Kakteenk. 18: 75. 1908) are unpublished names.

Cereus kageneckii Gmelin (Pfeiffer, Enum. Cact. 77. 1837), also, according to the Index Kewensis, *Cactus hageneckii* (De Candolle, Prodr. 3: 463. 1828) and *Cereus ochracanthus* (Pfeiffer, Enum. Cact. 78) were published as synonyms.

In the Engelmann Herbarium is a single specimen labeled "*Cereus multangularis*" with the following note: "Columnar, similar to *serpentinus*, coll. Germantown, Pa., October 27, 1869." We believe this plant is *Nyctocereus serpentinus*.

Dr. A. Hrdlička collected in March 1913, in the mountains southeast of Nasea, Peru, at an altitude of 5,000 to 7,000 feet, a curious plant which may represent the one referred here by Schumann. It is a low cespitose plant, rarely 2 feet high, with numerous low almost indistinct ribs, nearly hidden by the numerous spines; areoles approximate, 4 to 5 mm. apart, felted and spiny; spines 25 or more, brown or white with brown tips, the longest ones 12 mm. long; flower-buds scaly, woolly, and setose in their axils. Living specimens were sent to Washington, but these eventually died without flowering.

Cereus lecchii (Pfeiffer, Enum. Cact. 78. 1837; *Cactus lecchii* Colla and *C. lanuginosus aureus* Colla, Hort. Ripul. 25. 1825; *Echinocactus lecchii* Don in Sweet, Hort. Brit. ed. 3. 283. 1839) is referred here by Schumann. *Cereus lanuginosus aureus* (Pfeiffer, Enum. Cact. 78. 1837) was given as a synonym of *C. lecchii*. *Cactus lecchii* was illustrated by Colla in his Fourth appendix to the Hortus Ripulensis (Mem. Accad. Sci. Torino 35: pl. 2).

CEREUS LIMENSIS Salm-Dyck, Allg. Gartenz. 13: 353. 1845.

> *Echinocereus limensis* Rümpler in Förster, Handb. Cact. ed. 2. 824. 1885.
> *Cereus multangularis limensis* Maass, Monatsschr. Kakteenk. 15: 119. 1905.

Stems erect, thick, very green; ribs 12, obtuse, subrepand; areoles close together, oval, filled with yellow tomentum; spines acicular, setaceous, rigid, the central ones 8 to 10, divergent, yellowish red, one longer than the others; radial spines 20 to 25, reddish yellow above, white below.

The above is a free translation of the original.

This species is not determinable but was referred by Schumann to *Cereus multangularis*. *Echinocereus multangularis limensis* Lemaire (Rümpler in Förster, Handb. Cact. ed. 2. 824. 1885) was given as a synonym of *Echinocereus limensis*.

CEREUS LANGLASSEI, Monatsschr. Kakteenk. 14: 145. 1904. Mentioned as a seedling from Paris. Weingart (Monatsschr. Kakteenk. 29: 105. 1919) described the plant after it had made some growth and compared it with *C. eburneus* (*Lemaireocereus griseus*.)

CEREUS HORIZONTALIS Gillies in Sweet, Hort. Brit. ed. 3. 285. 1839. Described as horizontal with stems of 5 or 6 angles.

CEREUS AMBLYOGONUS G. Don in Sweet, Hort. Brit. ed. 3. 284. 1839. Described as "blunt angled" and introduced from South America.

CEREUS CAUDATUS Gillies in Sweet, Hort. Brit. ed. 3. 285. 1839. Described only as "tailed" and introduced from Chile in 1828.

CEREUS LONGIFOLIUS Karwinsky in Sweet, Hort. Brit. ed. 3. 286. 1839. Described as "long-leaved."

CEREUS DE LAGUNA Haage in Förster, Handb. Cact. 433. 1846. Said to be similar to *C. geometrizans* and *C. eburneus* and to be from Brazil.

CEREUS REGALIS Haworth in Sprengel, Syst. 2: 496. 1825. Described as erect, 9-ribbed, and with elongated yellow equal spines.

Cereus ovatus Don (Loudon, Hort. Brit. 195. 1830; *Cactus ovatus* Gillies) and *Cereus decorus* Loddiges (Voigt, Hort. Suburb. Calcutt. 62. 1845) were both introduced into India in 1840 but are not now known nor have they been described.

Cereus flavispinus Roezl in Morren (Belg. Hort. 24: 39. 1874), collected by Roezl probably in the high mountains above Lima, was never formally published.

The following names of *Cereus* we have been unable to refer to any of the species otherwise mentioned in this work:

Cereus aculeatus Förster, Handb. Cact. 433. 1846.
    albertinii Fobe, Monatsschr. Kakteenk. **18**: 175. 1908.
    atrovirens Förster, Handb. Cact. 433. 1846.
    concinnus Haage in Schumann, Gesamtb. Kakteen 167. 1897.
    damacaro Haage in Schumann, Gesamtb. Kakteen 167. 1897.
    incrassatus Link and Otto, Verh. Ver. Beförd. Gartenb. **6**: 432. 1830.
    jacquinii Rebut in Schumann, Gesamtb. Kakteen 167. 1897.
    karwinskii Haage in Schumann, Gesamtb. Kakteen 167. 1897.
    longipendunculatus Förster, Handb. Cact. 433. 1846.
    lormata Maass, Monatsschr. Kakteenk. **15**: 119. 1905.
    ophites Lemaire, Monatsschr. Kakteenk. **4**: 173. 1894.
    pruinatus, Monatsschr. Kakteenk. **11**: 181. 1901.
    robustus Schumann, Monatsschr. Kakteenk. **13**: 111. 1903.
    rogalli Schumann, Monatsschr. Kakteenk. **9**: 96. 1899.
    salpingensis Schumann, Monatsschr. Kakteenk. **11**: 181. 1901.
    schoenemannii Hildmann, Monatsschr. Kakteenk. **5**: 43. 1895.
    spathulatus Förster, Handb. Cact. 433. 1846.
    steckmannii Jacobi, Monatsschr. Kakteenk. **5**: 43. 1895.
    tellii, Monatsschr. Kakteenk. **5**: 43. 1895. A name from Hildmann's Catalogue.
    trichocentrus Förster, Handb. Cact. 433. 1846.
    verschaffeltii Haage in Schumann, Gesamtb. Kakteen 167. 1897.

2. MONVILLEA gen. nov.

Night-blooming cacti with long, slender, half-erect stems, often forming thickets; flowers borne toward the top of the stem, of medium size, without felt or spines; tube proper in typical species slender, tapering into a short throat; scales minute; outer perianth-segments greenish or pinkish; inner perianth-segments white or yellow; stamens white, not in definite rows but scattered over the throat; style slender, white, with linear stigma-lobes; flower-tube rigid after anthesis, withering on the ovary; scales on the ovary minute, their axils naked; fruit glabrous, red, plump, spineless; flesh of fruit white, juicy; seeds small, black.

Type species: *Cereus cavendishii* Monville.

The generic name commemorates M. Monville, a well-known student of this family. We recognize 7 species, all South American.

KEY TO SPECIES.

Flower-tube slender, straight; stamens and style more or less exserted.
  Ribs 6 to 10.
    Flowers white...1. *M. cavendishii*
    Flowers yellow..2. *M. insularis*
  Ribs 3 to 5.
    Plants erect, bluish green, more or less spotted; branches with 3 or 4 sharp ribs, these
      deeply serrate..3. *M. spegazzinii*
    Plants decumbent; branches with 4 or 5 rounded ribs.....................4. *M. phatnosperma*
Flower-tube short and stout, somewhat curved; stamens and style included.
  Spines subulate, often elongated.
    Fruit globular; flower strongly angled; flower-bud pointed.........................5. *M. diffusa*
    Fruit oblong; flower not strongly angled; flower-bud obtuse......................6. *M. maritima*
  Spines acicular, all very short..7. *M. amazonica*

1. Monvillea cavendishii (Monville).

    Cereus serpentinus splendens Salm-Dyck in Lemaire, Cact. Gen. Nov. Sp. 79. 1839.
    Cereus cavendishii Monville, Hort. Univ. **1**: 219. 1840.
    Cereus paxtonianus Monville in Salm-Dyck, Cact. Hort. Dyck. 1849. 211. 1850.
    Cereus splendens Salm-Dyck, Cact. Hort. Dyck. 1849. 214. 1850.
    Cereus saxicola Morong, Annals N. Y. Acad. Sci. **7**: 121. 1893.
    Cereus euchlorus Weber in Schumann, Gesamtb. Kakteen 84. 1897.
    Cereus rhodoleucanthus Schumann, Monatsschr. Kakteenk. **9**: 187. 1899.
    Eriocereus cavendishii Riccobono, Boll. R. Ort. Bot. Palermo **8**: 239. 1909.

In cultivation more or less branched at base, 1 to 3 meters high, suberect or clambering, green, 2 to 3 cm. in diameter; ribs 9 or 10, low and rounded; areoles small, about 1 cm. apart; spines acicular,

8 to 12, brown; flower 10 to 12 cm. long, the tube 5 to 6 cm. long; outer perianth-segments pinkish; inner perianth-segments white; ovary small, bearing a few very small scales, these broader than long, with minute brown chartaceous tips; fruit globular, 4 to 5 cm. in diameter.

Type locality: Carthagene.*

Distribution: Brazil, northern Argentina and Paraguay.

This is one of the best flowering species we have in cultivation. The flowers open at night and appear more or less abundantly from April to September.

The species was named for William Spencer Cavendish, Duke of Devonshire, who had a magnificent collection of plants at Chatsworth.

Cereus anguiniformis (Weingart, Monatsschr. Kakteenk. **18**: 6. 1908) and *C. saxicola anguiniformis* Riccobono (Boll. R. Ort. Bot. Palermo **8**: 252. 1909) probably belong here.

We have referred here *Cereus euchlorus,* which originally came from São Paulo, Brazil. We have specimens growing which were obtained under this name from M. Simon, of St. Ouen, Paris, in 1901.

FIG. 20.—Monvillea cavendishii.

FIG. 21.—Monvillea insularis. ×0.5.

Botanists have been much in doubt as to the relationship of this species. Schumann in his Monograph refers it along with *Cereus striatus* (now *Wilcoxia*) to his series *Tenuiores.* In his Nachträge, published in 1903, he places it with *Cereus obtusangulus* (now *Zygocactus*) in his series *Anomali,* but in his Keys, published about the same time, he again has it in the series *Tenuiores.* Berger places it in his subgenus *Piptanthocereus,* while Riccobono has recently transferred it to his genus *Eriocereus.*

Dr. Schumann discusses *Cereus lauterbachii* Schumann (Bull. Herb. Boiss. II. **3**: 250. 1903) in connection with this species, but does not point out how they differ.

Both the names *Cereus cavendishii* and *C. paxtonianus* were in general use until Schumann in 1897 suggested that the plants are the same. Sir Joseph Hooker in 1899 united them definitely under the name of *Cereus paxtonianus.*

Illustrations: Monatsschr. Kakteenk. **19**: 77, as *Cereus saxicola;* Monatsschr. Kakteenk. **13**: 12, as *Cereus rhodoleucanthus;* Curtis's Bot. Mag. **125**: pl. 7648, as *Cereus paxtonianus.*

*When first published it was stated that the species came from "Carthagene." Schumann says probably South America, possibly Brazil, while Morong collected it in central Paraguay.

M. E. Eaton del.

1. Top of stem of *Cereus validus*.
2. Top of flowering stem of same.
3. Top of flowering branch of *Monvillea cavendishii*.
4. Top of branch with fruit of same.

(Natural size.)

Plate III, figure 3, shows a flowering branch in the collection of the New York Botanical Garden; figure 4 shows the fruit from a plant in the same collection. Figure 20 is from a photograph taken in 1917 by Dr. Shafer at Catilegua, Argentina.

2. Monvillea insularis (Hemsley).

Cereus insularis Hemsley, Voyage of Challenger Bot. 1²: 16. 1884.

Creeping or clambering, forming a dense thicket, much branched; branches nearly cylindric, 2.5 to 3 cm. in diameter, 6-angled; spines 12 to 15, unequal, spreading, terete, yellow; flowers described as yellow, 12.5 cm. long; ovary and flower-tube bearing only a few minute scales, but no spines or hairs; flower-tube very slender; perianth-segments in several series; filaments and style protruding; stigma-lobes 13, radiating.

Type locality: St. Michael's Mount, off Brazil, 5° S. latitude.

Distribution: Known only from the type locality.

This plant is noteworthy as inhabiting an island on which no other cactus exists. It is the most eastern in natural distribution of all cactus species. So far as we are informed, it has never been in cultivation.

Illustration: Voyage of Challenger Bot. 1²: pl. 14, as *Cereus insularis.*

Figure 21 is copied from the plate above cited.

3. Monvillea spegazzinii (Weber).

Cereus spegazzinii Weber, Monatsschr. Kakteenk. 9: 102. 1899.
Cereus anisitsii Schumann, Monatsschr. Kakteenk. 9: 185. 1899.

Erect, strongly 3-angled or ribbed, bluish green, more or less spotted with white; ribs strongly undulate or serrate; spines on young branches brown to black, 3 at an areole, 5 mm. long, with broad conic bases; on old wood 6 at an areole, of these 5 radial, 1 central; bud and flower rigid and erect, but after anthesis abruptly reflexed; flowers 11 to 12 cm. long, narrow, funnelform; outer perianth-segments purplish, the inner nearly white, serrate above, acuminate.

Type locality: Near Resistencia, Chaco Territory, Argentina.

Distribution: Paraguay and northeastern Argentina.

Cereus marmoratus Zeissold (Cat. 1899), unpublished, is referred by Gürke (Monatsschr. Kakteenk. 18: 131. 1908) to *Cereus anisitsii;* Gürke (Monatsschr. Kakteenk. 16: 146. 1906) also refers *Cereus lindenzweigianus,* name only, to *C. anisitsii.*

Illustrations: Monatsschr. Kakteenk. 12: 193; Schumann, Gesamtb. Kakteen Nachtr. f. 5; Schelle, Handb. Kakteenk. f. 31; Rev. Hort. Belge 40: after 184*, as *Cereus spegazzinii;* Blühende Kakteen 2: pl. 107, as *Cereus anisitsii.*

FIG. 22.—Monvillea spegazzinii.

Figure 22 is from a photograph of the type specimen given to Dr. Rose by Dr. Spegazzini.

---

*In the Revue de L'Horticulteur for 1914 was published a number of plates, mostly of cacti grown by Frantz de Laet. These plates are not numbered and we have indicated their position by the page they follow. Most of them were reproduced in De Laet's Catalogue Général.

4. Monvillea phatnosperma (Schumann).

Cereus phatnospermus Schumann, Monatsschr. Kakteenk. **9**: 186. 1899.

Decumbent, 1 to 2 meters long; branches 4 or 5-ribbed, bright green, 2.5 cm. in diameter; ribs rounded, somewhat concave on the sides; spines brown, subulate; radial spines 5 or 6, spreading, 15 mm. long; central spines, when present, straight or somewhat curved, up to 2.5 cm. long; flowers white, 12 cm. long; ovary subnaked, narrow, cylindric, about 3 cm. long.

Type locality: Near Porongo, Paraguay.
Distribution: Paraguay.
The plant is known to us only from the description above cited.

5. Monvillea diffusa sp. nov.

Stems slender, 4 to 5 cm. in diameter, at first erect, afterwards with long arching branches, when growing in the open often forming thickets 2 to 5 meters in diameter; ribs high and thin, usually 8; areoles 2.5 to 3 cm. apart, gray-felted; radial spines 6 to 10, spreading, acicular, 6 to 12 mm. long; central spines 1 to 3, one usually much elongated, 2 to 3 cm. long, subulate, gray with black tips; flowers 7.5 cm. long, the tube strongly ribbed; scales on the flower-tube ovate, acute; ovary globose with elongated tubercles or ribs; scales on ovary minute, acute.

FIG. 23.—Monvillea diffusa. ×0.5.

Common on the hillsides of the Catamayo Valley in southern Ecuador.

Collected by J. N. Rose, A. Pachano, and George Rose, October 3, 1918 (No. 23325). Figure 23 shows a flower and young fruit from the type collection.

6. Monvillea maritima sp. nov.

Stems slender, 5 to 8 cm. in diameter, at first erect, sometimes 4 to 5 meters high, growing among shrubs and trees and often high-clambering, either simple or with few distant branches, these weak, ascending or drooping; ribs 4 to 6, somewhat undulating, the areoles borne in the depressions, 2 to 3 cm. apart; spines about 8, all gray, with black tips; central spines 1 or 2, one much longer and stouter, 5 to 6 cm. long; upper part of flower-bud nearly globular, merely acute at apex; flowers 6 cm. long; flower-tube faintly angled without, naked for about 3 cm. above the base of the style; ovary oblong, faintly angled, the scales broad with a minute scarious tip.

Common in the thickets along the coast of southern Ecuador near Santa Rosa where it was collected by J. N. Rose and George Rose, October 1918 (No. 23495).

The flowers of this species are similar to those of *M. diffusa*, but the two plants grow in very different situations and are of different habit. *M. diffusa* grows on the mountain-side of a very arid interior valley at an altitude of about 2,170 meters, while *M. maritima* is from a humid region near sea-level; the former grows in the open while *M. maritima* grows among bushes and trees.

7. Monvillea amazonica (Schumann).

Cereus amazonicus Schumann in Vaupel, Monatsschr. Kakteenk. **23**: 164. 1913.

At first erect, up to 5 meters long, not much branched; ribs 7, low, acute; areoles about 17 mm. apart; spines about 15, acicular, weak, 8 mm. long; flowers borne on the upper part of the stem but not at the tip, straight, 8 cm. long; areoles on ovary and flower-tube without hairs, bristles, or spines, subtended by minute scales; perianth-segments numerous, obovate, rounded above; ovary and fruit oblong, capped by the withering flower.

Type locality: Loreto, Peru.
Distribution: On the upper Amazon in eastern Peru.
This is evidently a remarkable species. It is known to us only from the description and illustration, and may very likely represent a distinct generic type.
Illustration: Monatsschr. Kakteenk. **23**: 165, as *Cereus amazonicus*.

MONVILLEA sp.

Stems slender, sometimes 3 to 4 meters high, nearly simple, rather weak and often supported by other plants, 3 to 5 cm. in diameter; ribs about 8, 1 cm. high; areoles 1 to 1.5 cm. apart; radial spines about 10, somewhat unequal, the longest about 1 cm. long; central spines 2 or 3, longer and stouter than the radial, usually about 2 cm. long, black at tip; flowers and fruit not seen.

Collected by J. N. Rose and George Rose near Guayaquil, Ecuador, August 11, 1918 (No. 22117).

This species was quite common in the flat country northwest of Guayaquil associated with a larger arborescent cactus, a species of *Lemaireocercus*, and at first was supposed to be its juvenile form. Unfortunately, no flowers or fruit were seen. Living specimens were brought back to the New York Botanical Garden, but these have not yet flowered. We are not certain of the generic position of this plant, but it so much resembles *Monvillea maritima* in habit that we suspect that its relationship is here.

3. CEPHALOCEREUS Pfeiffer, Allg. Gartenz. 6: 142. 1838.

Cephalophorus Lemaire, Cact. Aliq. Nov. xii. 1838. Not *Cephalophora* Cavanille. 1801.
Pilocereus Lemaire, Cact. Gen. Nov. Sp. 6. 1839.

Elongated cacti, various in habit, mostly columnar and erect, sometimes much branched with a short trunk or in one species with spreading and procumbent branches; in some species the flowering areoles develop an abundance of wool which confluently forms a dense mass called a pseudocephalium either at the top or on one side near the top; in others long wool or hairs grow from the areoles but a pseudocephalium is not formed; in others the flowers are produced in a circle at the top and the bristles and fruit afterwards form a collar at the base of the new growth; in other species neither wool nor hairs are produced in the flowering areoles; flowers nocturnal, short-campanulate to short-funnelform or pyriform, straight or curved; perianth persisting on the ripening fruit, except in one species; fruit usually depressed-globose, sometimes oblong; seeds black, smooth or tuberculate.

We know 48 species, distributed from southern Florida and northern Mexico to eastern Brazil and Ecuador. The type species is *Cactus senilis* Haworth, which is also the type of Lemaire's genera *Cephalophorus* and *Pilocereus*. The name *Cephalocereus* is from the Greek, signifying headed-cereus, with reference to the pseudocephalium of the typical species.

KEY TO SPECIES.

A. Flowering areoles confluent, forming a pseudocephalium.
  Pseudocephalium lateral.
    Ovary bearing few distant scales; areoles of the flower with tufts of short wool;
      plants simple, tall, columnar.
      Plant cylindric; top rounded; bristles of pseudocephalium twice as long as wool. 1. *C. senilis*
      Plant tapering to the apex; bristles of pseudocephalium little longer than the wool. 2. *C. hoppenstedtii*
    Ovary naked.
      Plant unbranched.. 3. *C. purpureus*
      Plant branched at the base.
        Ribs 12 to 17; flowers 6 to 7 cm. long................................. 4. *C. fluminensis*
        Ribs up to 23; flowers 4 cm. long..................................... 5. *C. dybowskii*
  Pseudocephalium terminal... 6. *C. macrocephalus*
AA. Flowering areoles not confluent, though sometimes close together, not forming a pseudocephalium.
  Flower-tube strongly bent about middle; areoles wholly without hairs; plant blue... 7. *C. pentaedrophorus*
  Flower-tube straight or a little curved at the base.
    Ribs 10 to 18; flowers red.
      Ribs 15 to 18; perianth-segments not reflexed.......................... 8. *C. polylophus*
      Ribs 8; perianth-segments reflexed.................................... 9. *C. euphorbioides*
    Ribs 4 to 13; flowers mostly whitish to purplish.
      Perianth falling away from the ovary by abscission......................10. *C. russelianus*
      Perianth withering-persistent (so far as known).
        Ribs strongly tubercled.
          Spines all brown, the radials widely spreading, the centrals stout, subulate..11. *C. gounellei*
          Spines all yellow, acicular, the radials only slightly spreading............12. *C. zehntneri*
        Ribs not tubercled.
          Flower-tube curved at base; areoles of the stem all densely long-woolly....13. *C. leucostele*
          Flower-tube straight; only flowering areoles, if any, long-woolly.

KEY TO SPECIES—continued.

AA. Flowering areoles not confluent, though sometimes close together, not forming a pseudocephalium—*continued.*
  Ovary well-developed; flower-tube little scaly or without scales.
    Flowering areoles without wool, or wool very short.
      Fruit oblong to ovoid...14. *C. smithianus*
      Fruit globose or depressed.
        Spines acicular; berry large.
          Perianth-segments rounded, acute, or mucronate.
            Plant grayish green; at least the perianth-segments rounded
              or mucronate.
              Outer perianth-segments rounded.......................15. *C. bahamensis*
              All perianth-segments rounded.........................16. *C. deeringii*
            Plant glaucous green when young, dull green when old; outer
              perianth-segments acute.
              Much branched, the branches ascending; ribs 10 to 13; style
                exserted...............................17. *C. robinii*
              Little branched, the branches nearly erect; ribs 9 or 10; style
                scarcely exserted; young growth very glaucous.18. *C. keyensis*
          Perianth-segments retuse....................................19. *C. monoclonos*
        Spines subulate; berry small..............................20. *C. scoparius*
    Flowering areoles definitely long-woolly.
      Ribs 5 to 13, separated by narrow valleys.
        Ribs 8 mm. high or higher.
          Plant light green to dark green.
            Spines short, subulate.................................21. *C. moritzianus*
            Spines slender, acicular.
              Wool of flowering areoles sparse, not matted............22. *C. arrabidae*
              Wool of flowering areoles mostly copious, matted.
                Wool of flowering areoles brown...................23. *C. urbanianus*
                Wool of flowering areoles white.
                  Plant bright green, shining.
                    Joints slender, dark green; wool short.............24. *C. nobilis*
                    Joints stout, light green; wool long...............25. *C. barbadensis*
                  Plant dull green, not shining.
                    Wool elongated, up to 5 to 7 cm. long............26. *C. millspaughii*
                  Wool short, 2 cm. long or less.
                    Ribs 1 to 2 cm. high.
                      Only the flowering areoles woolly.............27. *C. swartzii*
                      Both flowering and flowerless areoles of young
                        joints woolly..........................28. *C. polygonus*
                  Ribs 8 mm. high or less......................29. *C. gaumeri*
         Plant, at least young joints, blue or bluish green, glaucous.
          Young spines yellow.
            Flower 7 to 8 cm. long...............................30. *C. chrysacanthus*
            Flower 5 to 6 cm. long.
              Ribs 6 to 8.....................................31. *C. maxonii*
              Ribs mostly 9 to 12.
                Ribs 13.....................................32. *C. piauhyensis*
                Ribs 9 or 10.
                  Perianth-segments rounded or mucronate.
                    Young joints bright blue; ribs low.............33. *C. lanuginosus*
                    Young joints bluish green, glaucous; ribs high...34. *C. brooksianus*
                  Perianth-segments acute........................35. *C. royenii*
          Young spines brown or nearly black.
            Areoles approximate, their spines overlapping...........36. *C. robustus*
            Areoles separated, their spines not overlapping.
              Ribs 9 to 12.
                Wool short, 2 cm. long........................37. *C. cometes*
                Wool 10 cm. long.............................38. *C. leucocephalus*
              Ribs 7 to 9.
                Ribs strongly horizontally grooved below the areoles.
                Flowers rose-red............................39. *C. sartorianus*
                Flowers brown..............................40. *C. palmeri*
              Ribs not grooved..............................41. *C. tweedyanus*
         Color of plant and of young spines unknown.
          Mexican...42. *C. alensis*
          Colombian...43. *C. colombianus*
       Ribs only 5 to 6 mm. high; plant green.......................44. *C. purpusii*
      Ribs 4 to 6, separated by broad valleys.
        Glaucous, up to 10 meters high; flowers 6 to 8 cm. long........45. *C. catingicola*
        Bright green, up to 3 meters high; flowers about 5 cm. long.....46. *C. brasiliensis*
    Ovary very short and flat; flower-tube scaly.. ....47. *C. phaeacanthus*
AAA. Species not grouped..:...48. *C. ulei*

1. **Cephalocereus senilis** (Haworth) Pfeiffer, Allg. Gartenz. **6**: 142. 1838.

> *Cactus senilis* Haworth, Phil. Mag. **63**: 31. 1824.
> *Cactus bradypus* Lehmann, Ind. Sem. Hamburg 17. 1826.
> *Cereus senilis* De Candolle, Prodr. **3**: 464. 1828.
> *Cephalophorus senilis* Lemaire, Cact. Aliq. Nov. xii. 1838.
> *Pilocereus senilis* Lemaire, Cact. Gen. Nov. Sp. 7. 1839.
> *Echinocactus senilis* Beaton, Loudon's Gard. Mag. **15**: 550. 1839.
> *Echinocactus staplesiae* Tate, Loudon's Gard. Mag. **16**: 27. 1840.

Plants 6 to 10 or even 15 meters high, columnar, simple or rarely branched above, sometimes branched at base; ribs numerous; pseudocephalium developing on plants when 6 meters high, broadening above, rarely confined to one side but usually encircling the top of the plant; areoles closely set; the ones at base of old plants producing weak, gray bristles 2 to 3 dm. long, the ones in the pseudocephalium producing similar but shorter bristles intermixed with dense, tawny wool, 4 to 6 cm. long; flower, including the ovary, 5 cm. long, rose-colored; scales few on the tube; fruit obovoid, 2.5 to 3 cm. long, rose-colored, capped by the chartaceous base of the flower, bearing a few minute scales with hairs in their axils.

Type locality: Mexico.

Distribution: Hidalgo and Guanajuato, Mexico.

Cephalocereus senilis has long been considered a great curiosity and small plants are shipped in quantities to Europe. The young plants are covered with long white silky hairs resembling a beard and hence the name old man cactus, and similar names. Large plants are not often seen since the species grows in regions difficult of access. It is very common on limestone hills of eastern Hidalgo, where it is often the most conspicuous plant in the landscape. Large individuals are common here and are often 15 meters high. Very little wood-tissue is developed and the largest individuals can easily be cut down with a pick or small knife.

Salm-Dyck (Cact. Hort. Dyck. 1844. 24. 1845) named two varieties of *Pilocereus senilis: longispinus* and *flavispinus;* later (Cact. Hort. Dyck. 1849. 40, 186. 1850) he described *P. senilis longisetus,* saying nothing about *longispinus* and *flavispinus.*

Cereus bradypus Lehmann and *Melocactus bradypus* Lehmann (Steudel, Nom. ed. 2. 1: 333. 1840; 2: 122. 1841) are cited as synonyms of *Cereus senilis.*

Illustrations: Schumann, Gesamtb. Kakteen f. 40; Grässner, Haupt-Verz. Kakteen 2; Knippel, Kakteen pl. 1; Engler and Prantl, Pflanzenfam. 3[6a]: pl. opp. p. 180; f. 60; Safford, Ann. Rep. Smiths. Inst. 1908: f. 15, 16, all as *Cephalocereus senilis;* Förster, Handb. Cact. ed. 2. f. 91, 92; Monatsschr. Kakteenk. **4**: 124, 125; Rev. Hort. **61**: f. 139; **62**: f. 38, 39; Rümpler, Sukkulenten f. 78; Dict. Gard. Nicholson Suppl. f. 634, all as *Pilocereus senilis;* Monatsschr. Kakteenk. **1**: 32, as *Pilocereus senilis cristatus;* Nov. Act. Nat. Cur. **16**: pl. 12, as *Cactus bradypus.*

Figure 24 is from a photograph of a small plant in the collection of the New York Botanical Garden.

2. **Cephalocereus hoppenstedtii** (Weber) Schumann in Engler and Prantl, Pflanzenfam. 3[6a]: 181. 1894.

> *Pilocereus hoppenstedtii* Weber, Cat. Pfersdorff. 1864 (according to Schumann).
> *Pilocereus hagendorpi* Regel, Gartenflora **18**: 220. 1869.
> *Pilocereus lateralis* Weber, Dict. Hort. Bois 966. 1898.
> *Cereus hoppenstedtii* Berger, Rep. Mo. Bot. Gard. **16**: 70. 1905.

Slender, columnar, said sometimes to reach 10 meters in height, but in cultivation much lower, often bent or clambering, the apex tapering; ribs low, close together, 20 or more, the whole plant hidden under the numerous spines; areoles close together; radial spines 14 to 18, very short, white; central spines 5 to 8, the longest one sometimes 7.5 cm. long, usually reflexed, brownish; pseudo-cephalium at the top of the plant but to one side (said to be on the north side); flower described as 7.5 cm. long, whitish, with rosy tips, bell-shaped; fruit not known.

Type locality: Zapotitlan, near Tehuacán, Mexico.

Distribution: Southern Mexico.

This plant is clearly a close relative of *Cephalocereus senilis.*

So far as we know, the type has not been preserved. The species is sometimes cultivated, but it has never done well with us under glass. Mr. Berger was able to grow it at La Mortola, Italy.

There has long been considerable confusion regarding the characters of this species, partly because other cacti have been confused with it. For instance, the only specimens (several flowers) in the Engelmann Herbarium, so named, although from the region of this species, are those of a *Pachycereus*.

Illustrations: Knippel, Kakteen pl. 29; Grässner, Haupt-Verz. Kakteen 29. 1912; Möllers Deutsche Gärt. Zeit. **29**: 355. f. 10; Schelle, Handb. Kakteenk. f. 38, as *Pilocereus hoppenstedtii;* Bull. Soc. Acclim. France **52**: f. 15, as *Pilocereus lateralis.*

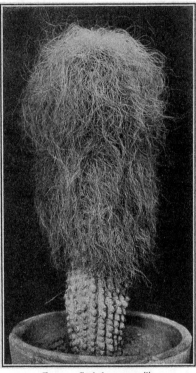

FIG. 24.—Cephalocereus senilis.

FIG. 25.—Cephalocereus purpureus.

3. Cephalocereus purpureus Gürke, Monatsschr. Kakteenk. **18**: 86. 1908.

Columnar, upright, unbranched, 3 meters high or more; ribs 12 to 15, broad, low, separated by narrow intervals, marked by upturned V-shaped depressions, one from the top of each areole; areoles large, longer than broad, white-woolly and spiny; radial spines 15 to 20, acicular, white, short, 1 cm long or less; central spines 8 to 10, the longer ones 5 cm. long, brown; pseudocephalium on the west side of the plant, confined to only a few of the ribs (3 to 7); flowers open at night, closing in the morning, 4 to 5 cm. long; tube and outer perianth-segments pinkish; inner perianth-segments white; stamens and style included; fruit small; seeds black, roughened, large at the top, narrowed at base.

Type locality: Serra do Sincorá, Bahia, Brazil, 800 to 1,200 meters altitude.
Distribution: Southern central Bahia, Brazil.

Photographs and an abundance of flowers and seed of this most interesting species were obtained by Dr. Rose from Dr. L. Zehntner, who had two plants growing in his garden at Joazeiro, Bahia.

Figure 25 is from a photograph of one of the plants above mentioned; figure 28 shows a spine-areole of its stem; figure 27 shows the flower.

4. Cephalocereus fluminensis (Miquel).

Cactus melocactus Vellozo, Fl. Flum. 205. 1825. Not Linnaeus, 1753.
Cereus fluminensis Miquel, Bull. Sci. Phys. Nat. Neerl. 1838: 48, 1838.
Pilocereus vellozoi Lemaire, Rev. Hort. 1862: 427. 1862.
Cephalocereus melocactus Schumann in Martius, Fl. Bras. 4²: 215. 1890.
Pilocereus melocactus Schumann, Monatsschr. Kaktecnk. 3: 20. 1893.
Cereus melocactus Berger, Rep. Mo. Bot. Gard. 16: 62. 1905.*

Growing generally in clumps, clambering over rocky cliffs; branches erect, spreading or pendent, 1 to 2 meters long; ribs 12 to 17, 1 to 1.5 cm. high, acute, separated by acute intervals; spines acicular, yellow, the longest ones 3 cm. long; pseudocephalium on one side of the branch, of a dense white felt, 2 to 3 cm. thick, intermixed with long yellow bristles, 4 to 7 cm. long; areoles close together, circular, with short white wool but with no long hairs; flowers 6 to 7 cm. long; style long-exserted; fruit bright red to purple, obovoid, 3 cm. long, naked, almost hidden in the mass of white wool of the pseudocephalium; seeds black, 1 mm. in diameter, tuberculate.

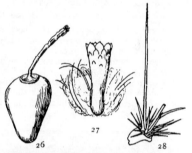

Type locality: On island in harbor of Rio de Janeiro, Brazil.

Distribution: On rocky cliffs and islands along Brazilian coast from Rio de Janeiro to Cabo Frio.

This plant was first collected and figured by Vellozo about 1790 and named *Cactus melocactus*, a name which had already been used by Linnaeus

Fig. 26.—Fruit of Cephalocereus fluminensis. ×0.7.
Fig. 27.—Flower of Cephalocereus purpureus. ×0.7.
Fig. 28.—Cluster of spines of same. ×0.7.

Fig. 29.—Cephalocereus fluminensis.

for another plant. Although this species is very common on all the rocky knolls and outcrops about the harbor of Rio de Janeiro, it has rarely been collected and no living or herbarium material was in the Washington and New York collections until Dr. Rose collected it in Brazil in 1915.

---

*Schumann (Martius, Fl. Bras. 4²: 216. 1890) erroneously refers this binomial to Vellozo.

Illustrations: Vellozo, Fl. Flum. **5**: pl. 20, as *Cactus melocactus;* Martius, Fl. Bras. 4²: pl. 43; Engler and Prantl, Pflanzenfam. 3⁶ᵃ: f. 65, B, as *Cephalocereus melocactus;* Monatsschr. Kakteenk. **3**: 25, as *Pilocereus melocactus.*

Figure 29 is from a photograph taken at Rio de Janeiro by Paul G. Russell in 1915; figure 26 shows the fruit as drawn by A. Löfgren.

Cereus ferox Haworth (Phil. Mag. **7**: 109. 1830) may be of this relationship. It is described as upright, stout, oblong, terete, 9 inches high, 2 inches in diameter, dark green; ribs about 18, densely covered with spreading yellow spines; radial spines about 6; central spines 4 or 5, one twice as long as the others, much stouter, up to an inch long. This species was introduced from Brazil by Loddiges, in whose collection it was seen and described by Haworth. It is stated to be near *Cereus multangularis.* Förster and Schumann did not know the species.

5. Cephalocereus dybowskii (Gosselin).

Cereus dybowskii Gosselin, Bull. Soc. Bot. France **55**: 695. 1908.

Stems much branched at the base, sending up many strict, usually simple branches 2 to 4 meters high, 8 cm. in diameter, almost hidden by the white cobwebby hairs of the areoles; ribs numerous, often 23, low; pseudocephalium on the west side of the plant, consisting of a mass of long white wool extending from the top of the branch downward sometimes for 5 to 6 dm.; spines yellow, the radials short, hidden in the white hairs, the central spines 2 or 3, porrect, acicular, 2 to 3 cm. long; flower opening at night, 4 cm. long; inner perianth-segments white, broad, short; stamens in 2 series; stamens in upper series with short filaments or none; stamens in lower series united at base into a short tube; style slender, cream-colored, 3.5 cm. long; stigma-lobes about 15, linear; fruit globular, naked, pinkish, 2.5 cm. in diameter; seeds black, roughened.

FIG. 30.—Cephalocereus dybowskii.

Type locality: Itumirin, Bahia, Brazil.

Distribution: Common on the dry hills in Bahia, where it forms dense thickets sometimes to the exclusion of all other plants.

Our description is based largely on the specimens collected by Dr. Rose at Barrinha, Bahia, in 1915 (No. 19785).

In Bahia this species is called cabeça branca or mandacaru de penacho.

Figure 30 is from a photograph taken at Barrinha, Bahia, by Paul G. Russell in 1915.

6. **Cephalocereus macrocephalus** Weber in Schumann, Gesamtb. Kakteen 197. 1897.

Pilocereus macrocephalus Weber, Dict. Hort. Bois 966. 1898.
Cereus macrocephalus Berger, Rep. Mo. Bot. Gard. **16**: 62. 1905.

Plant of great size, 10 to 16 meters high, with a very solid woody trunk 3 to 6 dm. in diameter, simple or with a few ascending branches; pseudocephalium not so conspicuous as in *Cephalocereus senilis;* ribs numerous (about 24), low, obtuse, pale green; radial spines about 12, spreading; central spines several, sometimes 6 cm. long; flowering areoles spineless but bearing white, stiff hairs or weak bristles; perianth about 5 cm. long, the tube bearing a few distant scales, the limb short, the outer segments rounded.

Type locality: Tehuacán, Mexico.

Distribution: Southern Puebla, Mexico.

Dr. Rose found this species very common on a single hill near Tehuacán, forming a forest of considerable size. The individual plants are often very large and the trunk is so stout and woody that one can not cut down the plants readily, as is the case with *Cephalocereus senilis* and some other species of this genus.

Illustrations: Contr. U. S. Nat. Herb. **10**: pl. 43, f. B; MacDougal, Bot. N. Amer. Des. pl. **15**; Nat. Geogr. Mag. **21**: 698; Möllers Deutsche Gärt. Zeit. **29**: 351. f. 6.

Plate 1 is from a photograph taken by Dr. MacDougal near Tehuacán in 1906.

7. **Cephalocereus pentaedrophorus** (Labouret).

Cereus pentaedrophorus Labouret, Monogr. Cact. 365. 1853.
Pilocereus polyedrophorus Lemaire, Rev. Hort. **1862**: 428. 1862.
Pilocereus pentaedrophorus Console in Schumann, Gesamtb. Kakteen 174. 1897.

Stems very slender, usually only 2 to 5 meters, rarely 7 or occasionally 10 meters high, 10 cm. in diameter or less, bluish, glaucous especially toward the growing tip; ribs usually 4 to 6, but occasionally as many as 8; areoles without wool, often large, separated by horizontal grooves or depressions; spines yellow, various as to size and number, usually 6 to 12, the longest often 4 cm. long; flowers 4 to 6 cm. long; perianth-tube bent near the middle; ovary and tube green, glabrous, rarely with a few minute scales; perianth-segments small, white; fruit depressed-globose, 3 cm. broad; pulp red, juicy.

Type locality: Moro-Queimado, Bahia, Brazil.

Distribution: Very common in the brush country of Bahia, Brazil, but not found in dry parts of that state.

Labouret (Monogr. Cact. 365. 1853) gives *Cereus pentalophorus* Labouret and *Cereus pentagonus glaucus* Morel as synonyms.

Fig. 31.—Cephalocereus pentaedrophorus.

Fig. 32.—Fruit of Cephalocereus pentaedrophorus. ×0.6.
Fig. 33.—Flower of same. ×0.6.

This species differs from its relatives in its smaller flowers and fruits, in the flower-tube being more or less curved or sometimes abruptly bent, and in the flowering areoles never producing long hairs or wool. The plant, although widely distributed in Bahia, is not found in the dry parts where other cacti are common, but prefers the borders of the deserts, growing usually as solitary individuals surrounded by bushes and small trees. The stems, which are erect and usually unbranched, project about the surrounding vegetation. They are of a vivid glaucous-blue color and thus in striking contrast to their surroundings.

Plate IV, figure 1, shows the top of a flowering plant in the collection of the New York Botanical Garden. Figure 31 is from a photograph taken in Bahia by Paul G. Russell in 1915; figure 32 shows a fruit collected by Dr. Rose at Machado Portella, Brazil; figure 33 shows a flower.

8. **Cephalocereus polylophus** (De Candolle) Britton and Rose, Contr. U. S. Nat. Herb. **12**: 419. 1909.

   Cereus polylophus De Candolle, Mém. Mus. Hist. Nat. Paris **17**: 115. 1828.
   Pilocereus polylophus Salm-Dyck, Cact. Hort. Dyck. 1844. 24. 1845.

Erect, with simple stems 10 to 13 meters high, green; ribs 15 to 18; areoles small, 1 cm. apart or less, bearing white felt but no wool; spines 7 or 8, yellow, straight, spreading; central spine single, longer than the others; flowers 4 to 5 cm. long, about 3 cm. broad at top, narrowly funnelform; free part of tube 6 to 8 mm. long with ridges down the inside; stamens included, inserted on the throat; filaments about 5 mm. long, red; inner perianth-segments probably red, broad and short, rounded at apex; ovary somewhat tuberculate; scales small, without felt, wool, or hairs in their axils; scales of flower-tube small, acute, spreading, with the tip reflexed.

Type locality: Mexico.
Distribution: Eastern Mexico.

In 1909 Dr. C. A. Purpus sent Dr. Rose a small plant labeled *Pilocereus polylophus* which is probably this species. It is now only about 30 cm. high and may be briefly described as follows: Ribs 14, strongly notched below the areoles; areoles white-felted; spines 3 to 6, at first brown, becoming white, acicular, about 1 cm. long.

The flower of this species is not typical for the genus. We have never seen it in bloom, but it did flower in the Missouri Botanical Garden, August 24, 1905, and our description is based on photographs, specimens, and notes made by Mr. C. H. Thompson at that time. The plant is known in trade also as *Cereus nickelsii* and is a shy bloomer in cultivation. The name occurs in the Monatsschrift für Kakteenkunde for 1910 (**20**: 27).

FIG. 34.—Cephalocereus polylophus.

Cereus angulosus Stieber (Schumann, Gesamtb. Kakteen 175. 1897) belongs here.
Illustration: Bull. U. S. Dept. Agr. Bur. Pl. Industr. **262**: pl. 9, as *Pilocereus polylophus*.

Figure 34 is from a photograph of the plant in flower at the Missouri Botanical Garden in 1905, copied from Bulletin No. 262 of the Bureau of Plant Industry.

PLATE IV

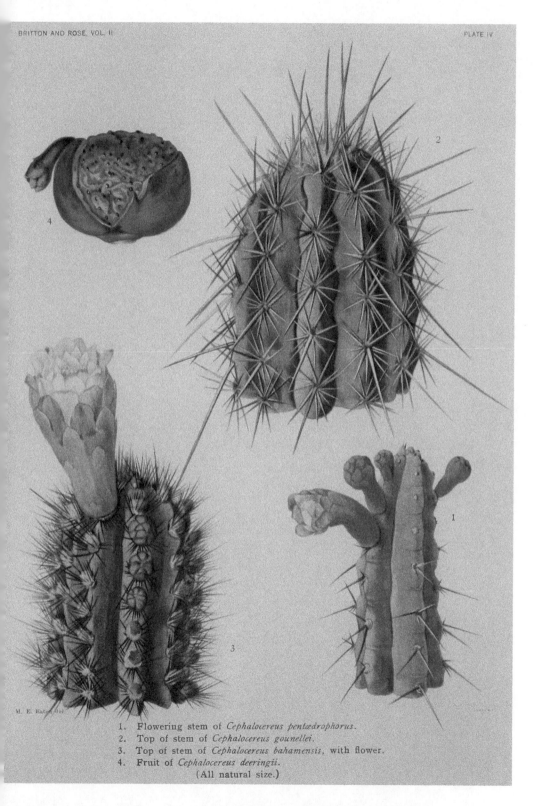

1. Flowering stem of *Cephalocereus pentædrophorus*.
2. Top of stem of *Cephalocereus gounellei*.
3. Top of stem of *Cephalocereus bahamensis*, with flower.
4. Fruit of *Cephalocereus deeringii*.
(All natural size.)

9. Cephalocereus euphorbioides (Haworth).

Cereus euphorbioides Haworth, Suppl. Pl. Succ. 75. 1819.
Cactus euphorbioides Sprengel, Syst. 2: 496. 1825.
Pilocereus euphorbioides Rümpler in Förster, Handb. Cact. ed.
2. 658. 1885.

Plant 3 to 5 meters high, columnar, usually simple; ribs 8, acute, somewhat crenate; areoles less than 1 cm. apart, white-felted; spines few, sometimes only 4 or 5, and then only 1 prominent, dark brown, porrect, about 1 cm. long; ovary 2 cm. long, spirally tuberculate; tubercles bearing triangular scales sparingly woolly in their axils, with 1 to 4 yellow spines; perianth-tube funnelform, campanulate, 4.5 cm. long; outer perianth-segments 15 mm. long, fleshy, reflexed, brown or reddish brown, the inner ones 2 cm. long, reflexed, rose-red; flowers diurnal.

Type locality: Not cited.
Distribution: The Index Kewensis says South America; Schumann says Brazil, not Mexico; Rümpler says Mexico and tropical America. Known to us only from cultivated specimens.

Rümpler refers here *Cereus conicus* Hort. Berol. (Pfeiffer, Enum. Cact. 97. 1837), which Pfeiffer states is from Mexico. *Cereus olfersii* Salm-Dyck (Hort. Dyck. 335. 1834) probably belongs here. The Theodosia B. Shepherd Company, in their Descriptive Catalogue for 1916, describes briefly *Cereus olfersii* from Brazil as follows: "A magnificent *Cereus*, exceedingly stout growth; color light blue; beautiful spines which are jet black and very long."

Cereus polylophus is very similar in its habit and flowers to this species. Although Haworth did not know its origin, it is usually stated to have come from Brazil. In habit it resembles *Cephalocereus fluminensis*.

Illustrations: Monatsschr. Kakteenk. **17**: 89, as *Pilocereus euphorbioides*; Rev. Hort. **57**: 279. f. 47, 48, as *Cereus olfersii*.

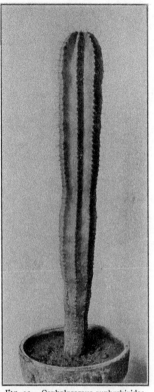

FIG. 35.—Cephalocereus euphorbioides.

Figure 35 is from a photograph of a plant in the New York Botanical Garden.

10. Cephalocereus russelianus (Otto) Rose, Stand. Cycl. Hort. Bailey 2: 715. 1914.

Cereus russelianus Otto in Salm Dyck, Cact. Hort.
Dyck. 1849. 201. 1850.
Pilocereus russelianus Rümpler in Förster, Handb.
Cact. ed. 2. 682. 1885.

Often tree-like, up to 8 meters high, with a much branched top and a definite woody trunk, 2 meters long and 2.5 dm. in diameter; branches elongated, nearly erect, dark green; ribs 4 to 6, stout, with prominent horizontal creases on the sides; areoles 1 to 2 cm. apart, large and circular, when young bearing white wool sometimes 1 to 1.5 cm. long; spines 8 to 14, at first dark brown but in age becoming gray except at the tips, 1 to 1.5 cm. long; flowers nocturnal, 7 to 9 cm. long, appearing from areoles anywhere on the branches or even from the base of the old trunk, cutting off after anthesis at the top of the ovary; top of unopened flower tuberculate; upper scales and outer perianth-segments broad, obtuse, thick, fleshy, pinkish; inner perianth-segments narrow, almost

FIG. 36.—Flower of Cephalocereus russelianus. ×0.7.
FIG. 37.—Fruit of same. ×0.7.

linear, cream-colored, erect; stamens numerous, included; ovary naked or nearly so, oblong, olive-green; fruit crowned by the persisting style, salmon-colored, about 6 cm. long, when fully mature splitting from top to bottom, exposing the white juicy pulp.

Type locality: La Guayra, Venezuela.

Distribution: Northern Venezuela and Colombia.

Collected by Dr. and Mrs. J. N. Rose in 1915, on the mountains between La Guayra and Carácas (No. 21828) and again near Puerto Cabello, Venezuela (No. 21859).

This species was also collected by William R. Maxon in April 1906, at Puerto Colombia, Colombia, and in June of the same year at the same place by H. Pittier, and by John G. Sinclair in 1914, at Santa Marta, Colombia. Mr. Maxon's plant was confused with *Cephalocereus colombianus,* based on material collected by Mr. Pittier from the State of Cauca.

Illustration: Möllers Deutsche Gärt. Zeit. **25**: 473. f. 5, No. 5, as *Pilocereus russelianus.*

Figure 38 is from a photograph taken by Mrs. J. N. Rose above Carácas, Venezuela, in 1916; figure 39 is from a photograph of the top of a plant collected by J. G. Sinclair at Santa Marta; figure 36 shows the flower of the Carácas plant and figure 37 its fruit.

FIG. 38.—Cephalocereus russelianus.   FIG. 39.—Cephalocereus russelianus.

11. Cephalocereus gounellei (Weber).

Pilocereus gounellei Weber in Schumann, Gesamtb. Kakteen 188. 1897.
Cereus setosus Gürke in Ule, Monatsschr. Kakteenk. **18**: 19. 1908. Not Loddiges. 1832.
Pilocereus setosus Gürke, Monatsschr. Kakteenk. **18**: 52. 1908.

Low, 1 to 2 or rarely 3 meters high, much branched and spreading to as much as 5 meters in diameter, often with a definite woody trunk up to 12 cm. in diameter; lower branches at first spreading or creeping, the tips ascending or even erect, pointed, the upper branches horizontal; ribs 10 or 11, stout, acute, more or less tuberculate; areoles large, 1.5 cm. in diameter; flowering areoles with many long white hairs covering the flower-buds; radial spines 15 to 24, widely spreading, brown; central spines 4 to 6, subulate, much stronger than the radials, sometimes 10 cm. long; perianth tubular to funnelform, white, 7 to 9 cm. long, glabrous, the limb shorter than the tube; stigma-lobes 15 to 18; ovary glabrous; fruit purplish, depressed-globose.

Type locality: Certão, Pernambuco, Brazil.

Distribution: Semi-arid parts of Pernambuco and Bahia, Brazil.

This species is very common in the dry parts of Bahia and Pernambuco, where it is known as chique-chique. The town Chique Chique on the São Francisco River takes its name from this plant. Several collections of this plant were made by Dr. Rose in Bahia in 1915 (Nos. 19945, 19846, and 19289).

FIG. 40.—Cephalocereus gounellei.

The perianth is relatively longer and narrower than that of other species of this genus.

Illustrations: Monatsschr. Kakteenk. 18: 21, as *Cereus setosus;* Vegetationsbilder 6: pl. 15, as *Pilocereus setosus.*

Plate IV, figure 2, shows the top of a plant collected by Dr. Rose near Joazeiro, Bahia, in 1915. Figure 40 is a nearby view of a good-sized plant taken by P. H. Dorsett in northern Bahia, Brazil, in 1914.

FIG. 41.—Flower of C. zehntneri. ×0.6.

12. Cephalocereus zehntneri sp. nov.

Low, much branched, and spreading; branches 3 to 4 cm. in diameter, about 9-ribbed, more or less tubercled; areoles 1 to 2 cm. apart, long-hairy when young; spines often 30 or more, only slightly spreading, all acicular and bright yellow, the centrals similar to the radials or a little longer, the longer ones 3 to 4 cm. long; flowering areoles producing with the flowers long tufts of white wool; flowers slender, tubular, 6 to 7 cm. long, white to light cream-colored; inner perianth-segments oblong, obtuse; style slender, glabrous, cone-shaped at base; ovary naked.

Collected by Leo Zehntner, from the Serra de Tiririca, Bahia, Brazil, November 1917. It is called chique-chique das pedras and is similar in habit to the one from Joazeiro described above. Dr. Zehntner says, however, that it prefers a rocky habitat while the common chique-chique is found on sandy ground and this statement is in accordance with Dr.

Rose's observations. Dr. Zehntner also says it differs from the latter in its more numerous, finer spines, which are of a light orange-yellow color. We find, too, that the radial spines are less spreading, while the centrals are much like the radials, acicular, and not stout-subulate as in the other species, and the flowers appear to be smaller.

Figure 41 shows a flower from the type plant.

13. Cephalocereus leucostele (Gürke).

Cereus leucostele Gürke, Monatsschr. Kakteenk. **18**: 53. 1908.

Plants normally simple, 2 to 5 meters tall, 4 to 8 cm. in diameter, the joints surrounded by peculiar bands or collars of long bristles; ribs 13 to 18, low; spines numerous, white, acicular, the lateral ones spreading, the central much longer, 3 to 5 cm. long; flowers borne in a mass of wool at top of plant but fruit becoming lateral by the prolongation of the stem; perianth slightly curved

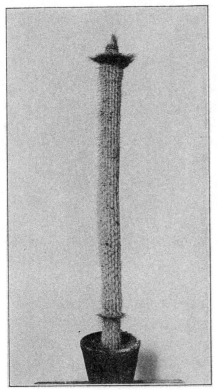

FIG. 42.—Cephalocereus leucostele.

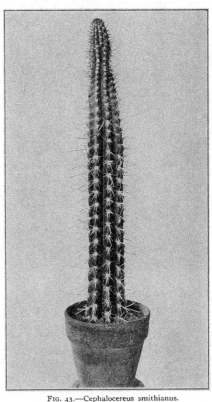

FIG. 43.—Cephalocereus smithianus.

downward, dull green, with a few small scales, 7 cm. long, opening in the early evening; perianth-segments short, waxy, white, tightly recurved; stamens numerous, included; filaments white, the upper cluster thickly set all over the long throat, very short; the lower cluster few, fixed at top of short tube proper, longer than the others, bent in just above their bases forming a knee and pressing against the style; space between the two clusters of stamens short but definite; perianth-tube proper 1.5 cm. long; style slender, white, pressed against the upper part of tube; anthers dehiscing soon after the flower expands, appressed against the tube; fruit smooth, longer than broad, 5 cm. long, bluish green, the rind thick, the pulp white; seeds black, tuberculate.

Type locality: Calderão, Bahia.

Distribution: Deserts of southern Bahia, Brazil.

A cutting, received from the Berlin Botanical Garden with the name *Cereus albispinus* Salm-Dyck, is strikingly similar to *Cephalocereus leucostele*.

In developing, the perianth carries flecks of wool with it from the dense white cushion at the areole; the perianth-tube bears several distant scales passing into the numerous outer, green, obtuse segments; inner perianth-segments about 25, ovate, white, acute, firm in texture, reflexed-spreading, about 12 mm. long; stigma-lobes pale yellow, slightly exserted when the perianth is fully expanded; stamens unequal in length. Dr. Rose collected living and herbarium specimens in Bahia in 1915 (No. 19902).

Illustration: Bot. Jahrb. Engler **40**: Beibl. **93**: pl. 5, as *Cereus leucostele*.

Figure 42 is from a photograph of a plant brought to the New York Botanical Garden from Machado Portella, Bahia, by Dr. Rose in 1915; figure 44 shows a flower of this plant and figure 45 its fruit.

FIG. 44.—Flower of Cephalocereus leucostele. ×0.7.
FIG. 45.—Fruit of same. ×0.7.

FIG. 46.—Flower of C. smithianus. ×0.7.
FIG. 47.—Fruit of same. ×0.7.

14. Cephalocereus smithianus sp. nov.

Stems weak and slender, 4 to 7 cm. in diameter, simple or much branched, erect or more or less clambering; ribs 9 to 11, low and rounded, sometimes constricted between the areoles; areoles rather large, felted; radial spines short, white, acicular, 1 cm. long or less; central spines several, nearly porrect, the longest ones 3 to 4 cm. long, at first black, in age black only at tips; flower 6 to 8 cm. long, 4 cm. broad across the mouth, with a short funnelform tube bearing a few broad ovate scales with reddish tips; inner perianth-segments short, rounded, white; ovary with a few minute scales; fruit ovoid, 3 to 4 cm. in diameter, red, splitting on one side when mature; areoles on the fruit each represented by a horizontal line 8 mm. long, subtended by a minute brown scale; pulp white; seeds black.

Collected by Dr. and Mrs. J. N. Rose just below Zig Zag, between La Guayra and Carácas, Venezuela, October 25, 1916 (No. 21889, type) and by Dr. Rose and Major C. C. Smith near Puerto Cabello, Venezuela, October 28 (No. 21852); also by Dr. Britton, Mr. W. G. Freeman, and Professor T. E. Hazen on Patos Island, Trinidad, a few miles from the Venezuelan Coast, March 13, 1920 (No. 532).

In form its flower is not quite typical of the genus.

This species is named for Major Cornelius C. Smith, U. S. Army, who accompanied and assisted Dr. Rose during some of his excursions in northern Venezuela in 1916.

Figure 43 is from a photograph of a plant brought by Dr. Rose to the New York Botanical Garden from Puerto Cabello, Venezuela, in 1916; figure 46 shows the flower and figure 47 the fruit, collected by Dr. Rose between Carácas and La Guayra.

15. Cephalocereus bahamensis Britton, Contr. U. S. Nat. Herb. 12: 415. 1909.

Cereus bahamensis Vaupel, Monatsschr. Kakteenk. 23: 23. 1913.

Plant 3 to 4 meters high, often 20 cm. thick at the base; branches divergent-ascending, 7 to 9 cm. thick, dull green, not pruinose, 10 or 11-ribbed, the ribs blunt or acutish, rather higher than wide; areoles 1 to 1.5 cm. apart; spines 15 to 20, acicular, radiately spreading and ascending, grayish brown to yellowish brown when old, 1 to 1.5 cm. long, the young ones yellowish with darker bases, the uppermost 2.5 to 3 cm. long; wool very short, shorter than the spines, or none; flower 5 to 6 cm. long, brownish outside, the tube bluish; inner perianth-segments creamy white, tinged with pink, acute; style pale greenish white, sometimes slightly exserted; fruit depressed-globose, 3 to 4 cm. in diameter.

Type locality: Frozen Cay, Berry Islands, Bahamas.

Distribution: Bahamas.

Illustration: Journ. N. Y. Bot. Gard. 11: f. 20.

Plate IV, figure 3, shows a cutting of the type plant which flowered in the New York Botanical Garden July 24, 1912. Figure 48 is from a photograph of the type plant in flower; figure 49 is from a photograph taken by Dr. Paul Bartsch on Andros Island.

FIG. 48.—Cephalocereus bahamensis.   FIG. 49.—Cephalocereus bahamensis.

16. Cephalocereus deeringii Small, Journ. N. Y. Bot. Gard. 18: 201. 1917.

Plant slender, often becoming 10 meters tall, the stem erect, simple or with few erect, short or elongated fastigiate branches which are ascending or erect and appressed to the main stem, the branches deep green, but sometimes rather light, usually 10-ribbed, sometimes 9-ribbed; areoles copiously short-hairy, the hairs rather persistent; spines acicular, 25 to 31 together, the longer ones 1 cm. long or more; flowers opening in the afternoon, about 6 cm. long, elongate-campanulate, light green without; outer perianth-segments obovate, obtuse, rounded, or emarginate; inner perianth-segments 9 to 11 mm. long, clawless, oval, rounded at the apex, erose, scarcely narrowed at the base; anthers less than 2 mm. long; fruit much depressed, 3.5 to 4 cm. in diameter, dark red; seeds about 2 mm. long, shining.

A clump of plants of *Cephalocereus deeringii* on Lower Matacumbe Key, Florida.

Type locality: Lower Matecumbe Key, Florida.

Distribution: Rocky hammocks, Lower Matecumbe Key, Florida.

The plant was named for Charles Deering, whose deep interest in the botanical exploration of Florida and in the preservation of its hammocks from destruction and its rare native plants from extermination, enabled Dr. Small to rediscover, study, and satisfactorily determine the relationship of this plant.

Plants similar to those from Upper and Lower Matecumbe Key have been collected on Umbrella Key, which is a few miles north of Lower Matecumbe, and these plants represent, without much doubt, the same species.

Illustration: Journ. N. Y. Bot. Gard. **18**: pl. 206.

Plate IV, figure 4, shows a fruit from the type plant; plate V is from a photograph of the type colony of plants on Lower Matecumbe Key, taken by Dr. Small in May 1917. Figure 50 shows the flower and figure 51 the fruit with withering persistent corolla.

FIG. 50.—Flower of C. deeringii. ×0.7.
FIG. 51.—Fruit of same. ×0.7.

17. Cephalocereus robinii (Lemaire).

Pilocereus robinii Lemaire, Illustr. Hort. **11**: Misc. 74. 1864.
Cephalocereus bakeri Britton and Rose, Contr. U. S. Nat. Herb. **12**: 415. 1909.
Cereus bakeri Vaupel, Monatsschr. Kakteenk. **23**: 23. 1913.

Plant 3 to 8 meters high, branching near and above the base; branches ascending, 7 to 10 cm. thick, dull green, bright glaucous green when young; ribs 10 to 13, acutish; areoles 1 to 1.5 cm. apart, bearing short wool; spines 15 to 20, acicular, 1 to 2.5 cm. long, yellow when young, becoming gray, the centrals hardly different from the radials; flowering areoles close together; flowers brownish green, 5 cm. long, 3 cm. broad at widest part of throat, constricted at top of tube proper, alliaceous in odor; tube green and slightly glaucous; ovary and lower

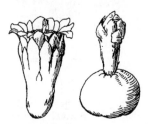

FIG. 52.—Flower of C. robinii. ×0.7.
FIG. 53.—Fruit of same. ×0.7.

FIG. 54.—Cephalocereus robinii.

part of tube with a few small scales; upper scales broadly ovate with bluish purple tips passing into greenish or cream-colored perianth-segments, the inner segments white; tube proper very short (1 cm. long or less); throat 2.5 cm. long, bearing stamens all over its surface; stamens white, included, the inner row appressed against the style; style creamy white, 6 cm. long, exserted beyond the perianth-segments; fruit 4 cm. in diameter, flattened above, dark wine-colored; seeds smooth, black, shining.

Type locality: Near Habana, Cuba.

Distribution: Coastal regions of Matanzas and Habana, Cuba.

This species was recorded by Grisebach as *Cereus royenii armatus.*

Illustrations: Journ. N. Y. Bot. Gard. **11**: 226, f. 28; Roig, Cact. Fl. Cub. pl. [3], f. 1, as *Cephalocereus bakeri.*

Figure 52 shows a flower of *Cephalocereus robinii,* and figure 53 its fruit; figure 54 is from a photograph of the plant obtained by Brother Léon at the type locality.

18. **Cephalocereus keyensis** Britton and Rose, Contr. U. S. Nat. Herb. **12**: 416. 1909.

> *Cereus keyensis* Vaupel, Monatsschr. Kakteenk. **23**: 23. 1913.

Plant 5 to 6 meters high, little branched, the branches almost erect, 5 to 6 cm. in diameter, the trunk up to 12 cm. thick; ribs 9 or 10, narrow, separated by deep grooves, bluish green, very glaucous; areoles 1 to 2 cm. apart, slightly elevated; spines about 15, acicular, yellow, diverging, 1.5 cm. long or less; wool very short, less than 1 mm. long, white, turning grayish; flowers brownish purple, narrowly campanulate, 6 cm. long, with a strong odor of garlic when opening in the late afternoon or evening, odorless the next morning; outer perianth-segments oblong-spatulate, bluntly pointed, the inner acutish; style scarcely exserted; fruit depressed-globose, reddish, 3.5 cm. thick, about 2 cm. high.

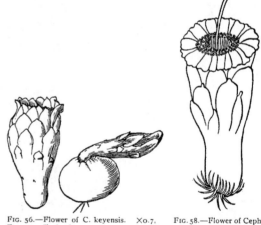

FIG. 55.—Cephalocereus keyensis.

FIG. 56.—Flower of C. keyensis.   ×0.7.
FIG. 57.—Fruit of same.   ×0.7.

FIG. 58.—Flower of Cephalocereus monoclonos.

Type locality: Hammock, Key West, Florida.

Distribution: Key West, Big Pine Key, and Boca Chica Key.

The plant is now very nearly exterminated on Key West, owing to the necessity for military purposes during the war with Germany of clearing the hammock in which it grew. Dr. Small succeeded in establishing it in flourishing masses in the cactus garden of Mr. Charles Deering at Buena Vista, Miami, Florida.

Illustration: Journ. N. Y. Bot. Gard. **10**: f. 25.

Figure 55 is from a photograph of the type plant taken by Marshall A. Howe; figure 56 shows its flower and figure 57 its fruit.

19. Cephalocereus monoclonos (De Candolle) Britton and Rose, Contr. U.S. Nat. Herb. **12**: 418. 1909.

Cereus monoclonos De Candolle, Prodr. **3**: 464. 1828.
Melocactus monoclonos Steudel, Nom. ed. 2. **2**: 122. 1841.

Stems simple, erect, tall, mostly 8-ribbed; ribs triangular in section, high, obtuse; spines 10 to 16, short, about equal, spreading; flowering areoles with only a few short hairs; flower-tube short and thick, bearing a few broad, pointed scales; perianth-segments white, numerous, spreading, retuse; stamens not exserted, numerous; style slender, long-exserted, with 5 or 6 stigma-lobes; fruit purple, globular, naked, thick-walled, with numerous shining seeds.

Type locality: Caribbean Islands, according to De Candolle.
Distribution: Probably Hispaniola.

There are probably two species of this genus on Hispaniola, although a half-dozen species have been described, two of them based on the same illustration. We believe that *C. monoclonos* is one of these. De Candolle, however, who first took up this species under *Cereus*, and who knew it only from Plumier's description and his somewhat conventionalized plate, gave its range as Caribbean Islands, but since most of Plumier's plates are supposed to be based on Hispaniola plants, it should be looked for on that island; however, explorations on both sides of the island in recent years have failed to bring it to light.

Illustration: Plumier, Pl. Amer. ed. Burmann, pl. 191.

Figure 58 is copied from Plumier's plate above cited.

20. Cephalocereus scoparius (Poselger) Britton and Rose, Contr. U. S. Nat. Herb. **12**: 419. 1909.

Pilocereus scoparius Poselger, Allg. Gartenz. **21**: 126. 1853.
Cereus scoparius Berger, Rep. Mo. Bot. Gard. **16**: 63. 1905.

"Arborescent; much branched, 20 to 25 feet high. Trunk a foot and more in diameter. Branches often very long, 2 to 3 inches in diameter. The younger branches, which have not yet borne flowers, are somewhat different from the older ones which have borne flowers. The former (younger) have 12 to 15 ribs. Ribs blunt, the furrows tolerably sharp. Areoles 8 to 12 lines apart, naked, somewhat thickened and protruding, close under the areole a strongly marked horizontal impression, through which the ribs appear serrated. Radial spines 5, somewhat bent downward, 2½ to 4 lines long. Central spine 1, stout, sharp, bent upward, blackish when young, later whitish, 1 inch long. The latter (older) flower-bearing branches are usually thinner, having 20 to 25 ribs; these are lower, blunter, and much closer together. Areoles very thick. Radial spines 5 to 7, central spine 1, all of the spines 10 to 15 lines long, bristle-like, brown. Flowers very sparse, small, almost campanulate (bell-shaped) reddish. Fruit red, of the size of a hazel-nut. Seeds large, black, shiny." This is a translation of the original description.

Type locality: Near La Soledad (near Vera Cruz, Mexico).
Distribution: Known only from the type locality.

Pilocereus sterkmannii Hortus is an unpublished name cited in the synonymy of *P. scoparius* by Schumann (Gesamtb. Kakteen 179. 1897).

We know the plant only from description. Schumann (Gesamtb. Kakteen 179) states that the flowering tops have up to 25 ribs.

21. Cephalocereus moritzianus (Otto).

Cereus moritzianus Otto in Pfeiffer, Enum. Cact. 84. 1837.
Pilocereus moritzianus Lemaire, Illustr. Hort. **13**: under pl. 469. 1866.

Tree-like, up to 10 meters high, sometimes with 50 ascending branches, green or bluish; ribs 7 to 10, obtuse, separated by acute intervals; areoles 10 to 12 mm. apart, all white-woolly at first; flowering areoles with tufts of wool 1 cm. long or longer; spines slender, at first brownish, rigid, straight, 1 to 3.5 cm. long; radial spines 6 to 8; central spines 3; flowers 5 cm. long, the outer perianth-segments broad, short, the inner white, obtuse; fruit red, depressed, naked, 4 to 5 cm. broad.

Type locality: La Guayra, Venezuela.
Distribution: Venezuela; northwestern mainland and Bocas Islands of Trinidad; Tobago.

This species is common above La Guayra and about Puerto Cabello, Venezuela, especially about the latter place, where it is the most common cactus seen, being abundant both on the hills and on the flats near the sea. Its branches are often overrun by orchids, vines, and bromeliads.

It is abundant on Monos, Chacachacare, and Patos Islands of Trinidad, inhabiting rocky hillsides and cliffs, and varying from slender, light green, and simple-stemmed in sunny situations to stout, dark green, and much branched in woodlands.

Cereus pfeifferi Parmentier (Pfeiffer, Allg. Gartenz. 5: 370. 1837) is referred to the synonymy of *Cereus moritzianus* by Labouret (Monogr. Cact. 344. 1853), who also states on the same page that in Monville's Catalogue is indicated the variety *C. moritzianus pfeifferi*. As *Cereus pfeifferi* is supposed to have originally come from Buenos Aires, it is more likely to be a true *Cereus.*

FIG. 59.—Flower of Cephalocereus moritzianus.
FIG. 60.—Fruit of same. Both ×0.7.

FIG. 61.—Cephalocereus moritzianus.

Figure 61 is from a photograph taken by Mrs. J. N. Rose near Puerto Cabello, Venezuela, in 1916; figure 59 shows the flower of this plant; figure 60 a fruit of same.

22. Cephalocereus arrabidae (Lemaire).

Pilocereus arrabidae Lemaire, Rev. Hort. **1862:** 429. 1862.
Cereus warmingii Schumann in Martius, Fl. Bras. **4²:** 204. 1890.
Pilocereus exerens Schumann in Engler and Prantl, Pflanzenfam. **3⁶ᵃ:** 181. 1894.
Cephalocereus exerens Rose, Stand. Cycl. Hort. Bailey **2:** 715. 1914.

Rather low but sometimes 3 meters high, often much branched at base, usually pale, somewhat glaucous; branches 6 to 10 cm. in diameter; ribs 6 to 8, high, obtuse; areoles rather close together, producing long hairs when young, but no tufts of hairs or wool at flowering time; spines 5 to 10, acicular to subulate, unequal, the longest up to 4 cm. long, brownish or sometimes yellowish; flowers 6 cm. long; inner perianth-segments white; fruit depressed, 6 cm. broad; seeds black, shining.

Type locality: Not cited.
Distribution: Along the sandy coast of Brazil.

The synonymy of this coastal species of Brazil is very complicated, for it has been confused with a Mexican species of uncertain relationship. An attempt is here made to account for the various names. Schumann took up the specific name *exerens* for it, basing it on *Cereus exerens,* an unpublished name of Link. *Pilocereus arrabidae* Lemaire seems to be the oldest definite name for the plant. This is not to be confused with *Cereus arrabidae* (Steudel, Nom. ed. 2. **1:** 333. 1840) as it has been in the Index Kewensis.

PLATE VI

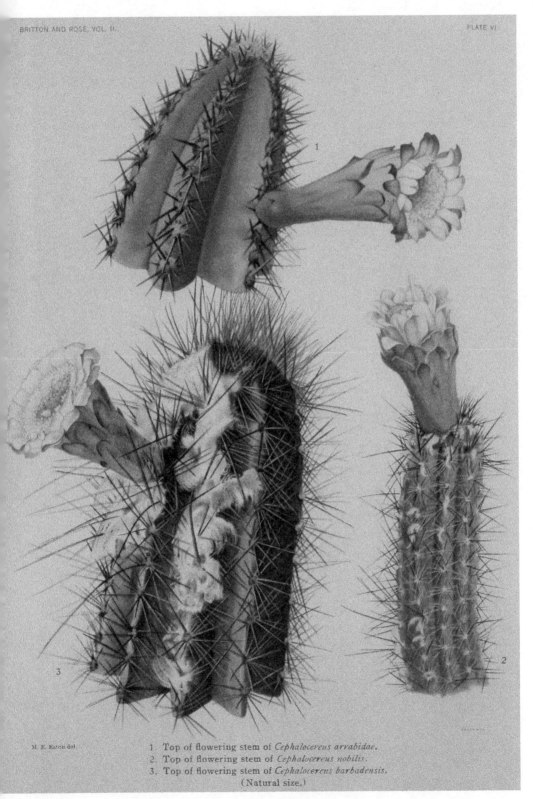

M. E. Eaton del.

1. Top of flowering stem of *Cephalocereus arrabidae*.
2. Top of flowering stem of *Cephalocereus nobilis*.
3. Top of flowering stem of *Cephalocereus barbadensis*.
(Natural size.)

Schumann refers here *Cereus virens* Pfeiffer (Enum. Cact. 99. 1837; *Pilocereus virens* Lemaire, Illustr. Hort. 13: Misc. 20. 1866), but Pfeiffer really did not propose a new name, although the plant he described may have been different from De Candolle's (Mém. Mus. Hist. Nat. Paris 17: 116. 1828), which came from Mexico, for the latter is definitely stated to have been sent by T. Coulter from there, and is described as a simple, light green, 5-ribbed plant; it may be a *Lemaireocereus*. Schumann refers *Cereus sublanatus* Salm-Dyck (Hort. Dyck. 337. 1834) here, but this reference is to be questioned. If the two are the same the name *sublanatus* must be taken up instead of *arrabidae*.

Cereus exerens Link (Pfeiffer, Enum. Cact. 99. 1837) was never described but given as a synonym of *Cereus virens*. *Cereus retroflexus* Pfeiffer (Allg. Gartenz. 3: 380. 1835) and *C. reflexus* Steudel (Nom. ed. 2. 1: 335. 1840) were given as synonyms of *C. tilophorus*. *Cereus ericomus*, given as a synonym of *Pilocereus exerens*, was given by Salm-Dyck (Cact. Hort. Dyck. 1849. 47. 1850) as a synonym of *C. virens*.

Illustrations: Schumann, Gesamtb. Kakteen f. 39, as *Pilocereus exerens;* Monatsschr. Kakteenk. 2: 41, as *Pilocereus virens;* Martius, Fl. Bras, 4²: pl. 40, as *Cereus macrogonus;* Vellozo, Fl. Flum. 5: pl. 18, as *Cactus hexagonus;* also pl. 19, as *Cactus heptagonus.*

FIG. 62.—Fruit of Cephalocereus arrabidae. ×0.7.

FIG. 63.—Cephalocereus arrabidae.

Plate VI, figure 1, shows a flowering joint of a plant brought by Dr. Rose to the New York Botanical Garden from Iguaba Grande, Brazil, in 1915. Figure 62 shows the fruit collected by Dr. Rose at Bahia in the same year; figure 63 is from a photograph taken by Paul G. Russell on Juparyba Island, Bay of Rio de Janeiro, Brazil, in the same year.

23. **Cephalocereus urbanianus** (Schumann) Britton and Rose, Contr. U. S. Nat. Herb. 12: 420. 1909.

Pilocereus urbanianus Schumann, Gesamtb. Kakteen 193. 1897.
Cereus urbanianus Berger, Rep. Mo. Bot. Gard. 16: 63. 1905. Not Gürke and Weingart. 1904.

Simple and columnar or branching at base, sometimes 4 meters high, 3 cm. in diameter; branches 4 to 5 cm. in diameter, woolly at apex; ribs 8 to 12, obtuse; spines 10 to 13, spreading, stiff but flexuous; central spines distinct from the radials; flowers on one side of the stem, the flowering areoles bearing long brown wool and bristle-like spines often 4 to 6 cm. long; flower 3 to 4 cm. long; ovary bearing small scales; fruit depressed, 3 cm. broad; seeds black, smooth, shining.

Type locality: Guadeloupe Island, West Indies.
Distribution: Guadeloupe, Martinique, and, apparently, Grenada.

We have not seen this species alive; it is based upon Père Duss's No. 3506, of which material is preserved in the herbaria of the New York Botanical Garden and the United States National Museum. Specimens from Woodlands, St. George's, Grenada (W. E Broadway, No. 1766) appear to be referable to this species.

24. **Cephalocereus nobilis** (Haworth) Britton and Rose, Contr. U. S. Nat. Herb. **12**: 418. 1909.

Cereus nobilis Haworth, Syn. Pl. Succ. 179. 1812.
Cactus strictus Willdenow, Enum. Suppl. 32. 1813. Not C. strictus Haworth. 1803.
Cactus haworthii Sprengel, Syst. **2**: 495. 1825.
Cactus niger Salm-Dyck in Sprengel, Syst. **2**: 495. 1825.
Cereus strictus De Candolle, Prodr. **3**: 465. 1828.
Cereus haworthii De Candolle, Prodr. **3**: 465. 1828.
Cereus aureus Salm-Dyck in De Candolle, Prodr. **3**: 465. 1828.
Cereus curtisii Otto in Pfeiffer, Enum. Cact. 81. 1837.
Cereus lutescens Salm-Dyck in Pfeiffer, Enum. Cact. 84. 1837.
Cereus violaceus Lemaire, Cact. Gen. Nov. Sp. 57. 1839.
Cereus nigricans Lemaire, Cact. Gen. Nov. Sp. 57. 1839.
Pilocereus curtisii Salm-Dyck, Cact. Hort. Dyck. 1844. 24. 1845.
Pilocereus consolei Lemaire, Rev. Hort. 1862: 427. 1862.
Pilocereus haworthii Console in Lemaire, Rev. Hort. 1862: 428. 1862.
Pilocereus nigricans Sencke in Lemaire, Illustr. Hort. **13**: Misc. 20. 1866.
Pilocereus lutescens Rümpler in Förster, Handb. Cact. ed. 2. 675. 1885.
Pilocereus strictus Rümpler in Förster, Handb. Cact. ed. 2. 687. 1885.
Pilocereus nobilis Schumann in Engler and Prantl, Pflanzenfam. 3⁶ᵃ. 181. 1894.
Pilocereus strictus consolei Schumann, Gesamtb. Kakteen 190. 1897.

Plant much branched and spreading, the ultimate branches slender, erect, green, shining when young, not at all glaucous, 8 to 10-ribbed; areoles about 1 cm. apart, at first producing only a little wool and this appressed against the ribs, but wool in flowering areoles very dense but short, white; spines up to 3.5 cm. long, acicular, at first yellow, soon brown; flower-buds obtuse or nearly truncate; flowers 4 to 6 cm. long; upper scales and outer perianth-segments broad, rounded at apex; inner perianth-segments purple; style exserted; fruit depressed-globose.

Type locality: West Indies.
Distribution: St. Christopher to Grenada.
The plant has escaped from cultivation on the island of St. Thomas, and has been grown at Hope Gardens, Jamaica.

As to the locality for *C. curtisii*, Pfeiffer (Enum. Cact. 81. 1837) gives Grenada, following Hooker, who originally published it as from Grenada, while Pfeiffer and Otto (Abbild. Beschr. Cact. **1**: pl. 11) give New Granada also as its original habitat.

Cereus aureus pallidior Salm-Dyck (Hort. Dyck. 63. 1834), given by name only, is referred by Pfeiffer (Enum. Cact. 84. 1837) as a synonym of *C. lutescens* Salm-Dyck.

Cereus mollis and *C. nigricans* (Pfeiffer, Enum. Cact. 83. 1837) and *C. mollis nigricans* (Labouret, Monogr. Cact. 349. 1853) were given as synonyms of *Cereus strictus*. *C. niger** Salm-Dyck (Observ. Bot. **3**: 4. 1822) and *C. niger gracilior* Salm-Dyck (Hort. Dyck. 63. 1834) may also belong here. *Cereus trichacanthus* (Salm-Dyck, Cact. Hort. Dyck. 1849. 46. 1850) was given as a synonym of *Cereus lutescens* and *Pilocereus trichacanthus* (Rümpler in Förster, Handb. Cact. ed. 2. 675. 1885) of *Pilocereus lutescens*. Here also, *Echinocereus trichacanthus*, only a name, is referred by the Index Kewensis.

Illustrations: Rep. Mo. Bot. Gard. **16**: pl. 4, f. 2, as *Cereus strictus;* Pfeiffer and Otto, Abbild. Beschr. Cact. **1**: pl. 11, as *Cereus curtisii;* Curtis's Bot. Mag. **59**: pl. 3125, as *Cereus royeni;* Krook, Handb. Cact. 92, as *Pilocereus.*

Plate VI, figure 2, shows a flowering branch of a plant in the collection of the New York Botanical Garden. Figure 64 is from a photograph of the plant growing on St. Thomas, taken by W. R. Fitch in 1913.

25. **Cephalocereus barbadensis** sp. nov.

Plant light green, tall, 3 to 6 meters high, with ascending or spreading columnar branches; ribs usually 8 or 9, high, separated by acute intervals; areoles 1 cm. apart; spines acicular, 1 to 4 cm.

---

* *Pilocereus niger* is different from *Cactus niger* Salm-Dyck and is not a synonym of *Cereus nobilis*, although it was referred to *P. strictus* by Schumann (Gesamtb. Kakteen 189. 1897). Neumann described *Pilocereus niger* (Rev. Hort. II. **4**: 289. 1845), a new species based on plants sent from Mexico by M. Ocampo. This species in the Index Kewensis and also in Schumann's Monograph is attributed to Poiteau, one of the editors of the Revue, but the article is signed by Neumann and, therefore, he should be made the author for the name. The Index Kewensis also makes it a synonym of *Cereus niger*, which it is not, nor should it be referred to this plant as it is by Schumann.

long, numerous, light brown; flowering areoles confined to one side of the branch and near its top; sometimes only on 3 ribs, producing abundant, long, white wool; flowers 5 to 6 cm. long; tube short and thick, greenish below, red above; perianth-segments numerous, light pink, spreading, obtuse; stamens scarcely exserted, dull yellowish white; style included; fruit red, subglobose, 3.5 cm. in diameter; seeds minute, shining, black.

Collected by J. N. Rose in company with W. Nowell, of the Imperial Department of Agriculture for the West Indies, on Barbados, September 30, 1915 (No. 21181). This species is found only on the exposed hills near the ocean on the eastern side of the island.

Plate VI, figure 3, shows the top of the type specimen in the New York Botanical Garden in flower in 1916. Figure 65 is from a photograph of the same plant; figure 66 is from a photograph taken near Boscobel, St. Andrew, February 9, 1919, communicated by Sir Francis Watts.

FIG. 64.—Cephalocereus nobilis.

FIG. 65.—Cephalocereus barbadensis.

26. Cephalocereus millspaughii Britton, Contr. U. S. Nat. Herb. **12**: 417. 1909.

Cereus millspaughii Vaupel, Monatsschr. Kakteenk. **23**: 23. 1913.

Stem branched, 2 to 6 meters high, 20 cm. thick at the base, the branches nearly erect, 8 to 12 cm. thick, pale grayish green, pruinose, 8 to 13-ribbed; ribs acutish, about as wide as high or a little wider; areoles 1 to 2 cm. apart; spines about 20, acicular, widely radiating, 1 to 2 cm. long, or at the flower-bearing (upper) areoles 3 to 7 cm. long, the old ones gray-brown, the young ones yellow or yellowish brown, with darker bases; upper areoles on one side of the plant with large tufts of whitish wool 5 to 7 cm. long, often as long as the spines or longer; flowers greenish, 6 to 7 cm. long; tube obconic with a spreading limb, 6 to 7 cm. broad, slightly purple, a little glaucous; scales on ovary and flower-tube few, small, acute; inner perianth-segments waxy, rigid, white, 1.5 to 2 cm. long; style white; fruit depressed-globose, about two-thirds as long as thick, about 4 cm. in diameter.

FIG. 66.—Cephalocereus barbadensis.

FIG. 67.—Cephalocereus millspaughii.

Type locality: Cave Cay, Exuma Chain, Bahamas.

Distribution: Bahamas; Cays of northern Cuba.

Figure 67 is from a photograph, taken by Marshall A. Howe in 1907 on the island Mariguana, Bahamas; figure 68 represents the fruit of the type specimen.

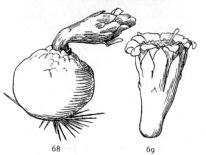

68   69

27. Cephalocereus swartzii (Grisebach) Britton and Rose, Contr. U. S. Nat. Herb. **12**: 420. 1909.

    Cereus swartzii Grisebach, Fl. Brit. W. Ind. 301. 1860.

FIG. 68.—Fruit of Cephalocereus millspaughii. ×0.7.
FIG. 69.—Flower of Cephalocereus royenii. ×0.7.

Tall, 2 to 7 meters high, often simple; branches obtuse at apex; ribs 10, obtuse, strongly indented between the areoles; spines 8 to 10, or in young plants 20 or more from an areole, the longer ones 2.5 cm. long, slightly spreading; flowers pinkish to greenish yellow, sometimes borne on all the ribs, usually near the tops of the branches, surrounded with masses of white hair and long bristles; perianth 5 to 6 cm. long, the inner perianth-segments obtuse; fruit depressed-globose, 3 cm. in diameter, perhaps larger.

Type locality: Jamaica.

Distribution: Southern side of Jamaica.

Cephalocereus swartzii, which is confined to the dry southern portions of Jamaica, has frequently been confused with *Lemaireocereus hystrix,* which is very commonly used as a hedge plant along the country roads about Kingston.

Schumann (Gesamtb. Kakteen 184. 1897) by mistake attributed the name *Pilocereus swartzii* to Grisebach.

Figure 70 is from a photograph obtained by Wm. Harris near Port Henderson, Jamaica.

28. Cephalocereus polygonus (Lamarck) Britton and Rose, Contr. U. S. Nat. Herb. **12**: 418. 1909.

> *Cactus polygonus* Lamarck, Encycl. **1**: 539. 1783.
> *Cereus polygonus* De Candolle, Prodr. **3**: 466. 1828.
> *Pilocereus plumieri* Lemaire, Rev. Hort. **1862**: 427. 1862.
> *Pilocereus schlumbergeri* Weber in Schumann, Gesamtb. Kakteen 186. 1897.
> *Pilocereus polygonus* Schumann, Gesamtb. Kakteen 196. 1897.

FIG. 70.—Cephalocereus swartzii.

Plants at first simple, but when old with large, much branched tops, 3 meters high or more; trunk erect, 1 to 1.5 meters long below the branches; branches elongated, erect or ascending, 5 to 13-ribbed; young growth, at least in some forms, very blue; ribs rather narrow, 2 cm. high or more, grooved on their sides; areoles closely set, often only 1 cm. apart, producing long tawny wool, longer than the short acicular spines; old areoles without wool, vigorous and producing very different spines from the new ones; first spines acicular or setaceous, 1 to 1.5 cm. long, yellow, becoming gray or darker by age; supplementary spines elongated, subulate, yellowish brown, 2 to 7 cm. long; flowering areoles very woolly; flowers 5 to 6 cm. long, white; perianth-segments rounded or somewhat acutish; fruit globular, 3 to 4 cm. in diameter; seeds numerous, small, 2 cm. long, smooth, shining.

Type locality: Santo Domingo.

Distribution: Dry parts of Hispaniola.

Illustration: Plumier, Pl. Amer. ed. Burmann, pl. 196.

Plate VII, figure 1, is from a photograph taken by Paul G. Russell near Azua, Santo Domingo, in 1913.

29. Cephalocereus gaumeri sp. nov.

Plant 6 meters high, light green, slender, often only 2 to 3 cm. in diam., but sometimes 6 cm., in diameter; ribs 8 or 9, 6 to 8 mm. high; areoles 6 to 10, bearing short felt and cobwebby hairs when young; flowering areoles bearing tufts of white wool 1 to 2 cm. long, 1 to 2 mm. apart; spines numerous, 15 to 25, acicular, 1 to 5 cm. long, yellowish brown when young; flowers "light green," 5 to 7 cm. long; scales on the ovary and lower part of the flower-tube few, minute, acute; scales on the upper part of the tube and outer perianth-segments broadly ovate, acute; inner perianth-segments oblong, acute; stamens included; style long-exserted; stigma-lobes 12; fruit depressed, brownish, somewhat ridged, 4.5 cm. long.

This species has been repeatedly collected by Dr. George F. Gaumer in Yucatan and has been distributed by him under various numbers. In 1918 he sent living plants to the New York Botanical Garden and these flowered the same year. This number (No. 23934) is made the type of the species.

Schott also collected this species in Yucatan and indicated it as a new species of *Cereus*, but this was never published. His sheet, now in the Field Museum of Natural History, bears drawings and paintings of the flowers and fruit.

30. **Cephalocereus chrysacanthus** (Weber) Britton and Rose, Contr. U. S. Nat. Herb. **12**:416. 1909.

Pilocereus chrysacanthus Weber in Schumann, Gesamtb. Kakteen 178. 1897.
Cereus chrysacanthus Orcutt, West Amer. Sci. **13**: 63. 1902.

Plant 3 to 5 meters high, branching near the base; branches erect or ascending, glaucous; ribs about 12; areoles about 1 cm. apart; spines 12 to 15, the longer ones 3 to 4 cm. long, at first golden yellow, becoming darker in age; flowers borne in definite zones on one side of the branch, accompanied by dense masses of long white hairs, nocturnal, 7 to 8 cm. long, rose-red; fruit smooth, reddish or purplish, about 3 cm. in diameter, the flesh red; seeds black.

Type locality: Near Tehuacán, Mexico.
Distribution: Puebla and Oaxaca, Mexico.
Illustration: MacDougal, Bot. N. Amer. Des. pl. 17, in part, as *Pilocereus chrysacanthus;* Möllers Deutsche Gärt. Zeit. **29**: 356. f. 12.

Plate VII, figure 2, is from a photograph taken by Dr. MacDougal near Esperanza, Mexico, in 1906.

31. **Cephalocereus maxonii** Rose, Contr. U. S. Nat. Herb. **12**: 417. 1909.

Cereus maxonii Vaupel, Monatsschr. Kakteenk. **23**: 23. 1913.

Plant 2 to 3 meters high, with few long branches, erect or nearly so, in mature plants the tops of the branches for about 30 cm. clothed with white hairs 4 to 5 cm. long; ribs 6 to 8, acute, pale blue and somewhat glaucous; areoles small; spines about 10, slender, yellow, the central single, 4 cm. long, all nearly hidden in flowering areoles by the long white hairs; flowers purple, 4 cm. long; ovary naked except for a few small scales; fruit 3.5 cm. broad, broader than high; seeds brownish, reticulate, with an oblique basal hilum.

FIG. 71.—Cephalocereus maxonii.   FIG. 72.—Cephalocereus piauhyensis.

Type locality: Near El Rancho, Guatemala.
Distribution: Guatemala.

This species, although discovered only a few years ago, has been repeatedly collected since and is now to be found in living collections. It is called organo in Guatemala.

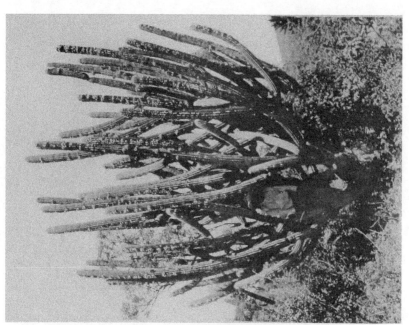

2. Large plant of *Cephalocereus chrysacanthus*.

1. Plants of *Cephalocereus polygonus*.

Illustration: Contr. U. S. Nat. Herb. **12:** pl. 64.

Figure 71 is from a photograph taken near Salama, Guatemala, by W. R. Maxon in 1905.

32. Cephalocereus piauhyensis (Gürke).

Cereus piauhyensis Gürke, Monatsschr. Kakteenk. **18:** 84. 1908.

Plant tree-like, 5 to 10 meters high; trunk woody, 3 to 5 dm. in diameter, with a smooth, nearly spineless bark; branches 20 to 100, slender, bluish green; ribs 13, low; areoles large, each flowering one bearing a tuft of long white hairs; spines numerous, yellowish brown, acicular, unequal, the longest 3 cm. long; flowers 3.5 to 4 cm. long, naked; fruit depressed, glaucous, 4 cm. broad, naked.

Type locality: Rocks of the Serra Branca, Piauhy, Brazil.

Distribution: On the dry hills in the caatinga along the São Francisco River in the States of Bahia and Piauhy, Brazil.

It resembles *Cephalocereus catingicola*, but has more slender branches, more ribs, smaller flowers, and smaller fruits. The trunk is woody, very heavy, and is often sawed into boards and used for making picture frames and the like. We have referred here the plant collected by Dr. Rose in Bahia without having seen the type of the species.

Figure 72 is from a photograph taken by Paul G. Russell east of Joazeiro, Bahia, Brazil, in 1915.

33. Cephalocereus lanuginosus (Linnaeus) Britton and Rose, Contr. U. S. Nat. Herb. **12:** 417. 1909.

Cactus lanuginosus Linnaeus, Sp. Pl. 467. 1753.
Cereus lanuginosus Miller, Gard. Dict. ed. 8. No. 3. 1768, as to name only.
Cereus crenulatus Salm-Dyck, Observ. Bot. 3: 6. 1822.
Cereus lanuginosus glaucescens Pfeiffer, Enum. Cact. 80. 1837.
Pilocereus crenulatus Rümpler in Förster, Handb. Cact. ed. 2. 655. 1885.
Pilocereus lanuginosus Rümpler in Förster, Handb. Cact. ed. 2. 672. 1885.

Often tall and tree-like, either nearly simple or much branched; branches elongated, 9 to 13-ribbed, bright blue, somewhat glaucous; ribs rounded when young, separated by acute intervals; spines acicular, light yellow when young; young areoles all woolly, the flowering ones bearing dense tufts of wool, but this not very long; flowering areoles confined to 2 to 4 ribs on the south side of the plant; flower-buds short, green, rounded at the apex; flowers opening in the early evening, 6 cm. long; outer perianth-segments short, green; inner perianth-segments ovate, white, short; stamens numerous, included; style rigid, white, slightly exserted; stigma-lobes white; fruit depressed, red, naked.

Type locality: Island of Curaçao.

Distribution: Curaçao, Aruba, Bonaire.

Cereus crenulatus gracilior Salm-Dyck (Hort. Dyck. 63. 1834) is only a mentioned name.

Illustrations: Loudon, Encyl. Pl. f. 6861, as *Cactus lanuginosus;* Rep. Mo. Bot. Gard. 16: pl. 4, f. 5, as *Cereus lanuginosus;* Hermann, Par. Botavus pl. 115, as *Cereus erectus*, etc; Monatsschr. Kakteenk. **12:** 56, as *Pilocereus lanuginosus.*

Figure 73 is from a photograph taken on Curaçao by Mrs. J. N. Rose in 1916.

34. Cephalocereus brooksianus Britton and Rose, Torreya **12:** 14. 1912.

Cereus brooksianus Vaupel, Monatsschr. Kakteenk. **22:** 66. 1912.

Plant 3 to 6 meters high, stout, much branched at base, bluish green, glaucous; ribs 8 or 9, obtuse; areoles closely set, in flowering specimens almost contiguous, bearing silky hairs when young and tufts of long white hairs at flowering ones; spines about 16, acicular, up to 3 cm. long, yellow, all somewhat similar, the upper ones in each areole ascending; flowers 5 to 6 cm. long, opening in the evening, odorless, somewhat flattened; tube stout, rigid, green, with only 2 or 3 small scales; inner perianth-segments about 10, rather rigid, broad, a little spreading; throat of flower wide; stamens very numerous, all included; filaments white, attached all over the long broad throat, 3 cm. long; tube proper very short, 8 mm. long or less; style white, rigid, 5 cm. long; ovary naked.

Type locality: Near Novaliches, about six miles south of Guantánamo, Cuba.

Distribution: Dry, rocky situations, provinces of Oriente and Santa Clara, Cuba.

Plate VIII, figure 1, is from a plant collected by Dr. Britton at Guantánamo Bay, Cuba, in 1909, which flowered at the New York Botanical Garden in 1913.

35. **Cephalocereus royenii** (Linnaeus) Britton and Rose, Contr. U. S. Nat. Herb. **12**: 419. 1909.

Cactus royenii Linnaeus, Sp. Pl. 467. 1753.
Cereus royenii Miller, Gard. Dict. ed. 8. No. 7. 1768.
Cereus fulvispinosus Haworth, Syn. Pl. Succ. 183. 1812.
Cactus fulvispinosus Sprengel, Syst. **2**: 496. 1825.
Cereus floccosus Otto in Pfeiffer, Enum. Cact. 81. 1837.
Cereus armatus Otto in Pfeiffer, Enum. Cact. 81. 1837.
Pilocereus floccosus Lemaire, Illustr. Hort. **13**: under pl. 470. 1866.
Cereus leiocarpus Bello, Anal. Soc. Esp. Hist. Nat. **10**: 276. 1881.
Pilocereus barbatus Rebut in Förster, Handb. Cact. ed. 2. 650. 1885.
Pilocereus royenii Rümpler in Förster, Handb. Cact. ed. 2. 682. 1885.
Pilocereus royeni armatus Salm-Dyck in Förster, Handb. Cact. ed. 2. 682. 1885.
Pilocereus strictus fouachianus Schumann, Gesamtb. Kakteen 190. 1897.
Pilocereus fulvispinosus Schumann, Gesamtb. Kakteen 196. 1897.
Pilocereus fouachianus Weber in Gosselin, Bull. Mus. Hist. Nat. Paris **10**: 386. 1904.
Cereus fouachianus Vaupel, Monatsschr. Kakteenk. **23**: 25. 1913.

FIG. 73.—Cephalocereus lanuginosus.　　　　FIG. 74.—Cephalocereus royenii.

Stout, 2 to 8 meters high or more, either branching near the base or with a short definite trunk up to 3 dm. in diameter; branches stout, erect or ascending, glaucous, green to blue; ribs 7 to 11, high; areoles close together; spines acicular, very variable, often only 1 cm. long, but sometimes 6 cm. long, yellow; young areoles bearing soft wool; flowering areoles producing tufts of long white hairs; flowers about 5 cm. long, greenish yellow to purplish; inner perianth-segments white, acute; fruit reddish or green, 5 cm. broad; pulp red; seeds black, shining.

Type locality: America, but no definite locality cited.

Distribution: Antigua to Anegada, St. Croix, St. Thomas, Culebra, Porto Rico, Mona, and Desecheo.

Philip Miller states that this species was sent to him from the British Islands of America in 1728. The combination *Cereus royenii* is generally credited to Haworth (1812), but it was first used by Miller in 1768, although the true *Cactus royenii* of Linnaeus may not be the one he actually described.

Cereus barbatus Wendland (Salm-Dyck, Cact. Hort. Dyck. 1844. 29. 1845) was given as a synonym of *Cereus floccosus*. *Cereus royenii armatus* Salm-Dyck (Walpers, Repert. Bot. 2: 276. 1843), and *C. royenii floccosus* Monville (Labouret, Monogr. Cact. 343. 1853) are given only as synonyms.

Cephalocereus fouachianus Quehl (Monatsschr. Kakteenk. 20: 39. 1910), name only, belongs here.

Cereus gloriosus (Pfeiffer, Enum. Cact. 80. 1837) was printed as a synonym.

Illustrations: Monatsschr. Kakteenk. 12: 56, as *Pilocereus royenii;* Journ. N. Y. Bot. Gard. 7: f. 4, this last as *Pilocereus* sp.

Figure 74 is from a photograph taken by Frank E. Lutz on Desecheo Island, in 1914; figure 69 represents a flower of the plant from Culebra Island, collected by Dr. Britton.

FIG. 75.—Cephalocereus robustus.

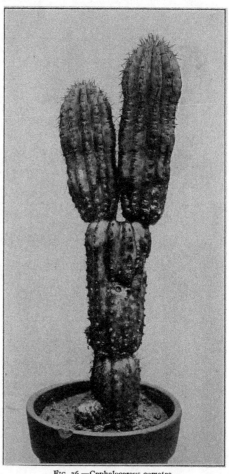

FIG. 76.—Cephalocereus cometes.

36. Cephalocereus robustus nom. nov.

Pilocereus ulei Schumann, Gesamtb. Kakteen Nachtr. 64. 1903. Not *Cephalocereus ulei* Gürke, 1908.
Cereus ulei Berger, Rep. Mo. Bot. Gard. 16: 70. 1905.

Tall, much branched, with a rather indefinite trunk, 3 to 7 meters high, pale whitish blue, roughish; ribs 8 or 9, high, separated by acute intervals; areoles closely set, with short dark spines and longer silky hairs; 3 of the ribs bearing flowers and their flowering areoles producing long, curly, white hairs, 5 to 6 cm. long; flower nocturnal, 5 cm. long, its tube proper 1 cm. long; perianth-segments acute, nearly white; stamens numerous, scattered all over the broad long throat, scarcely exserted; anthers purple; style slender, included; fruit 2 cm. in diameter; seeds minute, black, shining.

Type locality: Cabo Frio, Rio de Janeiro, Brazil.
Distribution: Coast of State of Rio de Janeiro, Brazil.

This species is common on the hills about Araruama Lake and near Cabo Frio, where it forms small forests and is the dominant feature of many landscapes. Dr. Rose and Señor Campos Porto obtained from São Pedro, near Cabo Frio, a living specimen (No. 20706) which flowered in the New York Botanical Garden in July 1916.

This plant is very unlike any of the other Brazilian species of this genus, of which there are at least three in the State of Rio de Janeiro.

Figure 75 is from a photograph of a specimen collected by Dr. Rose near São Pedro, Rio de Janeiro, in 1915.

37. Cephalocereus cometes (Scheidweiler) Britton and Rose, Contr. U. S. Nat. Herb. 12: 416. 1909.

Cereus cometes Scheidweiler, Allg. Gartenz. 8: 339. 1840.
Pilocereus jubatus Salm-Dyck in Förster, Handb. Cact. 356. 1846.
Cereus flavicomus Salm-Dyck, Cact. Hort. Dyck. 1849. 202. 1850.
Pilocereus flavicomus Rümpler in Förster, Handb. Cact. ed. 2. 658. 1885.

Erect, cylindric; ribs 12 to 15 (Schumann says 9 to 12), hardly tuberculate, obtuse; areoles close together, round; spines unequal, straight, spreading, 2 cm. long or less, flesh-colored or brownish, becoming gray; flowering areoles bearing masses of yellow hairs or wool, longer than the spines; neither the flowers nor the fruit known.

Type locality: Near San Luis Potosí, Mexico.
Distribution: State of San Luis Potosí, Mexico.

A small specimen in the New York Botanical Garden (No. 6710) has 12 ribs, with areoles bearing long white deciduous hairs and short spines, brownish at first, becoming gray.

Förster (Handb. Cact. 357. 1846) gave both *Pilocereus cometes* Mittler and *Cereus jubatus* Salm-Dyck, as synonyms of *P. jubatus.* See also Schelle, Handb. Kakteenk. 104. 1907.

Figure 76 is from a photograph of a plant in the collection of the New York Botanical Garden obtained from M. Simon of St. Ouen, Paris, France.

38. Cephalocereus leucocephalus (Poselger) Britton and Rose, Contr. U. S. Nat. Herb. 12: 417. 1909.

Pilocereus leucocephalus Poselger, Allg. Gartenz. 21: 126. 1853.
Pilocereus houlletii Lemaire, Rev. Hort. 1862: 428. 1862.
Pilocereus foersteri Lemaire, Illustr. Hort. 13: under pl. 472. 1866.
Cereus houlletii Berger, Rep. Mo. Bot. Gard. 16: 70. 1905.

Plants 2 to 5 meters high, branched below, the branches 3 to 15, erect or ascending; ribs usually 12, low; spines about 10 in each cluster, acicular, 12 to 20 mm. long; flowering areoles clustered on one side of the plant toward the top and producing an abundance of long white hairs (sometimes 4 to 10 cm. long); flowers and fruit not seen.

Type locality: Near Horcasitas, Sonora, Mexico.
Distribution: Sonora and southeastern Chihuahua, Mexico.

This species has been much misunderstood in recent years. The specific name *houlletii* is a clear synonym of the older name *leucocephalus.* Both were described as species of *Pilocereus* and based on plants from Sonora, Mexico, but no further Sonoran material being collected the name was transferred to an East Mexican species from Vera Cruz

(*Cephalocereus sartorianus*), and is sometimes applied to the Guatemalan species (*C. maxonii*). As a result of Dr. Rose's explorations in Sonora in 1910, additional material, both living and herbarium, was obtained, which enables us to reëstablish this species as Sonoran.

In 1908, Dr. Palmer sent photographs and specimens of a *Cephalocereus* from Batopilas, Chihuahua, which we believe may belong here.

Cereus foersteri (Sencke, Cat. 1861) and *Pilocereus marschalleckianus* (Zeissold, Cat. 1899) are given by Schumann as synonyms of this species. The latter is mentioned in Nicholson's Dictionary of Gardening (Suppl. 602. 1901) as having been introduced but very rare in cultivation.

Illustrations: Dict. Gard. Nicholson 3: f. 153; Gartenwelt 7: 291; Knippel, Kakteen pl. 29; Lemaire, Cact. f. 5, 6; Monatsschr. Kakteenk. 3: 145; 11: 76; 21: 37; 22: 133; Engler and Prantl, Pflanzenfam. 3⁶ᵃ: f. 59, A, B; Rev. Hort. 1862: f. 38 to 41; Förster, Handb. Cact. ed. 2. f. 89, 90; Rümpler, Sukkulenten f. 77; Rev. Hort. Belge 40: after 184, all as *Pilocereus houlletii;* Cact. Journ. 2: 5, as *P. houlletianus;* Rep. Mo. Bot. Gard. 16: pl. 4, f. 3, 4, as *Cereus houlletii.*

Figure 77 is from a photograph by E. Palmer, at Batopilas, Chihuahua, in 1908.

39. Cephalocereus sartorianus Rose, Contr. U. S. Nat. Herb. **12:** 419. 1909.

Plant 3 to 5 meters high or more, with nearly erect branches, these 7 to 10 cm. in diameter, bluish or bluish green; ribs (in the three individuals examined) 7, 2 cm. high, marked by a pair of grooves descending obliquely, one on each side, from each areole; areoles closely set, usually 1.5 cm. apart; radial spines at first 7 or 8, others apparently developing later; central normally one; all spines short, 1 cm. long or less, at first straw-colored, in age grayish; all areoles producing few or many cobwebby hairs; flowering areoles appearing on one side of the plant, in the specimen under observation on a single rib, and producing long white hairs 4 to 6 cm. long; flowers 6 to 8 cm. long, "dirty rose-red"; fruit red.

Type locality: State of Vera Cruz, Mexico.

Distribution: Vera Cruz, Mexico.

FIG. 77.—Cephalocereus leucocephalus.

In the original description, based on material sent by Dr. C. A. Purpus, we stated that the branches were "light or yellowish green, apparently not pruinose." The illustration in Blühende Kakteen referred to below, however, shows very blue and probably pruinose branches.

It seems to grow in thickets, and is very slender, with a few slender, nearly erect branches bearing large masses of wool at the top.

Illustration: Blühende Kakteen 2: pl. 79, as *Pilocereus houlletii.*

40. Cephalocereus palmeri Rose, Contr. U. S. Nat. Herb. **12:** 418. 1909.

Cereus victoriensis Vaupel, Monatsschr. Kakteenk. 23: 24. 1913.

Tall, 2 to 6 meters high, with 20 branches or more (often 5 to 8 cm. in diameter), dark green or when young glaucous and bluish; ribs 7 to 9, rounded on the edge, rather closely set, clothed from top downward for 20 to 30 cm. with long white hairs (4 to 5 cm. long) usually hiding the brown spines; radial spines 8 to 12, slender, the central one much longer than the others, 2 to 3 cm. long; areoles 1 cm. apart, scarcely woolly except toward the top; flowers 6 cm. long, somewhat tubular, purplish to brownish, the ovary without spines or hairs; fruit globular, about 6 cm. in diameter, naked but the surface somewhat warty; seeds black, shining, minutely pitted, 2 mm. long, oblique at bases.

Type locality: Near Victoria, Mexico.

Distribution: Eastern Mexico.

The spines of seedlings are yellow. This species flowered in the New York Botanical Garden in June 1918.

E. O. Wooton made a trip into eastern Mexico in 1919 and obtained a photograph of a large *Cephalocereus*, presumably this species. The plant was common on the coastal plain and extended the known range of this species northwards. Mr. Wooton's locality was on the Chamal Hacienda, about halfway between Matamoras and Tampico.

41. Cephalocereus tweedyanus sp. nov.

Sometimes only 1 to 2 meters high and much branched at base, or sometimes tall, 5 to 7 meters high and branched above, with a large woody trunk; branches 8 to 10 cm. in diameter, ascending or slightly spreading, bluish green when young, grayish green in age; ribs 7 to 9, obtuse; spines brown when young; radial spines several, 1.5 cm. long or less; central spines often solitary, porrect, 2 to 3 cm. long; flowering areoles bearing long white wool; flowers 7 cm. long; inner perianth-segments short, oblong, obtuse; scales and outer perianth-segments obtuse, purplish; fruit nearly globular, about 4 cm. in diameter, reticulated.

FIGS. 78 and 79.—Cephalocereus tweedyanus.

The species is based on two collections from widely separated localities in Ecuador, one being from the Pacific coast near sea-level, and the other from east of the coast range at an altitude of about 3,000 feet. The first was collected by J. N. Rose and George Rose in thickets near Santa Rosa, Province Del Oro, October 18, 1918 (No. 23494, type), and the other east of Ayapamba, same province, October 15, 1918 (No. 23454). This is the first species of *Cephalocereus* reported from Ecuador and is the most southern species known on the west coast of South America. It is dedicated to Mr. Andrew Mellick Tweedy, who assisted Dr. Rose in his Ecuadorean Expedition in 1918.

Figure 78 shows the type plant as it grows in thickets along the coast at Santa Rosa; figure 79 shows it as it grows in the open below Ayapamba, both from photographs by George Rose; figure 80 shows a flower and figure 81 a fruit collected by Dr. Rose near Ayapamba, Ecuador, in 1918.

42. Cephalocereus alensis (Weber) Britton and Rose, Contr. U. S. Nat. Herb. **12:** 415. 1909.

> *Pilocereus alensis* Weber in Gosselin, Bull. Mus. Hist. Nat. Paris **11:** 508. 1905.
> *Cereus alensis* Vaupel, Monatsschr. Kakteenk. **23:** 23. 1913.

Fig. 80.—Flower of C. tweedyanus. ×0.5.
Fig. 81.—Fruit of same. ×0.5.

Erect, sometimes 5 to 6 meters high, branching from the base; branches rather slender, spreading, 12 to 14-ribbed, the ribs somewhat tuberculate; spines 10 to 14, acicular, about 1 to 1.5 cm. long, brownish; flowering areoles on one side of the stem, developing white or yellowish hairs 5 cm. long; flowers light purple to purplish green; perianth-segments fleshy, usually rounded at apex; ovary nearly naked; fruit not known.

Type locality: Sierra del Alo, Mexico.

Distribution: Western Mexico.

The type of the species was collected by Léon Diguet and is preserved in the Museum of Paris, where it was studied by Dr. Rose in 1912. To this species we would refer specimens collected in Jalisco, Mexico, in 1892, by M. E. Jones.

Illustration: Bull. Soc. Acclim. France **52:** f. 16, as *Pilocereus alensis*.

43 Cephalocereus colombianus Rose, Contr. U. S. Nat. Herb. **12:** 416. 1909.

> *Cereus colombianus* Vaupel, Monatsschr. Kakteenk. **23:** 23. 1913.

Plant 5 to 6 meters high, more or less branched throughout, the branches nearly erect; ribs 8, obtuse; spines many, 25 at an areole or more, long and slender; wool of the areoles long and white, produced for 1 meter down from the top of the plant; flowers 7 cm. long, smooth, pale pink.

Fig. 82.—Cephalocereus colombianus.

Type locality: Venticas del Dagua, Colombia.

Distribution: Northwestern Colombia.

In our original description we referred here specimens from northern Colombia which we now include in *Cephalocereus russelianus*.

Illustrations: Contr. U. S. Nat. Herb. **12**: pl. 62, 63.

Figure 82 is from a photograph of the plant; figure 83 *a* shows a piece of the stem; figure 83 *b* a cross-section of the stem; figure 83 *c* a flower; and figure 83 *d* a flower-bud; all are from photographs of the type specimen by Henry Pittier.

44. Cephalocereus purpusii sp. nov.

Stems slender, 2 to 3 meters high, simple or more or less branched; branches green, erect, 3 to 4 cm. in diameter, usually simple; ribs 12, low, 5 to 6 mm. high, separated by narrow intervals; areoles closely set, 10 mm. apart or less on the lower part of the stem, but much closer together toward the top, on the young growth with long silky white hairs, but on old parts without hairs; spines acicular, swollen at base, 1 to 3 cm. long, bright yellow at first, in age gray.

Collected by Rose, Standley, and Russell at Mazatlan, Mexico, on the hills near the town overlooking the sea, March 31, 1910 (No. 13749, type), and also a short distance inland at Guadalupe, in thickets, April 18, 1910 (No. 14741).

This species differs from the other Mexican ones in having very slender stems. It is named for the veteran collector, Dr. C. A. Purpus, who writes that he collected the species several years earlier than above recorded. We have not, however, seen his specimens.

The plant is growing in the New York Botanical Garden, from Dr. Rose's collection at Mazatlan.

Fig. 83.—Cephalocereus colombianus.

45. Cephalocereus catingicola (Gürke).

Cereus catingicola Gürke, Monatsschr. Kakteenk. **18**: 54. 1908.

Tree-like, 3 to 10 meters high, with a short definite trunk and a large, much branched top, bluish green; ribs 4 or 5, separated by broad intervals; areoles large, woolly; spines yellow when young, numerous, unequal, the longest 3 cm. long; flowers 6 to 8 cm. long, 6 cm. broad when fully open, with a broad throat, opening in the evening, odorless; flower-tube short, about 1 cm. long, with broad scales near its top, these green with brownish margins; perianth-segments numerous, broad, short, white, stiff; anthers dehiscing soon after the flowers open; filaments short, the lower ones much longer than the upper one but all included, attached all over the throat; style stout, soon exserted, at first raised against the upper part of the throat, white; stigma-lobes at first white but pinkish the second day after anthesis; fruit broader than high, glaucous, 6 to 7 cm. broad, capped by the withered perianth; rind thick; pulp purple.

Type locality: In the caatinga of Bahia, Brazil.

Distribution: Common in the caatinga of Bahia.

Illustration: Monatsschr. Kakteenk. **18**: 55, as *Cereus catingicola*.

Plate VIII, figure 2, shows the top of a plant brought from Bahia by Dr. Rose in 1915.

PLATE VIII

A. E. Eaton del.

1. Top of flowering stem of *Cephalocereus brooksianus*.
2. Top of stem of *Cephalocereus catingicola*.
3. Top of stem of *Cephalocereus phaeacanthus*.
4. Flowering branch of *Leptocereus assurgens*.

(Natural size.)

46. Cephalocereus brasiliensis sp. nov.

Somewhat branching at base, 1 to 3 meters high, producing long, slender, weak branches, these at first erect, but soon spreading or reclining, bright green when old, but when young somewhat glaucous; branches at base nearly square in section, but toward the tip with 4, sometimes 5, prominent obtuse ribs; areoles very close together, with long white hairs longer than the spines; radial spines several, very short, brown, spreading, acicular; central spines generally solitary, porrect, 1 to 2 cm. long; flowers 5 cm. long; fruit globular.

Collected by Rose and Russell on the base of Corcovado, Rio de Janeiro, Brazil, July 10, 1915 (No. 20190). This plant is not uncommonly found with *Cephalocereus fluminensis*, but is not as abundant as that species.

In the open flats and valleys toward Cabo Frio, Brazil, similar plants occur, but they are stouter and usually erect; of these, flowers and fruit were not obtained (Rose, No. 20705).

Figure 84 is from a photograph taken by Paul G. Russell of the type plant.

Fig. 84.—Cephalocereus brasiliensis growing above Cephalocereus fluminensis.

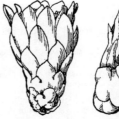

Fig. 85.—Flower of C. phaeacanthus. ×0.7.
Fig. 86.—Fruit of same. ×0.7.

47. Cephalocereus phaeacanthus (Gürke).

Cereus phaeacanthus Gürke, Monatsschr. Kakteenk. 18: 57. 1908.

Slender, usually branching at base, rarely branching above, more or less erect, often 4 meters high, the branches 4 to 9 cm. in diameter; ribs usually 13, low, narrow, bearing approximate areoles about 5 mm. apart, with acicular spines and small tufts of short white wool; spines numerous, when young yellowish brown, 1 to 1.5 cm. long; flowers 6 cm. long, slightly bent upward near the top of the tube, the limb 6 cm. broad when fully expanded; perianth-tube and ovary bearing several ovate scales; outer perianth-segments greenish brown; inner perianth-segments white, acute; upper series of stamens 2 cm. long; lower series of stamens 4 cm. long; filaments green; style white; fruit 1.5 cm. in diameter, smooth, somewhat tubercled; seeds 2 cm. long.

Type locality: Maracás, Bahia, Brazil.
Distribution: In thickets, State of Bahia, Brazil.

We have placed this species near the end of the genus, for it is very unlike the other species and may not be congeneric with them. It has very slender stems, low ribs, no long hairs at the flowering areoles, and a bent flower with a very small, flattened ovary.

Plate VIII, figure 3, shows the top of a plant brought by Dr. Rose from Toca da Onca, Brazil, in 1915. Figure 85 shows the flower, and figure 86 shows the fruit and withering perianth.

48. Cephalocereus ulei Gürke, Monatsschr. Kakteenk. **18**: 85. 1908.

"Trunk upright, strongly branched, columnar, several meters high, and in this instance 7 cm. in diameter. Ribs 18 to 20, blunt, separated by deep furrows from each other and rather deeply crenate, 8 to 9 mm. high, 7 to 8 mm. wide at the base, semi-elliptic and with rounded angles; areoles 10 to 12 mm. apart, elongated, 4 to 5 mm. in diameter, covered with gray wool, developing on one side of the crown of a branch into a stout, brownish, dirty yellow cephalium, with wool about 8 to 10 mm. long; radial spines 13 to 15, radiating, extending obliquely from the plant body, 10 to 12 mm. long; central spines 2 or 3, somewhat longer than the radial spines, up to 18 mm. long; all of the spines brown, not very sharp, elastic; flowers from the cephalium short, tubular, 45 mm. long, 17 to 20 mm. in diameter; ovary and tube thickly covered with lanceolate or narrowly triangular scales, 2 to 4 mm. long and bearing in their axils fascicles of short, closely lying reddish-brown hairs; petals white, lanceolate-spatulate, with short tips, the innermost 10 mm. long and 5 mm. wide; anthers arising from the upper part of the tube, not extending beyond the perianth; filaments 15 mm. long; pistil 27 mm. long, the stigma slightly exceeding, with 10 stigma-lobes 3 mm. long; fruit pear-shaped, 6 cm. long, 4 cm. in diameter; seeds black, shining, 1.5 mm. long.

"Of the hitherto known species of the genus *Cephalocereus* only one comes from Brazil, the rest from Mexico. The Brazilian species, *C. melocactus* Schumann, has only 12 ribs; 3 to 6 radial spines; red flowers, 3 cm. long; and through these characteristics differs from the here-described species." Translated by Paul G. Russell from Ule, Monatsschr. Kakteenk. **18**: 85. 1908.

Type locality: Serra do S. Ignacio, Bahia, Brazil.

Distribution: Known only from the type locality.

The plant is known to us only from the description and illustration. It would seem from these to be related to *C. dybowskii.*

Illustrations: Bot. Jahrb. Engler **40**: Beibl. **93**: pl. 9; Vegetationsbilder **6**: pl. 18.

PUBLISHED SPECIES, PERHAPS OF THIS GENUS.

The species of this genus have often been described under *Pilocereus*, while others have appeared under *Cereus.* There are also some species of *Cephalocereus* which we do not know, and these are all grouped here under the above heading.

CEPHALOCEREUS HERMENTIANUS (Monville) Britton and Rose, Contr. U. S. Nat. Herb. **12**: 416. 1909.

> *Cereus hermentianus* Monville, Illustr. Hort. **6**: Misc. 90. 1859.
> *Pilocereus hermentianus* Lemaire in Weber, Dict. Hort. Bois 965. 1898.

Upright, slender, 3 meters high, 5 to 7 cm. in diameter, branching; ribs about 19, rounded, shallow; areoles close together, round, with short brown wool and silky, persistent, hanging hairs; spines about 20, small, slender, yellowish; flowers 5 to 6 cm. long.

Type locality: Not cited.

Distribution: Haiti, according to Weber.

Monville did not know the origin of this species, but Weber assigned it to Haiti without question. We do not know any cactus from Hispaniola with 19 ribs, but further explorations may prove its occurrence there.

PILOCEREUS ALBISETOSUS (Haworth) Schumann, Gesamtb. Kakteen 196. 1897.

> *Cereus albisetosus* Haworth, Suppl. Pl. Succ. 77. 1819.

This certainly does not belong to this genus. It may be a *Selenicereus.* Evidently a low creeping plant, green, 5-angled, with areoles bearing brown wool and several white setaceous spines. It is a native of "Domingo," and is said to be similar to *Cereus reptans* It was introduced into England by A. B. Lambert in 1816.

PILOCEREUS VERHEINEI Rümpler in Förster, Handb. Cact. ed. 2. 690. 1885.

Columnar, simple so far as known, pale green, the apex covered with white wool, soon turning gray; ribs 12 or 13, 8 to 10 mm. high, obtuse; areoles 6 to 8 mm. apart, 2 to 3 mm. in diameter; spines yellowish at first, in age gray; radial spines 7 or 8, spreading, subulate, 1 to 1.5 cm. long; central spine solitary or wanting, 1 cm. long.

This species, recognized by Schumann as a good species of *Pilocereus*, we do not know. Its flowers and fruits are unknown and hence its exact place can not be determined. Its origin, too, is unknown and so far as we are aware it is not now in cultivation. The above description has been compiled.

PILOCEREUS GLAUCESCENS Labouret, Monogr. Cact. 279. 1853.

> *Pilocereus coerulescens* Lemaire, Rev. Hort. 1862: 427. 1862.
> *Pilocereus andryanus* Cels in Lemaire, Rev. Hort. 1862: 427. 1862, as synonym.

"Stem erect, at first simple, later probably becoming branched, dark bluish gray, glaucescent, bearing 10 rounded, blunt, inflated ribs; sinuses sharp, shallow, with age becoming effaced toward the base of the stem; areoles close together, almost confluent toward the base of the stem, rounded, with short almost black tomentum, furnished with hairs and very abundant, fine, undulating, weak, white bristles, especially on the areoles recently developed and toward the summit of the stem, rarer on the lower areoles; spines radiating, of different lengths and thicknesses, biserial, the exterior fine, divergent, inserted to the number of 5 or 6 on each side of the areole, the interior stouter, disposed irregularly in the center, to the number of 5 or 6 also, all of a dull yellow, brown at the base. The plant is 8 cm. in diameter by about 20 cm. in height; the bristles and the hairs of the areoles are about 1 to 2 cm. long; the interior spines, which are the strongest, are 10 to 15 mm. long; flowers unknown.

"The general aspect of this plant, which I believe is unique in Europe, resembles that of a *Pilocereus*. However, in the absence of definite and certain characters, it is not without doubt that we place it in the genus near *Pilocereus columna* and *chrysomalus*, as much for the long bristles of its areoles as for its branching stem. If later, however, its flower makes it a *Cereus*, its place would very probably be among the *Lanuginosi*, just after the *Coerulescentes*, and in th is event, which seems doubtful to me because of its many points of resemblance to *Pilocereus*, it would certainly constitute one of the most remarkable species of *Cerei*."

The above is taken from Labouret's monograph.

The Index Kewensis refers one of these names to *Cereus glaucescens* and the other to *Cereus coerulescens*, but doubtless in error, while the last is called *Cereus andryanus* by Schumann (Gesamtb. Kakteen 196). Lemaire says his plant came from Serra do Cipo, District Diamartina, Brazil.

PILOCEREUS ALBISPINUS (Salm-Dyck) Rümpler in Förster, Handb. Cact. ed. 2. 649. 1885.

> *Cereus albispinus* Salm-Dyck, Observ. Bot. 3: 5. 1822.
> *Cereus crenatus* Salm-Dyck in Labouret, Monogr. Cact. 341. 1853.
> *Pilocereus albispinus crenatus* Rümpler in Förster, Handb. Cact. ed. 2. 649. 1885.
> *Cereus serpentinus albispinus* Weingart, Monatsschr. Kakteenk. 18: 30. 1908.

Columnar, usually simple but sometimes branched at base; branches with 8 to 12 low, obtuse ribs, these dull green and woolly at apex; radial spines 8 to 13, spreading, white except at tip and there red; central spines 1 to 4; flowers and fruit not known.

Type locality: Curaçao, according to Schumann, but nothing like it was found there by Dr. Britton in 1913.

Distribution: Unknown. Rümpler says it is South America.

Cereus albispinus major Monville (Labouret, Monogr. Cact. 341. 1853) is undescribed.

The original publication of *Cereus acromelas* (Hortus, Berol. Ind. Cact. 1833) we have not seen. Pfeiffer refers it to *Cereus crenulatus* and Labouret to *C. albispinus.*

Cereus octogonus Otto (Allg. Gartenz. 1: 365. 1833) and *C. decagonus* (Pfeiffer, Enum. Cact. 85. 1837) are unpublished names for this species.

We have studied a living specimen of this plant which is growing in the New York Botanical Garden. Its flowers and fruit are not known. See note under *Cephalocereus leucostele;* see also Weingart's reference (Monatsschr. Kakteenk. 18: 30. 1908).

PILOCEREUS FLAVISPINUS Rümpler in Förster, Handb. Cact. ed. 2. 659.   1885.

    Cereus flavispinus Salm-Dyck, Observ. Bot. 3: 5.   1822.

These names were both referred by Schumann to *Pilocereus strictus*. The former is said to come from South America and the latter from tropical America. The specific name comes from *Cactus flavispinus* Colla (Hort. Ripul. 24) and probably applies to some Chilean plant.

CEREUS GHIESBREGHTII Schumann, Gesamtb. Kakteen 81.   1897.

    Columnar, simple or somewhat branched, short-jointed; joints nearly as broad as long; ribs 6 to 8, separated by broad intervals; radial spines 10 to 12, subulate, about 1.4 cm. long; central spines 2 to 4, 5 cm. long; flowers and fruit unknown.

Type locality: Mexico.

Distribution: Known only from type locality.

    A plant in the New York Botanical Garden so named suggests a small *Cephalocereus* and here we refer the species for the present. Schumann's illustration suggests a greenhouse seedling and may differ widely from the wild form.

    This is different from *Pilocereus ghiesbrechtii* Rümpler (Förster, Handb. Cact. ed. 2. 661. 1885) which Rümpler says (p. 662) is in the Paris Gardens as *Echinocactus ghiesbrechtii*. This is doubtless what Salm-Dyck (Allg. Gartenz 18: 395. 1850) described under that name, a species which has not been recognized by later students.

Illustration: Schumann, Gesamtb. Kakteen f. 16.

NOTES.

    A species of *Cephalocereus* with woolly areoles occurs at Tehuantepec, Mexico, as shown by a photograph obtained there by O. F. Cook and G. N. Collins of the United States Department of Agriculture.

    A species of *Cephalocereus*, with slender, deflexed, white spines, occurs at Coro, Venezuela, as shown by a plant brought by Dr. Rose to the New York Botanical Garden in 1916.

    A species of *Cephalocereus* inhabits the Serra de Borborema, Pernambuco, Brazil, as shown by a photograph received from A. Löfgren; his notes describe it as several meters high, with stout, erect branches, numerous low ribs, the yellow pseudocephalium on one side, elongated, the acicular spines yellow.

    A very peculiar plant which was collected by Luetzelburg near Bom Jesus, Bahia altitude about 1,700 meters, should probably be placed in this genus and next to *C. leucostele*. It is called the bottle cactus on account of its shape. A brief description follows

    Plant simple, short, and stubby, 10 cm. high; globular at first, in time lengthening from 20 to 40 cm. and becoming more or less bottle-shaped, the upper part being more slender and jointed; ribs 12 to 15, acute; areoles close together, arranged along the ribs; spines from the upper areoles white, elongated, and soft; flowers reddish, 8 to 9 cm. long, opening during the day.

4. ESPOSTOA gen. nov.

    Columnar plants with numerous low ribs and when flowering developing a pseudocephalium similar to that of some species of *Cephalocereus*; areoles strongly armed with spines, and bearing long white hairs; flowers small, short-campanulate, nearly hidden by the surrounding wool, probably opening at night; tube short; outer perianth-segments pinkish, the inner ones probably white; stamens and style short, included; scales on ovary and flower-tube small, acute, with long silky caducous hairs; fruit subglobose to broadly obovoid, smooth, its flesh pure white, slightly acid, very juicy, edible; seeds very small, black, shining.

    This genus resembles the typical species of *Cephalocereus*. Berger suggested that it was an *Oreocereus*, but this was before he had seen any flowers of the latter; we now know that there is much difference not only in the flowers but also in the fruit and seeds. It is named for Nicolas E. Esposto, a very keen botanist who is connected with the Escuela Nacional de Agricultura at Lima, Peru.

1. Espostoa lanata (HBK.).

Cactus lanatus Humboldt, Bonpland, and Kunth, Nov. Gen. et. Sp. 6: 68. 1823.
Cereus lanatus De Candolle, Prodr. 3: 464. 1828.
Pilocereus dautwitzii Haage, Gard. Chron. 1873: 7. 1873.
Pilocereus haagei Rümpler in Förster, Handb. Cact. ed. 2. 665. 1885.
Pilocereus lanatus Weber, Dict. Hort. Bois 965. 1898.
Cereus dautwitzii Orcutt, West Amer. Sci. 13: 63. 1902.
Cleistocactus lanatus Weber in Gosselin, Bull. Mens. Soc. Nice 44: 37. 1904.
Pilocereus lanatus haagei Jostmann, Monatsschr. Kakteenk. 21: 25. 1911.
Oreocereus lanatus Britton and Rose, Stand. Cycl. Hort. Bailey 4: 2404. 1916.

Plant simple, 2 to 4 meters high, sometimes with several strict branches or with a simple erect stem, 4 to 10 cm. in diameter, with many spreading branches at first nearly horizontal or curved upward and becoming erect near the tip, the tip hidden under a mass of hairs and brown bristles; ribs numerous, 20 to 25, low, 5 to 8 mm. high, rounded; areoles rather large, 5 to 6 mm. apart; radial spines numerous, acicular, 4 to 7 mm. long, brownish, intermixed with long white hairs; central spine solitary, yellow or brown to black, subulate, 2 to 5 cm. long; flowers borne on one side of the stem from a prominent pseudocephalium, 3.5 to 5 cm. long; scales on the tube many, triangular-lanceolate, acute, about 6 cm. long; fruit 3 to 4 cm. long, juicy, edible, white except the small pinkish scales; seeds 1 mm. broad.

Type locality: Near Rio Aranza and Guancabamba, Ecuador.

Distribution: On the dry hills of northern Peru and Ecuador, altitude 1,200 to 2,250 meters.

FIGS. 87 and 88.—Espostoa lanata.

In 1918, while in Ecuador, Dr. Rose attempted to reach the exact locality of Humboldt's *Cactus lanatus*, but was unsuccessful. In the Catamayo Valley somewhat north of Humboldt's station and in what is doubtless a part of the same desert he collected this species and upon this our description above is largely based. These plants are so different in habit from other plants collected by Dr. Rose in central Peru that we have been very much in doubt whether they should all be referred here or a part separated as a new species. That

FIG. 89.—Flower of Espostoa lanata. ×0.7.
FIG. 90.—Fruit of same. ×0.7.

there is more than one species in this genus has been further suggested since receiving a photograph from G. M. Dyott, taken at Chagual, on the west bank of the Marañon River, in northern Peru. In this photograph are shown several very striking cactus plants, perhaps of this genus, but very unlike any we have heretofore seen.

We have followed most recent writers in combining *Cereus dautwitzii* with *Cereus lanatus*, although we have not seen the type of either. We know, however, that *Cereus dautwitzii* came from Huancabamba, Peru, while *Cactus lanatus*, upon which *Cereus lanatus* was based, came from Guancabamba, Ecuador; the names, varying only in the initial letter, are different spellings for the same place. The northern boundary-line of Peru has pushed north since Humboldt visited this region; his station of Guancabamba is now in Peru instead of Ecuador.

The sweet, edible fruit is called soroco in southern Ecuador; it is also called piscol colorado, according to Humboldt.

FIG. 91.—Espostoa lanata.

The typical form was collected by J. N. Rose, A. Pachano, and George Rose in the Catamayo Valley, southern Ecuador, October 3, 1918 (No. 23326) and the other form was collected by Dr. and Mrs. Rose near Matucana, central Peru, altitude about 7,000 feet, July 9, 1914 (No. 18649). Dr. Rose also collected a living plant above Chosica (No. 18537) and herbarium specimens between Matucana and San Bartelome (No. 18748). Dr. W. H. Osgood has sent us photographs of a cactus which we would refer here. One was taken near Chilete, Peru, altitude 1,000 feet, and the other between Menocucho and Otuzco, Peru, altitude 3,000 feet.

Pilocereus haageanus (Monatsschr. Kakteenk. **6**: 96. 1896) is sometimes referred to but was never published.

Illustrations: Dict. Gard. Nicholson 3: f. 152; Fl. Serr. 21: pl. 2163; Förster, Handb. Cact. ed. 2. f. 87; Gard. Chron. 1873: f. 1; Knippel, Kakteen pl. 29, all as *Pilocereus dautwitzii;* Cact. Journ. 2: 4, as *Pilocereus dautwitzii cristatus,* Monatsschr. Kakteenk. 21: 23; 24: 131, both as *Pilocereus lanatus;* Monatsschr. Kakteenk. 19: 183, as *Pilocereus lanatus cristatus;* Monatsschr. Kakteenk. 21: 23; 23: 125, both as *Pilocereus lanatus haagei.*

Figure 87 is from a photograph taken by George Rose in southern Ecuador in 1918; figure 88 is from a photograph taken by Dr. Rose at Matucana, Peru, in 1914; figure 89 shows the flower and figure 90 the fruit of the plant photographed by him; figure 91 is from a photograph taken at the New York Botanical Garden of the plant obtained by Dr. Rose at Chosica, Peru, in 1914.

5. BROWNINGIA gen. nov.

Plants solitary, with an upright trunk, branching only at top, the branches spreading or drooping; ribs numerous, low; young and sterile plants formidably spined; flowering branches naked or bearing only weak bristle-like spines; flowers solitary at the areoles, nocturnal, large, with slightly curved tubes; stamens and style shorter than the perianth-segments; flowers nearly white; ovary and flower-tube covered with large, thin, fleshy scales, these naked in their axils; fruit slightly acid, yellow, becoming naked by the falling away of the scales; seeds black, strongly papillose.

This genus does not closely approach any other. In the thin scales of the ovary and flower-tube there is a hint of *Escontria* of Mexico, but the scales are not chartaceous and the flowers are otherwise different. The ovary and perianth perhaps most resemble those of *Hylocereus.*

It is named in honor of W. E. Browning, formerly director of the Instituto Ingles at Santiago, Chile, who for many years did efficient educational work in Chile, and who was the friend of all Americans who visited Santiago.

1. Browningia candelaris (Meyen).

Cereus candelaris Meyen, Allg. Gartenz. 1: 211. 1833.

Stems 3 to 5 meters high, with a simple trunk sometimes 3 dm. in diameter at base, tapering gradually upward; trunk when young strongly armed with many long spines, but when very old shedding the spines and in some cases becoming nearly naked; ribs 30 to 34, rounded, about 5 mm. high; branches from and near the top usually many, sometimes as many as 50, but sometimes as few as 3 to 6, in whorls or pseudo-whorls, slender, often spreading at right angles to the trunk, sometimes erect, or sometimes drooping and even touching the ground; areoles circular, usually about 1 cm. apart, 5 to 15 mm. in diameter and, when old, much elevated; spines of the trunk-areoles normally about 20, very unequal, the longest ones 6 to 10 cm. long, but sometimes 50 or more, the longest 15 cm. long, at first brownish, then gray or black; spines on flowering branches weak, yellow, sometimes bristle-like or even wanting; flower-buds globular, obtuse, covered with thin imbricating scales; flowers opening in the evening, closing in early morning, not fragrant, 8 to 12 cm. long, a little curved; scales on ovary and flower-tube large, numerous; throat of flower rather narrow, 3 to 4 cm. long, covered with filaments; tube proper 4 cm. long; inner perianth-segments narrow, about 2 cm. long, brown or rose-colored or the innermost pale rose to white; filaments cream-colored, numerous, the lower 3 cm. long, the upper 2 cm. long; style slender, 7 cm. long, cream-colored; stigma-lobes about 12, 4 to 5 mm. long, cream-colored; fruit said to be edible; seeds 2 mm. broad.

Type locality: On mountain slopes along the way from Tacna, Chile, to Arequipa, Peru, up to 9,000 feet (2,740 meters) altitude.

Distribution: Southern Peru and northern Chile.

The name, *Cactus candelaris* Meyen (Reise 2: 40. 1835), occurs in Meyen's narrative, where he states that it was first found in the Cordilleras of Tacna (now in Chile) in isolated examples, confined between 7,000 and 9,000 feet altitude. This plant is very conspicuous in the desert below Arequipa and was collected there by Dr. Rose in 1914 (No. 18794).

Figure 92 shows a flower collected by Dr. Rose below Arequipa, Peru, in 1914; figure 93 shows the young fruit and persistent withering perianth from the same plant; figure 94 is from a photograph taken by T. A. Corry near Arequipa, Peru, in 1917; the plant immediately in front is *Trichocereus fascicularis*.

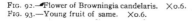

Fig. 92.—Flower of Browningia candelaris. ×0.6.
Fig. 93.—Young fruit of same. ×0.6.

Fig. 94.—Browningia candelaris, with Trichocereus fascicularis immediately in front of it.

6. STETSONIA gen. nov.

A tall, erect, much branched cactus, with strongly ribbed branches, the areoles felted and bearing several unequal stiff subulate spines; flowers funnelform, large, solitary at upper areoles; ovary oblong-globose, densely covered by small, broad, erose, ciliate, abruptly subulate-tipped, membranous scales; flower-tube cylindric, somewhat expanded above, bearing distant scales similar to those of the ovary; outer perianth-segments broad, green, obtuse, the inner oblong-oblanceolate, spreading, acute; stamens numerous, not exserted; anthers large, oblong; style rather stout; stigma-lobes many, linear.

Only the following species is known to us, a conspicuous plant of the Argentine deserts. The genus is dedicated to Francis Lynde Stetson, of New York.

1. Stetsonia coryne (Salm-Dyck).

Cereus coryne Salm-Dyck,* Cact. Hort. Dyck. 1849. 205. 1850.

Plants large and massive, 5 to 8 meters high, with a thick, short trunk up to 4 dm. in diameter and 4 to 6 dm. long, and many (100 or more) ascending or upright elongated branches; ribs 8 or 9, 1 to 1.5 cm. high, obtuse, more or less crenate; spines 7 to 9, unequal, the longest 5 cm. long, subulate; flowers 12 to 15 cm. long; inner perianth-segments white, spreading; fruit not known.

Type locality: Not cited.
Distribution: Northwestern Argentina.

Although this species has long been known in collections, it is usually represented by very small specimens and has been poorly described.

*Both Weber and Schumann make Otto the author of this name.   Salm-Dyck credits it to the Berlin Gardens.

A large plant of *Stetsonia coryne* in the desert of northern Argentina.

This tree-like cactus is native in the dry parts of northwestern Argentina, and occurs over a considerable area, growing with scattered shrubs and small trees on plains and low ridges. It is one of the most striking cacti in South America and often forms the dominant feature of the landscape on the high plains of northern Argentina.

In 1917, Dr. Shafer collected living specimens and flowers in Santiago del Estero, Argentina, which have enabled us to redescribe the species. Flowers were also collected by Wilhelm Bodenbender in 1905, but these were not accompanied by stems and were not at first associated with this species.

Illustrations: Monatsschr. Kakteenk. **3**: 177; **13**: 187; Schelle, Handb. Kakteenk. f. 17; Rev. Hort. Belge **40**: after 184, as *Cereus coryne.*

FIG. 95.—Flower of S. coryne. ×0.6.

FIG. 96.—Stetsonia coryne.

Plate IX is from a photograph contributed by Dr. Spegazzini. Figure 96 is from a photograph of flowering branches taken by Dr. Shafer at Santiago del Estero, Argentina, in 1917; figure 95 shows the flower of one of these branches.

7. ESCONTRIA Rose, Contr. U. S. Nat. Herb. **10**: 125. 1906.

Large and much branched plants; ribs few; spines all similar, arranged in peculiar pectinate clusters; flowers small, yellow, somewhat campanulate, one at an areole, diurnal; ovary globular, covered with imbricated chartaceous, translucent, persistent scales, their axils without spines or hairs; inner perianth-segments erect, narrow; stamens and style included; fruit globular, scaly, purple, fleshy, edible; seeds numerous, black, rugose, with a flattened, broad, basal hilum.

Type species: *Cereus chiotilla* Weber.

Only 1 species is definitely known. The genus commemorates Señor Don Blas Escontria, a distinguished Mexican, who died in 1906.

1. Escontria chiotilla (Weber) Rose, Contr. U. S. Nat. Herb. **10**: 126. 1906.

Cereus chiotilla Weber in Schumann, Gesamtb. Kakteen 83. 1897.

Plant 4 to 7 meters high; trunk very short; branches numerous, forming a compact top, weak and easily broken, bright green, not at all glaucous; ribs 7 or 8, acute; areoles close together, often confluent, elliptic; radial spines 10 to 15, rather short, often reflexed; central spines several, one much longer than the others, somewhat flattened, sometimes 7 cm. long, all light colored; flowers borne near the ends of the branches, including the ovary about 3 cm. long; inner perianth-segments yellow, acuminate; scales on ovary and flower-tube arranged in many overlapping series, ovate, 8 to 15 mm. long; fruit glabrous, about 5 cm. in diameter, scaly, edible.

Type locality: Mexico.
Distribution: Southern Mexico.

FIG. 97.—Flower of Escontria chiotilla. ×0.7.
FIG. 98.—Fruit of same. ×0.7.

The ripe fruit is sold in the market at Tehuacán under the name of geotilla or chiotilla and tuna. Dr. H. H. Rusby reports that the dried fruit, which tastes like gooseberries, is also sold in the markets.

This species was collected by Dr. A. Weber while connected with the French army in Mexico. Material was sent to Dr. Engelmann in 1864, but it was not described by him.

Illustrations: Bull. Soc. Acclim. France **52**: f. 5; Möllers Deutsche Gärt. Zeit.**29** : 445, as *Cereus chiotilla*; Contr. U. S. Nat. Herb. **10**: pl. 43, f. A; **12**: pl. 65.

Plate x is from a photograph taken by Dr. MacDougal at Tomellín, Mexico, in 1906. Figure 97 shows a flower and figure 98 a fruit collected in 1905 by Dr. Rose near Tehuacán.

8. CORRYOCACTUS gen. nov.

Stems columnar, usually very short, branching from the base, the branches stiff, more or less erect, strongly ribbed; areoles very spiny; flowers diurnal (?), rather large, with a broad, open throat, the tube proper very short; perianth-segments yellow or orange; filaments numerous, stiff, short, scattered all over the throat, much shorter than the segments; style short and stiff, with numerous stigma-lobes; ovary and flower-tube bearing numerous conspicuous areoles with brown or black wool and subtended by minute scales; fruit juicy, globular, covered with clusters of deciduous acicular spines; seeds small.

Type species: *Cereus brevistylus* Schumann.

A genus of three known species of similar habit and flowers, natives of Peru and Bolivia. The flowers have very short tubes, but are quite different from those of *Eulychnia*, to which Berger referred the only species he knew.

While the species are similar in a general way, they are individually different in habit, armament, and in shades of color and size of the flowers; their ranges do not overlap, as they are found in different regions and at different altitudes. One of them occurs in the coastal mountains of southern Peru, altitude 550 meters; one in the foothills of the Andes proper, altitude 2,300 meters; and one in the great valley of the Andes in Bolivia, altitude 3,650 meters.

The genus is named for T. A. Corry, chief engineer of the Ferrocarril del Sur of the Peruvian Corporation, who much facilitated our exploration of this region. It is rather remarkable that all three of these species are found along this very interesting railroad, which extends from the sea-level to an altitude of 16,000 feet.

KEY TO SPECIES.

Flowers very broad, 10 cm. wide, yellow...1. *C. brevistylus*
Flowers much narrower than the last, 4 to 7 cm. broad.
    Inner perianth-segments orange; longest spines 10 to 16 cm. long.......................2. *C. brachypetalus*
    Inner perianth-segments yellow; longest spines 5 to 7 cm. long..........................3. *C. melanotrichus*

1. Corryocactus brevistylus (Schumann).

Cereus brevistylus Schumann in Vaupel, Bot. Jahrb. Engler **50**: Beibl. **111**: 17. 1913.

Plants 2 to 3 meters high, usually much branched from the base, often forming large clumps, light green to almost yellow; ribs few, 6 or 7, very prominent; areoles 3 cm. apart, large, circular and elevated, with short, dense wool and spines; spines about 15, at first brownish, very unequal,

A plant of *Escontria chiotilla*, near Tehuacán, Mexico.

some less than 1 cm. long, some about 3 cm. long, and still others 20 to 24 cm. long; flower broadly funnelform, constricted just above the ovary, 9 cm. long, 10 cm. broad when fully expanded; throat 4.5 cm. broad at the top; perianth-segments bright yellow, oblong, spreading; filaments numerous, yellow; style short, thick, white, with numerous, slender, white stigma-lobes; scales of the ovary and flower-tube small, with brown wool, white bristles, and short spines; fruit globular, juicy, covered with numerous spine-clusters, these tardily deciduous.

Type locality: Yura, near Arequipa, Peru.

Distribution: In the mountains of southern Peru, altitude 2,000 to 3,300 meters.

This species is one of the three or four common cacti found on the hills and in the valleys both above and below Arequipa, and, while not the largest, is often the most abundant and conspicuous. Dr. Rose studied this plant in Peru in 1914, collecting living and herbarium specimens, including the very long spines described above (Nos. 18780, 18965).

Figure 99 is from a photograph taken near Arequipa, Peru, by T. A. Corry in 1917; figure 101 represents a flower collected by Dr. Rose near the same place in 1914.

Fig. 99.—Corryocactus brevistylus.

Fig. 100.—Corryocactus brachypetalus.

2. Corryocactus brachypetalus (Vaupel).

Cereus brachypetalus Vaupel, Bot. Jahrb. Engler 50: Beibl. 111: 16. 1913.

Plant 2 to 4 meters high, usually with many (sometimes 100 or more) strict branches from the base, forming a top 3 to 4 meters in diameter; ribs usually 7 or 8, somewhat prominent; areoles usually 2 cm. apart, large, 1 cm. in diameter or less, with short wool and spines; spines at first black with brown bases, about 20 at an areole, very unequal, most of them less than 1 cm. long, the longest ones 10 to 16 cm. long; flowers broadly funnelform, 4 to 6 cm. broad; throat 2 to 3 cm. broad at top; inner perianth-segments deep orange, 1 to 1.5 cm. long, the outer ones apiculate, the inner ones obtuse or truncate; filaments very short, 5 to 8 mm. long, yellow; style, including the slender stigma-lobes, 2 cm. long; areoles of the ovary and flower small, filled with black and white wool and nascent spines; fruit globular, 6 to 7 cm. in diameter, greenish yellow, covered with clusters of deciduous spines, juicy, said to be edible; seeds dull in color, 1.5 mm. long.

Type locality: Rocky sandy bottoms near Mollendo, southern Peru.
Distribution: Foothills of southern Peru, altitude 600 meters.

This plant is very abundant in the foothills of southern Peru. In many places it is the only conspicuous plant in this arid region, which in the dry season is otherwise almost devoid of plant life. In the shelter of these plants thousands of lizards live and, doubtless, feed upon the flowers. Dr. Rose collected the species in 1914 (No. 18810) at Posco, Peru, not far from the type locality.

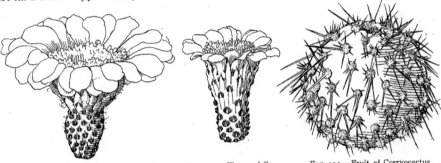

FIG. 101.—Flower of C. brevistylus. ×0.7.   FIG. 102.—Flower of C.   FIG. 103.—Fruit of Corryocactus
brachypetalus. ×0.7.   brachypetalus. ×0.7.

Figure 102 represents a flower and figure 103 a fruit, collected by Dr. Rose at Posco, Peru, in 1914; figure 100 is from a photograph taken near Posco, Peru, by T. A. Corry in 1918.

3. Corryocactus melanotrichus (Schumann).

Cereus melanotrichus Schumann, Gesamtb. Kakteen 71. 1897.*

Plant 1 to 2 meters high, forming small clumps with erect slender branches 3 to 4 cm. in diameter; ribs 7 or 8, much lower than in the other species; areoles 1 to 1.5 cm. apart, black or nearly so; spines 7 to 15, light yellow, subulate, somewhat unequal, the longest ones 5 to 7 cm. long; flowers broadly funnelform, 6 to 8 cm. long, 5 to 7 cm. broad; perianth-segments yellow; filaments much longer than in the other species; areoles of the flower with 1 to 5 long, black, bristle-like spines; fruit globular, 5 to 6 cm. in diameter, very juicy, covered with clusters of small acicular spines.

Type locality: Near La Paz, Bolivia.
Distribution: Central Bolivia, altitude 3,300 meters.

Plants of this species are much smaller than those of the other two and often form low thickets, growing on the barren hills in and about La Paz. The species was re-collected by Dr. Rose in 1914 (No. 18843).

9. PACHYCEREUS (Berger) Britton and Rose, Contr. U. S. Nat. Herb. **12:** 420. 1909.

Usually very large plants, more or less branched, with definite trunks, the stems and branches stout, columnar, ribbed; flowers diurnal, with rather short tubes; outer perianth-segments short, spatulate; stamens included, numerous, inserted along the throat; style included; ovary and flower-tube covered with small scales bearing felt and bristles in their axils; fruit large, bur-like, dry, usually densely covered with clusters of deciduous spines and bristles; seeds large and black.

Type species: *Cereus pringlei* S. Watson.

We recognize 10 species, all natives of Mexico, from northern Sonora to Yucatan. The name was first used by Berger as a subgenus of *Cereus* (Rep. Mo. Bot. Gard. **16:** 63. 1905); we agree with his limitation of the group, except by excluding *Cereus thurberi* Engelmann,

---

*This name occurred in print two years earlier, but without description, in the Memoirs Torrey Botanical Club (**4:** 207).

referred by us to *Lemaireocereus*, and by including *Cereus marginatus* De Candolle, placed by him in his subgenus *Stenocereus*.

The name *Pachycereus* is from the Greek and means thick-cereus, referring to the stout stems and branches.

KEY TO SPECIES.

```
A. Scales of ovary and perianth-tube fleshy or herbaceous.
    Wool of ovary-areoles copious, mostly longer than the scales.
        Perianth-tube broad; branches many-ribbed.
            Areoles of ovary and perianth-tube densely felted, but without long wool.
                Joints green or but slightly glaucous.
                    All areoles of the perianth-tube densely felted, the scales short.
                        Spines brown to gray or sometimes black................  1.  P. pringlei
                        Spines of young growth yellowish brown.................  2.  P. orcuttii
                    Upper areoles of perianth-tube little or scarcely felted, scales long.
                        Flowering areoles bearing many short, weak spines.........  3.  P. pecten-aboriginum
                        Flowering areoles bearing several acicular stiff spines.......  4.  P. gaumeri
                    Young growth glaucous, the bloom persistent as whitish streaks......  5.  P. grandis
                Areoles of ovary and perianth-tube bearing copious yellow-brown wool 1.5
                    to 2.5 cm. long.........................................  6.  P. chrysomallus
            Perianth-tube narrow; branches 5 to 7-angled..........................  7.  P. marginatus
        Wool of ovary-areoles sparse, shorter than the coriaceous scales...............  8.  P. ruficeps
AA. Scales of ovary and perianth-tube dry........................................  9.  P. lepidanthus
AAA. Species not grouped.......................................................  10. P. columna-trajani
```

1. Pachycereus pringlei (S. Watson) Britton and Rose, Contr. U. S. Nat. Herb. **12**:422. 1909.

> *Cereus pringlei* S. Watson, Proc. Amer.
> Acad. **20**: 368. 1885.
> *Cereus calvus* Engelmann in Coulter, Contr.
> U. S. Nat. Herb. **3**: 409. 1896.
> *Cereus titan* Engelmann in Coulter, Contr.
> U. S. Nat. Herb. **3**: 409. 1896.
> *Pilocereus pringlei* Weber, Dict. Hort. Bois
> 966. 1898, name only.
> *Pachycereus calvus* Britton and Rose, Contr.
> U. S. Nat. Herb. **12**: 420. 1909.
> *Pachycereus titan* Britton and Rose, Contr.
> U. S. Nat. Herb. **12**: 422. 1909.

Tree-like, up to 11 meters high, usually with a very short, thick trunk, sometimes 1 or even 2 meters long or more, often 6 dm. but sometimes 2 meters in diameter or more, very woody and in age naked; stem sometimes nearly simple but often with numerous, thick, upright branches, more or less glaucous, very spiny or in some forms nearly naked; ribs usually 11 to 15 but sometimes 17, obtuse; areoles, especially the flowering ones, very large, brown-felted, usually confluent or connected by a groove; spines variable as to length, abundance, and structure, usually more formidable in young plants than in old plants, often wanting in very old plants; spines on young growth 20 or more at an areole, 1 to 2 cm. long, white but with black tips, or on young plants sometimes 12 cm. long and black throughout; flower-bearing region of the branches extending from near the top downward sometimes for 2 meters, the areoles becoming broad and uniting, often spineless; flower-buds greenish; flowers 6 to 8 cm. long, the tube and ovary bearing small, acute scales, these nearly hidden by the mass of brown hairs produced in their axils; inner perianth-segments white, broad, spreading; fruit globular, covered with brown felt and bristles, dry; seeds large, edible.

FIG. 104.—Pachycereus pringlei.

Type locality: South of the Altar River, Sonora, Mexico.

Distribution: Sonora and Lower California.

This is a very interesting and important cactus in northwestern Mexico, often the dominant plant in the landscape. On the plain about Guaymas solitary plants, giants of the race, are seen, which are doubtless remnants of great forests which once covered this plain. In Lower California protected valleys and hillsides are now covered with forests made up almost entirely of this species. The natives call these plants cardon. They gather the wood for firewood and use it to make walking-canes, or in building their simple houses, especially for rafters and beams; the Yaquí Indians, especially, gather the seeds and make a kind of flour by crushing them, and this is made into tomales. It is common in western Sonora, on many of the islands in the Gulf of California, all along the east coast of Lower California, and along the west coast of Lower California as far north as Magdalena Bay. In this distribution we have included the two species *Cereus calvus* and *C. titan*, both of which were described from spine-clusters. They may or may not be specifically distinct from *P. pringlei*, but without further data it is best to refer them here.

Illustrations: Gard. and For. 2: f. 92; Monatsschr. Kakteenk. 18: 119; Rep. Mo. Bot. Gard. 16: pl. 1, f. 1 to 4; Schumann, Gesamtb. Kakteen f. 13; Schelle, Handb. Kakteenk. f. 19; MacDougal, Bot. N. Amer. Des. pl. 12, 13; Rep. U. S. Nat. Mus. 1897: pl. 6, as *Cereus pringlei;* Contr. U. S. Nat. Herb. 16: pl. 130; Stand. Cycl. Hort. Bailey 5: f. 2695.

Figure 104 is from a photograph taken at Magdalena Bay, Lower California.

2. **Pachycereus orcuttii** (K. Brandegee) Britton and Rose, Contr. U. S. Nat. Herb. 12: 422. 1909.

    Cereus orcuttii K. Brandegee, Zoe 5: 3. 1900.

"Stems erect, branching, bright green, reaching a height of 3 meters and a diameter of 15 cm., with hard woody center; ribs 14 to 18, about 1 cm. high; areoles round, about 6 mm. in diameter and about half that distance apart, densely covered with short, light gray wool; spines all slender, spreading, yellowish brown, irregularly 3-seriate; radials 12 to 20, about 12 mm. long, deficient above; intermediates about 10, one-third to more than twice as long, less spreading, one of the upper spines of this row usually stouter and darker, porrect, often reaching a length of 7 cm.; centrals about 5, porrect, spreading a little longer than the intermediates; flowers greenish brown, darker outside, diurnal, entire length about 4 cm.; petals short-apiculate; ovary densely covered with short scales, almost completely concealed by thick, rounded tufts of yellowish wool, in which are imbedded dark brown bristles 4 to 6 cm. long; stamens lining the upper half of the tube; style tips acute; fruit not known.

"The plant from which this description is drawn was obtained by Mr. C. R. Orcutt near Rosario, Baja California, in May 1886. It was brought to him by his guide, who found it off the trail some little distance. The cutting was planted in Mr. Orcutt's garden, and is now about 2 meters in height; has flowered but has formed no fruit. It is much the finest of the large *Cerei* of Baja California, being densely covered with bright yellow-brown spines."

Type locality: Rosario, Lower California.

Distribution: Known only from the type locality.

The above description and account are taken from Mrs. Brandegee's article in Zoe, June 1900. Dr. Rose saw the type plant in 1908 at San Diego, California, and at that time obtained a flower and bud from Mr. Orcutt. Afterwards Mr. Orcutt photographed the plant and a flower and sold the prints. The photograph has also been printed on cardboard and distributed in an advertisement for Orcutt's American plants. A set of these photographs is in the National Herbarium.

3. **Pachycereus pecten-aboriginum** (Engelmann) Britton and Rose, Contr. U. S. Nat. Herb. 12: 422. 1909.

    Cereus pecten-aboriginum Engelmann in S. Watson, Proc. Amer. Acad. 21: 429. 1886.

Tree-like, 5 to 10 meters high, with a trunk 1 to 2 meters high and 3 dm. in diameter, crowned with many erect branches; ribs 10 or 11; areoles 1 cm. in diameter or even less, extending downward in narrow grooves, in the flowering ones forming brownish cushions connecting with the areoles below, densely tomentose (grayish except in flowering ones, which are brownish or reddish); spines 8 to 12, 1 to 3 central, all short, usually 1 cm. long or less, but in some cases 3 cm. long, grayish with black tips; flowering areoles not much larger than the others; flowers 5 to 7.5 cm. long; ovary covered with dense soft hairs with only a few bristles or none; outer perianth-segments purple, succulent; inner ones white, fleshy; stamens very numerous; style with 10 linear stigma-lobes; fruit 6 to 7.5 cm. in diameter, dry, covered with yellow wool and long yellow bristles.

Type locality: Hacienda San Miguel, Chihuahua, Mexico.

Distribution: Chihuahua, Sonora, Colima, and Lower California.

Illustrations: Contr. U. S. Nat. Herb. **5**: f. 32; pl. 57, 58; Gard. and For. **7**: f. 54; Dict. Gard. Nicholson Suppl. f. 233, all as *Cereus pecten-aboriginum.*

Figures 105 and 106 are copied from the two plates first cited above.

4. Pachycereus (?) gaumeri sp. nov.

Plant slender, 2 to 7 meters high, erect, simple or few-branched; branches 4-angled or winged; ribs thin, 3 to 4 cm. high; areoles large, 1 to 2.5 cm. apart, brown-felted; spines several, slender, 1 to 3 cm. long, brownish; flowers yellowish green, 5 cm. long; scales of the ovary and flower-tube more or less foliaceous, drying black and thin, with brown felt in the areoles; scales on the ovary linear, puberulent; fruit not known.

This species is based on two collections, both made in Yucatan by George F. Gaumer, as follows: No. 23778 at Hodo, April 1917 (type), No. 648 at Port Silam, 1895. Dr. Gaumer writes of these numbers as follows:

"As to my No. 23778 I sent many fine specimens of flowers and several cross-sections of a moderately large plant to Dr. Millspaugh. It grows erect, has few branches, many flowers on each plant; it is very common at the senote Hodo where the most of the plants range from 6 to 10 ft. high; it is a delicate-looking Cactus of a light pea-green color, quite showy, the flowers are of a light green tinged with cream-color, they do not open out much but remain almost cylindrical. Living specimens were sent to Dr. Britton at Bronx Park. It blooms in May and is found about four leagues east of Izamal.

"648 was taken by myself at the port of Silam in 1894 and sent to Dr. Millspaugh. Only two plants were seen; one was about 10 ft. and the other 20 ft. high. It grows erect and the larger plant had but one branch. My son Geo. J. has failed to find it in the region of Progresso."

Since the above description was written, Dr. Gaumer has sent another plant (No. 23935) which we believe belongs here, although it differs somewhat from the other plants. A cutting was sent to the New York Botanical Garden which produced a bud in the spring of 1919, but this only partially developed. This plant may be described as follows:

Erect; ribs 5 to 7, separated by broad intervals; areoles 1 cm. apart; spines about 15, 2 to 3 cm. long, weak, gray in age; flower-bud acute, ovoid, covered with green imbricating scales.

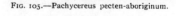

FIG. 105.—Pachycereus pecten-aboriginum.

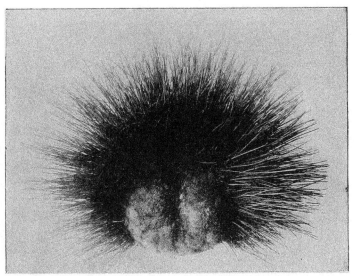

FIG. 106.—Fruit of Pachycereus pecten-aboriginum.

5. **Pachycereus grandis** Rose, Contr. U. S. Nat. Herb. **12**: 421. 1909.

  Cereus bergerianus Vaupel, Monatsschr. Kakteenk. **23**: 24. 1913.

Plant 6 to 10 meters high, either simple or much branched, the trunk sometimes a meter in diameter; branches, when present, columnar, generally simple, becoming erect almost from the first, with numerous constrictions, pale green, or when young glaucous, with some bloom which persists in streaks; ribs 9 to 11, acutish, high; sterile areoles circular, large, bearing white felt and subulate spines, 2 to 3 cm. apart, not running together, not extending below the spines as in *P. pecten-aboriginum;* old spines grayish to white with black tips; radial spines 9 or 10; central spines 3, the lower one longer, sometimes 6 cm. long, somewhat flattened; flowering areoles large, elliptic, bearing acicular or bristle-like spines; flowers rather small for the genus, about 4 cm. long; ovary and flower-tube bearing small, acuminate scales, their axils filled with downy hairs; fruit large, globular, dry, covered with long yellow bristles and yellow felt.

Type locality: On the pedregal near Cuernavaca, Mexico.

Distribution: Common in the State of Morelos, Mexico.

This plant is very common on the pedregal north of Cuernavaca, where it was first observed by Dr. Rose in 1906 (No. 11087), and is frequent on the hills south of Cuernavaca. Mr. Dowell, the cactus dealer in Mexico City, told Dr. Rose that he had exported plants to Europe, but whether they are now in the trade we do not know. A living specimen sent back by Dr. Rose has since been growing in the Washington Botanical Garden.

6. **Pachycereus chrysomallus** (Lemaire) Britton and Rose, Contr. U. S. Nat. Herb. **12**: 421. 1909.

  Pilocereus chrysomallus Lemaire, Fl. Serr. **3**: under pl. 242. 1847.
  Cereus chrysomallus Hemsley, Biol. Centr. Amer. Bot. **1**: 541. 1880.
  Cephalocereus chrysomallus Schumann in Engler and Prantl, Pflanzenfam. **3**6a: 182. 1894.
  Pilocereus fulviceps Weber in Schumann, Gesamtb. Kakteen 176. 1897.
  Cereus fulviceps Berger, Rep. Mo. Bot. Gard. **16**: 64. 1905.

Stem columnar, massive, at first simple, but in very old plants much branched, giving off hundreds of erect branches which form an almost compact cylinder up to 5 meters in diameter, becoming 12 to 18 meters high; branches glaucous green, 11 to 14-ribbed; flowering branches capped

M. E. Eaton del

Top of flowering plant of *Pachycereus chrysomallus*.
(Natural size.)

by dense masses of brownish wool; areoles approximate or even confluent; radial spines about 12, slender; centrals 3, 1 very long, sometimes 12 to 13 cm. long; flowers borne near the tops of the stems or branches, 6 to 7 cm. long; the bud, afterwards the flower, and finally the fruit, completely concealed in the long wool; ovary covered with small, pale, imbricated scales; flower-tube also covered with imbricated scales, but these larger and pinkish, pointed; flowers doubtless opening at night, but still expanded at 8 o'clock in the morning; tube proper 10 mm. long or less; throat funnelform, 3 cm. long; inner perianth-segments numerous, 1.5 to 3 cm. long, cream-colored; inner perianth-segments and stamens inflexed after anthesis, with the stiff outer perianth-segments pressed down upon them; stamens attached all over the throat, the innermost and lower row united at base and appressed against the style; filaments cream-colored; style stout, stiff, 7.5 cm. long, cream-colored; stigmalobes linear, erect, cream-colored.

Fig. 107.—Pachycereus chrysomallus.

Type locality: Mexico.

Distribution: Puebla and Oaxaca, Mexico.

This is one of the characteristic plants on the mesas around Tehuacán. When fully grown, it is a very large plant with many upright branches; the trunk and old branches are stout and woody, making it very difficult to obtain botanical specimens. In 1906 Dr. MacDougal and Dr. Rose shipped a very large plant to the New York Botanical Garden, which flowers annually and from which an abundance of flowers has been obtained.

Cereus militaris Audot (Rev. Hort. II. 4: 307. 1845) and *Pilocereus militaris* (Salm-Dyck, Cact. Hort. Dyck. 1849. 40. 1850, as synonym) probably belong here.

Illustrations: Contr. U. S. Nat. Herb. 10: pl. 18; MacDougal, Bot. N. Amer. Des. pl. 16; Nat. Geogr. Mag. 21: 699, as *Pilocereus fulviceps;* Contr. U. S. Nat. Herb. 12: pl. 66.

Plate xi illustrates the top of a flowering plant in the New York Botanical Garden brought from Tehuacán, Mexico, by Dr. MacDougal and Dr. Rose in 1906. Figure 107 is from a photograph taken by Dr. Rose near Tehuacán, in 1906; figure 108 shows the flower of this plant; and figure 109 a longitudinal section of the flower.

7. **Pachycereus marginatus** (De Candolle) Britton and Rose, Contr. U. S. Nat. Herb. **12**: 421. 1909.

Cereus marginatus De Candolle, Mém. Mus. Hist. Nat. Paris **17**: 116.   1828.
Cereus gemmatus Zuccarini in Pfeiffer, Enum. Cact. 96.   1837.

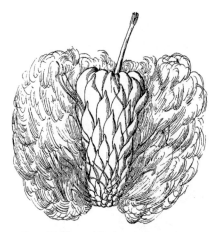

FIG. 108.—Flower of P. chrysomallus.   ✕0.7.   FIG. 109.—Longitudinal section of flower
                                                        of P. chrysomallus.   ✕0.7.

Stems 3 to 7 meters high, erect, usually simple; ribs 5 or 6 (7 in the original specimen), some-what acute when young, obtuse in age; areoles close together, usually confluent, their wool forming a dense white cushion along the ridge of each rib; spines at first 5 to 8 (1 central), in old areoles more numerous, 1 cm. long or less, but in flowering areoles often numerous, bristly and 2 cm. long; flowers and fruit usually closely set, one above the other, apparently only one at an areole, but recorded as often geminate, and appearing anywhere along the ribs from the top downward; flower funnelform, 3 to 4 cm. long including the ovary; tube and ovary more or less scurfy and with ovate scales sub-tending bunches of wool and small spines; fruit globular, about 4 cm. in diameter, not very fleshy, yellowish red within, covered with spines and wool which finally drop off; seeds numerous, black, and somewhat shining, rather large, 4 mm. long, the hilum depressed.

Type locality: Mexico.

Distribution: Hidalgo, Querétaro, and Guanajuato, and widely planted and naturalized throughout Mexico.

This species is commonly cultivated throughout central and southern Mexico as a hedge plant and when properly cared for forms an impene-trable barrier; it is there called organo.

Cereus cupulatus, Cereus incrustatus, and *Cereus mirbelii* are all re-ferred by Pfeiffer (Enum. Cact. 97. 1837) to this species. *Cereus incrus-tans* Steudel (Nom. ed. 2. **1**: 334. 1840) was only a garden name but was referred to this species by Steudel.

FIG. 110.—Flower of P. marginatus. Natural size.

Illustrations: Contr. U. S. Nat. Herb. **5**: pl. 59, 60; Bull. Soc. Acclim. France **52**: f. 8; Monatsschr. Kakteenk. **19**: 62; Reiche, Veg. Alred. Cap. Mex. f. 21, 22; Schumann, Gesamtb. Kakteen f. 17; U. S. Dept. Agr. Bur. Pl. Ind. Bull. **262**: pl. 6; Journ, Intern. Gard. Club **3**: 18, all as *Cereus marginatus;* Mo-natsschr. Kakteenk. **23**: 149, as *Cereus marginatus gibbosus;* Cact. Journ. **1**: 59; **2**: 169, as *Cereus gemmatus;* Schelle, Handb. Kakteenk. f. 22, as *C. marginatus gemmatus.*

Figure 110 shows a flower drawn from an herbarium specimen collected by Edward Palmer at San Luis Potosí, Mexico, in 1905; figure 111 is from a photograph of the plant used as a hedge near the City of Mexico.

Fig. 111.—Pachycereus marginatus used as a hedge plant in Mexico.

8. Pachycereus ruficeps (Weber).

Pilocereus ruficeps Weber in Gosselin, Bull. Mus. Hist. Nat. Paris 11: 509. 1905.
Cereus ruficeps Vaupel, Monatsschr. Kakteenk. 23: 27. 1913.

Stout, columnar, 15 meters high, from a simple trunk, 3 to 4 dm. in diameter, but branched above; branches erect; ribs about 26; young spines all reddish; radial spines 8 to 10, about 1 cm. long, rigid, grayish; central spines 1 to 3, the longest 4 to 5 cm. long, porrect or deflexed; flowers at the top of the plant, campanulate, 5 cm. long, the ovary and tube bearing small chartaceous scales, these with small tufts of felt and a few yellow bristles in their axils; stamens numerous, arranged in 2 series; style stout, light flesh-colored; stigma-lobes 7 to 9; fruit small, not edible; seeds small, brownish, shining.

Type locality: Near Tehuacán, Mexico.
Distribution: Oaxaca, Mexico.
This species has been described rather fully by Roland-Gosselin, but we are still in some doubt as to its relationship.
Dr. Rose collected flowers of it in 1905, but these were confused with specimens of *Cephalocereus macrocephalus*, which seems to indicate that the two species grow together.
When Dr. Rose was at the Museum of Paris in 1912 he was given a flower from the type collection made by M. Diguet.
Illustration: Bull. Soc. Acclim. France 52: 58. f. 17, as *Cereus ruficeps*.

9. Pachycereus lepidanthus (Eichlam).

Cereus lepidanthus Eichlam, Monatsschr. Kakteenk. **19:** 177. 1909.

Stems simple or with few stout branches, light green; ribs 7 to 9, rather low, separated by broad, rounded intervals; areoles about 1 cm. apart, small; radial spines about 10, slender, 1.5 cm. long or the longer ones 4 cm. long; the central ones stouter and somewhat flattened, 3 to 6 cm. long; flowers 7 cm. long, 2.5. cm. broad; perianth-segments arranged in 3 or 4 series, 2.5 cm. long, 8 mm. broad, below red, above sepia-brown, persisting on the fruit; ovary and flower-tube covered with membranous scales; fruit dry.

Type locality: Rancho San Agustin, Guatemala.

Distribution: Guatemala.

This plant resembles *Escontria chiotilla*, with which we at one time thought it was related, but it has very different areoles on the stems, while the areoles in the axils of the fruit scales, instead of being naked, are described as bearing felt and bristles, and the fruit as dry instead of juicy. We have studied living specimens of the plant both in the New York Botanical Garden and in the Cactus House at Washington, but none of these has flowered, and we know its flowers and fruits only from Eichlam's description above cited.

Illustration: Monatsschr. Kakteenk. **23:** 53, as *Cereus lepidanthus*.

10. Pachycereus columna-trajani (Karwinsky) Britton and Rose, Contr. U. S. Nat. Herb. **12:** 421. 1909.

Cereus columna-trajani Karwinsky in Pfeiffer, Enum. Cact. 76. 1837.
Cephalophorus columna-trajani Lemaire, Cact. Aliq. Nov. xii. 1838.
Pilocereus columna Lemaire, Cact. Gen. Nov. Sp. 9. 1839.
Pilocereus lateribarbatus Pfeiffer in Förster, Handb. Cact. ed. 2. 672. 1885.
Cephalocereus columna Schumann in Engler and Prantl, Pflanzenfam. 3^{aa}: 182. 1894.
Pilocereus columna-trajani Schumann, Gesamtb. Kakteen 198. 1897, as synonym.

Plants erect, stout, up to 15 meters high, 4.5 to 5 dm. in diameter, often simple; ribs many, green; areoles oblong, bearing brown felt; radial spines 8 to 10, 12 to 25 mm. long; central spines more elongated, sometimes 16 cm. long, deflexed; spines all rigid, white or horn-colored except the brown bases and tips, sometimes said to be soft and erect; flowers described as purple.

Type locality: San Sebastian, Puebla, Mexico.

Distribution: Puebla and Oaxaca, Mexico.

In 1906, Dr. Rose collected in the Tomellín Canyon in southern Mexico, not far from the type locality of this species, what appeared to him to be this species. It forms forests which cover the surrounding hills, but, unfortunately, no flowers or fruit could be procured.

Melocactus columna-trajani (Pfeiffer, Enum. Cact. 46. 1837) is usually referred to this species, but is not formally published at the place here cited.

Cereus lateribarbatus (Rev. Hort. **1862:** 427. 1862) belongs here, according to Lemaire.

Illustrations: Blanc, Cacti 77. f. 1715; Rev. Hort. **62:** 129. f. 40, as *Pilocereus columna-trajani;* Möllers Deutsche Gärt. Zeit. **29:** 354. f. 9; MacDougal, Bot. N. Amer. Des. pl. **22,** as *Pilocereus tetetzo;* Schelle, Handb. Kakteenk. f. 43, as *Cephalocereus columna-trajani.*

Plate xii is from a photograph taken by Dr. MacDougal at Tomellín Canyon, Mexico.

Cereus tetazo Coulter (Contr. U. S. Nat. Herb. **3:** 409. 1896; *Pilocereus tetetzo* Weber in Schumann, Gesamtb. Kakteen 175. 1897), which we first confused with *Pachycereus columna-trajani,* is not of this genus, for its ovary is glabrous and the fruit more or less fleshy and edible. Coulter, however, does state that it is closely related, if not identical, with one of the species of this genus, that is, *Pachycereus pecten-aboriginum.* It should be compared with *Cephalocereus macrocephalus.*

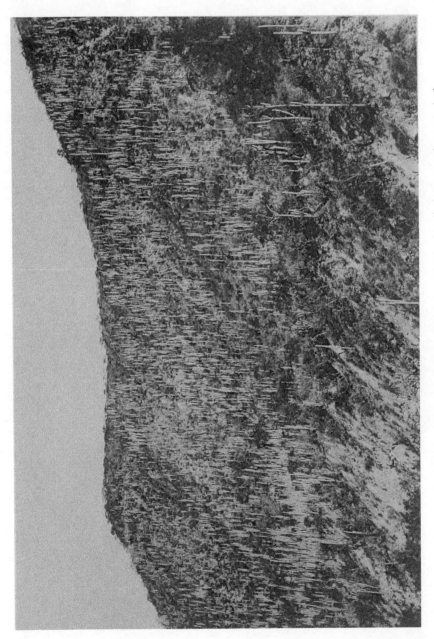

A mountain-side along Tomellin Canyon, Mexico, covered with *Pachycereus columna-trajani.*

10. LEPTOCEREUS (Berger) Britton and Rose, Contr. U. S. Nat. Herb. **12**: 433. 1909.

Arborescent, bush-like, vine-like, or diffusely branching cacti; joints with 3 to 8 prominent, thin, high, crenate ribs, without aërial roots; spines slender, acicular; flowers diurnal, small; ovary spiny; flower-tube short, campanulate, spiny; stamens very numerous, borne at the base of the throat, scarcely exserted; stigma-lobes a little exceeding the stamens; fruit globose to oblong, more or less spiny, fleshy; seeds numerous, black.

Type species: *Cereus assurgens* Grisebach.

This genus is composed of eight species, six of them Cuban, one Santo Domingan, and one Porto Rican. Some are weak and clambering; others develop woody trunks. The branches are strongly ribbed and armed with clusters of long acicular spines. The earliest species were referred to *Cereus*. *Leptocereus assurgens* and *L. quadricostatus*, the only species known to Schumann, were placed by him in different sections of the genus *Cereus*, the former in his series *Tortuosi* and the latter in his series *Oligogoni*. A. Berger in his treatment of *Cereus* proposed the subgenus *Leptocereus*, which we afterward raised to generic rank. *Cereus gonzalezii* and *C. tonduzii*, also referred here by Berger, we have referred to other genera.

The generic name is from the Greek, signifying thin-cereus, referring to the thin ribs.

KEY TO SPECIES.

```
Ultimate joints slender, 1 to 2 cm. thick.
    Vine-like, elongated; ultimate joints 1 cm. thick, 4 to 7-ribbed.........................1. L. weingartianus
    Ultimate joints 2 cm. thick, 6 to 8 ribbed.
        Tree-like, 5 meters high, the trunk 3 cm. in diameter at the base; flowers 3.5 cm. long,
            sparingly short-spiny......................................................2. L. leonii
        Prostrate, creeping; flowers 1.5 cm. long, densely long-spiny......................3. L. prostratus
Ultimate joints stout, 2 to 6 cm. thick.
    Fruit densely long-spiny.
        Bush-like, 1 to 3 meters high; ultimate joints 2 to 3 cm. thick; fruit 3 to 6 cm. long.
            Joints 4-ribbed; spines brown...............................................4. L. assurgens
            Joints 5 to 7-ribbed; spines yellow.........................................5. L. maxonii
        Tree-like, 5 to 6 meters high; ultimate joints 5 to 6 cm. thick; fruit 8 to 10 cm. long...6. L. arboreus
    Fruit sparingly short-spiny; tree-like species.
        Joints 5 to 7-ribbed, the ribs very broad; fruit 7 to 8 cm. long....................7. L. sylvestris
        Joints 3 to 8-ribbed, the ribs low; fruit 3 to 5 cm. long...........................8. L. quadricostatus
```

1. Leptocereus weingartianus (Hartmann).

Cereus weingartianus Hartmann in Dams, Monatsschr. Kakteenk. **14**: 155. 1904.

Roots in clusters, tuberous, thick; stems becoming terete and woody below, the branches creeping or climbing among shrubs and trees, sometimes to the height of 8 to 10 meters; stems 4 to 7-ribbed, at first slender and weak, 1.5 to 2 cm. in diameter; areoles 1 to 1.5 cm. apart, circular,

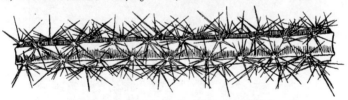

FIG. 112.—Part of branch of L. weingartianus. X0.7.

small, at first filled with short, whitish wool (afterwards disappearing) and acicular spines at first brownish or yellowish brown but in age becoming gray; radial spines 10 to 12, spreading; central spines a little stouter than the radials, a little spreading, often as many as 6, bulbose at base, 1 to 1.5 cm. long; flowers small, about 4 cm. long; fruit about 2 cm. long, covered with clusters of small, deciduous spines.

Type locality: Haiti.

Distribution: Hispaniola.

This species has heretofore been known only from the type material from "Haiti." In 1913 Dr. Rose collected an abundance of both living and herbarium material near

Azua, Santo Domingo, and a little later received living specimens from Father M. Fuertes, of Barahona, Santo Domingo. Dr. Rose's material showed for the first time the peculiar root system of this species. With it also were old flowers and fruit, heretofore unknown. The species is rare about Azua, only two stations being found in the lower foothills north of the town (No. 3941). Dr. Paul Bartsch collected specimens in Haiti in 1917 (No. 221).

Illustration: Monatsschr. Kakteenk. **14:** 155, as *Cereus weingartianus.*

Figure 112 shows part of a branch of a plant collected by Dr. Rose at Azua, Santo Domingo, in 1913.

2. Leptocereus leonii Britton and Rose, Torreya **12:** 15. 1912.

> *Cereus leonii* Vaupel, Monatsschr. Kakteenk. **22:** 66. 1912.

Plant up to 5 meters high, repeatedly branching, the rounded trunk 3 cm. in diameter at the base, the cortex scaly-roughened; ultimate branches about 1.5 cm. in diameter, slender, elongated, 6 to 8-ribbed; old areoles 1 to 1.5 cm. apart in vertical rows, bearing acicular spines; ribs crenate, with the areoles borne at the depressions; spines 6 to 12 at an areole, long, yellowish when young, gray when old, 2 to 9 cm. long; flower 3.5 cm. long, campanulate; inner perianth-segments pink, about 15, withering-persistent; tube of flower bearing scattered areoles each with 1 to 4 short spines or some of them spineless; fruit globose-ovoid, 2 cm. in diameter, with a few scattered spine-bearing areoles; seeds black.

Type locality: Sierra de Anafe, near Guayabal, Cuba.

Fig. 113.—Leptocereus leonii.

Distribution: On limestone rocks, Sierra de Anafe and Sierra de Guane, western Cuba.

The wood is very hard; the flowers appear from August to November.

At the type locality this tree-like species inhabits a steep rocky slope and cliff, difficult of access, growing as a colony.

Figure 113 is from a photograph of the type plant, obtained by Brother Léon, of the Colegio de la Salle, Habana, in whose honor the species was named.

3. Leptocereus prostratus sp. nov.

Plant prostrate, bright green, 7-ribbed, 1.5 to 2 cm. thick, the ribs scarcely crenate; areoles elevated, about 1 cm. apart; spines 15 to 20 at an areole, acicular, 1 to 2 cm. long, yellow when young, gray when old; ovary densely covered with yellow spines; perianth about 1.5 cm. long; fruit about 1.5 cm. in diameter.

On high, dry, exposed rocks, La Guira, north of Sumidero, Pinar del Rio, Cuba (Shafer, No. 13754, August 17, 1912).

Leptocereus prostratus is related to *L. leonii*, which differs in having an erect trunk, the ribs of the branches deeply crenate, the areoles depressed in the crenatures, and larger flowers and fruit.

4. Leptocereus assurgens (C. Wright) Britton and Rose, Contr. U. S. Nat. Herb. 12: 433. 1909.

Cereus assurgens C. Wright in Grisebach, Cat. Pl. Cub. 116. 1866.

Plant 2 to 3 meters high, not much branched, the ultimate joints 3 cm. in diameter or less; ribs 4; areoles 1 to 2.5 cm. apart; spines acicular, brown, 2 to 8 cm. long; flowers 4 to 5 cm. long; tube and ovary bearing scattered clusters of spines; inner perianth-segments short, numerous, spreading or even turned backward; stamens and style pale greenish white; fruit covered with clusters of short spines.

Type locality: Western Cuba.
Distribution: On limestone, near northern coast of Habana Province, Cuba.

Fig. 114.—Leptocereus assurgens.

This species was long known only from the collections of Charles Wright, but has been rediscovered by collectors connected with the New York Botanical Garden.

The name *Cereus pellucidus* (Pfeiffer, Enum. Cact. 108. 1837) belongs to this species or to some other Cuban member of the genus; the published description is not sufficiently complete to enable us to identify the plant more accurately.

Illustration: Schumann, Gesamtb. Kakteen f. 33, as *Cereus assurgens.*

Plate VIII, figure 4, shows a plant collected by Britton and Cowell at Cojimar, Cuba, in 1911, which flowered in the New York Botanical Garden in July 1915. Figure 114 is from a photograph obtained by Brother Léon at the same locality.

5. Leptocereus maxonii sp. nov.

Stems 1 to 1.5 meters high, more or less branched, erect or sometimes with recurved branches; ribs 5 or 7, usually 6, thin, 6 to 15 mm. deep, scalloped; areoles 1.5 to 2 cm. apart, circular; spines when young of a decided yellowish-brown color, dark brown or sometimes whitish in age, about 20 from an areole, needle-shaped, the longer ones 3 cm. long; flowers 5 to 6 cm. long; inner perianth-segments about 32, spreading at right angles to the tube, linear-oblong, yellowish green inside, the outer obtuse, the inner acute; stamens cream-colored; ovary and flower-tube densely covered with yellowish spines; immature fruit bur-like, 4 cm. long, densely covered with yellow or brownish spines.

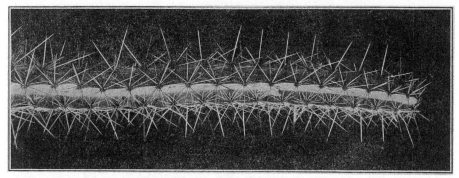

Fig. 115.—Leptocereus maxonii.

Collected by Wm. R. Maxon at Berraco, 8 miles east of Daiquiri, Cuba, April 13, 1907 (No. 4023), and by Britton and Cowell at the same locality, March 1912 (No. 12657, type).

This species differs from *L. assurgens* in habit, in having more ribs, and in the flowers and young shoots being covered with yellow spines and bristles instead of dark brown ones.

Figure 115 is from a photograph of a branch of the plant collected by Mr. Maxon as above cited.

6. Leptocereus arboreus Britton and Rose, Torreya 12: 15.   1912.

> *Cereus arboreus* Vaupel, Monatsschr. Kakteenk. 22: 65. 1912.

Plants up to 6 meters high, erect, much branched; joints 3 to 10 dm. long, 5 to 6 cm. in diameter, narrowed at base; ribs 4, narrow, thin, 1.5 to 2 cm. deep, somewhat depressed between the areoles; areoles 2.5 to 4 cm. apart or less; spines 10 or fewer, acicular, yellowish, becoming gray, radiating, the longer up to 5 cm. long; flower short, campanulate, 2 to 3 cm. long; inner perianth-segments short, spreading, greenish white to cream-colored; ovary and flower-tube very spiny; fruit ellipsoid, 8 to 10 cm. long, 5 to 6 cm. in diameter, its areoles bearing tufts of numerous light-yellow spines.

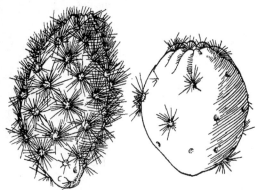

Fig. 116.—Fruit of Leptocereus arboreus.   ×0.7.
Fig. 117.—Fruit of Leptocereus sylvestris.   ×0.7.

Type locality: Punta Sabanilla, Santa Clara, Cuba.

Distribution: Near southern coast of the Province of Santa Clara, Cuba.

Plate XIII, figure 1, shows the plant collected by Britton, Cowell, and Earle at Castillo de Jagua, Cuba, in 1911, which flowered in the New York Botanical Garden in 1913. Figure 116 shows a fruit of the type specimen.

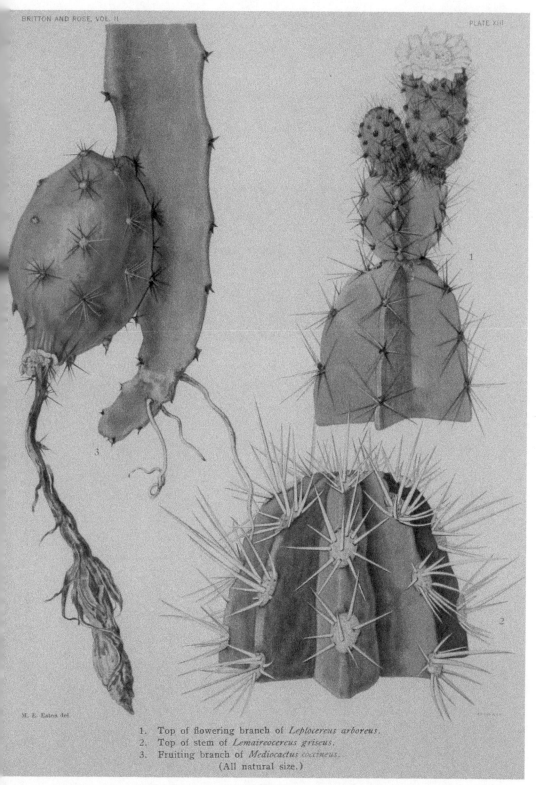

M. E. Eaton del.

1. Top of flowering branch of *Leptocereus arboreus*.
2. Top of stem of *Lemaireocereus griseus*.
3. Fruiting branch of *Mediocactus coccineus*.
(All natural size.)

7. Leptocereus sylvestris sp. nov.

Tree-like, up to 5 meters high; joints 2 to 3 cm. in diameter, 5 to 7-ribbed; ribs strongly crenate; areoles 1 to 1.5 cm. apart; spines light brown, long and acicular, the longest ones 9 cm. long; fruit subglobose, 7 to 8 cm. long, bearing clusters of short spines, these early deciduous.

Collected by Britton, Cowell, and Shafer in coastal woods, Ensenada de Mora, Province of Oriente, Cuba, March 20 to 29, 1912, No. 13060.

Figure 117 shows a fruit of the type specimen and figure 118 a branch.

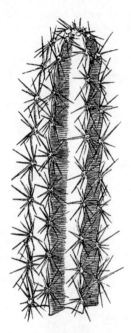

FIG. 118.—Top of branch of Leptocereus sylvestris. ×0.6.

FIG. 119.—Leptocereus quadricostatus.

8. Leptocereus quadricostatus (Bello) Britton and Rose, Contr. U. S. Nat. Herb. 16: 242. 1913.

Cereus quadricostatus Bello, Anal. Soc. Esp. Hist. Nat. 10: 276. 1881.

Plants erect or arching, up to 4 meters high, with numerous lateral, usually elongated branches, often forming thickets; branches dull, dark green, usually 4-ribbed, sometimes 3-ribbed, the ribs thin and low; spines acicular, 1 to 4 cm. long; flowers 4 cm. long, 2 cm. wide at the mouth; outer perianth-segments green; inner perianth-segments greenish white or yellowish white, truncate, the apex lacerate or erose; ovary and flower-tube bearing a few clusters of short spines; style and filaments greenish; fruit subglobose to obovoid, 3 to 5 cm. long, not very spiny, red.

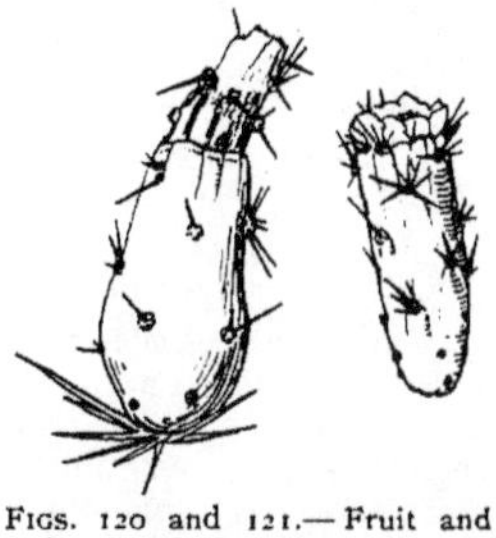

Type locality: Porto Rico.

Distribution: Southwestern Porto Rico.

This plant inhabits hillsides and plains in the dry southwestern part of Porto Rico, sometimes forming dense thickets, penetrable only by the use of the machete; it is known as sebucan.

FIGS. 120 and 121.—Fruit and flower of L. quadricostatus. ×0.7.

Figure 119 is from a photograph taken by Frank E. Lutz at Ensenada, near Guanica, Porto Rico, in 1915; figure 120 shows a fruit collected by Dr. Britton and Dr. Shafer at Guanica in 1913; figure 121 shows a flower from a plant at the same locality.

PUBLISHED SPECIES, PERHAPS OF LEPTOCEREUS.

CEREUS PANICULATUS De Candolle, Prodr. **3**: 466. 1828.

Cactus paniculatus Lamarck, Encyl. **1**: 540. 1783.

This has long been in doubt and is known only from imperfect description and illustration. Lamarck states that it is from Santo Domingo, in a region called cul-de-sac, and is based on Burmann's plate 192 of Plumier. It is apparently a *Leptocereus*, perhaps *L. weingartianus*.

11. **EULYCHNIA** Philippi, Fl. Atac. 23. 1860.

Stout, erect or procumbent and ascending, green cacti, usually with many branches, the branches parallel-ribbed, armed with spines; perianth white to pinkish, withering and persisting on the ovary; flowers single at the areoles, opening during both day and night, short and broad for the group, with an open throat, the tube very short if not wanting altogether; scales on ovary and flower-tube numerous, their axils usually with bristles or long hairs; filaments very short, covering the face of the throat; style short and thick; fruit globular, fleshy, somewhat acid, hardly edible; seeds small, dull black, containing endosperm (according to Mr. Söhrens).

Type species: *Eulychnia breviflora* Philippi.

This genus as here defined contains 4 species found along the coast and central valleys of the provinces of Aconcagua, Coquimbo, Atacama, Antofagasta, and Tarapaca, Chile.

To this group, treated as a subgenus of *Cereus* by Mr. Berger, has been referred a number of anomalous species which we place elsewhere; they are similar to this genus in the fact that they have very short flower-tubes, but in habit, fruit, and other characters they are quite distinct. These species will be discussed in this work under other genera.

The genus *Eulychnia* was first established in 1860 by Rudolph Philippi, who based it upon a single species, *E. breviflora*. In 1864 two other species, *E. acida* and *E. castanea*, were described, while the fourth is transferred by us from *Cereus*.

The plants are usually found on dry hills, and are often associated with other cacti and other desert plants. In many regions they form the dominant feature in the vegetation. At least two species are commonly used for fuel, and one (*E. acida*) is used for hedge fences.

The generic name is from the Greek, signifying a candlestick.

KEY TO SPECIES.

```
Areoles of the ovary and perianth-tube without stiff bristles.
    Areoles of the ovary and perianth-tube with long wool.
        Wool chestnut-brown; areoles of the joints small, little felted...........................1. E. spinibarbis
        Wool white; areoles of the joints large, approximate, densely felted......................2. E. iquiquensis
    Areoles of the ovary and perianth-tube with very short wool.................................3. E. acida
Areoles of the ovary and perianth-tube with stiff brownish bristles and short wool..............4. E. castanea
```

1. Eulychnia spinibarbis (Otto).

Cereus spinibarbis Otto in Pfeiffer, Enum. Cact. 86. 1837.
Cereus panoplaeatus Monville, Hort. Univ. **1**: 220. 1840.
Eulychnia breviflora Philippi, Fl. Atac. 24. 1860.
Echinocereus spinibarbis Schumann, Monatsschr. Kakteenk. **5**: 124. 1895.
Cereus breviflorus Schumann, Gesamtb. Kakteen Nachtr. 23. 1903.

Stems 2 to 4 meters high, much branched; branches 7.5 cm. in diameter; ribs 12 or 13; spines about 20 from an areole, usually 18 mm. long, but the longest one at times 15 cm. long; flowers 3 to 5 cm. long; scales on ovary and flower small, bearing in their axils long brown wool; outer perianth-segments short, acuminate; inner perianth-segments white to pinkish, oblong, 2 cm. long, acute; style short, 1.5 cm. long including the stigma-lobes; scales on the ovary small, their axils filled with long brown wool.

Type locality: Near Coquimbo, Chile.
Distribution: Along the coast of the province of Coquimbo, Chile.

The only difference Dr. Rose was able to observe in the field between this species and *E. acida* is the very woolly flower and fruit of *E. spinibarbis*. The flowers seemed to be identical in form, size, and color and at La Serena the two were growing side by side; the form with long silky wool on the flowers was never seen in the interior valleys, although thousands of flowers were there observed.

Cereus tortus (Förster, Handb. Cact. 391. 1846) may belong here.

Cereus chilensis breviflorus (Hirscht, Monatsschr. Kakteenk. 8: 159. 1898) doubtless belongs here, but has not been described.

Illustrations: Fl. Atac. pl. 2, f. A, as *Eulychnia breviflora;* Rep. Mo. Bot. Gard. 16: pl. 4, f. 1, as *Cereus breviflorus;* Schumann, Gesamtb. Kakteen f. 11, as *C. coquimbanus.*

Figure 122 is from a photograph of a plant brought by Dr. Rose to the New York Botanical Garden from the Botanical Garden of Santiago, Chile, under the name *Eulychnia breviflora*, in 1914.

2. Eulychnia iquiquensis (Schumann).

Cereus iquiquensis Schumann, Monatsschr. Kakteenk. 14: 99. 1904.

Plant 2 to 7 meters high, when old quite spineless below, but very spiny toward the top; trunk usually very short, 2 to 2.5 cm. in diameter, its outer layers pulpy and yellow, terete, with many branches from near the base, these nearly erect or more or less spreading and again branching; ribs 12 to 15, no broader at base than above, somewhat tuberculate, separated by acute intervals; areoles approximate, sometimes with only a very little space between them, 5 to 10 mm. in diameter, with short white wool, on many old stems and branches the areoles die and fall, leaving a row of indentations along the top of the rib; spines various, on vigorous sterile shoots about 12 to 15 at an areole, most of them about 1 cm. long, while 1 or 2 are very stout, porrect, elongated, and sometimes 12 cm. long; on flowering branches the spines numerous, soft and hair-like or some of them bristle-like; flowers borne near the tops of branches, 6 to 7 cm. long including the ovary; flower-buds globular, covered with long, white, silky hairs; inner perianth-segments white, short; fruit globular, 5 to 6 cm. in diameter, fleshy, said to be acid, densely clothed with white hairs; seeds not known.

FIG. 122.—Eulychnia spinibarbis.

Type locality: Iquique, Province of Tarapaca, Chile.

Distribution: On top and slopes of the coastal hills in the Provinces of Atacama, Antofagasta, and Tarapaca, Chile.

According to published records, this species is known only from the original collection made by Carlos Reiche in 1904, and has never been in cultivation. Dr. Rose collected living, herbarium, and formalin specimens in 1914 not only at Iquique, but also at Antofagasta. It grows only on the coastal hills, which at both towns come down almost to the sea or rise from a narrow coastal plain, and is not found on the pampas, which extend east of the coastal hills to the Andes. In both the Provinces of Antofagasta and Tarapaca, it is the most conspicuous plant seen, in fact it is the only woody plant met with on their western borders. It is called by the natives copado, and the old dead branches are carried to the towns and used for firewood.   The flowers begin to appear late in October; the fruit is eaten by animals, doubtless by birds, as all old fruits had large holes on one side, and no seeds remained.

Plate xv, figure 1, shows the top of plant collected by Dr. Rose at Antofagasta, 1914.

3. Eulychnia acida Philippi, Linnaea 33: 80.   1864.

Cereus acidus Schumann, Gesamtb. Kakteen Nachtr. 22.   1903.

Plant various in habit, usually 3 to 7 meters high, with a definite trunk 1 meter long and then more or less branching, forming a more or less rounded top, but sometimes without trunk, forming a low mass 1 meter high or less, with branches often procumbent or ascending; ribs 11 to 13, broad and low; spines various, nearly porrect, grayish in age but brownish when young, sometimes 20 cm. long; flowers 5 cm. long, turbinate, 13 cm. in circumference at top; ovary and tube covered with small, ovate, imbricating scales, fleshy at base but with acute, callous tips; limb somewhat oblique; inner perianth-segments at first pale rose-colored, then white, 20 to 22 mm. long; throat very short, covered with stamens; stamens white, 1 to 1.5 cm. long, included; style 2 cm. long, stiff, white, with 12 to 15 stigma-lobes; fruit fleshy, somewhat acid.

Type locality: Near Illapel and Choapa, Chile.
Distribution: From near Choapa to Copiapo, in western Chile.

This species is called tuna de cobado by the natives, according to Philippi.

This species was originally described from material obtained by Landbeck near Illapel and Choapa, but nothing of the type has been preserved in the Philippi herbarium at San-

tiago.   Dr. Rose, however, visited both Illapel and Choapa in 1914, and was able to decide definitely upon the species described by Philippi.   At both places *E. acida* was quite common, usually grow-ing with *Cereus chiloensis*, but from which it differs so much in habit and flowers that one is soon able to distinguish the two readily.

It is sometimes referred to as *Cereus chilensis acidus* (Monatsschr. Kakteenk. 8: 159. 1898), but the name has never been formally published.

Figure 123 shows a flower collected by Dr. Rose at Illapel, Chile, in 1914.

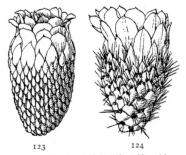

123   124

F<small>IG</small>. 123.—Flower of Eulychnia acida. X 0.7.
F<small>IG</small>. 124.—Flower of Eulychnia castanea. X 0.7.

4. Eulychnia castanea Philippi, Linnaea 33: 80.   1864.

Cereus castaneus Schumann, Gesamtb. Kakteen Nachtr. 22. 1903.

Forming dense thickets sometimes 20 meters broad; branches 6 to 8 cm. in diameter, spreading at base or decumbent, with ascending tips, reaching a height of 1 meter or less; ribs 9, 10, or 11, low and rounded; areoles about 1 cm. apart, large and circular; spines, when young, yellow with brown tips, gray or nearly white in age; radial spines 8 to 10, unequal but short, usually 5 to 20 cm. long; central spine 1, 6 to 10 cm. long, stout, porrect; flowers borne near the tips of the branches, 3 to 5 cm. long; ovary tuberculate, its numerous areoles with short brown wool and slender brown bristles 1 to 1.5 cm. long, resembling somewhat a chestnut bur; areoles subtended by minute scales each with a callous tip; inner perianth-segments 1 to 1.5 cm. long, broad, with mucronate tips, white or pinkish; fruit globular, said to be insipid, 5 cm. in diameter, fleshy, the small scales persistent, but nearly devoid of bristles except near the top, crowned by the withering perianth; seeds 1.5 mm. long, dull black.

Type locality: Near Los Molles, Province of Aconcagua, Chile.
Distribution: On bluffs near and facing the sea along the shores of Aconcagua from Los Molles to Los Vilos.

The history of this species, though short, is interesting.   It was collected by Land-beck, at Los Molles, Chile, in November 1862, and was described by Rudolph Philippi in 1864.   The type material, consisting of two flowers and a few bunches of spines, is preserved in the Museo Nacional de Santiago.   Unfortunately, the original material and labels had been mixed with other species, but Dr. Rose, who studied the Philippi collection in 1914, was able to make the separation, and through the kindness of the Director, brought back a flower and cluster of spines, which are now preserved in the United States National Herbarium in Washington. From 1862 to 1914 there is no record that this species has been seen by botanists. Dr. Rose, while exploring in Chile, after several efforts was finally successful in obtaining living, herbarium, and formalin material (No. 19393), and also a fairly good photograph.

Figure 124 shows a flower collected by Dr. Rose at the type locality in 1914.

12. LEMAIREOCEREUS Britton and Rose, Contr. U. S. Nat. Herb. **12**: 424. 1909.

Stenocereus Riccobono, Boll. R. Ort. Bot. Palmero **8**: 253. 1909.

Plants usually large, tall, and branching, but rarely low, nearly prostrate, simple, forming thickets; areoles rather large, felted; spines usually stout and numerous; flowers diurnal or in some species nocturnal, one at an areole, tubular-funnelform or campanulate, the short tube tardily separating with the style from top of the ovary; stamens numerous, borne in many rows all along the inner surface of the throat; ovary more or less tubercled, bearing scales felted in the axils, the areoles at first spineless or nearly so, soon developing a cluster of spines; fruit globular to oval, often edible, irregularly bursting when old, exposing the seeds, at first very spiny, but when ripe the spines are often deciduous; seeds many, black.

The genus commemorates Charles Lemaire (1801–1871), a distinguished French cactologist and horticulturist; it consists of about 21 species, distributed from southern Arizona and Cuba to Peru and Venezuela.

Type species: *Cereus hollianus* Weber.

KEY TO SPECIES.

```
A. Ribs 6 to 20, separated by deep intervals.
   B. Areoles with white, brown, or gray felt, not glandular.
      Spines slender, acicular to subulate.
         Spines not appressed to the joints, a central one usually evident.
            Ribs 6 to 12.
               Areoles borne on ribs, when these are crenate borne on elevations.
                  Joints green, not glaucous.
                     Flowers 10 cm. long; central spine long, reflexed.........  1. L. hollianus
                     Flowers 7 to 9 cm. long; central spine spreading or
                        ascending.............................  2. L. hystrix
                  Young growth glaucous, the bloom persistent as curved,
                        whitish streaks.
                     Spines subulate; plants relatively light green.
                        Ribs 8 to 10; young growth slightly glaucous......  3. L. griseus
                        Ribs 6 or 7; young growth definitely glaucous.
                           Spines terete, 5 cm. long or less.............  4. L. pruinosus
                           Spines flattened above, up to 8 cm. long.......  5. L. longispinus
                     Spines acicular; plants dark green....................  6. L. eichlamii
               Areoles borne in depressions of the crenate ribs.
                  Plants bright green.
                     Flowers greenish yellow to rose.
                        Ribs 9 to 12; flowers greenish yellow..............  7. L. chichipe
                        Ribs 7 to 9; flowers rose-colored..................  8. L. chende
                     Flowers white.......................................  9. L. godingianus
                  Plants glaucous, the bloom persistent as whitish streaks.
                     Ribs 6 to 8, bluntly acute........................ 10. L. aragonii
                     Ribs 8 to 12, rounded............................. 11. L. stellatus
            Ribs about 20............................................. 12. L. treleasei
         Spines usually all radial, appressed to the joints................ 13. L. deficiens
      Spines very stout, at first reddish brown or nearly black............ 14. L. weberi
   BB. Areoles with dark brown or black felt, glandular.
      Ribs 6 to 8.
         Scales of the ovary 2 mm. long or less........................... 15. L. queretaroensis
         Scales of the ovary 4 to 6 mm. long.............................. 16. L. montanus
      Ribs 12 to 17................................................... 17. L. thurberi
A. Ribs 3 to 7, separated by broad and shallow intervals.
   Areoles large, widely separated.
      Stems very stout, erect.
         Stems bluish gray; spines of fruit brown....................... 18. L. laetus
         Stems green; spines of fruit yellow........................... 19. L. cartwrightianus
      Stems slender, weak, usually 3 or 4-ribbed......................... 20. L. humilis
   Areoles small, nearly contiguous................................... 21. L. dumortieri
```

1. **Lemaireocereus hollianus** (Weber) Britton and Rose, Contr. U. S. Nat. Herb. **12**: 425. 1909.

Cereus hollianus Weber in Coulter, Contr. U. S. Nat. Herb. **3**: 411. 1896.
Cereus bavosus Weber in Schumann, Gesamtb. Kakteen 84. 1897.

Stem simple or branching only at base, 4 to 5 meters high; ribs 8 to 12, acute; areoles 1 to 3 cm. apart; spines at first bright red, but soon gray; radial spines about 12, very unequal, 1 to 3 cm. long, mostly spreading; centrals 3 to 5, swollen at base, unequal, the lower ones much longer than the others, sometimes 10 cm. long, strongly deflexed; flowers borne at the upper areoles, 10 cm. long, white; scales on ovary and flower-tube with lanate and bristly axils; fruit "as large as a goose egg," dark purple to red, covered with clusters of spines and bristles; seeds black, shining.

Type locality: Tehuacán, Puebla, Mexico.
Distribution: Puebla, Mexico.

This is a remarkable species, with unusually large fruit. It is called by the Mexicans bavoso.

The two names *C. hollianus* and *C. bavosus* are based on Weber's collection of 1864–66, and hence the latter is a synonym.

About the town of Sebastian in southern Puebla it is used as a hedge plant as well shown in our illustration.

Cereus brachiatus Galeotti (Salm-Dyck, Cact. Hort. Dyck. 1849. 195. 1850) must be very close to *L. hollianus*, if not identical, although Schumann did not believe they were the same; both came from near Tehuacán, Mexico. *Cereus militaris californicus* (Schumann, Gesamtb. Kakteen 85. 1897) is said to be a horticultural form of *Cereus bavosus*.

Illustrations: Contr. U. S. Nat. Herb. 10: pl. 19, as *Cereus hollianus;* Möllers Deutsche Gärt. Zeit. **29**: 438. f. 14, as *Cereus bavosus.*

Fig. 125.—Lemaireocereus hollianus.

Figure 125 is from a photograph by Dr. Rose at Sebastian, Puebla, Mexico, in 1905.

2. **Lemaireocereus hystrix** (Haworth) Britton and Rose, Contr. U. S. Nat. Herb. **12**: 425. 1909.

Cactus hystrix Haworth, Suppl. Pl. Succ. 73. 1819.
Cereus hystric Salm-Dyck, Observ. Bot. **3**: 7. 1822.
Echinocactus hystrix Haworth, Phil. Mag. **7**: 116. 1830.

Plant often 8 to 12 meters high and then with 10 to 50 erect branches; trunk short, often indefinite, sometimes 3 dm. in diameter; branches 7 to 10 cm. in diameter, with 9 or 10, rarely 12, ribs separated by V-shaped intervals; spines gray with brown tips, acicular, the radials about 10; central spines usually 3, one often longer than the others, often 4 cm. long; flower, including the ovary, 8 to 9 cm. long; tube 5 cm. long, broadly obconic, 3 cm. broad at mouth, spineless, purplish to dark green, bearing few short broad scales; inner perianth-segments white, spreading or recurved; stamens numerous, erect, white; style white, slender, club-shaped; ovary tuberculate, spineless, bearing small ovate scales; fruit 5 to 6 cm. long, longer than broad, scarlet, covered with clusters of deciduous spines, when mature breaking open and exposing the dark red pulp.

Type locality: West Indies.
Distribution: Dry parts of Cuba, Jamaica, Hispaniola, and the Porto Rican islands Desecheo and Cayo Muertos.

On the outskirts of Kingston, Jamaica, the stout branches are planted close together, forming a fence or an almost impenetrable hedge about fields, especially along the roadsides

The flowers of this species open at about 7 o'clock in the evening in Jamaica. The style is exserted from the tip of the bud several days before the flower opens, but it seems to be withdrawn before the flower is ready to expand. The flower-tube cuts off from

FIG. 126.—Lemaireocereus hystrix.

the ovary as is done in *Cereus*, except that the style comes off with the perianth. Numerous wasps visit the flowers to gather the nectar which oozes from the back of the scales on the flower-tube.

The plant is called Spanish dildos in Jamaica. On hillsides at the United States Naval Station, Guantánamo Bay, Cuba, this cactus occurs in great abundance, forming large colonies, individual plants differing much in the length of their spines, which in some are all less than 1 cm. long.

Illustrations: Journ. N. Y. Bot. Gard. **10:** f. *20*, as *Cereus hystrix;* Gard. Chron. II. **10:** 185. f. *37;* Möllers Deutsche Gärt. Zeit. **18:** 342, as *Cereus swartzii.*

Figure 126 is from a photograph taken by Marshall A. Howe near Guantánamo Bay, Cuba, in 1909; figure 127 shows a flower collected by William Harris in Jamaica and figure 128 a fruit from the same source.

Here is probably to be referred Descourtilz's plate 419 (Fl. Med. Antill., vol. 6), which he supposed to be *Cactus fimbriatus* Lamarck. This plate seems to be based largely on Burmann's plate of Plumier 195, f. 2. This latter plate was made the type of Haworth's *Cereus grandispinus* (Phil. Mag. **7:** 113. 1830; *Pilocereus grandispinus* Lemaire, Rev. Hort. **1862:** 427. 1862).

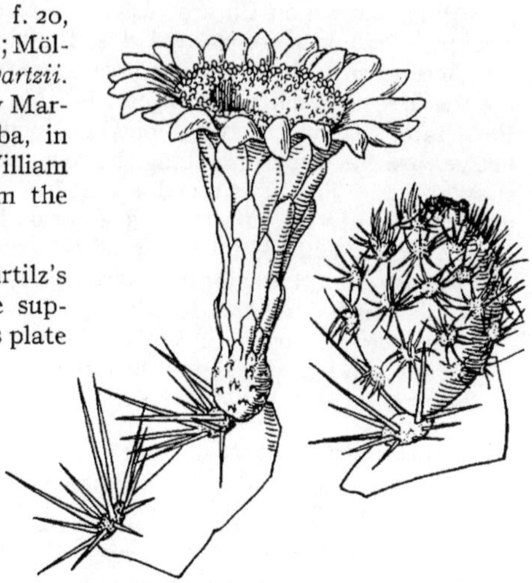

3. **Lemaireocereus griseus** (Haworth) Britton and Rose, Contr. U. S. Nat. Herb. **12:** 425. 1909.

FIG. 127.—Flower of Lemaireocereus hystrix. ×0.7.
FIG. 128.—Fruit of same.

Cereus griseus Haworth, Syn. Pl. Succ. 182. 1812.
Cereus eburneus Salm-Dyck, Observ. Bot. **3:** 6. 1822.
Cereus crenulatus griseus Salm-Dyck in Pfeiffer, Enum. Cact. 85. 1837.
Cereus eburneus polygonus Pfeiffer, Enum. Cact. 91. 1837.
Cereus resupinatus Salm-Dyck, Allg. Gartenz. **8:** 10. 1840.
Cereus gladiger* Lemaire, Hort. Univ. **6:** 60. 1845.

*At place cited, by error spelled *gladiiger*, which some have cited as *gladilger*, thus making another error.

Plant 8 meters high or less, sometimes branching at the base, sometimes with a definite trunk up to 3.5 dm. in diameter, smooth when old; branches 8 to 10-ribbed, more or less glaucous; spines acicular, gray, the longer ones 4 cm. long; flower-bud obtuse or rounded at apex, covered with overlapping scales, these obtuse and brown; flowers pinkish, 7 cm. long; inner perianth-segments white; style exserted before the flower opens; fruit subglobose, about 5 cm. in diameter, spiny, edible, the pulp red.

Type locality: South America, but no definite locality cited.

Distribution: Northern coast of Venezuela and adjacent islands; Curaçao; Aruba; Bonaire; Margarita; Patos Island, Trinidad; and now cultivated in many parts of tropical America for its delicious fruits.

Cereus polygonatus (Pfeiffer, Enum. Cact. 91. 1837) was given as a synonym of *C. eburneus polygonus.*

Cactus coquimbanus, a Chilean species, has sometimes been confused with this species.

Cereus gladiger, sometimes referred to Cels and sometimes to Lemaire as the author, seems to have come originally from Colombia.

In this species as well as in many others, abnormal forms occur, among which is *C. eburneus monstrosus* Salm-Dyck (De Candolle, Prodr. 3: 465. 1828).

Cereus enriquezii (Monatsschr. Kakteenk. 19: 92. 1909) was sent to Europe from Jalapa, Mexico, by Señor Murrilo. It is considered by W. Weingart to be *C. eburneus monstrosus.*

The common cultivated species of Mexico seems to belong here.

According to Boldingh, this cactus is known in the Dutch West Indies as daatoe, kadoesji, and jaatoe. It is widely grown on Curaçao Island as a hedge plant, where the branches are planted close together in rows.

According to Captain Lens, poor people of Curaçao use the fleshy branches as a vegetable. Mr. Harold G. Foss states that in the region of Coro, Venezuela, the natives use the wood in making the roofs and walls of their houses. The heart wood is split into two pieces and then tied to the rafters so as to form the support for the mortar and tiles. The wood is rich in potash, and the ash from it is shipped in large quantities to the United States for use as a fertilizer.

FIG. 129.—Lemaireocereus griseus.

Illustration: Contr. U. S. Nat. Herb. 12: pl. 67.

Plate XIII, figure 2, shows the top of a plant collected on Curaçao. Figure 129 is from a photograph taken by Mrs. J. N. Rose on the same island in 1916.

4. Lemaireocereus pruinosus (Otto).

Echinocactus pruinosus Otto in Pfeiffer, Enum. Cact. 54. 1837.
Cactus pruinosus Monville in Steudel, Nom. ed. 2. 1: 246.1840.
Cereus pruinosus Otto in Förster, Handb. Cact. 398. 1846.
Cereus laevigatus Salm-Dyck, Cact. Hort. Dyck. 1849. 204. 1850.

Plant usually tall, with a more or less definite trunk; ribs 5 or 6, very high, separated by broad intervals; spines few, the radial ones 5 to 7, brownish; central spine solitary, 3 cm. long; flowering areoles large, brown-felted; flowers about 9 cm. long; upper scales and outer perianth-segments 1 cm. long or less, rounded at apex; inner perianth-segments longer and thinner than the outer ones. ovary with numerous brown-felted areoles; fruit ovoid, spiny, 6 to 7 cm. long.

Type locality: Mexico.
Distribution: South-central Mexico.

This plant is certainly native in south-central Mexico, and distinguishable from the related cultivated *L. griseus* by fewer ribs, larger flowers, and ovoid fruit.

Cereus roridus (Pfeiffer, Enum. Cact. 54. 1837) was given as a synonym of *Echinocactus pruinosus*.

Cereus edulis Weber (Monatsschr. Kakteenk. **10**: 55. 1900) is another name for this species, never described.

Illustrations: Bull. Soc. Acclim. France **52**: f. 1, as *Cereus pruinosus;* Bradley, Hist. Succ. Pl. ed. 2. pl. 12, as *Cereus americanus octangularis;* Monatsschr. Kakteenk. **18**: 171, in part; **21**: 37; U. S. Dept. Agr. Bur. Pl. Ind. Bull. **262**: pl. 8; pl. 13, f. 2; MacDougal, Bot. N. Amer. Des. pl. 23; Journ. N. Y. Bot. Gard. **8**: f. 6, all these as *Cereus eburneus.*

Figure 130 shows a fruit collected by H. H. Rusby in Oaxaca in 1910.

Fig. 130.—Fruit of Lemaireocereus pruinosus. ×0.7.

5. Lemaireocereus (?) longispinus sp. nov.

Erect, rather stout, light green, the young growth more or less glaucous; ribs 6, broad at base, somewhat acute, more or less undulate; areoles borne at the tops of the undulations; radial spines about 10, spreading or even reflexed, acicular; central spine elongated, porrect, flattened above, up to 8 cm. long, gray; flowers and fruit unknown.

Collected by F. Eichlam in Guatemala in 1909.

Figure 131 is from a photograph of the type specimen in the collection of the New York Botanical Garden.

6. Lemaireocereus eichlamii nom. nov.

Cereus laevigatus guatemalensis Eichlam in Weingart, Monatsschr. Kakteenk. **22**: 182. 1912. Not *C. guatemalensis* Vaupel.

Cylindric, simple in cultivation, deep green except for narrow glaucous bands showing the commencement of new growth; ribs 8 to 10, rather broad and rounded, with acute intervals between; areoles large, brown-felted at first, soon gray; spines 4 to 6, acicular, nearly porrect, 2 cm. long or less; flower-buds obtuse; flower 6 to 7 cm. long; outer perianth-segments greenish purple, obtuse, with serrulate margins; inner perianth-segments purple, 10 to 15 mm. long, widely spreading or even rolled backward; tube proper 15 to 18 mm. long, ribbed within; tube funnelform, 2.5 cm. long, its surface covered with stamens; filaments unequal, white, numerous; style slender, white below, orange above, included; ovary tuberculate, each tubercle crowned by a minute scale; areoles on the ovary bearing brown felt but no spines.

Type locality: Guatemala.
Distribution: Guatemala.
Illustration: Monatsschr. Kakteenk. **22**: 183, as *Cereus laevigatus guatemalensis.*
Figure 132 shows a plant in the collection of the New York Botanical Garden.

7. Lemaireocereus chichipe (Gosselin).

Cereus chichipe Gosselin, Bull. Mus. Hist. Nat. Paris **11**: 507. 1905.
Cereus mixtecensis J. A. Purpus, Monatsschr. Kakteenk. **19**: 52. 1909.
Lemaireocereus mixtecensis Britton and Rose, Contr. U. S. Nat. Herb. **12**: 425. 1909.

Tree-like, up to 5 meters high, with a short trunk 8 to 10 dm. in diameter and a large very much branched top; branches 9 to 12-ribbed, undulate, acutish, 2 cm. high; areoles 1 to 1.5 cm.

apart; radial spines 6 or 7, 5 to 10 cm. long, grayish; central spine 1; flowers small, yellowish green; fruit spiny, globose, 2 to 2.5 cm. in diameter, red both within and without; seeds small, black, with a basal hilum.

Type locality: Cerro Colorado, near Tehuacán, Mexico.

Distribution: Puebla and Oaxaca, Mexico.

The plant is known as chichipe, or, according to Dr. C. A. Purpus, chichibe. The fruit, which is sold in the Mexican markets, like many other Mexican cactus fruits, has a different name from the plant; it is called chichituna.

Illustrations: Bull. Soc. Acclim. France **52**: f. 7, as *Cereus chichipe;* Monatsschr. Kakteenk. **19**: 53, as *Cereus mixtecensis.*

FIG. 131.—Lemaireocereus longispinus.   FIG. 132.—L. eichlamii.

8. Lemaireocereus chende (Gosselin).

Cereus chende Gosselin, Bull. Mus. Hist. Nat. Paris **11**: 506. 1905.
Cereus del moralii J. A. Purpus, Monatsschr. Kakteenk. **19**: 89. 1909.

Plant 5 to 7 meters high, with a short indefinite trunk, very much branched above, forming a large top; branches rather slender, ascending or erect; ribs 7 to 9, rather sharp; areoles on old branches 1.5 cm. apart, on young branches perhaps closer together; radial spines usually 5, the centrals when present a little longer than the radials, brown to bright yellow, in age grayish, acicular; flowers small, about 3 to 4 cm. long including the ovary; fruit said to be deep red, very spiny.

Type locality: In the Cerro Colorado, near Tehuacán, Mexico.

Distribution: Puebla and Oaxaca, Mexico.

According to Roland-Gosselin, the Mexican name for this species is chende; according to J. A. Purpus, chente; and according to Dr. Rose, chinoa. Dr. Rose collected this species between Tehuacán and Esperanza in 1912 (No. 11429) and Dr. C. A. Purpus sent a living plant to Washington in 1909 from San Luis, Oaxaca, Mexico. Dr. Purpus's plant has 7 acute ribs.

Illustrations: Bull. Soc. Acclim. France **52**: f. 6, as *Cereus chende*; Monatsschr. Kakteenk. **19**: 87, as *Cereus del moralii;* Contr. U. S. Nat. Herb. **12**: pl. 68, as *Lemaireocereus mixtecensis*, in error.

Figure 133 is from a photograph taken by Dr. MacDougal near Tehuacán in 1906.

Fig. 133.—Lemaireocereus chende.

9. Lemaireocereus godingianus sp. nov.

Large plant 3 to 10 meters high with a short, thick, woody trunk 2 to 5 dm. in diameter, becoming smooth; joints bright green when young, grayish afterwards; ribs 7 to 11; spines acicular, 2 to 4 cm. long, brownish when young; flowers large, white, 10 to 11 cm. long; tube proper 2 cm. long with walls 1 cm. thick or more; areoles on flower-tube and ovary closely set, large, bearing brown wool and yellow bristles; fruit large, 10 cm. long or more, covered with yellow spines.

Collected by J. N. Rose and George Rose at Huigra, Ecuador, August to November 1918 (No. 22127).

This species is very common on the dry hills both below and above Huigra, Ecuador, ranging from about 3,500 to 6,000 feet, where it is the most conspicuous plant in the landscape. It is associated with a *Furcraea*, several species of *Opuntia*, a *Bauhinia*, and a *Zanthoxylum*. It is frequently overrun by vines, such as species of *Passiflora* and *Ipomoea*.

It overlaps the lower range of an undescribed species of *Trichocereus* and has been frequently confused with that species. (See page 135.)

This plant is named for Dr. F. W. Goding, United States Consul-General at Guayaquil, Ecuador, a well-known entomologist, who assisted Dr. Rose in his botanical explorations in Ecuador.

Illustration: Smiths. Misc. Coll. **70:** f. 48, as giant cactus.

Figure 134 is from a photograph taken by George Rose at Huigra.

FIG. 134.—Lemaireocereus godingianus.

10. Lemaireocereus aragonii (Weber).

> *Cereus aragonii* Weber, Bull. Mus. Hist. Nat. Paris **8:** 456. 1902.

Columnar, 5 to 6 meters high, dark green with glaucous bands at intervals of growth; terminal branches about 3 meters long, 12 to 15 cm. in diameter; ribs 6 to 8, very large, 2 to 3 cm. high, rounded; areoles about 2 cm. apart, large, brown-felted; spines gray, about 8 to 10, but new ones developed from time to time, acicular, the radial ones about 1 cm. long, one of the centrals 2 to 3 cm. long; flowers 6 to 8 cm. long; ovary tuberculate, bearing clusters of spines; flesh of the fruit white; seeds large, black, 5 to 6 mm. long.

Type locality: Western Costa Rica.

Distribution: Costa Rica.

This cactus is used a good deal as a hedge plant in Costa Rica, much as is *Pachycereus marginatus* on the table-lands of Mexico. It is the only columnar cactus in Costa Rica. We have had living specimens of it in Washington since 1907, but they have never grown very much.

A cristate form of *Cereus aragonii* was named as a variety (*palmatus*) by Weber (Bull. Mus. Hist. Nat. Paris **8:** 456. 1902).

Illustrations: Boletin de Fomento Costa Rica **4:** 117; Iberica **48:** 339, both illustrations from the same source as the one used as figure 135.

Figure 135 is from a photograph taken by Otto Lutz at Tres Rios, Costa Rica, 1,350 meters altitude.

11. Lemaireocereus stellatus (Pfeiffer) Britton and Rose, Contr. U. S. Nat. Herb. **12:** 426. 1909.

> *Cereus stellatus* Pfeiffer, Allg. Gartenz. **4:** 258. 1836.
> *Cereus dyckii* Martius in Pfeiffer, Enum. Cact. 87. 1837.
> *Cereus tonelianus* Lemaire, Illustr. Hort. **2:** Misc. 63. 1855.
> *Stenocereus stellatus* Riccobono, Boll. R. Ort. Bot. Palermo **8:** 253. 1909.
> *Stenocereus stellatus tonelianus** Riccobono, Boll. R. Ort. Bot. Palermo **8:** 254. 1909.

Plant 2 to 3 meters high, branching at base, rarely branching above, pale bluish green; ribs 8 to 12, low, obtuse; radial spines 10 to 12; centrals several, often much longer than the others, some-

---

*Riccobono in error spells it "*tenellianus.*"

times 5 to 6 cm. long; areoles 1 to 2 cm. apart; flowers appearing at or near the top of the plant, red, small, narrowly campanulate, about 4 cm. long; ovary bearing small scales subtending wool and bristly spines; fruit red, spiny, globular, about 3 cm. in diameter; spines deciduous; seeds dull, pitted.

Type locality: Mexico.

Distribution: Southern Mexico.

The fruit is known in the markets as joconostle and sometimes as tuna.

The above description is drawn from Dr. Rose's specimens, which seem to represent *L. stellatus*, but the identification has not been confirmed by reference to the type specimen.

Cereus joconostle Weber (Schumann, Gesamtb. Kakteen 79. 1897) is known only as a synonym of this species.

Fig. 135.—Lemaireocereus aragonii.

Illustrations: Contr. U. S. Nat. Herb. 10: pl. 20; Rep. Mo. Bot. Gard. 16: pl. 3, f. 1 to 4; U. S. Dept. Agr. Bur. Pl. Ind. Bull. 262: pl. 12, as *Cereus stellatus;* Bull. Soc. Acclim. France 52: f. 3, as *Cereus dyckii;* Contr. U. S. Nat. Herb. 12: pl. 69.

Figure 136 is from a photograph taken by Dr. MacDougal at Tomellín, Mexico, in 1906.

12. **Lemaireocereus treleasei** Britton and Rose, Contr. U. S. Nat. Herb. 12: 426. 1909.

Cereus treleasei Vaupel, Monatsschr. Kakteenk. 23: 37. 1913.

Plant 5 to 7 meters high, simple or with a few strict branches; ribs about 20; areoles approximate with a peculiar V-shaped depression just above each one; spines rather short, yellowish; flowers pinkish, 4 to 5 cm. long, diurnal; scales on ovary and flower-tube subtending slender whitish bristles; fruit red, about 5 cm. in diameter, covered with clusters of deciduous spines; seeds black with a dull, rugose surface and a large oblique basal hilum.

Type locality: Road between Mitla and Oaxaca, Mexico.

Distribution: Oaxaca, Mexico.

In flower and fruit this much resembles *L. stellatus*, but has a different habit, more ribs, and different areoles. This plant is not common in the deserts about Oaxaca, but when it does occur is found in clumps. It is characterized by its strict elongated stems, which seldom branch.

Illustration: Contr. U. S. Nat. Herb. **12**: pl. 70.

Figure 137 is from a photograph taken by Dr. MacDougal at the type locality in 1906.

FIG. 136.—Lemaireocereus stellatus.

13. Lemaireocereus deficiens (Otto and Dietrich).

Cereus deficiens Otto and Dietrich, Allg. Gartenz. **6**: 28. 1838.
Cereus clavatus Otto and Dietrich, Allg. Gartenz. **6**: 28. 1838.
Cereus eburneus clavatus Fobe, Monatsschr. Kakteenk. **18**: 78. 1908.

A tall tree-like plant, with a more or less definite trunk and many stout erect branches, the old trunk often spineless; branches somewhat glaucous; ribs 7 or 8, very broad at base; areoles borne at the depressions on the ribs, large, white or brown-felted; spines about 8, grayish with black tips, more or less spreading, sometimes appressed, 1 to 1.5 cm. long, the clusters either with or without central ones, these, when present, 3 cm. long and a little flattened; flowers only 5 to 6 cm. long; ovary without spines, the areoles felted; fruit very spiny, edible, its flesh either red or white, juicy.

Type locality: Carácas, Venezuela.

Distribution: Central part of coast of Venezuela.

This species is common on all the hills about La Guayra, is less common in the mountains toward Carácas, and is also to be found along the coast at Puerto Cabello. About towns it is much used as a hedge plant.

Figure 138 is from a photograph taken by Mrs. J. N. Rose near Puerto Cabello, Venezuela, in 1916.

FIG. 137.—Lemaireocereus treleasei.

14. **Lemaireocereus weberi** (Coulter) Britton and Rose, Contr. U. S. Nat. Herb. **12**: 426. 1909.

Cereus weberi Coulter, Contr. U. S. Nat. Herb. 3: 410. 1896.
Cereus candelabrum Weber in Schumann, Gesamtb. Kakteen 106. 1897.

Plant very large, 10 meters high or more, with a trunk short but thick and often with hundreds of nearly erect branches arising from near the base, dark bluish green, slightly glaucous; ribs usually 10, rounded; areoles large; radial spines usually 6 to 12, spreading, more or less acicular, 1 to 2 cm. long; central spine usually up to 10 cm. long, solitary, flattened, often more or less deflexed, except those of the upper areoles, at first brown to blackish, much longer than the laterals; areoles white-felted; flowers 8 to 10 cm. long; scales on flower-tube narrow, thin, bearing long brown hairs in their axils; inner perianth-segments oblong, 2 cm. long; ovary globular, covered by the dense brown felt of its areoles; fruit oblong, edible, 6 to 7 cm. long, very spiny, the spine-clusters deciduous in ripening.

Type locality: A few miles south of Tehuacán, Puebla, Mexico.
Distribution: Puebla and Oaxaca, Mexico.

This plant is called cardon and candebobe.

Cereus belieuli and *C. pugionifer* are two garden names referred here by Schumann (Gesamtb. Kakteen 107. 1897).

Illustrations: Contr. U. S. Nat. Herb. 10: pl. 21; MacDougal, Bot. N. Amer. Des. pl. 21; Nat. Geogr. Mag. 21: 705; Journ. Intern. Gard. Club 3: 16; U. S. Dept. Agr. Bur. Pl. Ind. Bull. 262: pl. 11, all as *Cereus weberi;* Schelle, Handb. Kakteenk. f. 37; Schumann, Gesamtb. Kakteen f. 24; Möllers Deutsche Gärt. Zeit. 29: 352. f. 7; 353. f. 8, as *Cereus candelabrum;* Contr. U. S. Nat. Herb. 12: pl. 71.

Figure 139 is from a photograph taken by Dr. Rose at Tomellín, Mexico, in 1905; figure 140 shows clusters of spines and figure 141 a fruit collected by H. H. Rusby at Cuicatlan, Oaxaca, in 1910.

FIG. 138.—Lemaireocereus deficiens.

FIG. 139.—Lemaireocereus weberi.

15. **Lemaireocereus queretaroensis** (Weber) Safford, Ann. Rep. Smiths. Inst. **1908:** pl. 6, f. 2. 1909.

Cereus queretaroensis Weber in Mathsson, Monatsschr. Kakteenk. **1:** 27. 1891.
Pachycereus queretaroensis Britton and Rose, Contr. U. S. Nat. Herb. **12:** 422. 1909.

Plant 3 to 5 meters high, with a short woody trunk, much branched above; ribs 6 to 8, prominent, obtuse; areoles about 1 cm. apart, large, brown-woolly, very glandular; spines 6 to 10, at first red, becoming grayish in age, acicular, rather unequal, sometimes only 15 mm. long, at other times 5 cm. long; flowers 7 to 8 cm. long; ovary with many woolly areoles subtended by ovate scales 2 mm. long or less; fruit spiny, edible.

Type locality: Querétaro, Mexico.
Distribution: Central Mexico.

This species was formerly referred by us to the genus *Pachycereus*, but it has since been learned that the fruit is not dry, but juicy and edible, and therefore the plant is more properly a *Lemaireocereus*. Its peculiar glandular areoles are like those of *L. thurberi*, although otherwise the two species are quite different. This plant is said to be cultivated in Jalisco and Querétaro, Mexico, doubtless for its edible fruits, which are also called pitahaya. We have had the plant in cultivation in Washington since 1907, but it has made little or no growth.

Dr. Rose has collected the species at several localities in central Mexico, including the type locality (No. 11133).

Illustrations: Bull. Soc. Acclim. France **52**: 18. f. 2, as *Cereus queretaroensis;* Ann. Rep. Smiths. Inst. **1908**: pl. 6, f. 2.

Figure 142 shows the spine-bearing stem-areoles of an herbarium specimen collected by Dr. Rose near Querétaro, Mexico, in 1906.

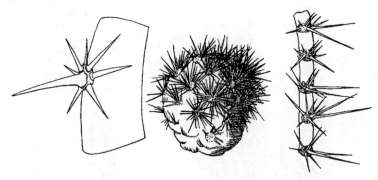

FIG. 140.—Cluster of spines of Lemaireocereus weberi. ×0.7.
FIG. 141.—Fruit of same. ×0.7.
FIG. 142.—Part of rib, showing spine-clusters of Lemaireocereus queretaroensis. ×0.7.

16. Lemaireocereus montanus sp. nov.

Tree-like, 6 to 7 meters high, with a definite smooth trunk 1 meter long or more, with few branches, at first spreading, then nearly erect; ribs few, usually 8, prominent areoles 1 to 1.5 cm. apart, large, filled with short brown wool; spines few, 6 or less, pale in color, rather stout, one of them longer, sometimes 3 cm. long; flowers 6 to 7 cm. long, opening during the day; outer perianth-segments purplish; scales on the ovary ovate, 4 to 6 mm. long, imbricated, acuminate, with erose margins.

This species was found well up on the side of Alamos Mountain, associated with *Lemaireocereus thurberi,* but usually at a higher altitude than that at which that species is generally found. It differs from *L. thurberi* in its habit, number of ribs, armament, and flowers. Like *L. thurberi* it has brown areoles, which are not found in any of the other species except *L. queretaroensis* of the table-land region of central Mexico.

Collected by Rose, Standley, and Russell above Alamos, Mexico, March 18, 1910 (No. 13039).

17. Lemaireocereus thurberi (Engelmann) Britton and Rose, Contr. U. S. Nat. Herb. **12**: 426. 1909.

Cereus thurberi Engelmann, Amer. Journ. Sci. II. **17**: 234. 1854.
Pilocereus thurberi Rümpler in Förster, Handb. Cact. ed. 2. 689. 1885.
Cereus thurberi littoralis K. Brandegee, Zoe **5**: 191. 1904.

Usually without a definite trunk, sending up from the base 5 to 20, or even more, erect or ascending branches 3 to 7 meters high, 15 to 20 cm. in diameter, the basal ones usually simple but occasionally with lateral branches, this doubtless being caused by injuries to the growing tips; ribs numerous, 12 to 17, rather low but sometimes 2 cm. high, rounded, separated by narrow intervals; areoles 10 to 15 or rarely 30 mm. apart, large, sometimes becoming 1 cm. in diameter, circular, brown-felted, more or less glandular, the whole areole becoming a wax-like mass; spines numerous, acicular to subulate, unequal, brownish to black, becoming gray in age, the longest ones sometimes 5 cm. long; flowers mostly borne near the top of the stem but sometimes 3 dm. below the top, 6 to 7.5 cm. long including the ovary, opening during the day; outer perianth-segments broad, reddish, imbricated, gradually passing into the scales on the tube; inner perianth-segments light purple with nearly white margins, widely spreading or even turned back at the apex, broad, obtuse; filaments short, numerous, erect, white, borne all over the throat, 2 to 2.5 cm. long; lower part of flower-tube or tube proper smooth within; ovary tuberculate, bearing small, ovate, acute scales, these with white and brown hairs in their axils; fruit globular, 4 to 7.5 cm. in diameter, edible, very spiny, but in age naked, olive without, crimson within; seeds black, shining, 1.8 to 2 mm. long.

Type locality: Canyon near the mountain pass of Bachuachi, Sonora, Mexico.

Distribution: Southern Arizona, in the Comobabi, Quijotoa, and Ajo Mountains, throughout western Sonora, and on both coasts of Lower California. The Index Kewensis says it is from New Mexico, doubtless an error for northern Mexico. In the cape region of Lower California a slender form is found which has been described as a variety.

The flowers, which appear from March to August, are followed by the large delicious fruit much prized by the native, who knows it as pitahaya or pitahaya dulce.

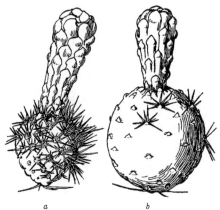

FIG. 143.—Lemaireocereus thurberi.

FIG. 144.—L. thurberi: *a*, flower; *b*, fruit. X0.7.

The species was named for George Thurber (1821–1890), one of the collectors on the first Mexican Boundary Survey.

The habit of branching just at the base is unusual in this genus, in which most of the species have definite, though often short, trunks.

This is the only species of *Lemaireocereus* which reaches the United States and is the only one found in northwestern Mexico or Lower California. Two other species were credited to Lower California in our former treatment (Contr. U. S. Nat. Herb. **12**: 425), but these we now refer to another genus (see pages 115, 116).

Whether the flowers open at night or during the day has been in dispute. Dr. Rose, who studied the species in Lower California, observed the flowers widely expanded at 2 o'clock on a bright sunny day. F. E. Lloyd, in a letter dated September 6, 1909, says, "I notice that what we have hitherto called *Cereus thurberi* is stated by you as having a day-blooming flower. You may recall that I made a special study with reference to this point at the Quijotoa Mountains and found it strictly night-blooming. The photograph which you have of the flower I made between 4 and 5 o'clock in the morning, just before sun-up."

Cereus thurberi monstrosus (E. Dams, Monatsschr. Kakteenk. **14**: 182. 1904) is not an unusual form.

Illustrations: Bull. Soc. Acclim. France **52**:f. 4; Cact. Mex. Bound. pl. 74, f. 15; Hornaday, Campfires on Des. and Lava opp. 68, 136; MacDougal, Bot. N. Amer. Des. pl. 8; Monatsschr. Kakteenk. **17**: 105, as *Cereus thurberi.*

Figure 143 is from a photograph taken by Dr. MacDougal at Torres, Sonora, in 1902; figure 144*a* shows a flower of the plant collected by F. E. Lloyd on the Quijotoa Mountains, Arizona, in 1906; and figure 144*b* shows a fruit of the same.

18. Lemaireocereus laetus (HBK.) Britton and Rose, Journ. N. Y. Bot. Gard. **20**: 157. 1919.

Cactus laetus Humboldt, Bonpland, and Kunth, Nov. Gen. et Sp. **6**: 68. 1823.
Cereus laetus De Candolle, Prodr. **3**: 466. 1828.

Plant 4 to 6 meters high, much branched, bluish gray but not glaucous; ribs 4 to 8, prominent; areoles 2 to 3 cm. apart; spines brown when young, becoming gray to nearly white in age, usually 1 to 3 cm. but sometimes 8 cm. long, subulate; flowers 7 to 8 cm. long; inner perianth-segments white, 2 cm. long; fruit green without, very spiny, splitting down the side when ripe, white within; pulp edible; seeds black.

Type locality: Near Sondorillo, formerly in Ecuador but now in Peru.
Distribution: Central Peru and southern Ecuador.

Dr. Rose found the species in Catamayo Valley in southern Ecuador, where it is very common (No. 23340); it was, however, seen only in this one locality in Ecuador. We also refer here the plant collected by Dr. and Mrs. Rose along the Rimac River below Matucana, Peru, July 9, 1914 (No. 18650).

FIG. 145.—Lemaireocereus laetus.

FIG. 146.—Lemaireocereus laetus.

We have referred here the plant from Catamayo, as it is the only wild one we know in this region which could possibly have been described as *Cactus laetus.* It is such a conspicuous plant that we do not believe Humboldt would have passed it by without some reference. Through the kindness of Dr. Charles Wood we were able to send to the Natural History Museum of Paris specimens of this Catamayo plant in order to have it compared, if possible, with Humboldt's type. M. Lecomte, however, informs us that no specimen can be found. It is of interest to note that in 1825 Sprengel, who redescribed this species in his edition of the Systema, placed it next to *Cereus eburneus* and questions whether it is not the same as *C. hystrix.*

The original description of *Cactus laetus* is very brief and unsatisfactory.

Figure 145 shows a plant, growing on the flats of the Catamayo Valley, southern Ecuador, photographed by George Rose in 1918; figure 146 is from a photograph taken by Mrs. J. N. Rose at Matucana, Peru; figure 147 shows its flower and figure 148 its fruit.

19. Lemaireocereus cartwrightianus sp. nov.

Plant 3 to 5 meters high, with woody trunk, much branched; branches consisting of short stout joints, 15 to 60 cm. long, 8 to 15 cm. in diameter; ribs 7 or 8; areoles large, brown-felted; young spines white, brown, black, or variegated, about 20, 1 to 2 cm. long, except on the old trunk and here 12 cm. long or more; flowers slender, 7 to 8 cm. long, opening in the early evening; outer perianth-segments narrow, 1 to 1.5 cm. long, reddish, erect; inner perianth-segments small, white except at the spreading tips; filaments numerous, short, included; fruit globular to oblong, 8 to 9 cm. long, covered with clusters of weak spines, deciduous when ripe, red without, white within.

Very common on the flats near Guaya-quil; collected by J. N. Rose, J. H. Burns, and George Rose, north of the city, August 11, 1918 (No. 21118). It is characterized by very narrow flowers.

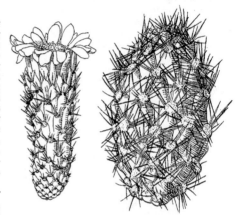

Fig. 147.—Flower of Lemaireocereus laetus. X0.6.
Fig. 148.—Fruit of same. X0.6.

Fig. 149.—Lemaireocereus humilis.

It is named for Mr. Alfred Cartwright who has for many years been connected with the British Consular Service at Guayaquil and who has aided many visiting scientists. Mr. Cartwright first described this plant to us and directed us to its peculiar habitat.

20. Lemaireocereus humilis sp. nov.

Stems weak, forming dense thickets, dark green, 1 to 4 meters long, about 4 cm. thick, usually with few branches or none; ribs 3 or 4, sometimes 6, more or less interrupted, little undulate; areoles borne in the depressions of the ribs, large, white-felted, bearing spines only in the lower part; spines 5 to 8, brown, becoming white, acicular, 1 to 2 cm. long; flowers greenish white, about 6 cm. long; outer perianth-segments linear-oblong, spreading; ovary with small scattered scales, at first without spines; fruit very spiny, spherical, 4 cm. long.

Fig. 150.—Flowering branch of Lemaireocereus humilis.

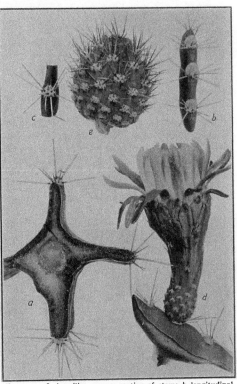

Fig. 151.—L. humilis: a, cross-section of stem; b, longitudinal section of rib; c, cluster of spines; d, flower; e, fruit.

Collected by H. Pittier at Venticas del Dagua, Dagua Valley, Western Cordillera of Colombia, altitude 700 to 1,000 meters, February 1906, and described from a plant collected by him (New York Botanical Garden, No. 34794) and from his field notes and detailed and habit photographs. It is called tuna colorado.

This plant is quite different from the other species in its slender stems with very few ribs and in its tendency to form dense thickets, but it has the characteristic flower and fruit of this genus.

Figure 149 is from a photograph taken by Henry Pittier at the type locality; figure 150 shows a flowering branch, and figure 151 shows details of the type.

21. Lemaireocereus dumortieri (Scheidweiler) Britton and Rose, Contr. U. S. Nat. Herb. **12**: 425. 1909.

Cereus dumortieri Scheidweiler, Hort. Belge **4**: 220. 1837.

Often tree-like, 6 to 15 meters high, the trunk proper short, 6 to 10 dm. long, 3 dm. in diameter or more, woody; branches many, erect almost from the first, with numerous constrictions, very pale bluish green or somewhat glaucous; ribs generally 6, sometimes 5 or 7, occasionally 9 on very old joints; areoles elliptic, approximate or often confluent, gray-felted; spines various in number and in length, 10 to 20 radials, 1 central or more, the longer ones often 4 cm. long, all at first straw-colored but in age blackened; flowers 5 cm. long, the tube and ovary bearing small ovate scales with bunches of felt and occasionally bristles in their axils, the limb about 2.5 cm. broad; fruit oblong, small, 3 to 4 cm. long, reddish within, not spiny, its areoles nearly contiguous, felted; seeds brownish, 1.5 mm. long, dull, roughened.

Type locality: Incorrectly given as Buenos Aires (see note below).
Distribution: Central Mexico.

Our description is drawn from numerous specimens collected by Dr. Rose in central Mexico. This is the plant which passes as *Cereus dumortieri* in collections, but from the description alone one can hardly be certain. It ranges over a considerable territory, but is never abundant, being found generally as large isolated individuals on the sides of rocky hills and cliffs.

Greenhouse plants much resemble *Pachycereus marginatus*, and both species have small flowers; but the wild plants are very unlike and the fruit and seeds differ widely.

Although Scheidweiler in his original description of this species referred it to "Buenos Ayres," he doubtless made a mistake, as he must have done in his reference of *Mammillaria obconella* in the same publication. The original description does not correspond to any known South American cactus, but does represent fairly well our central Mexican species which passes under this name. In 1845 the species was listed by Salm-Dyck (Cact. Hort. Dyck. 1844. 30) as from the Belgian Gardens (H. Belg.). In 1850 (Cact. Hort. Dyck. 1849. 210) he published an original description apparently based on the Belgian specimens; but evidently he had forgotten the older publication. Schumann and most writers since 1850 have assigned Prince Salm-Dyck as the author of this species. Weber (Dict. Hort. Bois 279. 1895) seems to have been the first botanist to refer the species to Mexico.

Fig. 152.—Fruit of Lemaireocereus dumortieri. ×0.8.

Fig. 153.—Lemaireocereus dumortieri.

Cereus anisacanthus De Candolle (Mém. Mus. Hist. Nat. Paris **17**: 116. 1828) is doubtfully referred here by Schumann. If it should prove to be the same, it would, of course, supplant the present name. Its two varieties, *ortholophus* and *subspiralis* (De Candolle, Mém. Mus. Hist. Nat. Paris **17**: 117. 1828), so far as we can determine, belong here also.

This species is anomalous in *Lemaireocereus*, having very small flowers and spineless fruit, but the scales of the ovary sometimes subtend bristles, if not spines, in their axils.

Illustration: Hort. Belge **4**: pl. 15, as *Cereus dumortieri*.

Plate xv, figure 2, shows the top of a plant brought by Dr. Rose from Cuernavaca, Morelos, Mexico, in 1906. Figure 152 shows the fruit of a plant from Hidalgo; figure 153 is from a photograph taken by him in Hidalgo, Mexico, in 1905.

SPECIES NOT GROUPED.

LEMAIREOCEREUS SCHUMANNII (Mathsson) Britton and Rose, Contr. U. S. Nat. Herb. **12**: 425. 1909.

Cereus schumannii Mathsson in Schumann, Monatsschr. Kaktecnk. **9**: 131. 1899.

Plants tall and stout, 15 meters high, with few branches; ribs 8, thick and high, very obtuse, somewhat pruinose; spines 6 or 7, radial, 1 central, all white with brown tips; flowers and fruit unknown.

Type locality: Honduras.

Distribution: Known only in cultivation.

Little is known regarding this species and from the brief description we are unable to place it definitely in our key. It may be only a form of *L. griseus* so widely cultivated in Mexico and Central America and is near *L. aragonii* and possibly not specifically distinct.

LEMAIREOCEREUS sp.

Cereus rigidispinus Monville, Hort. Univ. **1**: 223. 1840.

"Erect, stout, dark green, somewhat glaucous; ribs thick, rounded; sinuses open, deep, acute. Spines very strong and stiff, whitish, divaricate. Trunk 2¼ feet in diameter, having 7 ribs about 9 lines by 5 lines thick at the middle. Areoles 6 to 10 lines apart, a little sunken, subovate, a little convex, covered with a very short grayish nap, bearing 6 to 8 very unequal spines, the strongest, as well as the weakest, arising from no particular point, 3 to 13 lines long and ¼ to 1 line in diameter, all exceedingly stiff, whitish and black at the tip, sometimes 2 centrals or larger ones united along their entire length. Habitat: Mexico. Flowers and fruit unknown. This plant should be placed in Cat. Monv. between *Cer. hystrix* and *eburneus*. In spite of its peculiar appearance, it shows some similarities to them, especially to the latter." (Translated from De Monville, Hort. Univ. **1**: 223. 1840.)

Type locality: Mexico.

Schumann refers *Cereus hildmannii* Hortus (Gesamtb. Kakteen 57. 1897) here as a synonym.

LEMAIREOCEREUS sp.

Cereus conformis Salm-Dyck, Cact. Hort. Dyck. 1849. 203. 1850.

Stems erect, robust, 3 dm. high, 10 cm. in diameter, glaucous, green; ribs 7, crenate; areoles 18 mm. apart, orbicular, densely grayish, tomentose; radial spines 7 to 9, 6 to 8 mm. long; central spines 1 to 3, a little stouter than the radial; flowers and fruit unknown.

Type locality: Mexico.

It was sent from Mexico by Ehrenberg to the Berlin Botanical Garden in 1840, but has doubtless long since disappeared. Schumann did not know the species, but Weingart (Monatsschr. Kakteenk. **15**: 79. 1905) considers it identical with *Cereus aragonii*. It may be a *Lemaireocereus*, but we doubt its being *L. aragonii*, which is a native of Costa Rica.

LEMAIREOCEREUS sp.

Joints bright green, not at all glaucous; ribs about 2 cm. high, separated by V-shaped intervals; margin of ribs somewhat crenate with the areoles borne at the top of the crenations; radial spines about 8, 1 to 2 cm. long; central spine usually solitary, erect or porrect, sometimes 10 cm. long.

This plant was sent to the New York Botanical Garden by Dr. George F. Gaumer in 1918, but it has not yet flowered (New York Botanical Garden No. 46120). Dr. Gaumer has also sent from Yucatan two other plants which are of this relationship which we are unable to place. His No. 23941 has 7 ribs and numerous short spines; it did not live. His No. 23922 has 10 ribs and also short spines.

13. ERDISIA gen. nov.

Stems much branched at base, sometimes mainly subterranean, the branches slender, erect, ascending, or pendent; ribs few, crenate, with spiny areoles; flowers small, funnelform-campanulate, the tube short; throat short, funnelform, covered with stamens; outer perianth-segments obtuse or sometimes with acute tips; filaments numerous, white, about half the length of the inner perianth-segments; style stout, a half longer than the stamens; ovary tuberculate, bearing minute ovate scales with spines and felt in their axils; fruit juicy, small, globular, bearing clusters of deciduous spines; seeds numerous, minute.

The genus consists of 4 species, so far as known; *Cereus squarrosus* Vaupel is the type species. It is named in honor of Ellwood C. Erdis, who was in charge of the topographical work of the Yale University Peruvian Expedition, 1914.

The plants resemble in habit some of the bushy Cuban species of the genus *Leptocereus*. In the shape of the flowers, the spiny ovary, and the deciduous spines on the fruit, some of them suggest *Echinocereus*, but the habit is very different, and no *Echinocereus* is known to be of South American origin.

KEY TO SPECIES.

Stem and branches cylindric.
    Flowers bright red or scarlet; inner series of stamens not united..............................1. *E. squarrosa*
    Flowers yellow; inner series of stamens united into a tube.................................2. *E. philippii*
Branches clavate; stem more or less subterranean.
    Flowers yellow...3. *E. meyenii*
    Flowers purple..4. *E. spiniflora*

1. Erdisia squarrosa (Vaupel).

Cereus squarrosus Vaupel, Bot. Jahrb. Engler **50**: Beibl. **111**: 21. 1913.

Stems 1 to 2 meters long, 1 to 3 cm. in diameter; ribs 8 or 9; areoles 1 to 1.5 cm. apart; spines about 15, yellowish, very unequal, somewhat swollen at base, the longest ones 4 cm. long; flowers borne toward the ends of branches, 2.5 to 4 cm. long including the ovary, sometimes as much as 5 cm. broad; inner perianth-segments 1.5 cm. long; filaments 1 cm. long or less; style stout, 1.5 cm. long; fruit 1.5 to 2 cm. in diameter, juicy, spiny, the clusters of spines falling off early; seeds minute.

FIG. 154.—Erdisia squarrosa.

Type locality: Tarma, Department of Junin, Peru.
Distribution: The highlands of eastern Peru.

This species was collected by Dr. J. N. Rose below Cuzco, Peru, in 1914, when flowers and stems were obtained. Some of the living plants which were sent to the New York

Botanical Garden survived. It was first seen by him about 10 miles below Cuzco, along the railroad running to Juliaca, and was frequently observed a long distance below Cuzco, being easily recognized by its scarlet flowers, which in September were just appearing.

In June 1914, Ellwood C. Erdis collected living specimens 40 miles west of Cuzco, at 2,450 meters altitude, but these died. In November of the same year he again collected the species, this time in flower.

In May 1915, O. F. Cook and G. B. Gilbert collected the plant at Ollantaytambo, Peru, at an altitude of about 3,000 meters. These specimens were accompanied by both flowers and fruit, and some good habit and detail photographs were taken.

Figure 154 is from a photograph taken by O. F. Cook at Ollantaytambo, Peru, in 1915; figure 155 shows portions of the plant photographed.

2. Erdisia philippii (Regel and Schmidt).

Cereus philippii Regel and Schmidt, Gartenflora 31: 98. 1882.
Echinocactus philippii Schumann, Gesamtb. Kakteen 427. 1898.
Echinopsis philippii Nicholson, Dict. Gard. Suppl. 338. 1901.

Stems slender, cylindric; ribs 8 to 10, strongly tubercled; radial spines about 8, 10 to 12 mm. long; central spines much stouter and longer, 2.5 cm. long; flowers 4 cm. long, campanulate, yellow below, reddish above; outer segments ovate, acuminate; inner segments oblong, acute; stamens in two distinct series, the outer arising from the base of the segments, the inner series united into a tube around the style; style included; stigma-lobes very short; ovary globular, bearing clusters of acicular spines.

FIG. 155.—Erdisia squarrosa.

Type locality: Chile.

Distribution: Known only from the type collection.

This species has been described in turn under *Cereus*, *Echinocactus*, and *Echinopsis*, from all of which it is distinct. It is remarkable in having the lower series of stamens united into a tube.

Illustrations: Gartenflora 31: pl. 1079, f. 1, *a*, *b*, as *Cereus philippii*.

3. Erdisia meyenii nom. nov.

Cereus aureus Meyen, Allg. Gartenz. 1: 211. 1833. Not Salm-Dyck, 1828.
Cactus aureus Meyen, Reise 1: 447. 1834.
Echinocactus aureus Meyen in Pfeiffer, Enum. Cact. 68. 1837.
Cleistocactus aureus Weber in Gosselin, Bull. Mens. Soc. Nice 44: 39. 1904.

Stems subterranean, often forming large colonies sending up short, usually unjointed branches; joints 10 to 20 cm. long, 3 to 5 cm. in diameter, more or less clavate; ribs 5 to 8, high (1 cm. high or more), somewhat undulate; spines several, subulate, unequal, brown to blackish, the longest 5 to 6 cm. long; flowers small, about 4 cm. long, yellow; scales on ovary and flower-tube small, 3 to 4 mm. long, acute, bearing felt and spines in their axils; fruit 2 cm. in diameter, reddish.

Type locality: Cordilleras de Tacna, Chile (formerly Peru), 600 meters altitude.

Distribution: Northern Chile and near Arequipa, Peru.

Dr. Rose found this plant (No. 18801) very common on the dry hills just below Arequipa, Peru, growing mostly underground. The separated branches at first seemed to represent distinct plants; it is inconspicuous, its purple stems with black spines resembling a dead plant.

Meyen's Travels is usually cited as the original place of publication for this species, but it was published a year earlier as *Cereus aureus.* This last combination has usually been credited to Schumann (Gesamtb. Kakteen 124. 1897). The name assigned to this plant being a homonym, we have renamed it for its discoverer, Franz Julius Ferdinand Meyen (1804–1840), a celebrated traveler and writer.

Figure 156 shows a branch collected by Dr. Rose near Arequipa, Peru, in 1914.

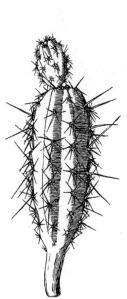

FIG. 156.—Branch of E. meyenii. ×0.5.   FIG. 157.—Erdisia spiniflora.

4. Erdisia spiniflora (Philippi).

Opuntia spiniflora Philippi, Linnaea **30:** 211. 1859.
Opuntia bicolor Philippi, Linnaea **33:** 83. 1864.
Opuntia clavata Philippi, Anal. Univ. Chile **41:** 722. 1872. Not Engelmann, 1848.
Cereus hypogaeus Weber in Regel, Gartenflora **31:** 165. 1882.
Echinocereus hypogaeus Rümpler in Förster, Handb. Cact. ed. 2. 784. 1885.
Eulychnia clavata Philippi in Engler and Prantl, Pflanzenfam. **3**ᵍᵃ**:** 185. 1894, as synonym.
Echinocereus clavatus Schumann, Monatsschr. Kakteenk. **5:** 123. 1895.

Underground stems slender, spineless, branching near the surface of the ground; branches somewhat clavate, becoming bronzed, 6-ribbed; spines all black at base, brown at tip; radial spines about 6, acicular, central spine solitary, porrect, slender; flowers probably purplish, 5 to 6 cm. long, with a rather broad throat; fruit fleshy, spiny; seeds not known.

Type locality: Near Aranas, Santiago, Chile.

Distribution: High mountains of Chile, near Santiago.

This species has been described under four specific names, and has been referred to four genera. We refer it to *Erdisia* on account of floral similarity to *E. squarrosa.*

Illustrations: Gartenflora **21**: pl. 721, f. 3, as *Opuntia clavata;* Gartenflora **31**: pl. 1085, as *Cereus hypogaeus.*

Figure 157 is a copy of the first illustration above cited.

PUBLISHED SPECIES, PERHAPS OF THIS GENUS.

CEREUS APICIFLORUS Vaupel, Bot. Jahrb. Engler **50**: Beibl. **111**: 15. 1913.

Prostrate or ascending, the stems 2 to 2.5 cm. thick, about 10, spreading; central spine solitary, 2 to 3 times as long as the radials; flowers numerous, forming a crest at the top of the plant, 4 cm. long; ovary terete, 1 cm. long, covered with numerous small lanceolate scales bearing brown felt and reddish-brown bristles in their axils.

Type locality: Valley of Puccha River, Department of Ancachs, Peru.

The author compares the flowers of this plant with those of *Cereus aureus* Meyen.

14. BERGEROCACTUS Britton and Rose, Contr. U. S. Nat. Herb. **12**: 435. 1909.

Low, much branched cactus, with stout, cylindric, spreading or ascending branches; ribs many, low; areoles approximate; spines many, yellow, acicular; flower small, pale yellow, with short tube and widely expanded limb; scales on ovary and flower-tube small, bearing felt and spines in their axils; perianth-segments small, obtuse; fruit globose, densely spiny; seeds obovate.

FIG. 158.—Bergerocactus emoryi.

The genus is monotypic; it is named in honor of Alwin Berger, author of an excellent discussion of the genus *Cereus*, who was long in charge of the garden of Sir Thomas Hanbury at La Mortola, Italy.

1. Bergerocactus emoryi (Engelmann) Britton and Rose, Contr. U. S. Nat. Herb. **12**: 435,* 474. 1909.

Cereus emoryi Engelmann, Amer. Journ. Sci. II. **14**: 338. 1852.
Echinocereus emoryi Rümpler in Förster, Handb. Cact. ed. 2. 804. 1885.

Branches 2 to 6 dm. long, 3 to 6 cm. in diameter, entirely covered with the dense spiny armament; ribs 20 to 25, very low, only a few millimeters high, somewhat tuberculate; spines 10 to 30, yellow to yellowish brown, acicular, 1 to 4 cm. long; flowers about 2 cm. long and about as broad when expanded; outer perianth-segments obovate, obtuse; inner perianth-segments oblong, about 1 cm. long.

Type locality: "About the boundary line" of California and Lower California.

Distribution: Near the coast of southwestern California and northwestern Lower California and adjacent islands.

Illustration: Engelmann, Cact. Mex. Bound. pl. 60, f. 1 to 4, as *Cereus emoryi.*

Figure 158 is from a photograph taken by E. O. Wooton on San Clemente Island, California, in 1912; figure 159 shows a dried flower collected by Le Roy Abrams at Tia Juana, San Diego County, California, in 1903.

Fig. 159.—Flower of B. emoryi. ×0.8.

15. LEOCEREUS gen. nov.

Stems long and slender, nearly terete, somewhat vine-like in habit; ribs numerous, but low and indistinct; areoles approximate, bearing acicular spines and felt, but no wool or hairs; flowers axillary, solitary, small, narrowly campanulate, with a short limb; ovary and flower-tube very scaly, the scales bearing numerous silky hairs and bristly spines in their axils; fruit small, globular; seeds black, shining, pitted.

In its narrow flower, in the hairs in axils of the scales on the ovary and tube, this genus suggests *Oreocereus*, but is very different in habit. The flower of *Leocereus* is different from that of *Nyctocereus* in its narrow throat, short perianth-segments, hairy and bristly areoles.

The genus is named for Señor A. Pacheco Leão, Director, Jardim Botanico, Rio de Janeiro, Brazil. The first of the 3 species here described is taken as the type.

KEY TO SPECIES.

Flowers 4 cm. long; spines yellowish brown..1. *L. bahiensis*
Flowers 6 to 7 cm. long; spines dark chestnut-brown.
    Axils of scales on ovary densely lanate; fruit villous...2. *L. melanurus*
    Axils of scales on ovary sparsely lanate; fruit nearly naked...3. *L. glaziovii*

1. Leocereus bahiensis sp. nov.

Somewhat branched, sometimes erect, sometimes clambering, up to 2 meters long, 1 to 1.5 cm. in diameter; ribs 12 to 14, low; areoles close together, circular, bearing white felt and spines; spines numerous, the central ones much longer than the radials, often 3 cm. long, acicular, yellowish, spreading; flowers 4 cm. long, densely woolly and spiny; inner perianth-segments small, white; fruit 10 to 12 mm. in diameter; seeds 1.5 mm. long.

Living and dried flowers of this species were obtained from the Horto Florestal, Joazeiro, Brazil, through Dr. L. Zehntner (No. 266, type); later Dr. Rose obtained more material at Barrinha, Bahia, Brazil, and living plants were sent to the New York Botanical Garden; in October 1917 Dr. Zehntner obtained fruit in the Chique-Chique district of Bahia.

---

*By typographical error *Bergerocereus.*

It is called, in Bahia, rabo de raposa and tail of the fox.

Figure 160 is from a photograph of a plant collected by Dr. Rose in Bahia in 1915; figure 161 shows a flower of an herbarium specimen received from Dr. Zehntner.

2. Leocereus melanurus (Schumann).

Cereus melanurus Schumann in Martius, Fl. Bras. 4²: 200. 1890.

Stems more or less cespitose from fibrous roots, slender, 1 meter long or more, 2 to 2.5 cm. in diameter; ribs 12 to 16, low, only 2 to 3 mm. high; branches 10 to 40 cm. long, often short-jointed, 2 to 2.5 cm. in diameter; areoles approximate, 2 to 5 mm. apart, white-felted when young; spines numerous, very unequal; lower radials about 20, white, bristle-like, 5 to 8 mm. long; upper radial spines and centrals about 15, all brown, stouter than the lower radials and a little longer, except that 1 or sometimes 2 of the centrals are much elongated (3 to 5 cm. long); flower narrow, 5 to 6 cm. long, somewhat enlarged above, appearing in December; flower-tube 4 to 5 cm. in diameter, covered with closely appressed hairs; perianth-segments narrow, erect, acute; seeds 1.5 mm. long, brownish.

Type locality: Serra de S. João del Ray, Brazil.

Distribution: Minas Geraes, Brazil.

The above description is drawn from the original of Schumann, supplemented by notes from specimens collected in Minas Geraes, Brazil, by Campos Porto and sent to Washington by Dr. A. Löfgren in 1917. These specimens differ considerably in habit from the plant as originally described, but since they come from near the type locality and have the same ribs and spines we believe we are justified in so referring them.

Illustration: Martius, Fl. Bras. 4²: pl. 39, as *Cereus melanurus*.

Figure 162 is copied from the illustration above cited.

Fig. 160.—Leocereus bahiensis.

3. Leocereus glaziovii (Schumann).

Cereus glaziovii Schumann in Martius, Fl. Bras. 4²: 200. 1890.

Stems erect, with somewhat spreading branches, 1.5 to 2 cm. in diameter; ribs 12, low; areoles a little longer than broad; spines 20 to 30, subulate, brownish, 1.5 to 2.5 cm. long; flowers 6 cm. long, funnelform; inner perianth-segments white, 2.5 to 3.5 cm. long, 5 mm. broad, acuminate; stamens included; scales of the ovary woolly in their axils; fruit narrowly oblong, 2 cm. long, 5 mm. in diameter; seeds small, black.

Type locality: Near Pico d'Itabira do Campo.

Distribution: Known only from the type locality.

Cereus glaziovii Schumann was placed by K. Schumann next to *C. melanurus* Schumann, and is probably congeneric with it; its flowers are similar, but the ovary and fruit are not spiny. It is known only from the collection made by Glaziou in the State of Minas Geraes, Brazil, near Pico d'Itabira do Campo.

161

162

Fig. 161.—Flower of L. bahiensis. ×0.9.

Fig. 162.—Flower of L. melanurus. ×0.9.

Leocereus ? sp.

Rootstock 1 to 2 dm. broad, flattened, shallow-seated; stems several, erect or ascending, unbranched, up to 1 meter long or more, 3 to 4 cm. in diameter; ribs 13 to 15, low, 3 to 4 mm. high; areoles close together, 3 to 4 mm. apart, brown-felted when young; spines yellowish, 20 or more, acicular, about 1 cm. long; flowers said to be tubular, 2.5 to 3 cm. long, somewhat hairy; perianth-segments white.

Collected by Campos Porto, on the Serra do Ouro Branco, Minas Geraes, Brazil, December 1916.

This plant was collected for *Cereus melanurus*, but it is too tall and stout and has different spines and smaller flowers. We have living specimens of this plant collected by Señor Porto, but they have not yet flowered in cultivation.

16. WILCOXIA Britton and Rose, Contr. U. S. Nat. Herb. 12: 434. 1909.

Plants usually low and weak, producing a cluster of dahlia-like roots; stems very slender, more or less branched, the branches often only the diameter of a lead pencil; ribs few and low; spines of all the areoles similar; flowers diurnal, funnelform-campanulate, red or purple, large for the size of the plant, only 1 from an areole, the tube rather short, its areoles bearing spines or bristles and wool; areoles of the ovary and fruit bearing spines or bristles and wool; seeds black; aril large, basal.

Type species: *Echinocereus poselgeri* Lemaire.

Four species, of Texas and Mexico, compose the genus as known.

The type species has been included in *Echinocereus*, but its habit is very unlike that genus, while the second and third species have been considered as belonging to *Cereus* proper.

The genus was named for General Timothy E. Wilcox, U. S. A., who for many years has been an enthusiastic student of plants.

KEY TO SPECIES.

```
Areoles on ovary and flower-tube bearing long bristles.
    Stems puberulent......................................................................1. W. viperina
    Stems glabrous.
        Corolla about 5 cm. long; tube indefinite; seeds dull; spine-clusters approximate, 3 to 5 mm. apart..2. W. poselgeri
        Corolla 10 to 12 cm. long; tube definite; seeds shining; spine-clusters distant, 7 to 15 mm. apart..3. W. striata
Areoles on ovary and lower part of flower-tube without long bristles...........................4. W. papillosa
```

1. Wilcoxia viperina (Weber) Britton and Rose, Contr. U. S. Nat. Herb. 16: 242. 1913.

Cereus viperinus Weber in Gosselin, Bull. Mus. Hist. Nat. Paris 10: 385. 1904.

Stems elongated, branching, the largest ones seen 1 cm. in diameter and becoming spineless; branches densely velvety-puberulent, 8 mm. in diameter or less; ribs about 8, inconspicuous; spines about 8, appressed, dark, about 5 mm. long; flowers red, about 3 cm. long; spines of ovary and corolla-tube black, bristle-like, intermixed with long white wool.

FIG. 163.—Sections of stem of W. viperina. × 0.8.

Type locality: Zapotitlan, Mexico.
Distribution: Southern Puebla, Mexico.

The type of this species was collected by L. Diguet and is now in the

FIG. 164.—Wilcoxia poselgeri.

Herbarium of the Museum of Paris, where it was examined by Dr. Rose in 1912. It is the same as C. A. Purpus's No. 3301 collected at the type locality in 1908 and distributed in his sets of specimens. It is called in Mexico organito de vibora.

The plant is remarkable among *Cereae* in having puberulent stems. We include it in *Wilcoxia*, but are uninformed as to the characters of the roots, which are tuberous in the other species.

Figure 163 shows pieces of the stem, from an herbarium specimen collected by C. A. Purpus at the type locality.

2. Wilcoxia poselgeri (Lemaire) Britton and Rose, Contr. U. S. Nat. Herb. **12:** 434. 1909.

> *Cereus tuberosus* Poselger, Allg. Gartenz. **21:** 135. 1853. Not Pfeiffer, 1837.
> *Echinocereus poselgeri* Lemaire, Cact. 57. 1868.
> *Echinocereus tuberosus* Rümpler in Förster, Handb. Cact. ed. 2. 783. 1885.
> *Cereus poselgeri* Coulter, Contr. U. S. Nat. Herb. **3:** 398. 1896.

Roots tuberous, black, several, near the surface of the ground; stems 60 cm. high or less, 6 to 10 mm. thick, with 8 to 10 inconspicuous ribs, the lower and older parts naked, spiny above, the spines almost hiding the ribs; radial spines 9 to 12, appressed, 3 to 5 mm. long, delicate, puberulent; central one ascending, black-tipped, about 1 cm. long, stouter than the radials; flowers purple or pink, 5 cm. long, spines of ovary and flower-tube intermixed with white hairs; perianth-segments linear, acuminate, about 2.5 cm. long, widely spreading or strongly recurved; style pale green; stigma-lobes slender, green; seeds pitted or rugose, 8 mm. long.

Type locality: Texas.

Distribution: Southern Texas and Coahuila.

This cactus does not grow well on its own roots in greenhouse cultivation, but gradually loses its vitality; we have had plants, however, to persist in cultivation for ten years. If grafted on cuttings of *Selenicereus pteranthus*, very vigorous plants can be developed, which will flower each year. It is sometimes called sacasil.

The flowers open in the afternoon, but close at night, opening and closing in this way for from 5 to 9 days. They have a pleasing odor.

Illustrations: Monatsschr. Kakteenk. **13:** 77; Knippel, Kakteen pl. 15; Blühende Kakteen **1:** pl. 38; Schelle, Handb. Kakteenk. f. 53, as *Echinocereus tuberosus;* Engelmann, Cact. Mex. Bound. pl. 59, f. 12; Goebel, Pflanz. Schild. **1:** pl. 4, f. 1; Blanc, Cacti 38. f. 348, 349, as *C. tuberosus.*

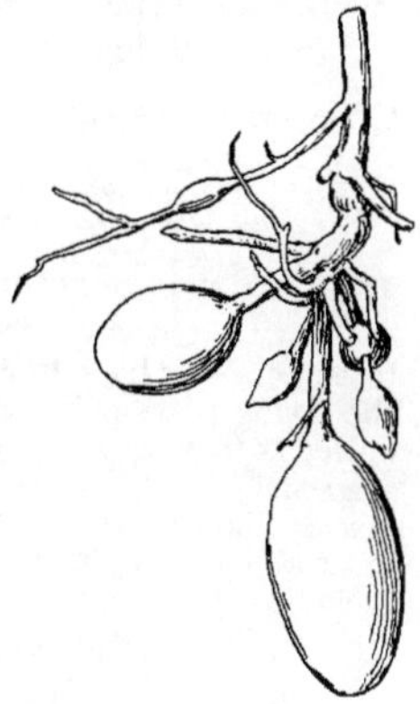

FIG. 165.—Cluster of tuberous roots of W. poselgeri. ×0.6.

Figure 164 is from a photograph of a flowering plant in the collection of the New York Botanical Garden; figure 165 shows the cluster of tuberous roots of a plant grown at Floral Park, New York, in 1890.

3. Wilcoxia striata (Brandegee) Britton and Rose, Contr. U. S. Nat. Herb. **12:** 434. 1909.

> *Cereus striatus* Brandegee, Zoe **2:** 19. 1891.
> *Cereus diguetii* Weber, Bull. Mus. Hist. Nat. Paris **1:** 319. 1895.

Roots brownish, deep-seated; stem vine-like, very slender, usually with 9 indistinct ribs, grayish; spines about 9, 1.5 to 3 mm. long, acicular, weak, appressed, brownish, the areoles rather distant; flowers 10 to 12 cm. long, purple, the areoles bearing slender, bristle-like spines and long wool; fruit pyriform, 3 to 4 cm. long, scarlet, spiny, the spines deciduous; seeds minutely pitted.

Type locality: San José del Cabo, Lower California.

Distribution: Lower California and Sonora, Mexico.

The natives call it pitayita, pitahayita, sacamatraca, saramatraca, and jaramataca.

This differs from the type species of the genus in its much larger, funnelform flowers.

4. Wilcoxia papillosa sp. nov.

Tap-root spindle-shaped, fleshy, 4 to 7 cm. long, 2 cm. in diameter, this giving off long fibrous roots; stems slender with few branches, 3 to 4 dm. long, perhaps longer, 3 to 5 mm. in diameter, glabrous, but the whole surface covered with minute papillæ; ribs low, indistinct, perhaps 3 to 5; areoles small, distant, 1 to 3 cm. long, white-woolly; spines in clusters of 6 to 8, minute, yellowish brown, bulbose at base, 1 to 3 mm. long; flowers scarlet, 4 to 5 cm. long; scales on the ovary and flower-tube small, linear-cuspidate, the lower ones naked or nearly so, those at the top of the tube with long white wool and several brown bristles (8 to 12 mm. long) in their axils; perianth-segments 2 cm. long; fruit probably spineless.

Collected by C. A. Purpus at Culiacan, Sinaloa, Mexico, October 1, 1904, and now deposited in the Herbarium of the University of California (No. 160654), and in the same State at Tinamaxtita, San Ignacio, altitude 1,340 meters, May 20, 1919, by a Mexican Commission which was studying the natural resources of Sinaloa (No. 848).

The plant is called cardoncillo.

17. PENIOCEREUS (Berger) Britton and Rose, Contr. U. S. Nat. Herb. **12**: 428. 1909.

Plants low, slender, from an enormous, fleshy, turnip-shaped root; stems and branches usually 4 or 5-angled, rarely 3 or 6-angled; spines of all the areoles similar; flowers very large for the size of the plant, funnelform, nocturnal, white, the outer perianth-segments tinged with red; tube of flower long, slender, with long hairs in the axils of the upper scales, but with clusters of spines on the lower part as also on the ovary; fruit spiny, ovoid, long-pointed, bright scarlet, fleshy, and edible; seeds black, rugose, with a large oblique hilum.

A monotypic genus of the southwestern United States and northern Mexico.

The generic name is from the Greek, signifying thread-cereus.

1. Peniocereus greggii (Engelmann) Britton and Rose, Contr. U. S. Nat. Herb. **12**: 428. 1909.

Cereus greggii Engelmann in Wislizenus, Mem. Tour North. Mex. 102. 1848.
Cereus pottsii Salm-Dyck, Cact. Hort. Dyck. 1849. 208. 1850.
Cereus greggii transmontanus Engelmann, Proc. Amer. Acad. 3: 287. 1856.
Cereus greggii cismontanus Engelmann, Proc. Amer. Acad. 3: 287. 1856.
Cereus greggii roseiflorus Kunze, Monatsschr. Kakteenk. 20: 172. 1910.

Root often very large, sometimes 6 dm. in diameter, weighing 60 to 125 pounds, usually 15 to 20 cm. long by 5 to 8 cm. in diameter; stems 3 dm. to 3 meters high, 2 to 2.5 cm. in diameter, the young parts pubescent; spines small, blackish; radials 6 to 9; central usually 1, sometimes 2; flower 15 to 20 cm. long, the tube slender and terminating in a short funnelform throat, covered with stamens; inner perianth-segments lanceolate, acute, 4 cm. long, spreading, or the outer ones reflexed; filaments erect, exserted; style slender, the stigma-lobes about 1 cm. long; fruit tuberculate, 12 to 15 cm. long, including the elongated beak.

Type locality: Near Chihuahua, Mexico.

Distribution: Western Texas, southern New Mexico and Arizona to Sonora, Chihuahua, and Zacatecas.

FIG. 166.—Peniocereus greggii.

FIG. 167.—Flower of Peniocereus greggii. ×0.5.
FIG. 168.—Fruit of same. ×0.5.

In the southwest it is called deerhorn cactus or night-blooming cereus.

The petals were first described as pale purple, but this was probably incorrect.

The species is found occasionally in valleys and on mesas in its range, but is never abundant. It is hard for the novice to find, as the short, dull-colored stems resemble dead sticks or the common sage bush, while the large flowers appear only at night.

Mrs. W. R. Kitt informs us that in cultivation this plant sometimes reaches a height of 6 feet. About Tucson, Arizona, it flowers usually between June 12 and 16 and many of the flowers appear on the same night everywhere throughout the desert. The flowers are extremely fragrant and collectors are thus guided when searching for the plants.

Illustrations: Gard. Chron. III. **34**: f. 43; Cact. Emory's Exped. 157. b 6; Förster, Handb. Cact. ed. 2. f. 14, 94; Monatsschr. Kakteenk. **5**: 150, 151; **14**: 135; Schumann, Gesamtb. Kakteen f. 18; Cact. Mex. Bound. pl. 63, 64; Schelle, Handb. Kakteenk. f. 23, as *Cereus greggii*; Cact. Mex. Bound. pl. 65, as *Cereus greggii transmontanus*; Contr. U. S. Nat. Herb. **12**: pl. 74, 75.

Figure 166 is from a photograph taken at night by F. E. Lloyd at Tucson, Arizona; figure 167 shows a flower and figure 168 a fruit collected by F. E. Lloyd near Tucson.

18. DENDROCEREUS gen. nov.

Tree-like, with a thick, upright, terete trunk crowned with numerous erect or pendent branches; branches 3 to 5-flanged; ribs thin and high, very spiny; areoles without long hairs; flowers nocturnal, broadly funnelform, the perianth finally falling from the ovary by abscission; tube of flower subcylindric, narrowed below, bearing short, often reflexed scales, the lower ones subtending short spines; perianth-segments numerous, spreading; stamens numerous, somewhat exserted; ovary with few areoles, these often bearing a few spines; fruit indehiscent, globular, naked, green, hard, with a very thick outer wall; seeds brownish, roughened, truncate at base.

A monotypic genus of Cuba. The name is from the Greek, meaning tree-cereus, this cactus being, in outline, more like a tree than any other.

1. Dendrocereus nudiflorus (Engelmann).

Cereus nudiflorus Engelmann in Sauvalle, Anal. Acad. Cienc. Habana **6**: 98. 1869.

Plant often 7 to 10 meters high, with a definite woody trunk and a very large, much branched top; trunk 1 meter long or more, up to 6 dm. in diameter, with a solid wood core, the bark close, grayish brown, armed with 3 to 5 rows of clusters of spines, sometimes borne on rounded knobs; spines pale gray, stout but acicular, 8 cm. long or less; branches dull green, when young weak, 3 to 5-winged, made up of numerous short joints, with a very slender woody axis, about 12 cm. thick; ribs or wings 4 to 7 cm. high, with low crenate margins; areoles 5 to 50 mm. apart, felted, on branches rather large, sometimes spineless, sometimes bearing 2 to 15 spines, these acicular, sometimes 4 cm. long, with black tips; flowers 10 to 12 cm. long, borne near the tops of the terminal joints, the wall of the flower-tube thick and firm; the flower-bud nearly erect, subcylindric, narrowed at base, with a few scattered areoles below the middle, ovoid-conic, blunt-pointed, viscid, shining, green streaked with brown; areoles of the ovary bearing tufts of white wool and usually 1 to 3 short black spines; outermost segments of the perianth triangular, reflexed; outer segments linear-oblong, greenish yellow, blunt, 2 to 3 cm. long, the inner narrowly oblong, white, 4 cm. long; stamens numerous, borne on the elongated throat, slightly exserted; style very thick, 5 to 6 mm. in diameter, entirely filling the tube proper, 2.5 cm. long; stigma-lobes numerous; fruit globular or longer than thick, sometimes pointed, 8 to 12 cm. long, smooth, greenish, naked, with a very thick tough rind 10 to 15 mm. thick; seeds 3 mm. long.

Type locality: Flats around Habana, Cuba.

Distribution: Coast of Habana, Matanzas, Santa Clara, and Oriente provinces, Cuba.

Dendrocereus nudiflorus is one of the most striking and interesting of all cacti. Many individuals have the general aspect of apple trees and one realizes that it is a cactus only by rather close observation. It grows in level ground, wherever observed by us, often densely surrounded by trees and bushes of various kinds. Dr. Howe's photograph, here reproduced (see fig. 169), was obtained only after cutting away a large number of bushes in order to place the camera.

The Cuban name for this plant is flor de copa.

Illustrations: Contr. U. S. Nat. Herb. **12**: pl. 49 to 51; Journ. N. Y. Bot. Gard. **10**: f. 19; Roig, Cact. Fl. Cub. pl. 2, as *Cereus nudiflorus.*

Plate XIV, figures 1 and 2, show branch and flower of the plant as it flowered at the New York Botanical Garden in 1911. Figure 169 is from a photograph taken by Marshall A. Howe at Guantánamo Bay, Cuba, in 1909; figure 170 shows a fruit collected by N. L. Britton and Percy Wilson at Punta Colorado, Cienfuegos Bay, Cuba, in 1910.

FIG. 169.—Dendrocereus nudiflorus.

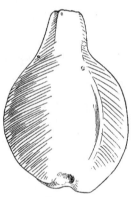

FIG. 170.—Fruit of D. nudiflorus.
×0.5.

19. MACHAEROCEREUS gen. nov.

Plants prostrate or low and bushy, often with long horizontal or prostrate stout branches, very spiny throughout; ribs low; areoles large, felted, and spiny; spines numerous, the centrals flattened and dagger-like; flowers diurnal, 1 at an areole, long, slender, funnelform, the perianth persisting on the fruit; stamens numerous, borne on the narrow elongated throat; ovary and lower part of flower-tube bearing many small scales, these subtending felted areoles which afterwards bear clusters of spines; fruit globular, edible when young, covered with clusters of spines, but when fully mature becoming naked; seeds dull black, somewhat punctate, acute on the back.

In its fruit this genus is nearest *Lemaireocereus,* to which we once referred its two species; the perianth, however, is much more elongated and more persistent; in habit and shape of spines the species are very different from any of *Lemaireocereus.*

Two species, natives of Lower California, are recognized, of which *Cereus eruca* Brandegee is the type.

The generic name is from the Greek, signifying dagger-cereus, with reference to the dagger-like spines.

KEY TO SPECIES.

Prostrate, the tips ascending; flowers yellow..1. *M. eruca*
Bushy, erect, 1 meter high or less; flowers purple...2. *M. gummosus*

M. E. Eaton del.

1. Part of branch of *Dendrocereus nudiflorus*.
2. Flowering branch of the same.
3. Flowering branch of *Nyctocereus guatemalensis*.
(Natural size.)

1. Machaerocereus eruca (Brandegee).

Cereus eruca Brandegee, Proc. Calif. Acad. II. **2**: 163. 1889.
Lemaireocereus eruca Britton and Rose, Contr. U. S. Nat. Herb. **12**: 425. 1909.

Prostrate, except the erect or ascending tips; branches 1 to 3 meters long, 4 to 8 cm. in diameter, usually simple, rooting on the under surface, dying at the older end and growing forward at the other; sometimes several plants starting as branches from a common parent as a center and first radiating out, then dying at the rear; ribs about 12; areoles large, 2 cm. apart; spines about 20, very unequal, pale gray, the outer ones terete, the inner ones stout and flatter, the longest about 3 cm. long; flowers 10 to 12 cm. long, described as yellow; tube about 10 cm. long, nearly 6 mm. in diameter; limb 4 to 6 cm. broad; ovary very spiny; fruit spiny, 4 cm. long; seeds black.

Type locality: Magdalena Island, Lower California.
Distribution: Lower California.

The plant is known in Lower California as chirinola and creeping devil cactus. Mr. Brandegee describes it as follows:

"Its manner of growth with uplifted heads and prominent reflexed spines gives the plants a resemblance to huge caterpillars."

While this resemblance is true of the plants when growing in the open, it is especially striking when the plant meets with some obstruction such as a log or large stone. Then it raises its head, crawls up one side and down the other, and finally by the dying of the rear virtually passes over the obstruction.

Fig. 171.—Machaerocereus eruca.

Mr. E. A. Goldman (Contr. U. S. Nat. Herb. **16**: 352, 353. 1916) speaks of it as follows:

"We first saw this remarkable cactus on the coastal plain near Santo Domingo, about 30 miles north of Matancita and here made a collection. From this point southward it was noted at intervals on the plains as far as the Llano de Yrais and on the lower and more sandy parts of Magdalena Island. The stems grow 1 to 3 meters in length and are nearly prostrate, and from this habit and their long whitish recurved spines have aptly been likened to huge caterpillars. The growing ends of the branches stand up from the ground, but progressive growth leaves the main body lying prostrate. The stems become rooted along the lower sides and gradually die behind, resulting in a slow progression of the living portion along the ground. Multiplication of individuals frequently results from the decay of connecting parts. In some places disconnected plants forming a hollow circle can be traced by the remains of dead trunks to a common center. The plants show a preference for soft parts of the coastal plain and grow usually in groups, often topping a slight eminence formed of wind-drifted material. These cactuses serving as a sand binder and preventing erosion tend to favor further accumulations. The desert foxes (*Vulpes macrotis devius*) of the region find congenial burrowing places among the procumbent trunks."

Illustrations: Monatsschr. Kakteenk. 5: 71; Proc. Calif. Acad. II. 2: pl. 7; Schumann, Gesamtb. Kakteen f. 29; Nat. Geogr. Mag. 22: 466, as *Cereus eruca;* Contr. U. S. Nat. Herb. 16: pl. 127, as *Lemaireocereus eruca.*

Figure 171 is from a photograph taken by E. A. Goldman at Santo Domingo, Lower California; figure 172 is from a photograph of a plant collected by C. R. Orcutt at Magdateria Bay, Lower California.

FIG. 172.—Machaerocereus eruca.

2. Machaerocereus gummosus (Engelmann).

> *Cereus gummosus* Engelmann in Brandegee, Proc. Calif. Acad. II. 2: 162. 1889.
> *Cereus cumengei* Weber, Bull. Mus. Hist. Nat. Paris 1: 317. 1895.
> *Cereus flexuosus* Engelmann in Coulter, Contr. U. S. Nat. Herb. 3: 411. 1896.
> *Lemaireocereus cumengei* Britton and Rose, Contr. U. S. Nat. Herb.
>   12: 424. 1909.
> *Lemaireocereus gummosus* Britton and Rose, Contr. U. S. Nat. 12:
>   425. 1909.

Erect or ascending, but usually not a meter high, or with long, spreading, sometimes prostrate, branches, the whole plant sometimes having a spread of 6 to 7 meters; branches 4 to 6 cm. in diameter; ribs usually 8, rarely 9, low and obtuse; areoles rather large, about 2 cm. apart; spines stout, the radials 8 to 12, somewhat unequal, about 1 cm. long; central spines 3 to 6, stout, flattened, one much longer than the others and about 4 cm. long; flowers 10 to 14 cm. long, the tube long and slender; inner perianth-segments 2 to 2.5 cm. long, purple; stamens about as long as the segments; fruit subglobose, 6 to 8 cm. in diameter, spiny; skin of fruit bright scarlet; pulp purple; seeds rugose, pitted, 2.5 mm. long.

Type locality: Lower California.

Distribution: Lower California and adjacent islands.

Dr. Rose, who visited Lower California in 1911, found this the most widely distributed there of all the cacti. He observed it at all stations visited on the main peninsula and on all the islands of the Gulf of California except Tiburon and Estaban. The plant is rather diverse in its habit; it often sends out long horizontal branches which take root and start other colonies. In habit it much resembles *Rathbunia alamosensis,* but is usually stouter and less gregarious. The

FIG. 173.—M. gummosus.

fruit is called pitahaya agre or pitahaya agria and is probably the most valuable fruit of Lower California. A fish poison is prepared by bruising the stems. The mashed pulp is then thrown into a running stream.

Fig. 175.—Flower of M. gummosus. ×0.6.

Fig. 174.—Machaerocereus gummosus.

Cereus gummatus, C. gumminosus, and *C. pfersdorffii* Hildmann (Schumann, Gesamtb. Kakteen 125. 1897) are only garden names of this species.

Illustrations: Grässner, Haupt-Verz. Kakteen 3; Monatsschr. Kakteenk. 13: 105, both as *Cereus gummosus;* Contr. U. S. Nat. Herb. 16: pl. 126 A, as *Lemaireocereus gummosus.*

Figure 173 is from a photograph of a plant collected by Dr. Rose at Santa Maria Bay in 1911; figure 174 is from a photograph taken by E. A. Goldman on Esperito Santo Island, Lower California, in 1906; figure 175 shows a flower drawn from an herbarium specimen obtained from C. R. Orcutt, collected in northern Lower California.

20. NYCTOCEREUS (Berger) Britton and Rose, Contr. U. S. Nat. Herb. 12: 423. 1909.

Erect or clambering, slender, sparingly branched cacti, with cylindric, ribbed stems and branches; ribs numerous, low; areoles each bearing a tuft of short white wool and small radiating acicular bristles or weak spines; flowers large, white, nocturnal; ovary bearing small scales, short or long wool, and tufts of weak spines or bristles; perianth funnelform, gradually expanding above, bearing scales and tufts of weak bristles below the middle, above the middle bearing narrowly lanceolate scales distant from each other and grading into the blunt outer perianth-segments; inner perianth-segments widely spreading, obtuse or acutish; stamens numerous, shorter than the perianth; style about as long as the stamens; fruit fleshy, scaly, spiny or bristly; seeds large, black.

Type species: *Cereus serpentinus* De Candolle.

Nyctocereus was considered by A. Berger a subsection of his subgenus *Eucercus* but his conception of it was of a complex, from which we would exclude all but three of the species which he referred to it. He speaks of certain forms in the type species which have smaller flowers and no fruit; this variation we have also noticed in *N. guatemalensis.*

The name is from the Greek, meaning night-cereus. Five species are here recognized, natives of Mexico and Central America.

KEY TO SPECIES.

Flower-tube longer than the limb..1. *N. serpentinus*
Flower-tube not longer than the limb.
    Flowers 4 to 7 cm. long.
        Spines acicular; ribs acute..2. *N. hirschtianus*
        Spines subulate; ribs obtuse..3. *N. guatemalensis*
    Flowers 9 cm. long or more.
        Perianth-segments long-acuminate...4. *N. neumannii*
        Perianth-segments acute or obtusish..5. *N. oaxacensis*

1. Nyctocereus serpentinus (Lagasca and Rodrigues) Britton and Rose, Contr. U. S. Nat. Herb. **12**: 423. 1909.

    Cactus serpentinus Lagasca and Rodrigues, Anal. Cienc. Nat. Madrid **4**: 261. 1801.
    Cactus ambiguus Bonpland, Descr. Pl. Rares 90. 1813.
    Cereus serpentinus De Candolle, Prodr. **3**: 467. 1828.
    Cereus ambiguus De Candolle, Prodr. **3**: 467. 1828.
    Cereus serpentinus stellatus Lemaire, Cact. Gen. Nov. Sp. 78. 1839.
    Cereus serpentinus splendens Salm-Dyck in Lemaire, Cact. Gen. Nov. Sp. 79. 1839.
    *Cereus splendens** Salm-Dyck, Cact. Hort. Dyck. 1849. 214. 1850.
    Echinocereus serpentinus Lemaire, Cact. 57. 1868.
    Echinocereus splendens Lemaire, Cact. 57. 1868.
    Cereus serpentinus albispinus† Weingart, Monatsschr. Kakteenk. **18**: 30. 1908.

Stems growing in a cluster or clump, at first erect, then clambering through bushes or over walls or, when without support, creeping or hanging, often 3 meters long, 2 to 5 cm. in diameter; ribs 10 to 13, low and rounded; areoles close together, felted and with acicular or bristle-like spines; spines about 12, white to brownish, the tips usually darker, the longest about 3 cm. long; flowers borne at the upper areoles, sometimes terminal, 15 to 19 cm. long, the limb 8 cm. broad; areoles on ovary and flower-tube bristly; inner perianth-segments white, spatulate, obtuse; fruit red, covered with deciduous spines, 4 cm. long; seeds black, 5 mm. long.

Type locality: Not cited; described from a garden plant.

Distribution: Mexico, probably native near the eastern coast.

Cereus serpentinus strictior Walpers (Repert. Bot. **2**: 278. 1843) is only a published name.

Cereus ambiguus strictior (Weingart, Monatsschr. Kakteenk. **19**: 9. 1909) seems never to have been published.

Cereus kalbreyerianus Wercklé (Monatsschr. Kakteenk. **17**: 38. 1907) is known only from its flowers, which, from the description, closely resemble those of *N. serpentinus* and it is said to resemble this species in its habit. It was found near Bogotá, Colombia.

Although Mexico is given as the home of this species, no wild specimens have been collected there in recent times; it is now widely cultivated in that country, or is half-wild in hedges or running over walls about yards. A. Berger (Rep. Mo. Bot. Gard. **16**: 75, 76, 1905) has this interesting note:

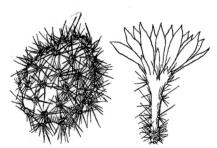

FIG. 176.—Fruit of Nycto- FIG. 177.—Flower of Nyctocereus serpentinus. ×0.7. cereus hirschtianus. ×0.7.

"*Cereus serpentinus* P. DC. possesses the largest seeds of *Cereus* known to me. There are only a few in each fruit, bedded in the crystalline red pulp. Several varieties of this species occur in gardens. There are two very pronounced forms at La Mortola. One has weaker and more serpentine stems, with smaller spines and smaller flowers. This never produces any fruit. The other form has stronger, upright stems with longer spines. Its flowers are remarkably larger and produce a great quantity of fruits. The former variety seems to have undeveloped stigmata, and it may prove to be the male plant. Similar cases of heterogamy are known in *Opuntia* and *Mammillaria*, but nothing of the kind has ever been shown in *Cereus*. This male form at La Mortola corresponds well with the figure in the Botanical Magazine, pl. 3566. Strictly terminal flowers, as shown in this plate, are also occasionally produced by our plant."

---

    *We have followed Weingart (Monatsschr. Kakteenk. **18**: 30. 1908) in referring this name here rather than to *Monvillea cavendishii*.
    †Weingart states that this and *Cereus albispinus* are identical with *Cereus splendens*.

PLATE XV

M. E. Eaton del.

1. Top of branch of *Eulychnia iquiquensis*.
2. Top of stem of *Lemaireocereus dumortieri*.
3. Part of flowering stem of *Nyctocereus serpentinus*.
(Natural size.)

Known in Mexico as junco or junco espinoso.

Illustrations: Link and Otto, Ic. Pl. Select. pl. 42, as *Cactus serpentinus;* Bonpland, Descr. Pl. Rares pl. 36; Van Geel, Sert. Bot. 3: pl. 17, the last two as *Cactus ambiguus;* Abh. Bayer. Akad. Wiss. München 19: pl. 2; Cact. Journ. 1: 59; Curtis's Bot. Mag. 64: pl. 3566; Dict. Gard. Nicholson 1: f. 410; Förster, Handb. Cact. ed. 2. f. 95; Gartenflora 31: pl. 1079, f. 2. c; Mém. Mus. Hist. Nat. Paris 17: pl. 12; Rep. Mo. Bot. Gard. 16: pl. 11, f. 1 to 3; Rümpler, Sukkulenten f. 65, as *Cereus serpentinus.*

Plate xv, figure 3, shows the flower of a plant in the collection of the New York Botanical Garden. Figure 176 shows the fruit collected in Mexico by H. H. Rusby in 1910.

2. Nyctocereus hirschtianus (Schumann) Britton and Rose, Contr. U. S. Nat. Herb. **12:** 424. 1909.

Cereus hirschtianus Schumann, Gesamtb. Kakteen 130. 1897.

Stems columnar, erect, slender, 10 mm. in diameter; ribs 10, somewhat acute, 3 mm. high; radial spines 7 to 9, slender, 4 to 5 mm. long; central spines 1 to 5, the lower one stouter and porrect; flowers probably white, 5 to 6 cm. long, funnelform; perianth-segments spreading, acute; stamens numerous, somewhat exserted; ovary and tube very spiny; fruit not known.

Type locality: Nicaragua.

Distribution: Known only from the type locality.

This species differs from *N. guatemalensis* in its habit, more slender stem, its spines, which are much more slender and delicate but not as long, and its smaller flowers. Weingart has written extensively (Monatsschr. Kakteenk. **23:** 108 to 111. 1913) about this species, reaching the conclusion that it and *N. guatemalensis* are the same. We have both types before us, and feel convinced that the species are distinct.

Illustration: Schumann, Gesamtb. Kakteen f. 31, as *Cereus hirschtianus.*

Figure 177 shows the flower of a cotype specimen in the herbarium of the United States National Museum.

3. Nyctocereus guatemalensis Britton and Rose, Contr. U. S. Nat. Herb. **16:** 240. 1913.

Cereus guatemalensis Vaupel, Monatsschr. Kakteenk. **23:** 86. 1913.

Stems half erect, arching, creeping, or even prostrate, 1 meter long or longer, 3 to 6 cm. in diameter; ribs 8 to 12, very low; radial spines about 10; central spines 3 to 6, usually much longer than the radials, the longer ones 3 to 4 cm. long; flowers very fragrant, 4 to 7 cm. long; ovary somewhat tuberculate, each tubercle crowned by an areole bearing a bunch of pinkish or brownish spines; outer perianth-segments brownish; inner perianth-segments lanceolate, acute, nearly white; stamens much shorter than the perianth, attached all along the surface of the wide throat; style stout, 3 cm. long; fruit about 2 cm. long, spiny; seeds black, shining, 3 mm. in diameter.

Type locality: El Rancho, Guatemala.

Distribution: Guatemala.

Illustrations: Monatsschr. Kakteenk. **19:** 167, as *Cereus hirschtianus;* Contr. U. S. Nat. Herb. **16:** pl. 70, 71.

Plate xiv, figure 3, shows a part of the type specimen, which flowered at the New York Botanical Garden in 1915. Figure 178 is from a photograph of another part of the type specimen.

4. Nyctocereus neumannii (Schumann) Britton and Rose, Contr. U. S. Nat. Herb. **12:** 424. 1909.

Cereus neumannii Schumann in Loesener, Bot. Jahrb. Engler **29:** 99. 1900.

Stems columnar, up to 1 meter long, 3 cm. in diameter, ascending or decumbent; ribs 13, somewhat crenate; spines 10 to 14, radials and centrals similar, acicular, up to 4 cm. long, grayish, brownish when young; flower 10 cm. long; ovary tuberculate, bearing felt and brown or reddish spines in its areoles; inner perianth-segments white, lanceolate, long-acuminate.

Type locality: Near Chiquitillo, Metagalpa, Nicaragua.

Distribution: Known only from the type locality.

The plant is known to us only from description.

5. Nyctocereus oaxacensis sp. nov.

Stems branching, slender, 2 to 3 cm. in diameter; ribs 7 to 10, rather low; areoles 10 mm apart; radial spines 8 to 12, 4 to 15 mm. long, slender, brownish; centrals 3 to 5; flowers 8 to 10 cm. long, "whitish inside, dirty purplish or reddish outside"; perianth-segments linear to oblong, rounded at apex; stamens not extending nearly as far as the perianth-segments; ovary densely covered with brownish bristly spines.

Collected by E. W. Nelson about Lagunas, Oaxaca, Mexico, altitude 255 meters, June 5, 1895 (No. 2543, type).

We refer here tentatively another specimen also collected by Mr. Nelson near Huilo-tepec, Oaxaca, altitude 30 meters, May 4 to 11, 1895 (No. 2585).

Fig. 178.—Nyctocereus guatemalensis, as it flowered in Washington.

21. BRACHYCEREUS gen. nov.

Stems low, forming candelabrum-like masses; branches numerous, cylindric; ribs many, low, with closely set areoles bearing felt and numerous acicular spines; flowers narrow-funnelform, bearing small scales which subtend large spiny areoles; outer perianth-segments lanceolate; inner perianth-segments very narrow, long-acuminate, described as possibly white but more likely yellow; filaments very short; ovary obliquely subglobose, bearing scattered spiny areoles; fruit ellipsoid, very spiny, but in age probably naked.

The name is from the Greek, meaning short-cereus.

Only one species is known, native of the Galápagos Islands.

1. Brachycereus thouarsii (Weber).

Cereus thouarsii Weber, Bull. Mus. Hist. Nat. Paris **5**: 312. 1899.
Cereus nesioticus Schumann in Robinson, Proc. Amer. Acad. **38**: 179. 1902.

Stems 6 to 10 dm. high; branches numerous, radiating and ascending, 3 to 5 cm. in diameter, entirely covered by a mass of yellow spines; ribs about 20, low, 3 mm. high; areoles 5 to 6 mm. apart; spines about 40, unequal, the longer ones about 3 cm. long, bristle-like; flower 7 cm. long; outer perianth-segments 1.5 cm. long, 2 mm. broad; inner perianth-segments longer than the outer, narrow; filaments 1 mm. long or less; fruit 2.5 to 4 cm. long, 1.3 cm. in diameter; seeds numerous, 1.2 mm. long, ellipsoid, brownish, slightly punctate.

Type locality: Charles Island, Galápagos.

Distribution: Albemarle, Abingdon, Chatham, James, Charles, and Tower Islands, Galápagos.

We have identified *Cereus thouarsii* Weber, by photographs of the specimens sent by Professor Agassiz to Dr. Engelmann, preserved at the Missouri Botanical Garden, and mentioned by Dr. Weber at the place of first publication.

Schumann says this species is a very peculiar one, "from its long, brown, non-pungent spines, which clothe the stem so densely that its surface is invisible. I have never before seen a species of the genus with such short filaments as in this. The petals are also uncommonly narrow."

Berger refers this species to his subsection *Nyctocereus*, with which it is probably most nearly related. It was named for Abel Aubert Du Petit-Thouars (1793–1864).

Illustration: Proc. Calif. Acad. Sci. IV. 1: pl. 5, as *Cereus nesioticus.*

Figure 179 shows the flower of the type specimen of *Cereus nesioticus* preserved in the Gray Herbarium; figure 180 shows the fruit of *Brachycereus thouarsii* collected by A. Stewart, preserved in the herbarium of the California Academy of Sciences.

Fig. 179.—Flower of B. thouarsii. ×0.8. Fig. 180.— Fruit of same species. ×0.8.

22. ACANTHOCEREUS (Berger) Britton and Rose, Contr. U. S. Nat. Herb. 12: 432. 1909.

Weak, elongated, many-jointed cacti, at first erect but soon clambering or trailing, the joints usually strongly 3-angled, sometimes 4 or 5-angled, in one species sometimes 7-angled, the seedlings and juvenile branches not as strongly angled, with more ribs and with different spines; areoles bearing short wool or felt and several stiff spines; flowers funnelform, nocturnal, 1 at an areole; flower-tube remaining rigid after anthesis, gradually drying and remaining on the ripe fruit, green, rather slender, expanded toward the summit, bearing a few areoles similar to those of the branches, subtended by small scales; limb somewhat shorter than the tube, widely expanded; outer perianth-segments narrowly lanceolate to linear, acuminate, green, shorter than the white, inner segments; stamens not extending as far as perianth-segments, attached all along the upper half of the tube or throat; style very slender, divided at the apex into several linear stigma-lobes; fruit spiny or naked, with a thick, dark-red skin breaking irregularly from top downward; flesh red; seeds numerous, black.

This genus has a wide distribution; its species are usually found at low altitudes in semiarid regions, especially about the Gulf of Mexico and the Caribbean Sea; although occurring on the coasts of Texas and Florida and recorded from Cuba, it has not been reported from any of the other larger Antilles, but is represented on the Venezuelan and Colombian coasts and also in Central America and Brazil. It is found not only on the east and west coasts of Mexico but also in the interior.

The type of this genus is based on the *Cactus pentagonus* of Linnaeus. Linnaeus in his Species Plantarum cites no definite habitat for it, while his description is very meager. His earlier reference in Hortus Cliffortianus (182. 1737), although somewhat fuller, is still uncertain. It is there stated that the ribs are 5, sometimes 6. Most of the species of this genus, especially those which would have been known in Linnaeus's time, usually have 3 ribs, occasionally 4, rarely 5. The young plants and the young growth, however, often have 5 and 6 ribs, which would account for variations in descriptions of the same species.

Curiously enough, the type species is one of the species of Linnaeus which Miller omits in his Gardener's Dictionary (1768).

Cereus pellucidus Pfeiffer (Enum. Cact. 108. 1837), which we formerly referred to this genus (Contr. U. S. Nat. Herb. 12: 432), following previous authors, is to be looked for in *Leptocereus*. Both Schumann and Berger regard this group as consisting of but a single species, the former placing it with *Cereus greggii* in his series *Acutangules*, and the latter in a subsection *Acanthocereus*; Pfeiffer, on the other hand, recognized several species as belonging to this group; we distinguish at least 7. The name is from the Greek, meaning thorn-cereus.

KEY TO SPECIES.

Ribs usually 3, rarely 4, thick.
  Joints 8 to 10 cm. wide, deeply crenate; spines very stout, subulate.
    Spines 1 to 6; perianth-tube about 7 cm. long...1. *A. horridus*
    Spines about 10, the outer 5 to 8, very short; perianth-tube about 12 cm. long..........2. *A. colombianus*
  Joints 2 to 8 cm. wide, low-crenate; spines slender.
    Spines well developed, subulate...3. *A. pentagonus*
    Spines short or none, when present acicular..4. *A. subinermis*
Ribs 4 to 7, mostly thin.
  Plants green.
    Spines up to 7 cm. long; ribs 3 to 5...5. *A. occidentalis*
    Spines 3 cm. long or less; ribs 5 to 7...6. *A. brasiliensis*
  Plants bluish white; joints 4-angled; spines 2 to 6, the longest 2 cm. long..................7. *A. albicaulis*

1. Acanthocereus horridus sp. nov.

Plants stout, the joints strongly 3-angled or 3-winged, the young growth 5 or 6-angled; wings with deep undulations; areoles large, 3 to 6 cm. apart; spines brown or blackish when young; radial spines 1 to 6, very short, conic, less than 1 cm. long; central spine usually 1, sometimes 2, often very stout and elongated, sometimes 8 cm. long; flower, including the ovary, 18 to 20 cm. long; tube 4 cm. long, including the funnelform throat 12 cm. long; throat 4 cm. broad at mouth; outer perianth-segments linear, brown or greenish, 6 cm. long; inner perianth-segments 3 to 4 cm. long; stamens white; style thick, cream-colored; fruit 3.5 cm. long, light red, glossy, covered with large areoles bearing white felt; skin thick, finally splitting as the fruit ripens; pulp red.

Collected in Guatemala by F. Eichlam in 1909 (New York Botanical Garden No. 34788). It has frequently flowered in cultivation, both at Washington and at New York.

Here we are disposed to refer E. W. Nelson's plant from San Juan Guichicovi, Oaxaca, Mexico, collected June 21 to 24, 1895 (No. 2729).

Figure 181 shows a part of a joint of the type specimen.

2. Acanthocereus colombianus sp. nov.

Erect, branching dichotomously, 2 to 3 meters high; joints about 9 cm. wide, strongly 3-winged; areoles large, 5 cm. apart; radial spines 5 to 8, very short, less than 5 mm. long; central spines 1 or 2, very stout, 4 to 5.5 cm. long; flower 25 cm. long, white, with a rather stout tube 12 cm. long, the gradually expanded throat 5 to 6 cm. long.

Collected by Francis W. Pennell and Henry H. Rusby near Calamar, Colombia, July 10, 1917 (No. 23, type), and by Herbert H. Smith near Bonda, Colombia, in 1898–1899 (No. 2423). According to Mr. Smith this species grows in dry forests and thickets at low altitudes; here it is known as pitahaya.* His

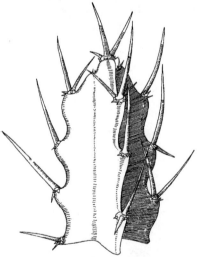

FIG. 181.—Part of joint of A. horridus. X0.4.

*Pitahaya is a well-known name in tropical America for many species of cacti, especially of *Cereus* and its relatives, of which there are various spellings, such as pitajaya, pitajuia, pitalla, pitaya, and pithaya. Several suffixes are sometimes used with it, as pitahaya agre, pitahaya agria, pitahaya de San Juan, and pitahaya dulce, and it has the diminutives pitayita and pitahayita.

plant comes from near the type locality of *Cactus pitajaya* Jacquin, but we refer that species to *A. pentagonus*, also found in northern Colombia.

The species is near *A. horridus*, but has a much longer flower-tube.

3. **Acanthocereus pentagonus** (Linnaeus) Britton and Rose, Contr. U. S. Nat. Herb. **12**: 432. 1909.

<blockquote>
Cactus pentagonus Linnaeus, Sp. Pl. 467. 1753.
Cactus pitajaya Jacquin, Enum. Pl. Carib. 23. 1761.
Cereus pentagonus Haworth, Syn. Pl. Succ. 180. 1812.
Cactus prismaticus Willdenow, Enum. Pl. Suppl. 32. 1813.
Cereus prismaticus Haworth, Suppl. Pl. Succ. 77. 1819.
Cereus pitajaya De Candolle, Prodr. 3: 466. 1828.
?*Cereus undulosus* De Candolle, Prodr. 3: 467. 1828.
?*Cactus undulosus* Kosteletzky, Allg. Med. Pharm. Fl. 4: 1393. 1835.
Cereus cognatus Pfeiffer, Enum. Cact. 106. 1837, as synonym.
Cereus acutangulus Otto in Pfeiffer, Enum. Cact. 107. 1837.
Cereus princeps Pfeiffer, Enum. Cact. 108. 1837.
Cereus ramosus Karwinsky in Pfeiffer, Enum. Cact. 108. 1837.
Cereus baxaniensis Karwinsky in Pfeiffer, Enum. Cact. 109. 1837.
Cereus variabilis Engelmann, Bost. Journ. Nat. Hist. 5: 205. 1845. Not Pfeiffer, 1837.
Cereus nitidus Salm-Dyck, Cact. Hort. Dyck. 1849. 211. 1850.
Cereus vasmeri Young, Fl. Texas 276. 1873.
Cereus dussii Schumann, Gesamtb. Kakteen 89. 1897.
Cereus sirul Weber in Gosselin, Bull. Mus. Hist. Nat. Paris 10: 384. 1904.
</blockquote>

Stem clambering, usually 2 to 3, sometimes 7 meters high, but when growing in the open more or less arched and rooting at the tips, then making other arches and thus forming large colonies; old trunk becoming nearly round, 5 cm. in diameter or more, covered with a thick mucilaginous, spineless cortex and a hard-wood axis with only a small pithy cavity; joints 3 to 8 cm. broad, 3 to 5-angled, low-crenate; juvenile growth nearly terete, with 6 to 8 low ribs, approximate areoles and numerous short acicular spines; areoles on normal branches 3 to 5 cm. apart; spines gray, acicular to subulate, various; radials at first 6 or 7, 1 to 4 cm. long; central spine often solitary, longer than the radials; spines of old areoles often as many as 12, of which several are centrals; flowers 14 to 20 cm. long; tube and ovary bearing conspicuous areoles with brown felt and several subulate spines; outer perianth-segments green; inner perianth-segments white, acuminate; fruit oblong, red, edible; cotyledons broadly ovate, 5 to 8 mm. long, thick, united at base, gradually passing below into the spindle-shaped hypocotyl.

Type locality: America, but no definite locality cited.

Distribution: Keys of southern Florida; coast of Texas, south along the eastern coast of Mexico to Gua-

FIG. 182.—Acanthocereus pentagonus.

FIG. 183.—Acanthocereus pentagonus.

temala and Panama; the coasts of Colombia and Venezuela and Guadeloupe. Introduced on St. Thomas and St. Croix. Recorded from Cuba.

As understood by us this species varies greatly in the relative thickness of its branches, in armament, and in the size of its flowers. Its geographical range is, in our conception, greater than that of most cacti.

Cereus baxaniensis ramosus (Salm-Dyck in Walpers, Repert. Bot. **2**: 277. 1843) is published only as a synonym. *Cereus arcuatus* Zuccarini (Monatsschr. Kakteenk. **14**: 55. 1904) from its description is of this relationship. It was originally collected at Totolapa, Mexico, by Zuccarini.

Cereus bajanensis Wercklé (Monatsschr. Kakteenk. **15**: 166. 1905) was never described but belongs here. *Cereus quadrangularis* Haworth (Syn. Pl. Succ. 181. 1812; *C. trigonus quadrangularis* Pfeiffer, Enum. Cact. 118. 1837; *Cactus quadrangularis* Loudon, Encycl. Pl. 412. f. 6876. 1829) may belong here, but Pfeiffer referred it with a question to *Cereus caripensis* De Candolle (Prodr. **3**: 467. 1828; *Cactus caripensis* Humboldt, Bonpland, and Kunth, Nov. Gen. et Sp. **6**: 66. 1823), but this species was referred by Schumann to the genus *Rhipsalis*.

Fig. 184.—Acanthocereus pentagonus in cactus plantation of Charles Deering at Buena Vista, Florida, May 1918.

Cereus undulatus Pfeiffer (Enum. Cact. 107. 1837), based on a specimen in the Dresden Garden, is usually referred to *Cereus acutangulus*, but was not described by Pfeiffer at the place here cited.

A specimen in the Berlin Garden also was called *Cereus undulatus* by D. Dietrich (Syn. Pl. **3**: 104. 1843) and described, but should be referred elsewhere. It is of quite different relationship, being very slender, dull green, 10-ribbed. The flowers are large, 12.5 cm. in diameter, white. Its native habitat is unknown.

Illustrations: Cact. Journ. **1**: 125; Cact. Mex. Bound. pl. 60, f. 5, 6, all these as *Cereus variabilis*; Monatsschr. Kakteenk. **13**: 158; Rev. Hort. Belge **40**: after p. 184; Tribune Hort. **4**: pl. 140, as *Cereus baxaniensis*.

Plate XVI, figure 1, shows a flower and part of a joint of a plant sent from the Berlin Botanic Garden to the New York Botanical Garden. Figure 182 is from a photograph

M. E. Eaton del.

1. Top of flowering branch of *Acanthocereus pentagonus*.
2. Top of flowering branch of *Acanthocereus subinermis*.
3. Top of a fruiting branch of *Acanthocereus subinermis*.
(All natural size.)

taken by Marshall A. Howe on Boot Key, Florida, in 1909; figure 183 shows the fruit and withering perianth of a specimen collected by Dr. Rose at Laredo, Texas, in 1906; figure 184 is from a photograph by J. K. Small of a plant in the cactus plantation of Charles Deering, Buena Vista, Miami, Florida, May 1918, originally brought from Sands Key in 1917.

4. Acanthocereus subinermis sp. nov.

Plants 1 meter high or higher; joints stout, 5 to 7 cm. broad, strongly 3 or 4-angled, bright green, somewhat shining, usually short; areoles 3 to 4 cm. apart; spines either wanting or short, when present 6 to 10 at an areole, acicular, usually less than 1.5 cm. long; flowers various in size, 15 to 22 cm. long; outer perianth-segments narrow, reddish, acute; inner perianth-segments white; areoles of ovary and flower-tube somewhat spiny; fruit globular to short-oblong, 4 cm. long, dull red.

Collected by J. N. Rose between Mitla and Oaxaca City, Mexico, September 6, 1908 (No. 11304). It has since been grown in Washington and in the New York Botanical Garden, where it has frequently flowered and fruited.

Plate XVI, figure 2, represents a flowering joint of the type specimen, and figure 3 shows its fruit.

5. Acanthocereus occidentalis sp. nov.

Stems rather weak, forming dense thickets; branches slender, 4 to 5 cm. in diameter, 3 to 5-angled, dull green, often bronzed; margins of ribs slightly sinuate; areoles 1 to 3 cm. apart, filled with short brown wool; spines numerous, nearly equal, yellowish, acicular, up to 7 cm. long; flowers 14 to 18 cm. long; fruit unknown.

Common on the western coast of Mexico, where it was frequently collected by Rose, Standley, and Russell at the following places: San Blas, Sinaloa, March 24, 1910 (No. 13431, type); Mazatlan, April 4, 1910 (No. 14050); Guadalupe, April 18, 1910 (No. 14752); and by Dr. Rose at Rosario in 1897 (No. 3170).

This species is widely separated geographically from the others of this genus, being confined to low thickets along the coast of Sinaloa, western Mexico.

Figure 185 shows part of a joint of a plant brought by Dr. Rose from Sinaloa in 1910.

FIG. 185.—Joint of Acanthocereus occidentalis. X0.5.

6. Acanthocereus brasiliensis sp. nov.

Stems weak, at first erect but soon prostrate or clambering over bushes, usually much branched at base, bright green, somewhat shining; ribs 5 to 7, high and thin, slightly undulate; areoles small, 2 to 4 cm. apart; spines numerous, acicular, white with brown tips, 3 cm. long or less; flowers about 15 cm. long; buds acuminate; outer perianth-segments linear; ovary bearing clusters of acicular spines; fruit globular, slightly tuberculate, 8 cm. in diameter, green, covered with clusters of acicular deciduous spines; pulp greenish white; seeds few, large, brownish.

Common in thickets in the subarid parts of Bahia, Brazil, where it was frequently observed by Dr. Rose in 1915; the type is from Machado Portella (No. 19903).

This species not only is out of the range of the preceding species of this genus, but is otherwise somewhat anomalous, for it normally has more ribs and these of different texture. The fruit, too, is much larger than that of the other species, is covered with deciduous spines, and has a greenish white pulp.

Figure 186 is from a photograph taken by Paul G. Russell at the type locality in 1915.

7. Acanthocereus (?) albicaulis sp. nov.

A low, weak plant, although erect at first, a meter high or less, afterward elongating and arching; branches few, usually sharply 4-angled, 1 to 3 cm. broad, bluish white, the margins only slightly undulate; areoles 2 to 3 cm. apart, small, brown-felted; spines 2 to 6, acicular, brown, swollen at base, unequal, the longest 2 cm. long; flowers and fruit unknown.

Collected near Barrinha, Bahia, Brazil, by Rose and Russell, June 8, 1915 (No. 19808).

This is a very distinct and remarkable plant. In the shape and color of the branches it suggests some species of *Hylocereus* such as *H. ocamponis*, but it is a true terrestrial and never develops aërial roots. It is inconspicuous, growing in the bushy flats, and easily overlooked. Numerous cuttings were sent to the New York Botanical Garden by Dr. Rose, but only one of these lived, and this has not yet made any new growth. It may not be of this genus, for it does not resemble closely any of the described species.

Figure 187 is from a photograph taken by Paul G. Russell in 1915 at the type locality.

FIG. 186.—Acanthocereus brasiliensis.   FIG. 187.—Acanthocereus (?) albicaulis.

DESCRIBED SPECIES, PERHAPS OF THIS GENUS.

CEREUS TENELLUS Salm-Dyck in Pfeiffer, Enum. Cact. 109. 1837.

Suberect, slender, 8 to 12 mm. in diameter; ribs 4 or 5, thin, compressed; areoles 8 to 10 mm. apart; spines setiform, brown, short, 6 to 8 mm. long; flowers and fruit unknown.

Type locality: Brazil.

This species is not known to us from the incomplete description.

Pfeiffer refers here as a synonym *C. candelabrius* (Enum. Cact. 109. 1837).

23. HELIOCEREUS (Berger) Britton and Rose, Contr. U. S. Nat. Herb. **12**: 433. 1909.

Stems usually weak, procumbent or climbing over rocks and bushes, in cultivation often bushy and erect; branches strongly angled or ribbed; ribs or angles usually 3 or 4, sometimes up to 7; spines of all areoles similar; flowers diurnal, large, funnelform, only 1 at an areole, usually scarlet, sometimes white; tube short but definite; inner perianth-segments elongated; stamens numerous, declined; ovary spiny.

Type species: *Cactus speciosus* Cavanilles.

Heliocereus was considered a subsection of *Cereus* by Berger and, as stated by him, the species are closely related, the chief differences being in the flowers; they are all confined to Mexico and Central America. We recognize 5 species.

The plants are easily propagated by cuttings, but it has been our experience that they are among the most difficult cacti to grow under glass. It is said, however, if plants are grown out of doors during the summer, they make strong branches and flower abundantly during the winter. *H. speciosus* has been much used in hybridizing with various species of *Epiphyllum*, resulting in many types, some of which are greatly admired, and for which new specific, varietal, and form names have been proposed.

The name is from the Greek, meaning sun-cereus.

KEY TO SPECIES.

```
Flowers red.
    Inner perianth-segments acuminate.
        Style not longer than the stamens.............................1. H. elegantissimus
        Style definitely longer than the stamens......................2. H. schrankii
    Inner perianth-segments apiculate, rounded or abruptly tipped.
        Perianth-segments apiculate or rounded........................3. H. speciosus
        Perianth-segments abruptly tipped.............................4. H. cinnabarinus
Flowers white.......................................................5. H. amecamensis
```

1. Heliocereus elegantissimus nom. nov.

> *Cereus coccineus* Salm-Dyck in Pfeiffer, Enum. Cact. 122. 1837. Not *C. coccineus* De Candolle, 1828.
> *Cereus speciosissimus coccineus* Rümpler in Förster, Handb. Cact. ed. 2. 773. 1885.
> *Cereus speciosus coccineus* Graebener, Monatsschr. Kakteenk. **19**: 137. 1909.
> *Heliocereus coccineus* Britton and Rose, Contr. U. S. Nat. Herb. **12**: 433. 1909.

Stems at first erect, low, 1 to 2 dm. high; branches often decumbent, light green, 3 to 5 cm. broad, mostly 3 or 4-angled; ribs strongly undulate; areoles large, 1.5 to 2 cm. apart, yellow-felted; spines acicular, short, 1 cm. long or less, the radial ones bristly and white, the inner ones stiff and recurved; flowers scarlet, 10 to 15 cm. broad; perianth-segments lanceolate, acuminate, 7 cm. long or less; ovary 3 to 4 cm. long, oblong, with a few scattered spreading scales; style red, slender, not longer than the stamens; stigma-lobes white.

Type locality: Mexico.
Distribution: Mexico.
Illustrations: Blühende Kakteen **2**: pl. 118; Monatsschr. Kakteenk. **5**: 135; Pfeiffer and Otto, Abbild. Beschr. Cact. **1**: pl. 15, all three as *Cereus coccineus*.

Plate XVII, figure 1, shows a flowering branch of a plant in the collection of the New York Botanical Garden.

2. Heliocereus schrankii (Zuccarini) Britton and Rose, Contr. U. S. Nat. Herb. **12**: 434. 1909.

> *Cereus schrankii* Zuccarini in Seitz, Allg. Gartenz. **2**: 244. 1834.

Stems ascending, branching; joints 1 to 2 cm. broad, 3 or 4-angled, somewhat winged, when young reddish, in age green; areoles 1.5 to 2 cm. apart, somewhat elevated; spines 6 to 8, acicular, white when young, yellowish brown in age; flowers dark red, large, 14 cm. broad; stamens numerous; style stout, red, longer than the stamens; stigma-lobes white; ovary oblong, 4 cm. long, spiny.

Type locality: Zimipan, Mexico.
Distribution: Known only from the type locality.

We know this plant only from descriptions and the cited illustration. It must be closely related to the preceding species and may not be specifically distinct from it.

Illustration: Pfeiffer and Otto, Abbild. Beschr. Cact. **1**: pl. 27, as *Cereus schrankii*.

3. **Heliocereus speciosus** (Cavanilles) Britton and Rose, Contr. U. S. Nat. Herb. **12**: 434. 1909.

Cactus speciosus Cavanilles, Anal. Cienc. Nat. Madrid **6**: 339. 1803.
Cactus speciosissimus Desfontaines, Mém. Mus. Hist. Nat. Paris **3**: 193. 1817.
Cereus bifrons Haworth, Suppl. Pl. Succ. 76. 1819.
Cereus speciosissimus De Candolle, Prodr. **3**: 468. 1828.
Cereus speciosus Schumann in Engler and Prantl, Pflanzenfam. 3⁰ₐ: 179. 1894. Not Sweet, 1826.

Stems clambering or hanging, strongly 3 to 5-ribbed, old parts bright green, young parts reddish; ribs strongly undulate; areoles distant, often 3 cm. apart, usually large, with felt and acicular spines; spines numerous, yellow or brownish in age, 1 to 1.5 cm. long; flowers scarlet, 15 to 17 cm. long, lasting for several days; perianth-segments oblong, 10 to 12 cm. long, with rounded, often apiculate tips; filaments weak, red; style little longer than the stamens; stigma-lobes white; ovary bearing scattered minute scales; fruit ovoid, 4 to 5 cm. long.

Type locality: Described from a garden plant.
Distribution: Central Mexico and reported from Central America.

Dr. Rose found this species very common on the pedregal near the City of Mexico. It there forms large masses, usually growing in the pot holes and at the mouths of dark caves, clambering over the rocks and occasionally giving off roots. Mr. Pringle found it at high elevations on the mountain ranges south of the City of Mexico.

Cereus speciosissimus grandiflorus (Pfeiffer, Enum. Cact. 122. 1837) is a hybrid with *Selenicereus grandiflorus*.

Cereus speciosissimus hansii Baumann (Förster, Handb. Cact. ed. 2. 773. 1885; *C. hansii* Haage in Förster, Handb. Cact. 428. 1846) is a hybrid with *Epiphyllum ackermannii*. *Cereus jenkinsoni* (Sweet, Hort. Brit. ed. 2. 237. 1830; *C. speciosissimus jenkinsonii* Pfeiffer, Enum. Cact. 121. 1837) is a hybrid obtained in 1824. *Cereus jenkinsonii verus* Haage (Förster, Handb. Cact. 429. 1846) is another hybrid. Here also belongs *Cereus speciosissimus lateritius* Pfeiffer (Enum. Cact. 121. 1837), which was earlier described and figured as *Cactus speciosissimus lateritius* (Edwards's Bot. Reg. **19**: pl. 1596. 1833) and, afterwards, as *Cereus lateritius* Salm-Dyck (Cact. Hort. Dyck. 1849. 53. 1850). The variety of *Cereus speciosissimus, albiflorus* (*Cereus albiflorus* Schumann, Gesamtb. Kakteen Nachtr. 54. 1903), though first mentioned in 1837, was without description, but was taken up and described along with *coccineus, hoveyi,* and *peacocki* by Rümpler (Förster, Handb. Cact. ed. 2. 772, 773) in 1885.

Cereus speciosissimus aurantiacus (Pfeiffer, Enum. Cact. 122. 1837; *C. aurantiacus* Förster, Handb. Cact. 428. 1846) is very briefly described.

The following are some of the hybrids of *Cereus speciosissimus* with *Epiphyllum phyllanthoides* which are listed by Walpers (Repert. Bot. **2**: 278. 1843): *bodii, bollwillerianus, bowtrianus, curtisii, eugenia, guillardieri, ignescens, kiardii, longipes, lothii, maelenii, mexicanus* Salm-Dyck, *roidii, sarniensis, superbus, unduliflorus, vandesii, vitellinus,* and *suwaroffii.* Some of these names had been previously used by Pfeiffer (Enum. Cact. 121. 1837) as varieties of this species, as follows: var. *curtisii, eugenia, guillardieri, ignescens, kiardii, lothii,* and *roydii.*

Among other named hybrids, Pfeiffer gave var. *devauxii* (*Cereus devauxii* Förster, Handb. Cact. 428. 1846). Förster (Handb. Cact. 428 to 431. 1846) also mentioned 66 hybrids with this species, among which are: *blindii* Haage, *colmariensis* Haage, *danielsii* Haage, *edesii* Booth, *elegans* Booth, *finkii* Salm-Dyck, *gebvillerianus* Haage, *gloriosus* Haage, *hitchensii* and its varieties *hybridus* and *speciosus, kampmannii* Haage, *kobii, latifrons, loudonii, macqueanus* Salm-Dyck, *maurantianus, merckii* Booth, *mittleri* Salm-Dyck, *muhlhausianus, peintneri* Haage, *rintzii* Salm-Dyck and the two varieties *roseus albus* and *roseus superbus, scidelii* Booth, *scitzii, selloii, smithii* (*Epiphyllum smithianum* Marnock, Floricult. Mag. **8**: pl. 13), *suwarowii,* and *triumphans.* In addition to these there are many hybrids with only an English name. There are also many quadrinomials.

PLATE XVII

M. E. Eaton del.

1. End of flowering branch of *Heliocereus elegantissimus*.
2. End of a flowering branch of *Heliocereus speciosus*.
3. A tip of a fruiting branch of *Harrisia portoricensis*.
(All natural size.)

Cereus setiger Haworth, Phil. Mag. **7**: 110. 1830, although said to have come originally from Brazil, probably belongs here. *Cereus aurantiacum superbus* Haage (Labouret, Monogr. Cact. 428. 1853), a hybrid of this species, is only mentioned.

Cereus josselinacus D. Gaillard (Rev. Hort. **5**: 56. 1841) is probably only a form.

Cereus serratus Weingart (Monatsschr. Kakteenk. **22**: 185. 1912) is of this relationship. Rother believed it was of hybrid origin and Weingart at first agreed, but afterwards considered it distinct.

Cereus mexicanus Lemaire (Förster, Handb. Cact. 430. 1846) is a hybrid of which *Heliocereus speciosus* is one parent.

Illustrations: Blühende Kakteen 1: pl. 17; Schumann, Gesamtb. Kakteen f. 36, as *Cereus speciosus;* Herb. Génér. Amat. **5**: pl. 351; Curtis's Bot. Mag. **49**: pl. 2306; Loddiges, Bot. Cab. **10**: pl. 924; Mém. Mus. Hist. Nat. Paris **3**: pl. 9; Edward's Bot. Reg. **6**: pl. 4; **28**: pl. 49; Loudon, Encycl. Pl. 410. f. 6857, as *Cactus speciosissimus;* Schelle, Handb. Kakteenk. f. 35, as *Cereus speciosissimus.*

Plate XVII, figure 2, shows a flowering joint of a plant in the collection of the New York Botanical Garden.

4. Heliocereus cinnabarinus (Eichlam).

> *Cereus cinnabarinus* Eichlam in Weingart, Monatsschr. Kakteenk. **20**: 161. 1910.

Stems erect or in time creeping and more or less rooting, very slender, 1 to 1.5 cm. in diameter; ribs few, sometimes only 3 or 4; areoles 2 to 3 cm. apart; spines about 10, bristle-like, 6 to 8 mm. long; flowers about 15 cm. long, the tube bent just above the ovary, more or less funnelform; outer perianth-segments narrow, acute, green; inner perianth-segments oblong to spatulate, sometimes 2.5 cm. broad, abruptly acuminate, somewhat erose toward the apex; style rose-colored; stigma-lobes 7, white.

Type locality: Vulcan Agua, Guatemala.

Distribution: Guatemala.

We know the plant from specimens collected by E. W. Nelson on the volcano of Santa Maria, altitude 2,600 to 3,800 meters, January 24, 1896 (No. 3719).

It is like *Heliocereus elegantissimus*, but with slenderer stems, lower ribs, weaker spines, and abruptly acuminate inner perianth-segments.

This must be a very beautiful species and, growing at such high altitudes in Guatemala, suggests the possibility of its cultivation in the open in certain parts of the United States.

5. Heliocereus amecamensis* (Heese) Britton and Rose, Contr. U. S. Nat. Herb. **12**: 433. 1909.

> *Cereus amecamensis* Heese† in Rother, Prakt. Ratgeb. **11**: 442. 1896.
> *Cereus amecaensis* Heese, Gartenwelt **1**: 317. 1897.

Plant pale green when young, similar to *H. speciosus* in habit and spines; ribs 3 to 5; flower 11 cm. long, 8 to 12.5 cm. in diameter; flower-tube 3.5 cm. long, 1 cm. in diameter, green, with green scales and whitish bristles; outer perianth-segments yellowish green, grading into oblanceolate white inner segments, 7 cm. long, 2 cm. wide; stamens white except the pale-green bases, attached all over the tube; anthers creamy white; style white, slightly exserted beyond the stamens, strongly curved down in the tube; stigma-lobes 11, linear, light creamy white; ovary cylindric, 6 mm. long.

Type locality: Amecameca, Mexico.

Distribution: Central Mexico.

This species has been introduced into Europe by Dr. C. A. Purpus, where it is now much cultivated.

Illustrations: Curtis's Bot. Mag. **135**: pl. 8277; Rother, Prakt. Ratgeb. **11**: 442; Garden **76**: 306, all as *Cereus amecamensis;* Blühende Kakteen **3**: pl. 157; Gard. Mag. **55**: 427; Gartenwelt **1**: 316, 317. f. 1 to 3, as *Cereus amecaensis.*

---

* Confusion of the type locality, Amecameca, with another Mexican town, Ameca, doubtless accounts for the two spellings of the name of this plant.
† Rother here spells this name *Hesse*, doubtless erroneously.

24. **TRICHOCEREUS** (Berger) Riccobono, Boll. R. Ort. Bot. Palermo **8**: 236. 1909.

Columnar plants, more or less branched; ribs few to numerous, either low or prominent, usually very spiny; flowers nocturnal, large, funnelform, the perianth either persistent or separating from the fruit by abscission; perianth-segments elongated; stamens numerous, filiform, arranged in two groups; stigma-lobes numerous; ovary and flower-tube bearing numerous scales, their axils bearing long hairs; fruit without bristles or spines, dull colored.

Type species: *Cereus macrogonus* Otto.

This genus consists of 19 species, confined to South America. It is based on the subgenus of the same name by Berger, but only 2 of Berger's species were transferred to it by Riccobono.

While the flowers of this genus suggest *Echinopsis*, we can not agree with Berger's suggestion that the genera might be united.

The name is from the Greek and signifies thread-cereus, referring to the hairy flower-areoles.

KEY TO SPECIES.

```
Stems more or less branched, usually erect.
    Limb of flower broad.
        Joints relatively slender, 5 to 9 cm. thick.
            Ribs transversely sulcate between the areoles.
                Tubercles prominent.....................................  1. T. thelegonus
                Tubercles not prominent.................................  2. T. thelegonoides
            Ribs not transversely sulcate between the areoles.
                Central spine solitary..................................  3. T. spachianus
                Central spines 4.......................................  4. T. lamprochlorus
        Joints stout.
            Ribs on old plants very numerous and the spines bristle-like..................  5. T. pasacana
            Ribs 4 to 17.
                Spines slender, 1 to 7 cm. long.
                    Ribs 4 to 9.
                        Spines yellow, at least when young; ribs 4 to 8..................  6. T. bridgesii
                        Spines brown from the first; ribs 6 to 8.
                            Plant dark green; spines few at each areole or wanting........  7. T. pachanoi
                            Plant light green; spines several at each areole.
                                Spines acicular, 2.5 cm. long or less..................  8. T. macrogonus
                                Spines subulate, up to 7 cm. long.
                                    Spines swollen at base; young growth green.........  9. T. cuzcoensis
                                    Spines not swollen at base; young growth very
                                        glaucous...................................10. T. peruvianus
                    Ribs 16 or 17.............................................11. T. chiloensis
                Spines very stout, formidable.
                    Spines dark brown.......................................12. T. coquimbanus
                    Spines yellow..........................................13. T. terscheckii
        Limb of flower narrow..........................................14. T. fascicularis
Stems usually simple, low, cespitose.
    Flowers red or yellow, short, more or less campanulate...................15. T. huascha
    Flowers elongated, funnelform, white.
        Tube longer than the limb.
            Ribs few, 9 to 11.............................................16. T. candicans
            Ribs 12 to 18.
                Stem slender, elongated....................................17. T. strigosus
                Stem stout, short ........................................18. T. shaferi
        Tube about the length of the limb.....................................19. T. schickendantzii
```

1. Trichocereus thelegonus (Weber).

Cereus thelegonus Weber in Schumann, Gesamtb. Kakteen 78. 1897.

Stems procumbent or sometimes with erect branches, elongated, 4 to 10 dm. long, dark green, cylindric, 4 to 7.5 cm. in diameter; ribs 12 or 13, broad and obtuse, divided into prominent, more or less distinctly 6-sided tubercles; areoles circular, felted; spines at first brown, some turning gray, others black; radial spines 6 to 8, acicular, somewhat spreading, 1 to 2 cm. long; central spine solitary, porrect, 2 to 4 cm. long; flowers white, about 20 cm. long, funnelform; outer perianth-segments greenish; axils of scales on flower-tube long-woolly; fruit about 5 cm. long, hairy, red, splitting on one side; seed black.

Type locality: Tucuman, Argentina.
Distribution: Northwestern Argentina.

This species has heretofore been unknown to us, but fine specimens, both living and for the herbarium, were obtained by J. A. Shafer at Tapia, Tucuman, near the type locality, February 9, 1917 (No. 98).

Illustration: Schumann, Gesamtb. Kakteen f. 14, as *Cereus thelegonus.*

Figure 188 is from a photograph of a flowering branch in the garden of Dr. Spegazzini at La Plata, Argentina; figure 189 is from a photograph of the wild plant taken by Dr. Shafer in 1917 at Tapia, Argentina.

2. Trichocereus thelegonoides (Spegazzini).

Cereus thelegonoides Spegazzini, Anal. Mus. Nac. Buenos Aires III. 4: 480. 1905.

More or less branched above; trunk 4 to 6 meters high, cylindric, 18 cm. in diameter; branches more or less curved, ascending, 5 to 8 cm. in diameter, obtuse at apex; ribs 15, low, obtuse, at first strongly tubercled by a strong depression between the areoles, but gradually disappearing in age; areoles small, circular, felted; spines 8 to 10, yellow or brownish, setaceous, short, 4 to 8 mm. long; flowers 20 to 24 cm. long, greenish without; inner perianth-segments oblanceolate, acute, white; scales on the ovary and flower-tube hairy in their axils.

FIG. 188.—Trichocereus thelegonus.

Type locality: Jujuy, Argentina.
Distribution: Northern Argentina.

Living specimens were brought from Argentina by Dr. Rose in 1915.

FIG. 189.—Trichocereus thelegonus.

3. Trichocereus spachianus (Lemaire) Riccobono, Boll. R. Ort. Bot. Palermo 8: 237. 1909.

Cereus spachianus Lemaire, Hort. Univ. 1: 225. 1840.
Echinocereus spachianus Rümpler in Förster, Handb. Cact. ed. 2. 827. 1885.
Cereus santiaguensis Spegazzini, Anal. Mus. Nac. Buenos Aires III. 4: 478. 1905.

Stem upright, at first simple, later profusely branching at the base; branches ascending parallel with the main stem, 6 to 9 dm. high by 5 to 6 cm. in diameter, columnar; ribs 10 to 15, obtuse, rounded; areoles about 1 cm. apart, large, covered with curly yellow wool, becoming white; radial spines 8 to 10, 6 mm. to 1 cm. long, spreading, stiff, sharp, amber-yellow to brown; central solitary, stronger and longer than the radials; all the spines later becoming gray; flowers about 20 cm. long by about 15 cm. in diameter, white.

Type locality: Argentina, but definite locality not cited.
Distribution: Western Argentina.
This species was named for Edward Spach (1801–1879).

Illustrations: Monatsschr. Kakteenk. **10**: 93; Rep. Mo. Bot. Gard. **16**: pl. 8, f. *2, 3*; Sunset Mag. July 1915, p. 166; Schelle, Handb. Kakteenk. f. 18, as *Cereus spachianus.*

Figure 190 is from a photograph of a plant in the collection of the New York Botanical Garden.

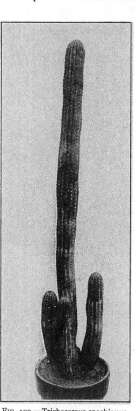

FIG. 190.—Trichocereus spachianus.

FIG. 191.—Trichocereus pasacana.

4. Trichocereus lamprochlorus (Lemaire).

Cereus lamprochlorus Lemaire, Cact. Aliq. Nov. 30. 1838.
Cereus nitens Salm-Dyck, Allg. Gartenz. **13**: 354. 1845.
Echinocereus lamprochlorus Rümpler in Förster, Handb. Cact. ed. 2. 831. 1885.
Echinopsis lamprochlora Weber, Dict. Hort. Bois 471. 1896, as synonym.

Columnar, simple or branching at base, 1.5 to 2 meters high, 7 to 8 cm. in diameter; ribs 10 to 17, low and rounded; radial spines 11 to 14, acicular to subulate, 8 to 10 mm. long; central spines 4, 2 cm. long; flowers funnelform, 20 to 24 cm. long; outer perianth-segments red; inner perianth-segments white, 2.5 cm. long, acuminate.

Type locality: Not cited.
Distribution: Northern Argentina and, according to Rümpler, Bolivia.

Cereus lamprochlorus salinicolus Spegazzini (Anal. Mus. Nac. Buenos Aires II. 4:286. 1902) from southern Argentina, may belong here, but it is much south of the range of this species; *Cereus chiloensis lamprochlorus* Monville (Labouret, Monogr. Cact. 326. 1853) is given as a synonym. *Echinocactus wangertii* (Labouret, Monogr. Cact. 326. 1853) has been referred here as a synonym.

The type specimens were without flowers and fruit. Afterward, Schumann referred to this species a plant collected by Otto Kuntze in Jujuy, Argentina, in October 1892. A specimen of this collection is now in the herbarium of the New York Botanical Garden, and has been used in drawing up the above description, together with plants and specimens obtained by Dr. Shafer at Andalgala, Argentina, in 1917 (No. 13). A cespitose plant with long procumbent stems is sometimes associated with this species, but whether conspecific with it or distinct we have been unable to ascertain.

Illustration: Monatsschr. Kakteenk. **26**: 60, as *Cereus lamprochlorus*.

Figure 192 is from a photograph of plants in flower, taken by Dr. Shafer in 1917.

Fig. 192 —Trichocereus lamprochlorus.

5. Trichocereus pasacana (Weber).

Pilocereus pasacana Rümpler in Förster, Handb. Cact. ed. 2. 678. 1885.
Cereus pasacana Weber, Monatsschr. Kakteenk. **3**: 165. 1893.

Plant often 6 to 10 meters high, sometimes less than 1 meter, usually either simple or with few branches and resembling a small *Carnegiea gigantea*, sometimes with a number of branches from the base, more or less club-shaped, 3 dm. in diameter near the top, when old spineless at base; ribs 20 to 38, low, 2 cm. high; areoles large, approximate, sometimes touching one another; spines numerous, rather variable on young plants; spines yellow, stiff, subulate, the longer ones 4 to 14 cm. long; on old plants, especially flowering ones, elongated, flexible, sometimes bristle-like, 10 to 12 cm. long, yellow or even white; flowers 10 cm. long, the ovary and tube covered with long brown hairs; fruit globular, about 3 cm. in diameter; seeds small, dull black.

Type locality: High valleys of cordilleras of Catamarca and Salta, Argentina.
Distribution: Argentina and Bolivia.

This species is very characteristic of the high plains of northern Argentina and Bolivia, sometimes growing in valleys, but usually along cliffs and on rocky hillsides, and often forms the most conspicuous plant in the landscape. The woody trunks are used for making goat corrals and rude huts. The fruit, which is said to be edible, is called pasacana.

Illustrations: Nov. Act. Soc. Sci. Upsal. IV. **1**: pl. 4; pl. 5, f. 1, as *Cereus pasacana*.

Figure 191 is from a photograph taken by Dr. Rose near Comanche, Bolivia, in 1914; figure 193 shows a flower and figure 194 a fruit collected by Dr. Shafer near Andalgala, Argentina, in 1916.

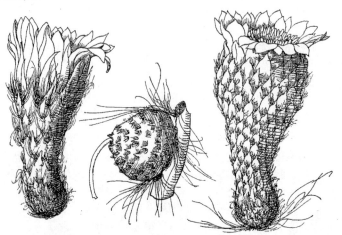

FIG. 193.—Flower of T. pasacana. ×0.6.   FIG. 194.—Fruit of T. pasacana. ×0.6.   FIG. 195.—Flower of T. candicans. ×0.6.

6. Trichocereus bridgesii (Salm-Dyck).

Cereus bridgesii Salm-Dyck, Cact. Hort. Dyck. 1849. 208.   1850.
Cereus lagenaeformis Förster, Hamb. Gartenz. 17: 164.   1861.
Cereus bridgesii brevispinus Schumann, Gesamtb. Kakteen 108.   1897.
Cereus bridgesii lageniformis Schumann, Gesamtb. Kakteen 108.   1897.
Cereus bridgesii longispinus Maass, Monatsschr. Kakteenk. 15: 119.   1905.
Cereus lasianthus Schumann in Rusby, Bull. N. Y. Bot. Gard. 4: 365.   1907, as hyponym.

Tall, 2 to 5 meters high, more or less branching, pale green, a little glaucous; branches 1 to 1.5 dm. in diameter, 4 to 8-ribbed; ribs obtuse, separated by broad but shallow intervals; areoles large, about 2 cm. apart; spines 2 to 6, yellowish, acicular to subulate, very unequal, sometimes 10 cm. long, not swollen at base; flowers large, 18 cm. long; flower-tube 5 to 6 cm. long; throat broad; inner perianth-segments oblong, perhaps white, 5 to 6 cm. long; scales on ovary and flower-tube small, sometimes only 3 to 4 mm. long, scattered, bearing numerous hairs in their axils; fruit scaly, long-hairy, 5 to 6 cm. long.

Type locality: Not cited.

Distribution: About La Paz, Bolivia, where it is frequently grown as a hedge plant or placed on the tops of walls for the protection of gardens.

Mr. Juan Söhrens reports a similar plant from northern Chile which may belong here, or it may be the little-known *Cereus arequipensis.*

The origin of this species is unknown, but since it was named for Bridges, who collected in Bolivia, it is probable that it came from that country.   Dr. Rose's specimens from Bolivia (No. 18842) closely resemble living plants so named from European collections, now represented in the New York Botanical Garden, so that we have no hesitancy in referring them here.

7. Trichocereus pachanoi sp. nov.

Plants tall, 3 to 6 meters high, with numerous strict branches, slightly glaucous when young, dark green in age; ribs 6 to 8, broad at base, obtuse, with a deep horizontal depression above the areole; spines often wanting, when present few, 3 to 7, unequal, the longest 1 to 2 cm. long, dark yellow to brown; flower-buds pointed; flowers very large, 19 to 23 cm. long, borne near the top of branches, night-blooming, very fragrant; outer perianth-segments brownish red; inner perianth-segments oblong, white; filaments long, weak, greenish; style greenish below, white above; stigma-lobes linear, yellowish; ovary covered with black curled hairs; axils of scales on flower-tube and fruit bearing long black hairs.

Collected by J. N. Rose, A. Pachano, and George Rose at Cuenca, Ecuador, September 17 to 24, 1918 (No. 22806, type).

This species is widely cultivated throughout the Andean region of Ecuador, where it is grown both as an ornamental and as a hedge plant. In some of the lateral valleys on the western slope of the Andes it appears to be native, as for instance above Alausí, but as it has doubtless long been cultivated it is impossible to be sure of its natural habitat.

It is known to the Ecuadoreans as agua-colla or giganton and has been passing in Ecuador under the names of *Cereus peruvianus* and *Cereus giganteus*. It is named for Professor Abelardo Pachano of the Quinta Normal at Ambato, Ecuador, who accompanied Dr. Rose in 1918 on his travels in the high Andes of Ecuador.

Fig. 196.—Trichocereus pachanoi.

This species belongs to the high Andes, ranging from 2,000 to 3,000 meters in altitude. In the Chanchan Valley it certainly comes down to about 2,000 meters and overlaps the upper range of *Lemaireocereus godingianus*, which differs from it greatly in habit and flowers. Different as the two plants are, Richard Spruce, keen botanist as he was, confused them, as the following quotation will show; the part in italics refers to the *Lemaireocereus:*

"The brown hill-sides began to be diversified by an arborescent Cactus, with polygonal stems and white dahlia-like flowers, *which, Briareus-like, threw wide into the air its hundred rude arms. Lower down, at about 6,000 feet, I saw specimens full 30 feet high and 18 inches in diameter.*"

Figure 196 shows the top of a large plant growing on the sides of a cliff on the outskirts of Cuenca, Ecuador, photographed by George Rose in September 1918.

8. Trichocereus macrogonus (Salm-Dyck) Riccobono, Boll. R. Ort. Bot. Palmero **8:** 236. 1909.

Cereus macrogonus Salm-Dyck, Cact. Hort. Dyck. 1849. 203. 1850.
Eriocereus tephracanthus Riccobono, Boll. R. Ort. Bot. Palermo **8:** 244. 1909.

Stem probably tall, stout, but in cultivation often slender, bluish green, especially on young growth; ribs usually 7, low and rounded, 1.5 cm. high, separated by acute intervals; areoles large, 1.5 to 2 cm. apart; spines several from an areole, acicular, brown; radial spines 5 to 8 mm. long; central spine about 2 cm. long; flowers probably large and white; fruit unknown.

Type locality: Not cited.

Distribution: South America, but not known definitely in the wild state.

This species is represented in the New York Botanical Garden by a live specimen from Kew, which we consider typical. Salm-Dyck described it from specimens growing in the Botanical Garden at Berlin, but did not know their origin. Schumann figured what he supposed to be it in the Flora Brasiliensis, referring it to Brazil; his plant is from the Province of Rio de Janeiro, collected by Glaziou, and is undoubtedly *Cephalocereus arrabidae.*

Cereus tetracanthus Labouret (Rev. Hort. IV. 4: 25. 1855) and *C. tephracanthus bolivianus* Weber (Schumann, Gesamtb. Kakteen 81. 1897) are probably of this relationship; both forms come from Bolivia. Rümpler (Förster, Handb. Cact. ed. 2. 712. 1885) says the former came from Chuquisaca, Bolivia. An earlier reference (Steudel, Nom. ed. 2. 1: 336. 1840), but of slightly different spelling, cites Link and Otto as authors of this name, but the species was not described. To one of these forms may belong the plant in the New York Botanical Garden (No. 6231), obtained from M. Simon, St. Ouen, Paris, in 1901, which is called *Cereus bolivianus.* The last name, first credited to Weber (Monatsschr. Kakteenk. 12: 21. 1902), is occasionally met in literature.

Cereus hempelianus Bauer (Monatsschr. Kakteenk. 17: 55. 1907) is, according to F. Fobe, only a stout, bluish-green variety of *C. macrogonus.*

9. Trichocereus cuzcoensis sp. nov.

Plants tall, 5 to 6 meters high, much branched, the branches somewhat spreading, light green when young; ribs 7 or 8, low and rounded; areoles rather close together, 1 to 1.5 cm. apart; spines numerous, often 12, very stout, rigid, sometimes 7 cm. long, swollen at base; flowers 12 to 14 cm. long, doubtless nocturnal but, sometimes at least, remaining open during the morning, fragrant; flower-tube green, 5 to 6 cm. long; inner perianth-segments oblong, white, 4 to 5 cm. long; filaments weak, declining on the lower side of the throat; scales on the ovary and flower-tube small, bearing a few long hairs in their axils; fruit not known.

Collected by J. N. Rose below Cuzco, Peru, September 1, 1914 (No. 19022).

10. Trichocereus peruvianus sp. nov.

Plant 2 to 4 meters high with numerous erect or ascending, stout branches, 15 to 20 cm. in diameter, glaucous when young; ribs 6 to 8, broad and rounded; areoles large, 2 to 2.5 cm. apart, brown-felted; spines brown from the first, about 10, unequal, some of them 4 cm. long, rigid and stout, not at all swollen at base; areoles on ovary and flower-tube hairy; mature flowers not seen but evidently large and probably white.

Collected by Dr. and Mrs. Rose near Matucana, Peru, altitude 2,100 meters, July 9, 1914 (No. 18658).

Fig. 197.—*Trichocereus peruvianus.*

This species resembles *T. bridgesii* but has stouter and darker spines. It is found on the western slopes of the Andes at a much lower altitude than that species.

Figure 197 is from a photograph taken by Mrs. J. N. Rose at Matucana, Peru, in 1914.

11. Trichocereus chiloensis (Colla).

Cactus chiloensis Colla, Mem. Accad. Sci. Torino **31**: 342. 1826.
Cereus chiloensis De Candolle, Prodr. **3**: 465. 1828.
Cereus chilensis Pfeiffer, Enum. Cact. 86. 1837.
Cereus panoplaeatus Monville, Hort. Univ. **1**: 220. 1840.
Cereus heteromorphus Monville, Hort. Univ. **1**: 221. 1840.
Cereus longispinus Salm-Dyck, Allg. Gartenz. **13**: 354. 1845.
Cereus pepinianus Lemaire in Salm-Dyck, Allg. Gartenz. **13**: 354. 1845.*
Cereus subuliferus Salm-Dyck, Allg. Gartenz. **13**: 354. 1845.
Cereus gilvus Salm-Dyck, Allg. Gartenz. **13**: 355. 1845.
Cereus quisco Remy in Gay, Fl. Chilena **3**: 19. 1847.
Cereus linnaei Förster, Hamb. Gartenz. **17**: 165. 1861.
Cereus funkii Schumann, Gesamtb. Kakteen 61. 1897.
Cereus chilensis pycnacanthus Schumann, Gesamtb. Kakteen 63. 1897.
Cereus chilensis zizkaanus Schumann, Gesamtb. Kakteen 63. 1897.
Cereus chilensis panhoplites Schumann, Gesamtb. Kakteen 63. 1897.
Cereus chilensis poselgeri Schumann, Gesamtb. Kakteen 63. 1897.
Cereus chilensis heteromorphus Schumann, Gesamtb. Kakteen 63. 1897.
Cereus chilensis polygonus Schumann, Gesamtb. Kakteen 63, 1897.

FIG. 198.—Trichocereus chiloensis.	FIG. 199.—Trichocereus chiloensis.

* *Cereus pepinianus* was described by Salm-Dyck in 1845 (Allg. Gartenz. **13**: 354. 1845) who there credits the name to Lemaire. Lemaire evidently had reported the name under some other genus, for in 1850 (Salm-Dyck, Cact. Hort. Dyck. 1849. 44, 197) Salm-Dyck redescribed the species, crediting himself with the name and citing "*Echinocactus pepinianus* Cat. Cels" as synonym. The name *Echinocactus pepinianus* Lemaire occurs first in 1846 (Förster, Handb. Cact. 347), but without description. Labouret in 1853 takes it up as *Echinocactus echinoides pepinianus* (Monogr. Cact. 178), with the statement that Salm-Dyck considered it synonymous with *Cereus pycnacanthus*. These two combinations in *Echinocactus*, while evidently referring to *Cereus pepinianus*, being without description, can not be properly referred here as synonyms. They are, however, both referred by Schumann to *Echinocactus pepinianus*. The plant which he describes, however, is different from *Cereus pepinianus*. If a good *Echinocactus*, it should be credited to Schumann, with the citation to his monograph (Gesamtb. Kakteen 420. 1898).

Stems rarely single, usually of several branches, sometimes of many, arising from near the base, starting nearly at right angles to the main trunk but soon erect, the tallest sometimes 8 meters high; ribs usually 16 or 17, low and broad, separated by narrow intervals, divided into large tubercles even when fully mature; radial spines when young light yellow with brown tips but soon becoming gray, 8 to 12, slightly spreading, often stout, 1 to 2 or even 4 cm. long; central spine single, porrect, often stout, 4 to 7 or even 12 cm. long; flowers 14 cm. long, outer perianth-segments white but tinged with red or brown; inner perianth-segments white, acuminate; style green below, cream-colored above; stigma-lobes cream-colored, about 18, 1.5 cm. long; fruit globular.

Type locality: Described from cultivated plants supposed to have come from Chile.

Distribution: On the hills in and about the great central valley of Chile, extending from Curico north to Puenta Colorado in the northern part of the province of Coquimbo.

While this plant shows considerable variation in its spines, we do not believe it possible to separate the species into varieties as Schumann has done.

Echinocactus jeneschianus Pfeiffer (Allg. Gartenz. **8**: 406. 1840) and *Echinocactus pepinianus echinoides* (Labouret, Monogr. Cact. 177. 1853) are referred to *Echinocactus echinoides* by Labouret.

Echinocereus chiloensis Console and Lemaire (Rev. Hort. **35**: 173. 1864) is only mentioned, but Lemaire later (Cact. 61. 1868) states that it is based on *Cereus chiloensis*, which definitely places it here.

Cereus chilensis funkianus (Schumann, Gesamtb. Kakteen 61. 1897) has never been formally published.

Cereus polymorphus (published as a synonym of *Opuntia polymorpha* in Förster, Handb. Cact. 472. 1846), referred here by Schumann, should doubtless go elsewhere, for it is said to come from Mendoza, Argentina. It may be a form of *Opuntia glomerata.*

Fig. 200.—Flower of T. chiloensis. ×0.5.

Fig. 201.—Trichocereus coquimbanus.

Cereus pycnacanthus Salm-Dyck (Allg. Gartenz. **13**: 355. 1845), and *Cereus panoplaeatus* Cels (Salm-Dyck, Cact. Hort. Dyck. **1849**. 44. 1850) published as a synonym of the former, were both referred to *Cereus chilensis* by Schumann, but they came from Bolivia and the description does not fit this species.

Cereus fulvibarbis Otto and Dietrich (Allg. Gartenz. **6**: 28. 1838; *Cereus chilensis ful-vibarbis* Salm-Dyck in Walpers, Repert. Bot. **2**: 276. 1843), said to have come from Chile, is referred to *Cereus chilensis* by Schumann, but it is described as having 10 to 13 ribs.

Cereus polymorphus G. Don (Loudon, Hort. Brit. 195. 1830) and *Cactus polymorphus* Gillies (published here as a synonym), referred to *Cereus chilensis* by Schumann, can not be identified from the meager description. It is said to have been introduced from Chile in 1827.

The following names belong here; they have not been accompanied by descriptions.

Cereus quintero Pfeiffer, Enum. Cact. 86. 1837.
    chilensis brevispinulus Salm-Dyck in Walpers, Repert. Bot. **2**: 276. 1843.
    spinosior Salm-Dyck in Förster, Handb. Cact. 377. 1846.
    flavescens Salm-Dyck, Cact. Hort. Dyck. 1849. 44. 1850.
    eburneus (Schumann, Gesamtb. Kakteen 63. 1897) based on *Eulychnia eburnea* Philippi, must belong here.
    linnaei Schumann, Gesamtb. Kakteen 63. 1897.
    quisco Weber in Hirscht, Monatsschr. Kakteenk. **8**: 110. 1898.

Cereus spinibarbis var. *minor* Monville and var. *purpureus* Monville (Labouret, Monogr. Cact. 334. 1853) have been referred here.

Cereus elegans Lemaire and *C. duledevantii* Lemaire (Illustr. Hort. **5**: Misc. 10. 1858), unpublished, doubtless were given to forms of this species. *Echinocactus pyramidalis* and *E. elegans* (Pfeiffer, Enum. Cact. 86. 1837) were given only as synonyms of *Cereus chilensis*.

Illustration: Engler and Drude, Veg. Erde **8**: pl. 19, as *Cereus chilensis*.

Figure 198 is from a photograph of a group of plants taken in Valparaiso, Chile, by Dr. Rose in 1914; figure 199 is from a photograph of a branch from the same group as grown in the New York Botanical Garden; figure 200 is from a drawing of a flower brought back by Dr. Rose from La Serena, Chile, in 1914.

12. Trichocereus coquimbanus (Molina).

Cactus coquimbanus Molina, Sagg. Stor. Nat. Chil. 170. 1782.
Cereus nigripilis Philippi, Fl. Atac. 23. 1860.
Cereus coquimbanus Schumann, Gesamtb. Kakteen 58. 1897.

Plant low, 1 meter high or more, or sometimes prostrate and forming dense thickets; branches 7 to 8 cm. in diameter, with 12 or 13 ribs; areoles large, circular, filled with short wool; spines about 20, very for: idable, often 7 to 8 cm. long; central spines several, 2 to 6 cm. long; flowers large, white, about 10 cm. long; inner perianth-segments acute; scales of ovary and tube subtending black hairs.

Fig. 202.—Trichocereus coquimbanus.

Type locality: Coquimbo to Paposo, Chile.

Distribution: Along the coast of the province of Coquimbo, Chile.

Cereus chilensis nigripilis (Hirscht, Monatsschr. Kakteenk. **8**: 159. 1898) doubtless belongs here.

Illustrations: Monatsschr. Kakteenk. **11**: 27; Schumann, Gesamtb. Kakteen Nachtr. f. 3, both as *Cereus nigripilis.*

Figure 201 is from a photograph of a plant brought by Dr. Rose from the Botanical Garden at Santiago, Chile, in 1914; figure 202 is from a photograph taken by Dr. Rose at Coquimbo, Chile, in 1914.

13. Trichocereus terscheckii (Parmentier).

Cereus terscheckii Parmentier in Pfeiffer, Allg. Gartenz. **5**: 370. 1837.
Cereus ferscheckii Parmentier, Hort. Belge **5**: 66. 1838 (*fide* Index Kewensis).
Cereus fulvispinus Salm-Dyck, Cact. Hort. Dyck. 1849. 46. 1850.
Pilocereus terscheckii Rümpler in Förster, Handb. Cact. ed. 2. 688. 1885.

At first columnar, in age becoming much branched, 10 to 12 meters high; trunk woody, up to 4.5 cm. in diameter; branches 1 to 2 dm. in diameter; ribs 8 to 14, prominent, 2 to 4 cm. high, obtuse; areoles large, 1 to 1.5 cm. in diameter, felted, 2 to 3 cm. apart; spines 8 to 15, subulate, yellow, up to 8 cm. long; flowers very large, 15 to 20 cm. long, 12.5 cm. broad; inner perianth-segments oblong, 7 cm. long, acute, white; scales on the ovary and flower-tube ovate, mucronate-tipped, their axils filled with long brown wool.

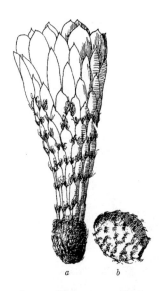

FIG. 203.—Trichocereus terscheckii.
a, flower; *b,* fruit. ×0.4.

FIG. 204.—Trichocereus terscheckii.

Type locality: Argentina, but no definite locality cited.

Distribution: Northern Argentina.

This is a very large cactus, called in Argentina cardon grande. It has frequently been confused with another species, *T. pasacana,* of the same region, but it is more branched, with fewer ribs, different spines, and larger flowers.

Figure 203*a* shows a flower and figure 203*b* a fruit, collected by Dr. Shafer near Salta, Argentina, in 1916; figure 204 is from a photograph taken by Dr. Shafer at Salta, Argentina, in 1917.

14. Trichocereus fascicularis (Meyen).

> *Cereus fascicularis* Meyen, Allg. Gartenz. 1: 211. 1833.
> *Cactus fascicularis* Meyen, Reise 1: 447. 1834.
> *Echinocactus fascicularis* Steudel, Nom. ed. 2. 1: 536. 1840.
> *Cereus weberbaueri* Schumann in Vaupel, Bot. Jahrb. Engler **50**: Beibl. 111: 22. 1913.

Fig. 205.—Trichocereus fascicularis.

Growing in large clusters made up of many slender, erect or ascending branches, 2 to 4 meters high; ribs about 16, low, rounded, separated by narrow intervals; areoles filled with tawny felt, closely set, large; spines numerous, at first yellowish to brown; radial spines acicular, often only 1 cm. long or less; central spines much stouter and often 4 cm. long; flowers 1 from an areole, 8 to 11 cm. long, slender, somewhat curved near the base; ovary and flower-tube bearing small ovate scales, their axils filled with long white and brown hairs; outer perianth-segments narrow, acute, passing into broader ones, simply mucronate, pinkish; inner perianth-segments thinner and a little broader than the outer ones, obtuse, 1.5 cm. long, greenish to brownish (not white); filaments numerous, slender, scattered over the narrow throat, somewhat exserted; style bulbose at base, slender, 7 cm. long, exserted; stigma-lobes short, greenish; lower part of tube or tube proper 1.5 cm. long, somewhat scabrous within; fruit globular, 3 to 4 cm. in diameter, yellowish to reddish, splitting open on one side and exposing the pulp; seeds black, shining, 2 mm. long, a little longer than broad, minutely punctate.

Type locality: Southern Peru.

Distribution: Mountains of southern Peru and northern Chile, at about 2,300 meters altitude. At Arequipa it is especially common, being found both above and below the city, where it was collected by Dr. Rose in 1914 (No. 18781).

This species, although recently described as new under the name of *Cereus weberbaueri*, is the one described by Meyen in 1833 as *Cereus fascicularis*. Meyen's description is very unsatisfactory, but he does describe the

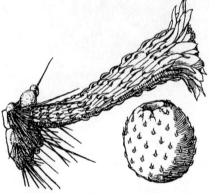

Fig. 206.—Flower of Trichocereus fascicularis. ×0.7.
Fig. 207.—Fruit of same. ×0.7.

Fig. 208.—Flower of Trichocereus huascha. ×0.7.
Fig. 209.—Fruit of same. ×0.7.

habit, number of ribs, and size of flowers, all of which answer fairly well to our plant. A translation of his brief description is as follows: Erect, 16-angled, 4 feet high, somewhat jointed (3 to 4 joints); spines 8 or 9, in a radiating circle; flowers 9 or 10, white, 3.5 inches long, at the ends of the branches.

The flowers of this species differ from those of typical *Trichocereus* in that they are very slender, bent near the base, and have short perianth-segments.

Figure 205 is from a photograph taken by Dr. Rose at Arequipa, Peru, in 1914; figure 206 shows the flower and figure 207 the fruit of the plant photographed.

15. Trichocereus huascha (Weber).

Cereus huascha Weber, Monatsschr. Kakteenk. **3**: 151. 1893.
Cereus huascha flaviflorus Weber, Monatsschr. Kakteenk. 3: 151. 1893.

More or less cespitose, forming clumps 8 to 20 dm. broad; stems 8 to 16 dm. high, cylindric, 4 to 5 cm. in diameter; ribs 12 to 18, low, rounded; areoles approximate, often only 5 to 7 mm. apart; spines numerous, acicular, unequal, the longest often 5 to 6 cm. long, yellowish to brown; flowers very variable in color and size, red to yellow, 7 to 10 cm. long, broadly funnelform; scales on the ovary bearing long brown hairs.

Type locality: Yacutala, Catamarca, Argentina.
Distribution: Northern Argentina.

Cereus huascha flaviformis Weber (Monatsschr. Kakteenk. **3**: 136. 1893) is only a name.

Figure 208 shows a flower and figure 209 a fruit collected by Dr. Shafer near Andalgala, Argentina, in 1916; figure 210 is from a photograph of the plant from which the flowers and fruit were taken.

Fig. 210.—Trichocereus huascha.

16. Trichocereus candicans (Gillies).

Cereus candicans Gillies in Salm-Dyck, Hort. Dyck. 335. 1834.
Cereus candicans tenuispinus Pfeiffer, Enum. Cact. 91. 1837.
Cereus gladiatus Lemaire, Cact. Aliq. Nov. 28. 1838.
Cereus candicans robustior Salm-Dyck, Cact. Hort. Dyck. 1849. 43. 1850.
Echinocereus candicans Rümpler in Förster, Handb. Cact. ed. 2. 832. 1885.
Echinocereus gladiatus Rümpler in Förster, Handb. Cact. ed. 2. 833. 1885.
Echinopsis candicans Weber, Dict. Hort. Bois 471. 1896, as synonym.
Cereus candicans courantii Schumann, Gesamtb. Kakteen 70. 1897.
Cereus candicans gladiatus Schumann, Gesamtb. Kakteen 70. 1897.

Cespitose, forming large clumps often 1 to 3 meters in diameter; joints erect or spreading, 6 dm. long or less, 14 cm. in diameter or less, rounded at apex; ribs 9 to 11, low, rounded or obtuse; areoles large, white-felted when young, 2 to 3 cm. apart; spines subulate, brownish yellow, more or less mottled; radial spines 10 or more, more or less spreading, unequal, the longest 4 cm. long;

central spines several, the longest nearly 10 cm. long; flowers very large, funnelform, very fragrant, showy, 15 cm. long; scales on flower-tube ovate, acuminate, bearing long hairs in their axils; inner perianth-segments white, oblong; fruit globose to ellipsoid, splitting on one side.

Type locality: Not cited, but doubtless Mendoza, Argentina.

Distribution: Mendoza and northward, Argentina,

Schumann describes 3 varieties, all apparently from Mendoza, which we have merged into the species. Plants as seen in the field show even greater variation than is called for in Schumann's descriptions, but they all evidently grade into one another.

Cereus montezumae Hortus (Pfeiffer, Enum. Cact. 91. 1837, as synonym), *C. dumesnilianus* Haage (Schumann, Monatsschr. Kakteenk. 4: 172. 1894, as a probable variety of *C. candicans*), *C. dumesnilianus* Monville (Weber, Dict. Hort. Bois 279. 1894, as synonym), *Echinopsis dumesniliana* Cels (Schumann, Gesamtb. Kakteen 69. 1897, as synonym; *C. candicans dumesnilianus* Zeissold, Monatsschr. Kakteenk. 3: 140. 1893), and *Echinocereus candicans tenuispinus* Pfeiffer (Förster, Handb. Cact. ed. 2. 833. 1885) are usually referred here. *Echinocactus candicans* (Pfeiffer, Enum. 91. 1837) is a synonym only.

Cereus candicans spinosior Salm-Dyck (Walpers, Repert. Bot. 2: 276. 1843), undescribed, belongs here.

Schumann refers *Echinocactus auratus* Pfeiffer (Abbild. Beschr. Cact. 2: under pl. 14. 1846 to 1850) and its synonym *Echinopsis aurata* Salm-Dyck (Cact. Hort. Dyck. 1849. 39. 1850) to *Cereus candicans*, but this can not be, for the descriptions are very different. The former was described as depressed, 12 to 15 inches in diameter, only 4 to 5 inches high, and with 28 ribs. The type locality was Bellavista, Chile. It should be compared with *Eriosyce sandillon* and its relatives.* *Echinopsis dumeliana* Cels (Salm-Dyck, Cact. Hort. Dyck. 1849. 39. 1850) is given as a synonym only; it is doubtless the name referred to by Schumann, but with different spelling.

Figure 195 shows a flower collected by Dr. Rose near Córdoba, Argentina, in 1915.

17. Trichocereus strigosus (Salm-Dyck).

Cereus strigosus Salm-Dyck, Hort. Dyck. 334. 1834.
Cereus intricatus Salm-Dyck, Cact. Hort. Dyck. 1849. 194. 1850.
Echinocereus strigosus Lemaire in Förster, Handb. Cact. ed. 2. 826. 1885.
Echinocereus strigosus spinosior Rümpler in Förster, Handb. Cact. ed. 2. 827. 1885.
Echinocereus strigosus rufispinus Rümpler in Förster, Handb. Cact. ed. 2. 827. 1885.
Echinocereus intricatus Rümpler in Förster, Handb. Cact. ed. 2. 830. 1885.
Cereus strigosus intricatus Weber in Schumann, Gesamtb. Kakteen 68. 1897.
Cereus strigosus longispinus Maass, Monatsschr. Kakteenk. 15: 119. 1905.

Cespitose, forming clumps 2 to 10 dm. in diameter, the branches usually simple, erect, or ascending, sometimes 6 dm. high, 5 to 6 cm. in diameter, very spiny; ribs 15 to 18, very low, 4 to 5 mm. high, obtuse; areoles circular, rather large, approximate, 4 to 8 mm. apart, densely white-felted when young; spines numerous, very variable as to color and length, either white, yellowish, or pinkish to nearly black, 1 to 5 cm. long, acicular; flowers white, large, 20 cm. long, funnelform, the scales on the ovary and tube with long silky hairs in their axils; seeds black, glossy, about 2 mm. long; hilum basal but oblique.

Type locality: Not cited.

Distribution: Western Argentina.

This species is very common in the deserts of the Province of Mendoza, especially about the city of Mendoza, and in the mountain valleys farther to the west. The first specimens were doubtless sent out through Chile, for before the railroads this was the most accessible route out from Mendoza.

Cereus myriophyllus Gillies (Allg. Gartenz. 1: 365. 1833), given by Schumann as a synonym of this species, was never described and the name was referred here originally with

---

* We have found that *Echinocactus ceratistes* Otto, one of the synonyms of *Eriosyce*, originally came from Bellavista, Chile, also.

doubt. *C. strigosus spinosior* Salm-Dyck (Cact. Hort. Dyck. 1844. 27. 1845) and *C. strigosus rufispinus* (Monatsschr. Kakteenk. 7: 184. 1897) also belong here, and, perhaps, *C. spinibarbis flavidus* (Labouret, Monogr. Cact. 325. 1853).

Illustration: Rep. Mo. Bot. Gard. **16**: pl. 8, f. 1, as *Cereus strigosus.*

Figure 211 is from a photograph taken by Dr. Shafer at Andalgala, Argentina, in 1916.

Fig. 211.—Trichocereus strigosus.

18. Trichocereus shaferi sp. nov.

Cespitose, cylindric, 3 to 5 dm. high, 10 to 12.5 cm. in diameter, light green; ribs about 14, 10 to 15 mm. high; areoles approximate, 5 to 7 mm. apart, white-felted when young; spines about 10, acicular, 12 mm. long or less, light yellow; flowers from the top of plant, 15 to 18 cm. long; tube slender; outer perianth-segments linear; inner segments probably white; scales of the ovary and flower-tube bearing long brown hairs.

Collected by J. A. Shafer in wooded ravine, altitude 1,800 meters, near San Lorenzo, Salta, Argentina, January 11, 1917 (No. 44).

19. Trichocereus schickendantzii (Weber).

Echinopsis schickendantzii Weber, Dict. Hort. Bois 473. 1896.

Plants simple or cespitose, slender, 15 to 25 cm. high, 6 cm. in diameter, dark green, shiny; ribs 14 to 18, low, 5 mm. high, somewhat crenate; spines yellowish, flexible, 5 to 10 mm. long; radial spines at first 9, in age more numerous; central spines 2 to 8; flower-bud pointed, covered with black wool; flowers funnelform, several from the top of the plant, inodorous, 20 to 22 cm. long; scales on the ovary and flower-tube with hairy axils; inner perianth-segments acute, oblong, white; fruit edible, agreeable.

Type locality: Tucuman, Argentina.

Distribution: Northwestern Argentina.

Spegazzini thinks it is not an *Echinopsis* but a *Cereus*, although he leaves it under the former genus. The flowers are not those of a true *Cereus.*

Weber (Dict. Hort. Bois 473. 1896) gives *Cereus schickendantzii* Weber as a synonym.

Illustration: Monatsschr. Kakteenk. **15**: 125, as *Echinopsis schickendantzii.*

PUBLISHED SPECIES, PERHAPS REFERABLE TO TRICHOCEREUS.

CEREUS AREQUIPENSIS Meyen, Allg. Gartenz. **1**: 211. 1833.

This may be a *Trichocereus*; we give a translation of Meyen's account of it:

"The number of cacti here, as well as in the southern provinces of Peru, is unusually large, and only a few of them are known in our greenhouses, also it is very difficult to transport them to us, as many of them die in the trip around Cape Horn. *Cactus candelaris*, which we first found in the Cordilleras of Tacna, appears here also, in isolated examples, and its distribution appears to be sharply confined between 7,000 and 9,000 feet altitude. However, close upon its heels comes another *Cereus* which surpasses it in beauty; it is 8-angled and reaches a height of 20 to 35 feet; upon its ribs appear at regular distances hairy areoles, from which protrude the spine clusters and the long white flowers. There is no more beautiful plant in this remarkable family, and we name it *Cereus arequipensis*."—Meyen, Reise **2**: 41. 1835.

CEREUS ATACAMENSIS Philippi, Fl. Atac. 23. 1860.

Usually simple and columnar, 6 meters high or more, 5 to 7 dm. in diameter, containing a thick woody cylinder; ribs numerous, very spiny; areoles 1.2 cm. in diameter, filled with brown wool; spines numerous, sometimes 30 to 40, often very slender, 10 cm. long.

Type locality: Mines of "San Bartolo."
Distribution: Province of Atacama, Chile.

The original specimen came from the desert of northern Chile, not far from the Bolivian line; it is not unlikely that the plant extends into Bolivia and northern Argentina. Indeed, Dr. Rose found a large woody section in the Museo Nacional del Santiago bearing this name and coming from Argentina. This material, supplemented by the illustration by Reiche from Chile and by Fries (Nov. Act. Soc. Sci. Upsal. IV. **1**: pl. 4, 5. 1905) from Argentina, suggests the probability of *Trichocereus pasacana* belonging here.

In the Museo Nacional del Santiago are two very interesting wood sections of this species. One from Atacama is 1.55 meters long and 41 cm. in diameter, while the other, from Argentina, is 3 meters long and 44 cm. in diameter, with a hollow center 22 cm. in diameter. All that is left of the type material in the Philippi herbarium are two clusters of spines, these very long, slender, numerous, and brown.

Illustration: Engler and Drude, Veg. Erde 8: pl. 9, as *Cereus atacamensis*.

CEREUS ERIOCARPUS Philippi, Anal. Mus. Nac. Chile 1891[2]: 27. 1891.

Stems large, erect, simple below, with small branches above, 5 to 6 cm. in diameter, the upper part densely covered with white curly hairs; ribs 27 to 29; areoles very close together, 14 mm. in diameter, with grayish tomentum intermixed with straight spines 11 cm. long, grading into stiff bristles 4 cm. long; expanded flowers unknown; ovary 22 mm. in diameter, densely covered with white hairs.

Type locality: Calcalhuay, altitude 12,000 feet (3,700 meters).
Distribution: Province of Tarapaca, Chile.

We know the plant only from description and from fragments of the type specimen.

CEREUS MALLETIANUS Cels in Schumann, Gesamtb. Kakteen 120. 1897.

"Stem upright, cylindric, somewhat crooked, slightly constricted above, hardly sunken in at the crown, exceeded by a brownish yellow thick tuft of spines which can not be seen under the wool, up to 4 cm. in diameter, bluish green; ribs 17, separated by sharp shallow furrows, hardly 4 mm. high, rounded and lightly sinuate, disappearing at the base; areoles 6 to 8 mm. apart, circular, 3 to 3.5 mm. in diameter, covered with short yellow wool, later turning gray, which gradually disappears; radial spines about 30, radiating horizontally, the inner spreading, needle-like, so thickly intertwined that they surround the entire body, the inner pair the longest, measuring 10 mm.; central spines 4, in an upright cross, sometimes more, since they are not sharply distinguishable from the radial spines, the lowermost, sometimes, however, the uppermost, the longest, measuring up to

2 cm., this one is yellowish brown, darker above; the remaining spines are yellowish when young, then become white, almost translucent, finally they turn gray and are knocked off." (Translation of Schumann's description.)

Type locality: Not definitely cited.
Distribution: Andes, South America.

Echinopsis catamarcensis Weber, Dict. Hort. Bois 471.   1896.
    Echinocactus catamarcensis Spegazzini, Anal. Mus. Nac. Buenos Aires III. **4**: 500.   1905.

Stems simple, ellipsoid to shortly columnar, up to 1 meter high, grayish green; ribs 13 to 17, high, somewhat undulate; radial spines 10, pale brown, subulate, somewhat curved; central spines 4, arranged in a single perpendicular row, somewhat curved; flowers supposed to be yellow.

Type locality: Catamarca, Argentina.
Distribution: Argentina.
Weber gives *Cereus catamarcensis* (Dict. Hort. Bois 471) as a synonym of this species.

25. JASMINOCEREUS gen. nov.

Stems upright and tall with a definite cylindric trunk and a much branched top; ribs numerous, low; areoles circular, bearing felt and spines; flowers slender, salverform or perhaps funnelform, the slender tube narrowly cylindric, the limb broad, spreading; inner perianth-segments narrow, yellow or brownish; stamens and style exserted; ovary bearing small spreading scales with small tufts of wool in their axils; fruit oblong, smooth, except the small scarious scales, these naked in their axils; seeds minute, black.

A monotypic genus of the Galápagos Islands.   The name signifies jasmine-like cereus, with reference to the flowers.

Fig. 212.—Jasminocereus galapagensis.

1. **Jasminocereus galapagensis** (Weber).

        Cereus galapagensis Weber, Bull. Mus. Hist. Nat. Paris **5**: 312.   1899.
        Cereus sclerocarpus Schumann in Robinson, Proc. Amer. Acad. **38**: 179.   1902.

Tall, often 8 meters high or more; trunk large, cylindric, 15 to 30 cm. in diameter; branches spreading, very stout, composed of many short joints, about 14 cm. in diameter; ribs 15 to 18, low, about 1 cm. high, separated by broad, rounded intervals; areoles rather close together, 1 cm. apart or less, bearing brown felt; spines various as to length, sometimes only 1 cm. long, sometimes 8 cm. long, usually slender, sometimes bristle-like, often 10 or more at an areole; flowers various in size, 5 to 11 cm. long, "chocolate-brown with yellow stripes"; outer perianth-segments spatulate, 2 to 3 cm. long; inner perianth-segments linear, about 2.4 to 3 cm. long; ovary terete, scaly; scales few, 1 to 1.5 mm. long, ovate, acute; stigma-lobes 11; flower-tube about twice as long as the segments; fruit greenish, short-oblong, 5 cm. long, 4 cm. in diameter, with a thin tough rind, palatable.

Type locality: St. Charles Island, Galápagos.
Distribution: Various islands of the Galápagos group.

This species and the other cacti of the Galápagos Islands were discovered by Charles Darwin in 1835. He associated this species with *Cereus peruvianus*, which it resembles only in its large cylindric trunk. The various collectors who have since visited these islands have noted this striking plant, but little material has been collected and even to-day our knowledge is very limited.

W. Botting Hemsley has written most interestingly of the cactus flora of the islands (Gard. Chron. III. **24**: 265, 266. 1898; **27**: 177, 178. 1900).

Mr. Alban Stewart, who made extensive collections in the Galápagos Islands in 1905 and 1906, discusses the cacti in considerable detail (Proc. Calif. Acad. Sci. IV. **1**: 107 to 115). He recognizes two columnar arborescent species under the names *Cereus galapagensis* Weber and *Cereus sclerocarpus* Schumann, and indicates that they may be distinguished by habit characters, but remarks particularly on the great variability of the flowers of both.

Illustrations: Proc. Calif. Acad. Sci. IV. **1**: pl. 6, 16, as *Cereus sclerocarpus*; Gard. Chron. III. **27**: 185. f. 61, as *Cereus* sp.; Wolf, Geographia y Geologia del Ecuador f. 41, pl. 11; Bull. Mus. Comp. Zool. **23**: pl. 16, 20.

Figure 212 is from a photograph of the plant in its natural habitat on Charles Island, Galápagos, contributed by the United States Fish Commission; figures 213 and 214 show flowers drawn from an herbarium specimen in the collection of the California Academy of Sciences, collected by Alban Stewart (No. 2097) in 1905 and 1906.

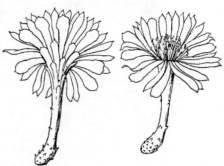

FIGS. 213 and 214.—Flowers of *J. galapagensis*. ×0.6.

26. HARRISIA Britton, Bull. Torr. Club **35**: 561. 1908.

Eriocereus Riccobono, Boll. R. Ort. Bot. Palermo **8**: 238. 1909.

Night-flowering cacti with slender, branched stems, the branches fluted or angled, each areole with several acicular spines; flowers borne singly at areoles near the ends of the branches, funnel-form, large, with a cylindric, scaly tube as long as the limb or longer; buds globose, ovoid or obovoid, the scales subtending areoles which bear tufts of long or short hairs, persistent or sometimes deciduous as the flower expands; outer perianth-segments mostly pink or greenish, linear to lanceolate; inner perianth-segments white or pinkish; stamens shorter than the perianth; ovary and young fruit tubercled; style somewhat longer than the stamens; fruit globose to obovoid-globose, spineless or spiny, but with mostly deciduous scales, the corolla withering-persistent; seeds numerous, small.

The genus is named in honor of William Harris, superintendent of Public Gardens and Plantations of Jamaica, distinguished for his contributions to the knowledge of the flora of that island.

We recognize 17 species, distributed from Florida and the Bahamas and the Greater Antilles to Argentina. The type species is *Cereus gracilis* Miller.

The vegetative characters of the first 9 species here recognized, natives of Florida and the West Indies, are very much alike; their showy yellow or orange-red fruits are edible. The young stem-areoles are subtended by subulate small deciduous leaves in several species.

FIG. 215.—Harrisia eriophora.

KEY TO SPECIES.

A. Fruit yellow or orange-red, not splitting (*Euharrisia*).
    B. Plants erect.
        Hairs of the flower-areoles white.
           Perianth-segments entire.
               Hairs of the flower-areoles copious, 1 to 1.5 cm. long.
                  Fruit yellow... 1. *H. eriophora*
                  Fruit orange-red.. 2. *H. fragrans*
               Hairs of the flower-areoles few and short.
                  Flower-buds depressed-truncate; fruit yellow.............. 3. *H. portoricensis*
                  Flower-buds pointed.
                      Flower-buds obovoid, short-pointed; color of fruit un-
                        known.. 4. *H. nashii*
                      Flower-buds ovoid, very long-pointed; fruit yellow....... 5. *H. brookii*
           Perianth-segments denticulate.
               Fruit yellow.. 6. *H. gracilis*
               Fruit orange-red... 7. *H. simpsonii*
         Hairs of the flower-areoles tawny or brown.
            Hairs of the flower-areoles 1 to 1.5 cm. long; color of fruit unknown;
               spines up to 6 cm. long............................... 8. *H. fernowi*
           Hairs of flower-areoles 7 mm. long or less; fruit yellow; spines much
               shorter....................................... 9. *H. aboriginum*
    BB. Plants prostrate and pendent on rocks.....................................10. *H. earlei*
AA. Fruit red, often splitting (*Eriocereus*).
    Joints several-ribbed or subterete.
        Ribs of the joints prominent.
           Ribs not tubercled.
               Plants bright green...11. *H. tortuosa*
               Plants bluish green...12. *H. pomanensis*
           Ribs of old joints strongly tubercled.
               Central spine 1, much longer than radial spines....................13. *H. martinii*
               Spines of nearly the same length ....14. *H. adscendens*
        Ribs of the joints low and broad.......................................15. *H. platygona*
    Joints 3 to 5-angled.
        Scales of the perianth-tube copiously woolly in the axils....................16. *H. bonplandii*
        Scales of the perianth-tube scarcely woolly in the axils........................17. *H. guelichii*

M. E. Eaton del.

1. Tip of a flowering branch of *Harrisia eriophora*.
2. Fruiting branch of the same.
(Natural size.)

1. Harrisia eriophora (Pfeiffer) Britton, Bull. Torr. Club **35**: 562. 1908.

(?) *Cereus cubensis* Zuccarini in Seitz, Allg. Gartenz. **2**: 244. 1834.
Cereus eriophorus Pfeiffer, Enum. Cact. 94. 1837.

Plant 3.5 dm. high or less, the young joints bright green, the main stem 4 cm. in diameter or more, the branches nearly as thick, erect or ascending, 8 or 9-ribbed, the ribs prominent, the depressions between them rather deep; areoles 2 to 4 cm. apart; spines 6 to 9, the longer ones 2.5 to 4 cm. long, light brown with nearly black tips; buds ovoid, sharp-pointed, their scales subtending tufts of bright white-woolly hairs 1 to 1.5 cm. long; flowers 12 to 18 cm. long; scales of the tube lanceolate, acuminate, appressed, 1 to 1.5 cm. long, subtending long white hairs; outer perianth-segments pale pink outside, the outermost greenish; inner segments pure white, tipped with a hair-like cusp 5 mm. long; filaments white; anthers oblong, yellow; pistil cream-colored; fruit subglobose, yellow, about 6 cm. in diameter, edible.

Type locality: Cuba.
Distribution: Central and western Cuba and Isle of Pines.

The names *Cereus eriophorus laeteviridis* and *C. repandus laetevirens* (Salm-Dyck, Hort. Dyck. 335. 1834), both unpublished, may belong here.

The flower-buds, copiously covered with bright white wool, are conspicuous.

Plants grown in the Habana Botanical Garden, formerly referred to *Cereus undatus* (Bull. Torr. Club **35**: 564), apparently belong to this species.

Illustration: Pfeiffer and Otto, Abbild. Beschr. Cact. **1**: pl. 22, as *Cereus eriophorus*.

Plate XVIII, figure 1, shows the flower of a plant from Mariel, Cuba, painted at the New York Botanical Garden, July 12, 1912; figure 2 shows a fruiting branch of a plant sent by C. F. Baker in 1907. Figure 215 is from a photograph taken by C. S. Gager at Mariel, Cuba, in 1910.

FIG. 216.—Harrisia fragrans.

2. Harrisia fragrans Small, sp. nov.

Plants 5 meters tall or less, the stems erect, reclining or clambering prominently, 10 to 12-ridged, the ridges more or less depressed between the areoles, the grooves rather deep and sharp; areoles about 2 cm. apart; spines acicular, 9 to 13 in each areole, mostly grayish and yellowish at the tip, one of each areole longer than the others, mostly 2 to 4 cm. long; young buds copiously white-hairy; flowers 12 to 20 cm. long, odorous; ovary bearing subulate or lanceolate-subulate scales subtending long white hairs; scales of the flower-tube few and remote, subulate, slenderly acuminate, not turgid, with a tuft of long white hairs in each axil; outer perianth-segments very narrowly linear, slenderly acuminate; inner perianth-segments white or pinkish, spatulate, caudate-tipped; fruit obovoid or globose, about 6 cm. in diameter, dull red, with tufts of long hairs persistent with the scale-bases.

Coastal sand-dunes, Brevard and St. Lucie Counties, Florida. Type collected by John K. Small on sand-dunes 6 miles south of Fort Pierce, December 1917.

Plants, taken to the cactus plantation of Mr. Charles Deering at Miami, Florida, in 1917, flowered and fruited in April 1918.

Plate XIX, figure 1, shows a flowering top of a plant from an island east of Malabar, Florida, brought to the New York Botanical Garden by Dr. Small in 1903; figure 2 shows its fruit. Figure 216 is from a photograph of the plant in the cactus garden of Mr. Charles Deering, Miami, Florida, taken by Dr. Small.

3. Harrisia portoricensis Britton, Bull. Torr. Club 35: 563. 1908.

Cereus portoricensis Urban, Symb. Ant. 4: 430. 1910.

Plant slender, 2 to 3 meters high, little branched, the branches nearly erect, 3 to 4 cm. thick, 11-ribbed, the ribs rounded, the depressions between them shallow; areoles 1.5 to 2 cm. apart; spines 13 to 17, grayish white to brown with dark tips, the longer ones 2.5 to 3 cm. long; bud obovoid, depressed-truncate, its areoles with many curled white hairs 6 mm. long or less; flower about 1.5 dm. long; outer perianth-segments pinkish green inside, the inner white; scales of the flower-tube lanceolate, appressed, 1.5 cm. long, the areoles loosely hairy, the hair completely deciduous in flakes; fruit ovoid to globose, yellow, tubercled, becoming smooth or nearly so, 4 to 6 cm. in diameter.

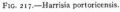

FIG. 217.—Harrisia portoricensis.

FIG. 218.—Harrisia nashii.

Type locality: Near Ponce, Porto Rico.

Distribution: Type locality and vicinity, and on the islands Mona and Desecheo.

Plate XVII, figure 3, shows a fruiting branch of the plant from the type locality, painted at the New York Botanical Garden in 1914. Figure 217 is from a photograph taken at the type locality by Delia W. Marble in 1913.

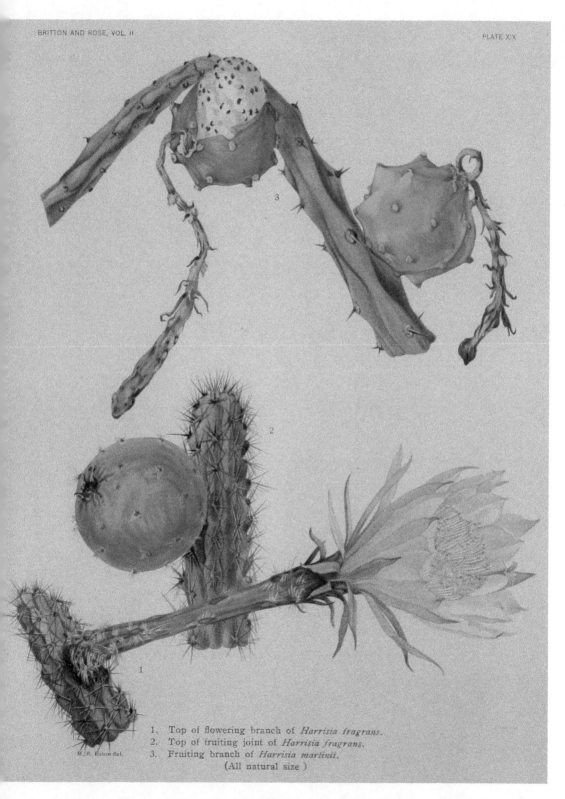

M. E. Eaton del.

1. Top of flowering branch of *Harrisia fragrans*.
2. Top of fruiting joint of *Harrisia fragrans*.
3. Fruiting branch of *Harrisia martinii*.
(All natural size)

4. Harrisia nashii Britton, Bull. Torr. Club **35**: 564. 1908.

>?*Cereus divergens* Pfeiffer, Enum. Cact. 95. 1837.
>*Cereus nashii* Vaupel, Monatsschr. Kakteenk. **23**: 27. 1913.

Slender, erect, 2 to 3 meters high; branches widely divergent, light green, 3 to 4 cm. thick, 9 to 11-ribbed, the ribs rounded; areoles 2 to 2.5 cm. apart; spines only 3 to 6, gray, the longer ones 15 mm. long; bud narrowly obovoid, obtuse, very short-pointed, its scales subtending many curled white hairs 6 mm. long or less; flower 1.6 to 2 dm. long; scales of the flower-tube linear, acuminate, 1.5 cm. long, subtending a few hairs; fruit ellipsoid, 6 to 8 cm. long, 4 to 5 cm. thick, very strongly tubercled, at least when immature, the conic tubercles 6 to 8 mm. high.

Type locality: Between Gonaives and Plaisance, Haiti.
Distribution: Arid parts of Hispaniola.
Cereus divergens Pfeiffer is known only from the description of a sterile plant.
Cereus divaricatus De Candolle (Prodr. **3**: 466. 1828; *Cactus divaricatus* Lamarck, Encycl. **1**: 540. 1783; *Pilocereus divaricatus* Lemaire, Rev. Hort. **1862**: 427. 1862) is based upon Plumier's plate 193, which can not be certainly associated with any known cactus.
Figure 218 is from a photograph by Paul G. Russell near Azua, Santo Domingo, in 1913.

Here perhaps is to be referred *Cactus fimbriatus* Lamarck (Encycl. **1**: 539. 1783; *Cereus fimbriatus* De Candolle, Prodr. **3**: 464. 1828; *Pilocereus fimbriatus* Lemaire, Rev. Hort. **1862**: 427. 1862) and *Cereus serruliflorus* Haworth (Phil. Mag. **7**: 113. 1830), both of which were based on Burmann's plate of Plumier (pl. 195, f. 1, A, B, C, and D), found along the coast of Haiti under the name of la bande du sud. *Cactus fimbriatus* Descourtilz (Fl. Med. Antill. ed. 2. **6**: 160. pl. 419), which refers to the same plate of Plumier, is really based upon pl. 195, f. 2, of Burmann, and is probably a *Lemaireocereus*.

5. Harrisia brookii Britton, Bull. Torr. Club **35**: 564. 1908.

>*Cereus brookii* Vaupel, Monatsschr. Kakteenk. **23**: 24. 1913.

Plant 5 meters high, much branched, light green; branches 3 to 4 cm. thick, 10-ribbed, the ribs sometimes prominent, with deep depressions between them; areoles about 2 cm. apart; spines 9 to 12, the longer ones 2 to 2.5 cm. long; young upper spines of areoles brown, others white; bud ovoid, prominently long-pointed, its scales with few curled white hairs 7 to 10 mm. long; fruit yellowish, ellipsoid or subglobose, about 8 cm. in diameter, rounded at both ends, the tubercles very low, with tips only 1.5 mm. high, the linear scales persistent.

Type locality: Clarence Town, Long Island, Bahamas.
Distribution: Long Island, Bahamas.
Figure 219 shows a fruit of the type plant; figure 220 shows a flower-bud of the same.

FIG. 219.—Fruit of Harrisia brookii. ×0.6.
FIG. 220.—Flower-bud of same. ×0.6.

6. Harrisia gracilis (Miller) Britton, Bull. Torr. Club **35**: 563. 1908.

>*Cereus gracilis* Miller, Gard. Dict. ed. 8, No. 8. 1768.
>*Cactus gracilis* Weston, Bot. Univers. **1**: 33. 1770.
>*Cereus repandus* Haworth, Syn. Pl. Succ. 183. 1812. Not *Cactus repandus* Linnaeus, 1753.
>*Cereus subrepandus* Haworth, Suppl. Pl. Succ. 78. 1819.
>*Cereus undatus* Pfeiffer, Enum. Cact. 94. 1837. Not Haworth, 1830.
>*Harrisia undata* Britton, Bull. Torr. Club **35**: 564. 1908.
>*Eriocereus subrepandus* Riccobono, Boll. R. Ort. Bot. Palermo **8**: 243. 1909.

Plant much branched, often 7 meters high, dark green, its branches rather slender, somewhat divergent, 9 to 11-ribbed, the ribs rounded, the depressions between them rather shallow; areoles 1.5 to 2 cm. apart; spines 10 to 16, whitish with black tips, the longer 2 to 2.5 cm. long; bud oblong-ovoid, short-pointed, its scales subtending a few straight white hairs 8 to 12 mm. long; corolla 2 dm. long, the scales of its tube greenish brown, narrowly lanceolate, abruptly bent upward near the base, acuminate, about 2 cm. long, subtending a few hairs, the outer perianth-segments pale brown, the inner white, denticulate (or sometimes entire?); fruit depressed-globose, yellow, about 5 cm. long,

6 to 7 cm. thick, the base flat, the top bluntly pointed, strongly tubercled when young, the tubercles low-conic, about 4 mm. high, about 1.5 cm. from tip to tip, bearing a deciduous triangular-lanceolate scale 6 to 8 mm. long, becoming confluent, the fruit finally smooth or nearly so, yellow.

Type locality: British Islands of America.

Distribution: Jamaica.

The following names were referred to *Cereus repandus* as synonyms by Schumann:

Cereus tinei Todaro (Ind. Sem. Hort. Panorm. 39. 1857; *C. cossyrensis* Tineo in Todaro, Ind. Sem. Hort. Panorm. 39. 1857), said to have come from Brazil, and *Cereus erectus* Pfeiffer (Enum. Cact. 95. 1837), stated definitely to have come from Mexico.

Illustrations: Trew, Pl. Select. pl. 14, as *Cereus* etc.; Loudon, Encycl. Pl. 411. f. 6862; Edwards's Bot. Reg. 4: pl. 336, as *Cactus repandus*; De Candolle, Mém. Mus. Hist. Nat. Paris 17: pl. 13, as *Cereus repandus*; Pfeiffer and Otto, Abbild. Beschr. Cact. 1: pl. 23, as *Cereus undatus*.

FIG. 221.—Harrisia gracilis.

FIG. 222.—Flower of Harrisia gracilis.

Plate xx, figure 1, shows a fruiting branch of a plant in the the New York Botanical Garden. Figure 221 is from a photograph taken in Jamaica, contributed by William Harris; figure 222 is copied from the last illustration above cited.

7. Harrisia simpsonii Small, sp. nov.

Plants up to 6 meters high, erect, reclining, or spreading, simple or more or less branched; ribs 8 to 10; areoles 1 to 2 cm. apart; spines 7 to 14, gray when mature, 1 to 2.5 cm. long; buds white-hairy; flowers 12 to 17 cm. long; scales of the ovary lanceolate-subulate, subtending few white hairs 10 mm. long or less; scales of the flower-tube lanceolate, distant; outer perianth-segments linear; inner perianth-segments spatulate, acute or acuminate, erose-denticulate; fruit depressed-globose, orange-red, 4 to 6 cm. in diameter.

PLATE XX

M. E. Eaton del.

1. Part of fruiting branch of *Harrisia gracilis*.
2. Top of flowering branch of *Harrisia martinii*.
(Natural size.)

Found on Hammocks, Keys of Florida, and southern mainland coast. Type from between Cape Sable and Flamingo, collected by John K. Small, November 29, 1916.

The species is dedicated to Charles Torrey Simpson, naturalist, long resident in Florida.

Flowers of a plant from Pumpkin Key, grown at the cactus garden of Mr. Charles Deering, Miami, Florida, and at the New York Botanical Garden, have the flower-tube little, if any, longer than the limb; the stems of this plant and its fruit are not different from those of the type, but are smaller, about 2 meters high.

Figure 223 is from a photograph taken by Mr. C. L. Pollard on Key Largo, Florida.

FIG. 223.—Harrisia simpsonii.

FIG. 224.—Harrisia taylori.

8. Harrisia fernowi Britton, Bull. Torr. Club **35**: 562. 1908.

　　Cereus pellucidus Grisebach, Cat. Pl. Cub. 116. 1866. Not *C. pellucidus* Otto, 1837.

Plant 2.5 to 3 meters high; branches slender, about 2.5 cm. thick, light green, 9-ribbed, the ribs not prominent, the depressions between them shallow; areoles about 2 cm. apart; spines 8 to 11, light brown with blackish tips, the longer ones 6 cm. long; bud subglobose-ovoid, its scales subtending and rather densely covered with tawny, curled woolly hairs 1 cm. long; flower nearly 2 dm. long, its ovary and tube bearing oblong-lanceolate, acute scales 1 to 2 cm. long, subtending tufts of long brown hairs; outer perianth-segments linear, acuminate, the inner white, spatulate, entire, short-acuminate.

　　Type locality: Between Rio Grande and Rio Ubero, Oriente, Cuba.

　　Distribution: Dry parts of Oriente Province, Cuba.

Plate XXIV, figure 1, shows a flowering branch of the type plant from a painting made at the New York Botanical Garden, July 9, 1912.

8a. Harrisia taylori Britton, Bull. Torr. Club **35**: 565. 1908.

　　Cereus taylori Vaupel, Monatsschr. Kakteenk. **23**: 37. 1913.

Plant light green, branched above, 1.5 to 2 meters high, the branches divaricate-ascending, rather stout, 4 to 5 cm. thick, 9-ribbed, the ribs rounded, the depressions between them rather deep;

areoles 2 to 3 cm. apart; spines 9 to 12, the longer 3 to 5 cm. long, ascending; bud globose-ovoid, short-pointed, its scales with sparse curled grayish-white wool, 3 to 6 mm. long.

Type locality: Sea-beach between Rio Grande and Rio Ubero, Oriente, Cuba.

Distribution: Known only from the type locality.

This plant was collected near the type locality of the preceding species; specimens of the two appeared to be different when first studied, but subsequent observations indicate that they may not be distinct; additional evidence is needed to determine this question.

Figure 224 is from a photograph of the type plant in its natural environment.

9. Harrisia aboriginum Small, sp. nov.

Plants 6 meters high or less, erect or reclining, simple or branched; ribs 9 to 11, rounded; areoles 1.5 to 3 cm. apart; spines 7 to 9, acicular, mostly 1 cm. long or less, sometimes longer, gray with brown tips when mature, pink when young; flower-buds densely brown-hairy; flowers slightly odorous, about 15 cm. long; scales of the ovary and flower-tube lanceolate, subtending short brown hairs; outer perianth-segments linear, acuminate, the inner oblanceolate, white, caudate-acuminate, erose-denticulate; fruit globular, yellow, 6 to 7.5 cm. in diameter.

On shell-mounds, western coast of Florida, north of the Ten Thousand Islands to Tampa Bay. Type collected by John K. Small on Terra Ceia Island, April 1919.

The type plants were found growing in shell heaps formed by the aborigines, whence the specific name.

10. Harrisia earlei sp. nov.

Pendent and prostrate on limestone rocks, 2 to 3 meters long, dark green, the old stems nearly or quite terete, 4 to 6 cm. in diameter and smooth, the younger branches 2 to 3 cm. in diameter, 5 to 7-angled, with spine-bearing areoles 2 to 4 cm. apart; spines gray, acicular, 5 to 8 at each areole, the longer 4 to 5 cm. long, ascending; flowers about 2 dm. long, the slender greenish tube about as long as the limb; ovary about 1 cm. in diameter, tubercled, bearing short subulate leaves, the areoles with short, white hairs; perianth-tube bearing distant, linear, acuminate scales 1 to 3 cm. long, the areoles with white hairs 1 to 1.5 cm. long; outer perianth-segments linear, greenish, acuminate, the inner somewhat broader, white, acute or acuminate; fruit yellow, depressed-globose, tubercled when young, nearly smooth when old, 6 to 7 cm. in diameter.

Limestone rocks, province of Pinar del Rio, Cuba. Type from San Diego de los Baños, August 31, 1910, collected by Britton, Earle, and Gager (No. 6667).

In habit and vegetative characters intermediate between typical *Harrisiae* and *Eriocerei*.

11. Harrisia tortuosa (Forbes).

Cereus tortuosus Forbes, Allg. Gartenz. **6**: 35. 1838.
Cereus arendtii Hildmann and Mathsson in Schumann, Monatsschr. Kakteenk. **4**: 173. 1894.
Eriocereus tortuosus Riccobono, Boll. R. Ort. Bot. Palermo **8**: 245. 1909.

Stem at first erect but soon arching, the slender, bright green branches 2 to 4 cm. thick; ribs few, usually 7, low, rounded, sometimes tuberculate, bright green; spines 6 to 10, subulate, the central one longer than the radials; flowers 12 to 15 cm. long; scales of the ovary and flower-tube ovate, about 1 cm. long, acute, bearing hair in their axils; outer perianth-segments narrow, dull colored; inner perianth-segments broader than the outer, acute, white to pink; stamens scarcely exserted; stigma-lobes green; fruit globular, tuberculate, red, 3 to 4 cm. in diameter, its areoles bearing a few short spines.

Type locality: Buenos Aires, Argentina.

Distribution: Argentina.

Riccobono gives as a synonym of this species *Cereus atropurpureus* (Hocay, Cacteen-cult. 91). Under this name it is also briefly described in the Theodosia B. Shepherd Company's Descriptive Catalogue for 1916.

Cereus davisii (Monatsschr. Kakteenk. **14**: 166. 1904) is an unpublished name; a specimen in the Succulent House at Kew indicates that it is related to *H. tortuosa*.

Illustrations: Monatsschr. Kakteenk. **14**: 89. f. b; Rep. Mo. Bot. Gard. **16**: pl. 9, f. 1; U. S. Dept. Agr. Bur. Pl. Ind. Bull. **262**: pl. 7, all as *Cereus tortuosus.*

Plate XXI, figure 1, shows a flowering branch, figure 2 a fruiting branch, both from plants in the collection of the New York Botanical Garden.

12. Harrisia pomanensis (Weber).

Cereus pomanensis Weber in Schumann, Gesamtb. Kakteen 136. 1897.

Often prostrate or arched, bluish green and glaucous; ribs 4 to 6, rounded, obtuse; radial spines 6 to 8, 1 cm. long; central spine solitary, 1 to 2 cm. long; spines all subulate, when young white or rose-colored; flowers 15 cm. long; outer perianth-segments linear, acute; inner perianth-segments oblong, acutish, probably white; stigma-lobes numerous, linear; scales on ovary and flower-tube ovate, acute.

Type locality: Poman, Catamarca, Argentina.
Distribution: Northwestern Argentina.

There is a living specimen of this species in the New York Botanical Garden (No. 39517). The stem is 4-angled, 2 cm. broad, and light green. The small areoles are 2 cm. apart and the acicular spines are less than 5 mm. long. The plant has not yet flowered.

Cereus bonplandii pomanensis Weber (Schumann, Gesamtb. Kakteen 137. 1897) is given as a synonym of this species. *C. pomanensis grossei* (Graebener, Monatsschr. Kakteenk. **19**: 137. 1909) is only a mentioned name.

Illustrations: Rep. Mo. Bot. Gard. **16**: pl. 7, f. 5, 6, both as *Cereus pomanensis.*

Figure 225 is from a photograph of a flowering branch in the collection of Dr. Spegazzini at La Plata, Argentina.

13. Harrisia martinii (Labouret).

Cereus martinii Labouret, Ann. Soc. Hort. Haute Garonne. 1854.
Eriocereus martinii Riccobono, Boll. R. Ort. Bot. Palermo **8**: 241. 1909.
Cereus martinii perviridis Weingart, Monatsschr. Kakteenk. **24**: 72. 1914.

Plant much branched, clambering, 2 meters long or longer; old stems terete, spineless; young stems vigorous, about 2 cm. thick, pointed, 4 or 5-angled; areoles with a stout central spine 2 to 3 cm. long, straw-colored with a black tip and a row of short radials, sometimes half as long as the central one; flower about 2 dm. long; outer perianth-segments narrow, becoming pinkish, acuminate; inner perianth-segments broader, short-acuminate, white or tinged with pink; style green; ovary tuberculate; scales on ovary ovate, acuminate, on tube similar, becoming more elongate above, all with brown felt in their axils; fruit red, 3.5 cm. long, bearing small scales, the flowers withering-persistent.

Type locality: Not cited.
Distribution: Argentina.

Cereus monacanthus Cels, not Lemaire, is not listed in the Index Kewensis, but it is cited by Schumann (Gesamtb. Kakteen 142. 1897) as a synonym of this species, quoting Cels, Catalogue, 1853. Here may belong *Pilocereus monacanthus* Lawrence in Loudon, Gard. Mag. **17**: 319. 1841.

A plant of this species in the Kew collection is said by Mr. Weingart to be *Cereus regelii* (Monatsschr. Kakteenk. **20**: 33. 1910).

Illustrations: Amer. Gard. **11**: 569; Cycl. Amer. Hort. Bailey **1**: f. 304 (both fruits spineless); Rep. Mo. Bot. Gard. **16**: pl. 10, f. 1, 2; Rev. Hort. **94**: f. 123 to 125, all as *Cereus martinii.*

Plate XIX, figure 3, represents a fruiting branch, and plate XX, figure 2, a flowering branch, both painted from plants in the collection of the New York Botanical Garden.

14. Harrisia adscendens (Gürke).

Cereus adscendens Gürke, Monatsschr. Kakteenk. **18**: 66. 1908.

At first erect, becoming much branched and bushy or sometimes with long clambering branches 5 to 8 meters long, 2 to 5 cm. thick; ribs 7 to 10, low, rounded, broken up into elongated tubercles;

trunk 2 to 4 cm. in diameter, with a woody cylinder, its center coarsely pithy; areoles large, rounded, subtended by small definite leaves like those of *Opuntia;* spines usually 10, stout, 2 to 3 cm. long, swollen at base, when young brownish or yellowish with brown tips; flowers 15 to 18 cm. long, opening at night; perianth-segments white; ovary bearing lanceolate acute scales with long hairs in their axils; fruit red, globular, tuberculate, 5 to 6 cm. in diameter, spineless, bearing scales and felt at the areoles, when mature splitting down on one side; flesh white, juicy; seeds large, black, 3 mm. long.

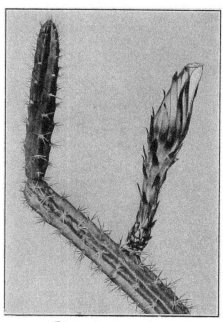

Fig. 225.—Harrisia pomanensis.   Fig. 226.—Harrisia adscendens.

Type locality: Near Tambury, Bahia, Brazil.

Distribution: In the subarid parts of the state of Bahia, Brazil.

Dr. Rose found this very common in Bahia, Brazil, either growing as a low bush in the open or clambering through bushes (No. 19730).

Illustration: Monatsschr. Kakteenk. **18:** 67, as *Cereus adscendens.*

Figure 226 is from a photograph taken by Paul G. Russell at Barrinha, Bahia, in 1915.

15. Harrisia platygona (Otto).

Cereus platygonus Otto in Salm-Dyck, Cact. Hort. Dyck. 1849. 199.   1850.
Eriocereus platygonus Riccobono, Boll. R. Ort. Bot. Palermo **8:** 242.   1909.

At first erect, but soon spreading; branches slender, 2 cm. in diameter or more, nearly terete, the 6 to 8 ribs flat or hardly elevated, separated only by shallow, narrow depressions, pale green or somewhat bronzed; spines 12 to 15, setaceous, very short, the longest only 12 mm. long; flowers 12 cm. broad; flower-tube 10 cm. long, bearing scales; ovary tuberculate, bearing scales, these woolly in their axils; stigma-lobes 14, linear.

Type locality: Not cited.

Distribution: Not known, probably South America.

M. E. Eaton del.

1. Flowering branch of *Harrisia tortuosa*.
2. Fruiting branch of *Harrisia tortuosa*.
(Natural size.)

This species has only once been reported as flowering, and then by Riccobono; our description of the flowers is based on his. We have studied a small plant in the collection of the New York Botanical Garden.

Illustration: Schumann, Gesamtb. Kakteen f. 19, as *Cereus platygonus.*

16. **Harrisia bonplandii** (Parmentier).

> *Cereus bonplandii* Parmentier in Pfeiffer, Enum. Cact. 108. 1837.
> *Cereus balansaei* Schumann in Martius, Fl. Bras. 4²: 210. 1890.
> *Eriocereus bonplandii* Riccobono, Boll. R. Ort. Bot. Palermo 8: 238. 1909.

Stems slender and weak, at first erect, up to 3 meters high or more, sometimes procumbent, arching or clambering, 3 to 8 cm. in diameter, strongly 4-angled; areoles 2 cm. apart; spines 6 to 8, acicular, the longest 4 cm. long, when young red, in age gray; flowers 15 to 22 cm. long, white, closing soon after sunrise; filaments numerous, borne almost to the base of the tube; style included; stigma-lobes numerous; fruit edible, globular, 4 to 6 cm. in diameter, red, bearing large scales with hairs in their axils, spineless, splitting on the side and exposing the white flesh and black seeds.

FIG. 227.—Harrisia bonplandii.

Type locality: Brazil.

Distribution: Paraguay, Argentina, and Brazil.

This species is widely cultivated, but under different names, one of which is *Cereus acutangulus.* The only specimens from wild plants which we have seen were collected by Thomas Morong at Trinidad, Paraguay, and by J. A. Shafer at Ascencion, Paraguay, and at Salta, Argentina. *Cereus bonplandii brevispinus* (Maass, Monatsschr. Kakteenk. 15: 119. 1905) is only mentioned, but Mr. Weingart says it is identical with the hybrid *Cereus jusbertii.*

Schumann's treatment of *Cereus balansaei* is confusing. In the Gesamtbeschreibung der Kakteen (p. 136) he refers it to *Cereus bonplandii.* In the Nachträge (p. 45) he puts the Balansa specimen (No. 2504, type) here, but not the name, while in his Keys of the Monograph of Cactaceae (p. 17) he recognizes *C. balansaei* as well as *C. bonplandii*, referring to the former the Argentine species *C. pomanensis*.

Cereus rhodocephalus Lemaire (Cact. Gen. Nov. Sp. 79. 1839) is cited as a synonym of *Cereus bonplandii.*

We do not know *Cereus ureacanthus* Förster, (Hamb. Gartenz. 17:166. 1861); it is recorded as originally from Peru. Förster thought it might come next to *Cereus bonplandii*, but no species of this relationship have heretofore been reported from Peru.

Illustrations: Rep. Mo. Bot. Gard. **16:** pl. 10, f. 3, 4, both as *Cereus bonplandii.*

Plate xxiv, figure 2, represents a fruiting branch of a plant in the collection of the New York Botanical Garden. Figure 227 is from a photograph taken by Dr. Shafer at Salta, Argentina, in 1917.

17. Harrisia guelichii (Spegazzini).

Cereus guelichii Spegazzini, Anal. Mus. Nac. Buenos Aires III. **4:** 482. 1905.

FIG. 228.—Harrisia guelichii.

Branching, high-climbing on trees, up to 25 meters long, the branches 3 to 5 cm. thick, 3 or 4-angled; ribs acute, undulate; radial spines 4 or 5; central spine 1, stouter than the radials; flowers large, green without; scales on the ovary and flower-tube prominent, nearly naked in their axils; fruit globular, strongly tuberculate, spineless, red, 4 to 4.5 cm. in diameter; pulp white, very sweet, edible.

Type locality: In the Chaco, Argentina.

Distribution: Argentina.

We have a living specimen of this species brought by Dr. Rose from Argentina in 1915; from Dr. Spegazzini's description this must be the most elongated cactus known.

Illustration: Monatsschr. Kakteenk. **19:** 19, as *Cereus guelichii.*

Figure 228 is from a photograph of a plant grown in the garden of Dr. Spegazzini, La Plata, Argentina.

PUBLISHED SPECIES, PERHAPS OF THIS GENUS.

CEREUS JUSBERTII Rebut in Schumann, Gesamtb. Kakteen 137. 1897.

Eriocereus jusbertii Riccobono, Boll. R. Ort. Bot. Palermo **8:** 240. 1909.

Somewhat erect, from the first more or less branched; ribs 6, usually low, with broad intervals; spines very short, the centrals a little longer than the radials; flowers funnelform; inner perianth-segments white; stigma-lobes numerous, linear, about 12, green; scales on ovary and tube with long hairs in their axils.

This plant, now common in living collections, is generally believed to be a hybrid. Berger says, "According to repeated assurances of Abbé Beguin, it is a hybrid between an *Echinopsis* and a *Cereus* raised by him."

Illustrations: Blühende Kakteen **2:** pl. 78; Schumann, Gesamtb. Kakteen f. 32; Möllers, Deutsche Gärt. Zeit. **26:** 305.

CEREUS AREOLATUS Mühlenpfordt in Schumann, Gesamtb. Kakteen 100. f. 20. 1897.

Cleistocactus areolatus Riccobono, Bol. R. Ort. Bot. Palermo 8: 264. 1909.

Described as columnar, somewhat branching, with 12 low, acutish ribs; ribs divided into tubercles by transverse lines running down from the areoles; radial spines 9 or 10, acicular; central spines 2 to 4, stouter, subulate; flowers and fruit unknown.

The above name was published in a garden catalogue in 1860, while the plant was listed as *Cereus dumesnilianus* Labouret in Gruson's Catalogue.

This cactus is described from plants which are supposed to have come from the Andes of South America. The species is recognized by Schumann in his monograph and is placed in his series *Graciles* after *Cereus platygonus*. It has been in cultivation in the Berlin Botanical Garden and at La Mortola. From the latter source Dr. Rose obtained a specimen in 1914. This plant may be described as follows:

Ribs 15, low, rounded, with a deep horizontal groove just above the areoles; spines yellowish brown, the 6 to 8 radials acicular, spreading, about 1 cm. long; the central subulate, 2 cm. long, porrect.

CEREUS MAGNUS Haworth, Phil. Mag. 7: 109. 1830.

This species has not been definitely identified. Haworth says it was procured from the captain of a French vessel, who obtained it from Santo Domingo. He describes it as a yard high, with 12 ribs and a very large white flower 6 inches long and open day or night. This does not correspond to any cactus known from Hispaniola. Pfeiffer suggests that it might be a form of *C. eyriesii*, that is an *Echinopsis*. In its large flower, open both day and night, it does agree with that genus.

Cereus microsphaericus Schumann (Fl. Bras. 4²: 196. 1890) and *C. damazioi* Schumann (Monatsschr. Kakteenk. 13: 63. 1903; 28: 62. 1918) are of this alliance. Both come from near Rio de Janeiro, Brazil.

27. BORZICACTUS Riccobono, Boll. R. Ort. Bot. Palermo 8: 261. 1909.

Low, slender cacti, erect or procumbent; ribs usually numerous but sometimes as few as 9, usually low and rounded; spines acicular or in some species subulate; areoles usually approximate, in some species producing wool with the flowers; flowers diurnal, orange to scarlet (in one species said to be white) solitary, narrow; tube-proper very short, smooth within; throat very narrow below, expanded above; limb somewhat spreading; axils of scales on ovary and flower-tube bearing long silky hairs; stamens long and slender, slightly exserted; fruit small, globular, edible.

Type species: *Borzicactus ventimigliae* Riccobono.

This genus is perhaps nearest *Rathbunia* of Mexico, but is of different habit and usually with different spines. The flowers are of much the same shape, but with a different limb, some of the stamens originating near the base of the flower-tube, while the areoles of the ovary and flower bear long silky hairs.

The plants are found in the mountains and hills of Ecuador, Peru, and northern Chile, where they have a remarkable development. The indications are that there are still other species to be referred here. It was named in honor of Professor Antonio Borzi, director of the Botanical Garden of Palermo, Italy. Eight species are here described.

KEY TO SPECIES.

Flowers red.
  Base of throat bearing a mass of hairs within.
    Ribs few, 8 to 11, prominent...1. B. sepium
    Ribs many, low..2. B. morleyanus
  Base of throat naked within.
    Flowers pinkish, not as narrow as in the next species, their areoles very hairy.........3. B. icosagonus
    Flowers dark red, very narrow, their areoles not very hairy......................4. B. acanthurus
Flowers white..5. B. decumbens
Not grouped.................................. {6. B. humboldtii
                                      {7. B. plagiostoma
                                        {8. B. aurivillus

1. **Borzicactus sepium** (HBK.).

    Cactus sepium Humboldt, Bonpland, and Kunth, Nov. Gen. et Sp. **6**: 67.   1823.
    Cereus sepium De Candolle, Prodr. **3**: 467.   1828.
    Cleistocactus sepium Weber in Gosselin, Bull. Mens. Soc. Nice **44**: 36.   1904.
    Borzicactus ventimigliae Riccobono, Boll. R. Ort. Bot. Palermo **8**: 262.   1909.
    Cereus ventimigliae Vaupel, Monatsschr. Kakteenk. **23**: 13.   1913.

    Stem slender, simple, columnar, 1.5 meters high, about 4 cm. thick; ribs 8 to 11, crenate, obtuse; areoles 1.5 to 2 cm. apart; radial spines 8 to 10, slender, spreading, 5 to 10 mm. long; central spine solitary, about 2 cm. long; spines all dark red with yellowish bases when young, gray in age; flowers somewhat zygomorphic, about 4 cm. long, 3 cm. broad; scales on ovary and flower-tube woolly in their axils; outer perianth-segments lanceolate, erect, scarlet; inner perianth-segments cuneate, red; pistil slightly exceeding the stamens; stigma-lobes 10, short, greenish; fruit globular, 2 cm. in diameter; flesh of fruit white; seeds numerous.

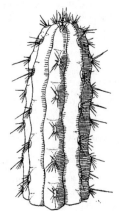

FIG. 229.—Top of plant of Borzicactus sepium. X0.6.

    Type locality: Near Riobamba, at foot of Chimborazo, Ecuador.

    Distribution: Dry hills along the interandean valley of Ecuador from San Antonio to Riobamba.

    The plant blooms from July to September, while the flowers are said to remain open for 48 hours.

    A careful examination of the description of Humboldt's *Cactus sepium* convinces us that it is the same as *Borzicactus ventimigliae*. Not only are the two descriptions similar, but the two types came from the high Andes of Ecuador and a plant sent by Mr. Riccobono from Palermo as *B. ventimigliae* is the same as one sent from the Berlin Botanical Garden as *Cereus sepium*. Dr. Rose, when in Ecuador in 1918, visited Riobamba, but did not see this species there; but he did find it a little north on the hills about Ambato (No. 22389). He also saw what he took to be this species between Ambato and Quito, and, again, collected the species at San Antonio, north of Quito (No. 23557).

    The fruit is eaten at Ambato and doubtless elsewhere and is known as muyusa.

    Figure 229 shows the top of a plant obtained by Dr. Rose from the Botanical Garden at Palermo, Italy.

2. **Borzicactus morleyanus** sp. nov.

    Plant low, growing in clumps, prostrate or with erect branches, sometimes hanging over cliffs or ascending and leaning against rocky banks for support, 4 to 6 cm. in diameter; ribs 13 to 16, low, obtuse, divided into tubercles by V-shaped creases above the areoles; areoles circular, 1 cm. apart or less; spines numerous, 15 to 20, bristly or somewhat acicular, brown, unequal, the longer ones 2.5 cm. long; flowers narrow, 5 to 6 cm. long, slightly oblique; perianth-segments spreading, acute; stamens exserted; filaments purple above, white or tinged with pink below, erect; style cream-colored; stigma-lobes 10, cream-colored.

    Very common at Sibambe, Ecuador, where it was collected by J. N. Rose and George Rose, August 29, 1918 (No. 22431, type), and above Huigra, August 28, 1918 (No. 22426).

    Here may belong Dr. Rose's plant (No. 22829) from Cuenca, although it has somewhat different spines and perhaps more ribs on the stem.

    It is named for Mr. Edward Morley, of Huigra, Ecuador, who greatly aided Dr. Rose in his explorations in Ecuador in 1918.

    Figure 230 shows the top of a flowering stem, and figure 231 shows the type, photographed by George Rose.

3. **Borzicactus icosagonus** (HBK.).

    Cactus icosagonus Humboldt, Bonpland, and Kunth, Nov. Gen. et Sp. **6**: 67.   1823.
    Cereus icosagonus De Candolle, Prodr. **3**: 467.   1828.
    Cereus isogonus Schumann, Gesamtb. Kakteen 102.   1897.
    Cleistocactus icosagonus Weber in Gosselin, Bull. Mens. Sci. Nice **44**: 34.   1904.

Plants small, procumbent or ascending, 2 to 6 dm. long, 3 to 5 cm. in diameter; ribs 18 to 20, low; areoles approximate; spines bright yellow, numerous, acicular, 1 cm. long or less; flower-buds covered with white wool or hairs; flowers near the end of the branches, 7 to 8 cm. long, pinkish to orange; flower-tube naked at base; perianth-segments oblong, acute, apiculate; scales on the ovary and flower bearing long white and brown hairs in their axils.

Type locality: Near Nabón, Ecuador.

Distribution: In the vicinity of Nabón, Ecuador.

This species has long been a puzzle and so far as we know the only record of its having been previously collected is that of the type at Nabón. Dr. Rose visited Nabón in 1918, where he found this species very abundant on the dry hills. Its range is very circumscribed, for it does not extend very far either north or south of Nabón. It is readily distinguished from the other species of the genus seen in Ecuador in its very dense mass of short yellow spines and its larger lighter-colored flowers. These flowers are very attractive and it is to be hoped that some of the living material sent to New York may produce flowers. The plants, however, had to be carried for a long distance by pack train before being shipped to New York and did not arrive in very good condition. Dr. Rose's plant from Nabón was collected September 25, 1918 (No. 23029). We have tentatively referred here his plant from Tablón de Oña, collected September 27, 1918 (No. 23130), but it has smaller flowers.

Illustration: Schumann, Gesamtb. Kakteen f. 21, as *Cereus isogonus.*

FIG. 230.—Borzicactus morleyanus.   FIG. 231.—Borzicactus morleyanus.

4. Borzicactus acanthurus (Vaupel).

Cereus acanthurus Vaupel, Bot. Jahrb. Engler 50: Beibl. 111: 13. 1913.

Plants low, spreading and procumbent, with the tips ascending, sometimes sprawling over the edge of a cliff, with long hanging branches, 1 to 3 dm. long and 2 to 4 cm. in diameter; ribs 15 to 18, very low, rounded, separated by narrow acute intervals; areoles small, approximate; flowering areoles producing tufts of white wool about the flowers; flowers scarlet; tube slender, straight or a little curved, 4 to 5 cm. long; limb about 2.5 cm. broad; inner perianth-segments spreading, acute; filaments white below, scarlet above; style rose-colored, longer than the filaments; stigma-lobes green; fruit globular, 2 cm. in diameter.

Type locality: Matucana, Peru.

Distribution: On the low hills and in the narrow valleys near Lima and along the Rimac River to the east of Lima.

Observed June 1839 at San Cristobal near Lima by A. T. Agate, of the Wilkes' Exploring Expedition. Agate's painting of it is preserved in the Library of the Gray Herbarium.

Plate xxv, figure 3, shows a flowering plant collected at the type locality by Dr. Rose in 1914 which flowered in the New York Botanical Garden in the same year.

5. Borzicactus decumbens (Vaupel).

Cereus decumbens Vaupel, Bot. Jahrb. Engler **50**: Beibl. **111**: 18. 1913.

Plant cespitose, procumbent or ascending, forming small clumps; branches slender, 3 to 4 cm. in diameter; ribs numerous, 20, low, almost hidden under the spines, the intervals acute; areoles close together, about 5 mm. apart; radial spines very numerous, about 30, acicular, short, 5 to 8 mm. long, yellowish; central spines usually 5, much longer and stouter than the radials, often 2 to 3 cm. long, subulate; flower 8 cm. long, with a slender cylindric tube gradually expanded into the throat, the limb about 5 cm. broad; perianth-segments described as white, oblong to oblanceolate.

Type locality: Rocky sandy bottoms, Mollendo, Peru.

Distribution: On hills, southwestern Peru, and northwestern Chile.

The type of this species was first collected by Weberbauer in 1902 on the hills about Mollendo, and here Dr. Rose collected living and herbarium specimens in 1914. Old flowers and fruits were obtained, but no flowers have appeared on the living specimens in the New York Botanical Garden.

Three collections made by Dr. Rose in southern Peru are referred here tentatively. One is from near Arequipa, altitude about 7,000 feet, the second is from near Posco, altitude about 2,000 feet, and the third is from hills above Mollendo, altitude about 200 feet, as mentioned above. This is an unusually wide range for a species in this region. The plants themselves show considerable variation, suggesting that more than one species is involved. Until fresh flowers have been obtained it seems best to recognize only the one species.

FIG. 232.—Borzicactus decumbens.

FIG. 233.—Flower of Borzicactus decumbens. ×0.7.

Figure 232 is from a photograph taken by Dr. Rose near Arequipa, Peru, showing this plant in the foreground at the base of a ledge; figure 233 shows a flower collected by Juan Söhrens near Tacna, Chile.

PLATE XXII

M. E. Eaton del.

Top of flowering plant of *Carnegiea gigantea*.
(Natural size)

6. Borzicactus humboldtii (HBK.).

> *Cactus humboldtii* Humboldt, Bonpland, and Kunth, Nov. Gen. et Sp. 6: 66. 1823.
> *Cereus humboldtii* De Candolle, Prodr. 3: 467. 1828.
> *Cleistocactus humboldtii* Weber in Gosselin, Bull. Mens. Soc. Nice 44: 33. 1904.

Procumbent, cylindric; ribs 10 to 12, low, somewhat tuberculate; spines setose, rigid; flowers red, about 7 cm. long; flower-tube elongated; the scales bearing long greenish gray hairs; perianth-segments lanceolate, acute, red; filaments slender, glabrous; style much longer than the perianth.

Type locality: Between Sondorillo and San Felipe, Peru.

Distribution: Northern Peru and probably southern Ecuador.

The type locality when this species was collected by Humboldt was located in Ecuador, but it is now in northern Peru.

Dr. Rose while collecting in Ecuador in 1918 did not reach Peru, but he found in southern Ecuador near Loja and again in the Catamayo Valley a species of *Borzicactus* which seemed to correspond to *Cactus humboldtii*.

7. Borzicactus plagiostoma (Vaupel).

> *Cereus plagiostoma* Vaupel, Bot. Jahrb. Engler 50: Beibl. 111: 20. 1913.

Columnar, erect, or suberect, about 1 meter high, attenuated and rounded at apex; ribs 15, low; areoles close together, orbicular; spines numerous, nearly black; flowers numerous, cylindric but somewhat zygomorphic; ovary bearing many small, ovate, acuminate scales with black felt in axils.

Type locality: San Miguel, Department of Cojamarca, Peru.

Distribution: Peru.

Said to resemble *Cleistocactus baumannii*, but the relationship is doubtless with the species which we have referred to *Borzicactus*. It is known to us only from description and illustrations.

Illustrations: Monatsschr. Kakteenk. 24: 165, 167, as *Cereus plagiostoma*.

8. Borzicactus aurivillus (Schumann). (See Appendix, p. 226.)

PUBLISHED SPECIES, PERHAPS OF THIS GENUS.

CLEISTOCACTUS CHOTAENSIS Weber, Bull. Mens. Soc. Nice 44: 47. 1904.

> *Cereus chotaensis* Vaupel, Monatsschr. Kakteenk. 23: 25. 1913.

Plant 2 meters high; flowers 5 cm. long, orange-colored; limb 2.5 cm. broad; scales on the ovary bearing long black hairs; stamens as long as the perianth-segments.

Type locality: On the Rio Chota, Peru.

According to Weber this species is similar to one of the so-called species of *Cereus* collected by Humboldt from this same general region.

CEREUS SERPENS (HBK.) De Candolle, Prodr. 3: 470. 1828.

> *Cactus serpens* Humboldt, Bonpland, and Kunth, Nov. Gen. et Sp. 6: 68. 1823.
> *Cleistocactus serpens* Weber in Gosselin, Bull. Mens. Soc. Nice 44: 39. 1904.

Stems creeping; branches somewhat angled; areoles 6-angled, spiny; spines 1 to 3.5 cm. long; flowers tubular, 5 cm. long, flesh-colored; scales few, the upper ones spreading, glabrous; the lower ones hirsute; inner perianth-segments 8 to 12, lanceolate, acute, arranged in 2 or 3 series; stamens a little shorter than the perianth-segments; ovary ovate; stigma-lobes 8.

Type locality: Dry barren hills, banks of Rio Guancabamba, near Sondorillo, Ecuador, now Peru.

Distribution: Known only from the type locality.

This species was originally described from Bonpland's manuscript notes and no specimens are extant. The type locality is definitely given and it should be re-collected and positively identified. Kunth, who referred all of Humboldt and Bonpland's plants to *Cactus*, questioned its belonging to the subgenus *Cereus*, while De Candolle, although referring it to *Cereus*, asks if it may not be an *Opuntia*. In the original description the areoles are described as 6-angled, which suggests a cylindric *Opuntia* with angled tubercles rather than areoles.

28. CARNEGIEA Britton and Rose, Journ. N. Y. Bot. Gard. **9**: 187. 1908.

A large, columnar cactus with stout, erect, many-ribbed stems and branches, the areoles felted and spiny, the spines of flowering and sterile areoles different; flowers borne singly at the uppermost areoles, diurnal, funnelform-campanulate, the stout tube nearly cylindric, expanded above into the throat; scales on tube few, broadly ovate to oblong, acute, bearing small tufts of felt in their axils; inner perianth-segments white, short, widely spreading or somewhat reflexed when fully expanded; ovary oblong, covered with scales similar to those of the tube; stamens very numerous,* about three-fourths as long as the inner perianth-segments; stigma-lobes 12 to 18, narrowly linear, reaching a little above the stamens; fruit an oblong, ellipsoid, or somewhat obovoid berry splitting down from the top into 2 or 3 sections, containing red pulp and bearing small distinct ovate scales, its areoles spineless or bearing a few short spines; seeds small, very numerous, black and shining; embryo hooked; cotyledons incumbent; endosperm wanting.

A monotypic genus of the southwestern United States and Sonora. It is dedicated to Andrew Carnegie (1835–1919), distinguished philanthropist and patron of science.

1. Carnegiea gigantea (Engelmann) Britton and Rose, Journ. N. Y. Bot. Gard. **9**: 188. 1908.

Cereus giganteus Engelmann in Emory, Mil. Reconn. 159. 1848.†
Pilocereus engelmannii Lemaire, Illustr. Hort. **9**: Misc. 97. 1862.
Pilocereus giganteus Rümpler in Förster, Hand. Cact. ed. 2. 662. 1885.

Stem simple and upright, up to 12 meters high, or with one or two lateral branches, or sometimes with 8 to 12 branches, the branches 3 to 6.5 dm. in diameter; ribs 12 to 24, obtuse, 1 to 3 cm. high; areoles about 2.5 cm. apart or nearly contiguous on the upper part of the plant, densely brown-felted; spines of two kinds, those at the top of flowering plants acicular, yellowish brown, porrect, those of sterile plants and on the lower parts of flowering plants more or less subulate, the central ones stouter than the radials, often 7 cm. long; flowers 10 to 12 cm. long, sometimes nearly as broad as long when fully expanded; tube about 1.5 cm. long, green, its scales broad and short, white-felted in their axils; throat about 3 cm. long, covered with numerous white stamens; style stout, 5 to 6 cm. long, white or cream-colored; ovary somewhat tuberculate, bearing scales with woolly axils; ovules numerous; berry red or purple, obtuse, 6 to 9 cm. long, edible, its few, distant scales ovate, 2 to 4 mm. long, with or without 1 to 3 short acicular spines in their axils.

Type locality: Along the Gila River, Arizona.
Distribution: Arizona, southeastern California, and Sonora, Mexico.

The size of the giant cactus is usually overestimated, for it is generally stated to be from 15 to 24.4 meters high, while the tallest plants actually measured are not over 12 meters high. Dr. MacDougal reports weighing a plant which was approximately 5.5 meters high, which weighed nearly 770 kilograms. There are a number of Mexican and South American species which are taller and which would weigh more than *Carnegiea gigantea; Lemaireocereus weberi* must be many times heavier.

Although this species was not described until 1848, it seems to have been known to the early missionaries in California and Mexico (about 1540). It is referred to by Humboldt, according to Engelmann, in his work on New Spain (2: 225). According to Dr. Mac-Dougal, the first Anglo-Saxon observation of *Carnegiea gigantea* was made by J. O. Pattee in 1825.

---

* Dr. Charles E. Bessey (Science n. s. 40: 680. 1914) reports that he had the stamens in one flower counted, and found that there were 3,482, while one ovary contained 1,980 ovules.

† It is usually stated that this species was published on page 158, this even being the reference given by Engelmann himself. Emory's report, in which this species was described, was printed at least twice the same year and about the same date, once as a Senate Document (Executive Document No. 7) and once as a House Document (Executive Document No. 41). In the former *Cereus giganteus* occurs on page 159 and in the latter on page 158. There has been considerable speculation and much difference of opinion as to which edition was published first, but we have recently come into possession of Emory's personal copy of the Senate Document No. 7 marked "with manuscript corrections by the author." From this copy the type of the other edition was set up.

This is sometimes called pitahaya, but it is more generally known in the Southwest by the Indian name of sahuaro or saguaro.* The ripe fruit is much used by the Indians.

While the fruit of this cactus sometimes bears short spines, we have not observed spines in the areoles of the ovary, and presume that they develop during the growth of the berry, as they are known to do in some other cacti.

FIG. 234.—Carnegiea gigantea.

Papago Saguaro, one of the United States National Monuments, is named for this plant. This monument, consisting of over 2,000 acres of desert land, is situated about 9 miles east of Phoenix, Arizona, where there is a wonderful display of *Carnegiea gigantea* on the rocky hillsides.

The sahuaro is the State flower of Arizona.

Dr. Forrest Shreve has contributed the following account of the sahuaro:

"The geographical range of the sahuaro extends from the headwaters of the Yaqui River in southern Sonora, northward to the southern edge of the Colorado Plateau. In Sonora it is rarely

*The following are some of the other published spellings of this name: suaharo, suguaro, suwarrow, suwarro, and zuwarrow.

found more than 150 miles inland from the coast of the Gulf of California, and in southern Arizona its range follows approximately the contour of 3,500 feet on the east and north, and the lower course of the Colorado River on the west. It is found in California only in three restricted localities on the Colorado River and reaches its northern limit on that stream at a point about 40 miles north of the mouth of the Bill Williams Fork.

"The occurrence of the sahuaro is by no means continuous throughout this area, for it is never found in deep alluvial soil and is relatively rare on the nearly level plains in the drainages of the Altar, Santa Cruz, and Gila rivers. It is extremely abundant on coarse detrital soils adjacent to the larger and smaller mountains and is very common wherever there is rock in place, ascending the mountains in diminishing numbers to an elevation of about 4,500 feet. The absence of the sahuaro from alluvial soils is undoubtedly related to the adverse conditions of soil aëration in these areas, and possibly to the lack of good mechanical support.

"The localities in which the sahuaro reaches its greatest size and abundance are the uppermost portions of the slopes adjacent to small mountain ranges and hills, particularly where there is a southern or southwestern exposure. In localities of this sort throughout southwestern Arizona, it reaches a height of 30 to 35 feet, which is very seldom exceeded. Individuals of this size are freely branched and often have a gross weight of as much as 6 to 8 tons. In the vicinity of Tucson branching begins on attaining a height of about 15 feet, but on the edges of the range of this cactus branching individuals are relatively uncommon and the maximum size is rarely reached.

"The flowers of the sahuaro are borne at the crown of the main trunk and the lateral branches, usually appearing in May, while the fruit matures some weeks in advance of the summer rainy season. The small seeds are borne in great profusion, but are eaten by birds and ants so rapidly that the crop is seriously decimated before the requisite conditions for germination occur. The seeds germinate readily at the high temperatures of the summer rainy season, but the growth of the seedlings is extremely slow, so that the end of the second year finds them only one-fourth of an inch in height, and at an age of 8 to 10 years they are still less than 4 inches high. The growth continues to be slow up to a height of 3 feet or more, so that individuals of that size are approximately 30 years of age. After reaching this size the growth rate is rapidly accelerated until it reaches a maximum of about 4 inches per year. The largest individuals are 150 to 200 years of age.

"The sahuaro appears to suffer from very few diseases and natural enemies, the greatest decimation in its numbers being occasioned by mechanical agencies. When struck by lightning or wounded in any other manner during the dry season, it recovers very rapidly by the formation of a heavy callus over the wounded spot. If it is wounded in the rainy season, however, bacterial decay sets in very rapidly and a large plant may be destroyed in less than a week as a result of a small wound. The nests made in them by woodpeckers are always lined by heavy callus and appear to occasion no permanent injury.

"The roots of the sahuaro are shallowly placed and widely extended, often reaching a distance of 50 to 60 feet from the base of the plant. The woody tissue may be compared to a series of bamboo fishing rods arranged parallel to each other in the form of a cylinder. These woody rods increase in thickness with the age of the plant, so that they form a very substantial framework at the base while they taper at the summit to slender elastic rods. The fleshy tissue is found both within and outside the circle of the woody rods and the water content of these two regions appears to be the same. Determinations made near the top of the plant indicate that there is 98 per cent of water on the basis of the wet weight. There are great fluctuations in the water content of the tissue from season to season and it has been shown that large quantities of water are taken up during the rainy seasons, particularly in the summer, and that this water is gradually lost during the dry seasons, particularly in May and June. The sahuaro, like many other cacti, is able by reason of its external form to adjust its size to these fluctuations in volume.

"This plant is an extremely useful one to the aborigines of its natural range. The heavy rods are used as construction material in building houses and enclosures, and the fruit and seeds are used for making both food and drink by the Papago and Pima Indians."

Illustrations: Amer. Bot. **20**: 87; Journ. N. Y. Bot. Gard. **9**: f. 32; pl. 49 to 52; Nat. Geogr. Mag. **21**: 651; Safford, Ann. Rep. Smiths. Inst. **1908**: f. 20; Shreve, Veg. Des. Mt. Range pl. 3 B, 4, 5 to 8; St. Nicholas **42**: 366. Amer. Gard. **11**: 451, 528; Ann. Rep. Bur. Amer. Ethn. **26**: pl. 8, f. b; pl. 9; Bull. Torr. Club **32**: pl. 3, 4; Cact. Journ. **2**: 84, 130; Cact. Mex. Bound. pl. 61, 62; Curtis's Bot. Mag. **118**: pl. 7222; Cycl.

FIG. 235.—Fruit of Carnegiea gigantea. X0.6.

The giant cactus, *Carnegeia gigantea*, near Tucson, Arizona.

Amer. Hort. Bailey 1: f. 413; Emory, Mil. Reconn. pl. opp. 72; Fl. Serr. 10: pl. 977 A; 15: pl. 1600; Gard. Chron. III. 45: f. 69; Gartenflora 31: 217; Hornaday, Camp-fires on Des. and Lava opp. 42, 68, 72, 82, 154; Lumholtz, New Trails in Mex. opp. 48; Monatsschr. Kakteenk. 10: 187; Bot. Wheeler Surv. frontispiece; Nat. Geogr. Mag. 21: 711; Orcutt, Cact. 5; Plant World 9: f. 46; 11⁵: f. 2; 11¹⁰: f. 2 to 4; Rümpler, Sukkulenten f. 63; Sargent, Man. Trees N. Amer. f. 558; Dict. Gard. Nicholson Suppl. f. 231; Garden 1: 263; Vegetationsbilder 4: pl. 40. B; pl. 41, 42; Garten-Zeitung 3: 58. f. 15; MacDougal, Bot. N. Amer. Des. pl. 48, 54 to 56, mostly as *Cereus giganteus;* Nat. Geogr. Mag. 27: 85, as *Cactus;* Förster, Handb. Cact. ed. 2. 663. f. 88, as *Pilocereus giganteus;* Journ. Intern. Gard. Club 3: 17.

Plate XXII shows the top of a plant, brought to the New York Botanical Garden by Dr. MacDougal in 1903, in flower June 1912; plate XXIII is from a photograph taken by Dr. MacDougal near Tucson, Arizona. Figure 234 is from a photograph also taken by Dr. MacDougal, 60 miles west of Tucson, showing a single plant; figure 235 shows the fruit collected by Dr. MacDougal, near Tucson, in 1905.

29. BINGHAMIA gen. nov.

Bushy, more or less branched cacti, the stout branches many-ribbed; ribs low, usually very spiny; flowers white, solitary at an areole, funnelform-campanulate, opening at night, of medium size, the tube straight and stout; style exserted; stamens weak and reclining on the underside of tube; scales on ovary and tube small, narrow, bearing a few hairs in their axils but no spines; fruit turgid, juicy, globular, crowned by the withering-persistent flower; seeds black, small.

We recognize 2 species in this genus, inhabitants of western Peru; it is dedicated to Hiram Bingham, Director of the Yale University Expedition to Peru, 1914–1915. The type species is *Cephalocereus melanostele* Vaupel.

KEY TO SPECIES.

Upper areoles of the flowering plant long-bristly, bearing spines.................................1. *B. melanostele*
Upper areoles bearing acicular spines similar to those of the lower...............................2. *B. acrantha*

1. Binghamia melanostele (Vaupel).

Cephalocereus melanostele Vaupel, Bot. Jahrb. Engler 50: Beibl. 111: 12. 1913.

Much branched at base, the 10 to 12 branches strict, usually only 1 meter high; ribs 18 to 22 (perhaps sometimes more), low, close together; areoles approximate, circular, bearing short white and yellow spines; spines very numerous, diverse, those on sterile branches stiff and pungent, the

Fig. 236.—Binghamia melanostele.

Fig. 237.—Binghamia acrantha.

central and longer ones sometimes 3 cm. long, those on old and flowering branches numerous, when young brownish, in age nearly white, all weak, bristle-like, 3 to 8 cm. long, hardly pungent; flowers 4 to 5 cm. long, white; scales on ovary and tube minute, numerous, bearing tufts of white wool in their axils; immature fruit sometimes longer than broad; mature fruit either globular or a little depressed, red, said to be edible, bearing scattered minute areoles with small tufts of wool; pulp white; seeds numerous, black.

Type locality: Near Chosica, Peru, at 800 meters altitude.

Distribution: Mountains of western Peru.

The top of the flowering plant is made up of a compact mass of long white or yellowish bristle-like spines from one side of which the flowers appear, and this F. Vaupel has termed a lateral cephalium.

Plate XXIV, figure 3, shows the top of a sterile plant brought by Dr. Rose from the type locality in 1914. Figure 236 is from a photograph taken by Dr. Rose at Santa Clara, Peru, in 1914; figure 238 shows the fruit of the plant photographed at Santa Clara.

 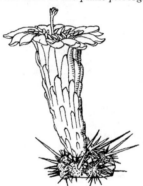 

FIG. 238.—Fruit of B. melanostele. ×0.6.   FIG. 239.—Flower of Binghamia acrantha. ×0.7.   FIG. 240.—Fruit of Binghamia acrantha. ×0.7.

2. Binghamia acrantha (Vaupel).

Cereus acranthus Vaupel, Bot. Jahrb. Engler **50**: Beibl. **111**: 14. 1913.

Stems 1 to 3 meters high, much branched at base, the branches usually erect, 5 to 8 cm. thick; ribs 12 to 14, low, somewhat tuberculate above, but on older parts with mere constrictions; areoles large, approximate, felted and spiny; felt at first yellow, then brown, finally black; spines at first yellow, numerous, short, and spreading, except the 1 or 2 centrals, which are stouter, 3 to 4 cm. long, porrect or reflexed; flowers opening in the early evening; flower 6 to 7 cm. long, gradually tapering upward from base, about 2.5 cm. in diameter at the top; scales on ovary and flower-tube small, acute, with small tufts of wool in their axils; upper scales and outer perianth-segments mauve; limb 4 to 5 cm. broad when fully expanded; inner perianth-segments usually white, sometimes greenish, oblong, obtuse, 2 to 2.5 cm. long; style cream-colored, much exserted; stigma-lobes greenish; fruit red, its pulp white, edible, slightly acid.

Type locality: Santa Clara, east of Lima, Peru.

Distribution: Very common on the hills above Lima, from Santa Clara to Matucana.

This is one of the most common species in central Peru, being especially abundant on the hillsides and in the narrow valleys between the hills, but not extending down into the broad valleys. It often forms dense thickets. In the lower parts of its range, where the fogs are abundant, especially about Santa Clara, the branches are often covered with lichens and tillandsias.

Our specimens of flowers were obtained by bringing in fully developed buds and allowing them to open; these began to open about 6 o'clock in the evening and were fully expanded at 9.

PLATE XXIV

M. E. Eaton del.

1. Top of flowering branch of *Harrisia fernowii*.
2. Flowering branch of *Harrisia bomplandii*.
3. Top of branch of *Binghamia melanostele*.
(Natural size.)

The name *Pilocereus acranthus* was proposed by Schumann (see plate 5 B of Engler and Drude, Veg. Erde 12: 1911), but was never published.

Illustration: Engler and Drude, Veg. Erde 12: pl. 5 B, as *Pilocereus acranthus.*

Figure 237 is from a photograph taken by Dr. Rose at the type locality in 1914; figure 239 shows the flower, and figure 240 the fruit of the plants photographed.

30. RATHBUNIA Britton and Rose, Contr. U. S. Nat. Herb. 12: 414. 1909.

Rather slender cacti, simple or bushy, the stems and branches weak, erect or bent; ribs few, 4 to 8, prominent; spines subulate, those of the flowering areoles not differing from the others; flowers diurnal, scarlet, solitary, usually at the upper areoles, narrowly tubular, the tube bearing distant long scales and united with it except at the tip, elongated, at first straight, or in age somewhat curved, the limb more or less oblique; perianth-segments short, spreading or reflexed; filaments exserted; style slender, exserted beyond the tube; stigma-lobes narrow; ovary with small scales bearing short felt and sometimes spines in their axils; fruit capped by the withered flower, spiny or becoming smooth, globular; seeds of the typical species black, compressed, minutely pitted, with a large basal oblique hilum.

This genus commemorates Dr. Richard Rathbun (1852–1918), Assistant Secretary of the Smithsonian Institution in charge of the United States National Museum, a well-known authority on marine invertebrates.

Type species: *Cereus sonorensis* Rünge.

We here include 2 species, natives of western Mexico.

KEY TO SPECIES.

Ribs 5 to 8; flowers 4 to 10 cm. long...1. *R. alamosensis*
Ribs 4; flowers 12 cm. long..2. *R. kerberi*

1. Rathbunia alamosensis (Coulter) Britton and Rose, Contr. U. S. Nat. Herb. 12: 415. 1909.

Cereus alamosensis Coulter, Contr. U. S. Nat. Herb. 3: 406. 1896.
Cereus sonorensis Rünge in Schumann, Monatsschr. Kakteenk. 11: 135. 1901.
Rathbunia sonorensis Britton and Rose, Contr. U. S. Nat. Herb. 12: 415. 1909.
Cereus pseudosonorensis Gürke, Monatsschr. Kakteenk. 20: 147. 1910.

Columnar, 2 to 3 meters high, at first erect but generally finally bent or curved, 8 cm. thick or less, rooting at or near the tip and thus forming new plants; ribs 5 to 8, obtuse; radial spines about 11 to 18, spreading, straight, whitish; centrals 1 to 4, much stouter than the radials, 3 to 5 cm. long, porrect or ascending; flowers scarlet, 4 to 10 cm. long; scales on ovary small, acute or obtuse, with a small tuft of felt and a few bristle-like spines in the axils, those on the flower-tube with a tuft of felt and sometimes with a spine; tube-proper 1.5 cm. long; style nearly white; stigma-lobes 6, cream-colored; ovary tuberculate; fruit red, globular, 3 to 4 cm. in diameter, naked or bearing scattered clusters of 5 or 6 white acicular spines.

Type locality: Near Alamos, Sonora, Mexico.

Distribution: Southern Sonora, Sinaloa, and Tepic, Mexico.

The plant grows in large clusters sometimes 8 meters in diameter; its flowers are various in size, and the perianth-limb is apparently quite variable in the degree of obliquity. In Mexico the plant is called cina.

Cereus simonii Hildmann (Monatsschr. Kakteenk. 5: 43. 1895), an unpublished name, according to Schumann and Gürke, belongs here. Schumann at one time described this plant as *Cereus stellatus*, a very different plant from southern Mexico which we have described elsewhere as *Lemaireocereus stellatus* (see page 92, *ante*).

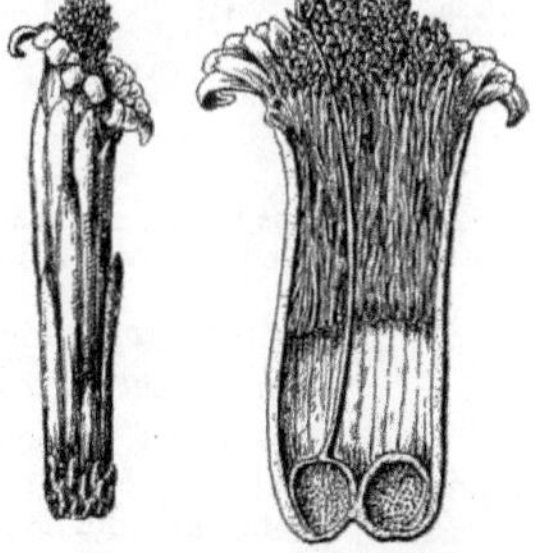

FIG. 241.—Flower of Rathbunia alamosensis. ×0.7.
FIG. 242.—Flower of same, cut open. ×0.7.

Illustrations: Blühende Kakteen 3: pl. 122; Monatsschr. Kakteenk. 11: 135; Rep. Mo. Bot. Gard. 16: pl. 3, f. 5, all as *Cereus sonorensis;* Schumann, Gesamtb. Kakteen Nachtr. f. 4, as *Cereus stellatus.*

Plate xxv, figure 1, shows the top of a plant received from the Missouri Botanical Garden in 1904, which flowered in the New York Botanical Garden, June 23, 1915; figure 2 shows a flowering piece of a plant sent to the New York Botanical Garden from Guaymas, Mexico, by Dr. Rose in 1910. Figures 241 and 242 show flowers of a plant collected by Dr. MacDougal at Torres, Sonora, in 1902.

2. **Rathbunia kerberi** (Schumann) Britton and Rose, Contr. U. S. Nat. Herb. **12**: 415. 1909.

> *Cereus kerberi* Schumann, Gesamtb. Kakteen 89. 1897.
> *Cleistocactus kerberi* Gosselin, Bull. Mens. Soc. Nice 44: 33. 1904.

Columnar, somewhat branched, 2 meters high; ribs 4, compressed; radial spines about 16, subulate; central spines 4, stouter than the radials, 4.5 cm. long; flowers 12 cm. long; outer perianth-segments linear-lanceolate, rose-colored, reflexed; stamens exserted; scales on the ovary lanate in the axils.

Type locality: On Volcano of Colima, Mexico.

Distribution: Known only from the type locality.

Dr. Rose saw flowers of this plant in the herbarium of the Botanical Garden at Berlin in 1912 and noted that it was a *Rathbunia;* otherwise it is known to us only from description. In transferring it to *Rathbunia* (*loc. cit.*) we associated specimens with it from Sinaloa and Tepic, Mexico, which now appear better referable to *Rathbunia alamosensis*, although the flowers are longer than in typical specimens (8 to 10 cm. long) and somewhat curved.

31. ARROJADOA gen. nov.

Stems low, much branched, cylindric; roots fibrous; ribs numerous, low, straight; areoles close together, bearing small acicular spines; flowers diurnal, borne in a pseudocephalium at the top of stem or branch, small, red or pink, resembling in color and size that of a large *Cactus* (*Melocactus*), nearly cylindric, the tube short; perianth-segments in several rows, short, erect; stamens and style included; fruit a small, oblong, naked, juicy berry; seeds small, black.

This is a peculiar genus, with no very close allies. The original reference of its two species to *Cereus* is not warranted by any taxonomic considerations, for the structure, origin, and shape of the flowers and fruit are quite different. In size and form the flower is similar to *Lophocereus*, but here the resemblance ends. Its terminal pseudocephalium is most characteristic, for instead of remaining as a permanent crown of the plant it forms a lateral collar for the new joint which is projected through its center.

The name is in honor of Dr. Miguel Arrojado Lisboa, the present superintendent of Estrada de Ferro Central de Brazil, to whom Brazil is indebted for the extensive botanical exploration of the semiarid regions made a few years ago.

The genus contains 2 species, of which *Cereus rhodanthus* is selected as the type.

KEY TO SPECIES.

Branches short and thick, 2 to 4 cm. in diameter...1. *A. rhodantha*
Branches long and slender, 1 to 1.5 cm. in diameter..2. *A. penicillata*

1. Arrojadoa rhodantha (Gürke).

> *Cereus rhodanthus* Gürke, Monatsschr. Kakteenk. **18**: 69. 1908.

Low, 1 to 2 meters long, at first erect, afterwards branching and clambering; joints short, cylindric, 2 to 4 cm. in diameter; ribs 10 to 13, low; areoles small, approximate, usually less than 1 cm. apart; spines at first brown, in age white, the central ones similar to the radials except a little longer, when young accompanied by some long cobwebby hairs; bristles at the tops of the joints long, brown; flowers solitary at the upper areoles, forming in clusters of 12 to 14 at the tops of branches, pink, rigid, 3 to 4 cm. long; ovary and lower part of tube naked; uppermost scales and perianth-segments similar, obtuse; stamens numerous, included; fruit red, oblong to obovate, about 2 cm. long.

Type locality: Caatinga de São Raimundo, Piauhy, Brazil.

M. E. Eaton del.

1. Flowering branches of *Rathbunia alamosensis*.
2. Flowering branches of *Rathbunia alamosensis*.
3. Top of flowering branch of *Borzicactus acanthurus*.
4. Top of stem of *Arrojadoa rhodantha*.
        (All natural size.)

Distribution: Arid parts of Bahia and Piauhy, Brazil.

Plate xxv, figure 4, shows the flowering top of a plant obtained by Dr. Rose near Joazeiro, Brazil, in 1915, which flowered soon afterward in the New York Botanical Garden; plate xxvii, figure 1, shows a fruiting branch of a plant collected by Dr. Rose near Salgada, Bahia, in 1915.

2. Arrojadoa penicillata (Gürke).

Cereus penicillatus Gürke, Monatsschr. Kakteenk. **18**: 70. 1908.

Plant slender, 1 to 2 meters high, much branched, often bushy, the branches 1 to 1.5 cm. in diameter; ribs usually 10, low; areoles small, close together; spines several; radial spines short, spreading; central spines longer, often 2 to 3 cm. long; pseudocephalium at the top of the joint 2 to 3 cm. in diameter, made up of long brown bristles and white wool; flowers 6 to 20 in a cluster, dark pink, 3 cm. long; fruit small, a little longer than broad, 1.5 cm. long, smooth, without scales, purplish, juicy; seeds numerous, black.

Type locality: Calderão, Bahia, Brazil.
Distribution: State of Bahia, Brazil.

Figure 243 is from a photograph taken by Paul G. Russell at Machado Portella, Bahia, in 1915.

FIG. 243.—Arrojadoa penicillata.

32. OREOCEREUS (Berger) Riccobono, Boll. R. Ort. Bot. Palermo **8**: 258. 1909.

Plants forming large clusters, usually low, erect, ascending or even prostrate, without a cephalium, but the areoles developing long white hairs, especially toward the tips of old branches, the stout stems and branches strongly ribbed; ribs strongly armed with spines; flowers slender, elongated, somewhat curved, diurnal; tube nearly cylindric, slightly expanded upward, the limb short, spreading, somewhat oblique, the inner perianth-segments dark red, narrow; filaments numerous, slender, exserted, attached all over the throat; anthers narrow, red; style long, exserted, with short green stigma-lobes; ovary and flower-tube bearing small narrow scales, with long black and white hairs in their axils; fruit globular, spineless, dry, dehiscing (like *Echinocactus*) by a basal opening; seeds numerous, dull black, with a large truncated hilum.

The name is from the Greek, signifying mountain-cereus. The genus is monotypic, in so far as known to us. The following species inhabits the Andes:

1. Oreocereus celsianus (Lemaire) Riccobono, Boll. R. Ort. Palermo **8**: 259. 1909.

Pilocereus celsianus Lemaire in Salm-Dyck, Cact. Hort. Dyck. 1849. 185. 1850.
Pilocereus fossulatus Labouret, Rev. Hort. IV. **4**: 24. 1855.
Pilocereus bruennowii Haage in Förster, Handb. Cact. ed. 2. 651. 1885.
Pilocereus fossulatus gracilis Rümpler in Förster, Handb. Cact. ed. 2. 661. 1885.
Pilocereus fossulatus pilosior Rümpler in Förster, Handb. Cact. ed. 2. 661. 1885.
?*Pilocereus kanzleri* Haage in Förster, Handb. Cact. ed. 2. 671. 1885.
Pilocereus celsianus lanuginosior Salm-Dyck in Schumann, Gesamtb. Kakteen 180. 1897.
Pilocereus celsianus gracilior Schumann, Gesamtb. Kakteen 180. 1897.
Pilocereus celsianus williamsii Schumann, Gesamtb. Kakteen 180. 1897.
Pilocereus celsianus bruennowii Schumann, Gesamtb. Kakteen 180. 1897.
Cleistocactus celsianus Weber in Gosselin, Bull. Mens. Soc. Nice **44**: 44. 1904.
Cereus celsianus Berger, Rep. Mo. Bot. Gard. **16**: 64. 1905.
?*Pilocereus straussii* Heese, Gartenflora **56**: 410. 1907.
?*Cereus straussii* Vaupel, Monatsschr. Kakteenk. **23**: 37. 1913.
Oreocereus celsianus bruennowii Britton and Rose, Stand. Cycl. Hort. Bailey **4**: 2404. 1916.

A bushy cactus, about 1 meter high, the slender branches either prostrate or ascending below, erect above, about 8 cm. thick; ribs about 10, obtuse, 1 cm. high, more or less broken up into tubercles; areoles bearing long hairs and several stout yellow spines sometimes over 5 cm. long; flowers borne near the tops of the stems, slender, 7 to 9 cm. long; limb 2 to 3 cm. broad; scales of the perianth-tube narrowly lanceolate, long-acuminate, 5 to 6 mm. long, much shorter than the hairs; inner perianth-segments linear to linear-oblong, acutish, the outer obtuse; fruit about 3 cm. in diameter, essentially smooth when mature, the basal pore about 5 mm. in diameter.

Type locality: Mountains of Bolivia.

Distribution: Bolivia, southern Peru, and northern Chile.

Pilocereus celsianus fossulatus Labouret (För-ster, Handb. Cact. ed. 2. 660. 1885) was given by Rümpler as a synonym of *P. fossulatus*. *Pilocereus foveolatus* Labouret (Rev. Hort. 1862: 428. 1862) was given by Lemaire as a synonym of *P.*

FIG. 244.—Oreocereus celsianus.

FIG. 245.—Oreocereus celsianus.

celsianus. Pilocereus williamsii Lemaire (Rev. Hort. 1862: 428. 1862), only a name, is usually referred here.

Illustrations: De Laet, Cat. Gen. f. 50, No.7; Knippel, Kakteen pl. 28; Wiener, Ill. Gart. Zeit. 29: f. 22, No. 7, all as *Pilocereus celsianus;* Cact. Journ. 2: 5; Gard. Chron. 1873: f. 197, both as *Pilocereus fossulatus;* Dict. Gard. Nicholson 3: f. 151; Förster, Handb. Cact. ed. 2. f. 86, both as *Pilocereus bruennowii;* Monatsschr. Kakteenk. 14: 169, as *Pilocereus celsianus bruennowii;* Rep. Mo. Bot. Gard. 16: pl. 2, as *Cereus celsianus;* Gartenflora 56: f. 49, as *Pilocereus straussii;* Monatsschr. Kakteenk. 24: 131, as *Pilocereus celsianus lanuginosior.*

Figure 244 is from a photograph taken by Dr. Rose above Arequipa, Peru, in 1914; figure 245 is from a photograph of a joint; figure 246 shows the flower, and figure 247 the immature fruit of the plant photographed.

FIG. 246.—Flower of Oreocereus celsianus. ×0.7.

FIG. 247.—Fruit of same. ×0.7.

PUBLISHED SPECIES, PERHAPS NEAR OREOCEREUS CELSIANUS.

CEREUS MONVILLEANUS Weber in Schumann, Gesamtb. Kakteen 67. 1897.

Cleistocactus monvilleanus Weber in Gosselin, Bull. Mens. Soc. Nice 44: 45. 1904.

Columnar, branching; ribs 19, obtuse, somewhat sinuate; radial spines about 20, setaceous or acicular.

Distribution: Uncertain. Perhaps Peru, Bolivia, or Ecuador.

According to Weingart, this species is near *Cereus aurivillus* and, if so, it is a *Borzicactus*. So far as we are aware, its flowers are unknown. We have never seen specimens of it.

33. FACHEIROA gen. nov.

Trunk short, with numerous slender, erect or ascending branches; ribs numerous, spiny; flowers borne in a pseudocephalium, this densely brown or red-felted; flowers small, the ovary and flower-tube covered with long silky brown or red hairs; tube-proper short, smooth within; throat short, not hairy at base, bearing numerous short, included stamens; inner perianth-segments short, white; fruit small, globular, greenish, and gelatinous within; seeds black, tuberculate, with a large basal hilum.

Dr. Zehntner states that the habit of this plant is like *Cereus squamosus*, but that the plants differ in the manner of producing their flowers. The flowers, although about the same size, show that the two species are generically different. The generic name is from the common Brazilian one used for a number of the cacti, this one being called facheiro preto da Serra de Cannabrava. Only one species is known.

1. Facheiroa publiflora sp. nov.

Erect, 1.5 to 5 meters high, much branched; trunk short, 10 to 12 cm. in diameter; branches slender, elongated, 5 to 7 cm. in diameter, at first light green, in age grayish green; ribs about 15, low, 5 to 6 mm. high; areoles 1 cm. apart, brown-felted; spines brownish, all acicular; radial spines 10 to 12; central spines 3 or 4, somewhat longer than the radials, often 2 to 2.5 cm. long; pseudocephalium extending from the top downward for 2 dm. or more, 2 to 4 cm. broad, composed of a dense mass of short brown or red hairs; flowers 3 to 3.5 cm. long; tube-proper about 1 cm. long, smooth within; inner perianth-segments orbicular, 3 to 4 mm. in diameter; style slender, glabrous; scales on ovary and flower-tube small, 2 to 6 mm. long, greenish, glabrous, obscured by the long hairs from the axils of other scales; fruit about 2 cm. in diameter, hairy; seeds 1.5 mm. long.

Collected by Leo Zehntner on the Serra de Cannabrava (Chique-Chique district) Bahia, Brazil, October 1917.

34. CLEISTOCACTUS Lemaire, Illustr. Hort. 8: Misc. 35. 1861.

Slender, erect or clambering cacti, with numerous low ribs and approximate areoles; flowers slender, tubular, the perianth withering-persistent on the fruit; perianth-segments small, erect, red to green; stamens and style exserted; ovary and flower-tube with numerous appressed scales bearing long hairs or wool in their axils; fruit small, globular, highly colored, becoming naked; pulp white; seeds black, slightly punctate.

Type species: *Cereus baumannii* Lemaire.

Berger recognizes only 1 species, but mentions 3 of *Cereus* (*C. hyalacanthus, C. laniceps*, and *C. parviflorus*) which may belong here, while Roland-Gosselin recognizes 14 species. 16 species have been described in the genus. We recognize 3 species. The name is from the Greek, signifying closed-cactus, referring to the unexpanded limb of the flower.

KEY TO SPECIES.

Flowers red or green.
    Flower-tube bent; inner perianth-segments red...1. *C. baumannii*
    Flower-tube straight; inner perianth-segments green...................................2. *C. smaragdiflorus*
Flowers orange-yellow...3. *C. anguinus*

1. **Cleistocactus baumannii** Lemaire, Illustr. Hort. 8: Misc. 35. 1861.

> *Cereus baumannii* Lemaire, Hort. Univ. 5: 126. 1844.
> *Cereus colubrinus* Otto in Förster, Handb. Cact. 409. 1846.
> *Cereus tweediei* Hooker in Curtis's Bot. Mag. 76: pl. 4498. 1850.
> *Aporocactus baumannii* Lemaire, Illustr. Hort. 7: Misc. 68. 1860.
> *Aporocactus colubrinus* Lemaire, Illustr. Hort. 7: Misc. 68. 1860.
> *Cleistocactus colubrinus* Lemaire, Illustr. Hort. 8: Misc. 35. 1861.
> *Cereus baumannii colubrinus* Schumann, Gesamtb. Kakteen 133. 1897.
> *Cereus baumannii flavispinus* Schumann, Gesamtb. Kakteen 133. 1897.
> *Cleistocactus baumannii colubrinus* Riccobono, Boll. R. Ort. Bot. Palermo 8: 266. 1909.
> *Cleistocactus baumannii flavispinus* Riccobono, Boll. R. Ort. Bot. Palermo 8: 266. 1909.

Somewhat branching at base, 2 meters high or more, 2.5 to 3.5 cm. in diameter, dark green; ribs 12 to 16, low; areoles approximate, brown or black-felted; spines acicular, 15 to 20, white, yellow, or brown, 4 cm. long or less; flower orange to scarlet, 5 to 7 cm. long, narrow, 1 cm. in diameter, curved, with oblique limb; scales on ovary and flower-tube ovate, acute; perianth-segments short and broad, acute; stamens numerous, shortly exserted, appressed against the upper part of the flower-tube; fruit 1 to 1.5 cm. in diameter, red with white pulp.

Type locality: Not cited.

Distribution: Argentina; reported also from Paraguay and Uruguay.

Cereus subtortuosus Hortus (Förster, Handb. Cact. 409. 1846) was given as a synonym of *Cereus colubrinus.* *Cereus colubrinus flavispinus* Salm-Dyck (Cact. Hort. Dyck. 1844. 32. 1845) seems never to have been described though Schumann takes it up under *C. baumannii* and attributes it to Salm-Dyck. Förster in his Handbuch refers it as a synonym of *C. colubrinus.*

According to Weingart, *C. grossei* (Monatsschr. Kakteenk. 18: 8. 1908) is only a variety of this species, while *C. anguiniformis* (Monatsschr. Kakteenk. 18: 6. 1908) is true *C. baumannii.*

Illustrations: Blühende Kakteen 1: pl. 57; Monatsschr. Kakteenk. 13: 139; Rep. Mo. Bot. Gard. 16: pl. 9, f. 2 to 5; pl. 12, f. 2, all as *Cereus baumannii;* Curtis's Bot. Mag. 76: pl. 4498; Fl. Serr. 6: pl. 559; Loudon, Encycl. Pl. ed. 3. f. 19394, all as *Cereus tweediei.*

Plate XXVII, figure 2, shows a flowering top of a plant in the New York Botanical Garden.

2. **Cleistocactus smaragdiflorus** (Weber).

> *Cereus smaragdiflorus* Weber, Dict. Hort. Bois 281. 1894.
> *Cereus baumannii smaragdiflorus* Weber in Schumann, Gesamtb. Kakteen 134. 1897.

Stems slender, 2 to 2.5 cm. in diameter; ribs low, 12 to 14; radial spines numerous, acicular; central spines porrect, several, stouter, the longer ones 2 cm. long, yellowish to dark brown; flowers small, 4 to 5 cm. long, straight, a little constricted above the ovary, the tube and ovary red; upper scales on flower-tube and outer perianth-segments with a long mucro; perianth-segments small, green, acute to mucronate; filaments included; style slightly exserted; stigma-lobes 5 to 8; fruit globose, 1.5 cm. in diameter; seeds small, black.

Type locality: Not cited.

Distribution: Provinces of Jujuy, Salta, Catamarca, and La Rioja, Argentina.

Fig. 248.—Cleistocactus smaragdiflorus.

We have known little of this species until quite recently. In 1917 Dr. Shafer collected on a dry sandy bank at Caligua, Jujuy, a plant (No. 69) which was sent to the New York Botanical Garden, where it flowered while this volume was going through the press.

The flowers are so different from the typical species that there is some doubt in our minds whether it is a true *Cleistocactus*.

Through the kindness of Dr. Juan A. Dominguez, director of the Museo Farma Cologico at Buenos Aires, Dr. Rose was permitted to bring to the United States certain critical specimens for detail study. Among these plants were flowers of a *Cleistocactus*, straight and regular but much larger than those of *C. smaragdiflorus*. Unfortunately, only flowers were preserved. These may be described as follows: flowers 6 to 7 cm. long, straight; outer perianth-segments apiculate; inner perianth-segments oblong, obtuse or rounded; stamens and style exserted. The plant was collected by Fritz Claren in Jujuy, Department of Santa Catalina, altitude 3,400 to 4,300 meters, in 1901 (No. 11576). It is probably an undescribed species, but it deserves further study.

The name *Cereus colubrinus smaragdiflorus* Weber (Dict. Hort. Bois 281. 1894), without formal publication, is implied, but the name was not actually used until later (Monatsschr. Kakteenk. 15: 122. 1905).

Illustrations: Blühende Kakteen 2: pl. 87; Monatsschr. Kakteenk. 15: 123, both as *Cereus smaragdiflorus*.

Figure 248 is from a photograph of an Argentine specimen communicated by Dr. Spegazzini as typical.

3. Cleistocactus anguinus (Gürke).

Cereus anguinus Gürke, Monatsschr. Kakteenk. 17: 166. 1907.

Branches decumbent; ribs 10 or 11, low; radial spines 18 to 22, grayish but brownish at base and apex, slender; central spines 1 or 2, stouter than the radials, yellowish; flowers somewhat one-sided, tubular, 7 cm. long, orange-yellow, 7.5 cm. long; stamens exserted.

Type locality: Paraguay.
Distribution: Paraguay.

We have studied a small plant in the collection of the New York Botanical Garden received from the Berlin Botanical Garden in 1914; vegetatively this resembles *Cleistocactus baumannii*. We also refer here a plant collected by J. A. Shafer at Paraguavi, Paragnay, March 21, 22, 1917 (No. 144).

PUBLISHED SPECIES, KNOWN TO US ONLY FROM DESCRIPTION.

CLEISTOCACTUS LANICEPS (Schumann) Gosselin, Bull. Mens. Soc. Nice 44: 32. 1904.

Cereus laniceps Schumann, Gesamtb. Kakteen 93. 1897.

Upright, 4 meters high or less; branches 5 cm. thick; ribs 9, blunt; areoles large, 6 mm. in diameter or more; spines usually 3 at an areole in a vertical row, subulate, gray, about 1.5 cm. long; flowers from a single rib, 3.5 cm. long; ovary spherical, 5 mm. long, covered with subulate scales, these bearing copious brown wool in their axils; fruit red, woolly, 1 cm. in diameter. It was collected near Tunari, Bolivia, at 1,300 meters altitude.

CLEISTOCACTUS PARVISETUS (Otto) Weber in Gosselin, Bull. Mens. Soc. Nice 44: 46. 1904.

Cereus parvisetus Otto in Pfeiffer, Enum. Cact. 79. 1837.

Originally described from a Brazilian specimen grown at Berlin, as follows: Simple, slender, 12 to 15 mm. in diameter, erect, with 12 angles; ribs somewhat compressed; areoles close together, white; upper spines 4 or 5, brown; lower spines 6 to 8, white, hair-like.

According to Schumann this species is found in the Serra da Lapa, Minas Geraes, Brazil. We do not know its relationship, although Weber thought it was a *Cleistocactus*. It was introduced only once, probably by Riedel, and is not now in cultivation.

This comes from the same region as the species of *Leocereus* and should be compared with plants of that genus.

CLEISTOCACTUS HYALACANTHUS (Schumann) Gosselin, Bull. Mens. Soc. Nice 44: 33.   1904.

Cereus hyalacanthus Schumann, Gesamtb. Kakteen 101.   1897.

This is described as upright, less than 1 meter high; ribs 20, low, obtuse; areoles elliptic; spines in clusters of 25 or more, the longest 2 cm. long, acicular, white, puberulent; flowers somewhat curved, 3 to 3.5 cm. long; ovary covered with numerous scales bearing copious brown wool in their axils.

It is known only from specimens collected by Otto Kuntze in the Province of Jujuy, Argentina.

CLEISTOCACTUS PARVIFLORUS (Schumann) Gosselin, Bull. Mens. Soc. Nice 44: 32.   1904.

Cereus parviflorus Schumann, Gesamtb. Kakteen 100.   1897.

Described as columnar, 2 to 3 meters high; branches 3 cm. in diameter; ribs 12, deeply marked by transverse furrows; radial spines 5 to 7, the longest one 4 mm. long, subulate, dark yellow; flowers from only a single rib, one above another, 2.5 to 3 cm. long; ovary covered with short, oblong to triangular scales bearing in their axils felt; fruit yellow, 1 cm. in diameter.

Collected near Parotani, Bolivia, by Otto Kuntze.

35. ZEHNTNERELLA gen. nov.

Tall and slender, much branched at base; ribs numerous, very spiny; flowers scattered along the upper part of the stem, 1 from an areole, perhaps night-blooming, very small; tube short but definite, about the length of the throat; base of throat filled with a ring of long white hairs; inner perianth-segments minute, white; ovary and flower-tube covered with small scales, their axils filled with hairs; fruit small, globular; seeds minute, tuberculately roughened, brownish to blackish, with a large basal slightly depressed hilum.

Named for Dr. Leo Zehntner, formerly of the Horto Florestal, Joazeiro, Brazil, who has furnished us specimens and valuable information concerning many of the cacti from this region.   It is a great pleasure to name a genus for this very keen observer, who has done such valuable work in Brazil, often under very trying circumstances. It is based upon a plant which Dr. Rose collected with him on the hills east of Joazeiro, Bahia, June 4, 1915 (No. 19760).   Our plant may be the same as *Cereus squamosus* Gürke (Monatsschr. Kakteenk. 18:70. 1908). A photograph of this species was reproduced (Bot. Jahrb. Engler 40: Beibl. 93: pl. 10) and this resembles *Zehntnerella squamulosa*, but the detailed description of *Cereus squamosus* does not wholly agree with it, and we have been unable to examine the type specimen of *Cereus squamosus*.

FIG. 249.—Zehntnerella squamulosa.

1. Zehntnerella squamulosa sp. nov.

Trunk, when present, 15 to 20 cm. in diameter, but usually a cluster of branches arising from the base; branches usually strict, 4 meters long or more, 5 to 7 cm. in diameter, covered with a mass of spines; ribs 17 to 20, low, close together; areoles circular, small; spines 10 to 15, acicular, chestnut-brown, the longest ones 3 cm. long; flowers small, 3 cm. long; inner perianth-segments oblong, 4 mm. long; lower scales on the ovary ovate, apiculate, 1 to 4 mm. long, the upper ones becoming oblong, all glabrous, the hairs in the axils white; fruit about 2 cm. in diameter, crowned by the withered perianth; seeds 1 mm. long.

This plant was common in a restricted rocky out-crop east of Joazeiro, called the Serra do Atoleiro, where flowers and photographs were obtained by Dr. Rose, June 4, 1915 (No. 19760, type) and ripe fruit was collected at the same locality by Dr. Zehntner in October 1917.

This species is called facheiro preto in Bahia.

Figure 249 is from a photograph taken by Paul G. Russell in Bahia, Brazil, in 1915; figure 250 shows a flower of the plant photographed.

FIG 250.—
Flower of
Z. squamulosa. Natural size.

36. **LOPHOCEREUS** (Berger) Britton and Rose, Contr. U. S. Nat. Herb. **12**: 426. 1909.

Columnar, stout cacti, the stems simple or with few branches, or much branched at base; ribs few; areoles on lower part of stem very different from the upper ones; flowering areoles large, felted, developing long bristles standing out at right angles to the axils of the stem; flowers usually several at each areole, small, funnelform, with short narrow tubes, nocturnal, beginning to open at about six o'clock at night and by eight or nine fully expanded, but closed the following morning, odorless; outer perianth-segments greenish; inner perianth-segments pink; stamens short, included; fruit small, red, globular, when mature bursting irregularly, glabrous or with a few spines and some felt in axils of lower scales; seeds numerous, small, black, shining, with a depressed basal hilum.

Type species: *Cereus schottii* Engelmann.

Three species have been recognized in this genus, all from the same floral region, which we now regard as reducible to one.

The generic name is from the Greek, signifying crested-cereus, with reference to the bristly top of the flowering stem.

1. **Lophocereus schottii** (Engelmann) Britton and Rose, Contr. U. S. Nat. Herb. **12**: 427. 1909.

Cereus schottii Engelmann, Proc. Amer. Acad. **3**: 288. 1856.
Pilocereus schottii Lemaire, Rev. Hort. **1862**: 428. 1862.
Cereus sargentianus Orcutt, Gard. and For. **4**: 436. 1891.
Pilocereus sargentianus Orcutt in Schumann, Monatsschr. Kakteenk. **2**: 76. 1892.
Cereus palmeri Engelmann in Coulter, Contr. U. S. Nat. Herb. **3**: 401. 1896.
Cereus schottii australis K. Brandegee, Zoe **5**: 3. 1900.
Lophocereus australis Britton and Rose, Contr. U. S. Nat. Herb. **12**: 427. 1909.
Lophocereus sargentianus Britton and Rose, Contr. U. S. Nat. Herb. **12**: 427. 1909.

Usually branching only at base, forming large clumps sometimes with as many as 50 or even 100 upright or ascending stems, 1 to 7 meters high; ribs usually 5 to 7, but sometimes 9, separated by broad intervals; bristles of the flowering areoles numerous, straight, finely acicular, gray, 6 cm. long or less; flowerless areoles smaller, little felted, with 3 to 7 short subulate spreading radial spines swollen at base and 1 or 2 central ones a little longer and stouter; flowers 3 to 4 cm. long; style, stigma-lobes, and filaments whitish; fruit 2 to 3 cm. in diameter, usually naked, rarely spiny; seeds 2.5 mm. long.

Type locality: In Sonora, toward Magdalena, Mexico.

Distribution: Southern Arizona, Sonora, and Lower California.

As with many other columnar cacti, this is sometimes used for fences. It is usually called sinita, with various spellings.

This species is remarkable among cacti on account of the long bristle-like spines, which develop at the ends of the flowering branches, giving the plant the appearance of bearing terminal brushes. This modification of the spines from the flowering areoles is similar to the changes we see in certain other genera such as *Cephalocereus, Arrojadoa*, and to a less extent in *Carnegiea*, on account of which both this species and *Carnegiea gigantea* have been referred by some authors to the genus *Pilocereus*.

Lophocereus schottii inhabits parts of western Mexico and southern Arizona, which have great aridity, but it usually grows in colonies and in this way seems to withstand the rigor of the desert. Its range is more extensive than that of most cacti and it shows considerable variability. Three species of *Lophocereus* have been described, but appear to be merely geographical races of this one.

Illustrations: MacDougal, Bot. N. Amer. Des. pl. 8; Cact. Mex. Bound. pl. 74, f. 16, as *Cereus schottii;* Schumann, Gesamtb. Kakteen f. 37, 38, as *Pilocereus schottii;* Orcutt, Gard. and For. 4: f. 69, as *C. sargentianus;* Monatsschr. Kakteenk. 5: 86, as *P. sargentianus.*

FIG. 251.—Lophocereus schottii.

Figure 251 is from a photograph obtained by Edward Palmer near Guaymas, Sonora; figure 252 shows a section through the upper part of a flowering stem collected by Dr. Rose at Abreojos Point, Lower California, in 1911; figure 253 shows a flower of a plant brought by Dr. MacDougal from Arizona to the New York Botanical Garden in 1902.

37. MYRTILLOCACTUS Console, Boll. R. Ort. Bot. Palermo 1: 8.   1897.

Large cacti, usually with short trunks and large, much branched tops, the stout, few-ribbed branches nearly erect, all the areoles bearing the same kind of spines; flowers diurnal, very small, several, sometimes as many as 9 at an areole, with very short tubes and widely spreading perianth-segments; ovary bearing a few minute scales with tufts of wool in their axils, spineless; fruit small, globular, edible; seed very small, black, with basal hilum.

Type species: *Cereus geometrizans* Martius.

This genus has no very close allies. We have grouped it with *Lophocereus* and the following genus, because they likewise have more than 1 flower from an areole, but otherwise little else in common. The small flowers somewhat resemble orange flowers, having scarcely any tubes; the short stamens are almost entirely exserted. The fruits are small berries.

We know 4 closely related species, natives of Mexico and Guatemala. The name is from the Greek, signifying berry-cactus, referring to the small fruit.

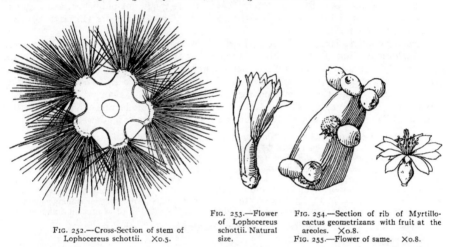

FIG. 253.—Flower of Lophocereus schottii. Natural size.

FIG. 252.—Cross-Section of stem of Lophocereus schottii. ×0.5.

FIG. 254.—Section of rib of Myrtillocactus geometrizans with fruit at the areoles. ×0.8.
FIG. 255.—Flower of same. ×0.8.

KEY TO SPECIES.

Young branches very blue; central spine elongated, reflexed, dagger-like........................1. *M. geometrizans*
Young branches green; central spine not dagger-like.
    Spines usually 3 to 5, ascending, with no definite central spine or, when present, very short..2. *M. cochal*
    Spines 6 or more, with definite central spine.
        Radial spines 5; fruit oblong, 10 to 15 mm. long....................................3. *M. schenckii*
        Radial spines more than 5; fruit globular, 6 mm. in diameter........................4. *M. eichlamii*

1. **Myrtillocactus geometrizans** (Martius) Console, Boll. R. Ort. Bot. Palermo 1: 10. 1897.

    Cereus geometrizans Martius in Pfeiffer, Enum. Cact. 90. 1837.
    Cereus pugioniferus Lemaire, Cact. Aliq. Nov. 30. 1838.
    Cereus gladiator Otto and Dietrich, Allg. Gartenz. 6: 34. 1838.
    Cereus geometrizans pugioniferus Salm-Dyck, Cact. Hort. Dyck. 1849. 48. 1850.
    Cereus geometrizans quadrangularispinus Lemaire in Labouret, Monogr. Cact. 367. 1853.

Tree-like, with a short definite trunk crowned by a large, much branched top; branches often a little curved, bluish green, usually 5 or 6-ribbed, 6 to 10 cm. in diameter, very blue when young; ribs 2 to 3 cm. high, rounded; areoles 2 to 3 cm. apart; radial and central spines very different, almost filling the areoles; radial spines usually 5, rarely 8 or 9, usually short, 2 to 10 mm. long, but sometimes 3 cm. long, more or less turned backward, a little flattened radially but swollen at base; central spine elongated, dagger-shaped, flattened laterally, 1 to 7 cm. long and sometimes 6 mm. broad; flowers appearing from the upper part of the areole, 2.5 to 3.5 cm. broad, the limb 3 to 4 times as long as the tube; perianth-segments oblong, 1.5 cm. long; stamens numerous, erect, exserted; fruit ellipsoid to subglobose, edible, purplish or bluish, 1 to 2 cm. long.

Type locality: Mexico.
Distribution: San Luis Potosí to Oaxaca, Mexico.

This cactus is very common on the Mexican tableland. The fruits, known as garrambullas, are to be found in all the Mexican markets, and are eaten both fresh and dried; the dried fruits very much resemble raisins in appearance and are used in much the same way.

The name *Cereus pugioniferus quadrangulispinus* Lemaire is given by Förster (Handb. Cact. 395. 1846) as a synonym for *C. pugioniferus; Cereus geometrizans quadrangulispinus* Lemaire is given only by name by Salm-Dyck (Cact. Hort. Dyck. 1849. 48. 1850); Labouret (Monogr. Cact. 366. 1853) gives *Cereus gladiator geometrizans* Monville (Cat. 1846) as a synonym for *C. geometrizans pugioniferus* Salm-Dyck; *Cereus arrigens* Monville and *C. gladiger* Lemaire are both given by Labouret (Monogr. Cact. 367. 1853) as synonyms for the variety *Cereus geometrizans quadrangularispinus* Lemaire.

Cereus aquicaulensis (Pfeiffer, Enum. Cact. 90. 1837) is not published, but is given only as a synonym of this species. *Cereus quadrangulispinus* Lemaire (Linnaea 19: 363. 1846) is only a name.

Cereus garambello Haage in Förster (Handb. Cact. 433. 1846), unpublished, belongs here.

Illustrations: Boll. R. Ort. Bot. Palermo 1: f. 1 to 4; Contr. U. S. Nat. Herb. 12: pl. 72; Safford, Ann. Rep. Smiths. Inst. 1908: pl. 9, f. 2; pl. 11, f. 1. Ber. Deutsch. Bot. Ges. 15: pl. 2, f. 1; Schumann, Gesamtb. Kakteen f. 23, these last two as *Cereus geometrizans*.

Plate XXVI, figure 1, is from a photograph taken by Dr. MacDougal near Tehuacán, Mexico, in 1908. Figure 254 shows a section of a rib and the small fruits of a plant collected by Edward Palmer at San Luis Potosí in 1905 and figure 255 shows its flower.

2. **Myrtillocactus cochal** (Orcutt) Britton and Rose, Contr. U. S. Nat. Herb. 12: 427. 1909.

Cereus cochal Orcutt, West Amer. Sci. 6: 29. 1889.
Cereus geometrizans cochal K. Brandegee, Zoe 5: 4. 1900.

Plant 1 to 3 meters high, much branched; trunk short, woody, sometimes 3 dm. in diameter; ribs 6 to 8, obtuse, separated by shallow intervals; spines grayish to black; radial spines 5, short; central spines when present 2 cm. long; flowers open night and day, 2.5 cm. long and fully as broad; perianth-segments usually 16, light green, the outer ones tinged with purple, oblong; filaments white; stigma-lobes 5 or 6, white; fruit edible, slightly acid, globular, 12 to 18 mm. in diameter, red.

Type locality: Todos Santos Bay, Lower California.

Distribution: Lower California.

This species is called cochal by the Indians of Lower California, who use the stems for firewood and are said to eat the fruit.

Illustration: Monatsschr. Kakteenk. 5: 74, as *Cereus cochal*.

3. **Myrtillocactus schenckii** (J. A. Purpus) Britton and Rose, Contr. U. S. Nat. Herb. 12: 427. 1909.

Cereus schenckii J. A. Purpus, Monatsschr. Kakteenk. 19: 38. 1909.

Tree-like, 3 to 5 meters high, with a very stout trunk and many short ascending branches, dark green; areoles circular, crowded with black felt, about 5 mm. apart; radial spines 6 to 8, straight, 5 to 12 mm. long, black or brownish; central spine 1, usually 2 cm. long, sometimes 5 cm. long; fruits small, oblong, 10 to 15 mm. long, naked; seeds black, pitted.

Type locality: Sierra de Mixteca, Puebla, Mexico.

Distribution: Puebla and Oaxaca, Mexico.

In habit and fruit this species is very similar to the well-known *Cereus geometrizans*, but differs from it greatly in color of stem and in the areoles and spines.

Illustrations: Monatsschr. Kakteenk. 19: 39, as *Cereus schenckii;* Contr. U. S. Nat. Herb. 12: pl. 73.

Plate XXVI, figure 2, shows a photograph taken by Dr. D. T. MacDougal between Mitla and Oaxaca City in 1906.

4. **Myrtillocactus eichlamii** sp. nov.

Branches strongly 6-angled, deep green or slightly glaucous; ribs obtuse; areoles 2 cm. apart, large, circular, with grayish wool at time of flowering; radial spines 5, bulbose at base; central spine 1, a little longer than the radials; flower-buds dark purple; outer perianth-segments greenish with red tips; inner perianth-segments creamy white, about 10, spreading almost at right angles to the tube; stamens numerous, pale, somewhat spreading; style white, a little longer than the stamens; flowers fully open at half past nine o'clock in the morning, deliciously fragrant; fruit small, globular, 6 mm. in diameter, wine-colored, naked except a few small scales.

1. *Myrtillocactus geometrizans*, Tehuacán, Mexico.
2. *Myrtillocactus schenckii*, near Mitla, Mexico.

This species is described from a plant sent by the late Federico Eichlam, in 1909, from Guatemala, which flowered in Washington, April 1910. It differs from *Myrtillocactus geometrizans* in its larger, greener branches and different armament.

Figure 256 shows a flower and two fruits of the type plant.

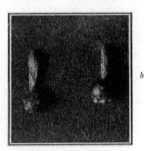

a   *b*

FIG. 256.—Myrtillocactus eichlamii.   Flower and fruits.   Natural size.

38. NEORAIMONDIA gen. nov.

A very stout cactus, the stems branched at base; branches erect, columnar, few-ribbed, the ribs separated by broad intervals, very spiny; areoles brown-felted, enormously developed, thick, the flowering ones sometimes elongated and branched, forming cephalium-like masses, these spineless but with ridges of short brown felt; flowers 2 on the areoles or solitary, funnelform, the tube stout, longer than the limb; perianth-segments oblong; scales of the ovary and flower-tube with short brown wool in their axils; fruit ellipsoid to globular, covered with globular areoles with short brown wool and short spines; seed dull black with pitted surface and a depressed hilum.

A monotypic genus of western Peru. It is named in honor of Antonio Raimondi (1825–1890), the great geographer and naturalist of Peru.

Berger says of the plant, which he referred to the subgenus *Eulychnia* (Rep. Mo. Bot. Gard. **16**: 68. 1905):

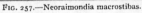

FIG. 257.—Neoraimondia macrostibas.

FIG. 258.—Flower and enlarged areole of N. macrostibas.   X0.7.

"But I am quite aware that in this form the subgenus *Eulychnia* is more artificial than natural. For instance, *C. macrostibas* A. Berger differs greatly from the rest, especially by its enlarged and prolonged flowering areoles.   But the material at hand is so scanty that I must refrain from any further statement."   "*Cereus macrostibas* A. Berger was originally described by Schumann as a *Pilocereus*, and as such it is another heterogeneous form of this conglomerate genus."

1. Neoraimondia macrostibas (Schumann).

Pilocereus macrostibas Schumann, Monatsschr. Kakteenk. **13**: 168.   1903.
Cereus macrostibas Berger, Rep. Mo. Bot. Gard. **16**: 69.   1905.

Plant with many branches arising from near the base, 2 to 4 meters high; areoles 1 to 2 cm. apart, very large, rarely less than 1 cm. broad, often globular or on old plants elongating into subcylindric spur-like bodies 10 cm. long or less; spines 12 or more at an areole, very unequal, the central ones often elongated on the old part of the stem, sometimes 24 cm. long; flowers 2.5 to 4 cm. long; inner perianth-segments about 1 cm. long; filaments numerous, short, included, white; style short, white; stigma-lobes pinkish; fruit sometimes 7 cm. in diameter, purple, the brown-woolly areoles finally falling off as little balls; pulp red, edible; seeds numerous.

Type locality: Near Mollendo, Peru.

Distribution: Throughout western Peru.

In 1914, Dr. Rose studied the plant in its native habitat and collected complete specimens. It is one of the most remarkable of all cacti in its very stout, few-ribbed branches, immense brown areoles and greatly elongated spines, these, perhaps, the longest of any. These areoles doubtless produce flowers year after year and the indication is that the largest of these areoles must be of great age. The plant itself must grow very slowly, for it is found only on the

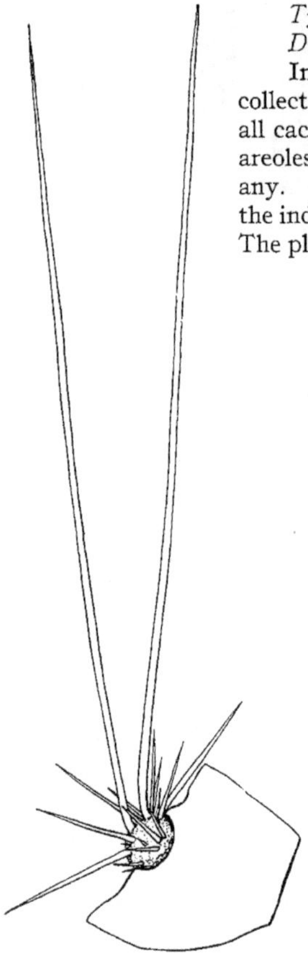

FIG. 259.—Cluster of spines of Neoraimondia macrostibas. ×0.6.

FIG. 260.—Neoraimondia macrostibas.

borders of the barren Peruvian deserts, where its water supply is very meager. The elongated spines (24 cm. long) are the longest we have seen in any cacti, although *Cereus jamacaru* is reported to have spines 30 cm. long, but the longest we have measured were only 19 cm. long.

The unusual specific name given to this plant probably refers to the peculiar areoles.

Illustrations: Monatsschr. Kakteenk. **13**: 168, 169, both as *Pilocereus macrostibas.*

Figure 257 is from a photograph taken by C. H. T. Townsend near Chosica, Peru; figure 258 shows a flower and young fruit on the much enlarged areole, collected by Dr. Rose near Arequipa in 1914; figure 259 shows a cluster of spines obtained by Dr. Rose at Chosica, Peru; figure 260 is from a photograph of a top of the Chosica plant brought by Dr. Rose to the New York Botanical Garden.

Subtribe 2. HYLOCEREANAE.

Elongated, vine-like, climbing, trailing or pendent, branched cacti, the stems and branches angled, ribbed, fluted, or rarely flat, the joints emitting aërial roots, the areoles usually spiny; flowers mostly large and white, rarely red or pink; perianth-limb regular, or in *Aporocactus* more or less oblique; fruit a fleshy berry, often large.

We group the species known to us in 9 genera.

KEY TO GENERA.

A. Joints angled, ribbed, winged, or fluted.
 Ovary and fruit covered with large foliaceous scales, their axils neither spiny, hairy, nor bristly;
  flowers mostly large, nocturnal; stems and branches 3-angled or 3-winged.
   Perianth-tube elongated; flowers very large, their scales naked in the axils.......1. *Hylocereus* (p. 183)
   Perianth-tube scarcely any; flowers small, some of their scales with tufts of short
    hairs and occasional bristles in the axils.......................2. *Wilmattea* (p. 195)
 Ovary and fruit not bearing large foliaceous scales, their axils spiny, hairy, or bristly.
  Flowers elongate-funnelform, very large, mostly nocturnal, the tube and ovary
   usually bearing scales, hairs, or spines.
    Stems ribbed, fluted, or angled...3. *Selenicereus* (p. 196)
    Stems winged.
     Areoles of ovary and flower-tube bearing felt and spines subtended by
      short scales; flowers nocturnal........................4. *Mediocactus* (p. 210)
     Areoles of ovary and flower-tube bearing long hairs; flowers diurnal.....5. *Deamia* (p. 212)
  Flowers short-funnelform or funnelform-campanulate.
   Perianth-limb regular, the tube stout; flowers white.
    Tube of the flower bearing short foliaceous scales; areoles of the tubercu-
     late ovary bearing long hairs............................6. *Weberocereus* (p. 214)
    Areoles of flower-tube and of non-tuberculate ovary bearing short black
     spines..7. *Werckleocereus* (p. 216)
   Perianth-limb somewhat oblique, the tube slender; flowers pink...........8. *Aporocactus* (p. 217)
AA. Joints flat..9. *Strophocactus* (p. 221)

1. HYLOCEREUS (Berger) Britton and Rose, Contr. U. S. Nat. Herb. **12:** 428. 1909.

Climbing cacti, often epiphytic, with elongated stems normally 3-angled or 3-winged, and branches emitting aërial roots, the areoles bearing a tuft of felt and several short spines, or spineless in one species; areoles on seedlings and juvenile growths often bearing bristles; flowers very large, nocturnal, funnelform, the limb as broad as long, and as long as the tube or longer; ovary and tube bearing large foliaceous scales but no spines, felt, wool, or hairs; outer perianth-segments similar to the scales on the tube, but longer; petaloid perianth-segments narrow, acute or acuminate, mostly white, rarely red; stamens very many, in two series, equaling or shorter than the style; style cylindric, rather stout and thick, the linear stigma-lobes numerous, simple or branched; fruit spineless but with several or many persistent foliaceous scales mostly large and edible; seeds small, black; cotyledons large, flattened above, thick, ovate, acute, connate at base.

We know 18 species, natives of the West Indies, Mexico, Central America, and northern South America. Most of them are closely related, having similar stems, flowers, and fruits.

Type species: *Cactus triangularis* Linnaeus.

The name is from the Greek, meaning forest-cereus.

KEY TO SPECIES.

A. Areoles spiniferous; ribs not deeply crenate.
 B. Stems bluish or more or less whitened or gray.
  Margin of joints horny.
   Spines short, conic.
    Outer perianth-segments acuminate, as long as the inner, white ones.... 1. *H. guatemalensis*
    Outer perianth-segments acute, much shorter than inner, golden-tipped ones. 2. *H. purpusii*
   Spines acicular, slender.
    Outer perianth-segments linear-lanceolate, acuminate................. 3. *H. ocamponis*
    Outer perianth-segments oblong-lanceolate, obtuse.................... 4. *H. bronxensis*
  Margin of joints not horny; spines few, conic.
   Branches slender, 4 cm. thick or less, scarcely crenate.
    Stigma-lobes entire.. 5. *H. polyrhizus*
    Stigma-lobes bifid.. 5a. *H. venezuelensis*
   Branches stout, 5 to 10 cm. thick................................... 6. *H. costaricensis*

KEY TO SPECIES—continued.

A. Areoles spiniferous; ribs not deeply crenate—*continued*.
  BB. Stems bright green.
    Margin of stems horny.
      Ribs of stem broad, thin, crenate.................................... 7. *H. undatus*
      Ribs of stem thick, scarcely crenate................................. 8. *H. cubensis*
    Margin of stems not horny.
      Stigma-lobes branched or forked.
        Spines several; margins of stem nearly straight; stigma-lobes branched.. 9. *H. lemairei*
        Spine one; margins of stem undulate; stigma-lobes forked (at least some-
          times)...10. *H. monacanthus*
      Stigma-lobes (so far as known) entire.
        Perianth-segments red or reddish purple.
          Ribs thin, almost wing-like; perianth-segments linear..............11. *H. stenopterus*
          Joints angular, not winged; inner perianth-segments oblanceolate...12. *H. extensus*
        Inner perianth-segments white.
          Scales on the ovary few and scattered.........................13. *H. napoleonis*
          Scales on the ovary brown, large, imbricated.
            Joint-angles strongly tubercled............................14. *H. trigonus*
            Joint-angles scarcely tubercled or not at all.
              Joints somewhat crenate...............................15. *H. triangularis*
              Joints not crenate....................................16. *H. antiguensis*
AA. Areoles without spines; ribs very deeply crenate.................................17. *H. calcaratus*

1. Hylocereus guatemalensis (Eichlam).

Cereus trigonus guatemalensis Eichlam, Monatsschr. Kakteenk. **21**: 68. 1911.

Stems high-climbing, 3 to 5 meters long, stout, mostly 3-angled, 2 to 7 cm. broad, the basal parts often narrow or nearly terete; joints beautifully glaucous or in time becoming more or less greenish; ribs low-undulate, the margins horny, not at all glaucous; areoles 2 cm. apart or less; spines 2 to 4, 2 to 3 mm. long, conic, dark, but on seedlings numerous and bristle-like; flowers 3 dm. long; outer perianth-segments rose-colored, acuminate; inner perianth-segments lanceolate, acute, white; style yellow, 7 mm. in diameter; stigma-lobes 25, entire; fruit 6 to 7 cm. in diameter, covered with large scales; seeds black.

Type locality: Guatemala.
Distribution: Guatemala.

We have grown plants from seeds; the seedlings are erect, 4-angled, the spines numerous, the bristles white, the cotyledons 15 mm. long.

Illustration: Monatsschr. Kakteenk. **23**: 155, as *Cereus trigonus guatemalensis*.

Figure 261 shows a joint of a plant sent to the New York Botanical Garden from Fiscal, Guatemala, by C. C. Deam.

FIG. 261.—Joint of H. guatemalensis. X0.5.

2. Hylocereus purpusii (Weingart).

Cereus purpusii Weingart, Monatsschr. Kakteenk. **19**: 150. 1909.

Stems bluish, climbing, elongate, epiphytic; ribs 3 or 4, with horny margins only slightly undulate; areoles small; spines 3 to 6, short; flowers large, 25 cm. long and nearly as broad when fully expanded; outer perianth-segments narrow, purplish; middle perianth-segments golden; inner perianth-segments broad, white except at the golden tips.

Type locality: Near Tuxpan, Mexico.
Distribution: Lowlands of western Mexico.

We have grown this plant but have not seen the flowers, our description of them being founded on that of Mr. Weingart.

Illustrations: Monatsschr. Kakteenk. **22**: 26, 27, both as *Cereus purpusii*.

3. Hylocereus ocamponis (Salm-Dyck) Britton and Rose, Contr. U. S. Nat. Herb. **12**: 429. 1909.

Cereus ocamponis Salm-Dyck, Cact. Hort. Dyck. 1849. 220. 1850.

Stems strongly 3-angled, at first bright green, soon glaucous, dull bluish green in age; ribs rather deeply undulate, their margins with a horny, brown border; areoles 2 to 4 cm. apart, borne near the

M. E. Eaton del

1. End of fruiting branch of *Arrojadoa rhodantha*.
2. Top of plant of *Cleistocactus baumannii*.
3. Flower on branch of *Hylocereus stenopterus*.
(All natural size.)

bottom of each undulation; spines 5 to 8, acicular, 5 to 12 mm. long; flowers 25 to 30 cm. long and fully as broad; outer perianth-segments narrow, long-acuminate, greenish, spreading or reflexed; inner perianth-segments oblong, acuminate, white; style stout; stigma-lobes linear, entire, green; ovary covered with imbricated, ovate, acute, purplish-margined scales.

Type locality: Mexico or Colombia.

Distribution: Mexico?

The above flower description is drawn from New York Botanical Garden specimens which bloomed in 1912 (No. 6170). The species is known to us from cultivated plants only.

Mr. Weingart is strongly of the opinion that *Cereus napoleonis* Graham is the same as this species and, if so, this name should be used. He states that *C. napoleonis* was described from an old plant, while the other species was described from young plants, which, he thinks, would account for the differences in the descriptions. We believe, however, that the two species are distinct and that *C. napoleonis* is much nearer *H. triangularis.*

Illustration: Monatsschr. Kakteenk. 23: 29, as *Cereus ocamponis.*

Plate xxviii shows a flowering joint of a plant in the collection of the New York Botanical Garden.

A species related to *H. ocamponis* but probably distinct was collected by T. S. Brandegee on rocks of Cerro Colorado, Sinaloa, Mexico, November 1904. Mr. Brandegee states that it is also epiphytic on trees. Rose, Standley, and Russell collected the same species at Villa Union, near Mazatlan, in 1910, but, although we have had it in our collections ever since, it has not yet flowered.

4. Hylocereus bronxensis sp. nov.

Joints strongly 3-angled, dull grayish green, 3 to 4 cm. broad; ribs strongly undulate, the margins horny and brown; areoles 2 to 3 cm. apart; spines about 10, acicular, brown in age, about 6 mm. long; flowers 25 cm. long; outer perianth-segments broad, ovate, obtuse or rounded; inner perianth-segments oblong, rounded at apex, more or less apiculate, but not long-acuminate; scales on the ovary broad; stigma-lobes (perhaps) bifid.

Described from specimens which flowered in the New York Botanical Garden (No. 9722) June 28, 1912. The plant was obtained from G. E. Barre in 1902, but its origin is otherwise unknown. It is related to *Hylocereus ocamponis* but its flowers are quite different from those of that species.

5. Hylocereus polyrhizus (Weber).

Cereus polyrhizus Weber in Schumann, Gesamtb. Kakteen 151. 1897.

Slender vines, sometimes only 3 to 4 cm. thick, normally 3-angled, at first green or purplish, but soon becoming white and afterwards green again; ribs or wings comparatively thin although in age becoming more turgid; margin nearly straight, obtuse, not horny; spines 2 to 4, rather stout, brownish, 2 to 4 mm. long, sometimes accompanied by two white hairs or bristles which finally drop off; young flower-buds globular, purple; flowers 2.5 to 3 dm. long or longer, strongly fragrant; outer perianth-segments linear-lanceolate, more or less reddish, especially at the tips; inner perianth-segments nearly white; stigma-lobes rather short, yellowish, entire; ovary covered with approximate ovate scales, with red or deep purple margins; fruit scarlet, oblong, 10 cm. long.

Type locality: Colombia.

Distribution: Colombia and Panama.

The original description of *Cereus polyrhizus* was, apparently, based on the juvenile state of the species for the branches are described as 5-angled; Weingart (Monatsschr. Kakteenk. 22: 106) associates the plant with the group in which we place it, and plants sent to the New York Botanical Garden in 1901 by M. Simon of St. Ouen, Paris, who had in his collection many cacti described by Weber, are, apparently, the same as others since obtained from Panama and Colombia; perhaps also from Ecuador.

We have referred here the 3 specimens collected by Dr. Rose in Ecuador although we are not sure that these are even conspecific. They all grow in very diverse habitats; only one was seen in flower. No. 22116 was found growing closely appressed to the trunk of a tree to which it was so tightly attached that it was with difficulty that specimens were obtained. The locality was on the edge of the mangrove swamp near Guayaquil. In the same region were seen other plants, presumably of the same species but these were clambering from tree to tree high up in their tops and far out of reach. No. 23342 was in a very peculiar habitat for a *Hylocereus*. It came from the edge of the Catamayo Valley, a hot semiarid region. Its stems were very stout, almost woody, and were spread out all over the top of a small tree. No flowers or fruit were seen and only a single plant was observed. The branches were nearly 10 cm. broad and the brown spines were usually 4 in a cluster and nearly 1 cm. long. On the other hand No. 23396 was found in a habitat very suitable for a *Hylocereus;* this was in a tree along a stream east of Portovelo in southern Ecuador; the plant was in flower but almost out of reach so that it was with difficulty we obtained a single flower. The following brief notes are based on our field observations:

Stems 3-angled, whitish; flowers 31 cm. long, fragrant; outermost segments short, purple; outer scales oblong, orange-red; inner perianth-segments white, tinged with pink; stamens yellow; scales on the ovary oblong, acute, dull green, with purple margins.

5a. Hylocereus venezuelensis sp. nov. (See Appendix, p. 226.)

6. Hylocereus costaricensis (Weber) Britton and Rose, Contr. U. S. Nat. Herb. **12**: 428. 1909.

    Cereus trigonus costaricensis Weber, Bull. Mus. Hist. Nat. Paris **8**: 457. 1902.

Vigorous vines, perhaps the stoutest of the genus, sometimes 10 cm. broad, normally 3-angled, at first green or purplish, but soon becoming white and afterwards green or gray; ribs or wings comparatively thin although in age becoming more turgid; margin rather variable, either straight or somewhat undulate, obtuse, never horny; spines 2 to 4, short, rather stout, brownish, usually accompanied by two white hairs or bristles which finally drop off; young flower-buds globular, purple; flowers 3 dm. long or more, strongly fragrant; outer perianth-segments narrow, more or less reddish, especially the tips; inner perianth-segments pure white; stigma-lobes rather short, yellowish, entire; ovary covered with closely set scales, these having deep purple margins; fruit scarlet, oblong, 10 cm. long.

FIG. 262.—Ovary of Hylocereus costaricensis transformed into branch. ×0.94.

Type locality: Costa Rica.
Distribution: Costa Rica.

    This species was originally described as a variety of *Cereus trigonus*, but it has much stouter blue stems and is otherwise different. It grows well in cultivation and frequently flowers. The very young areoles on the stem produce an abundance of nectar which runs down the stem in large sticky drops.
reduced leaves from the lower areoles.

    Figure 262 represents an arrested flower transformed into a branch showing scales or

M. E. Eaton del.

Flower and end of branch of *Hylocereus ocamponis*. × 0.7.

7. **Hylocereus undatus** (Haworth) Britton and Rose in Britton, Flora Bermuda 256. 1918.

Cactus triangularis aphyllus Jacquin, Stirp. Amer. 152. 1763.
Cereus triangularis major De Candolle, Prodr. 3: 468. 1828.
Cereus undatus Haworth, Phil. Mag. 7: 110. 1830.
Cereus tricostatus Gosselin, Bull. Soc. Bot. France 54: 664. 1907.
Hylocereus tricostatus Britton and Rose, Contr. U. S. Nat. Herb. 12: 429. 1909.

Stem long, clambering over bushes and trees or creeping up the sides of walls; ribs mostly 3, broad, thin, green; margin usually strongly undulate, more or less horny in age; areoles 3 to 4 cm. apart; spines 1 to 3, small, 2 to 4 mm. long; flowers up to 29 cm. long or more; outer perianth-segments yellowish green, all turned back, some strongly reflexed; inner perianth-segments pure white, erect, broad, oblanceolate, entire, with apiculate tips; filaments slender, cream-colored; stigmalobes as many as 24, slender, entire, cream-colored; style stout, 7 to 8 mm. in diameter, cream-colored; fruit oblong, 10 to 12 cm. in diameter, red, covered with large foliaceous scales, or nearly smooth when mature, edible; seeds black.

Type locality: China, evidently in cultivation.

Distribution: Common throughout the tropics and subtropics; often found as an escape and widely cultivated.

This species has long been known in cultivation under the name of *Cereus triangularis*, and it is to be regretted that the name *triangularis* can not be retained, but the plant which Linnaeus described as *Cactus triangularis* came from Jamaica. The latter is now well known to botanists but it has never been much cultivated, while *H. undatus* is grown all over the world and grows half-wild in all tropical countries. It is the best known of all the night-blooming cereuses and has one of the largest flowers. It makes a beautiful hedge plant; in Honolulu there is a hedge about Punahou College which is half a mile long and is said to produce 5,000 flowers in a single night.

Cereus undatus was described by Haworth from plants sent from China; he says it is similar to *C. triangularis*, but twice as large. Pfeiffer afterwards made it his variety *major* of *C. triangularis*, which Schumann referred doubtfully to *C. napoleonis*.

In the New York Botanical Garden herbarium are specimens of a *Hylocereus* collected on Martinique in 1884 by Père Duss (No. 904), which have the horny-margined ribs and large white flowers of this species. From this island Jacquin in 1763 described a variety *aphyllus* of *Cactus triangularis* from the mountain forests, which may very likely be this species, in which case Martinique may be the home of this widely cultivated plant.

FIG. 263.—Hylocereus undatus.

Two forms of this species are common in Yucatan. One is called chacoub; it has white flowers except that the perianth-segments have purple edges and tips; the fruit is globular and reddish purple. The other form called zacoub has white flowers and oblong and creamy-white fruit; these fruits are considered among the most desirable in Yucatan and are often to be found in the markets for sale.

Illustrations: Safford, Ann. Rep. Smiths. Inst. **1908**: pl. 6, f. 1, as *Hylocereus tricostatus;* Martius, Fl. Bras. 4²: pl. 42; Engler and Prantl, Pflanzenfam. 3⁶ᵃ: f. 57, A, B;

Edwards's Bot. Reg. 21: pl. 1807; Gard. Mag. 55: 689, all as *Cereus triangularis;* Curtis's Bot. Mag. 44: pl. 1884; Loudon, Encycl. Pl. f. 6870, as *Cactus triangularis;* Ann. Rep. Smiths. Inst. 1917: pl. 10; Scientific Monthly 5: 287, as night-blooming cereus; Britton, Flora Bermuda f. 278.

Plate xxx shows a flowering joint of a plant brought by Dr. Small from southern Florida to the New York Botanical Garden in 1903, where it has since bloomed every year;

FIG. 264.—Hylocereus undatus.

plate xxxii, figure 1, shows a fruiting joint of a plant in the same collection brought from Tehuacán, Mexico, by Dr. MacDougal and Dr. Rose in 1906. Figure 263 is from a photograph taken by Paul G. Russell at Machado Portella, Bahia, Brazil, in 1915; figure 264 is from a photograph by A. S. Hitchcock, 1918, showing a hedge of night-blooming cereus on a wall at Punahou College, Honolulu; the picture was taken early in the morning; the preceding evening the hedge was viewed by hundreds of people. The plant, in Honolulu, comes in full flower only once or twice a year and is then a marvelous sight.

FIG. 265.—Hylocereus cubensis. ×0.66.

8. Hylocereus cubensis sp. nov.

Stems slender, much elongated, freely rooting, 3-angled, dull green, 2 to 4 cm. in diameter; margin of joints scarcely crenate, becoming horny; spines 3 to 5, black, conic, 2 to 3 mm. long; flowers large, white, about 20 cm. long; ovary bearing large leafy scales; fruit a little longer than broad, 10 cm. long, reddish.

M. E. Eaton del.

Flower on short branch of *Hylocereus monacanthus*.
(Natural size.)

Collected by Brother Léon on a wall, Jata Hills, near Guanabacoa, Province of Habana, Cuba, July 14, 1913 (No. 3719). Living specimens were introduced into the New York Botanical Garden which flowered in September 1917. We are disposed to refer here J. A. Shafer's No. 13931 from lime rocks at Portales, Province of Pinar del Rio, Cuba. A plant from the Isle of Pines sent to us by O. E. Jennings probably belongs here, but the poor specimen which we have seen does not enable us to definitely refer it to this species.

Figure 265 shows a section of a branch of the type specimen.

9. Hylocereus lemairei (Hooker) Britton and Rose, Contr. U. S. Nat. Herb. **12:** 428. 1909.

*Cereus lemairei** Hooker in Curtis's Bot. Mag. **80:** pl. 4814. 1854.
Cereus trinitatensis Lemaire and Herment, Rev. Hort. IV. **8:** 642. 1859.

A somewhat slender, high-climbing vine; joints 3-angled, freely rooting on one side, 2 to 3 cm. in diameter, plain green; margins with slight elevations at the areoles; areoles 2 to 2.5 cm. apart; spines usually 2, very short, swollen at base, brownish; flower-buds elongated, acuminate; flower about 27 cm. long; tube, including ovary, 15 cm. long; scales on ovary and lower part of the tube ovate, dark green, with the margins and tips deep purple; scales on upper part of the tube much elongated, but marked like the lower ones; outer perianth-segments about 20, 12 cm. long, 1 cm. wide or less; edges slightly upturned, widely spreading or reflexed, yellowish green, sometimes a little purplish at the tip and the inner one somewhat rose-colored at the base; inner perianth-segments about 15, mostly oblanceolate, 3.5 cm. broad at the widest portion, acute, the lower portion pinkish, above nearly pure white; filaments cream-colored, about three-fourths the length of the inner perianth-segments; style thick, nearly as long as the inner perianth-segments; stigma-lobes cleft to the middle and the branches often notched at tip; flower somewhat odorous, not very pleasing; fruit purple, oblong, 6 to 7 cm. long, when mature splitting down the center almost to the base into 2 nearly equal parts and exposing the white flesh and black seeds.

Type locality: Not cited.
Distribution: Trinidad and Tobago. Perhaps also Surinam.

The above description was based upon specimens sent by Mr. Wm. Broadway in 1907 from Trinidad, which flowered in the New York Botanical Garden in July and August 1912 (No. 27689). Our reference (Contr. U. S. Nat. Herb. **12:** 428) of this species to Antigua and doubtfully to Culebra and Porto Rico, in which we followed previous authors, can not be supported by specimens in our collections.

This is a very beautiful species which has long been in cultivation, but the native home of which, until recently, has not been known. In 1909, Mr. Broadway sent specimens from Trinidad which soon flowered, enabling us to identify it definitely. Sir Joseph Hooker, under *Cereus lemairei* in Curtis's Botanical Magazine, volume 80, plate 4814, says, "Nothing is positively known of its native country; but it happens that I have in my possession a drawing made in Antigua, undoubtedly of this species; so that it is probably a native of that island." A copy of this drawing is now in the United States National Herbarium, and shows quite a different species from *Cereus lemairei*, and may represent the *Hylocereus* collected in the spring of 1913 on Antigua by Dr. Rose (No. 3297), of which we have both herbarium and living specimens,

FIG. 266.—Stigma-lobes of H. lemairei. ×0.7.

but the drawing is without stem and Dr. Rose's specimens were without flowers; however, it may be that Hooker's drawing is of a flower of the commonly cultivated *H. undatus.*

This is one of the few species of cacti having bifid stigma-lobes.

Illustration: Curtis's Bot. Mag. **80:** pl. 4814, as *Cereus lemairei.*

Plate XXXI is from Mr. Broadway's Trinidad plant which flowered in the New York Botanical Garden. Figure 266 shows its style and stigma-lobes.

Cereus lemoinei (Möllers Deutsche Gärt. Zeit. **6:** 92. 1891) may be only a misspelling of this name.

10. **Hylocereus monacanthus** (Lemaire).

Cereus monacanthus Lemaire, Hort. Univ. **6**: 60. 1845.

Stems green, 3-angled, the margins undulate; areoles remote, about 3 cm. apart, tomentose; spines usually single, sometimes 2, rigid, much swollen at base; flowers funnelform, large, 28 cm. long, 17 cm. broad; ovary and tube covered with large scales; outer perianth-segments narrow, greenish; inner perianth-segments oblong-ovate; filaments numerous, about 200, 8 to 9 cm. long, white but rose-colored at base; style thick, exserted, yellow; stigma-lobes numerous, spreading.

Type locality: Colombia.

Distribution: Colombia and Panama.

This species was first introduced by Cels and published in 1845. It was again introduced by Wercklé in 1905 and fully described by Weingart in 1911. Both Dr. Weber and Dr. Schumann considered it to be a variety of *Cereus martinii.*

A flower observed at the New York Botanical Garden September 6, 1918, and a plant brought by Dr. M. A. Howe from the Urava Islands in 1912, showed 2-forked stigma-lobes, the forks 2 to 3 mm. long; other flowers, previously observed, showed simple stigma-lobes.

Plate XXIX shows a branch of the plant collected by Dr. Howe, on Urava Island, Bay of Panama, in 1912, which flowered in the New York Botanical Garden in 1915.

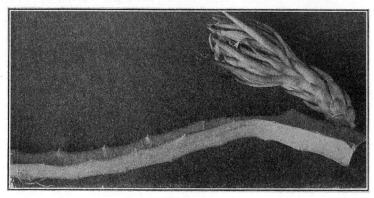

FIG. 267.—Hylocereus stenopterus.

11. **Hylocereus stenopterus** (Weber) Britton and Rose, Contr. U. S. Nat. Herb. **12**: 429. 1909.

Cereus stenopterus Weber, Bull. Mus. Hist. Nat. Paris **8**: 458. 1902.

A weak vine, not rooting freely from the sides, at least in cultivation, the joints 4 cm. broad, light green in color, not glaucous; ribs 3, thin; areoles slightly elevated; spines 1 to 3, small, yellow; flower 10 to 12 cm. long, opening at night, closing very early (completely closed at 9 a. m.); tube short, about 2 cm. long; perianth-segments all similar, reddish purple, linear, about 7 cm. long; stamens short, exserted; style white, thick, much exserted; stigma-lobes white, when closed forming an ovoid acuminate cluster; scales on ovary and flower-tube orbicular or the upper ones narrowly ovate, green, with purple margins.

Type locality: Vallée de Tuis, Costa Rica.

Distribution: Costa Rica, Central America.

This species is common in Costa Rica, and in recent years has been widely distributed by several Costa Rican collectors; it grows well under glass, and flowers frequently. It is the only *Hylocereus* in cultivation with red flowers except *H. extensus.*

Plate XXVII, figure 3, is from a plant obtained by Mr. William R. Maxon in San José, Costa Rica, in 1906, which flowered at the New York Botanical Garden. Figure 267 is from a photograph of a specimen which flowered in Washington from specimens received from the New York Botanical Garden in 1910 (No. 22197).

PLATE XXX

M. E. Eaton del.

Flower near end of branch of *Hylocereus undatus.* × 0.7.

12. Hylocereus extensus (Salm-Dyck).

Cereus extensus Salm-Dyck in De Candolle, Prodr. 3: 469. 1828.

Creeping and probably often climbing, bearing the usual aërial roots of the genus; joints green, rather slender, 1.5 cm. in diameter, 3-sided, the obtuse angles not at all winged; areoles remote, small, woolly and often setose; spines 2 or 3, rarely 4, very short and stout, dark brown, 1 to 2 mm. long; flowers large and handsome; tube green, cylindric; scales of the ovary ovate; scales of the tube rather short, becoming elongated above and passing into the narrow outer perianth-segments, greenish yellow, tipped and margined with red; inner perianth-segments oblong to obovate, acute, rose-red; style thick, longer than the stamens; stigma-lobes linear, entire; fruit not known.

Type locality: Not cited.

Distribution: Trinidad, according to Curtis's Botanical Magazine.

The above description is based on the figure and description found in Curtis's Botanical Magazine as below cited. This may or may not belong to the plant described by De Candolle (Prodr. 3: 469), for he describes the radial spines as 10 to 12, pilose and white, and the centrals as 2 to 4, small, rigid, and yellow; it is hardly the *Cereus extensus* of Pfeiffer (Enum. Cact. 119), where the inner perianth-segments are said to be white and obtuse.

Cereus subsquamatus Pfeiffer (Allg. Gartenz. 3: 380. 1835) is referred here by Pfeiffer.

Illustration: Curtis's Bot. Mag. 70: pl. 4066, as *Cercus extensus*.

13. Hylocereus napoleonis (Graham) Britton and Rose, Contr. U. S. Nat. Herb. 12: 429. 1909.

Cereus napoleonis Graham in Curtis's, Bot. Mag. 63: pl. 3458. 1836.

Stems much branched, light green, the joints with 3 acute angles and concave sides; angles tuberculate, with repand intervals, not at all horny; areoles about 4 cm. apart; spines 4 or 5, rigid, about 9 mm. long, with swollen bases; flowers 20 cm. long and nearly as broad; tube 7.5 cm. long, green, bearing a few subappressed, deep red scales, gradually enlarging upward; outer perianth-segments yellow, lanceolate, linear; inner perianth-segments pure white, spatulate-lanceolate, crenate at apex; stamens numerous, yellow; pistil stout; stigma-lobes numerous, entire.

Type locality: Unknown; described from a cultivated plant.

Distribution: West Indies and southern Mexico, according to Schumann; but we know it definitely only from the original illustration.

The origin of this species has long been in doubt. It was described by Graham at the time it flowered in the botanical garden at Edinburgh. The plant had then been in cultivation for about ten years, having been sent by a Mr. McKay of Clapton, but without any record of its source. It is possible that this species should be referred to the true *H. triangularis*, although Pfeiffer states in the most emphatic terms that they are very distinct. According to Loudon (Gard. Dict. 2: 65. 1827) *Cactus napoleonis* occurs in a list of new plants offered by L. C. Noisette, a nurseryman in Paris. This was about nine years before the name was published in the Botanical Magazine.

Cereus triangularis major Salm-Dyck, Allg. Gartenz. 4: 80. 1836 and *Cactus napoleonis* Hortus, unpublished names, are often given as synonyms.

Cereus lanceanus (G. Don in Sweet, Hort. Brit. ed. 3. 285. 1839), *C. inversus*, and *C. schomburgkii* are names of garden plants which are referred to this relationship by Förster (Handb. Cact. 422. 1846).

Plants from Santo Domingo resemble the original illustration in armament. We have these in cultivation, both at Washington and at New York, but they have not flowered (Rose, Nos. 3734, 3839, and 4147). Boldingh (Fl. Ned. West Ind. 296) records the plant from Aruba.

Pfeiffer (Enum. Cact. 117. 1837) referred here Burmann's plate of Plumier (pl. 199, f. 2) which is perhaps the best disposal to make of it. The fruit, however, has spiny areoles and in this respect resembles *Acanthocereus pentagonus*. Gosselin considered it an undescribed species which he called *Cereus plumieri* (Gosselin, Bull. Soc. Bot. France, 54: 668. 1907).

Illustrations: Curtis's Bot. Mag. 63: pl. 3458; Loudon, Encycl. Pl. ed. 2. f. 17363, both as *Cereus napoleonis;* (?) Plumier, Pl. Amer. ed. Burmann, pl. 200, f. 1, as *Cactus* etc.

14. **Hylocereus trigonus** (Haworth) Safford, Ann. Rep. Smiths. Inst. **1908**: 556.   1909.

> ?*Cactus triangularis foliaceus* Jacquin, Stirp. Amer. 152.   1763.
> *Cereus trigonus* Haworth, Syn. Pl. Succ. 181.   1812.
> *Cereus venditus* Paulsen, Journ. Bot. **56**: 235.   1918.

Stems slender, 2 to 3 cm. broad, clambering over bushes or rocks, sometimes 10 meters long, deep green; joints 3-angled, the margin of the ribs not horny, strongly undulate, the areoles borne on the tops of the undulations; spines usually 8, 4 to 7 mm. long, stiff, at first greenish, soon dark brown; accessory spines or bristles usually 2; perianth large; ovary bearing large foliaceous scales; fruit oblong or oblong-obovoid, red, 10 cm. long, becoming nearly smooth.

Type locality: Not cited.

Distribution: Hispaniola, Porto Rico, Vieques, Culebra, St. Jan, St. Thomas, Tortola, Virgin Gorda, and St. Croix.   Recorded by Boldingh (Fl. Ned. West Ind. 297) from St. Eustatius, Saba, and St. Martin.

This species, although known to Plumier and illustrated by Burmann (1750–1760), was not taken up as a species until 1812, when it was described by Haworth. In 1803 Haworth had described it as a variety of *Cactus triqueter* (Misc. Nat. 189), but had said it was twice the size. *Cereus venditus* Paulsen is based upon the juvenile form of this species from a plant collected on the Island of St. Jan.

Illustrations: Safford, Ann. Rep. Smiths. Inst. **1908**: pl. 12. Plumier, Pl. Amer. ed. Burmann, pl. 200, f. 2, as *Cactus* etc.; Contr. U. S. Nat. Herb.

FIG. 268.—Hylocereus trigonus.

8: pl. 25, as *Cereus* sp.;? Jacquin, Stirp. Amer. pl. 181, f. 65, as *Cactus triangularis foliaceus;* Loudon, Encycl. Pl. f. 6872, as *Cactus trigonus.*

Plate XXXVI, figure 1, represents a fruiting joint of a Porto Rican plant in the collection of the New York Botanical Garden.   Figure 268 is from a photograph taken by F. E. Lutz near Arecibo, Porto Rico.

15. **Hylocereus triangularis** (Linnaeus) Britton and Rose, Contr. U. S. Nat. Herb. **12**: 429.   1909.

> *Cactus triangularis* Linnaeus, Sp. Pl. 468.   1753.
> *Cereus compressus* Miller, Gard. Dict. ed. 8. No. 10.   1768.
> *Cereus triangularis* Haworth, Syn. Pl. Succ. 180.   1812.

High-clambering or creeping vines, sharply 3-angled, 3 to 4 cm. broad, giving off numerous long aërial roots; margin not horny, nearly straight or slightly elevated at the areoles; areoles about 2 cm. apart; principal spines 6 to 8, acicular, but with swollen bases; flowers 20 cm. long or more; outer perianth-segments linear-lanceolate, acuminate, 6 to 8 cm. long, longer than the inner segments; inner perianth-segments white, oblong; scales on the ovary and flower-tube oblong, green, 2 to 5 cm. long; fruit red.

Type locality: Jamaica.

Distribution: Very common on rocks and trees along the coast of Jamaica.

Plants of *H. triangularis* were collected by John F. Cowell in Panama, probably not native there, however.

Cereus triangularis pictus De Candolle (Prodr. **3**: 468) is said to have yellow or yellow and green joints, with spines often setiform, not rigid.

M. E. Eaton del. Flower on short branch of *Hylocereus lemairei*. ✕ 0.7.

Salm-Dyck (Cact. Hort. Dyck. 1849. 220. 1850) described *C. triangularis uhdeanus*, based upon a cultivated Mexican plant. It is described with 4 to 6 radial spines and 1 central, yellow, minute. Salm-Dyck was uncertain whether it was a garden variety or a distinct species.

Cereus anizogonus Salm-Dyck (Cact. Hort. Dyck. 1849. 52. 1850) was given as a synonym of *Cereus triangularis*.

Miller, who first published *Cereus compressus* distinguished it from *C. triangularis*, but based it upon Plukenet's illustration (Opera Bot. 1: pl. 29, f. 3), which Linnaeus referred to *Cactus triangularis*, and which we believe represents the Jamaican plant. Martyn in a later edition of Miller's Gardeners' Dictionary refers Miller's *Cereus compressus* to *Cactus pentagonus* (?), which seems hardly correct. The Index Kewensis refers *Cereus compressus* to Mexico. *Cephalocereus compressus* (Monatsschr. Kakteenk. Index, vol. 1 to 20. 36. 1912) belongs here.

Illustrations: Plukenet, Opera Bot. 1: pl. 29, f. 3, as *Cereus erectus cristatus*; Bradley, Hist. Succ. Pl. ed. 2. pl. 3, as *Cereus americanus triangularis* etc.

Figure 269 shows a joint of a plant collected by Dr. Britton near Mandeville, Jamaica, in 1907.

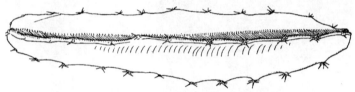

FIG. 269.—Joint of Hylocereus triangularis. ×0.5.

16. Hylocereus antiguensis sp. nov.

Stems high-clambering, forming great masses in the crotches of high trees or covering the tops of low trees; joints 2 to 4 cm. thick, 3-angled, rarely 4-angled; margins of ribs not horny, hardly undulate; areoles 2.5 to 3.5 cm. apart; principal spines 2 to 4, about 6 mm. long or less, accessory ones or bristles 2 to 5; flowers 14 cm. long; outer perianth-segments linear; inner perianth-segments yellow, at least drying so, broader than the outer segments; flower-tube bearing linear acute scales.

This species is nearest *H. trigonus*, but the margins of the ribs are very different. The description is based on specimens collected by Dr. Rose in Antigua (No. 3297), of which we have both living and herbarium specimens. It flowered in the New York Botanical Garden in 1916.

Figure 270 is from a photograph taken by Paul G. Russell on Antigua in 1913.

17. Hylocereus calcaratus (Weber) Britton and Rose, Contr. U. S. Nat. Herb. 12: 428. 1909.

Cereus calcaratus Weber, Bull. Mus. Hist. Nat. Paris 8: 458. 1902.

A climbing vine, the joints 4 to 6 cm. wide, strongly 3-winged, green, the margin divided into numerous prominent lobes; areoles small, from the upper angles of the marginal lobes, spineless but bearing 2 to 4 small, white bristles.

Type locality: Port Limón, Costa Rica.
Distribution: Costa Rica.

Neither flowers nor fruit were known to Dr. Weber when he described the plant; we have had it for a number of years and it has not yet flowered with us. It is very unlike the other species of *Hylocereus*, having very peculiar stems and no spines, and it may not be of this genus.

Figure 271 shows a joint of a plant, which was obtained by W. R. Maxon, in cultivation at San José, Costa Rica, in 1906.

HYLOCEREUS sp.

Dr. J. A. Samuels collected a species of this genus in the forest plantation, La Poule, Surinam, April 24, 1916 (No. 305), which is the first record we have of this genus being found in Dutch Guiana. Dr. Samuels's plants are juvenile ones, at least in part, as the spines on some specimens are represented by 10 to 12 spiny bristles at each areole; other branches which are more mature are less than 2 cm. broad, 3-angled, with the areoles only 1 to 1.5 cm. apart; the margin of the rib is almost straight; spines 3 to 5, brown, 2 to 3 mm. long. The specimens are without flowers or fruit. While reading the last proof, specimens have been received from Gerold Stahel, of Paramaribo, Surinam, which lead us to believe that the plant from that country is *Hylocereus lemairei*. Here probably belongs the plant from Surinam which Linnaeus called *Cactus triangularis* (Amoen. Acad. **8**: 257. 1785).

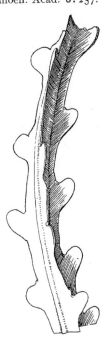

FIG. 270.—Hylocereus antiguensis.

FIG. 271.—Joint of H. calcaratus. ×0.5

HYLOCEREUS sp.

A species, apparently of this genus. It is a long, clambering plant running over and through tops of shrubs and trees and sometimes killing them, with strongly 3-angled joints, the margins of the ribs rather thick, hardly undulate; the areoles 5 to 6 cm. apart, with 6 to 8 subulate spines, the longer 12 to 15 cm. long. It was collected by E. A. Goldman at Carrizal, Vera Cruz, May 25, 1901 (No. 697). Its flowers and fruit are unknown.

HYLOCEREUS sp.

Branches slender, 3-angled, or sometimes nearly terete, 2 to 3 cm. broad, dull green, sometimes perhaps glaucous; margin not horny; areoles often distant, sometimes 5 cm. apart, borne on prominent and more or less reflexed knobs; spines brown, 2 to 4, stout, conic.

Collected by Dr. Rose near Puerto Cabello, Venezuela, in 1916 (No. 21870).

This very peculiar plant we have not been able to refer definitely to any of the above species. It suggests in a way the other *Hylocereus*, which Dr. Rose also obtained in Venezuela (No. 21835) and which we have described as new in the Appendix. It, however, has not yet flowered in the New York Botanical Garden where it is now being grown.

PUBLISHED SPECIES, PERHAPS OF THIS GENUS, KNOWN TO US ONLY FROM DESCRIPTION.

CEREUS RADICANS De Candolle, Prodr. 3: 468. 1828.

This species has not been well understood since its original description. It is probably either a *Selenicereus* or a *Hylocereus*. It was described as prostrate, light green in color, 3 or 4-angled, with rigid, slender, brown spines, of which 6 to 9 are radials and 1 is central. De Candolle refers it to Tropical America; Pfeiffer (Enum. Cact. 114), who redescribed it in 1837, refers it to Tropical America and the Antilles. Schumann did not know it, but referred it to South America.

Cereus reptans Salm-Dyck (De Candolle, Prodr. 3: 468. 1828) is an unpublished name which was first mentioned under *C. radicans*, while *Cereus reptans* Willdenow is referred by De Candolle to *Cereus pentagonus*.

CEREUS HORRENS Lemaire, Hort. Univ. 6: 60. 1845.

Climbing and rooting; ribs 3, prominent, strongly tubercled; areoles distant, bearing copious white down; spines 5 to 7, whitish, variable, stout, very long.

This species seems to have been lost. Its flowers and fruit are unknown as is also its origin. It is probably a *Hylocereus*.

2. WILMATTEA gen. nov.

A climbing cactus, epiphytic and rooting along the sides of the joints, slender, with few short spines; flowers solitary at the areoles (in one case 2 flowers seen), small for the tribe, nocturnal, with a narrow limb and with a very short tube; ovary small, covered with ovate, imbricating, reddish scales, each subtending a small areole filled with felt and occasionally with 1 bristle or more, perhaps sometimes naked; filaments and style short.

One species is known, native of Guatemala and Honduras. The genus is named in honor of Mrs. T. D. A. Cockerell (Wilmatte P. Cockerell) in recognition of her many discoveries of rare plants and animals in Central America.

In habit this plant resembles a slender-stemmed species of *Hylocereus* while the flower and ovary bear similar scales and this led us at one time to consider it as a species of that genus. The flowers, however, are so much smaller with scarcely any tube and bearing felt and bristles in their axils that we now regard it as generically distinct.

1. Wilmattea minutiflora.

Hylocereus minutiflorus Britton and Rose, Contr. U. S. Nat. Herb. 16: 240. 1913.
Cereus minutiflorus Vaupel, Monatsschr. Kakteenk. 23: 86. 1913.

A slender, high-climbing vine, the joints 3-angled, deep green, the angles sharp but not winged, not horny-margined; areoles 2 to 4 cm. apart; spines usually 1 to 3, minute, brownish; flowers only 5 cm. long, opening at night, rarely remaining open until 9 o'clock in the morning, very fragrant; flower-tube only 10 mm. long, or even less; outer perianth-segments linear, red on the midvein and at the tip, 3 to 4 cm. long; inner perianth-segments very narrow, acute, white; stamens white, about 1 cm. long, borne in a series at the base of the inner perianth-segments; scales on the ovary sometimes bearing bristles in their axils, sometimes naked, oblong to ovate, purple or greenish at base; style white, 2 cm. long, thick; stigma-lobes white.

Type locality: Near Lake Izabel, Guatemala.
Distribution: Guatemala and Honduras.

A cutting of the plant developed 3 thin wings 10 mm. wide, the areoles producing 2 to 5 long white hairs but no spines. In all the young joints 5 to 8 wings started, but all but

3 soon dropped out. In some cases the joints are nearly terete at base, or in cultivation develop long terminal shoots which are nearly terete.

This species was first collected by R. H. Peters in 1907. It was again collected by Mrs. T. D. A. Cockerell at Quirigoa in 1912, who sent living plants to Washington which flowered September 27, 1917, and in 1920 Harry Johnson sent us living specimens from Guatemala. In 1916 Francis J. Dyer sent from Honduras what seems to be this species.

Illustration: Contr. U. S. Nat. Herb. 16: pl. 69, as *Hylocereus minutiflorus.*

Plate XXXII, figure 2, shows a flowering branch of the type specimen, which was collected by R. H. Peters in Guatemala in 1907. Figure 272 shows a flowering joint of the type specimen, photographed in Washington.

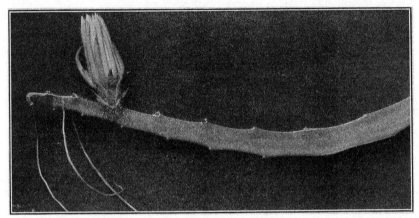

Fig. 272.—Wilmattea minutiflora.   ×0.6.

3. **SELENICEREUS** (Berger) Britton and Rose, Contr. U. S. Nat. Herb. 12: 429.   1909.

Slender, trailing, climbing or clambering, elongated cacti, the joints ribbed or angled, irregularly giving off aërial roots; areoles small, sometimes elevated on small knobs, bearing small spines or in one species spineless; flowers large, often very large, nocturnal; flower-tube elongated, somewhat curved; scales of ovary and flower-tube small, usually with long felt, hairs and bristles in their axils; upper scales and outer perianth-segments similar, narrow, greenish, brownish, or orange; inner perianth-segments broad, white, usually entire; filaments elongate, weak, numerous, in two clusters distinctly separated, one cluster forming a circle at top of flower-tube, the other scattered over the ong, slender throat; style elongated, thick, often hollow; stigma-lobes slender, numerous, entire; ruit large, reddish, covered with clusters of deciduous spines, bristles, and hairs.

Type species: *Cactus grandiflorus* Linnaeus.

The name is from the Greek and signifies moon-cereus, the plants being night-blooming.

All the species are clambering vines with aërial roots, and in the tropics often reach the tops of high trees; where there are no trees or shrubs, they trail over rocks and walls. Most of them have very large flowers; in fact, one of the largest flowered species of the family (*S. macdonaldiae*) belongs here. Several of the species, such as *S. hamatus*, *S. grandiflorus*, *S. macdonaldiae*, and *S. pteranthus* (better known as *Cereus nycticalus*), have long been favorites with amateurs. In our studies of the genus we have had several hundred growing plants under observation, representing all the species, and specimens of all have bloomed. The species of the genus range from southern Texas through eastern Mexico, Central America, the West Indies and along the northern coast of South America, while one species has been reported from Argentina. Sixteen species are here recognized.

PLATE XXXII

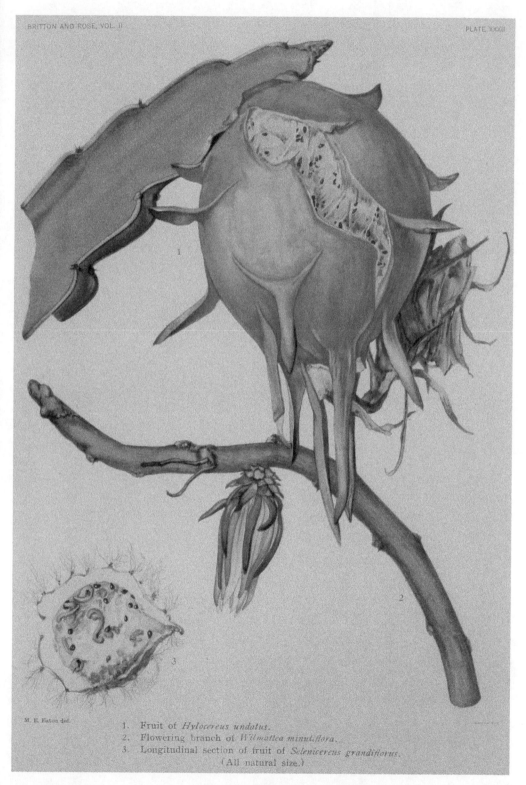

M. E. Eaton del.

1. Fruit of *Hylocereus undatus*.
2. Flowering branch of *Wilmattea minutiflora*.
3. Longitudinal section of fruit of *Selenicereus grandiflorus*.
(All natural size.)

KEY TO SPECIES.

Areoles of the ovary and flower-tube bearing long hairs.
  Branches ribbed, angled or subterete, not spurred.
    Areoles of the branches borne on the ribs or angles.
      Spines of the branch-areoles acicular.
        Hairs of flower-areoles tawny or whitish.............................. 1. *S. grandiflorus*
        Hairs of flower-areoles bright white.
          Branches 4 or 5-angled; stem-areoles without bristly hairs.
            Spines brown... 2. *S. urbanianus*
            Young spines yellow....................................... 3. *S. coniflorus*
          Branches 7 to 10-ribbed; stem-areoles with many bristly hairs.......... 4. *S. hondurensis*
      Spines of the branch-areoles short, conic.
        Branches 9 or 10-ribbed; branch-areoles with many appressed hairs......... 5. *S. donkelaarii*
        Branches 4 to 6-ribbed; young branch-areoles with few long hairs.
          Stems stout, 3 to 5 cm. thick.............................. 6. *S. pteranthus*
          Stems slender, 1.5 to 3 cm. thick.
            Hairs of flower-areoles white.
              Ribs not tubercled................................ 7. *S. kunthianus*
              Ribs tubercled.................................... 8. *S. brevispinus*
            Hairs of flower-areoles brown............................ 9. *S. boeckmannii*
      Areoles of the branches borne on prominent knobs....................10. *S. macdonaldiae*
    Branches with a stout, deflexed spur under each areole....................11. *S. hamatus*
Areoles of flower-tube and ovary without long hairs.
  Areoles of the branches spiniferous.
    Spines of the branch-areoles acicular....................12. *S. vagans*
    Spines of the branch-areoles short, conic.
      Ribs 7 or 8, obtuse; spines from the areoles on the ovary 1 to 3.................13. *S. murrillii*
      Ribs 4 to 6, acute; spines from the areoles on the ovary 10 or more.............14. *S. spinulosus*
  Areoles of the branches unarmed.
    Ribs prominent, 3 to 5; flowers white....................15. *S. inermis*
    Ribs low, 6 to 12; flowers red....................16. *S. wercklei*

1. Selenicereus grandiflorus (Linnaeus) Britton and Rose, Contr. U. S. Nat. Herb. **12:** 430. 1909.

    Cactus grandiflorus Linnaeus, Sp. Pl. 467. 1753.
    Cereus grandiflorus Miller, Gard. Dict. ed. 8. No. 11. 1768.
    Cereus grandiflorus affinis Salm-Dyck, Cact. Hort. Dyck. 1849. 51, 216. 1850.

Stems clambering, often 2.5 cm. in diameter, green or bluish green; ribs usually 7 or 8, sometimes fewer, low, separated by broad, rounded intervals; spines acicular, various, 1 cm. long or less, yellowish brown or brownish, in age gray, intermixed with the numerous whitish hairs; flower-buds covered with tawny hairs; flowers about 18 cm. long; outer perianth-segments narrow, salmon-colored; inner perianth-segments white, acute, entire; style often longer than the inner perianth-segments; fruit ovoid, 8 cm. long.

Type locality: Jamaica.
Distribution: Jamaica and Cuba. Widely planted and escaped from cultivation in tropical America.

We have observed seeds germinating within the fruit of this species.

S. grandiflorus and perhaps some of its allies are used medicinally as a heart tonic. In the trade the plant is sometimes called *Cactus mexicanus*.

Cereus scandens minor Boerhaave (Arendt, Monatsschr. Kakteenk. **1:** 82. 1891) probably refers to this species, as does also *C. grandiflorus* var. *minor* Salm-Dyck and var. *spectabilis* Karwinsky (Förster, Handb. Cact. 415. 1846). *C. grandiflorus uranos* Riccobono (Boll. R. Ort. Bot. Palermo **8:** 249. 1909), doubtless the same as *Cereus uranos* Hortus, is said to be but a form of this species. The *C. uranus nycticalus* mentioned (Monatsschr. Kakteenk. **3:** 117. 1893) is a hybrid and is like *C. grandiflorus callicanthus* Rümpler (Förster, Handb. Cact. ed. 2. 750. 1885; *Cereus callicanthus* Monatsschr. Kakteenk. **3:** 109. 1893). *Cereus grandiflorus viridiflorus* (Monatsschr. Kakteenk. **6:** 80. 1896) is a garden hybrid. The varieties *haitiensis* and *ophites* (Monatsschr. Kakteenk. **13:** 183. 1903) may belong to related species rather than to *Selenicereus grandiflorus*. This is probably true of the varieties *grusonianus* and *mexicanus* listed in Haage and Schmidt Catalogues. *Cereus maximiliani*, *C. grandiflorus maximiliani*, and *C. nycticalus maximiliani* are doubtless hybrids with one of the common cultivated species, perhaps *S. grandiflorus*, as believed by Berger. *Cereus schmidtii* (Monatsschr. Kakteenk. **4:** 189. 1894) may be *Cereus grandiflorus schmidtii*

(Berger, Hortus Mortolensis 70. 1912). *Cereus grandiflorus barbadensis* is also given by Berger.

The flowers of *S. grandiflorus* are almost identical with those of the 8 following species, which differ essentially only in vegetative characters and armament.

Illustrations: Andrews, Bot. Rep. **8**: pl. 508; De Candolle, Pl. Succ. **1**: pl. 52; Descourtilz, Fl. Med. Antill. **1**: pl. 65; Loddiges, Bot. Cab. **17**: pl. 1625; Loudon, Encycl. Pl. f. 6873, as *Cactus grandiflorus;* Cact. Journ. **1**: 125; Curtis's Bot. Mag. **62**: pl. 3381; Dict. Gard. Nicholson **1**: f. 407; Gartenflora **53**: 68, 401; Schumann, Gesamtb. Kakteen f. 34; Miller, Icones pl. 90; Rümpler, Sukkulenten f. 69; Monatsschr. Kakteenk. **10**: 60; Cycl. Amer. Hort. Bailey **1**: f. 414, all as *Cereus grandiflorus;* Trew, Pl. Ehret. pl. 31, 32, as *Cereus gracilis scandens* etc.; Cact. Journ. **1**: 79, as *Cereus grandiflorus major.*

Plate XXXII, figure 3, shows a section of the fruit of a plant in the New York Botanical Garden sent from Cuba by C. F. Baker in 1907, with germinated seeds within; plate XXXIII, figure 1, shows a flowering branch, figure 2 shows the tip of a branch, and figure 3 its fruit.

2. **Selenicereus urbanianus** (Gürke and Weingart) Britton and Rose, Contr. U. S. Nat. Herb. **16**: 242. 1913.

> *Cereus urbanianus* Gürke and Weingart, Notizbl. Bot. Gart. Berlin **4**: 158. 1904.
> *Selenicereus maxonii* Rose, Contr. U. S. Nat. Herb. **12**: 430. 1909.
> *Cereus roseanus* Vaupel, Monatsschr. Kakteenk. **23**: 27. 1913.
> *Cereus paradisiacus* Vaupel, Monatsschr. Kakteenk. **23**: 37. 1913.

Stems light green, but often becoming deep purple throughout, often 3 cm. in diameter; ribs 4 or 5, rarely 3 or 6, rather prominent but less so on the older branches; areoles small, white; spines 1 cm. long or less, brownish; reflexed bristles or hairs from the lower part of the areoles several, white, longer than the spines; flowers 20 to 30 cm. long; uppermost scales and outer perianth-segments narrow, brown to orange, paler within; inner perianth-segments spatulate to oblanceolate, the upper part more or less serrated, the very broad apex sometimes apiculate or entire and acuminate, pure white; stamens and style yellowish green, longer than the inner perianth-segments; flower-tube 17 cm. long, reddish brown, its areoles and those of the ovary bearing long, white hairs.

Type locality: Haiti.

Distribution: Cuba and Hispaniola.

Plants collected by Dr. John K. Small, escaped from cultivation near Halendale, Florida, are, apparently, referable to this species.

Illustrations: Monatsschr. Kakteenk. **16**: 137, as *Cereus urbanianus;* Blühende Kakteen **3**: pl. 153, 154, as *Cereus paradisiacus.*

Plate XXXIV shows a flowering branch of a plant collected by N. L. Britton and J. F. Cowell at El Cobre, Cuba, in 1912.

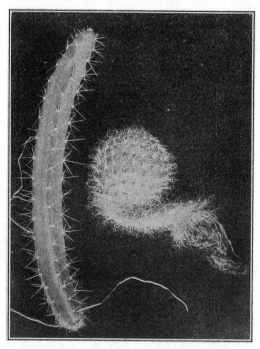

FIGS. 273 and 274.—Branch and fruit of Selenicereus coniflorus.

PLATE XXXIII

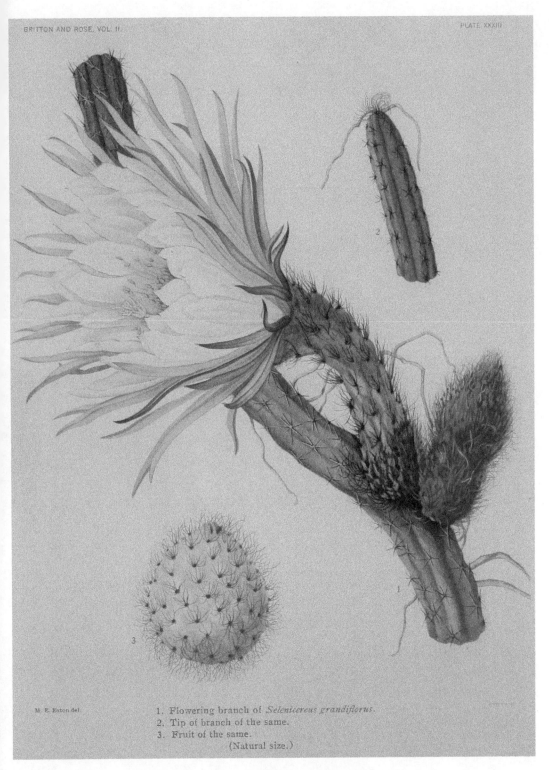

M. E. Eaton del.

1. Flowering branch of *Selenicereus grandiflorus*.
2. Tip of branch of the same.
3. Fruit of the same.
(Natural size.)

3. **Selenicereus coniflorus** (Weingart) Britton and Rose, Contr. U. S. Nat. Herb. **12**: 430. 1909.

Cereus coniflorus Weingart, Monatsschr. Kakteenk. **14**: 118. 1904.
Cereus nycticalus armatus Schumann, Gesamtb. Kakteen 147. 1897.
Selenicereus pringlei Rose, Contr. U. S. Nat. Herb. **12**: 431. 1909.
Cereus jalapaensis Vaupel, Monatsschr. Kakteenk. **23**: 26. 1913.

Stems high-climbing, giving off numerous aërial roots, pale green, becoming purplish along the ribs, 5 or 6-ribbed; intervals between the ribs either depressed or shallow; margins of the ribs slightly wavy to strongly knobby; spines acicular, pale yellow, the radials 4 to 6, with 1 central, porrect, 1 to 1.5 cm. long; bristles from the lower part of areoles, 2, reflexed; buds globular, covered with white hairs; flowers about 22 to 25 cm. long; outer perianth-segments linear, light orange or bronzed to lemon-yellow; inner perianth-segments pure white, apiculate; filaments greenish; style much shorter than the inner perianth-segments; stigma-lobes greenish yellow; scales on the ovary and flower-tube linear, reddish, their axils bearing white hairs and spines; fruit globose, about 6 cm. in diameter.

Type locality: Not cited; based on cultivated plants.
Distribution: Eastern Mexico, especially Vera Cruz.

This is a vigorous climbing vine, flowering freely in cultivation. It is often known in collections as *Cereus nycticalus armatus*. Living material was collected by Dr. Rose in Mexico in 1905, where he learned it was being gathered in large quantities and shipped to the United States, as *Cereus grandiflorus*, to be manufactured into medicine.

Illustration: Schumann, Gesamtb. Kakteen f. 35, as *Cereus nycticalus.*

Plate xxxv is from a plant collected by Dr. Rose in Mexico in 1905, which flowered at the New York Botanical Garden, May 8, 1913. Figure 273 shows a joint and figure 274 a fruit of a plant in the collection of the United States Department of Agriculture.

4. **Selenicereus hondurensis** (Schumann) Britton and Rose, Contr. U. S. Nat. Herb. **12**: 430. 1909.

Cereus hondurensis Schumann in Weingart, Monatsschr. Kakteenk. **14**: 147. 1904.

Climbing and clambering, 1.5 cm. in diameter, green, becoming in winter deep purple; ribs 7 to 10, low; areoles 6 to 10 mm. apart; spines rather short, 5 to 7 mm. long, but acicular, usually surrounded by numerous much longer white hairs or bristles, especially conspicuous on young branches; flowers 20 cm. long or more; outermost perianth-segments brownish and linear, the outer ones linear and acuminate, yellow; inner perianth-segments pure white, 10 cm. long, 10 to 15 mm. broad; scales on ovary and flower-tube linear, bearing numerous long bristly hairs in their axils; fruit not known.

Type locality: Honduras.
Distribution: Honduras and Guatemala.

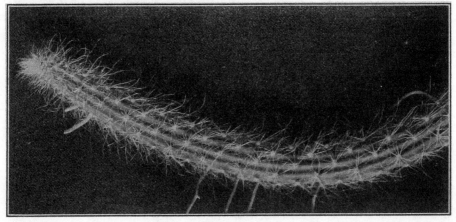

Fig. 275.—Tip of branch of Selenicereus hondurensis. Natural size.

This species has long passed as *Cereus kunthianus* and is the plant described by Schumann under that name.

Figure 275 is from a photograph of a branch of a plant collected by O. F. Cook at Panzos, Guatemala, in 1907.

5. Selenicereus donkelaarii (Salm-Dyck).

Cereus donkelaarii* Salm-Dyck, Allg. Gartenz. **13**: 355. 1845.

Stems elongated, creeping or ascending, 8 meters long or more, slender, about 1 cm. thick; ribs 9 or 10, obtuse, often indistinct; spines in clusters of 10 to 15, the radials 3 to 4 mm. long, setaceous, appressed; central spine 1 or several, 1 to 2 mm. long; flowers 18 cm. long, the slender tube 6 to 7 cm. long; outer perianth-segments reddish, linear; inner perianth-segments white, entire, 6 to 8 cm. long, about 1 cm. wide, acuminate; stamens and style nearly white above, greenish below; fruit unknown.

Type locality: Not cited.
Distribution: Yucatan, Mexico.

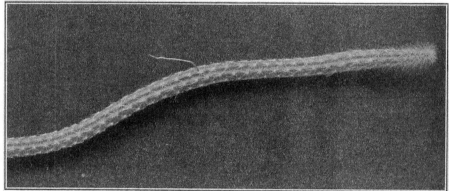

FIG. 276.—Selenicereus donkelaarii.

This species has long been known only from cultivated plants. Schumann reports it as from Brazil but this we are now disposed to question since it has recently been discovered by Dr. George Gaumer in Yucatan growing in dense forests, and we now have living specimens from his collections. We now find that Major E. A. Goldman collected it some years ago in Yucatan but it was not recognized at that time. Goldman's plant grows in dense patches on Cantay Island, collected April 22, 1901 (No. 661).

Figure 276 represents a sterile branch of the plant as grown in the collection of the United States Department of Agriculture.

6. Selenicereus pteranthus (Link and Otto) Britton and Rose, Contr. U. S. Nat. Herb. **12**: 431. 1909.

Cereus pteranthus Link and Otto, Allg. Gartenz. **2**: 209. 1834.
Cereus nyclicallus† Link in A. Dietrich, Verh. Ver. Beförd. Gartenb. **10**: 372. 1834.
Cereus brevispinulus Salm-Dyck, Hort. Dyck. 339. 1834.

Stems stout, often 3 to 5 cm. in diameter, bluish green to purple, strongly 4 to 6-angled; ribs of young branches sometimes 2 to 3 mm. high; spines 1 to 4, 1 to 3 mm. long, dark, conic; flowers 25 to 30 cm. long, very fragrant; the tube and throat 13 cm. long, swollen above, 5 cm. in diameter; outer perianth-segments linear, 12 cm. long; inner perianth-segments white, spatulate-oblong, 3 to 4 cm. broad above, acuminate; filaments numerous, greenish to cream-colored, the upper row reaching forward, upturned near the tip, 6 cm. long; lower stamens elongated, unequal, 8 to 12 cm.

---

*The species was originally spelled in the Allgemeine Gartenzeitung *Cereus donkelaarii* but was indexed in the same book as *Cereus donkelarii*. It is also written *Cereus donkelaeri*.
†Although the usual spelling of this name is with one l, it was originally spelled by Link as it is here.

PLATE XXXIV

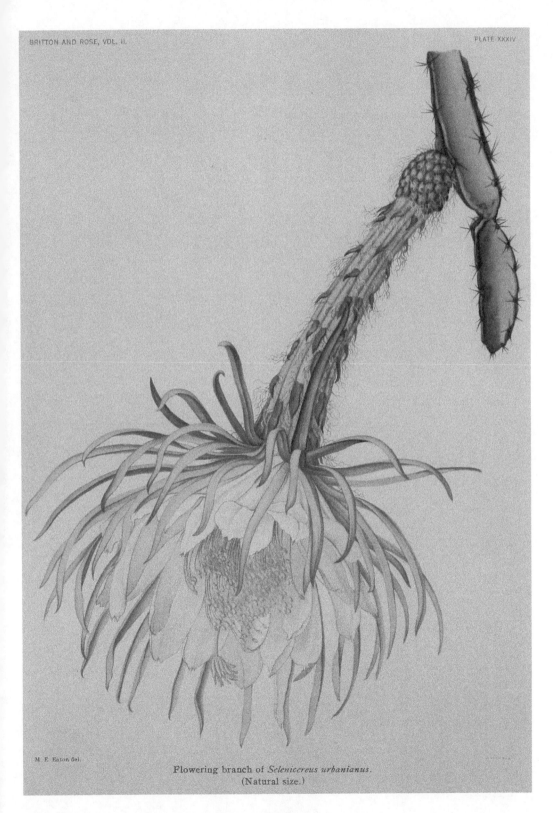

M. E. Eaton del.

Flowering branch of *Selenicereus urbanianus.*
(Natural size.)

long, weak, reclining on the lower side of the flower-tube and attached along the inner face of the tube for 7 to 8 cm.; tube-proper about 2 cm. long, yellow within; style 20 cm. long, yellowish green, bronzed above, thick but weak; stigma-lobes numerous, linear; ovary covered with long white silky hairs and bristles, 10 to 12 mm. long; fruit globular, red, 6 to 7 cm. in diameter.

Type locality: Mexico.

Distribution: Mexico; known to us only from cultivated specimens or from plants escaped from gardens.

Cercus antoinii (Pfeiffer, Enum. Cact. 114. 1837) is known only as a synonym of *Cereus nycticallus.* *Cereus rosaceus,* first mentioned by De Candolle (Prodr. 3: 471. 1828), is only a garden name which Pfeiffer (Enum. Cact. 114. 1837) referred to *C. nycticallus.*

Cercus peanii Beguin first mentioned in Rebut's Catalogue (Monatsschr. Kakteenk. 4: 173. 1894) has never been formally published. According to Weber, it is a hybrid of which *Cercus nycticallus* is one of the parents. *Cereus nothus* (Monatsschr. Kakteenk. 4:173. 1894), grown by Gruson but never described, is, according to Schumann, *Cereus pterogonus. Cereus nothus* Wendland (Schumann, Gesamtb. Kakteen 143. 1897), however, he says is a hybrid.

Several varieties of this species have been named, most of which doubtless belong here; at least the following do: *C. nycticalus gracilior* Haage (Förster, Handb. Cact. 416. 1846), *C. nycticalus maximiliani* (Arendt, Monatsschr. Kakteenk. 1: 58. 1891), and *C. nycticalus viridior* Salm-Dyck (Cact. Hort. Dyck. 1849. 51, 216. 1850). It has frequently been used by gardeners in making hybrids, especially with *S. grandiflorus* and *Heliocereus speciosus.* This is a common plant in conservatories.

Illustrations: Amer. Garden 11: 471; Dict. Gard. Nicholson 1: f. 408; Lemaire, Cact. f. 11; Rümpler, Sukkulenten f. 70, 71; Verh. Ver. Beförd. Gartenb. 10: pl. 4, all as *Cereus nycticallus.*

Plate xxxviii, figure 1, shows a fruiting branch of a plant obtained from J. E. Barre in 1901, which flowered in the New York Botanical Garden in 1915.

7. Selenicereus kunthianus (Otto) Britton and Rose, Contr. U. S. Nat. Herb. **12**: 430. 1909.

    Cereus kunthianus Otto in Salm-Dyck, Cact. Hort. Dyck. 1849. 217. 1850.

Stems elongate, weak, bluish green or purplish, 1 to 2.5 cm. thick; ribs 5 to 10, low; spines only 1 to 2 mm. long, 7 to 9; flowers 24 cm. long, the slender tube about 12 cm. long; outer perianth-segments numerous, linear, shorter than the white inner ones; axils of scales on ovary and flower-tube with long silky hairs; fruit unknown.

Type locality: Not cited.

Distribution: Known only in cultivation; said to have come from Honduras.

We are basing our determination of this species on a plant sent under this name to Dr. Rose from the Berlin Botanical Garden (1909); this has 5-angled stems. The original description of the species calls for 7-angled to 10-angled stems, however. There may be this amount of variability in the stems, or there may be two species involved.

Figure 277 shows a branch of a plant in the collection of the New York Botanical Garden, received from the Berlin Botanical Garden.

8. Selenicereus brevispinus sp. nov.

Stems rather stout, climbing or clambering, 2 to 3 cm. thick, in cultivation somewhat branching, light green, the growing branches tipped with white hairs; ribs 8 to 10, separated by narrow intervals, undulating, with knobby areoles; areoles circular, with short tawny felt; spines about 12, conic, stiff, about 1 mm. long, the 3 or 4 centrals thicker than the somewhat curved or hooked radials; bristles from the lower parts of the areoles, 6 or more, longer than the spines, hair-like; flower-

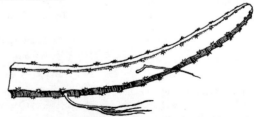

FIG. 277.—Branch of S. kunthianus. ×0.5.

buds covered with long white hairs; flower 25 cm. long; outer perianth-segments narrow, in 2 or 3 series, brown, or inner series yellowish, acuminate, 8 to 9 cm. long; inner perianth-segments shorter and broader than the outer, pure white, entire, acute; filaments numerous, included; style not projecting beyond the stamens, 17 to 18 cm. long; stigma-lobes linear, about 20; scales on the ovary and tube spreading, 4 to 6 mm. long; fruit not known.

Collected by Dr. J. A. Shafer on Cayo Romano, Cuba, in 1909 (No. 2811).

This species is clearly distinct from *S. boeckmannii*. Both flowered May 2, 1915, in Washington, when decided differences were observed in the color of the hairs on the flower-tube and in the color of the outer perianth-segments.

Figure 278 is from a photograph of a branch of the type plant.

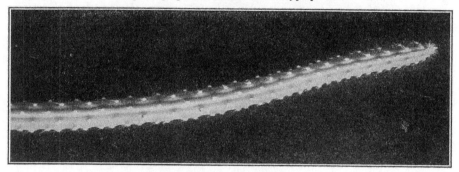

Fig. 278.—Selenicereus brevispinus.

9. **Selenicereus boeckmannii** (Otto) Britton and Rose, Contr. U. S. Nat. Herb. **12**: 429.   1909.

   Cereus boeckmannii Otto in Salm-Dyck, Cact. Hort. Dyck. 1849. 217.   1850.
   Cereus irradians Lemaire, Illustr. Hort. **11**: Misc. 74.   1864.
   Cereus eriophorus Grisebach, Cat. Pl. Cub. 116.   1866.   Not Pfeiffer, 1837.
   Cereus vaupelii Weingart, Monatsschr. Kakteenk. **22**: 106.   1912.

Stems light green, 1 to 2 cm. in diameter, strongly angled; ribs 3 to 8, slightly if at all undulating; areoles at first brownish but white in age; spines and hairs in the areoles at first purplish, the spines 3 to 6, becoming yellowish, 2 mm. long or less; flowers not fragrant, 24 to 39 cm. long; outer perianth-segments and scales linear, brownish; inner perianth-segments oblanceolate, 10 cm. long by 3 cm. broad at widest place, pure white; tube and throat 14 cm. long, bearing scattered, short, linear, acute, reddish scales, their axils bearing long brown silky hairs and brown bristles; filaments greenish, long, slender, and weak; style greenish, about 4 mm. in diameter; ovary strongly tuberculate; fruit globular, 5 to 6 cm. in diameter.

Type locality: Not cited.

Distribution: Cuba, Hispaniola, and eastern Mexico; introduced into the Bahamas.

Illustration: Roig, Cact. Fl. Cub. pl. 3, f. 3.

Plate XXXVI, figure 2, shows a specimen collected by J. A. Shafer on Cayo Guayaba, Cuba, in 1909, which flowered in the New York Botanical Garden, May 14, 1913; figure 3 is from a specimen collected in Cuba by Dr. Britton, which flowered and set fruit in 1915.

Fig. 279.—Piece of branch of S. macdonaldiae.   ×0.5.

10. **Selenicereus macdonaldiae** (Hooker) Britton and Rose, Contr. U. S. Nat. Herb. **12**: 430.   1909.

   Cereus macdonaldiae Hooker in Curtis's Bot. Mag. **79**: pl. 4707.   1853.

The old stems always terete, 10 to 15 mm. in diameter; younger stems somewhat 5-angled, giving off aërial roots, with rather prominent, flattened tubercles 1 to 5 cm. apart, 2 to 3 mm. high; spines

PLATE XXXV

M. E. Eaton del.

Flower on branch of *Selenicereus coniflorus.*
(Natural size.)

several, 2 mm. long or less; flowers 30 to 34 cm. long; outer perianth-segments and upper scales linear, yellow, the outermost scales red or brownish; inner perianth-segments pure white, 10 mm. long, oblanceolate, 2 to 3 cm. broad at widest point, acute; tube proper 12 cm. long, clothed with small scales bearing brown hairs and spines in their axils; fruit oblong, about 8 cm. long.

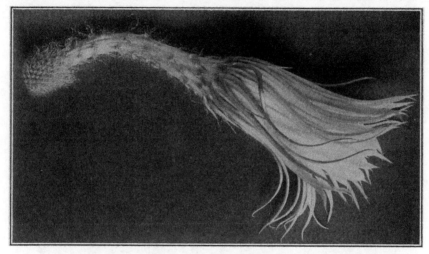

FIG. 280.—Selenicereus macdonaldiae.

Type locality: Cited as Honduras.

Distribution: According to Dr. Spegazzini (Anal. Mus. Nac. Buenos Aires III. 4:484), it is found in Uruguay and Argentina, and he thinks that Maldonado, near Montevideo, is the type locality; we know only plants in cultivation.

Cereus donatii (Schumann, Monatsschr. Kakteenk. 13:185. 1903), first listed in Haage and Schmidt's Catalogue, seems to belong here.

Cereus grusonianus Weingart (Monatsschr. Kakteenk. 15: 54. 1905) is apparently a race of this species, judging from the description and from small plants at the New York Botanical Garden.

Illustrations: Cact. Journ. 2: 135; Curtis's Bot. Mag. 79: pl. 4707; Fl. Serr. 9: pl. 896, 897; Cassell's Dict. Gard. 1: 194; Monatsschr. Kakteenk. 14:57, all as *Cereus macdonaldiae*; Contr. U. S. Nat. Herb. 12: pl. 76.

Figure 280 shows a flower of a plant in the collection of the United States Department of Agriculture; figure 279 shows a piece of a branch from a plant in the New York Botanical Garden; figure 281 shows a fruiting branch.

11. Selenicereus hamatus (Scheidweiler) Britton and Rose, Contr. U. S. Nat. Herb. 12: 430. 1909.

Cereus hamatus Scheidweiler, Allg. Gartenz. 5: 371. 1837.
Cereus rostratus Lemaire, Cact. Aliq. Nov. 29. 1838.

FIG. 281.—Selenicereus macdonaldiae.

Stem bright green, long and clambering, the branches strongly 4-angled, rarely 3-angled, about 1.5 cm. thick; areoles with spines and black wool, remote, at the upper edges of knobby projections, these often forming obtuse, deflexed spurs about 1 cm. long; spines on juvenile plants bristle-like, white, on old branches fewer, stouter, brown or black; flower 20 to 25 cm. long; upper scales dark green, tinged with red; outer perianth-segments pale green, narrow, about 8 cm. long; inner perianth-segments broad, white; flower-tube 10 cm. long, 22 mm. in diameter, its areoles long-hairy; filaments, style, and stigma-lobes yellow.

Fig. 282.—Selenicereus hamatus.

Type locality: Mexico.

Distribution: Southern and eastern Mexico.

According to the Index Kewensis *Cereus rostratus* occurs on the island of Antigua, but Dr. Rose was unable to find it there in 1913.

Fig. 283.—Part of branch of S. hamatus. ×0.5.

This species is common in cultivation in greenhouses and is occasionally seen in yards and patios in Mexico. Although we have seen no wild specimens, it seems to be common along the eastern coast of Mexico, probably in the wooded regions.

M. E. Eaton del.

1. Fruit of *Hylocereus trigonus*.
2. Flower of *Selenicereus boeckmannii*.
3. Fruit of *Selenicereus boeckmannii*.
(All natural size.)

Illustrations: Monatsschr. Kakteenk. **9**: 23; Rep. Mo. Bot. Gard. **16**: pl. 11, f. 4, 5; Schumann, Gesamtb. Kakteen Nachtr. f. 7; Blühende Kakteen **3**: pl. 161, 162; Wildeman, Icon. Select. **3**: pl. 103, all as *Cereus hamatus;* Möllers Deutsche Gärt. Zeit. **14**: 340; De Laet, Cat. Gen. f. 30, as *Cereus rostratus;* Rev. Hort. Belge **40**: after 184, as *Cereus kostratus;** Bull. Brooklyn Inst. Arts and Sci. **5**: 236, 237 (2 figures).

Figure 282 is from a photograph of a flower, taken at the New York Botanical Garden on the evening of October 10, 1910; figure 283 shows a part of a branch.

FIG. 284.—Selenicereus vagans.

12. **Selenicereus vagans** (K. Brandegee).

Cereus vagans K. Brandegee, Zoe **5**: 191. 1904.
Cereus longicaudatus Weber in Gosselin, Bull. Mus. Hist. Nat. Paris **10**: 384. 1904.

Stems creeping over rocks, often forming large clumps, more or less rooting, 1 to 1.5 cm. in diameter; ribs about 10, low; areoles 1 to 1.5 cm. apart; spines acicular, numerous, less than 1 cm. long, brownish yellow; flower 15 cm. long; tube, including throat, about 9 cm. long, slightly curved, brownish, with small scattered scales bearing clusters of 5 to 8 acicular spines in their axils; throat narrow, 5 cm. long; outer perianth-segments linear, brownish to greenish white, 6 cm. long; inner

*Doubtless error for *rostratus.*

perianth-segments white, oblanceolate, 6 cm. long, with short acuminate tips, the margins undulate or toothed, especially above; stamens numerous, weak; filaments white or white with greenish bases; style greenish or greenish with cream-colored upper part, slender; stigma-lobes 12, linear; ovary covered with acicular spines.

Type locality: Mazatlan, Mexico.

Distribution: Western coast of Mexico.

Illustration: Pamphlet descriptive of Carnegie Institution of Washington, seventh and eighth issues, p. 23 (reproduced here on p. 239).

Figure 284 shows a flower of a plant which bloomed at the National Botanical Garden, Washington, D. C., in 1905; figure 285a shows a tip of shoot and 285b a shoot with flower-bud, from specimens grown at the United States Department of Agriculture.

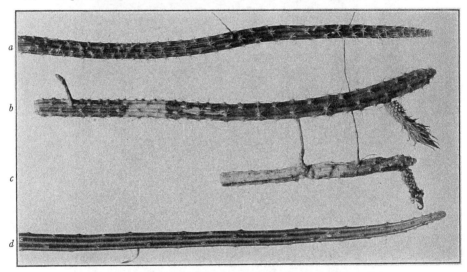

FIG. 285.—*a* and *b*, Selenicereus vagans; *c* and *d*, Selenicereus murrillii.   ×0.66.

13. Selenicereus murrillii sp. nov.

A very slender vine, 6 meters long or more, 8 mm. in diameter, freely giving off long slender aërial roots, dark green with the ribs more or less purplish, the scaly leaves at tips of branches minute, pinkish; ribs 7 or 8, low, obtuse, separated by low broad intervals; areoles 1 to 2 cm. apart, small, bearing white wool and minute spines; spines 5 or 6, minute, the two lower ones longer and reflexed, 1 to 2 cm. long; the other spines conic, greenish to black; flower-buds small, oblong, long-acuminate; flower opening at night, 15 cm. long, 15 cm. broad from tip to tip of the outer perianth-segments; tube and throat 6 cm. long, purplish green without, narrowly funnelform, bearing a few slightly elevated areoles, these white-felted and bearing one or two minute spines, the scales on the tube minute but those on the throat lanceolate, 3 to 10 mm. long and widely spreading even on the flower-buds; tube-proper smooth within; throat about 2 cm. long, covered with stamens; outer perianth-segments 12 to 14, greenish yellow or the outer ones purplish on the back, widely spreading, linear to linear-lanceolate, acute; inner perianth-segments pure white except the outermost ones and these greenish, together forming a campanulate corolla; segments broadly spatulate, 4 to 5 cm. long, obtuse; stamens numerous, slender, weak and somewhat declining on the perianth-segments, cream-colored; style slender, weak, cream-colored; stigma-lobes 9, linear, cream-colored; ovary bearing numerous rather large areoles, these white-felted and with 1 to 3 short spines but no long hairs.

Collected by Dr. W. A. Murrill, near Colima, Mexico, in 1910 (No. 31802 N. Y. B. G.). Although we have had it growing in Washington and New York for more than eight years, we have obtained but one flower. It grows vigorously, giving off many long aërial roots, soon reaching the top of the greenhouses. It has occasionally made small flower-buds, but these soon fall. Toward the last of May 1918, plants in Washington began to develop numerous flower-buds and gave every promise of an abundance of flowers, but a very hot spell occurred the first of June when the thermometer in the greenhouse rose to 114° Fahrenheit, and all the buds but one were killed. The plant, doubtless, needs half-shade conditions. Now that we have studied a mature flower we feel justified in referring this plant to *Selenicereus*, although it does not belong with the typical forms. The flower-bud and flower are similar to those of *S. vagans*. The flower itself in its bell-shaped perianth of short white segments, in its funnel-shaped flower-tube bearing scattered areoles, and in its ovary with short stubby spines resembles very much species of *Acanthocereus* but in habit and other respects it is very different.

Figure 285c shows a branch with young flower-buds, 285d a terminal shoot.

14. Selenicereus spinulosus (De Candolle) Britton and Rose, Contr. U. S. Nat. Herb. **12**: 431. 1909.

Cereus spinulosus De Candolle, Mém. Mus. Hist. Nat. Paris **17**: 117. 1828.

Stems clambering, 2 to 4 meters long, 1 to 2 cm. in diameter, producing numerous aërial roots, light green, somewhat shining, usually angled but sometimes nearly terete; ribs 4 to 6, or sometimes more; spines very short, yellowish or becoming blackish; radial spines 5 or 6, with 2 reflexed bristles at the base of the areole; central spine 1, rarely 2, on juvenile branches more numerous and more acicular, white; flower 12 to 14 cm. long; its tube about 5 cm. long, with a few clusters of small spines; outer perianth-segments narrowly oblong, 5 to 6 cm. long, acute, spreading; inner perianth-segments pinkish to white, narrowly oblong, acute; stamens white, attached along the inner surface of the throat; stigma-lobes white; ovary covered with clusters of spines similar to those on the tube.

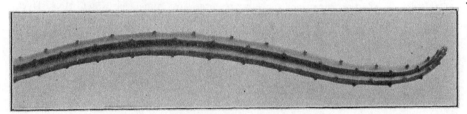

FIG. 286.—Selenicereus spinulosus. ×0.66.

Type locality: Mexico.

Distribution: Eastern Mexico to southeastern Texas.

Illustration: Blühende Kakteen **1**: pl. 53, as *Cereus spinulosus*.

Plate XXXVIII, figure 2, shows a flowering branch of a specimen obtained by Dr. Rose from Texas in 1900, which flowered in the New York Botanical Garden, April 9, 1912. Figure 286 shows a growing shoot from a plant obtained by Dr. E. Palmer at Victoria, Mexico, in 1907.

15. Selenicereus inermis (Otto).

Cereus inermis Otto in Pfeiffer, Enum. Cact. 116. 1837.
Cereus karstenii Salm-Dyck, Cact. Hort. Dyck. 1849. 218. 1850.

Creeping or clambering over rocks and bushes, deep green, the branches 1 to 2.5 cm. thick, 3 to 5-ribbed or angled, the ribs compressed, acute, undulate; old branches naked but young branches bearing setae from the small areoles; areoles remote, sometimes 6 cm. apart, when young each borne on a knob or elevation terminating in a subtending tip or scale; flower just before opening 15 cm. long, with a long acuminate tip, nocturnal; outer perianth-segments linear-oblong, 9 to 10 cm. long, 8 to 10 mm. broad, yellowish green, but more or less purplish at base; inner perianth-segments oblong, 8 to 9 cm. long, pure white except the pinkish bases; filaments numerous, slender, weak, white; style very thick, hollow, 7 mm. in diameter, pinkish, 15 cm. long; stigma-lobes numerous, greenish, 12 mm. long; flower-tube green, 8 cm. long, cylindric, 1.5 cm. in diameter, bearing a few scattered areoles, these brown-felted and with a cluster of 10 to 15 brown acicular spines, 1 cm. long or less and each subtended by an ovate linear scale; areoles on ovary closely set with clusters of brown acicular spines but no hairs.

Type locality: La Guayra, Venezuela.

Distribution: Venezuela and Colombia.

Flower description drawn from flower opening in the New York Botanical Garden in June 1917, on specimen obtained from M. Simon in Paris, 1905.

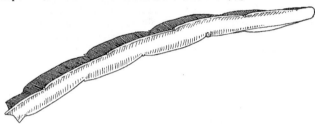

FIG. 287.—Joint of Selenicereus inermis. X0.5.

Cereus karstenii was sent by Hermann Karsten from Colombia and was described as near *Cereus inermis*, but twice as slender. We find, however, that true *Cereus inermis*, especially in cultivation, becomes elongated and slender. In the Jardin des Plantes, Paris, Dr. Rose found specimens labeled *Cereus karstenii* which proved to be only slender forms of *S. inermis*. In 1916 Dr. Rose collected *S. inermis* at its type locality and obtained fruit of this species for the first time. *C. inermis laetevirens* Salm-Dyck (Cact. Hort. Dyck. 1849. 51. 1850) is only a name.

Figure 287 shows a joint of a plant collected by Dr. Rose between Carácas and La Guayra, Venezuela, in 1916.

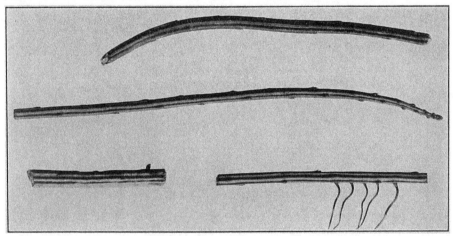

FIG. 288.—Branches of Selenicereus wercklei. X0.66.

16. Selenicereus wercklei (Weber).

Cereus wercklei Weber, Bull. Mus. Hist. Nat. Paris **8**: 460. 1902.

Epiphytic, slender, much branched, freely rooting, the young growth producing small swollen knobs at the areoles tipped by small red scale-like leaves; branches pale green, 5 to 15 cm. in diameter, nearly terete, with 6 to 12 faint ribs; areoles minute, each bearing a small tuft of felt subtended by a small scale but no spines; flower 15 to 16 cm. long, bright red; outer perianth-segments narrow, greenish, spreading; inner perianth-segments oblong; flower-tube narrow; style green at base, pink in the middle, nearly white above; ovary spiny; fruit ovoid, yellow, bearing clusters of brown spines at the areoles.

Type locality: Cerro Mogote, near Miravalles, Costa Rica.

Distribution: Costa Rica.

It resembles some species of *Rhipsalis* in its epiphytic habit and in its long, slender, naked branches, but not in its flower. In its naked stems, large flowers, and spiny fruit it resembles *S. inermis*, but differs from it in its many low ribs. We have had this plant under observation for a number of years but it has flowered only once. We have seen a second flower which Mr. Otón Jiménez brought us in alcohol from Costa Rica in 1919. He states that even in Costa Rica the plant rarely flowers.

Figure 288 shows branches of *Selenicereus wercklei* from a plant grown in Washington which was sent from Costa Rica by O. Jiménez; figure 289 shows a plant which flowered in the New York Botanical Garden in 1918.

Fig. 289.—Flowering plant of Selenicereus wercklei.

DESCRIBED SPECIES, PERHAPS OF THIS GENUS.

CEREUS ACANTHOSPHAERA Weingart, Monatsschr. Kakteenk. **24**: 81. 1914.

"Dark green, climbing in trees and hanging down, branched at the base; branches uniform, 3 to 7 meters long, with narrow, short, equal, rectangular joints and 4 or 5 compressed-winged ribs; sinuses acute; areoles small, with scanty shining tomentum; spines acicular, 1 to 3, diverging, short, and brownish above; flower unknown; fruit large, round, pendent, yellowish green, pilose and very spiny, crowned by the rotting perigon."

Type locality: On Rio de Santa Maria, State of Vera Cruz, Mexico.

Distribution: Known only from the type locality, and to us only from the description.

It may be a near relative of *Deamia testudo*. In Mexico the two plants are found in the same river valley.

CEREUS HUMILIS De Candolle, Prodr. **3**: 468. 1828.

Plant low, 2.5 cm. in diameter, with spreading, elongated, rooting branches; ribs 4 or 5, strongly compressed, repand; areoles 8 mm. apart, bearing white felt or nearly naked; spines 4 to 8 mm. long; radial spines 8 to 12, setaceous, white; central spines 3 or 4, stouter than the radial, straw-colored.

This species was described by De Candolle in 1828, who stated that the flowers and the country from which it came were unknown. Salm-Dyck had also sent it to him as a new species, under the name of *Cereus gracilis*. In 1837, Pfeiffer redescribed the species, adding the variety *minor* Pfeiffer (Enum. Cact. 115), which latter he described as having fasciculate, slenderer branches and subsetaceous spines. He gave as synonyms of this variety *Cereus mariculi* Hortus and *C. myriacaulon* Martius, sometimes misspelled *nyriacaulon*.

Lemaire in 1839 listed the species and also described the variety *major* Lemaire (Cact. Gen. Nov. Sp. 80. 1839), which he stated to be three times as stout as the species. To the variety he referred *C. rigidus* Lemaire, but this he seems never to have described.

In 1913 Weingart sent Dr. Rose a cutting labeled *Cereus rigidus* which is still growing in the Cactus House of the U. S. Department of Agriculture but it has never flowered. It gives off aërial roots and otherwise looks like a *Selenicereus* but is clearly distinct from any of our described species. The stem is slender, about 8 mm. in diameter, strongly 5-angled; areoles closely set, about 8 mm. apart; spines small, acicular, the centrals a little stouter than the radials, bulbose at base and yellowish brown in color. Weingart's plant proves to be the same as No. 6791 received from M. Simon of St. Ouen, Paris, under the name *Cereus pentagonus*, at the New York Botanical Garden.

Salm-Dyck in 1845 listed the varieties *rigidior* and *myriacaulon.** The latter name he published in 1850 (Cact. Hort. Dyck. 1849. 22, 222), when he states that the species has short spreading branches about 7.5 cm. long, while the variety is even shorter, slenderer, and often appressed to the ground. He would refer here *Cereus pentalophus radicans* De Candolle (Mém. Mus. Hist. Nat. Paris **17**: 117. 1828).

Several of the West Indian species of *Selenicereus* are known to us to develop very little for long periods after commencing growth; we suspect that the name *Cereus humilis* was based on a plant in that condition.

CEREUS MAYNARDII Paxton, Bot. Mag. **14**: 75. 1847.

> *Cereus grandiflorus speciosissimus* Pfeiffer, Enum. Cact. 113. 1837.
> *Cereus grandiflorus hybridus* Haage in Förster, Handb. Cact. 415. 1846.
> *Cereus grandiflorus maynardii* Paxton, Rev. Hort. III. **1**: 285. 1847.
> *Cereus fulgidus*† Hooker in Curtis's Bot. Mag. 96: pl. 5856. 1870.
> *Cereus grandiflorus ruber* Rümpler in Förster, Handb. Cact. ed. 2. 751. 1885.

Stems bright green, 3 or 4-angled, 3.5 cm. in diameter; spines about 9 in each cluster, acicular, 12 to 18 mm. long, straw-colored, with brown tips; flowers 15 to 18 cm. broad; flower-tube 7.5 to 10 cm. long, bearing small red scales with hairs in their axils; flower parts in several series, scarlet; stamens numerous, shorter than the inner perianth-segments; style elongate; stigma-lobes numerous, linear, white.

This is known to be of hybrid origin, being a cross between *Selenicereus grandiflorus* and *Heliocereus speciosus*.

The publication of the combination *Cereus maynardii* has been only incidental and is attributed to both Paxton and Lemaire. As it is named for Viscountess Maynard, it should have been spelled *maynardae*.

Illustrations: Paxton's Bot. Mag. **14**: pl. opp. 75, as *Cereus grandiflorus maynardi;* Fl. Serr. **3**: pl. 233, 234, as *Cereus grandifloro-speciosissimus maynardii;* Curtis's Bot. Mag. **96**: pl. 5856, as *C. fulgidus;* Deutsche Gärt. Zeit. **9**: 276, as *C. hybridus.*

4. MEDIOCACTUS gen. nov.

A more or less epiphytic cactus, usually growing in trees, with long procumbent branches; branches usually 3-winged, slender, producing aërial roots, the areoles short-spiny; flowers large, funnelform, nocturnal, the tube bearing distant scales; inner perianth-segments white; ovary tuberculate, its felted and spiny areoles subtended by small scales; fruit oblong, red, its areoles felted and spiny.

In habit and flowers this plant much resembles *Hylocereus*, but differs from it in its tuberculate ovary and in the felted and spine-bearing areoles of the fruit, which resemble those of *Selenicereus*.

The genus has 2 species, so far as known to us, the type being *Cereus coccineus* Salm-Dyck. Its name implies intermediate characters as it suggests both *Hylocereus* and *Selenicereus*.

---

* Walpers (Repert. Bot. **2**: 278. 1843) gave this variety as a synonym of this species.
† *Cereus fulgens* (Monatsschr. Kakteenk. **6**: 190. 1896) is a misspelling.

KEY TO SPECIES.

Flowers 25 to 30 cm. long; eastern coast of South America..................................1. *M. coccineus*
Flowers 38 cm. long; western Andes...2. *M. megalanthus*

1. Mediocactus coccineus (Salm-Dyck).

Cereus coccineus Salm-Dyck in De Candolle, Prodr. 3: 469. 1828.
Cereus setaceus Salm-Dyck in De Candolle, Prodr. 3: 469. 1828
Cereus setaceus viridior Salm-Dyck, Hort. Dyck. 65. 1834.
Cereus lindbergianus Weber in Schumann, Gesamtb. Kakteen 151. 1897.
Cereus lindmanii Weber in Schumann, Gesamtb. Kakteen 163. 1897.
Cereus hassleri Schumann, Monatsschr. Kakteenk. 10: 45. 1900.

FIG. 290.—Mediocactus coccineus.

Stems usually climbing on trees, sometimes clambering over rocks or walls, developing many aërial roots, the joints pale green, various, sometimes 8 cm. broad, often only 2 cm. broad; angles usually 3, but sometimes 4 or even 5 on the same plant; young areoles 5 to 10 mm. apart, bearing brown felt and 10 to 15 white, radial, deciduous bristles followed by several spines; areoles of mature branches 2 to 3 cm. apart; spines at first pinkish, then brown or yellowish brown, conic, 1 to 2 mm. long, more or less swollen at base, usually only 2 or 3, sometimes more, rarely only 1; flowers 25 to 30 cm. long; outer perianth-segments linear, green, widely spreading; inner perianth-segments erect, broader than outer, upper margins serrate; style exserted, yellow; stigma-lobes about 16, linear, entire, yellow; the fruit somewhat pointed, 7 cm. long, edible, strongly tuberculate when young, its areoles bearing a cluster of spines 1 to 2 cm. long; flesh white; seeds black.

Type locality: Brazil.

Distribution: Argentina to Brazil.

All writers on the *Cactaceae*, including Salm-Dyck, are agreed that the *Cereus coccineus* described by De Candolle (Prodr. 3: 469. 1828) is different from the plant after-

FIG. 291.—Mediocactus coccineus.

wards described by Salm-Dyck under that name. This name of De Candolle has priority of place over *Cereus setaceus* and is, therefore, adopted by us for this well-known plant of eastern South America. The name *coccineus* was evidently given because the flowers were supposed to be red but it would very properly apply to the color of the fruit.

A plant was found growing on a garden wall, half-wild, at Cali, Cauca Valley, Colombia, December 1905, by H. Pittier, but we do not know it to be a native of Colombia.

Cereus prismaticus Salm-Dyck (De Candolle, Prodr. 3:469. 1828. Not Haworth. 1819) is doubtless a *Mediocactus;* if really of South American origin, as stated by Schumann, it is probably *M. coccineus.*

Illustrations: Pfeiffer and Otto, Abbild. Beschr. Cact. 1: pl. 16, as *Cereus setaceus;* Vellozo, Fl. Flum. 5: pl. 24, as *Cactus triangularis.*

Plate XIII, figure 3, shows a fruiting branch and plate XXXVII a flowering branch of plants in the collection of the New York Botanical Garden. Figure 290 is from a photograph taken by Paul G. Russell at Nichteroy, Brazil, in 1915; figure 291 is from a photograph of a branch bearing young fruit collected by H. Pittier from a half-wild plant at Cali, Cauca Valley, Colombia, in 1905, possibly referable to the following species.

2. Mediocactus megalanthus (Schumann).

Cereus megalanthus Schumann, Bot. Jahrb. Engler 40: 412. 1907.

Growing in trees, forming masses of long pendent branches; branches often only 1.5 cm. broad, rooting freely, 3-angled; margin of angles only slightly undulating; spines 1 to 3, yellowish, 2 to 3 mm. long, when young associated with several white bristles; flowers very large, 38 cm. long, white; inner perianth-segments 11 cm. long, 3.5 cm. broad; stamens numerous; stigma-lobes numerous.

Type locality: Near the town of Tarapoto, Department of Loreto, eastern Peru.

Distribution: Andes of Peru and possibly Colombia and Bolivia.

This species was very briefly described at place cited above and had been previously illustrated (see below). Vaupel (Notizbl. Bot. Gart. Berlin 5: 284. 1913) has published an extended account which enables us to refer the plant definitely to this genus.

In 1914 Mr. Weingart sent Dr. Rose a cutting of this species but it has grown little since, although it has developed long aërial roots. It was briefly described in a Kew Bulletin (Kew Bull. Misc. Inf. 1914; App. 61. 1914).

The plant seems to have one of the largest flowers known among cacti and, according to Vaupel, is rivaled only by *Selenicereus urbanianus.*

A specimen similar to *Cereus megalanthus* was collected growing in trees by R. S. Williams at Charopampa, Bolivia, September 27, 1901 (No. 881). Mr. Williams says that his plant was many yards in length. It is without flowers or fruit.

Illustration: Karsten and Schenck, Vegetationsbilder 2: pl. 5, as *Cereus megalanthus.* Figure 292 is a reproduction of the illustration above cited.

5. DEAMIA gen. nov.

An elongated cactus, clambering over or pendent from rocks or climbing and growing on bark of living trees, the joints usually broadly 3-winged, but sometimes 5 to 8-ribbed or winged, clinging by aërial roots; spines of the areoles numerous, acicular, or in juvenile forms bristly; flowers diurnal, very large, the tube slender, elongated; throat funnelform; inner perianth-segments yellowish white; stamens numerous, slender, attached all over the throat; style rather slender; scales on ovary and tube very small, bearing 3 to 5 long brown bristles in their axils; stigma-lobes linear, entire; fruit not known.

A monotypic genus of Mexico, Central America, and Colombia, dedicated to Charles C. Deam, a diligent botanical collector, who sent the plant to us from Guatemala.

PLATE XXXVII

M. E. Eaton del.

Flower of *Mediocactus coccineus*. × 0.7.

1. **Deamia testudo** (Karwinsky).

> *Cereus testudo* Karwinsky in Zuccarini, Abh. Bayer, Akad. Wiss. München **2**: 682.   1837.
> *Cereus pterogonus* Lemaire, Cact. Gen. Nov. Sp. 59.   1839.
> *Cereus pentapterus* Otto in Salm-Dyck, Cact. Hort. Dyck. 1849. 221.   1850.
> *Cereus miravallensis* Weber, Bull. Mus. Hist. Nat. Paris **8**: 459.   1902.
> *Selenicereus miravallensis* Britton and Rose, Contr. U. S. Nat. Herb. **12**: 431.   1909.

Stems and joints various, 3 to 10 cm. broad, or perhaps even more; ribs thin, wing-like, 1 to 3 cm. high; areoles 1 to 2 cm. apart or on juvenile growth much closer together; spines spreading, 10 or more, 1 to 2 cm. long, brownish; flowers 28 cm. long, with a long slender tube 10 cm. long expanding into a broad throat nearly as long as the tube; inner perianth-segments linear-oblong, acuminate, 8 to 10 cm. long; stamens numerous; style slender, long, 24 to 25 cm. long; stigma-lobes linear, numerous; scales on ovary 1 mm. long or less; hairs on ovary and flower-tube brown, 1 to 3 cm. long.

Type locality: Mexico.
Distribution: Southern Mexico to Colombia.

Vaupel (Blühende Kakteen **3**: pl. 150. 1913) doubtfully refers here *Cereus pentagonus* Vellozo, both described and figured by Vellozo (Fl. Flum. **5**: pl. 22. text. ed. Netto 195). Vellozo's plate, however, represents *Cereus pernambucensis*.

This species, although described as *Cereus testudo* in 1837, has long been passing in collections as *Cereus pterogonus*, a later name. It has a rather wide range and there is considerable variation in stems and flowers. It needs more detailed observation than it has yet received.

FIG. 292.—Mediocactus megalanthus.   FIG. 293.—Deamia testudo.

Illustrations: Blühende Kakteen **3**: pl. 150; Curtis's Bot. Mag. **89**: pl. 5360, both as *Cereus pterogonus.*

Figure 293 is from a photograph taken by E. A. Goldman near Carrizal, Vera Cruz, Mexico, in 1901; figure 294 shows branches from a plant sent from Costa Rica in 1911.

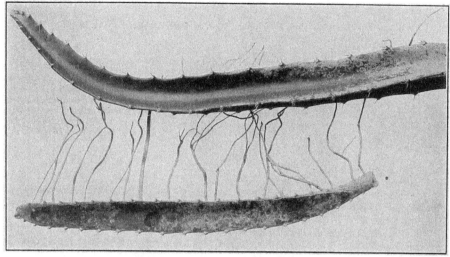

FIG. 294.—Branches of Deamia testudo. ×0.66.

6. WEBEROCEREUS Britton and Rose, Contr. U. S. Nat. Herb. **12**: 431. 1909.

Epiphytic cacti, with slender, climbing or hanging stems and branches, these terete, angled or rarely flattened, emitting aërial roots, the areoles bearing a tuft of felt and sometimes several weak acicular bristles or spines; flowers pink, rose-colored or white, nocturnal, short-funnelform or funnelform-campanulate; ovary tuberculate, its areoles bearing weak filiform bristles or stiff hairs, the lower part of the flower-tube with a few similar areoles, the upper part with a few foliaceous scales; outer perianth-segments reflexed-spreading, blunt, linear-oblong, the inner ones narrower; ovary hairy or bristly; areoles of the fruit hairy.

Type species: *Cereus tunilla* Weber.

Three species are here recognized, two from Costa Rica and one from Panama. They are all rather insignificant plants, growing in trees as does *Rhipsalis;* the seedlings and juvenile growths are similar to those of species of that genus, but the large flowers and fruits are quite different.

The genus was named for Dr. Albert Weber (1830–1903) of Paris, who gave much attention to the cacti.

KEY TO SPECIES.

Inner perianth-segments pinkish.
    Branches usually strongly angled...1. *W. tunilla*
    Branches terete or slightly angled...2. *W. biolleyi*
Inner perianth-segments white...3. *W. panamensis*

1. Weberocereus tunilla (Weber) Britton and Rose, Contr. U. S. Nat. Herb. **12**: 432. 1909.

    Cereus tunilla Weber, Bull. Mus. Hist. Nat. Paris **8**: 460. 1902.
    Cereus gonzalezii Weber, Bull. Mus. Hist. Nat. Paris **8**: 460. 1902.

Stems climbing, 5 to 12 mm. in diameter, usually strongly 4-angled, rarely 2, 3, or 5-angled, but in juvenile forms nearly terete; spines 6 to 12, stiff, swollen at base, yellowish at first, soon brown, 6 to 8 mm. long; flowers 5 to 6 cm. long, pinkish; outer perianth-segments linear, brownish, spreading or reflexed; inner perianth-segments oblong, acute, pink; filaments and style included, pinkish; stigma-lobes whitish; ovary strongly tubercled; tubercles bearing several yellow bristles.

Type locality: Near Tablón, southwest of Cartago, Costa Rica.
Distribution: Costa Rica.
Illustration: Curtis's Bot. Mag. **144**: pl. 8779, as *Cereus tunilla*.

M. E. Eaton del.

1. Fruiting branch of *Selenicereus pteranthus*.
2. Flowering branch of *Selenicereus spinulosus*.
3. Flowering branch of *Weberocereus panamensis*.
(All natural size.)

Plate XXXIX, figure 1, shows a flowering branch of a plant obtained by Wm. R. Maxon in Costa Rica in 1906, which flowered in the New York Botanical Garden in May 1913.

2. Weberocereus biolleyi (Weber) Britton and Rose, Contr. U. S. Nat. Herb. **12**: 431. 1909.

> *Rhipsalis biolleyi* Weber, Bull. Mus. Hist. Nat. Paris **8**: 467. 1902.
> *Cereus biolleyi* Weber in Schumann, Gesamtb. Kakteen Nachtr. 60. 1903.

Branches long, slender, and flexuous, climbing over or hanging from branches of trees, 4 to 6 mm. in diameter, terete or slightly angled, in juvenile plants often flattened or 3-winged, usually spineless but occasionally bearing 1 to 3 yellow spines from an areole; areoles small, remote; flowers 3 to 5 cm. long; all perianth-segments oblong, obtuse, the inner pinkish; ovary tuberculate, hairy.

Type locality: Vicinity of Port Limón, Costa Rica.
Distribution: Costa Rica.

The branches are often only 4 mm. in diameter and spineless. When cuttings are made from these branches queer juvenile forms develop. In one case a flat, thin, 2-edged branch 10 mm. broad was produced with closely set areoles filled with white, bristle-like hairs; from the same cuttings a similar branch was developed, but 3-angled, like a juvenile *Hylocereus.*

Plate XXXIX, figure 2, is from a plant collected at Zent, Costa Rica, by H. Pittier, which flowered in the New York Botanical Garden, July 18, 1913.

3. Weberocereus panamensis sp. nov.

Stems 1 to 2 cm. broad, strongly 3-angled or some joints flat; margins acute, indented; areoles small, each hidden beneath a small thick scale, sometimes bearing 1 to 3 short weak spines; flower 4 to 7 cm. long; outer perianth-segments and inner scales yellowish green, erect; inner perianth-segments white, oblong; tube proper smooth and white within; throat 1 cm. long; stamens included; filaments white, a part attached to the lower face of the throat and a part to the upper margin; style white, included; stigma-lobes white (in wild state said to be purple); ovary tuberculate, green, with spreading scales, each subtending 4 to 8 long white hairs; fruit red, 2 to 3 cm. in diameter, tubercled, at least when young.

FIG. 295.—Weberocereus panamensis.

Collected in forest thickets along the Rio Fato, Province of Colon, Panama, July 1911, by H. Pittier (No. 3903) and flowered first in Washington in 1913.

Plate XXXVIII, figure 3, is from the type specimen, which flowered in the New York Botanical Garden, September 20, 1915. Figure 295 shows a fruiting branch collected by Mrs. D. D. Gaillard at Lake Gatun, Panama, in 1913.

PUBLISHED SPECIES, KNOWN TO US ONLY FROM DESCRIPTION.

Cereus estrellensis Weber (Monatsschr. Kakteenk. **15**: 167. 1905) is, according to C. Wercklé, similar to *Cereus nycticallus* but weaker and more spiny. The stems are 6-angled; the flowers are small, rosy to salmon-colored, and nocturnal. It is of Costa Rican origin, but is known to us only from this brief characterization and may belong to our genus *Weberocereus*.

7. WERCKLEOCEREUS Britton and Rose, Contr. U. S. Nat. Herb. **12**: 432. 1909.

Epiphytic, climbing cacti, the 3-angled branches emitting aërial roots, their areoles bearing short bristles or very weak spines and a tuft of felt; flowers short-funnelform, the tube rather stout; ovary and flower-tube bearing many areoles, each with several nearly black, acicular spines and a tuft of short black felt, subtended by minute scales; outer perianth-segments lanceolate, acutish, narrow; inner perianth-segments broader; stamens many; style about as long as the longer stamens, with several linear stigma-lobes; berry globose, its areoles spiny.

Two species are known, 1 in Costa Rica and 1 in Guatemala; both are in cultivation. The genus is dedicated to C. Wercklé, a Costa Rican collector.

Type species: *Cereus tonduzii* Weber.

In habit the plants resemble species of *Hylocereus*, but the flowers are very different.

KEY TO SPECIES.

Flowers 8 cm. long or less; stem-areoles at most bristly...1. *W. tonduzii*
Flowers 10 cm. long or more; stem-areoles with weak but definite spines............................2. *W. glaber*

1. Werckleocereus tonduzii (Weber) Britton and Rose, Contr. U. S. Nat. Herb. **12**: 432. 1909.

Cereus tonduzii Weber, Bull. Mus. Hist. Nat. Paris **8**: 459. 1902.

Stems rather stout, bushy-branched, the joints 3-angled, rarely 4-angled, deep green, not at all glaucous, climbing by aërial roots; margins of ribs nearly straight; areoles small, felted, without spines, but sometimes with weak bristles; flowers 8 cm. long or less, areoles of the ovary and tube bearing clusters of dark spines and short black wool; outer perianth-segments brownish, oblong, 1 to 2 cm. long; inner perianth-segments oblong, creamy white, 2.5 cm. long; stamens exserted; style longer than the stamens; berry globose, citron-yellow, its apex umbilicate, its flesh white.

Type locality: Copey, near Santa Maria de Dota, Costa Rica.
Distribution: Costa Rica.

In greenhouse cultivation some plants are remarkably floriferous and very conspicuous when in bloom.

Plate XXXIX, figure 3, shows part of a plant which flowered in the New York Botanical Garden, March 30, 1908. Figure 296 is from a photograph of a plant in the same collection.

2. Werckleocereus glaber (Eichlam) Britton and Rose, Addisonia **2**: 13. 1917.

Cereus glaber Eichlam, Monatsschr. Kakteenk. **20**: 150. 1910.

Stems slender, 3-angled, about 2 cm. broad, pale green and slightly glaucous, climbing by aërial roots; margins somewhat knobby, the areole borne on the upper part of the knob, small, 3 to 4 cm. apart; spines 2 to 4, short, 1 to 3 mm. long, acicular, but with swollen bases; flower 10 cm. long or more, the ovary and tube bearing clusters of yellow to brown acicular spines; inner perianth-segments white, oblanceolate, acute, somewhat serrate; style pale yellow, weak, resting on the under side of the flower-tube; stigma-lobes white; fruit not known.

Type locality: Western coast of Guatemala.
Distribution: Guatemala.

In habit this species much resembles *Wilmattea minutiflora*, also from Guatemala, but its flower characters are quite different.

Illustration: Addisonia **2**: pl. 47.

Plate XXXIX, figure 4, is from a specimen obtained by Dr. Rose from Guatemala, which flowered in the New York Botanical Garden, April 14, 1915.

PLATE XXXIX

M. E. Eaton del.

1. Flowering branch of *Weberocereus tunilla*.
2. Flowering branch of *Weberocereus biolleyi*.
3. Flowering branch of *Werckleocereus tonduzii*.
4. A flower of *Werckleocereus glaber*.
(All natural size.)

FIG. 296.—Werckleocereus tonduzii.

8. APOROCACTUS Lemaire, Illustr. Hort. 7: Misc. 67. 1860.

Slender, vine-like cacti, creeping or clambering, sending out aërial roots freely, day-blooming; flowers rather small, one at an areole, funnelform, pink to red, the tube nearly straight, or bent just above the ovary, the limb somewhat oblique; outer perianth-segments linear, spreading or recurved, scattered; inner perianth-segments broad, more compact than the outer perianth-segments; stamens exserted, in a single, somewhat 1-sided cluster; filaments attached all along the throat; tube proper about the length of the narrow throat; fruit globose, small, reddish, setose; seeds few, reddish brown, obovate.

We recognize 5 species, the typical one being *Cactus flagelliformis* Linnaeus.

This genus, first described by Lemaire, included not only the typical *A. flagelliformis*, but also *Cereus baumannii* and *C. colubrinus*, but the next year he withdrew the last two species. The name has never come into very general use, in spite of the good generic characters. The geographical distribution of the genus is uncertain. Three of the species are known to grow wild in Mexico, while *A. flagelliformis*, also common in Mexico, was very early introduced into Europe as from South America.

The name is from the Greek, signifying impenetrable cactus, of no obvious application.

KEY TO SPECIES.

Flowers strongly bent just above the ovary.
    Branches very slender; ribs 7 or 8...1. *A. leptophis*
    Branches stouter; ribs 10 to 12.
        Outer perianth-segments narrow; inner perianth-segments apiculate.................2. *A. flagelliformis*
        Outer perianth-segments oblong; inner perianth-segments acuminate.................3. *A. flagriformis*
Flowers nearly straight.
    Inner perianth-segments acute...4. *A. conzattii*
    Inner perianth-segments acuminate..5. *A. martianus*

1. Aporocactus leptophis (De Candolle) Britton and Rose, Contr. U. S. Nat. Herb. **12**: 435.   1909.

Cereus leptophis De Candolle, Mém. Mus. Hist. Nat. Paris **17**: 117.   1828.
Cereus flagelliformis leptophis Schumann, Gesamtb. Kakteen 143.   1897.

Often creeping; branches cylindric, 8 to 10 mm. thick, rather strongly 7 or 8-ribbed; ribs obtuse, somewhat repand; areoles velvety, with 12 or 13 rigid setaceous spines; flower-tube curved just above the ovary; perianth-segments narrowly oblong, 2 to 3 cm. long, about 6 mm. wide.

Type locality: Mexico.
Distribution: Mexico.
Illustrations: De Candolle, Mém. Cact. pl. 12; Förster, Handb. Cact. ed. 2. f. 96; Rümpler, Sukkulenten f. 68, all as *Cereus leptophis.*

Plate XL, figure 1, shows a flowering plant in the collection of the New York Botanical Garden.   Figure 297 is reproduced from the first illustration above cited.

2. Aporocactus flagelliformis (Linnaeus) Lemaire, Illustr. Hort. **7**: Misc. 68.   1860.

Cactus flagelliformis Linnaeus, Sp. Pl. 467.   1753.
Cereus flagelliformis Miller, Gard. Dict. ed. 8. No. 12.   1768.
Cereus flagelliformis minor Salm-Dyck in Pfeiffer, Enum. Cact. 111.   1837.*

Stems at first ascending or erect, but weak and slender or pendent, 1 to 2 cm. in diameter; branches often prostrate and creeping or even pendent; ribs 10 to 12, low and inconspicuous, a little tuberculate; areoles 6 to 8 mm. apart; radial spines 8 to 12, acicular, reddish brown; central spines 3 or 4, brownish with yellow tips; flowers 7 to 8 cm. long, opening for 3 or 4 days, crimson; outer perianth-segments narrow, more or less reflexed; inner perianth-segments broader, only slightly spreading; fruit globose, small, 10 to 12 mm. in diameter, red, bristly; pulp yellowish.

Type locality: At first supposed to be from South America.

Distribution: Reported from Mexico, Central America, and South America; nowhere known to us in the wild state.

Said to have been introduced from Peru in 1690, but, presumably, originally from Mexico. The species is widely cultivated in all tropical countries. It is very common in Mexico to see this plant about the houses of the poorer Mexicans, often planted in the end of a cow's horn and hung on the side of the house. This species, too, has cristate forms.

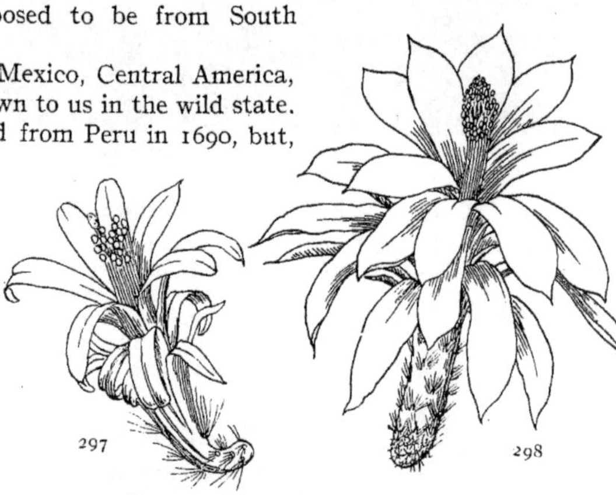

The plant is known as the rat-tail cactus and is much grown as a window plant. In Mexico the dried flowers are used

FIG. 297.—Flower of Aporocactus leptophis.
FIG. 298.—Flower of Aporocactus flagriformis.

as a household remedy and sometimes are sold in the drug markets under the name of flor de cuerno.

This species is recorded by Grisebach, citing Sloane and Swartz, as found in trees, in Jamaica along the coast, but it is not known to occur on that island at the present time. Sloane's description better applies to *Selenicereus grandiflorus.*

Cereus smithii Pfeiffer (Enum. Cact. 111. 1837) is a generic hybrid produced by adding the pollen of this species to one of the species of *Heliocereus* and was made by an English gardener, Mr. Mallison. It is figured in Curtis's Botanical Magazine for 1841 (**67**: pl. 3822) and in Edwards's Botanical Register (**19**: pl. 1565) and it was said to be one of the best hybrids which had yet been produced. The flower is nearly regular with scarcely any

Cereus minor (Weingart, Monatsschr. Kakteenk. **18**: 49.   1908) doubtless refers to the variety *minor* given above.

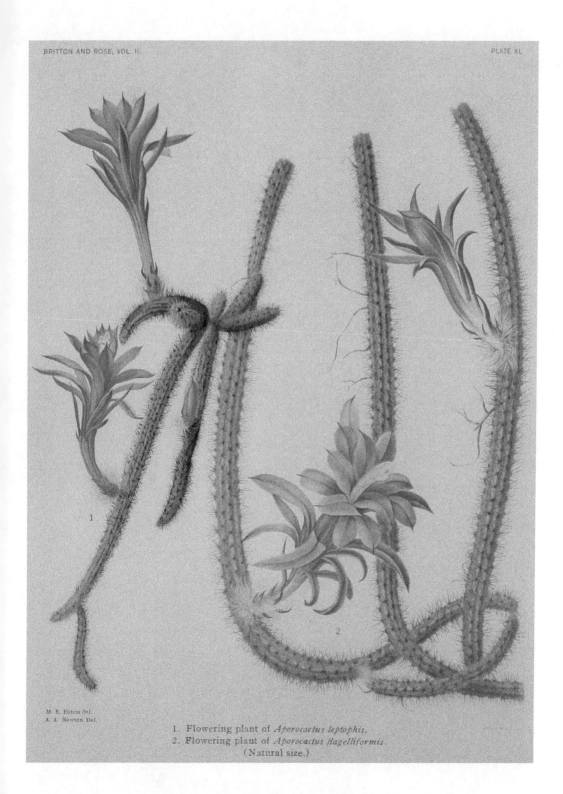

M. E. Eaton del.
A. A. Newton Del.

1. Flowering plant of *Aporocactus leptophis*.
2. Flowering plant of *Aporocactus flagelliformis*.
(Natural size.)

tube and with nearly erect filaments; the stem is weak and creeping, with about 6 angles; it is somewhat stouter than is *Aporocactus flagelliformis*. *Cereus mallisonii** (Pfeiffer, Enum. Cact. 111. 1837), *C. flagelliformis mallisonii* (Walpers, Repert. Bot. 2: 278. 1843), and *C. flagelliformis smithii* (Walpers, Repert. Bot. 2: 278. 1843) are other names for this same hybrid and this must also be *C. flagelliformis speciosus* Salm-Dyck (Cact. Hort. Dyck. 1849. 50. 1850) since it is based on the same illustration. While botanists generally refer the name, as we have above, to Pfeiffer, it was fully described and figured by Link and Otto (Verh. Ver. Beförd. Gartenb. 12: 134. pl. 1. 1837). While they announce that their plant was obtained by Mallison, as we state above, their illustration shows a very different flower from the one figured in the Botanical Register and suggests that it was from a different plant, although doubtless produced from the same parents. The flower differs from the other not only in its color but also in its narrower, more elongated tube. *Cereus crimsonii* (Pritzel, Icones 246. 1855) was also based on the plate in the Botanical Register (19: pl. 1565. 1833) and must represent this same hybrid.

Cereus aurora (Monatsschr. Kakteenk. 16: 81. 1906) is also of hybrid origin. According to E. Golz, one of its parents is some species of *Echinopsis*.

Cereus ruferi and *C. ruferi major* (Monatsschr. Kakteenk. 16: 10. 1906) are said to be hybrids of which *C. flagelliformis* is one of the parents.

Cereus moennighoffii Fischer (Monatsschr. Kakteenk. 15: 143. 1905) is a hybrid between this species and *C. martianus*. Other hybrids with *Cereus martianus* and *Epiphyllum ackermannii* have been reported.

Cereus vulcan (Monatsschr. Kakteenk. 16: 10. 1906) is a hybrid of *A. flagelliformis;* its other parent is unknown. It is illustrated by Rümpler (Sukkulenten f. 67).

There are several unpublished names which are referred to this species, among which are varieties *funkii, nothus, scotii,* and *smithii,* all in Walpers (Repert. Bot. 2: 278. 1843).

Illustrations: Safford, Ann. Rep. Smiths. Inst. 1908: f. 18; Stand. Cycl. Hort. Bailey 1: f. 237. Curtis's Bot. Mag. 1: pl. 17; De Candolle, Pl. Succ. Hist. 2: pl. 127; DeTussac, Fl. Antill. 2: pl. 28; Mag. Bot. and Gard. Brit. and For. 1: pl. 14, f. 4; Loudon, Encycl. Pl. f. 6875, as *Cactus flagelliformis;* Baillon, Hist. Pl. 9: f. 52, 53; Cact. Journ. 1: 10; Förster, Handb. Cact. ed. 2. f. 5; Martius, Fl. Bras. 4²: pl. 41, f. 2, all as *Cereus flagelliformis;* Rümpler, Sukkulenten f. 66, as *Cereus flagelliformis minor;* Trew, Pl. Select. pl. 30, as *Cereus.*

Plate XL, figure 2, shows a flowering plant in the collection of the United States Department of Agriculture.

3. **Aporocactus flagriformis** (Zuccarini) Lemaire in Britton and Rose, Contr. U. S. Nat. Herb. **12:** 435. 1909.

      Cereus flagriformis Zuccarini in Pfeiffer, Enum. Cact. 111. 1837.

At first erect and rather stout, afterwards creeping and very much branched; branches green, 10 to 24 mm. in diameter; ribs 11, very low, obtuse, somewhat tuberculate; areoles small, 4 to 6 mm. apart; radial spines 6 to 8, 4 mm. long, acicular, horn-colored; central spines 4 or 5, shorter than the radials but stouter, brown; flowers dark crimson, 10 cm. long, 7.5 cm. broad or more; flower-tube 3 cm. long or more; perianth-segments in 3 series, the series well separated; inner perianth-segments oblong, 10 mm. broad, acuminate; stamens red, erect, exserted; stigma-lobes 6, white.

Type locality: San José de l'Oro, Oaxaca, Mexico.

Distribution: Mexico.

This species seems not to have appeared in collections for a long time. As the type locality is known one would suppose it might have been reintroduced. We have repeatedly tried to have it re-collected but so far have failed; in putting forth this effort we have succeeded in discovering another species which is described below as new.

The binominal *A. flagriformis* appeared in Lemaire, Les Cactées, page 58, 1868, but it is not formally published at that place.

**Cactus mallisonii* is credited by the Index Kewensis to Loudon's Encyclopedia (Suppl. 1. 1202. 1840), but it appears there under *Cereus.*

Illustrations: Pfeiffer and Otto, Abbild. Beschr. Cact. 1: pl. 12; Engler and Prantl, Pflanzenfam. 3⁶ᵃ: f. 58, both as *Cereus flagriformis.*

Figure 298 is copied from the first illustration above cited.

4. Aporocactus conzattii sp. nov.

Creeping, clambering, or hanging from a support, developing aërial roots here and there; stems 12 to 25 mm. in diameter; ribs 8 to 10, rather prominent, low-tuberculate; aeroles 3 to 4 mm. apart; spines 15 to 20, acicular, light brown, unequal, the longest 12 mm. long; buds nearly erect, covered with brown acicular spines or bristles; flowers 8 to 9 cm. long; tube nearly straight, red, bearing a few ovate scales, their axils short-woolly and with a few bristle-like spines; limb slightly oblique, with a wide mouth; upper inner perianth-segments arching forward, the lower ones somewhat re-flexed, all narrow, 6 to 7 mm. broad, acute, brick-red; stamens and style shorter than the perianth-segments but exserted from the throat of the flower, long, connivent, nearly white; style slender, stiff, white, nearly 6 cm. long, not extending beyond the filaments; tube proper 2 to 2.5 cm. long; throat about 1 cm. long, narrow, bearing stamens all over its surface; filaments numerous, white; flower open for 2 days, remaining open at night.

FIG. 299.—Aporocactus conzattii. ×0.66.

Collected by Professor C. Conzatti in 1912 on Cerro San Felipe, Oaxaca, Mexico, and flowered in Washington, first on February 17, 1916, and again in 1917.

This species is near *Aporocactus martianus,* but the inner perianth-segments are not long-acuminate, and the flowers are smaller (at least than those shown in Hooker's illustration). It is a very valuable introduction for greenhouse culture.

Figure 299 is from a photograph of the type plant which flowered in Washington, in 1918; figure 300 shows a flower of the same.

5. Aporocactus martianus (Zuccarini).

Cereus martianus Zuccarini, Flora 15²: Beibl. 66. 1832.
Eriocereus martianus Riccobono, Boll. R. Ort. Bot. Palermo **8:** 240. 1909.

Stems rather stout, somewhat branched, 15 to 18 mm. in diameter; ribs about 8, low, obtuse; areoles 12 mm. apart; spines 6 to 8, acicular to bristle-like; flowers a deep rose-color, 8 to 10 cm. long; outer perianth-segments narrowly lanceolate, acuminate; inner perianth-segments similar but long-acuminate; style long exserted; fruit globular, 2 cm. in diameter, greenish, spiny.

Type locality: Mexico.

Distribution: Central Mexico.

This species is usually associated with *Aporocactus flagelliformis*, but, owing to its more regular flower, Berger was disposed to refer it elsewhere. It is known to us only from description and illustrations.

Illustrations: Blühende Kakteen **2**: pl. 65; Curtis's Bot. Mag. **66**: pl. 3768; Monatsschr. Kakteenk. **9**: 105; Rep. Mo. Bot. Gard. **16**: pl. 12, f. 1, as *Cereus martianus.*

Figure 301 is copied from Curtis's Botanical Magazine plate 3768.

The following description is based on a plant of this relationship, but it differs from figured specimens in the broader perianth-segments, shorter flower-tube, and red filaments. It was obtained from the Theodosia B. Shepherd Company, Ventura, California, and flowered in Washington, D. C., in 1916.

FIG. 300.—Flower of Aporocactus conzattii.
FIG. 301.—Flower of Aporocactus martianus.

Ribs 5 to 7, separated by broad intervals, somewhat undulate; areoles about 1 cm. apart, circular, bearing white wool and spines; spines about 10, acicular, yellow; flower-bud acute; flowers dark red, 9 cm. long, the tube shorter than the limb; tube-proper about 2 cm. long; throat 1.5 cm. long; scales on the ovary numerous, narrow, their axils with white wool and clusters of spines; axils of upper scales naked; outer perianth-segments narrow, acute; inner perianth-segments broadly lanceolate, 1.5 cm. broad, acute, carmine; stamens erect, not quite as long as the inner perianth-segments; filaments carmine; style carmine, weak, about as long as the filaments; stigma-lobes white; fruit not known.

The only flower seen opened in the early morning and was still open at half-past one.

9. **STROPHOCACTUS** Britton and Rose, Contr. U. S. Nat. Herb. **16**: 262. 1913.

As epiphytic cactus, the stems twining and climbing by aërial roots emitted along the midnerve, thin, broad, flat, somewhat branching, the margins bearing numerous closely-set areoles; spines numerous, acicular; flowers large, red, narrowly funnelform, nocturnal; ovary and flower-tube with numerous hairs and bristles in the axils of the scales; perianth cutting off from the ovary as in *Cereus;* fruit ovoid, with a truncate apex, its areoles bristly; seeds black, ear-shaped, with an open hilum.

Mr. Berger proposed a subgenus *Phyllocereus* for this plant, supposing it represented a connecting link between *Phyllocactus* and *Cereus.* Its flat stems are like those of *Epiphyllum;* its flower is most like that of *Selenicereus.* Berger's name, while appropriate, could not be used because of the *Phyllocereus* Miquel (Bull. Sci. Phys. Nat. Neerl. **1**12. 1839). The name is from the Greek, referring to the twisting or turning of the stem about trees.

A monotypic genus of the Amazonian forests, still rare in collections.

1. **Strophocactus wittii** (Schumann) Britton and Rose, Contr. U. S. Nat. Herb. **16**: 262. 1913.

. *Cereus wittii* Schumann, Monatsschr. Kakteenk. **10**: 154. 1900.

A thin, very flat plant, often 1 dm. broad, growing appressed to trunk of trees; joints broad, leaf-like, 3 or 4 times as long as broad, rounded at base and apex, with a stout central vein and nearly entire margin; areoles small, closely set along the margin of the joints, 6 to 8 mm. apart, bearing tufts of wool and bristles besides the spines; spines numerous, acicular, yellowish brown, 12 mm. long or less; flowers elongated, large, 25 cm. long; tube elongated, tapering upward, only about half as thick at top as at base; limb short; outer perianth-segments linear, about 10, nearly twice as long as the inner ones; inner perianth-segments narrowly oblanceolate, acute or acuminate; filaments not extending beyond the inner perianth-segments; limb short; fruit 2.5 to 3.5 cm. long, spiny.

Type locality: In the swampy woods near Manaos, Brazil.

Distribution: Very abundant and widely distributed in the swampy forests of the Amazon, Brazil.

We have in our collection a part of the type material.

The following account of this very re-markable plant is from the pen of Karl Schumann and was published in the Gardeners' Chronicle in 1901 p. 78:

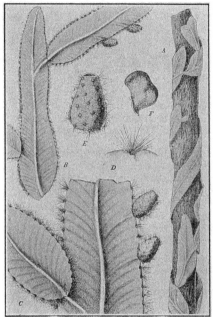

FIG. 302.—Strophocactus wittii.

"Among the numerous novelties which the last decade of the past century brought to Europe, the above named is surely one of the most inter-esting for both the amateur and the professional cultivator. I received this curious plant through the kindness of Mr. N. H. Witt, of Manaos, Erlado do Amazonas, Brazil. He told me long before he was able to send specimens that a climb-ing species of a genus he was not able to determine, grew in the swampy forest, or Igape, on the Amazon river. Closely appressed to the stems of the trees, and fixed to them by numerous roots, in the region of the yearly inundation, there creeps a cactus with the habit of a *Phyllocactus*, but armed with very sharp spines. It is so closely connected with the plant on which it grows that one must look carefully to distinguish it.

"When I had the specimen in my hand after it was taken out, I did not at all know how to class it. I was not able even to indicate the genus. It could not belong to *Phyllocactus*, however much the form of the leaf-like joints suggested that genus. Perhaps it might be a very abnormal species of *Rhipsalis*, but the flowers or fruits being absent, the question could not be answered.

"Last autumn I was fortunate enough to get, by the aid of Mr. N. H. Witt, plentiful specimens of the plant. After having carefully examined it, I found two fruits of ovoid form as large as a pigeon's egg, beset with very sharp prickles. This organ had all the characteristics of the genus *Cereus*, and I could now name the species, and did so in honour of the finder, *Cereus wittii*. The species is very interesting, because it is the 'missing link' between the genera *Phyllocactus* and *Cereus*. The form of the joints is perfectly typical of the former; the characteristics of the fruits and spines are those of a *Cereus*.

"Some days ago I received a notice from Dr. E. Ule, a botanist, whom I had sent from Manaos to the river Furnà, a tributary stream on the right side of the Amazon—that he had found a peculiar cactus in the upper part of the swampy forest, densely appressed to the tree-stems. His further description of the plant informed us that *C. wittii* is widely distributed. He told me that the older joints of *C. wittii* turn from green to a beautiful wine-red or purple colour, a peculiarity which I had also seen on the plants we cultivate in the Royal Botanic Garden of Berlin."

Illustrations: Gard. Chron. III. 29: f. 17; Monatsschr. Kakteenk. 10: 155; 12: 139; 15: 25; Schumann, Gesamtb. Kakteen Nachtr. f. 6, all as *Cereus wittii*; Contr. U. S. Nat. Herb. 16: pl. 84.

Figure 302 is a copy of the plate above cited (Monatsschr. Kakteenk. 10: 155)

APPENDIX.

We bring together here records of some species described in Germany during the war period, 1916–1918, cited from periodicals only recently received in the United States, together with a few supplementary observations upon other species described in this volume.

Cereus hexagonus. (See page 4, *ante.*)

Dr. Britton has recently studied this species on the western mainland of Trinidad and the small islands, Gasparee, Monos, and Chacachacare, adjacent. Here it inhabits rocky hillsides, attaining a height up to 15 meters; planted individuals observed were considerably taller. At St. Joseph large numbers of young plants up to 4 meters tall were seen growing upon branches of saman trees, evidently germinated from seeds carried by birds from the fruit of large planted specimens nearby, an interesting illustration of induced epiphytic habit of a typically saxicolous plant. Repeated field observations showed that this *Cereus* is usually 4-ridged when young, becoming 6-ridged later in life, many plants bearing some joints 4-ridged, some 6-ridged.

Illustration: Loudon, Encycl. Pl. 410. f. 6854, as *Cactus hexagonus.*

Cereus chalybaeus. (See page 16, *ante.*)

Cereus beysiegelii (Monatsschr. Kakteenk. 29: 48. 1919) is an abnormal form, similar to *Cereus peruvianus monstruosus*, which Mr. W. Weingart says looks like *Cereus chalybaeus* on account of its black spines and turquoise-green skin. Its origin is unknown.

Fig. 303.—Cereus grenadensis.

Fig. 304.—Section of flowering branch of C. grenadensis.

23. Cereus grenadensis sp. nov. (See page 18, *ante.*)

Tall, much branched, up to 7 meters high, the trunk short, sometimes 2.5 dm. in diameter, the branches grayish green, erect-ascending, about 7 cm. in diameter, 7 to 9-ribbed, the ribs about 1 cm. high, transversely grooved above each areole; areoles about 1 cm. apart, borne in slight depressions of the ribs, gray-pulverulent; spines about 17, subulate, straight, brownish or gray, the largest about 2 cm. long, the shortest about 3 mm, the central one often twice as long as any of the others; flowers many, borne towards the ends of the branches, about 7 cm. long, short-funnelform, open in the early morning, the buds rounded; outer perianth-segments with broad purple rounded or apiculate tips, the few inner ones rounded, purplish; ovary oblong, with a few naked areoles; stamens many, not exserted; immature fruit green, ellipsoid, 3 to 4 cm. long.

223

Collected on island of Grenada, British West Indies, by N. L. Britton and T. E. Hazen, February 24, 1920. Type from a slope on the harbor of St. George's.

As observed on the date of collection, this cactus is abundant about the harbor of St. George's and a conspicuous element of the vegetation; it was also studied on hills elsewhere in the southern part of the island, but only the type plant was seen in bloom. The species is closely related to *Cereus repandus* Miller of Curaçao, differing in its shorter spines, somewhat smaller, purple flowers, continuous unconstricted branches and transversely grooved ribs, and also to *Cereus margaritensis* Johnston of Margarita, from which it differs by straight spines, somewhat larger flowers, and grooved ribs. The fruit was said by negroes to be edible when ripe. It is called dildo, a common West Indian name for the tall-branching, cereus-like cacti.

Figure 303 shows the type plant; figure 304 shows one of its branches photographed by T. E. Hazen.

Cereus. (See page 21, *ante*.)

Cephalocereus californicus Hortus is credited by the Index Kewensis to Schumann (Engler and Prantl, Pflanzenfam. 3⁶ᵃ: 182. 1894), although it is not quite certain whether Schumann intended to list this name under *Cephalocereus* or as *Cereus californicus*. The *Cereus californicus* Nuttall we have already referred to *Opuntia serpentina* (see 1: 58, *ante*).

Cereus chlorocarpus De Candolle (Prodr. 3: 466. 1828; *Cactus chlorocarpus* Humboldt, Bonpland, and Kunth, Nov. Gen. et Sp. 6: 67. 1823) which originally came from the Peruvian and Ecuadorean boundary has not been identified. It is evidently not a true *Cereus*.

Cereus columnaris Loddiges (Voigt, Hort. Suburb. Calcutt. 61. 1845) is said to have been introduced into suburbs of Calcutta in 1840. Otherwise it is unknown. This name may apply to *Cereus hexagonus* (L.) Miller.

Cereus flavispinus hexagonus Salm-Dyck (Hort. Dyck. 63. 1834) is only a name.

Cereus geminisetus Reichenbach (Terscheck, Suppl. Cact. Verz. 3) we know only from Walpers's (Repert. Bot. 2: 340. 1843) brief description of a sterile plant of unknown origin.

Cereus heteracanthus Tweedie (Sweet, Hort. Brit. ed. 3. 284. 1839) was described simply as a variable-spined *Cereus*.

Cereus ictidurus (Hort. Univ. 1: 224. 1839), called the martin's-tail-cereus, reported as soon to be figured and described, we do not know.

Cereus zizkaanus or *C. ziczkaanus* (Montasschr. Kakteenk. 5: 44. 1895) is figured in the Gardeners' Chronicle for 1873 (75. f. 15) where it is referred to *Cereus eburneus* with a question. It is said to have come from Gruson's garden. This is doubtless the same as *Cereus chilensis zizkaanus*, sometimes spelled *zizkeanus* (see page 137, *ante*).

Pilocereus pfeifferi, sometimes credited to Otto, occurs frequently in German cactus works, but we have seen no description. The name is not found in the Index Kewensis or in Schumann's Monograph. Dr. Rose saw a living specimen in the Berlin Botanical Garden labeled "*Pilocereus pfeifferi*, Mexico" which he noted at the time as near *Lemaireocereus treleasei*.

Monvillea cavendishii. (See page 21, *ante*.)

Related to this species is the following which we know only from description:

CEREUS CHACOANUS Vaupel, Monatsschr. Kakteenk. **26**: 121. 1916.

Erect, 2 to 4 meters high, 6 cm. in diameter; ribs 8; spines 9 or 10; central spine solitary, 6 cm. long; flowers funnelform, 15 cm. long; outer perianth-segments rose-colored; inner perianth-segments white; fruit subglobose to ovoid, 3 cm. long.

Type locality: Gran Chaco, Paraguay.
Distribution: Paraguay.

Cephalocereus hoppenstedtii. (See page 27, *ante*.)

A wonderful display of this plant is shown in the photograph taken by C. A. Purpus near the type locality in 1912. A mountainside is shown with many of the plants which form the conspicuous objects in the landscape.

Cereus hoogendorpii (Monatsschr. Kakteenk. **4**: 80. 1894) and *Pilocereus hoogendorpii* (Schumann in Engler and Prantl, Pflanzenfam. 3⁰⁴: 181. 1894), only names, are the same as this plant, according to Schumann.

Espostoa lanata. (See page 61, *ante*.)

Our attention has been called to a paper by Vincenzo Riccobono (Bull. R. Ort. Firenze IV. **4**: 94. 1919) on the first flowering of *Pilocereus dautwitzii* in Europe and a flower of the plant has also been sent us by Riccobono. This seems to be the same plant as the one collected by Dr. Rose in southern Ecuador in 1918.

Illustration: Gartenflora **22**: 115, as *Pilocereus dautwitzii*.

Lemaireocereus hystrix. (See page 86, *ante*.)

  Cereus olivaceus, Lemaire, Rev. Hort. IV. **8**. 643. 1859.

The plant upon which *Cereus olivaceus* was based came from Santo Domingo.

Lemaireocereus griseus. (See page 87, *ante*.)

Both *Cereus eburneus* Salm-Dyck and *Cactus eburneus* Link (Enum. Hort. Berol. **2**: 22) were published in 1822 and to both *Cactus peruvianus* Willdenow (Enum. Hort. Berol. Suppl. 32. 1813) was referred. Willdenow's plant, from the description, suggests a *Cephalocereus* but is referred to *Cereus eburneus* by the Index Kewensis. Link also refers it to Hortus Dyckensis and to Haworth (Syn. Pl. Succ. 179), while Salm-Dyck's description indicates that he had a plant before him different from Willdenow's. The *Cereus eburneus* described by Pfeiffer (Enum. Cact. 90) was certainly a complex, a part coming from Curaçao and a part from Chile. For this reason, doubtless, Schumann (Gesamtb. Kakteen 59, 108) has referred both names to *Cereus coquimbanus* and *Cereus eburneus*.

Leocereus. (See page 108, *ante*.)

Cereus oligolepis Vaupel (Notizbl. Bot. Gart. Berlin **5**: 285. 1913) we know only from description. It is evidently not a *Cereus* but it suggests one of the species of *Leocereus* and comes from Campo der Serra do Mel on the Rio Surumu, northern Brazil, the region where these plants are found. It may be briefly described as follows: Plant 1 meter high; ribs 5, 1 cm. high; areoles 1 cm. apart; radial spines 8 to 10, 5 mm. long; central spine 1, 2 cm. long; ovary bearing small scales.

Cereus xanthochaetus Reichenbach (Terscheck. Suppl. Cact. Verz. 4) we know only from the description of Walpers (Repert. Bot. **2**: 340. 1843). He describes it as follows: Erect, light green; ribs 7, nearly continuous, compressed above, obtuse; areoles yellowish tomentose; spines 21, slender, yellowish, straight, the upper spines longer than the others.

Heliocereus schrankii. (See page 127, *ante*.)

Related to this species is *Cereus ruber* (Weingart, Monatsschr. Kakteenk. **15**: 22. 1905). The flowers are described as orange-yellow, passing into scarlet. It is said to come from Brazil, but no species of *Heliocereus* are known from South America. Weingart (Monatsschr. Kakteenk. **29**: 57. 1919) expresses his belief that *C. ruber* is of hybrid origin.

Trichocereus pasacana. (See page 133, *ante*.)

Of this relationship the following species are known only from descriptions:

CEREUS TACAQUIRENSIS Vaupel, Monatsschr. Kakteenk. **26**: 122. 1916.

  Columnar, 2.5 meters high; ribs low, about 1 cm. high, obtuse; spines numerous, setiform; hardly pungent, unequal, the longest 8 cm. long; flowers large, white, 20 cm. long, funnelform; inner perianth-segments oblong-spatulate; stamens in 2 series, shorter than the perianth-segments.

Type locality: Tacaquira, Bolivia.
Distribution: Southern Bolivia.

CEREUS TARIJENSIS Vaupel, Monatsschr. Kakteenk. **26:** 123. 1916.

Columnar, 1.5 meters high, 2.5 dm. in diameter; areoles broadly elliptic to oval; radial spines 10 to 13, stout, pungent, unequal, reddish brown; central spine solitary, 7 cm. long; flower 10 cm. long; outer perianth-segments lanceolate; inner perianth-segments spatulate.

Type locality: Escayache, near Tarijo, Bolivia.
Distribution: Southern Bolivia.

8. Borzicactus aurivillus (Schumann). (See page 163, *ante.*)

Cereus aurivillus Schumann, Monatsschr. Kakteenk 13: 67. 1903.

Cylindric, 2.5 dm. high or more, 2.5 cm. in diameter; ribs 17, crenate; areoles closely set, only 5 to 7 mm. apart, elliptic, bearing yellow curly wool; spines 30 or more, nearly equal, short, colorless except the yellow bases; flower from near the top of the plant, somewhat zygomorphic, 6 cm. long; inner perianth-segments obtuse.

Type locality: Probably Peru.
Distribution: Andes of Peru.
Illustrations: Monatsschr. Kakteenk. **29:** 7, 9, as *Cereus aurivillus.*

Oreocereus celsianus. (See page 171, *ante.*)

Restudy of *Pilocereus straussii* may show that it is specifically distinct from *Oreocereus celsianus.* The name *Cereus straussii* was really published by Heese in Gartenflora (**62:** 383) in 1907, although the illustration accompanying it bears the legend, *Pilocereus straussii.*

Illustrations: Möllers Deutsche Gärt. Zeit. **25:** 483. f. 15, as *Pilocereus celsianus bruennowii;* Schelle, Handb. Kakteenk. 100. f. 39, as *Pilocereus celsianus;* Gartenflora **62:** 383. f. 55, as *Pilocereus straussii.*

Cleistocactus baumannii. (See page 174, *ante.*)

Of this relationship is the following:

CEREUS TUPIZENSIS Vaupel, Monatsschr. Kakteenk. **26:** 124. 1916.

Slender, 2 to 3 meters high; ribs unknown; areoles large, oval; spines 15 to 20, subulate, pungent, reddish brown; central spines 2, one above the other, 4.5 cm. long; flower tubular, somewhat zygomorphic, 8 cm. long, pale salmon-colored; outer perianth-segments small; inner perianth-segments broader than the outer; stigma-lobes 8, 4 mm. long.

Type locality: Tupiza, Bolivia.
Distribution: Southern Bolivia.

5a. Hylocereus venezuelensis sp. nov. (See page 186, *ante.*)

Vines rather slender, climbing, bluish, 3-angled, the joints 3 to 4 cm. broad; margin of ribs not horny; spines 2 or 3, short, stubby, brown to black; flowers very fragrant, large, 2.5 dm. long; scales on ovary and perianth-tube green with purple margins; inner perianth-segments large, oblong, white above, pink below; stigma-lobes cream-colored, deeply cleft.

Collected by J. N. Rose near Valencia, Venezuela, in 1917 (No. 21835).

We were at first disposed to refer this plant to *H. polyrhizus* but when it flowered in the New York Botanical Garden in June 1920, it produced a flower strikingly different in its stigma-lobes, which are deeply cleft as in *H. lemairei.* In *H. polyrhizus* the stigma-lobes, so far as we know, are always entire. According to W. Weingart, a keen student of these plants, *H. lemairei* and *H. monacanthus* are the only two species he knows with bifid stigma-lobes; they may also occur in *H. bronxensis.*

INDEX.

(Pages of principal entries in heavy-face type.)

Acanthocereus, 1, 2, 15, **121–126,** 207
  albicaulis, 122, **125,** 126
  brasiliensis, 122, **125,** 126
  colombianus, **122**
  horridus, **122,** 123
  occidentalis, 122, **125**
  pentagonus, 122, **123,** 124, 191
  subinermis, 122, **125**
Acutangules, 122
Agua-colla, 135
Anomali, 22
Aporocactus, 183, **217–221**
  baumannii, 174
  colubrinus, 174
  conzattii, 217, **220,** 221
  flagelliformis, 217, **218,** 219, 221
  flagriformis, 217, 218, **219**
  leptophis, 217, **218**
  martianus, 217, **220,** 221
Arrojadoa, 2, **170, 171,** 178
  penicillata, 170, **171**
  rhodantha, **170**
Azureae, 4
Bande du sud, 151
Bauhinia, 92
Bavoso, 86
Bergerocactus, 2, **107, 108**
  emoryi, 107, **108**
Bergerocereus, 108
Binghamia, 2, **167–169**
  acrantha, 167, **168**
  melanostele, **167,** 168
Borzicactus, 2, **159–164,** 173
  acanthurus, 159, **161**
  aurivillus, 159, 163, **226**
  decumbens, 159, **162**
  humboldtii, 159, **163**
  icosagonus, 159, **160**
  morleyanus, 159, **160,** 161
  plagiostoma, 159, **163**
  sepium, 159, **160**
  ventimigliae, 159, 160
Bottle cactus, 60
Brachycereus, 2, **120, 121**
  thouarsii, **120,** 121
Breebee, 18
Bromeliads, 42
Browningia, 2, **63, 64**
  candelaris, **63,** 64
Cabeça branca, 30
Cactaceae, 1, 211
Cactanae, 1
Cacti, 1, 2, 3, 8, 9, 19, 21, 23, 25, 28, 32, 65, 67, 77, 82,
  113, 116, 117, 121, 122, 124, 127, 145, 147, 159,
  165, 166, 167, 169, 173, 176, 177, 178, 182, 183,
  185, 189, 196, 212, 214, 216, 217, 224
Cactus, 1, 5, 9, 19, 23, 25, 40, 42, 58, 61, 62, 64, 65, 70,
  71, 72, 76, 87, 88, 90, 92, 102, 107, 111, 113,
  115, 124, 125, 135, 140, 147, 150, 151, 153,
  158, 159, 163, 164, 165, 166, 167, 170, 172, 179,
  181, 191, 192, 195, 210, 212, 221, 222, 224
  abnormis, 12

Cactus—*continued.*
  ambiguus, 118, 119
  aureus, 105
  bradypus, 27
  candelaris, 63, 145
  caripensis, 124
  chiloensis, 137
  chlorocarpus, 224
  coquimbanus, 88, 139
  divaricatus, 151
  eburneus, 225
  euphorbioides, 33
  fascicularis, 141
  fimbriatus, 87, 151
  flagelliformis, 217, 218, 219
  flavispinus, 60
  fulvispinosus, 50
  gracilis, 151
  grandiflorus, 196, 197, 198
  haworthii, 44
  heptagonus, 43
  hexagonus, 3, 4, 5, 43, 223
  humboldtii, 163
  hystrix, 86
  icosagonus, 160
  jamacaru, 8
  kageneckii, 20
  laetus, 99
  lanatus, 61, 62
  lanuginosus, 49
  lanuginosus aureus, 20
  lecchii, 20
  mallisonii, 219
  melocactus, 29, 30
  mexicanus, 197
  multangularis, 19
  napoleonis, 191
  niger, 44
  octogonus, 4
  ovatus, 20
  paniculatus, 82
  pentagonus, 15, 121, 123, 193
  peruvianus, 11, 13, 225
  pitajaya, 15, 123
  polygonus, 47
  polymorphus, 139
  prismaticus, 123
  pruinosus, 88
  quadrangularis, 124
  repandus, 17, 151, 152
  royenii, 50
  senilis, 25, 27
  sepium, 160
  serpens, 163
  serpentinus, 118, 119
  speciosissimus, 128, 129
  speciosissimus lateritius, 128
  speciosus, 127, 128
  strictus, 44
  tetragonus, 9, 14
  triangularis, 183, 187, 188, 192, 193, 194, 212
  triangularis aphyllus, 187

227

Cactus—*continued.*
    triangularis foliaceus, 192
    trigonus, 192
    triqueter, 192
    undulosus, 123
Cactuses, 116
Candebobe, 96
Cardon, 70, 96
Cardon grande, 140
Cardoncillo, 112
Carnegiea, 2, **164–167,** 178
    gigantea, 133, **164,** 165, 166, 178
Cephalocereus, 1, 3, 13, **25–60,** 178, 224, 225
    alensis, 26, **55**
    arrabidae, 26, **42,** 43, 136
    bahamensis, 26, **38**
    bakeri, 39, 40
    barbadensis, 26, **44,** 45, 46
    brasiliensis, 26, **57**
    brooksianus, 26, **49**
    californicus, 224
    catingicola, 26, 49, **56**
    chrysacanthus, 26, **48**
    chrysomallus, 72
    colombianus, 26, 34, **55,** 56
    columna, 76
    columna–trajani, 76
    cometes, 26, 51, **52**
    compressus, 193
    deeringii, 26, **38,** 39
    dybowskii, 25, **30,** 58
    euphorbioides, 25, **33**
    exerens, 42
    fluminensis, 25, **29,** 33, 57
    fouachianus, 51
    gaumeri, 26, **47**
    gounellei, 25, **34,** 35
    hermentianus, **58**
    hoppenstedtii, 25, **27,** 225
    keyensis, 26, **40**
    lanuginosus, 18, 26, **49,** 50
    leucocephalus, 26, **52,** 53
    leucostele, 25, **36,** 37, 59, 60
    macrocephalus, 25, **31,** 75, 76
    maxonii, 26, **48,** 53
    melanostele, 167
    melocactus, 29, 30, 58
    millspaughii, 26, **45,** 46
    monoclonos, 26, 40, 41
    moritzianus, 26, **41,** 42
    nobilis, 26, **44,** 45
    palmeri, 26, **53**
    pentaedrophorus, 25, **31**
    phaeacanthus, 26, **57**
    piauhyensis, 26, 48, **49**
    polygonus, 26, **47**
    polylophus, 25, **32**
    purpureus, 25, **28,** 29
    purpusii, 26, **56**
    robinii, 26, **39,** 40
    robustus, 26, 51, **52**
    royenii, 26, 46, **50**
    russelianus, 25, **33,** 34, 56
    sartorianus, 26, **53**
    scoparius, 26, **41**
    senilis, 25, **27,** 28, 31
    smithianus, 26, 36, **37**

Cephalocereus—*continued.*
    swartzii, 26, **46,** 47
    tweedyanus, 26, **54,** 55
    ulei, 26, 52, **58**
    urbanianus, 26, **43**
    zehntneri, 25, **35**
Cephalophora, 25
Cephalophorus, 25
    columna-trajani, 76
    senilis, 27
Cereeae, **1,** 111
Cereanae, 1
Cerei, 59, 70
Cereus, 1, **3–21,** 33, 41, 42, 47, 58, 59, 68, 77, 82, 87, 105,
    108, 110, 118, 122, 127, 144, 145, 147, 152, 158,
    163, 164, 170, 173, 192, 219, 221, 222, 223, 224,
    225
    abnormis, 12
    acanthosphaera, **209**
    acanthurus, 161
    acidus, 84
    acranthus, 168
    acromelas, 59
    aculeatus, 21
    acutangulus, 123, 124, 157
    adscendens, 155, 156
    aethiops, 4, 16, **17,** 18
    affinis, 14
    alacriportanus, 4, **6,** 7
    alamosensis, 169
    albertinii, 21
    albiflorus, 128
    albisetosus, 58
    albispinus, 37, 59, 118
    albispinus major, 59
    alensis, 55
    amazonicus, 24
    ambiguus, 118
    ambiguus strictior, 118
    amblyogonus, **20**
    amecaensis, 129
    amecamensis, 129
    americanus octangularis, 89
    americanus triangularis, 193
    andryanus, 59
    anguiniformis, 22, 174
    anguinus, 175
    angulosus, 32
    anisacanthus, 102
    anisacanthus ortholophus, 102
    anisacanthus subspiralis, 102
    anisitsii, 23
    anizogonus, 193
    antoinii, 201
    apiciflorus, **107**
    aquicaulensis, 180
    aragonii, 92, 103
    aragonii palmatus, 92
    arboreus, 80
    arcuatus, 124
    arendtii, 154
    areolatus, **159**
    arequipensis, 134, **145**
    argentinensis, 4, **11,** 12
    armatus, 50
    arrabidae, 42, 43
    arrigens, 180

Cereus—*continued*.
  assurgens, 77, 79
  atacamensis, **145**
  atropurpureus, 154
  atrovirens, 21
  aurantiacum superbus, 129
  aurantiacus, 128
  aureus, 44, 105, 106, 107
  aureus pallidior, 44
  aurivillus, 173, 226
  aurora, 219
  azureus, 4, **15**, 16
  azureus seidelii, 15
  bahamensis, 38
  bajanensis, 124
  bakeri, 39
  balansaei, 157, 158
  barbatus, 51
  baumannii, 173, 174, 217
  baumannii colubrinus, 174
  baumannii flavispinus, 174
  baumannii smaragdiflorus, 174
  bavosus, 86
  baxaniensis, 123, 124
  baxaniensis ramosus, 124
  belieuli, 96
  beneckei, **18**
  beneckei farinosus, 18
  bergerianus, 72
  beysiegelii, 223
  bifrons, 128
  biolleyi, 215
  boeckmannii, 202
  bolivianus, 136
  bonariensis, 7
  bonplandii, 157, 158
  bonplandii brevispinus, 157
  bonplandii pomanensis, 155
  brachiatus, 86
  brachypetalus, 67
  bradypus, 27
  brandii, 14
  breviflorus, 82, 83
  brevispinulus, 200
  brevistylus, 66
  bridgesii, 134
  bridgesii brevispinus, 134
  bridgesii lageniformis, 134
  bridgesii longispinus, 134
  brookii, 151
  brooksianus, 49
  caesius, 4, 13, **15**
  calcaratus, 193
  californicus, 224
  callicanthus, 197
  calvescens, 11
  calvus, 69, 70
  candelabrius, 126
  candelabrum, 95, 96
  candelaris, 63
  candicans, 142, 143
  candicans courantii, 142
  candicans dumesnilianus, 143
  candicans gladiatus, 142
  candicans robustior, 142
  candicans spinosior, 143
  candicans tenuispinus, 142
  caripensis, 124

Cereus—*continued*.
  castaneus, 84
  catamarcensis, 146
  catingicola, 56
  cauchinii, 8
  caudatus, **20**
  cavendishii, 21, 22
  celsianus, 171, 172
  chalybaeus, 4, **16**, 17, 223
  chakoanus, **224**
  chende, 90, 91
  chichipe, 89, 90
  chilensis, 137, 138, 139
  chilensis acidus, 84
  chilensis breviflorus, 83
  chilensis brevispinulus, 139
  chilensis eburneus, 139
  chilensis flavescens, 139
  chilensis fulvibarbis, 139
  chilensis funkianus, 138
  chilensis heteromorphus, 137
  chilensis linnaei, 139
  chilensis nigripilis, 140
  chilensis panhoplites, 137
  chilensis polygonus, 137
  chilensis poselgeri, 137
  chilensis pycnacanthus, 137
  chilensis quisco, 139
  chilensis spinosior, 139
  chilensis zizkaanus, 137, 224
  chilensis zizkeanus, 224
  chiloensis, 84, 137, 138
  chiloensis lamprochlorus, 133
  chiotilla, 65, 66
  chlorocarpus, 224
  chotaensis, 163
  chrysacanthus, 48
  chrysomallus, 72
  cinnabarinus, 129
  clavatus, 94
  coccineus, 127, 210, 211, 212
  cochal, 180
  coerulescens, 17, 59
  coerulescens fulvispinus, 17
  coerulescens landbeckii, 17
  coerulescens longispinus, 17
  coerulescens melanacanthus, 17
  coeruleus, 17
  cognatus, 123
  colombianus, 55
  colubrinus, 174, 217
  colubrinus flavispinus, 174
  colubrinus smaragdiflorus, 175
  columna-trajani, 76
  columnaris, 224
  colvillii, 14
  cometes, 52
  compressus, 192, 193
  concinnus, 21
  conformis, 103
  conicus, 33
  coniflorus, 199
  coquimbanus, 83, 139, 225
  coracare, 10
  coryne, 64, 65
  cossyrensis, 152
  crenatus, 59
  crenulatus, 49, 59

Cereus—*continued.*
  crenulatus gracilior, 49
  crenulatus griseus, 87
  crimsonii, 219
  cubensis, 149
  cumengei. 116
  cupulatus, 74
  curtisii, 44
  damacaro, 21
  damazioi, 159
  dautwitzii, 61, 62
  davisii, 154
  dayamii, 4, 11
  de laguna, 20
  decagonus, 59
  decandollii. 13
  decorus, 20
  decumbens, 162
  deficiens, 94
  del moralii, 90, 91
  devauxii, 128
  diguetii, 111
  divaricatus, 151
  divergens, 151
  donatii, 203
  donkelaarii, 200
  donkelaerii, 200
  donkelarii, 200
  duledevantii, 139
  dumesnilianus, 143, 159
  dumortieri, 102
  dussii, 123
  dybowskii, 30
  dyckii, 92, 93
  eburneus, 20, 87, 89, 99, 103, 224. 225
  eburneus clavatus, 94
  eburneus monstrosus, 88
  eburneus polygonus, 87, 88
  edulis, 89
  elegans, 139
  emoryi, 108
  enriquezii, 88
  erectus, 49, 152
  erectus cristatus, 193
  erectus maximus, 13
  ericomus, 43
  eriocarpus, 145
  eriophorus, 149, 202
  eriophorus laeteviridis, 149
  eruca, 114, 115, 116
  estrellensis, 216
  euchlorus, 21, 22
  euphorbioides, 33
  exerens, 42. 43
  extensus. 191
  eyriesii, 159
  farinosus, 18
  fascicularis. 141
  fercheckii, 140
  fernambucensis, 14
  ferox, 30
  fimbriatus, 151
  flagelliformis, 218, 219
  flagelliformis funkii, 219
  flagelliformis leptophis, 218
  flagelliformis mallisoni, 219
  flagelliformis minor, 218, 219
  flagelliformis nothus, 219

Cereus—*continued.*
  flagelliformis scotii, 219
  flagelliformis smithii, 219
  flagelliformis speciosus, 219
  flagriformis, 219, 220
  flavescens, 20
  flavicomus, 52
  flavispinus, 20, 60
  flavispinus hexagonus, 224
  flexuosus, 116
  floccosus, 50, 51
  fluminensis, 29
  foersteri, 53
  forbesii, 7
  formosus, 14
  fouachianus, 50
  fulgens, 210
  fulgidus, 210
  fulvibarbis, 139
  fulviceps, 72
  fulvispinosus, 50
  fulvispinus, 140
  funkii, 137
  galapagensis, 146, 147
  garambello, 180
  gemmatus, 74
  geminisetus, 224
  geometrizans, 20, 179, 180
  geometrizans cochal, 180
  geometrizans pugioniferus, 179, 180
  geometrizans quadrangularispinus, 179, 180
  geometrizans quadrangulispinus, 180
  ghiesbreghtii, 60
  giganteus, 135, 164, 167
  gilvus, 137
  glaber, 216
  gladiator, 179
  gladiator geometrizans, 180
  gladiatus, 142
  gladiger, 87, 88, 180
  gladiiger, 87
  gladilger, 87
  glaucescens, 59
  glaucus, 8, 15
  glaucus speciosus, 14
  glaziovii, 109
  gloriosus, 51
  gonzalezii, 77, 214
  gracilis, 19, 147, 151, 209
  gracilis scandens, 198
  grandifloro-speciosissimus maynardii, 210
  grandiflorus, 197, 198, 199
  grandiflorus affinis, 197
  grandiflorus barbadensis, 198
  grandiflorus callicanthus, 197
  grandiflorus grusonianus, 197
  grandiflorus haitiensis, 197
  grandiflorus hybridus, 210
  grandiflorus major, 198
  grandiflorus maynardii, 210
  grandiflorus maximiliani, 197
  grandiflorus mexicanus, 197
  grandiflorus minor, 197
  grandiflorus ophites, 197
  grandiflorus ruber, 210
  grandiflorus schmidtii, 197
  grandiflorus speciosissimus, 210
  grandiflorus spectabilis, 197

Cereus—*continued*.
  grandiflorus uranos, 197
  grandiflorus viridiflorus, 197
  grandis, 14
  grandis gracilior, 14
  grandis ramosior, 14
  grandispinus, 87
  greggii, 112, 113, 122
  greggii cismontanus, 112
  greggii roseiflorus, 112
  greggii transmontanus, 112, 113
  grenadensis, 4, 18, **223**
  griseus, 87
  grossei, 174
  grusonianus, 203
  guatemalensis, 89, 119
  guelichii, 158
  gummatus, 117
  gumminosus, 117
  gummosus, 116, 117
  haageanus, 19
  haematuricus, 8
  hamatus, 203, 205
  hankeanus, 3, 7, 8
  hansii, 128
  hassleri, 211
  haworthii, 44
  hempelianus, 136
  hermannianus, 17
  hermentianus, 58
  heteracanthus, 224
  heteromorphus, 137
  hexagonus, **4**, 5, 9, 13, 223, 224
  hexangularis, 14
  hildmannii, 103
  hildmannianus, 4, **6**
  hirschtianus, 119
  hollianus, 85, 86
  hondurensis, 199
  hoogendorpii, 225
  hoppenstedtii, 27
  horizontalis, **20**
  horrens, **195**
  horribarbis, 8
  horridus, 5, 9
  houlletii, 52, 53
  huascha, 142
  huascha flaviflorus, 142
  huascha flaviformis, 142
  humboldtii, 163
  humilis, **209**, 210
  humilis major, 210
  humilis minor, 209
  humilis myriacaulon, 210
  humilis rigidior, 210
  hyalacanthus, 173, 176
  hybridus, 210
  hypogaeus, 106, 107
  hystrix, 86, 87, 99, 103
  icosagonus, 160
  ictidurus, 224
  incrassatus, 21
  incrustans, 74
  incrustatus, 74
  inermis, 207, 208
  inermis laetevirens, 208
  insularis, 23
  intricatus, 143

Cereus—*continued*.
  inversus, 191
  iquiquensis, 83
  irradians, 202
  isogonus, 160, 161
  jacquinii, 21
  jalapaensis, 199
  jamacaru, 4, 5, 6, **8**, 9, 15, 182
  jamacaru caesius, 15
  jamacaru glaucus, 9
  jenkinsoni, 128
  jenkinsonii verus, 128
  joconostle, 93
  josselinaeus, 129
  jubatus, 52
  jusbertii, 157, **158**
  kageneckii, 20
  kalbreyerianus, 118
  karstenii, 5, 207, 208
  karwinskii, 21
  kerberi, 170
  keyensis, 40
  kostratus, 205
  kunthianus, 200, 201
  labouretianus, 8
  laetevirens, 8
  laetevirens caesius, 15
  laetus, 99
  laevigatus, 88
  laevigatus guatemalensis, 89
  lagenaeformis, 134
  lamprochlorus, 132, 133
  lamprochlorus salinicolus, 133
  lamprospermus, 4, **10**
  lanatus, 61, 62
  lanceanus, 191
  landbeckii, 17
  langlassei, **20**
  laniceps, 173, 175
  lanuginosus, 49
  lanuginosus aureus, 20
  lanuginosus glaucescens, 49
  lasianthus, 134
  lateribarbatus, 76
  lateritius, 128
  lauterbachii, 22
  lecchii, 20
  leiocarpus, 50
  lemairei, 189
  lemoinei, 189
  leonii, 78
  lepidanthus, 76
  lepidotus, 4, 5, 6
  leptophis, 218
  leucostele, 36, 37
  limensis, **20**
  lindbergianus, 211
  lindenzweigianus, 23
  lindmanii, 211
  linnaei, 137
  lividus, 8, 9
  lividus glaucior, 9
  longicaudatus, 205
  longifolius, **20**
  longipedunculatus, 21
  longispinus, 137
  lormata, 21
  lutescens, 44

Cereus—*continued.*
  macdonaldiae, 202, 203
  macrocephalus, 31
  macrogonus, 43, 130, 136
  macrostibas, 181, 182
  magnus, **159**
  malletianus, **145**
  mallisonii, 219
  margaritensis, 4, **18**, 224
  marginatus, 69, 74
  marginatus gemmatus, 74
  marginatus gibbosus, 74
  mariculi, 209
  marmoratus, 23
  martianus, 219, 220, 221
  martinii, 155, 190
  martinii perviridis, 155
  maximiliani, 197
  maxonii, 48
  maynardae, 210
  maynardii, **210**
  megalanthus, 212
  melanacanthus, 17
  melanotrichus, 68
  melanurus, 109, 110
  melocactus, 29
  mendory, 17
  mexicanus, 129
  microsphaericus, **159**
  militaris, 73
  militaris californicus, 86
  millspaughii, 45
  minor, 218
  minutiflorus, 195
  miravallensis, 213
  mirbelii, 74
  mixtecensis, 89, 90
  moennighoffii, 219
  mollis, 44
  mollis nigricans, 44
  monacanthus, 155, 190
  monoclonos, 13, 41
  monstrosus, 12
  monstrosus minor, 12
  monstruosus, 12
  montezumae, 143
  monvilleanus, **173**
  moritzianus, 41, 42
  moritzianus pfeifferi, 42
  multangularis, **19**, 20, 30
  multangularis albispinus, 20
  multangularis limensis, 20
  multangularis pallidior, 19
  multangularis prolifer, 20
  multangularis rufispinus, 2 i
  myriacaulon, 209
  myriophyllus, 143
  nanus, 19
  napoleonis, 185, 187, 191
  nashii, 151
  nesioticus, 120, 121
  neumannii, 119
  nickelsii, 32
  niger, 44
  niger gracilior, 44
  nigricans, 44
  nigripilis, 139, 140
  nigrispinus, 17

Cereus—*continued.*
  nitens, 132
  nitidus, 123
  nobilis, 44
  northumberlandia, 4
  northumberlandianus, 4
  nothus, 201
  nudiflorus, 113, 114
  nycticallus, 200, 201, 216
  nycticalus, 196, 199
  nycticalus armatus, 199
  nycticalus gracilior, 201
  nycticalus maximiliani, 197, 201
  nycticalus viridior, 201
  nyriacaulon, 209
  obtusangulus, 22
  obtusus, 4, 13, 14, **15**, 16
  ocamponis, 184, 185
  ochracanthus, 20
  octogonus, 59
  olfersii, 33
  oligolepis, 225
  olivaceus, 225
  ophites, 21
  orcuttii, 70
  ovatus, 20
  pachyrhizus, 4, **10**
  palmeri, 177
  paniculatus, **82**
  panoplaeatus, 82, 137, 138
  paradisiacus, 198
  paraguayensis, 6, 7
  parviflorus, 173, 176
  parvisetus, 175
  pasacana, 133
  paxtonianus, 21, 22
  peanii, 201
  pecten-aboriginum, 70, 71
  pellucidus, 79, 122, 153
  penicillatus, 171
  pentaedrophorus, 31
  pentagonus, 123, 195, 210, 213
  pentagonus glaucus, 31
  pentalophorus, 31
  pentalophus radicans, 210
  pentapterus, 213
  pepinianus, 137
  perlucens, 4, **13**
  pernambucensis, 4, **14**, 15, 213
  perotetti, 9
  perrottetianus, 4, 6
  perrottetianus, 4, 6
  peruvianus, 3, 4, 5, **11**, 13, 135, 147
  peruvianus alacriportanus, 6
  peruvianus brasiliensis, 13
  peruvianus cristatus, 12
  peruvianus monstrosus, 12, 13
  peruvianus monstrosus minor, 13
  peruvianus monstruosus, 223
  peruvianus monstruosus nanus, 12
  peruvianus spinosus, 13
  peruvianus tortuosus, 12
  peruvianus tortus, 12
  pfeifferi, 42
  pfersdorffii, 117
  phaeacanthus, 57
  phatnospermus, 24
  philippii, 105
  piauhyensis, 49

Cereus—*continued*.
pitajaya, 13, 14, 15, 123
plagiostoma, 163
platygonus, 11, 156, 157, 159
plumieri, 191
polychaetus, 17
polygonatus, 88
polygonus, 47
polylophus, 32, 33
polymorphus, 138, 139
polyrhizus, 185
pomanensis, 155, 158
pomanénsis grossei, 155
portoricensis, 150
poselgeri, 111
pottsii, 112
princeps, 123
pringlei, 68, 69, 70
prismaticus, 123, 212
prismatiformis, 14
pruinatus, 21
pruinosus, 88, 89
pseudosonorensis, 169
pteranthus, 200
pterogonus, 201, 213
pugionifer, 96
pugioniferus, 179, 180
pugioniferus quadrangulispinus, 180
purpusii, 184
pycnacanthus, 137, 138
quadrangularis, 124
quadrangulispinus, 180
quadricostatus, 81
queretaroensis, 96, 97
quintero, 139
quisco, 137
radicans, 195
ramosus, 123
reflexus, 43
regalis, 20
regelii, 155
repandus, 4, 17, 18, 151, 152, 224
repandus laetevirens, 149
reptans, 58, 195
resupinatus, 87
retroflexus, 43
rhodanthus, 170
rhodocephalus, 158
rhodoleucanthus, 21, 22
rigidispinus, 103
rigidus, 210
robustus, 21
rogalli, 21
roridus, 89
rosaceus, 201
roseanus, 198
rostratus, 203, 204, 205
royeni, 44, 50
royenii armatus, 40, 51
royenii floccosus, 51
ruber, 225
ruferi, 219
ruferi major, 219
ruficeps, 75
russelianus, 33
salpingensis, 21
santiaguensis, 131

Cereus—*continued*.
sargentianus, 177, 178
saxicola, 21, 22
saxicola anguiniformis, 22
scandens minor, 197
schenckii, 180
schickendantzii, 144
schmidtii, 197
schoenemannii, 21
schomburgkii, 191
schottii, 177, 178
schottii australis, 177
schrankii, 127
schumannii, 103
sclerocarpus, 146, 147
scoparius, 41
seidelii, 15
senilis, 27
sepium, 160
serpens, 163
serpentinus, 20, 117, 118, 119
serpentinus albispinus, 59, 118
serpentinus splendens, 21, 118
serpentinus stellatus, 118
serpentinus strictior, 118
serratus, 129
serruliflorus, 151
setaceus, 211, 212
setaceus viridior, 211
setiger, 129
setosus, 34, 35
simonii, 169
sirul, 123
smaragdiflorus, 174, 175
smithii, 218
sonorensis, 169
spachianus, 131, 132
spathulatus, 21
speciosissimus, 128, 129
speciosissimus albiflorus, 128
speciosissimus aurantiacus, 128
speciosissimus blindii, 128
speciosissimus bodii, 128
speciosissimus bollwillerianus, 128
speciosissimus bowtrianus, 128
speciosissimus coccineus, 127, 128
speciosissimus colmariensis, 128
speciosissimus curtisii, 128
speciosissimus danielsii, 128
speciosissimus devauxii, 128
speciosissimus edesii, 128
speciosissimus elegans, 128
speciosissimus eugenia, 128
speciosissimus finkii, 128
speciosissimus gebvillerianus, 128
speciosissimus gloriosus, 128
speciosissimus grandiflorus, 128
speciosissimus guillardieri, 128
speciosissimus hansii, 128
speciosissimus hitchensii, 128
speciosissimus hitchensii hybridus, 128
speciosissimus hitchensii speciosus, 128
speciosissimus hoveyi, 128
speciosissimus ignescens, 128
speciosissimus jenkinsonii, 128
speciosissimus kampmannii, 128
speciosissimus kiardii, 128

THE CACTACEAE.

Cereus—*continued*.
speciosissimus kobii, 128
speciosissimus latifrons, 128
speciosissimus lateritius, 128
speciosissimus longipes, 128
speciosissimus lothii, 128
speciosissimus loudonii, 128
speciosissimus macqueanus, 128
speciosissimus maelenii, 128
speciosissimus maurantianus, 128
speciosissimus merckii, 128
speciosissimus mexicanus, 128
speciosissimus mittleri, 128
speciosissimus muhlhausianus, 128
speciosissimus peacocki, 128
speciosissimus peintneri, 128
speciosissimus rintzii, 128
speciosissimus roidii, 128
speciosissimus roseus albus, 128
speciosissimus roseus superbus, 128
speciosissimus roydii, 128
speciosissimus sarniensis, 128
speciosissimus seidelii, 128
speciosissimus seitzii, 128
speciosissimus selloi, 128
speciosissimus smithii, 128
speciosissimus superbus, 128
speciosissimus suwaroffii, 128
speciosissimus suwarowii, 128
speciosissimus triumphans, 128
speciosissimus unduliflorus, 128
speciosissimus vandesii, 128
speciosissimus vitellinus, 128
speciosus, 128, 129
speciosus albiflorus, 128
speciosus coccineus, 127
spegazzinii, 23
spinibarbis, 82
spinibarbis flavidus, 144
spinibarbis minor, 139
spinibarbis purpureus, 139
spinosissimus, 11
spinulosus, 207
splendens, 21, 118
squamosus, 173, 176
squarrosus, 104
steckmannii, 21
stellatus, 92, 93, 169
stenogonus, 4, **9**, 10, 11
stenopterus, 190
straussii, 171, 226
striatus, 22, 111
strictus, 44
strigosus, 143, 144
strigosus intricatus, 143
strigosus longispinus, 143
strigosus rufispinus, 144
strigosus spinosior, 144
subintortus, 19
subintortus flavispinus, 19
sublanatus, 43
subrepandus, 151
subsquamatus, 191
subtortuosus, 174
subuliferus, 137
surinamensis, 13
swartzii, 46, 87

Cereus—*continued*.
tacaquirensis, 225
tarijensis, 226
taylori, 153
tellii, 21
tenellus, **126**
tenuis, **19**
tephracanthus bolivianus, 136
terscheckii, 143
testudo, 213
tetazo, 76
tetracanthus, 136
tetragonus, 3, 4, 7, **9**, 14
tetragonus major, 9
tetragonus minor, 14
tetragonus ramosior, 9
thalassinus, 5, 9
thalassinus quadrangularis, 5
thelegonoides, 131
thelegonus, 130, 131
thouarsii, 120, 121
thurberi, 68, 97, 98, 99
thurberi littoralis, 97
thurberi monstrosus, 98
tilophorus, 43
tinei, 152
titan, 69, 70
tonduzii, 77, 216
tonelianus, 92
tortuosus, 154, 155
tortus, 83
treleasei, 93
triangularis, 187, 188, 192, 193
triangularis major, 187, 191
triangularis pictus, 192
triangularis uhdeanus, 193
trichacanthus, 44
trichocentrus, 21
tricostatus, 187
trigonodendron, **19**
trigonus, 186, 192
trigonus costaricensis, 186
trigonus guatemalensis, 184
trigonus quadrangularis, 124
trinitatensis, 189
tuberosus, 111
tunilla, 214
tupizensis, 226
tweediei, 174
ulei, 52
undatus, 149, 151, 152, 187
undulatus, 124
undulosus, 123
uranos, 197
uranus nycticalus, 197
urbanianus, 43, 198
ureacanthus, 158
vagans, 205
validus, 4, **7**
variabilis, 4, **13**, 14, 123, 124
variabilis glaucescens, 14
variabilis gracilior, 14
variabilis laetevirens, 14
variabilis micracanthus, 14
variabilis obtusus, 14
variabilis ramosior, 14
variabilis salm-dyckianus, 14

Cereus—*continued*.
 wismeri, 123
 vaupelii, 202
 venditus, 192
 ventimigliae, 160
 verschaffeltii, 21
 victoriensis, 53
 violaceus, 44
 viperinus, 110
 virens, 43
 vulcan, 219
 warmingii, 42
 weberbaueri, 141
 weberi, 95, 96
 weingartianus, 77, 78
 wereklei, 208
 wittii, 221, 222
 xanthocarpus, 4, 10
 xanthochaetus, 225
 ziczkaanus, 224
 zizkaanus, 224
Chacoub, 187
Chende, 91
Chente, 91
Chichibe, 90
Chichipe, 90
Chichituna, 90
Chinoa, 91
Chiotilla, 66
Chique-chique, 35
Chique-chique das pedras, 35
Chirinola, 115
Cina, 169
Cleistocactus, 2, 173–176
 anguinus, 173, **175**
 areolatus, 159
 aureus, 105
 baumannii, 163, 173, **174**, 175, 226
 baumannii colubrinus, 174
 baumannii flavispinus, 174
 celsianus, 171
 chotaensis, **163**
 colubrinus, 174
 humboldtii, 163
 hyalacanthus, **176**
 icosagonus, 160
 kerberi, 170
 lanatus, 61
 laniceps, **175**
 monvilleanus, 173
 parviflorus, **176**
 parvisetus, **175**
 sepium, 160
 serpens, 163
 smaragdiflorus, 173, **174**, 175
Cochal, 180
Coerulescentes, 3, 59
Compresso-costati, 3
Copado, 83
Corryocactus, 2, **66–68**
 brachypetalus, 66, **67**, 68
 brevistylus, **66**, 67, 68
 melanotrichus, 66, **68**
Coryphanthanae, 1
Creeping devil cactus, 115
Daatoe, 88
Deamia, 183, **212–214**
 testudo, 209, **213**, 214

Deerhorn cactus, 113
Dendrocereus, 2, **113, 114**
 nudiflorus, **113**, 114
Dildo, 224
Echinocactanae, 1
Echinocactus, 1, 105, 137, 171
 auratus, 143
 aureus, 105
 candicans, 143
 catamarcensis, 146
 ceratistes, 143
 echinoides, 138
 echinoides pepinianus, 137
 elegans, 139
 farinosus, 19
 fascicularis, 141
 ghiesbrechtii, 60
 hystrix, 86
 jeneschianus, 138
 lecchii, 20
 pepinianus, 137
 pepinianus echinoides, 138
 philippii, 105
 pruinosus, 88, 89
 pyramidalis, 139
 senilis, 27
 staplesiae, 27
 wangertii, 133
Echinocereanae, 1
Echinocereus, 3, 104, 110
 candicans, 142
 candicans tenuispinus, 143
 chiloensis, 138
 clavatus, 106
 emoryi, 108
 flavescens, 20
 gladiatus, 142
 hypogaeus, 106
 intricatus, 143
 lamprochlorus, 132
 limensis, 20
 multangularis, 19
 multangularis limensis, 20
 multangularis pallidior, 19
 poselgeri, 110, 111
 serpentinus, 118
 spachianus, 131
 spinibarbis, 82
 splendens, 118
 strigosus, 143
 strigosus rufispinus, 143
 strigosus spinosior, 143
 trichacanthus, 44
 tuberosus, 111
Echinopsis, 3, 105, 130, 144, 158, 159, 219
 aurata, 143
 candicans, 142
 catamarcensis, **146**
 dumeliana, 143
 dumesniliana, 143
 lamprochlora, 132
 philippii, 105
 schickendantzii, 144
Epiphyllanae, 1
Epiphyllum, 127, 221
 ackermannii, 128, 219
 phyllanthoides, 128
 smithianum, 128

Erdisia, 2, 104–107
    meyenii, 104, 105, 106
    philippii, 104, 105
    spiniflora, 104, 106
    squarrosa, 104, 105, 107
Eriocerei, 154
Eriocereus, 22, 147, 148
    bonplandii, 157
    cavendishii, 21
    jusbertii, 158
    martianus, 220
    martinii, 155
    platygonus, 156
    subrepandus, 151
    tephracanthus, 136
    tortuosus, 154
Eriosyce, 143
    sandillon, 143
Escontria, 2, 63, 65, 66
    chiotilla, 65, 66, 76
Espostoa, 2, 60–63
    lanata, 61, 62, 225
Eucereus, 3, 117
Euharrisia, 148
Eulychnia, 2, 66, 82–85, 181
    acida, 82, 83, 84
    breviflora, 82, 83
    castanea, 82, 84
    clavata, 106
    eburnea, 139
    iquiquensis, 82, 83
    spinibarbis, 82, 83
Euphorbia hystrix, 19
Facheiro preto, 177
Facheiro preto da Serra de Cannabrava, 173
Facheiroa, 2, 173
    publiflora, 173
Flor de copa, 114
Flor de cuerno, 218
Formosi, 3, 13
Furcraea, 92
Garrambullas, 179
Geotilla, 66
Giant cactus, 92, 164
Giganton, 135
Graciles, 159
Gymnocalycium, 1
Harrisia, 1, 2, 147–159
    aboriginum, 148, 154
    adscendens, 148, 155, 156
    bonplandii, 148, 157
    brookii, 148, 151
    earlei, 148, 154
    eriophora, 148, 149
    fernowi, 148, 153
    fragrans, 148, 149
    gracilis, 148, 151, 152
    guelichii, 148, 158
    martinii, 148, 155
    nashii, 148, 150, 151
    platygona, 148, 156
    pomanensis, 148, 155, 156
    portoricensis, 148, 150
    simpsonii, 148, 152, 153
    taylori, 153
    tortuosa, 148, 154
    undata, 151
Harrisiae, 154

Heliocereus, 2, 127–129, 218, 225
    amecamensis, 127, 129
    cinnabarinus, 127, 129
    coccineus, 127
    elegantissimus, 127, 129
    schrankii, 127, 225
    speciosus, 127, 128, 129, 201, 210
Hexagonae, 4
Hylocereanae, 1, 183
Hylocereus, 1, 63, 126, 183–195, 210, 215, 216
    antiguensis, 184, 193, 194
    bronxensis, 183, 185, 226
    calcaratus, 184, 193, 194
    costaricensis, 183, 186
    cubensis, 184, 188
    extensus, 184, 190, 191
    guatemalensis, 183, 184
    lemairei, 184, 189, 194, 226
    minutiflorus, 195, 196
    monacanthus, 184, 190, 226
    napoleonis, 184, 191
    ocamponis, 126, 183, 184, 185
    polyrhizus, 183, 185, 226
    purpusii, 183, 184
    stenopterus, 184, 190
    triangularis, 184, 185, 191, 192, 193
    tricostatus, 187
    trigonus, 184, 192, 193
    undatus, 184, 187, 188, 189
    venezuelensis, 183, 186, 226
Ipomoea, 92
Jaatoe, 88
Jaramataca, 111
Jasminocereus, 2, 146, 147
    galapagensis, 146, 147
Joconostle, 93
Junco, 119
Junco espinoso, 119
Kadoesji, 18, 88
La bande de sud, 151
Lanuginosi, 59
Lemaireocereus, 2, 25, 43, 69, 85–103, 114, 135, 151
    aragonii, 85, 92, 93, 103
    cartwrightianus, 85, 100
    chende, 85, 90, 91
    chichipe, 85, 89
    cumengei, 116
    deficiens, 85, 94, 96
    dumortieri, 85, 102
    eichlamii, 85, 89, 90
    eruca, 115, 116
    godingianus, 85, 91, 92, 135
    griseus, 20, 85, 87, 88, 89, 103, 225
    gummosus, 116, 117
    hollianus, 85, 86
    humilis, 85, 100, 101
    hystrix, 46, 85, 86, 87, 225
    laetus, 85, 99, 100
    longispinus, 85, 89, 90
    mixtecensis, 89, 91
    montanus, 85, 97
    pruinosus, 85, 88, 89
    queretaroensis, 85, 96, 97
    schumannii, 103
    stellatus, 85, 92, 93, 94, 169
    thurberi, 85, 96, 97, 98
    treleasei, 85, 93, 95, 224
    weberi, 85, 95, 96, 97, 164

Leocereus, 2, 108–110, 175, 225
    bahiensis, 108, 109
    glaziovii, 108, 109
    melanurus, 108, 109
Leptocereus, 2, 77–82, 104, 122
    arboreus, 77, 80
    assurgens, 77, 79, 80
    leonii, 77, 78, 79
    maxonii, 77, 80
    prostratus, 77, 79
    quadricostatus, 77, 81
    sylvestris, 77, 80, 81
    weingartianus, 77, 82
Lophocereus, 3, 170, 177, 178, 179
    australis, 177
    sargentianus, 177
    schottii, 177, 178, 179
Machaerocereus, 2, 114–117
    eruca, 114, 115, 116
    gummosus, 114, 116, 117
Mammillaria, 3, 118
    obconella, 102
Mandacaru, 9
Mandacaru de boi, 9
Mandacaru de penacho, 30
Martin's-tail-cereus, 224
Mediocactus, 183, 210–212
    coccineus, 211, 212
    megalanthus, 211, 212, 213
Melocactus, 170
    bradyus, 27
    columna-trajani, 76
    monoclonos, 41
Monvillea, 1, 21–25
    amazonica, 21, 24
    cavendishii, 21, 22, 118, 224
    diffusa, 21, 24
    insularis, 21, 22, 23
    maritima, 21, 24, 25
    phatnosperma, 21, 24
    spegazzinii, 21, 23
Muyusa, 160
Myrtillocactus, 3, 178–181
    cochal, 179, 180
    eichlamii, 179, 180, 181
    geometrizans, 179, 181
    schenckii, 179, 180
Neoraimondia, 3, 181–183
    macrostibas, 181, 182
Night-blooming cereus, 113, 187, 188
Nyctocereus, 1, 2, 108, 117–120, 121
    guatemalensis, 117, 118, 119, 120
    hirschtianus, 118, 119
    neumannii, 118, 119
    oaxacensis, 118, 120
    serpentinus, 20, 118
Obtusae, 4
Old man cactus, 27
Oligogoni, 3, 77
Opuntia, 3, 92, 118, 156, 164
    bicolor, 106
    clavata, 106, 107
    glomerata, 138
    pestifer, 19
    polymorpha, 138
    serpentina, 224
    spiniflora, 106
Orchids, 42

Oreocereus, 2, 60, 108, 171–173
    celsianus, 171, 172, 173, 226
    celsianus bruennowii, 171
    lanatus, 61
Organito de vibora, 111
Organo, 48, 74
Pachycereus, 2, 28, 68–76, 96
    calvus, 69
    chrysomallus, 69, 72, 73, 74
    columna-trajani, 69, 76
    gaumeri, 69, 71
    grandis, 69, 72
    lepidanthus, 69, 76
    marginatus, 69, 74, 75, 92, 102
    orcuttii, 69, 70
    pecten-aboriginum, 69, 70, 71, 72, 76
    pringlei, 69, 70
    queretaroensis, 96
    ruficeps, 69, 75
    titan, 69
Pasacana, 133
Passiflora, 92
Peniocereus, 2, 112, 113
    greggii, 112
Peruvianae, 4
Phyllocactus, 3, 221, 222
Phyllocereus, 221
Pilocereus, 3, 13, 25, 44, 51, 52, 58, 59, 178, 181
    acranthus, 169
    albisetosus, 58
    albispinus, 59
    albispinus crenatus, 59
    alensis, 55
    andryanus, 59
    arrabidae, 42
    barbatus, 50
    bruennowii, 171, 172
    celsianus, 171, 172, 226
    celsianus bruennowii, 171, 172, 226
    celsianus fossulatus, 172
    celsianus gracilior, 171
    celsianus lanuginosior, 171, 172
    celsianus williamsii, 171
    chrysacanthus, 48
    chrysomallus, 59, 72
    coerulescens, 59
    columna, 59, 76
    columna-trajani, 76
    cometes, 52
    consolei, 44
    crenulatus, 49
    curtisii, 44
    dautwitzii, 61, 63, 225
    dautwitzii cristatus, 63
    divaricatus, 151
    engelmannii, 164
    euphorbioides, 33
    exerens, 42, 43
    fimbriatus, 151
    flavicomus, 52
    flavispinus, 60
    floccosus, 50
    foersteri, 52
    fossulatus, 171, 172
    fossulatus gracilis, 171
    fossulatus pilosior, 171
    fouachianus, 50
    foveolatus, 172

Pilocereus—*continued.*
  fulviceps, 72, 73
  fulvispinosus, 50
  ghiesbrechtii, 60
  giganteus, 164, 167
  glaucescens, **59**
  gounellei, 34
  grandispinus, 87
  haageanus, 62
  haagei, 61
  haworthii, 44
  hagendorpii, 27
  hermentianus, 58
  hoogendorpii, 225
  hoppenstedtii, 27, 28
  houlletianus, 53
  houlletii, 52, 53
  jubatus, 52
  kanzleri, 171
  lanatus, 61, 63
  lanatus cristatus, 63
  lanatus haagei, 61, 63
  lanuginosus, 49
  lateralis, 27, 28
  lateribarbatus, 76
  leucocephalus, 52
  lutescens, 44
  macrocephalus, 31
  macrostibas, 182
  marschalleckianus, 53
  melocactus, 29, 30
  militaris, 73
  monacanthus, 155
  moritzianus, 41
  niger, 44
  nigricans, 44
  nobilis, 44
  pasacana, 133
  pentaedrophorus, 31
  pfeifferi, 224
  plumieri, 47
  polyedrophorus, 31
  polygonus, 47
  polylophus, 32
  pringlei, 69
  repandus, 17. 18
  robinii, 39
  royenii, 50, 51
  royenii armatus, 50
  ruficeps, 75
  russelianus, 33, 34
  sargentianus, 177, 178
  schlumbergeri, 47
  schottii, 177, 178
  scoparius, 41
  senilis, 27
  senilis cristatus, 27
  senilis flavispinus, 27
  senilis longisetus, 27
  senilis longispinus, 27
  setosus, 34, 35
  sterkmannii, 41
  straussii, 171, 172, 226
  strictus, 44, 60
  strictus consolei, 44
  strictus fouachianus, 50
  swartzii, 47
  terscheckii, 140

Pilocereus—*continued.*
  tetetzo, 76
  thurberi, 97
  trichacanthus, 44
  ulei, 52
  urbanianus, 43
  vellozoi, 29
  verheinei, **58**
  virens, 43
  williamsii, 172
Piptanthocereus, 3, 22
  azureus, 15
  beneckei, 18
  chalybaeus, 16
  forbesii, 7
  hankeanus, 7
  labouretianus, 7
  jamacaru, 8
  jamacaru caesius, 15
  jamacaru cyaneus, 8
  jamacaru glaucus, 8
  validus, 7
Piscol colorado, 62
Pitahaya, 96, 98, 122, 165
  agre, 117, 122
  agria, 117, 122
  de San Juan, 122
  dulce, 98, 122
Pitahayita, 111, 122
Pitajaya, 122
Pitajuia, 122
Pitalla, 122
Pitaya, 122
Pitayita, 111, 122
Pithaya, 122
Rabo de raposa, 109
Rat tail cactus, 218
Rathbunia, 2, 159, **169, 170**
  alamosensis, 117, **169**, 170
  kerberi, 169, **170**
  sonorensis, 169
Repandae, 4
Rhipsalidanae, 1
Rhipsalis, 3, 124, 209, 214, 222
  biolleyi, 215
Sacamatraca, 111
Sacasil, 111
Saguaro, 165
Sahuaro, 165, 166
Saman, 223
Saramatraca, 111
Sebucan, 81
Selenicereus, 1, 58, 183, 195, **196–210**, 221
  boeckmannii, 197, **202**
  brevispinus, 197, **201**, 202
  coniflorus, 197, 198, **199**
  donkelaari, 197, **200**
  grandiflorus, 128, 196, **197**, 198, 201, 210, 218
  hamatus, 196, 197, **203**, 204
  hondurensis, 197, **199**
  inermis, 197, **207**, 208, 209
  kunthianus, 197, **201**
  macdonaldiae, 196, 197, **202**, 203
  maxonii, 198
  miravallensis, 213
  murrillii, 197, **206**
  pringlei, 199
  pteranthus, 111, 196, 197, **200**

Selenicereus—*continued*.
  spinulosus, 197, **207**
  urbanianus, 197, **198**, 212
  vagans, 197, **205**, 206, 207
  wercklei, 197, **208**, 209
Sinita, 177
Soroco, 62
Spanish dildos, 87
Stenocereus, 69, 85
  stellatus, 92
  stellatus tenellianus, 92
  stellatus tonelianus, 92
Stetsonia, 2, **64, 65**
  coryne, **64**, 65
Strophocactus, 183, **221, 222**
  wittii, **221**, 222
Suaharo, 165
Suguaro, 165
Suwarro, 165
Suwarrow, 165
Tail of the fox, 109
Tenuiores, 22
Tortuosi, 77
Trichocereus, 2, 92, **130–146**
  bridgesii, 130, **134**, 136
  candicans, 130, 134, **142**
  chiloensis, 130, **137**, 138
  coquimbanus, 130, 138, **139**
  cuzcoensis, 130, **136**
  fascicularis, 64, 130, **141**
  huascha, 130, 141, **142**
  lamprochlorus, 130, **132**, 133
  macrogonus, 130, **136**
  pachanoi, 130, **134**, 135

Trichocereus—*continued*.
  pasacana, 130, 132, **133**, 134, 140, 145, 225
  peruvianus, 130, **136**
  schickendantzii, 130, **144**
  shaferi, 130, **144**
  spachianus, 130, **131**, 132
  strigosus, 130, **143**, 144
  terscheckii, 130, **140**
  thelegonoides, 130, **131**
  thelegonus, **130**, 131
Tuna, 66, 93
  colorado, 101
  de cobado, 84
Weberocereus, 183, **214–216**
  biolleyi, 214, **215**
  panamensis, 214, **215**
  tunilla, **214**
Werckleocereus, 183, **216, 217**
  glaber, **216**
  tonduzii, **216**, 217
Wilcoxia, 2, 22, **110–112**
  papillosa, 110, **112**
  poselgeri, 110, **111**
  striata, 110, **111**
  viperina, **110**
Wilmattea, 183, **195, 196**
  minutiflora, **195**, 196, 216
Zacoub, 187
Zanthoxylum, 92
Zehntnerella, 2, **176, 177**
  squamulosa, 176, **177**
Zuwarrow, 165
Zygocactus, 22